AF323673

COMPUTATIONAL METHODS IN NONLINEAR ANALYSIS

Efficient Algorithms, Fixed Point Theory and Applications

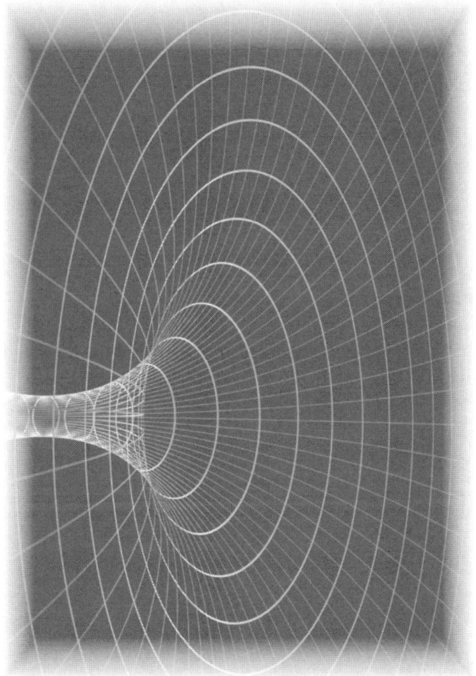

COMPUTATIONAL METHODS IN NONLINEAR ANALYSIS

Efficient Algorithms, Fixed Point Theory and Applications

Ioannis K Argyros *(Cameron University, USA)*

Saïd Hilout *(Poitiers University, France)*

NEW JERSEY · LONDON · SINGAPORE · BEIJING · SHANGHAI · HONG KONG · TAIPEI · CHENNAI

Published by

World Scientific Publishing Co. Pte. Ltd.

5 Toh Tuck Link, Singapore 596224

USA office: 27 Warren Street, Suite 401-402, Hackensack, NJ 07601

UK office: 57 Shelton Street, Covent Garden, London WC2H 9HE

Library of Congress Cataloging-in-Publication Data
Argyros, Ioannis K.
   Computational methods in nonlinear analysis : efficient algorithms, fixed point theory and applications /
by Ioannis K. Argyros (Cameron University, USA) & Saïd Hilout (Poitiers University, France).
    pages cm
  Includes bibliographical references and index.
  ISBN 978-981-4405-82-9 (hardcover : alk. paper)
  1. Nonlinear theories--Data processing. 2. Mathematics--Data processing. I. Hilout, Saïd. II. Title.
  QA427.A738 2013
  515'.355--dc23

                      2013005325

British Library Cataloguing-in-Publication Data
A catalogue record for this book is available from the British Library.

Printed in Singapore by World Scientific Printers.

The first author dedicates this book to his mother Anastasia, father
Konstantinos, sons Christopher, Gus and Michael and wife Diana
The second author dedicates to Aïcha, Lina, Nassim and Nouria

Preface

Many problems from applied sciences, computer graphics, engineering, optimization, economics, physics and other disciplines can be brought in the form of equations or variational inequalities using mathematical modeling. These equations or variational inequalities can be for examples: vectors (systems of linear or nonlinear algebraic equations); functions (difference, differential, integral equations); real or complex numbers (single algebraic equations with single unknowns); linear and nonlinear complementarity problems; optimality conditions for nonlinear programming. The field of computational sciences gives a lot of opportunity to researchers to solve these equations and has seen a considerable development in mathematics. The solutions of such equations can rarely be found in closed form. That is why most solution methods for these equations are iterative. The practice of numerical analysis for finding solutions is essentially connected to variants of Newton's method.

In 1669, Isaac Newton inaugurated his method through the use of numerical examples to solve equations, but did not use the current iterative expression. Later, in 1690, Raphson introduced Newton's method or the also called Newton–Raphson method. Newton's method is currently and undoubtedly the most popular one-point iterative procedure for generating a sequence approximating the solution of equation. In 1818, Fourier proved that the method converges quadratically in a neighborhood of the root, while Cauchy in 1829 and 1847 provided the multi-dimensional extension of Newton method. Kantorovich in 1948 published an important paper extending Newton's method for functional spaces.

There are usual concepts connected with iterative methods. The first concerns the evaluation of the function at each iterate to ensure that the iterates remain in the domain. In general, it is impossible to find the exact set of all initial data for which a given process is well defined and we restrict ourselves to giving conditions which guarantee that an iteration sequence is well defined for certain specific initial guesses. The secondly basic connection concerns the convergence of the sequences generated by a process and the question of whether their limit points are, in fact, solutions of the equation. The study about convergence matter of iterative methods is usually centered on two types: semi-local and local convergence analysis. The semi-local convergence matter is, based on the information around an initial point, to give criteria ensuring the convergence of Newton's method; while the local one is, based on the information around a solution, to find estimates of the radii of convergence balls. There is a plethora of studies on the weakness and/or extension of the hypothesis made on the underlying operators. The economy of the entire operations and the question of how fast a given sequence will converge is also a basic connection to iterative methods. Another concept affects the best chosen method, algorithm, or software program to solve a specific type of problem and its descriptions of when a given algorithm or method succeeds or fails. Our main objective is to expand the applicability of existing iterative procedures or introduce new ones. This is being achieved by using more precise majorizing sequences than before. This approach leads to weaker sufficient convergence conditions, tighter error bounds on the distances involved and a more precise information on the location of the solutions.

This Book adopts an updated scientific approach that combines recent results in numerical methods for nonlinear equations and variational inequalities with applications in various fields of optimization, economics, control theory, engineering, linear/nonlinear differential equations, partial differential equations and physics. We present the recent results on the convergence analysis in both finite dimensional and infinite dimensional spaces. Our attention has also been paid in studying iterative procedure on manifolds since there are many numerical problems posed on manifolds that arise naturally in many contexts. The book also provides comparison between various investigations made in recent years in the field of computational sciences and connects numerical analysis with functional analysis, theory of operators and their applications. Although the book is of a theoretical nature, with optimization and weakening of existing hypotheses considerations. Each chapter contains several new theoretical results and important applications in engineering, in dynamic systems, in input-output systems, in the solution of nonlinear and linear differential equations and optimization problems. The applications appear in the form of examples or study cases or they are implied since our results improve earlier ones that have already been applied in concrete problems. Note that we have endeavored to make the main text as self-contained as possible, to prove all results in full detail. In order to make the study useful as a reference source, we have complemented each chapter with a set of remarks, comments and corollaries in which literature citations are given, other related results are discussed and various possible extensions of the results of the text are indicated. Therefore we list numerous conjectures and open problems as well as alternative models which need to be explored. The book also contains abundant and updated bibliography in the field of computational sciences.

This book is intended for researchers and practitioners in applied computational sciences, mathematical programming, engineering, optimization, mathematical economics, senior undergraduate students and graduate students. The goal is to introduce these powerful concepts and techniques at the earliest possible stage. The reader is assumed to have had basic knowledge in numerical analysis, computational linear algebra, theory of operators, functional analysis and computer programming.

Our goal in chapter 1 is to find weaker convergence criteria for Newton's method than in earlier studies. It turns out that our sufficient convergence conditions are weaker and the error bounds are tighter than in earlier studies for many interesting cases. In this chapter new results on Newton–Kantorovich theory are presented. In section 1 we use hypotheses on the first Fréchet derivative of involved operator. These new results are illustrated by several numerical examples, for which the older convergence criteria do not hold but for which our weaker convergence criteria are satisfied. In section 2 we present a two-point Newton-like method to approximate a locally unique solution of a nonlinear equation containing a non-differentiable term. In section 3 we provide new local and semi-local convergence results for Newton's method. The sufficient convergence conditions used in this section do not include the Lipschitz constant usually associated with Newton's method. In section 4 we present a local convergence analysis for Newton's method under a weak majorant condition. Our results provide under the same information a larger radius of convergence and tighter error estimates on the distances involved than before in earlier studies. Special cases and applications are also provided for validating the theoretical results. we present a local convergence analysis for the continuous analog of Newton's method. The radius of convergence is larger, the error bounds tighter and under the same or weaker hypotheses than in earlier studies. In section 5 the concept nondiscrete mathematical induction inaugurated by Potra and Ptaák is used. We extend the applicability of Newton's method for approximating a solution of nonlinear equation using weaker hypotheses. Throughout chapter 1 we present an illustrative example involving a differential equation containing a Green's kernel, a nonlinear integral equation of Chandrasekhar-type and cubic polynomial equation to show that different theorems are applicable in some situations in which the other are not applicable.

In chapter 2 we study special conditions and present some new results for convergence analysis of Newton's method in a Banach space setting. In section 1 $w^\star$-conditioned second Fréchet-derivative is explored. This way we can handle equations, where the usual Lipschitz-type conditions are non verifiable. It turns out that a similar result using ω-conditioned hypotheses can provide usable error estimates indicating only linear convergence for Newton's method. We provide in section 2 an existence and uniqueness result for nonlinear equations involving regularly smooth operators, under weaker hypotheses. Our approach extends the applicability and improves the optimality of Newton-type methods using the same information, under weaker hypotheses. Moreover, the information on the location of the solution is at least as precise as in earlier studies. We use in section 3 the special Smale's α-theory by introducing the notion of the center γ_0-condition. We expand the applicability of Newton's method. Numerical examples and applications are also provided in each section of this chapter.

Chapter 3 exhibits convergence analysis of Newton-type methods in order to approximate a zero of a mapping on special spaces as Lie groups, Hilbert space and Banach space with convergence structure. Section 1 presents Newton's method to approximate a zero of a mapping from a Lie group into its Lie algebra. Under the same computational cost and weaker convergence conditions than before. Precise information on the location of the solution is obtained in the semi-local case and larger radius of convergence is established in the local case. Note that tighter error bounds on the distances involved is obtained in both cases. In section 2 we present new semi-local convergence results for continuous modified Newton-type methods to solve nonlinear operator equations in a real Hilbert space setting. Convergence analysis is presented in section 3 to approximate a solution of equations on Banach space with convergence structure. In sections 4 to 6, for approximating zeros of a vector field on Riemannian manifolds, we present a semi-local convergence analysis of a bilinear operator free third order method, Shamanskii-type method and Traub-type method, respectively. A characterization of the convergence under Kantorovich-type conditions, error estimates and applications are also given in these three sections.

Chapter 4 presents a recent developments on Secant method. Secant method uses divided differences operator and is an attractive replacement of Newton's method studied in chapters 1 and 2. This method uses a consistent approximation operator of the Fréchet derivative and is an alternative method of Newton's method. Note also that the Secant method is of convergence order $1.6180339887\cdots$, whereas Newton's method is of order at most 2. It is a self-correcting like Newton's method and it is of high efficiency. In section 1 we extend the applicability of the method of chords in some cases. The error bounds are tighter and the information on the location of the solution at least as precise under the same information as before in earlier studies. Application and examples are also provided in this section. Section 2 exhibits the convergence analysis of Secant-type method using the concept nondiscrete mathematical induction in order to improve error bounds. A local convergence analysis for an efficient Secant-type method is given in section 3 using both the Lipschitz continuous and center-Lipshchitz continuous divided differences of order one. An estimate of the radius of the convergence ball is provided. In section 4 we use our concept of recurrent functions to provide a new semi-local convergence analysis for secant-like methods. Our convergence criteria and sufficient convergence conditions are tighter than in earlier studies. A directional Secant-type methods in Euclidean space is presented in section 5. Using weaker hypotheses and motivated by optimization considerations, a new existence convergence results are established. A numerical example to implement the method is also presented in this section. We unify in section 6 the semi-local convergence analysis of Secant-type methods under more general Lipschitz-type conditions. We present very general majorizing sequences and we extend the applicability of Secant-type methods. Our analysis includes the computation of the bounds on the limit points of the majorizing sequences involved. As in chapters 1 to 3 we present in each section of chapter 4 some remarks, examples and applications of our theoretical results.

Chapter 5 develops Gauss–Newton methods for solving nonlinear equations and convex composite optimization problems. Section 1 concerns the local convergence analysis of the iteratively regularized Gauss–Newton method for solving ill-posed problems in a Hilbert space setting. In contrast to earlier studies with Lipschitz conditions, we only use the weaker center Lipschitz conditions. Our analysis expands the applicability of this method. A semi-local convergence analysis of the Gauss–Newton method for convex composite optimization is given in section 2. We use the concept of quasi-regularity. The convergence analysis is presented under L-average Lipschitz and generalized convex majorant conditions, respectively. Our results extend the applicability of Gauss-Newton method. Section 3 presents Gauss–Newton algorithm to find local minimum of penalized nonlinear least square problems. We develop local convergence analysis from more general Lipschitz-type conditions than used in earlier studies. Our results, even in special cases, provide a larger attraction ball and tighter error estimates. In section 4 we establish a local convergence analysis of inexact Gauss–Newton like methods using more precise majorant conditions. We also provide a clearer relationship between the majorant function and the associated least squares problem.

Chapters 6, 7 and 8 contain recent results on Halley's, Chebyshev's and Broyden's methods, respectively. For Halley's method, we present semi-local and local convergence under convex majorants. Kantorovich-type and Smale-type results are considered as applications and special cases. We finish chapter 6 by a unified approach to generating majorizing sequences for multi-point iterative procedures (Traub-type method). In chapter 7 we present in section 1 semi-local convergence analysis for directional Chebyshev-type methods in finite dimensional space using recurrent relations and Newton–Kantorovich-type hypotheses. We give in section 2 of chapter 7 a semi-local convergence theorem for a new family of iterative methods obtained as a combination of the well-known Secant method and Chebyshev's method. We give a very general convergence result that allow the application of these methods to non-differentiable problems. We finish chapter 7 by finer and precise majorizing sequences for Chebyshev's method. In chapter 8 we present a new results on Broyden's method using ω-type conditions. Examples and applications are also presented in chapters 6 to 8.

The goal of chapter 9 is to give a new convergence analysis of some Newton-like methods. In section 1 we give an estimate of the convergence radius of the well-known modified Newton's method for multiple zeros, when the involved function satisfies a Hölder and center-Hölder continuity conditions. We provide in section 2 a semi-local convergence analysis for a Newton-like method under weak conditions. In particular, we only assume that the Gateaux derivative of the operator involved is hemicontinuous. Section 3 exhibits a study on the radius of convergence for some cubically convergent Newton-type method. A comparison is given between the radii of this method and Newton's method. In section 4 we present a fast two-step Newton-like method. This method unifies earlier methods such as Newton-like, Chebyshev-Secant, Chebyshev-Newton, Steffensen, Stirling's and other single or multi-step methods. We unify in section 5 the semi-local convergence analysis of Newton-like methods under more general Lipschitz-type conditions and precise majorizing sequences. Section 6 contains a new result for collection of hybrid methods combining Newton's method with frozen derivatives and a family of high order iterative schemes. We establish in section 7 the local and sem-ilocal convergence of the relaxed Newton's method, that is Newton's method with a relaxation parameter. We give a new Kantorovich-like theorem. We also obtain in this section the recurrent sequence that majorizes the one given by the method and we characterize its convergence by a result that involves the relaxation parameter. We use a new technique that allows us on the one hand to generalize and on the other hand to extend the applicability of the result given initially by Kantorovich.

In the last chapter 10 we present some new results of convergence analysis for Newton–Tikhonov methods, in order to solve ill-posed operators problems in Hilbert space. In section 1 we expand the applicability of Newton–Tikhonov method by using more precise majorizing sequence. This way we provide a semi-local convergence analysis for this method with some advantages over earlier studies. These advantages are obtained, since we use the more precise center-Lipschitz condition instead of the Lipschitz condition for the computation of the upper bounds. We also study the semi-local convergence of the simplified Newton–Tikhonov method. Section 2 develops two-step directional Newton method to solve ill-posed problems under weak conditions. An iteratively regularized projection method, which converges quadratically, has been considered in section 3 for obtaining stable approximate solution of nonlinear ill-posed operator equations. We assume in this section that only a noisy data are available. Under weaker convergence condition, a choice of the regularization parameter using an adaptive selection and a stopping rule for the iteration index using a majorizing sequence are also presented in this section.

Note that we propose in the end of each chapter exercises to apply all obtained theoretical results.

Ioannis K. Argyros and Saïd Hilout

Contents

Chapter 1

Newton's Methods

Newton's method is currently and undoubtedly the most popular one-point iterative procedure for solving nonlinear equations. We present in this chapter a local as well as a semi-local convergence analysis of Newton's method in a Banach space setting.

1.1 Convergence under Lipschitz Conditions

In this section we are concerned with the problem of approximating a locally unique solution $x^\star$ of equation

$$F(x) = 0, \tag{1.1}$$

where F is a Fréchet-differentiable operator defined on a closed and convex subset $\mathcal{D}$ of a Banach space $\mathcal{X}$ with values in a Banach space $\mathcal{Y}$.

The basic idea of Newton's method is linearization. Starting from an initial guess $x_0 \in \mathcal{D}$, we can have the linear approximation of $F(x)$ in the neighborhood $\mathcal{D}_0$ of x_0 ($\mathcal{D}_0 \subseteq \mathcal{D}$): $F(x_0 + w) \simeq F(x_0) + F'(x_0)\, w$ and solve the resulting linear equation $F(x_0) + F'(x_0)\, w = 0$, leading to the recurrent Newton's method as follows

$$x_{n+1} = x_n - F'(x_n)^{-1} F(x_n) \qquad \text{for each} \quad n = 0, 1, 2, \cdots. \tag{1.2}$$

Here, $F'(x)$ denotes the Fréchet-derivative of F at $x \in \mathcal{D}$. We refer the reader to Refs. [106, 120, 163] for recent results on Newton's method (see also Refs. [16, 104, 112, 116, 161, 140, 155, 444] and the references therein).

Concerning the semi-local convergence of Newton's method, one of the most important results is the celebrated Kantorovich theorem (see Ref. [290]) for solving nonlinear equations. This theorem provides a simple and transparent convergence criterion for operators with bounded second derivatives F'' or the Lipschitz continuous first derivatives.

Let $U(x, r)$ and $\overline{U}(x, r)$ stand, respectively, for the open and closed ball in $\mathcal{X}$ with center x and radius $r > 0$. Let also $\mathcal{L}(\mathcal{X}, \mathcal{Y})$ stand for the space of bounded linear operators from $\mathcal{X}$ into $\mathcal{Y}$. We shall use the following $(\mathcal{A}_s)$ and $(\mathcal{B}_s)$ hypotheses to study the semi-local convergence and the $(\mathcal{A}_l)$ and $(\mathcal{B}_l)$ hypotheses to study the local convergence of Newton's method.

1

Hypotheses $(\mathcal{A}_s)$ are

$(\mathcal{A}_{s1})$ $F : \mathcal{D} \subseteq \mathcal{X} \longrightarrow \mathcal{Y}$ is Fréchet-differentiable. There exists $x_0 \in \mathcal{D}$ such that $F'(x_0)^{-1} \in \mathcal{L}(\mathcal{Y}, \mathcal{X})$ and

$$\| F'(x_0)^{-1} F(x_0) \| \leq \eta \quad \text{for some} \quad \eta > 0.$$

$(\mathcal{A}_{s2})$ There exists $L > 0$ such that

$$\| F'(x_0)^{-1} (F'(x) - F'(y)) \| \leq L \| x - y \| \quad \text{for each} \quad x, y \in \mathcal{D}.$$

$(\mathcal{A}_{s3})$ There exists $L_0 > 0$ such that

$$\| F'(x_0)^{-1} (F'(x) - F'(x_0)) \| \leq L_0 \| x - x_0 \| \quad \text{for each} \quad x \in \mathcal{D}.$$

$(\mathcal{A}_{s4})$ There exist $H > 0$, $K > 0$ such that

$$\| F'(x_0)^{-1} (F'(x_0 + \theta (x_1 - x_0)) - F'(x_0)) \| \leq H \theta \| x_1 - x_0 \|$$
$$\text{for each} \quad \theta \in [0, 1]$$

and

$$\| F'(x_0)^{-1} (F'(x_1) - F'(x_0)) \| \leq K \| x_1 - x_0 \|,$$

where $x_1 = x_0 - F'(x_0)^{-1} F(x_0)$.

We have that

$$(\mathcal{A}_{s2}) \Longrightarrow (\mathcal{A}_{s3}) \Longrightarrow (\mathcal{A}_{s4}), \quad K \leq H \leq L_0 \leq L$$

hold and L/L_0 can be arbitrarily large (see Refs. [106, 120, 163]). Note that constants H, L_0 and L are determined by computations at the initial data. Moreover, in practice the computation of L requires the computation of L_0 and H. Hence $(\mathcal{A}_{s3})$ or $(\mathcal{A}_{s4})$ are not additional hypotheses to $(\mathcal{A}_{s2})$.

Hypotheses $(\mathcal{B}_s)$ are

$(\mathcal{B}_{s1})$ $F : \mathcal{D} \subseteq \mathcal{X} \longrightarrow \mathcal{Y}$ is Fréchet-differentiable. There exist $A \in \mathcal{L}(\mathcal{X}, \mathcal{Y})$, $x_0 \in \mathcal{D}$ such that $F'(x_0)^{-1} \in \mathcal{L}(\mathcal{Y}, \mathcal{X})$, $A^{-1} \in \mathcal{L}(\mathcal{Y}, \mathcal{X})$ and

$$\| F'(x_0)^{-1} F(x_0) \| \leq \eta \quad \text{for some} \quad \eta > 0.$$

$(\mathcal{B}_{s2})$ There exists $\ell > 0$ such that

$$\| A^{-1} (F'(x) - F'(y)) \| \leq \ell \| x - y \| \quad \text{for each} \quad x, y \in \mathcal{D}.$$

$(\mathcal{B}_{s3})$ There exists $\ell_0 > 0$ such that

$$\| A^{-1} (F'(x) - A) \| \leq \ell_0 (\| x - x_0 \| + \| x - x_1 \|) \quad \text{for each} \quad x \in \mathcal{D},$$

where $x_1 = x_0 - F'(x_0)^{-1} F(x_0)$.

$(\mathcal{B}_{s4})$ There exists $h > 0$ such that

$$\| A^{-1} (F'(x_0 + \theta (x_1 - x_0)) - F'(x_0)) \| \leq h \theta \| x_1 - x_0 \|$$
$$\text{for each} \quad \theta \in [0, 1].$$

$(\mathcal{B}_{s}5)$ There exists $K_1 > 0$ such that

$$\| A^{-1}(F'(x_1) - A) \| \le K_1 \| x_1 - x_0 \| .$$

Note that

$$(\mathcal{B}_{s2}) \Longrightarrow (\mathcal{B}_{s4}), \quad (\mathcal{B}_{s3}) \Longrightarrow (\mathcal{B}_{s5}), \quad \ell \le \ell_0 \quad \text{and} \quad K_1 \le \ell_0$$

hold. Some possible choices for operator A are given by $A = F'(x_0)$ or $A = [x_0, x_1; F]$ where $[x, y; F]$ is divided into difference of F of order one at points $x, y \in \mathcal{D}$ (see Ref. [106]).

The Kantorovich hypothesis for the semi-local convergence of Newton's method using $(\mathcal{A}_{s1})$ and $(\mathcal{A}_{s2})$ (see Refs. [106, 290]) is given by

$$C_\star = 2 L \eta \le 1. \tag{1.3}$$

In a series of papers (see Refs. [100, 98, 104, 112, 116, 136, 140, 155]), respectively, we provided using $(\mathcal{A}_{s1})$–$(\mathcal{A}_{s3})$ the weaker conditions

$$C_1 = L_1 \eta \le 1, \tag{1.4}$$

$$C_2 = L_2 \eta \le 1, \tag{1.5}$$

$$C_3 = L_3 \eta \le 1, \tag{1.6}$$

where

$$L_1 = L_0 + L, \quad L_2 = \frac{1}{4} (L + 4 L_0 + (L^2 + 8 L_0 L)^{1/2})$$

and

$$L_3 = \frac{1}{4} (4 L_0 + (L L_0 + 8 L_0^2)^{1/2} + (L_0 L)^{1/2}).$$

If $L_0 = L$, then (1.4)–(1.6) coincide with (1.3). Otherwise, i.e. if $L_0 < L$, then

$$C_\star < 1 \Longrightarrow C_1 < 1 \Longrightarrow C_2 < 1 \Longrightarrow C_3 < 1 \tag{1.7}$$

but not vice versa. We also have that

$$\left. \begin{array}{lll} \dfrac{C_1}{C_\star} \longrightarrow \dfrac{1}{2}, & \dfrac{C_2}{C_\star} \longrightarrow \dfrac{1}{4}, & \dfrac{C_2}{C_1} \longrightarrow \dfrac{1}{2} \\[2ex] \dfrac{C_3}{C_\star} \longrightarrow 0, & \dfrac{C_3}{C_2} \longrightarrow 0, & \dfrac{C_3}{C_1} \longrightarrow 0 \end{array} \right\} \text{ as } \quad \frac{L_0}{L} \longrightarrow 0. \tag{1.8}$$

In this section we present convergence conditions for Newton's method using $(\mathcal{A}_s)$ if $A = F'(x_0)$ which are weaker than (1.6) or can be weaker than (1.6) under the $(\mathcal{B}_s)$ conditions. These hypotheses also lead to tighter majorizing sequences and a more precise information on the location of the solution.

Hypotheses $(\mathcal{A}_l)$ are

$(\mathcal{A}_{l1})$ $F : \mathcal{D} \subseteq \mathcal{X} \longrightarrow \mathcal{Y}$ is Fréchet-differentiable. There exists $x^\star \in \mathcal{D}$ such that $F'(x^\star)^{-1} \in \mathcal{L}(\mathcal{Y}, \mathcal{X})$ and $F(x^\star) = 0$.

$(\mathcal{A}_{l2})$ There exists $L > 0$ such that

$$\| F'(x^\star)^{-1} (F'(x) - F'(y)) \| \le L \| x - y \| \quad \text{for each} \quad x, y \in \mathcal{D}.$$

$(\mathcal{A}_{l3})$ There exists $L_0 > 0$ such that

$$\| F'(x^\star)^{-1} (F'(x) - F'(x^\star)) \| \le L_0 \| x - x^\star \| \quad \text{for each} \quad x \in \mathcal{D}.$$

Note that

$$(\mathcal{A}_{l2}) \Longrightarrow (\mathcal{A}_{l3}), \quad L_0 \le L$$

hold and L/L_0 can be arbitrarily large (see Ref. [106]).

Hypotheses $(\mathcal{B}_l)$ are

$(\mathcal{B}_{l1})$ $F : \mathcal{D} \subseteq \mathcal{X} \longrightarrow \mathcal{Y}$ is Fréchet-differentiable. There exist $A \in \mathcal{L}(\mathcal{X}, \mathcal{Y})$, $x^\star \in \mathcal{D}$ such that $A^{-1} \in \mathcal{L}(\mathcal{Y}, \mathcal{X})$ and $F(x^\star) = 0$.

$(\mathcal{B}_{l2})$ There exists $\ell > 0$ such that

$$\| A^{-1} (F'(x) - F'(y)) \| \le \ell \| x - y \| \quad \text{for each} \quad x, y \in \mathcal{D}.$$

$(\mathcal{B}_{l3})$ There exists $\ell_0 > 0$ such that

$$\| A^{-1} (F'(x) - A) \| \le \ell_0 (\| x - x_0 \| + \| x - x^\star \|)$$

for each $x \in \mathcal{D}$ and some $x_0 \in \mathcal{D}$.

Some possible choices for operator A are given by $A = F'(x^\star)$ or $A = [x_0, x^\star; F]$. The radius of convergence for Newton's method given by Traub (see Ref. [418]) using $(\mathcal{A}_{l1})$ and $(\mathcal{A}_{l2})$ is

$$R_0 = \frac{2}{3\,L}. \tag{1.9}$$

The radius of convergence given by Ref. [100] using $(\mathcal{A}_{l1})$–$(\mathcal{A}_{l3})$ is

$$R_1 = \frac{2}{2\,L_0 + L}. \tag{1.10}$$

Note that if $L_0 < L$, then we have that

$$R_0 < R_1 \tag{1.11}$$

and

$$\frac{R_1}{R_0} \longrightarrow 3 \quad \text{as} \quad \frac{L_0}{L} \longrightarrow 0. \tag{1.12}$$

Here, we show how to enlarge R_1 using the $(\mathcal{B}_l)$ hypotheses. We also provide tighter error estimates on the distances $\| x_{n+1} - x^\star \|$.

It is convenient for us to define certain parameters and quadratic polynomials. Let $H > 0$, $K > 0$, $L_0 > 0$, $L > 0$ and $\eta > 0$. Define parameters

$$\alpha = \frac{2\,L}{L + (L^2 + 8\,L_0\,L)^{1/2}}$$

$$t_0 = 0, \quad t_1 = \eta, \quad t_2 = t_1 + \frac{H\,(t_1 - t_0)^2}{2\,(1 - K\,t_1)},$$

$$\alpha_0 = \frac{L\,(t_2 - t_1)}{2\,(1 - L_0\,t_2)}$$

and quadratic polynomials

$$P_1(t) = (L\,H + 2\,\alpha\,L_0\,(H - 2\,K))\,t^2 + 4\,\alpha\,(L_0 + K)\,t - 4\,\alpha,$$

$$P_2(t) = L_0\,(H - 2\,(1 - \alpha)\,K)\,t^2 + 2\,(1 - \alpha)\,(L_0 + K)\,t - 2\,(1 - \alpha).$$

Note that $\alpha \in (0,1)$ and the discriminants Δ_1, Δ_2 of P_1, P_2, respectively, are such that

$$\Delta_1 = 16\,\alpha\,(\alpha\,(L_0 - K)^2 + (L + 2\,\alpha\,L_0)\,H) > 0$$

and

$$\Delta_2 = 4\,(1 - \alpha)\,((1 - \alpha)\,(L_0 - K)^2 + 2\,L_0\,H) > 0.$$

Moreover, define parameters

$$\eta_1 = \begin{cases} \dfrac{1}{L_0 + K} & \text{if } L\,H + 2\,\alpha\,L_0\,(H - 2\,K) = 0 \\ \text{positive root of } P_1 & \text{if } L\,H + 2\,\alpha\,L_0\,(H - 2\,K) > 0 \\ \text{small positive root of } P_1 & \text{if } L\,H + 2\,\alpha\,L_0\,(H - 2\,K) < 0 \end{cases}$$

$$\eta_2 = \begin{cases} \dfrac{1}{L_0 + K} & \text{if } H - 2\,(1 - \alpha)\,K = 0 \\ \text{positive root of } P_2 & \text{if } H - 2\,(1 - \alpha)\,K > 0 \\ \text{small positive root of } P_2 & \text{if } H - 2\,(1 - \alpha)\,K < 0. \end{cases}$$

Furthermore, note that $\eta_1 = \eta_2$, since by the definition of α, it follows that

$$L = \frac{2\,L_0\,\alpha^2}{1 - \alpha}, \quad L\,H + 2\,\alpha\,L_0\,(H - 2\,K) = \frac{2\,L_0\,\alpha}{1 - \alpha}\,(H - 2\,(1 - \alpha)\,K)$$

and

$$P_1(t) = \frac{2\,L_0\,\alpha}{1 - \alpha}\,P_2(t).$$

Suppose that

$$C_4 = \eta_1^{-1}\,\eta \le 1. \tag{1.13}$$

Then, it can easily be seen that

$$\alpha_0 \le \alpha \le 1 - \frac{L_0\,(t_2 - t_1)}{1 - L_0\,t}. \tag{1.14}$$

Indeed, the left-hand side inequality of (1.14) reduces to

$$P_1(\eta) \le 0,$$

whereas the inequality on the right of (1.14) reduces to

$$P_2(\eta) \leq 0.$$

These inequalities are true by the choice of η_1 and η_2, respectively.

Next, we need some auxiliary results on majorizing sequences for Newton's method.

Lemma 1.1.1. *Let $H > 0$, $K > 0$, $L > 0$, $L_0 > 0$ and $\eta > 0$. Suppose that (1.13) holds. Then, saclar sequence $\{t_n\}$ given by*

$$t_0 = 0, \quad t_1 = \eta, \quad t_2 = t_1 + \frac{H\,(t_1 - t_0)^2}{2\,(1 - K\,t_1)},$$
$$t_{n+2} = t_{n+1} + \frac{L\,(t_{n+1} - t_n)^2}{2\,(1 - L_0\,t_{n+1})} \quad for\ each \quad n = 1, 2, \cdots, \tag{1.15}$$

is well defined, increasing, bounded from above by

$$t^{\star\star} = \eta + \left(1 + \frac{\alpha_0}{1 - \alpha}\right) \frac{H\,\eta^2}{2\,(1 - K\,\eta)} \tag{1.16}$$

and converges to its unique least upper bound $t^\star$ which satisfies

$$t_2 \leq t^\star \leq t^{\star\star}. \tag{1.17}$$

Moreover, the following estimates hold:

$$0 < t_{n+2} - t_{n+1} \leq \alpha_0\,\alpha^{n-1} \frac{H\,\eta^2}{2\,(1 - K\,\eta)} \quad for\ each \quad n = 1, 2, \cdots \tag{1.18}$$

and

$$t^\star - t_n \leq \frac{\alpha_0\,(t_2 - t_1)}{1 - \alpha}\,\alpha^{n-2} \quad for\ each \quad n = 2, 3, \cdots. \tag{1.19}$$

Proof. We use the Mathematical Induction to prove that

$$0 < \frac{L\,(t_{k+1} - t_k)}{2\,(1 - L_0\,t_{k+1})} \leq \alpha \tag{1.20}$$

holds for each $k = 1, 2, \cdots$. Estimate (1.20) is true for $k = 1$ by the definition of η_1 and P_1. Then, we have by (1.15)

$$0 < t_3 - t_2 \leq \alpha_0\,(t_2 - t_1) \Longrightarrow t_3 \leq t_2 + \alpha_0\,(t_2 - t_1)$$
$$\Longrightarrow t_3 \leq t_2 + (1 + \alpha_0)\,(t_2 - t_1) - (t_2 - t_1)$$
$$\Longrightarrow t_3 \leq t_1 + \frac{1 - \alpha_0^2}{1 - \alpha_0}\,(t_2 - t_1) < t^{\star\star}$$

and for $k = 2, 3, \cdots$

$$t_{k+2} \leq t_{k+1} + \alpha_0\,\alpha^{k-1}\,(t_2 - t_1)$$
$$\leq t_k + \alpha_0\,\alpha^{k-2}\,(t_2 - t_1) + \alpha_0\,\alpha^{k-1}\,(t_2 - t_1)$$
$$\leq t_1 + (1 + \alpha_0\,(1 + \alpha + \cdots + \alpha^{k-1}))\,(t_2 - t_1)$$
$$= t_1 + (1 + \alpha_0\,\frac{1 - \alpha^k}{1 - \alpha})\,(t_2 - t_1) \leq t^{\star\star}.$$

Assume that (1.20) holds for all natural integers $n \leq k$. Then, we get by (1.15) and (1.20) that

$$0 < t_{k+2} - t_{k+1} \leq \alpha_0 \, \alpha^{k-1} \, (t_2 - t_1) \leq \alpha^k \, (t_2 - t_1)$$

and

$$t_{k+2} \leq t_1 + (1 + \alpha_0 \, \frac{1 - \alpha^k}{1 - \alpha}) \, (t_2 - t_1) \leq t_1 + \frac{1 - \alpha^{k+1}}{1 - \alpha} \, (t_2 - t_1) < t^{\star\star}.$$

Evidently estimate (1.20) is true, if k is replaced by $k + 1$ provided that

$$\frac{L}{2} \, (t_{k+2} - t_{k+1}) \leq \alpha \, (1 - L_0 \, t_{k+2})$$

or

$$\frac{L}{2} \, (t_{k+2} - t_{k+1}) + \alpha \, L_0 \, t_{k+2} - \alpha \leq 0$$

or

$$\frac{L}{2} \, \alpha^k \, (t_2 - t_1) + \alpha \, L_0 \left(t_1 + \frac{1 - \alpha^{k+1}}{1 - \alpha} \, (t_2 - t_1) \right) - \alpha \leq 0. \tag{1.21}$$

Estimate (1.21) motivates us to define recurrent functions $\{f_k\}$ on $[0, 1)$ by

$$f_k(s) = \frac{L}{2} \, (t_2 - t_1) \, s^{k+1} + s \, L_0 \, (1 + s + s^2 + \cdots + s^k) \, (t_2 - t_1) - (1 - L_0 \, t_1) \, s.$$

We need a relationship between two consecutive functions f_k. We get that

$$\begin{aligned}
f_{k+1}(s) &= \frac{L}{2} \, (t_2 - t_1) \, s^{k+2} + s \, L_0 \, (1 + s + s^2 + \cdots + s^{k+1}) \, (t_2 - t_1) \\
&\quad - (1 - L_0 \, t_1) \, s \\
&= \frac{L}{2} \, (t_2 - t_1) \, s^{k+2} + s \, L_0 \, (1 + s + s^2 + \cdots + s^{k+1}) \, (t_2 - t_1) \\
&\quad - (1 - L_0 \, t_1) \, s - \frac{L}{2} \, (t_2 - t_1) \, s^k \\
&\quad - s \, L_0 \, (1 + s + s^2 + \cdots + s^k) \, (t_2 - t_1) + (1 - L_0 \, t_1) \, s + f_k(s).
\end{aligned}$$

Therefore, we deduce that

$$f_{k+1}(s) = f_k(s) + \frac{1}{2} \, (2 \, L_0 \, s^2 + L \, s - L) \, s^k \, (t_2 - t_1). \tag{1.22}$$

Estimate (1.21) is satisfied, if

$$f_k(\alpha) \leq 0 \quad \text{holds for each} \quad k = 1, 2, \cdots. \tag{1.23}$$

Using (1.22) we obtain that

$$f_{k+1}(\alpha) = f(\alpha) \quad \text{for each} \quad k = 1, 2, \cdots.$$

Let us now define function f_∞ on $[0, 1)$ by

$$f_\infty(s) = \lim_{k \to \infty} f_k(s). \tag{1.24}$$

Then, we have by (1.24) and the choice of α that

$$f_\infty(\alpha) = f_k(\alpha) \quad \text{for each} \quad k = 1, 2, \cdots.$$

Hence, (1.23) is satisfied, if

$$f_\infty(\alpha) \leq 0. \tag{1.25}$$

Using (1.21) we get that

$$f_\infty(\alpha) = \left(\frac{L_0}{1-\alpha}(t_2 - t_1) + L_0 t_1 - 1\right)\alpha. \tag{1.26}$$

It then, follows from the definition of η_2, P_2 and (1.26) that (1.25) is satisfied. The induction is now complete. Hence, sequence $\{t_n\}$ is increasing, bounded from above by $t^{\star\star}$ given by (1.16) and as such it converges to its unique least upper bound $t^\star$ which satisfies (1.17). The proof of Lemma 1.1.1 is complete. $\qquad\square$

Next, first we present a semi-local convergence result relating majorizing sequence $\{t_n\}$ with Newton's method and hypotheses $(\mathcal{A}_s)$. Secondly, we use hypotheses $(\mathcal{B}_s)$

Theorem 1.1.1. *Suppose that hypotheses $(\mathcal{A}_s)$, hypotheses of Lemma 1.1.1 and $\overline{U}(x_0, t^\star) \subseteq \mathcal{D}$ hold, where $t^\star$ is given in Lemma 1.1.1. Then, sequence $\{x_n\}$ generated by Newton's method is well defined, remains in $\overline{U}(x_0, t^\star)$ and converges to a solution $x^\star \in \overline{U}(x_0, t^\star)$ of equation $F(x) = 0$. Moreover, the following estimates hold*

$$\| x_{n+1} - x_n \| \leq t_{n+1} - t_n \tag{1.27}$$

and

$$\| x_n - x^\star \| \leq t^\star - t_n \quad for\ each \quad n = 0, 1, 2, \cdots, \tag{1.28}$$

where sequence $\{t_n\}$ is given in Lemma 1.1.1. Furthermore, if there exists $R \geq t^\star$ such that

$$\overline{U}(x_0, R) \subseteq \mathcal{D} \quad and \quad L_0\,(t^\star + R) < 2,$$

then, the solution $x^\star$ of equation $F(x) = 0$ is unique in $\overline{U}(x_0, R)$.

Proof. We use Mathematical Induction to prove that

$$\| x_{k+1} - x_k \| \leq t_{k+1} - t_k \tag{1.29}$$

and

$$\overline{U}(x_{k+1}, t^\star - t_{k+1}) \subseteq \overline{U}(x_k, t^\star - t_k) \quad for\ each \quad k = 1, 2, \cdots. \tag{1.30}$$

Let $z \in \overline{U}(x_1, t^\star - t_1)$. Then, we obtain that

$$\| z - x_0 \| \leq \| z - x_1 \| + \| x_1 - x_0 \| \leq t^\star - t_1 + t_1 - t_0 = t^\star - t_0,$$

which implies $z \in \overline{U}(x_0, t^\star - t_0)$. Note also that

$$\| x_1 - x_0 \| = \| F'(x_0)^{-1} F(x_0) \| \leq \eta = t_1 - t_0.$$

Hence, estimates (1.29) and (1.30) hold for $k = 0$. Suppose these estimates hold for $n \leq k$. Then, we have that

$$\| x_{k+1} - x_0 \| \leq \sum_{i=1}^{k+1} \| x_i - x_{i-1} \| \leq \sum_{i=1}^{k+1} (t_i - t_{i-1}) = t_{k+1} - t_0 = t_{k+1}$$

and

$$\| x_k + \theta\, (x_{k+1} - x_k) - x_0 \| \leq t_k + \theta\, (t_{k+1} - t_k) \leq t^{\star}$$

for all $\theta \in (0, 1)$. Using Lemma 1.1.1 and the induction hypotheses, we get in turn that

$$\| F'(x_0)^{-1} (F'(x_{k+1}) - F'(x_0)) \| \leq M \| x_{k+1} - x_0 \| \leq M(t_{k+1} - t_0) \leq M t_{k+1} < 1, \tag{1.31}$$

where

$$M = \begin{cases} K & \text{if } k = 0 \\ L_0 & \text{if } k = 1, 2, \cdots. \end{cases}$$

It follows from (1.31) and the Banach lemma on invertible operators (see Refs. [106, 163, 290]) that $F'(x_{m+1})^{-1}$ exists and

$$\| F'(x_{k+1})^{-1} F'(x_0) \| \leq (1 - M \| x_{k+1} - x_0 \|)^{-1} \leq (1 - M\, t_{k+1})^{-1}. \tag{1.32}$$

Using (1.2), we obtain the approximation

$$\begin{aligned} F(x_{k+1}) &= F(x_{k+1}) - F(x_k) - F'(x_k)\, (x_{k+1} - x_k) \\ &= \int_0^1 (F'(x_k + \theta\, (x_{k+1} - x_k)) - F'(x_m))\, (x_{k+1} - x_k)\, d\theta. \end{aligned} \tag{1.33}$$

Then, by (1.33) we get in turn

$$\begin{aligned} \| &F'(x_0)^{-1} F(x_{k+1}) \| \\ &\leq \int_0^1 \| F'(x_0)^{-1} (F'(x_k + \theta\, (x_{k+1} - x_k)) - F'(x_k)) \| \| x_{k+1} - x_k \|\, d\theta \\ &\leq M_1 \int_0^1 \| \theta\, (x_{k+1} - x_k) \| \| x_{k+1} - x_k \|\, d\theta \leq \frac{M_1}{2}\, (t_{k+1} - t_k)^2, \end{aligned} \tag{1.34}$$

where

$$M_1 = \begin{cases} H & \text{if } k = 0 \\ L & \text{if } k = 1, 2, \cdots. \end{cases}$$

Moreover, by (1.2), (1.32), (1.34) and the induction hypotheses we get that

$$\begin{aligned} \| x_{k+2} - x_{k+1} \| &= \| (F'(x_{k+1})^{-1} F'(x_0))\, (F'(x_0)^{-1} F(x_{k+1})) \| \\ &\leq \| F'(x_{k+1})^{-1} F'(x_0) \| \| F'(x_0)^{-1} F(x_{k+1}) \| \\ &\leq \frac{\frac{M_1}{2}\, (t_{k+1} - t_k)^2}{1 - M\, t_{k+1}} = t_{k+2} - t_{k+1}. \end{aligned}$$

Hence, we showed that (1.29) holds for all $k \geq 0$. Furthermore, let $z \in \overline{U}(x_{k+2}, t^\star - t_{k+2})$. Then, we have that

$$\| z - x_{k+1} \| \leq \| z - x_{k+2} \| + \| x_{k+2} - x_{k+1} \|$$
$$\leq t^\star - t_{k+2} + t_{k+2} - t_{k+1} = t^\star - t_{k+1}.$$

That is $z \in \overline{U}(x_{k+1}, t^\star - t_{k+1})$. The induction for (1.29) and (1.30) is now complete. Lemma 1.1.1 implies that sequence $\{t_n\}$ is a complete sequence. It follows from (1.29) and (1.30) that $\{x_n\}$ is also a complete sequence in a Banach space $\mathcal{X}$ and as such it converges to some $x^\star \in \overline{U}(x_0, t^\star)$ (since $\overline{U}(x_0, t^\star)$ is a closed set). By letting $k \longrightarrow \infty$ in (1.34) we get $F(x^\star) = 0$. Estimate (1.28) is obtained from (1.27) (see Refs. [106, 163, 290]) by using standard majorization techniques. The proof for the uniqueness part has been given in Ref. [140] (see also the proof of Theorem 1.1.2 for the uniqueness part). The proof of Theorem 1.1.1 is complete. $\qquad\square$

Remark 1.1.1.

(a) The limit point $t^\star$ can be replaced by $t^{\star\star}$ given in closed form by (1.16).
(b) It follows from Lemma 1.1.1 that the smaller H, K, L_0 and L are, the weaker sufficient convergence conditions for sequence $\{t_n\}$ and consequently Newton's method $\{x_n\}$ will be. In particular, we have that $C_3 \leq 1 \implies C_4 \leq 1$ since $H \leq L_0 \leq L$ but not necessarily vice versa unless if $H = L_0$.

Next, we present a result on a majorizing sequence for Newton's method under conditions $(\mathcal{B}_s)$, Lemma 1.1.1 and sequence $\{s_n\}$.

Lemma 1.1.2. *Let $h > 0$, $K_1 > 0$, $\ell_0 > 0$, $\ell > 0$ and $\eta > 0$. Suppose that (1.13) holds for $H = h$, $K = K_1$, $L = \ell/(1 + \ell_0\,\eta)$ and $L_0 = 2\,\ell_0/(1 + \ell_0\,\eta)$. Then, saclar sequence $\{s_n\}$ given by*

$$s_0 = 0, \quad s_1 = \eta, \quad s_2 = s_1 + \frac{h\,(s_1 - s_0)^2}{2\,(1 - K_1\,s_1)},$$
$$s_{n+2} = s_{n+1} + \frac{\ell\,(s_{n+1} - s_n)^2}{2\,(1 - \ell_0\,(2\,s_{n+1} - s_1))} \quad \text{for each} \quad n = 1, 2, \cdots,$$

is well defined, increasing, bounded from above by

$$s^{\star\star} = \eta + \left(1 + \frac{\overline{\alpha_0}}{1 - \alpha}\right) \frac{h\,\eta^2}{2\,(1 - K_1\,\eta)}$$

and converges to its unique least upper bound $s^\star$ which satisfies $s_2 \leq s^\star \leq s^{\star\star}$, where

$$\overline{\alpha_0} = \frac{L\,(s_2 - s_1)}{2\,(1 - L_0\,s_2)}.$$

Moreover, the following estimates hold:

$$0 < s_{n+2} - s_{n+1} \leq \overline{\alpha_0}\,\alpha^{n-1}\,\frac{h\,\eta^2}{2\,(1 - K_1\,\eta)} \quad \text{for each} \quad n = 1, 2, \cdots$$

and

$$s^\star - s_n \leq \frac{\overline{\alpha_0}\,(s_2 - s_1)}{1 - \alpha}\,\alpha^{n-2} \quad \text{for each} \quad n = 2, 3, \cdots.$$

Proof. Simply use Lemma 1.1.1 for $H = h$, $K = K_1$, $L = \ell/(1 + \ell_0\,\eta)$ and $L_0 = 2\,\ell_0/(1 + \ell_0\,\eta)$. The proof of Theorem 1.1.2 is complete. $\qquad\square$

Next, we present the following semi-local convergence result for Newton's method using majorizing sequence $\{s_n\}$ under the $(\mathcal{B}_s)$ conditions.

Theorem 1.1.2. *Suppose that hypotheses $(\mathcal{B}_s)$, hypotheses of Lemma 1.1.2 and $\overline{U}(x_0, t^\star) \subseteq \mathcal{D}$ hold, where $s^\star$ is given in Lemma 1.1.2. Then, sequence $\{x_n\}$ generated by Newton's method is well defined, remains in $\overline{U}(x_0, s^\star)$ and converges to a solution $x^\star \in \overline{U}(x_0, s^\star)$ of equation $F(x) = 0$. Moreover, the following estimates hold*

$$\| x_{n+1} - x_n \| \le s_{n+1} - s_n \ \text{ and } \ \| x_n - x^\star \| \le s^\star - s_n \quad \text{for each} \quad n = 0, 1, \cdots,$$

where sequence $\{s_n\}$ is given in Lemma 1.1.2. Furthermore, if there exists $\varsigma \ge s^\star$ such that

$$\overline{U}(x_0, \varsigma) \subseteq \mathcal{D} \quad \text{and} \quad \ell_0\,(s^\star + \varsigma) < 2,$$

then, the solution $x^\star$ of equation $F(x) = 0$ is unique in $\overline{U}(x_0, \varsigma)$.

Proof. Using $(\mathcal{B}_s)$ instead of $(\mathcal{A}_s)$ conditions, we arrive as in the proof of Theorem 1.1.1 at

$$\| F'(x_1)^{-1} A \| \le \frac{1}{1 - K_1\,s_1}$$

$$\| F'(x_{k+1})^{-1} A \| \le \frac{1}{1 - \ell_0\,(2\,s_{k+1} - s_1)} \quad \text{for each} \quad k = 1, 2, \cdots,$$

$$\| A^{-1} F(x_1) \| \le \frac{h}{2}\,(s_1 - s_0)^2,$$

$$\| A^{-1} F(x_{k+1}) \| \le \frac{\ell}{2}\,(s_{k+1} - s_k)^2 \quad \text{for each} \quad k = 1, 2, \cdots,$$

leading to

$$\begin{aligned}
\| x_{k+2} - x_{k+1} \| &= \| (F'(x_{k+1})^{-1} A)\,(A^{-1} F'(x_{k+1})^{-1}) \| \\
&\le \| F'(x_{k+1})^{-1} A \|\,\| A^{-1} F'(x_{k+1})^{-1} \| \\
&\le s_{k+2} - s_{k+1} \quad \text{for each} \quad k = 1, 2, \cdots.
\end{aligned}$$

The rest of the proof follows exactly as in Theorem 1.1.1 until the uniqueness part. To show the uniqueness of the solution, let $y^\star \in \overline{U}(x_0, \varsigma)$ and $F(y^\star) = 0$. Then, we have in turn the following using $(\mathcal{B}_{s3})$

$$\begin{aligned}
&\left\| A^{-1} \int_0^1 (F'(y^\star + t\,(x^\star - y^\star)) - A)\,dt \right\| \\
&\le \int_0^1 \ell_0\,(\| y^\star + t\,(x^\star - y^\star) - x_0 \| + \| y^\star + t\,(x^\star - y^\star) - x_1 \|)\,dt \\
&\le \ell_0 \int_0^1 \ell_0\,((1 - t)\,\| y^\star - x_0 \| + t\,\| x^\star - x_1 \| + (1 - t)\,\| y^\star - x_0 + x_0 - x_1 \| + \\
&\quad t\,\| x^\star - x_1 \|)\,dt \le \frac{\ell_0}{2}\,(\varsigma + s^\star + \varsigma + \eta + s^\star - \eta) < 1.
\end{aligned}$$

If we denote by $\mathbb{Q} = \int_0^1 F'(y^\star + t\,(x^\star - y^\star))\,dt$, then the inverse of $\mathbb{Q}$ exists. Hence, from the identity $0 = F(x^\star) - F(y^\star) = \mathbb{Q}(x^\star - y^\star)$. We deduce that $x^\star = y^\star$. The proof of Theorem 1.1.2 is complete. $\qquad\square$

We also present useful and obvious generalizations of Lemma 1.1.1 and Lemma 1.1.2, respectively.

Lemma 1.1.3. *Let $H > 0$, $K > 0$, $L > 0$, $L_0 > 0$ and $\eta > 0$. Suppose $L \geq L_0$ and there exists a minimum integer $N > 1$ such that iterates t_i $(i = 0, 1, \cdots, N - 1)$ given by (1.15) are well defined,*

$$t_i < t_{i+1} < \frac{1}{L_0} \quad \text{for each} \quad i = 0, 1, \cdots, N - 2 \tag{1.35}$$

and

$$t_N \leq \frac{1}{L_0}\left(1 - (1 - L_0\,t_{N-1})\,\alpha\right). \tag{1.36}$$

Then, the following assertions hold

$$L_0\,t_N < 1, \tag{1.37}$$

$$t_{N+1} \leq \frac{1}{L_0}\left(1 - (1 - L_0\,t_N)\,\alpha\right), \tag{1.38}$$

$$\alpha_{N-1} \leq \alpha \leq 1 - \frac{L_0\,(t_{N+1} - t_N)}{1 - L_0\,t_N}, \tag{1.39}$$

sequence $\{t_n\}$ given by (1.15) is well defined, increasing, bounded from above by

$$t^{\star\star} = t_{N-1} + \left(1 + \frac{\alpha_{N-2}}{1 - \alpha}\right)(t_N - t_{N-1}) \leq t_{N-1} + \frac{t_N - t_{N-1}}{1 - \alpha}$$

and converges to its unique least upper bound $t^\star$ which satisfies $t_{N-1} \leq t^\star \leq t^{\star\star}$, where α_n is given by

$$\alpha_n = \frac{L\,(t_{n+2} - t_{n+1})}{2\,(1 - L_0\,t_{n+2})}.$$

Moreover, the following estimates hold:

$$0 < t_{N+n} - t_{N+n-1} \leq \alpha_{N-2}\,\alpha^{n-2}\,(t_{N+1} - t_N) \leq \alpha^{n-1}\,(t_{N+1} - t_N)$$
$$\text{for each} \quad n = 1, 2, \cdots.$$

Lemma 1.1.4. *Let $h > 0$, $K_1 > 0$, $\ell > 0$, $\ell_0 > 0$ and $\eta > 0$. Set $H = h$, $K = K_1$, $L = \ell/(1 + \ell_0\,\eta)$ and $L_0 = 2\,\ell_0/(1 + \ell_0\,\eta)$. Suppose $\ell \geq \ell_0$ and there exists a minimum integer $N > 1$ such that iterates s_i $(i = 0, 1, \cdots, N - 1)$ given in Lemma 1.1.2 are well defined,*

$$s_i < s_{i+1} < \frac{1}{L_0} \quad \text{for each} \quad i = 0, 1, \cdots, N - 2$$

and

$$s_N \leq \frac{1}{L_0} \left(1 - (1 - L_0\, s_{N-1})\, \alpha\right).$$

Then, the following assertions hold

$$\ell_0\, s_N < 1, \quad s_{N+1} \leq \frac{1}{L_0} \left(1 - (1 - L_0\, s_N)\, \alpha\right), \quad \alpha_{N-1} \leq \alpha \leq 1 - \frac{L_0\, (s_{N+1} - s_N)}{1 - L_0\, s_N},$$

sequence $\{s_n\}$ *given in Lemma 1.1.2 is well defined, increasing, bounded from above by*

$$s^{\star\star} = s_{N-1} + \left(1 + \frac{\alpha_{N-2}}{1 - \alpha}\right) (s_N - s_{N-1}) \leq s_{N-1} + \frac{s_N - s_{N-1}}{1 - \alpha}$$

and converges to its unique least upper bound $s^\star$ *which satisfies* $s_{N-1} \leq s^\star \leq s^{\star\star}$, *where* α_n *is given by*

$$\alpha_n = \frac{L\, (s_{n+2} - s_{n+1})}{2\, (1 - L_0\, s_{n+2})}.$$

Moreover, the following estimates hold:

$$0 < s_{N+n} - s_{N+n-1} \leq \alpha_{N-2}\, \alpha^{n-2}\, (s_{N+1} - s_N) \leq \alpha^{n-1}\, (s_{N+1} - s_N)$$
$$\text{for each} \quad n = 1, 2, \cdots.$$

Remark 1.1.2. If $N = 2$, Lemmas 1.1.3, 1.1.4 reduce to Lemmas 1.1.1 and 1.1.2, respectively. Clearly, Lemma 1.1.3 can replace Lemma 1.1.1 in Theorem 1.1.1 and Lemma 1.1.4 can replace Lemma 1.1.2 in Theorem 1.1.2.

Remark 1.1.3. Concerning the majorizing sequences under (1.3) or (1.4) or (1.5) or (1.6) or (1.13) conditions, we note that the Kantorovich majorizing sequence $\{w_n\}$ (using (1.3)) (see Refs. [290, 338, 342, 343, 356, 359, 369, 442–444]) is defined by

$$w_0 = 0, \ w_1 = \eta,$$
$$w_{n+2} = w_{n+1} + \frac{L\, (w_{n+1} - w_n)}{2\, (1 - L\, w_{n+1})} \quad \text{for each} \quad n = 0, 1, 2, \cdots.$$

The majorizing sequence $\{v_n\}$ (using (1.4) or (1.5)) (see Ref. [106]) is given by

$$v_0 = 0, \ v_1 = \eta,$$
$$v_{n+2} = v_{n+1} + \frac{L\, (v_{n+1} - v_n)}{2\, (1 - L_0\, v_{n+1})} \quad \text{for each} \quad n = 0, 1, 2, \cdots,$$

whereas the majorizing sequence $\{u_n\}$ (using (1.6)) (see Ref. [106]) is given by

$$u_0 = 0, \ u_1 = \eta, \ u_2 = u_1 + \frac{L_0\, (u_1 - u_0)^2}{2\, (1 - L_0\, u_1)},$$
$$u_{n+2} = u_{n+1} + \frac{L\, (u_{n+1} - u_n)}{2\, (1 - L_0\, u_{n+1})} \quad \text{for each} \quad n = 1, 2, 3, \cdots.$$

According to Lemma 1.1.1, a simple inductive argument shows that for $L_0 < L$

$$t_n < u_n < v_n < w_n \quad \text{for each} \quad n = 2, 3, \cdots,$$

$$t_{n+1} - t_n < u_{n+1} - u_n < v_{n+1} - v_n < w_{n+1} - w_n \quad \text{for each} \quad n = 1, 2, \cdots$$

and

$$t^\star \le u^\star = \lim_{n \to \infty} u_n \le v^\star = \lim_{n \to \infty} v_n \le w^\star = \lim_{n \to \infty} w_n.$$

These advantages hold with $\{s_n\}$ replacing $\{t_n\}$ provided that for $H = h$, $K = K_1$, $L = \ell/(1 + \ell_0\,\eta)$ and $L_0 = 2\,\ell_0/(1 + \ell_0\,\eta)$, $L_0 < L$ or $H < L_0$ or $K < L_0$. These estimations justify the claims made at the introduction of this section.

If we simply employ identity

$$x_{n+1} - x^\star = -(F'(x_n)^{-1}\,A)\,A^{-1} \int_0^1 (F'(x^\star + t\,(x_n - x^\star)) - F'(x_n))\,(x_n - x^\star)\,dt$$

instead of

$$x_{n+1} - x^\star = -(F'(x_n)^{-1}\,F'(x_n))\,F'(x_n)^{-1} \int_0^1 (F'(x^\star + t\,(x_n - x^\star)) - F'(x_n))\,(x_n - x^\star)\,dt$$

used in Ref. [100] and $(\mathcal{B}_l)$ instead of $(\mathcal{A}_l)$ hypotheses, we arrive at the following local convergence result for Newton's method.

Theorem 1.1.3. *Under the $(\mathcal{B}_l)$ hypotheses further assume that $\overline{U}(x^\star, R_2) \subseteq \mathcal{D}$ or $\overline{U}(x^\star, R_3) \subseteq \mathcal{D}$, where*

$$R_3 = \frac{2\,(1 - \ell_0 \parallel x_0 - x^\star \parallel)}{\ell + 4\,\ell_0} \quad \text{and} \quad R_2 = \frac{2}{\ell + 6\,\ell_0}.$$

Then, if $x_0 \in U(x^\star, R_2)$ or $U(x^\star, R_3)$, sequence $\{x_n\}$ generated by Newton's method is well defined, remains in $\overline{U}(x^\star, R_2)$ or $\overline{U}(x^\star, R_3)$ for each $n = 0, 1, \cdots$ and converges to $x^\star$. Moreover, the following estimates hold

$$\parallel x_{n+1} - x^\star \parallel \le \frac{\ell \parallel x_n - x^\star \parallel^2}{2\,(1 - \ell_0\,(\parallel x_n - x_0 \parallel + \parallel x_n - x^\star \parallel))}.$$

Remark 1.1.4. The error estimates in Ref. [418] using only $(\mathcal{A}_{l1})$ and $(\mathcal{A}_{l2})$ are given by

$$\parallel x_{n+1} - x^\star \parallel \le \frac{L \parallel x_n - x^\star \parallel^2}{2\,(1 - L \parallel x_n - x_0 \parallel)},$$

whereas the ones by us in Ref. [100] using $(\mathcal{A}_{l1})$–$(\mathcal{A}_{l3})$ are given by

$$\parallel x_{n+1} - x^\star \parallel \le \frac{L \parallel x_n - x^\star \parallel^2}{2\,(1 - L_0 \parallel x_n - x_0 \parallel)}.$$

Clearly, the latter estimate are tighter than the former if $L_0 < L$. Note also that if

$$\frac{\ell}{1 - 2\,\ell_0\,R_2} < L \quad \text{or} \quad \frac{\ell}{1 - 2\,\ell_0\,R_3} < L$$

and

$$\frac{\ell}{1 - 2\,\ell_0\,R_2} < L_0 \quad \text{or} \quad \frac{\ell}{1 - 2\,\ell_0\,R_3} < L_0,$$

then, the new estimates are tighter. We have provided a numerical example (Example 1.1.1 (local case)), where our results compare favorably to Refs. [100, 383, 384, 418, 420].

Recently, in Ref. [161] we presented a new technique of finding bounds for limit points in closed form. In this section on the one hand we present a new way for generating sufficient convergence conditions for Newton's method and on the other hand we develop a new technique for finding upper and lower bounds of limit points. In order for us to acheive these goals we first simplify sequence $\{t_n\}$. Let $H = L$, $K = L_0$, $b = L/L_0$ and $r_n = L_0 t_n$. Then, we can replace $\{t_n\}$ by sequence $\{r_n\}$ defined by

$$r_0 = 0, \ r_1 = L_0 \eta, \quad r_{n+1} = r_n + \frac{b\,(r_n - r_{n-1})^2}{2\,(1 - r_n)} \quad \text{for each} \quad n = 1, 2, \cdots . \quad (1.40)$$

Moreover, define sequence $\{\beta_n\}$ by setting $\beta_n = 1 - r_n$ for each $n = 0, 1, 2, \cdots$. That is sequence $\{\beta_n\}$ is defined by

$$\beta_0 = 1, \ \beta_1 = 1 - L_0 \eta, \quad \beta_{n+1} = \beta_n - \frac{b\,(\beta_n - \beta_{n-1})^2}{2\,\beta_n} \quad \text{for each} \quad n = 1, 2, \cdots .$$
$$(1.41)$$

Furthermore, let $\gamma_n = 1 - (\beta_n/\beta_{n-1})$. Then, sequence $\{\gamma_n\}$ is defined by

$$\gamma_1 = L_0 \eta, \ \gamma_{n+1} = \frac{b}{2} \left(\frac{\gamma_n}{1 - \gamma_n} \right)^2 \quad \text{for each} \quad n = 1, 2, \cdots . \quad (1.42)$$

Next, we study the convergence of sequence $\{r_n\}$.

Lemma 1.1.5. *Suppose that*

$$\delta_1 := L_0 \eta \leq a := \frac{4}{4 + b + (8\,b + b^2)^{1/2}}. \quad (1.43)$$

Then, the following assertions hold

(i) If $\delta_1 = a$, then $\gamma_n = \delta_1$ and if $\delta_1 < a$, then sequence $\{\gamma_n\}$ decreasingly converges to zero.

(ii) If $\delta_1 = a$, then $r_n = 1 - (1 - \delta_1)^n$, sequence $\{r_n\}$ is increasing and $\lim\limits_{n \to \infty} r_n = 1$. If $\delta_1 < a$, then sequence $\{r_n\}$ is increasing and converges to its unique least upper bound $r^\star$ which satisfies $r^\star \in (0, 1)$.

(iii) If $\delta_1 < a$, then for $\gamma_1 = L_0 \eta$ and $\gamma_2 = \dfrac{b}{2} \left(\dfrac{L_0 \eta}{1 - L_0 \eta} \right)^2$, we have the following bounds for $r^\star$:

$$1 - \exp\left(-2 \left(\frac{\gamma_1}{2 - \gamma_1} + \frac{\gamma_2}{2 - \gamma_2} \right) \right) \leq r^\star \leq 1 - \exp \frac{-\gamma_1}{(1 - \gamma_1)\,(1 - q_1\,\gamma_1)}, \quad (1.44)$$

where

$$q_1 = \frac{b}{2\,(1 - \gamma_1)^2}.$$

Proof.

(i) We have by the definition of sequence $\{\gamma_n\}$ that $\gamma_n > 0$ for each $n = 1, 2, \cdots$. It also follows from (1.42) and (1.43) that $\gamma_2 \leq \gamma_1 \leq a$. Let us assume that $\gamma_k \leq \gamma_{k-1} \leq a$ for all $k \leq n$. We must show that $\gamma_{k+1} \leq \gamma_k$ or equivalently

$$\frac{b}{2}\left(\frac{\gamma_k}{1 - \gamma_k}\right)^2 \leq \gamma_k \quad \text{or} \quad \gamma_k \leq a,$$

which is true by the induction hypothesis. Hence, sequence $\{\gamma_n\}$ is nondecreasing, bounded below by 0 and as such it converges to its unique maximum lower bound $\gamma^\star$. Then, by letting $n \to \infty$ in (1.42), we get that $\gamma^\star = 0$ or $\gamma^\star = a$. If $\delta_1 = a$, then $\gamma^\star = a = \gamma_n$ for each $n = 0, 1, 2, \cdots$. However, if $\delta_1 < a$, then $\gamma^\star = 0$, since if $a = \gamma^\star \leq \gamma_k < \gamma_{k-1} < \cdots < \gamma_2 < \gamma_1 < a$, leads to a contradiction. We need a more precise relationship between two consecutive iterates γ_k in the case $\gamma^\star = 0$ and $\delta_1 < a$. We have that

$$q_1\,\gamma_1 = \frac{b}{2}\,\frac{\gamma_1}{(1 - \gamma_1)^2}\,\frac{\gamma_1}{\gamma_1} = \frac{b}{2}\left(\frac{\gamma_1}{1 - \gamma_1}\right)^2\frac{1}{\gamma_1} < \frac{\gamma_1}{\gamma_1} = 1,$$

since $\dfrac{b}{2}\left(\dfrac{\gamma_1}{1 - \gamma_1}\right)^2 < \gamma_1$. We also have that $0 \leq \gamma_2 < \gamma_1$. Let us assume that $0 \leq \gamma_k < \gamma_{k-1} < \cdots < \gamma_2 < \gamma_1$. Using (1.42) and $q_1\,\gamma_1 < 1$, we get that $\gamma_{k+1} < q_1\,\gamma_k^2 < q_1\,\gamma_1\,\gamma_k < \gamma_k$. Hence, $\{\gamma_k\}$ is a decreasing sequence that satisfies

$$\gamma_{k+1} < q_1\,\gamma_k^2 < q_1^{1+2}\,\gamma_{k-1}^{2^2} < \cdots < \frac{1}{q_1}\,(q_1\,\gamma_1)^{2^k}. \tag{1.45}$$

(ii) If $\delta_1 = a$, then $\beta_n = (1 - \delta_1)^n$ and $r_n = 1 - (1 - \delta_1)^n$. Hence, since $\delta_1 = a$ then $0 < \delta_1 < 1$, we conclude that sequence $\{r_n\}$ is increasing and $\lim_{n \to \infty} r_n = 1$. If $\delta_1 < a$, we use the following equalities

$$\beta_n = (1 - \gamma_n)\,\beta_{n-1} = (1 - \gamma_n)\,(1 - \gamma_{n-1})\,\beta_{n-2}$$
$$= \cdots = (1 - \gamma_n)\cdots(1 - \gamma_1)\,\beta_0 = (1 - \gamma_n)\cdots(1 - \gamma_1).$$

Hence, we conclude that sequence $\{r_n\}$ converges to a limit $r^\star$. In fact, $\beta_n > (1 - \delta_1)^n > 0$. Then, $r_n = 1 - \beta_n < 1$ and sequence $\{r_n\}$ is bounded above by 1. As $1 - r_n > 0$ for each $n = 0, 1, 2, \cdots$ by (1.42), we deduce that $\{r_n\}$ is an increasing sequence. Consequently, $\{r_n\}$ converges to a limit $r^\star$.

(iii) Let us denote $\lim_{n \to \infty} \beta_n = \beta^\star = 1 - r^\star$. Then, we get in turn that

$$\beta^\star = \prod_{n=1}^{\infty}(1 - \gamma_n) \iff \ln\frac{1}{\beta^\star} = \sum_{n \geq 1}\ln\frac{1}{1 - \gamma_n}.$$

Using the following bounds of $\ln t$ for $t > 1$:

$$2\,\frac{t - 1}{t + 1} \leq \ln t \leq \frac{t^2 - 1}{2\,t},$$

we obtain that

$$\ln\frac{1}{\beta^\star} \leq \sum_{n \geq 1}\ln\frac{\gamma_n\,(2 - \gamma_n)}{2\,(1 - \gamma_n)} \leq \frac{1}{1 - \gamma_1}\sum_{n \geq 0}\gamma_{n+1}.$$

Using (1.45) and the inequality $2^n \geq n+1$ for each $n = 0, 1, \cdots$, we deduce that

$$\ln \frac{1}{\beta^\star} \leq \frac{1}{q_1 \left(1 - \gamma_1\right)} \sum_{n \geq 0} \left(q_1 \, \gamma_1\right)^{2^n} \leq \frac{1}{q_1 \left(1 - \gamma_1\right)} \sum_{n \geq 1} \left(q_1 \, \gamma_1\right)^n = \frac{\gamma_1}{\left(1 - \gamma_1\right) \left(1 - q_1 \, \gamma_1\right)}.$$

Since $r^\star = 1 - \beta^\star$, we have the upper bound in (1.44). To obtain the lower bound for $\ln(1/\beta^\star)$, we use the following inequalities

$$\ln \frac{1}{\beta^\star} \geq 2 \sum_{n \geq 1} \frac{\gamma_n}{2 - \gamma_n} > \frac{2 \gamma_1}{2 - \gamma_1} + \frac{2 \gamma_2}{2 - \gamma_2}.$$

Then, we deduce the lower bound given in (1.44). The proof of Lemma 1.1.5 is complete.
$\square$

Then, we have the following result on the convergence of sequence $\{t_n\}$.

Corollary 1.1.1. *Suppose that (1.43) holds. Then, sequence $\{t_n\}$ is increasing and converges to its unique least upper bound $t^\star = r^\star/L_0$, where $r^\star$ is defined in Lemma 1.1.5.*

Next, we present obvious and useful extentions respectively, of Lemma 1.1.5 and Corollary 1.1.1.

Lemma 1.1.6. *Suppose that there exists an integer $N \geq 1$ such that*

$$t_1 < t_2 < \cdots < t_N < \frac{1}{L_0} \tag{1.46}$$

and

$$\delta_N = L_0 \left(t_N - t_{N-1}\right) \leq a, \tag{1.47}$$

where a is given in (1.43). Then, the following assertions hold

(i) If $\delta_N = a$, then $\gamma_m = \delta_N$ for each $m = N, N+1, \cdots$ and if $\delta_N < a$, then sequence $\{\gamma_n\}$ decreasingly converges to zero.

(ii) If $\delta_N = a$, then $r_m = 1 - (1 - \delta_N)^m$ for each $m = N, N+1, \cdots$, sequence $\{r_n\}$ is increasing and $\lim\limits_{n \to \infty} r_n = 1$. if $\delta_N < a$, then sequence $\{r_n\}$ is increasing and converges to its unique least upper bound $r^\star$ which satisfies $r^\star \in (0, 1)$.

(iii) If $\delta_N < a$, then for $\gamma_N = L_0 \left(t_N - t_{N-1}\right)$ and $\gamma_{N+1} = \dfrac{b}{2} \left(\dfrac{\gamma_N}{1 - \gamma_N}\right)^2$, we have the following bounds for $r^\star$:

$$1 - \exp\left(2 \left(\frac{\gamma_N}{2 - \gamma_N} + \frac{\gamma_{N+1}}{2 - \gamma_{N+1}}\right)\right) \leq r^\star \leq 1 - \exp \frac{-\gamma_N}{\left(1 - \gamma_N\right) \left(1 - q_N \, \gamma_N\right)},$$

where

$$q_N = \frac{b}{2 \left(1 - \gamma_N\right)^2}.$$

Corollary 1.1.2. *Suppose that (1.46) and (1.47) hold. Then, sequence $\{t_n\}$ is increasing and converges to its unique least upper bound $t^\star = r^\star/L_0$.*

Remark 1.1.5. If $N = 1$ and $L_0 = L$, then $b = 1$ and (1.47) reduces to (1.3). But if $N = 1$, $H = L$ and $K = L_0$, then (1.47) reduces to (1.5). Moreover, if $N = 2$, $H = L_0$ and $K = L_0$, (1.47) reduces to (1.6). Furthermore, if $N = 2$, then (1.47) reduces to (1.13). Finally, if $N > 2$, (1.47) can be improved even further to obtain even weaker sufficient convergence conditions. Clearly, Lemma 1.1.6 can replace Lemma 1.1.1 in Theorem 1.1.1. Similar observations can follow for majorizing sequence $\{s_n\}$.

We finish this section by some examples validating the theoretical results.

Example 1.1.1. Semi-local case. Let $\mathcal{X} = \mathcal{Y} = \mathbb{R}$, $x_0 = 1$ and $\mathcal{D} = \overline{U}(1, 1 - a)$ for $a \in (0, .5)$. Define function F on $\mathcal{D}$ by

$$F(x) = x^3 - a. \tag{1.48}$$

Then, using (1.48), $(\mathcal{A}_{s1})$ and $(\mathcal{A}_{s2})$, we get that $\eta = (1 - a)/3$ and $L = 2\,(2 - a)$. The Newton–Kantorovich condition (1.3) is violated, since $C_\star = 4\,(1-a)\,(2-a)/3 > 1$ for each $a \in (0, .5)$. Hence, there is no guarantee under the Kantorovich theorem that sequence $\{x_n\}$ converges to $x^\star$. Using $(\mathcal{A}_{s3})$, we get that $L_0 = 3 - a$. Our hypothesis (1.6) holds for $a = [.4271907643, .5)$. Hence, (1.6) is violated, say for $a = .427$. However, hypotheses of Lemma 1.1.1 are satisfied, since

$$H = 2.191, \quad K = 1.809, \quad \alpha = .533840671, \quad L_0 = 2.573,$$

$$L = 3.146, \quad \eta = .191, \quad x_1 = .809, \quad x^\star = \sqrt[3]{.427} = .753024821,$$

$$P_1(t) = 2.972711379\,t^2 + 9.357159281\,t - 2.135362684$$

and

$$\eta_1 = .213698201 > \eta.$$

Hypotheses of Lemma 1.1.2 are also satisfied, since we have that for $A = (F(x_1) - F(x_0))/(x_1 - x_0)$,

$$|A| = |x_1^2 + x_1\,x_0 + x_0^2| = 2.463481, \quad 1/|A| = .405929658,$$

$$\ell = 6\,(2 - a)/|A| = 3.831164112, \quad \ell_0 = 4.146/|A| = 1.682984362$$

$$|3\,x^2 - A| = |(2\,x + x_0)\,(x - x_0) + (x + x_0 + x_1)\,(x - x_1)|$$
$$\leq (2\,(1 - a) + 3)\,|x - x_0| + (3 - a + x_1)\,|x - x_1|,$$

$$|(F'(x_0 + \theta\,(x_1 - x_0)) - F'(x_0))/A| = 3\,|2\,x_0 + \theta\,(x_1 - x_0)|/|A|$$
$$= 2.668175642 = h = K,$$

$$K_1 = K = |2\,x_1 + x_0|/|A| = 1.062723845,$$

$$\alpha = .521719197, \quad L_0 = 2.547178244, \quad L = 2.899212285,$$

$$P_2(t) = 9.178082746\, t^2 + 4.136947794\, t - 1.146$$

and condition (1.13) becomes

$$\eta_1 = .193740868 > \eta.$$

Hence, hypotheses of Theorem 1.1.1 and Theorem 1.1.2 are satisfied.

Local case. Let $\mathcal{D} = \overline{U}(x^\star, .5)$, $x_0 = .7$, where $x^\star = .753024821$. Then, using (1.48), the $(\mathcal{A}_l)$ and $(\mathcal{B}_l)$ hypotheses with $A = (F(x^\star) - F(x_0))/(x^\star - x_0)$, we have that

$$A = 1.584163756, \quad 1/A = .631247872,$$

$$3\, x^2 - A = (2\, x + x_0)\, (x - x_0) + (x + x^\star + x_0)\, (x - x^\star),$$

$$|2\, x + x_0| = |2\, (x - x^\star) + 2\, x^\star + x_0| \le 3.206049642,$$

$$|x + x^\star + x_0| = |(x - x^\star) + 2\, x^\star + x_0| \le 2.706049642.$$

We also have that

$$\ell_0 = .170818807, \quad \ell = 4.745815511,$$

$$L_0 = 3.537717036 \quad and \quad L = 4.419479122.$$

Therefore, we obtain that

$$R_0 = .150847339 < R_1 = .173990005 < R_2 = .346576702 < R_3 = .365049107.$$

Hence, the new convergence radii R_2 and R_3 given in Theorem 1.1.3 are larger than R_0 and R_1.

Example 1.1.2. Let $\mathcal{C}[0, 1]$ stand for the space of continuous functions defined on interval $[0, 1]$ and be equipped with the max-norm. Let also $\mathcal{X} = \mathcal{Y} = \mathcal{C}[0, 1]$ and $\mathcal{D} = U(0, r)$ for some $r > 1$. Define F on $\mathcal{D}$ by

$$F(x)(s) = x(s) - y(s) - \mu \int_0^1 \mathcal{G}(s, t)\, x^3(t)\, dt, \quad x \in \mathcal{C}[0, 1], \ s \in [0, 1].$$

$y \in \mathcal{C}[0, 1]$ is given, μ is a real parameter and the Kernel G is the Green's function defined by

$$\mathcal{G}(s, t) = \begin{cases} (1 - s)\, t \ if \ t \le s \\ s\, (1 - t) \ if \ s \le t. \end{cases}$$

Then, the Fréchet-derivative of F is defined by

$$(F'(x)\, (w))(s) = w(s) - 3\, \mu \int_0^1 \mathcal{G}(s, t)\, x^2(t)\, y(t)\, dt, \quad w \in \mathcal{C}[0, 1], \ s \in [0, 1].$$

Table 1.1   Comparison table of conditions (1.3), (1.4), (1.5) and (1.6)

r	$\mu^\star$	$C_\star$	C_1	C_2	C_3
2.09899	.9976613778	1.007515200	.9837576000	.9757536188	.9639223786
2.19897	.9831766058	1.055505600	1.007752800	.9915015816	.9678118280
2.29597	.9698185659	1.102065600	1.031032800	1.006635036	.9715205068
3.095467	.8796313211	1.485824160	1.222912080	1.127023800	1.000082409

Let us choose $x_0(s) = y(s) = 1$ and $|\mu| < 8/3$. Then, we have that

$$\| \mathcal{I} - F'(x_0) \| < \frac{3}{8}\,|\mu|, \quad F'(x_0)^{-1} \in \mathcal{L}(\mathcal{Y}, \mathcal{X}),$$

$$\| F'(x_0)^{-1} \| \le \frac{8}{8 - 3\,|\mu|}, \quad \eta = \frac{|\mu|}{8 - 3\,|\mu|}, \quad L_0 = \frac{12\,|\mu|}{8 - 3\,|\mu|},$$

$$L = \frac{6\,r\,|\mu|}{8 - 3\,|\mu|} \quad \text{and} \quad C_\star = \frac{12\,r\,|\mu|^2}{(8 - 3\,|\mu|)^2}.$$

Denote by $\mu^\star$ the positive root of equation $3\,(4\,r - 3)\,t^2 + 48\,t - 64 = 0$. Notice that if $\mu > \mu^\star$, then $C_\star > 1$. Hence the Newton–Kantorovich condition is not satisfied. Let us choose for example $r = 3$. Then, we obtain $\mu^\star = .888889$. In Table 1.1, we pick some values of r for $\mu = 1$, so we give the corresponding values of $\mu^\star$ and we compare the C conditions. We have chosen $H = K = L_0$ in this example. Therefore, for all values of r in Table 1.1, the C_4 conditions given by (1.13) coincide with the C_3 conditions (see the last column of Table 1.1). Hence, Table 1.1 shows that our conditions are always better than the Newton–Kantorovich conditions $C_\star$ (see the third column of Table 1.1).

Next we provide two more examples where $L_0 < L$.

Example 1.1.3. Let $x_0 = 0$. Define the scalar function F by $F(x) = d_0\,x + d_1 + d_2 \sin e^{d_3\,x}$, where d_i, $i = 0, 1, 2, 3$ are given parameters. It can easily be seen that for d_3 large and d_2 sufficiently small, $\dfrac{L_0}{L}$ can be arbitrarily small.

Example 1.1.4. Let $\mathcal{X}$ and $\mathcal{Y}$ as in Example 1.1.2. Consider the following nonlinear boundary value problem (see Ref. [120])

$$\begin{cases} u'' = -u^3 - \gamma\,u^2 \\ u(0) = 0, \quad u(1) = 1. \end{cases}$$

It is well known that this problem can be formulated as the integral equation

$$u(s) = s + \int_0^1 \mathcal{T}(s, t)\,(u^3(t) + \gamma\,u^2(t))\,dt, \tag{1.49}$$

where $\mathcal{T}$ as $\mathcal{G}$ in Example 1.1.2. We observe that

$$\max_{0 \le s \le 1} \int_0^1 |\mathcal{T}(s, t)|\,dt = \frac{1}{8}.$$

Then, problem (1.49) is in the form (1.1), where $F : \mathcal{D} \longrightarrow \mathcal{Y}$ is defined by

$$[F(x)](s) = x(s) - s - \int_0^1 \mathcal{T}(s,t)\,(x^3(t) + \gamma\,x^2(t))\,dt.$$

Set $u_0(s) = s$ and $\mathcal{D} = U(u_0, R_0)$. It is easy to verify that $U(u_0, R_0) \subset U(0, R_0+1)$, since $\| u_0 \| = 1$. If $2\,\gamma < 5$, the operator F' satisfies conditions of Theorem 1.1.1, with

$$\eta = \frac{1+\gamma}{5-2\,\gamma}, \quad L = \frac{\gamma + 6\,R_0 + 3}{4\,(5-2\,\gamma)}, \quad L_0 = \frac{2\,\gamma + 3\,R_0 + 6}{8\,(5-2\,\gamma)}.$$

Note that $L_0 < L$ (see also Fig. 1.1).

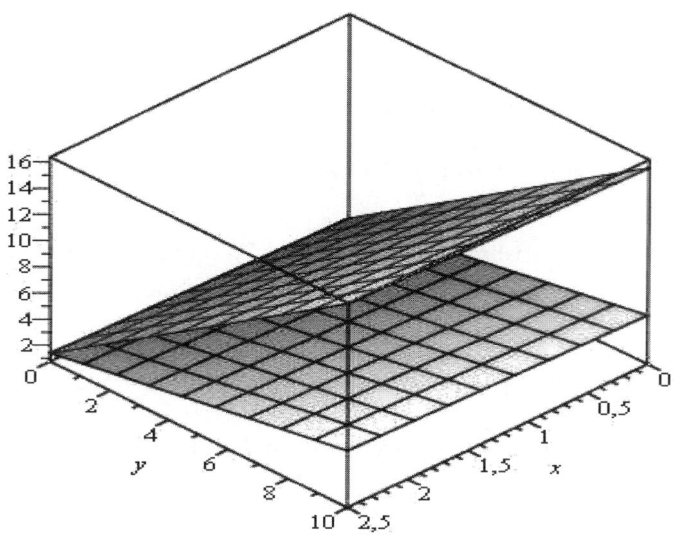

Fig. 1.1   Functions L_0 and L in 3d with respect to (γ, R_0) in $(0, 2.5) \times (0, 10)$. L is also above L_0.

Example 1.1.5. Let $\mathcal{X}$ and $\mathcal{Y}$ as in Example 1.1.2. Let $\zeta \in [0,1]$ be a given parameter. Consider the cubic integral equation

$$u(s) = u^3(s) + \lambda\,u(s) \int_0^1 \mathbb{K}(s,t)\,u(t)\,dt + y(s) - \zeta. \tag{1.50}$$

Nonlinear integral equations of the form (1.50) are considered Chandrasekhar-type equations (see Refs. [106, 120]) and they arise in the theories of radiative transfer, neutron transport and in the kinetic theory of gases. Here, the kernel $\mathbb{K}(s,t)$ is a continuous function of two variables $(s,t) \in [0,1] \times [0,1]$ satisfying

Table 1.2   Comparison table of conditions (1.3), (1.4), (1.5) and (1.6)

λ	ζ	$C_\star$	C_1	C_2	C_3
.97548	.954585	.4895734208	.4851994045	.4837355633	4815518345
.8457858	.999987	.4177974405	.4177963046	.4177959260	.4177953579
.3245894	.815456854	.5156159025	.4967293568	.4903278739	.4809439506
.3569994	.8198589998	.5204140737	.5018519741	.4955632842	.4863389899
.3789994	.8198589998	.5281518448	.5093892893	.5030331107	.4937089648
.458785	.5489756	1.033941504	.9590659445	.9332478337	.8962891928

(i) $0 < \mathbb{K}(s,t) < 1$,

(ii) $\mathbb{K}(s,t) + \mathbb{K}(t,s) = 1$.

The parameter λ is a real number called the albedo for scattering; $y(s)$ is a given continuous function defined on $[0,1]$ and $x(s)$ is the unknown function sought in $\mathcal{C}[0,1]$. For simplicity, we choose

$$u_0(s) = y(s) = 1 \quad \text{and} \quad \mathbb{K}(s,t) = \frac{s}{s+t} \quad \text{for all} \quad (s,t) \in [0,1] \times [0,1] \quad (s+t \neq 0).$$

Let $\mathcal{D} = U(u_0, 1 - \zeta)$ and define the operator F on $\mathcal{D}$ by

$$F(x)(s) = x^3(s) - x(s) + \lambda\, x(s) \int_0^1 \mathbb{K}(s,t)\, x(t)\, dt + y(s) - \zeta \quad \text{for all} \quad s \in [0,1].$$
$$(1.51)$$

Then every zero of F satisfies equation (1.50). We obtain using (1.51) and Ref. [106] that

$$[F'(x)\, v]\,(s) = \lambda\, x(s) \int_0^1 \mathbb{K}(s,t)\, v(t)\, dt + \lambda\, v(s) \int_0^1 \mathbb{K}(s,t)\, x(t)\, dt +$$
$$3\, x^2(s)\, v(s) - I(v(s)).$$

Therefore, the operator F' satisfies conditions of Theorem 1.1.1, with

$$\eta = \frac{|\lambda|\, \ln 2 + 1 - \zeta}{2\,(1 + |\lambda|\, \ln 2)}, \quad L = \frac{|\lambda|\, \ln 2 + 3\,(2 - \zeta)}{1 + |\lambda|\, \ln 2} \quad \text{and} \quad L_0 = \frac{2\,|\lambda|\, \ln 2 + 3\,(3 - \zeta)}{2\,(1 + |\lambda|\, \ln 2)}.$$

It follows from our main results that if one condition of our h conditions holds, then problem (1.50) has a unique solution near u_0. This assumption is weaker than the one given before using the Newton–Kantorovich hypothesis. Note also that $L_0 < L$ for all $\zeta \in [0,1]$ (see also Fig. 1.2).

Next, we pick some values of λ and ζ such that all hypotheses are satisfied, so we can compare the C conditions (see Table 1.2).

Finally, we provide two examples in the local case where $L_0 < L$.

Example 1.1.6. Let $\mathcal{X} = \mathcal{Y} = \mathbb{R}^3$, $\mathcal{D} = \overline{U}(0,1)$ and $x^\star = (0,0,0)$. Define function F on $\mathcal{D}$ for $w = (x,y,z)$ by

$$F(w) = \left(e^x - 1, \frac{e-1}{2}\, y^2 + y, z\right). \tag{1.52}$$

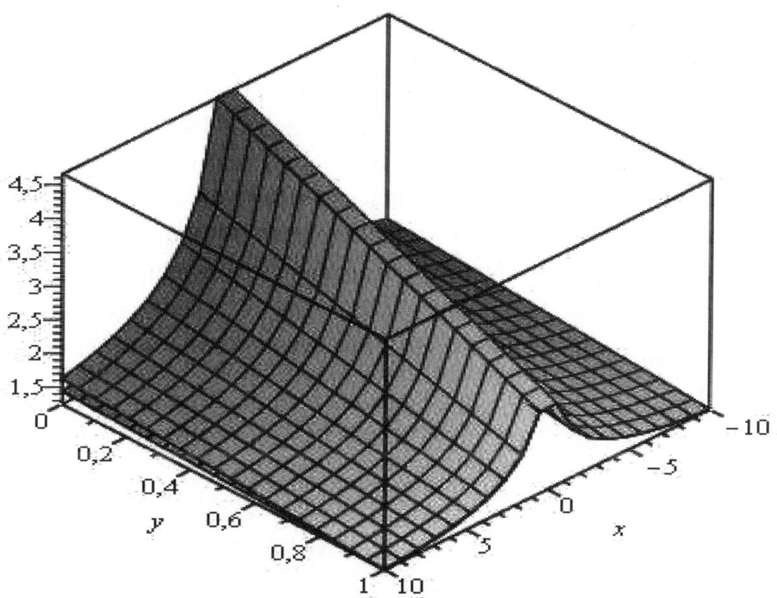

Fig. 1.2   Functions L_0 and L in 3d with respect to (λ, ζ) in $(-10, 10) \times (0, 1)$. L is above L_0.

Then, the Fréchet derivative of F is given by

$$F'(w) = \begin{pmatrix} e^x & 0 & 0 \\ 0 & (e-1)\,y + 1 & 0 \\ 0 & 0 & 1 \end{pmatrix}.$$

Notice that we have $F(x^\star) = 0$, $F'(x^\star) = F'(x^\star)^{-1} = \operatorname{diag}\{1, 1, 1\}$ and $L_0 = e - 1 < L = e$.

Example 1.1.7. Let $\mathcal{X}$ and $\mathcal{Y}$ as in Example 1.1.2. Let $\mathcal{D} = \overline{U}(0, 1)$. Define function F on $\mathcal{D}$ by

$$F(h)(x) = h(x) - 5 \int_0^1 x\,\vartheta\,h(\vartheta)^3\,d\vartheta. \tag{1.53}$$

Then, we have that

$$F'(h[u])(x) = u(x) - 15 \int_0^1 x\,\vartheta\,h(\vartheta)^2\,u(\vartheta)\,d\vartheta \quad \text{for all } u \in \mathcal{D}.$$

Using (1.53) we see that hypotheses of Theorem 1.1.3 hold for $x^\star(x) = 0$, where $x \in [0, 1]$, $L = 15$ and $L_0 = 7.5$.

1.2   Convergence under Generalized Lipschitz Conditions

In this section, we use a two-point Newton-like method to approximate a locally unique solution of a nonlinear equation containing a non-differentiable term. Using more precise majorizing sequences than in earlier studies such as Refs. [21, 100, 106, 443, 444], we present a tighter semi-local and local convergence analysis and weaker convergence criteria. This way we expand the applicability of these methods. Numerical examples are provided where the old convergence criteria do not hold but the new convergence criteria are satisfied. We are concerned with the problem of approximating a locally unique solution $x^\star$ of equation

$$F(x) + G(x) = 0, \tag{1.54}$$

where $F : \overline{U}(w, R) \longrightarrow \mathcal{Y}$ is Fréchet-differentiable around $x_0 \in U(w, R)$ and $G : \overline{U}(w, R) \longrightarrow \mathcal{Y}$ is continuous.

In Refs. [100, 106, 134, 163] the single-step Newton-like method defined by

$$
\begin{aligned}
&x_0 \text{ is a starting point in } U(w, R) \\
&x_{n+1} = x_n - A(x_n)^{-1} \left(F(x_n) + G(x_n) \right) \quad \text{for each} \quad n = 0, 1, \cdots
\end{aligned}
\tag{1.55}
$$

as well as the two-point Newton-like method given by

$$
\begin{aligned}
&y_{-1}, y_0 \text{ are starting points in } U(w, R) \\
&y_{n+1} = y_n - A(y_n, y_{n-1})^{-1} \left(F(y_n) + G(y_n) \right) \quad \text{for each} \quad n = 0, 1, \cdots
\end{aligned}
\tag{1.56}
$$

to generate respectively sequences $\{x_n\}$ and $\{y_n\}$ approximating $x^\star$, where $A(x)$, $A(v, w)$ belong in $\mathcal{L}(\mathcal{X}, \mathcal{Y})$. If $A(x) = F'(x)$ for each $x \in U(w, R)$, we obtain the Krasnosel'skii-Zincenko iteration (see Refs. [134, 163, 443, 444]). Moreover, if $G(x) = 0$ for each $x \in U(w, R)$, we obtain Newton's method (see Refs. [106, 134, 163, 290]) given by

$$
\begin{aligned}
&x_0 \text{ is a starting point in } U(w, R) \\
&x_{n+1} = x_n - F'(x_n)^{-1} F(x_n) \quad \text{for each} \quad n = 0, 1, \cdots .
\end{aligned}
\tag{1.57}
$$

If $A(x, y) = F'(x) + [x, y; G]$ where $[x, y; G]$ is the divided difference operator of order one for G, we obtain a Secant-type method studied in Refs. [106, 183]. Several other choices of operators A, F and G in (1.55)–(1.57) are given in Refs. [16, 444].

In this section, using more precise majorizing sequences we provide convergence criteria and a tighter semi-local and local convergence analysis for single and two-point Newton-like methods (1.55) and (1.56) than in Refs. [100, 443, 444]. This way we expand the applicability of these methods.

Let $R > 0$ be a constant and $r \in [0, R]$. Suppose there exist $w \in \mathcal{X}$ such that $A(w)^{-1} \in \mathcal{L}(\mathcal{Y}, \mathcal{X})$ and for any $x, y \in \overline{U}(w, r)$, $\theta \in [0, 1]$ the following conditions hold

$$\| A(w)^{-1} \left(A(x) - A(w) \right) \| \leq g_0(\| x - w \|) + \beta \tag{1.58}$$

and

$$
\begin{aligned}
&\| A(w)^{-1} \left((F'(x + \theta (y - x)) - A(x)) (y - x) + G(y) - G(x) \right) \| \\
&\leq (g_1(\| x - w \| + \theta \| y - x \|) - g_2(\| x - w \|) + g_3(r) + \gamma) \| y - x \|,
\end{aligned}
\tag{1.59}
$$

where $g_0(r)$, $g_1(r+\bar{r})-g_2(r)$ $(\bar{r} \geq 0)$, $g_2(r)$, $g_3(r)$ are non-decreasing and continuous functions for each r in $[0,R]$, $[0,R]^2$, $[0,R]$, $[0,R]$, respectively, $g_i(0) = 0$ for $i = 0,1,2,3$ and constants β, γ which satisfy $\beta \geq 0$, $\gamma \geq 0$.

Hypotheses (1.58) and (1.59) were used in Ref. [100] to provide a semi-local convergence analysis for single-step Newton-like methods. The majorizing sequence $\{t_n\}$ for $\{x_n\}$ was given by

$$t_0 = r_0 \in [r, R], \; t_1 = r_0 + \eta \text{ for some } \eta \geq 0$$
$$t_{n+2} = t_{n+1} + \delta_n (t_{n+1} - t_n) \quad \text{for each} \quad n = 0, 1, \cdots, \tag{1.60}$$

where

$$\delta_n = \frac{\displaystyle\int_0^1 (g_1(t_n + \theta (t_{n+1} - t_n)) - g_2(t_n) + \gamma) \, d\theta + g_3(t_{n+1})}{1 - \beta - g_0(t_{n+1})} \quad \text{for each} \quad n = 0, 1, \cdots. \tag{1.61}$$

Under the sufficient convergence conditions of [Theorem 3] in Ref. [100], we showed convergence of $\{x_n\}$ to $x^\star$ with the following error bounds

$$\| x_{n+1} - x_n \| \leq t_{n+1} - t_n \tag{1.62}$$

and

$$\| x_{n+1} - x^\star \| \leq t^\star - t_n, \tag{1.63}$$

where

$$t^\star = \lim_{n \longrightarrow \infty} t_n.$$

Next, we show how to improve these results. Let $x_0 \in \mathcal{X}$. Suppose that $A(x_0)^{-1} \in \mathcal{L}(\mathcal{Y}, \mathcal{X})$ and for $x_0, x_1 \in \overline{U}(w, r)$ with $x_1 = x_0 - A(x_0)^{-1} (F(x_0) + G(x_0))$, $\theta \in [0, 1]$, the following conditions hold

$$\| A(x_0)^{-1} (A(x_1) - A(x_0)) \| \leq g_0^0(\| x_1 - x_0 \|) + \beta_0 \tag{1.64}$$

and

$$\| A(x_0)^{-1} ((F'(x_0 + \theta (x_1 - x_0)) - A(x_0)) (x_1 - x_0) + G(x_1) - G(x_0)) \| \tag{1.65}$$
$$\leq (g_1^0(\theta \| x_1 - x_0 \|) - g_2^0(0) + g_3^0(\| x_1 - x_0 \|) + \gamma_0) \| x_1 - x_0 \|,$$

where g_i^0 $(i = 0, 1, 2, 3)$, β_0 and γ_0 are as g_i $(i = 0, 1, 2, 3)$, β and γ, respectively. We shall denote by $(\mathcal{C}_1)$ conditions (1.58), (1.59), (1.64) and (1.65). Let us define sequence $\{s_n\}$ by

$$s_0 = r_0 \in [r, R], \; s_1 = r_0 + \eta, \; s_2 = s_1 + \delta (s_1 - s_0)$$
$$s_{n+2} = s_{n+1} + \alpha_n (s_{n+1} - s_n) \quad \text{for each} \quad n = 0, 1, \cdots, \tag{1.66}$$

where

$$\delta = \frac{\displaystyle\int_0^1 (g_1^0(s_0 + \theta (s_1 - s_0)) - g_2^0(s_0) + \gamma_0) \, d\theta + g_3^0(s_1)}{1 - \beta_0 - g_0^0(s_1)}$$

and

$$\alpha_n = \frac{\int_0^1 \left(g_1(s_n + \theta\,(s_{n+1} - s_n)) - g_2(s_n) + \gamma \right) d\theta + g_3(s_{n+1})}{1 - \beta - g_0(s_{n+1})} \qquad \text{for each} \quad n = 1, 2, \cdots .$$

(1.67)

Suppose that for each $r \in [0, R]$, $r_1 \in [0, R - r]$, the following conditions hold

$$g_1^0(r + \bar{r}) - g_2^0(r) \le g_1(r + \bar{r}) - g_2(r),$$

(1.68)

$$g_0^0(r) \le g_0(r), \quad g_1^0(r) \le g_1(r),$$

(1.69)

$$g_2^0(r) \le g_2(r), \quad g_3^0(r) \le g_3(r),$$

(1.70)

$$\beta_0 \le \beta$$

(1.71)

and

$$\gamma_0 \le \gamma.$$

(1.72)

Then, a simple inductive argument shows that for each $n = 0, 1, \cdots$

$$s_n \le t_n,$$

(1.73)

$$s_{n+1} - s_n \le t_{n+1} - t_n$$

(1.74)

and

$$s^\star = \lim_{n \to \infty} s_n \le t^\star.$$

(1.75)

Next, we first provide sufficient conditions for the convergence of sequence $\{s_n\}$. Then, we show $\{s_n\}$ is a majorizing sequence for $\{x_n\}$. It is convenient for us to define functions p_n^N, q_n^N, p_∞^N on $[0, 1)$ for each fixed $N = 1, 2, \cdots$ and each $n = N, N + 1, \cdots$ by

$$p_n^N(t) = \int_0^1 \left(g_1\left(\frac{1 - t^{n-1}}{1 - t}(s_{N+1} - s_N) + s_N + \theta\, t^{n-1}(s_{N+1} - s_N)\right) - \right.$$
$$\left. g_2\left(\frac{1 - t^{n-1}}{1 - t}(s_{N+1} - s_N) + s_N\right) \right) d\theta + \gamma + g_3\left(\frac{1 - t^n}{1 - t}(s_{N+1} - s_N) + s_N\right) +$$
$$t\, g_0\left(\frac{1 - t^n}{1 - t}(s_{N+1} - s_N) + s_N\right) - t\,(1 - \beta),$$

(1.76)

$$q_n^N(t) = p_{n+1}^N(t) - p_n^N(t)$$

(1.77)

and

$$p_\infty^N(t) = g_1\left(\frac{s_{N+1} - s_N}{1 - t} + s_N\right) - g_2\left(\frac{s_{N+1} - s_N}{1 - t} + s_N\right) +$$
$$\gamma + g_3\left(\frac{s_{N+1} - s_N}{1 - t} + s_N\right) + t\, g_0\left(\frac{s_{N+1} - s_N}{1 - t} + s_N\right) - t\,(1 - \beta).$$

(1.78)

Then, we can show the following results on majorizing sequences for single-step Newton-like methods.

Lemma 1.2.1. *Suppose*

$$g_0(s_1) < 1 - \beta_0 \tag{1.79}$$

and there exists $\alpha \in (0,1)$ such that

$$0 \leq \alpha_1 \leq \alpha, \tag{1.80}$$

$$q_n^1(\alpha) \geq 0 \tag{1.81}$$

and

$$p_\infty^1(\alpha) \leq 0. \tag{1.82}$$

Then, sequence $\{s_n\}$ given by (1.66) is well defined, non-decreasing, bounded from above by

$$s_1^{\star\star} = s_1 + \frac{s_2 - s_1}{1 - \alpha} \tag{1.83}$$

and converges to its unique least upper bound $s^\star \in [0, s_1^{\star\star}]$. Moreover, the following estimates hold

$$0 \leq s_{n+2} - s_{n+1} \leq (s_2 - s_1)\,\alpha^n \quad for\ each \quad n = 1, 2, \cdots . \tag{1.84}$$

Proof. We have by (1.66) and (1.79) that s_2 is well defined and $s_2 \geq s_1$. We use mathematical induction to prove that

$$0 < \alpha_k \leq \alpha \quad \text{for each} \quad k = 1, 2, \cdots . \tag{1.85}$$

Estimate (1.85) holds for $k = 1$ by (1.80). Then, we have by (1.66) and (1.80) that

$$\begin{aligned}
0 \leq s_3 - s_2 \leq \alpha\,(s_2 - s_1) &\implies s_3 \leq s_2 + \alpha\,(s_2 - s_1) \\
&\implies s_3 \leq s_2 + (1 + \alpha)\,(s_2 - s_1) - (s_2 - s_1) \\
&\implies s_3 \leq s_1 + \frac{1 - \alpha^2}{1 - \alpha}(s_2 - s_1) < s_1^{\star\star}.
\end{aligned} \tag{1.86}$$

Assume that (1.85) holds for all natural integer $n \leq k$. Then, we get by (1.66) and (1.85) that

$$0 \leq s_{k+2} - s_{k+1} \leq s^k\,(s_2 - s_1) \tag{1.87}$$

and

$$s_{k+2} \leq s_1 + \frac{1 - \alpha^{k+1}}{1 - \alpha}\,(s_2 - s_1) < s_1^{\star\star}. \tag{1.88}$$

Evidently estimate (1.85) is true, if k is replaced by $k + 1$ provided that

$$p_k^1(\alpha) \leq 0 \quad \text{for each} \quad k = 1, 2, \cdots . \tag{1.89}$$

We have by (1.78) and (1.81) that

$$p_k^1(\alpha) \leq p_{k+1}^1(\alpha) \quad \text{for each} \quad k = 1, 2, \cdots . \tag{1.90}$$

Define function

$$p_\infty^1(t) = \lim_{k \longrightarrow \infty} p_k^1(t). \tag{1.91}$$

Using (1.77) and letting $n \longrightarrow \infty$ we see that p_∞^1 is given by (1.78) for $N = 1$. Estimate (1.89) holds by (1.91) and (1.82). The induction is now complete. Hence, sequence $\{s_n\}$ is increasing, bounded from above by $s_1^{\star\star}$ given by (1.83) and it converges to its unique least upper bound $s^\star \in [0, s_1^{\star\star}]$. The proof of Lemma 1.2.1 is complete. $\square$

Remark 1.2.1. The conclusions of Lemma 1.2.1 hold if (1.81) and (1.82) are replaced by

$$q_n^1(\alpha) \leq 0 \tag{1.92}$$

and

$$p_1^1(\alpha) \leq 0. \tag{1.93}$$

In this case, we have $p_{k+1}^1(\alpha) \leq p_k^1(\alpha) \leq p_1^1(\alpha) \leq 0$.

We have the following useful and obvious extension of Lemma 1.2.1.

Lemma 1.2.2. *Suppose there exists a minimum natural integer $N > 1$ and $\alpha \in (0,1)$ such that (1.81), (1.82),*

$$s_1 \leq s_2 \leq \cdots \leq s_{N+1},$$

$$0 < \alpha_N \leq \alpha,$$

$$g_0(s_{N+1}) < 1 - \beta$$

and

$$g_0\left(\frac{s_{N+1} - s_N}{1 - t}\left(s_{N+1} - s_N\right) + s_N\right) < 1 - \beta$$

hold. Then, sequence $\{s_n\}$ given by (1.66) is well defined, non-decreasing, bounded from above by

$$s_N^{\star\star} = s_{N-1} + \frac{s_N - s_{N-1}}{1 - \alpha}$$

and converges to its unique least upper bound $s_N^\star$ which satisfies $s_N^\star \in [0, s_N^{\star\star}]$. Moreover, the following estimates hold

$$0 \leq s_{N+n} - s_{N+n-1} \leq \left(s_{N+1} - s_N\right)\alpha^n \quad for \ each \quad n = 1, 2, \cdots.$$

Next, we present results for computing upper bounds on the limit point $t^\star$ using sequence $\{t_n\}$. Set

$$\nu_i(t) = \frac{g_i(t)}{1 - \beta} \quad \text{for all} \quad i = 0, 1, 2, 3 \quad \text{and} \quad \gamma_1 = \frac{\gamma}{1 - \beta}.$$

Then, iteration (1.60) can be written in the form

$$\begin{aligned}
&t_0 = r_0 \in [r, R], \ t_1 = r_0 + \eta \\
&t_{n+2} = t_{n+1} + \delta_n^1\left(t_{n+1} - t_n\right) \quad \text{for each} \quad n = 0, 1, \cdots,
\end{aligned} \tag{1.94}$$

where

$$\delta_n^1 = \frac{\displaystyle\int_0^1 \left(\nu_1(t_n + \theta\left(t_{n+1} - t_n\right)) - \nu_2(t_n) + \gamma_1\right) d\theta + \nu_3(t_{n+1})}{1 - \nu_0(t_{n+1})} \quad \text{for each} \quad n = 0, 1, \cdots.$$

Assume for the rest of this section that ν_4 (i.e. g_4) is strictly increasing on $[0, R]$. Define function $\chi(s, t)$ on $[0, t] \times [s, R]$ for each $t \in [0, R]$ by

$$\chi(s, t) = \left(\int_0^1 \left(\nu_1(s + \theta(t - s)) - \nu_2(t - s) + \gamma_1 \right) d\theta + \nu_3(t - s) \right) (t - s).$$

The result for upper bounds on the limit point $t^\star$ using sequence $\{t_n\}$ is as follows.

Lemma 1.2.3. *Let* $\lambda \in [0, \nu_4^{-1}(1)]$. *Let* f, ν_4 *be differentiable functions on* $[0, \nu_4^{-1}(1)]$. *Suppose* f *has a zero in* $[\lambda, \nu_4^{-1}(1)]$. *Denote by* ϱ *the smallest zero of function* f *in* $[\lambda, \nu_4^{-1}(1)]$. *Define functions* φ *and* g *on* $[0, \nu_4^{-1}(1))$ *by*

$$\varphi(t) = \frac{f(t)}{1 - \nu_4(t)} \quad and \quad g(t) = t + \varphi(t).$$

Moreover, suppose

$$g'(t) > 0 \quad for \ each \quad t \in [\lambda, \varrho].$$

Then, function g *is strictly increasing and bounded from above by* ϱ.

Proof. Function φ is well defined on $[\lambda, \varrho]$ with the possible exception when $\varrho = \nu_4^{-1}(1)$; but the L'Hospital theorem implies that f admits a continuous extension on the interval $[\lambda, \varrho]$. The function g is strictly increasing, since $g'(t) > 0$ on $[\lambda, \varrho]$. Therefore, we have for each $t \in [\lambda, \varrho]$ that

$$g(t) = t + \varphi(t) \leq \varrho + \varphi(\varrho) = \varrho.$$

The proof of Lemma 1.2.3 is complete. $\qquad\square$

Lemma 1.2.4. *Suppose that hypotheses of Lemma 1.2.3 hold. Define functions* ϕ *and* ψ *on* $\mathcal{I} := [\lambda, t] \times [s, \varrho]$ *for each* $t \in [\lambda, \varrho]$ *by*

$$\phi(s, t) = t + \frac{\chi(s, t)}{1 - \nu_4(t)}$$

and

$$\psi(s, t) = \begin{cases} \chi(s, t) - f(t) & if \ t \neq \varrho \\ 0 & if \ t = \varrho. \end{cases}$$

Moreover, suppose that

$$\psi(s, t) \leq 0 \quad for \ each \quad (s, t) \in \mathcal{I}.$$

Then, the following assertion holds

$$\phi(s, t) \leq g(t) \quad for \ each \quad (s, t) \in \mathcal{I}.$$

Proof. The result follows immediately from the definition of functions g, ϕ, ψ and the hypothesis of Lemma. That completes the proof of Lemma 1.2.4. $\qquad\square$

Lemma 1.2.5. *Let $N = 0, 1, 2, \cdots$ be fixed. Under the hypotheses of Lemma 1.2.4 with $\lambda = t_N$, further suppose that*

$$t_1 \leq t_2 \leq \cdots \leq t_N \leq t_{N+1} \leq \varrho$$

and

$$f(t_{N+1}) \geq 0.$$

Then, sequence $\{t_n\}$ generated by (1.94) is nondecreasing, bounded by ϱ and converges to its unique least upper bound $t^\star$ which satisfies $t^\star \in [t_N, \varrho]$.

Proof.   We can write that

$$t_{n+1} = \phi(t_{n-1}, t_n).$$

Then, we get that

$$t_{N+2} = \phi(t_N, t_{N+1}) \leq g(t_{N+1}) \leq \varrho.$$

That completes the proof of Lemma 1.2.5. $\square$

Remark 1.2.2.

(a) Hypotheses of Lemma 1.2.1 or Lemma 1.2.2 are satisfied in the interesting case of Newton's method (see Refs. [106, 290]). Set $\beta_0 = \gamma_0 = \beta = \gamma = 0$, $g_0(t) = L_0\, t$, $g_1(t) = g_2(t) = L\,t$ and $g_3(t) = 0$. Then, we get by (1.76) and (1.77) that

$$g_n^N(t) = \frac{1}{2}\left(2\,L_0\,t^2 + L\,t - L\right) t^n \left(s_{N+1} - s_N\right)$$

and

$$p_\infty^N(t) = t\left(\frac{L_0\left(s_{N+1} - s_N\right)}{1 - t} + s_N\right).$$

Set

$$\alpha = \frac{2\,L}{L + \sqrt{L^2 + 8\,L\,L_0}}.$$

Then, hypotheses of Lemma 1.2.1 are satisfied provided that

$$h_3 = L_3\,\eta \leq \frac{1}{2}, \tag{1.95}$$

where

$$L_3 = \frac{1}{8}\left((L_0\,L)^{1/2} + 4\,L_0 + (8\,L_0^2 + L_0\,L)^{1/2}\right).$$

Moreover, hypotheses of Lemma 1.2.2 become

$$s_1 \leq s_2 \leq \cdots \leq s_N < \frac{1}{L_0}$$

and

$$0 < \alpha_N \leq \alpha \leq 1 - \frac{L_0\left(s_{N+1} - s_N\right)}{1 - L_0\,s_N}.$$

Another interesting choice but not necessarily the best possible is given by the following scheme. Choose:

$$f(t) = \frac{\dfrac{L_0}{2}\,t^2 - t + \dfrac{L_3}{L_0}\,(t_N - t_{N-1})}{1 - L_0\,t} \quad \text{and} \quad \lambda = t_N \text{ for each } N = 1, 2, \cdots.$$

Suppose that

$$h_3^N = L_3\,(t_N - t_{N-1}) \le \frac{1}{2}.$$

Set

$$\varrho_3^N = \frac{2\,(t_N - t_{N-1})\,L_3}{L_0\,(1 + (1 - 2\,L_3\,(t_N - t_{N-1}))^{1/2})}$$

and

$$r_3 = \left(1 + \frac{L_0\,\eta}{2\,(1 - \alpha)\,(1 - L_0\,\eta)}\right)\eta.$$

Moreover, suppose

$$\psi(t_N, t) \le 0 \quad \text{for each} \quad t = g(t_N) \in [t_N, \varrho_3^N].$$

Then, ϱ_3^N are well defined. Elementary computations now show that all hypotheses of Lemma 1.2.5 are satisfied for $\varrho = \varrho_3^N$.

Moreover, the following estimates hold

$$\frac{2\,(t_N - t_{N-1})}{1 + (1 - 2\,L_3\,(t_N - t_{N-1}))^{1/2}} \le \varrho_3^N \le \frac{1}{L_0}$$

and

$$\frac{L_3}{L_0} \ge 1.$$

Furthermore, if

$$h_\star = L\,\eta \le \frac{1}{2}$$

and

$$(L_0^2 + L\,L_3)\,(t_N - t_{N-1}) \le L_0\,(1 - (1 - 2\,h_3^N)^{1/2}\,(1 - 2\,h_\star)^{1/2}),$$

then, we obtain that

$$\varrho_3^N \le \varrho^\star = \frac{2\,(t_N - t_{N-1})}{1 + (1 - 2\,h_\star)^{1/2}}.$$

We also have that for sufficiently small $t_N - t_{N-1}$, ϱ_3^N is smaller than r_3.

(b) Hypotheses of Lemma 1.2.5 are satisfied in the Kantorovich case (see Refs. [106, 134, 163, 290]). Indeed, set $\beta = \gamma = 0$, $g_0(t) = g_1(t) = g_2(t) = L\,t$ and $g_3(t) = 0$. Then, we must have that

$$h_\star \le \frac{1}{2} \quad \text{and} \quad \varrho = \frac{1 - \sqrt{1 - 2\,h_\star}}{L}$$

for

$$f(t) = \frac{L}{2}\,t^2 - t + \eta.$$

(c) Cartesian product $\mathcal{I}$ can be replaced by the more practical $\mathcal{J} = [\lambda, \varrho]^2$ in Lemma 1.2.5.

We can now present the semi-local convergence results for single-step Newton-like method. The proofs are omitted since they can be found in [Theorem 3] of Ref. [100] by simply replacing hypotheses of [Lemma 2] in Ref. [100] by those of Lemma 1.2.1 or Lemma 1.2.2 or Lemma 1.2.5 of this section.

Theorem 1.2.1. *Suppose that (1.58), (1.59) and hypotheses of Lemma 1.2.5 hold. Moreover, suppose that there exists $x_0 \in \overline{U}(w, r_0)$ such that*

$$\| A(x_0)^{-1} \left(F(x_0) + G(x_0) \right) \| \leq \eta. \tag{1.96}$$

Then, sequence $\{x_n\}$ generated by single-step Newton-like method is well defined, remains in $\overline{U}(w, t^\star)$ for each $n = 0, 1, \cdots$ and converges to a solution $x^\star$ of equation $F(x) + G(x) = 0$. Moreover, the following estimates hold for each $n = 0, 1, \cdots$

$$\| x_{n+1} - x_n \| \leq t_{n+1} - t_n \quad and \quad \| x_n - x^\star \| \leq t^\star - t_n \leq \varrho - t_n.$$

Furthermore, the solution $x^\star$ is unique in $\overline{U}(x_0, t^\star)$ if

$$\int_0^1 \left(g_1(t^\star + \theta\, t^\star) - g_2(t^\star) \right) d\theta + g_3(t^\star) + g_0(t^\star) + \beta + \gamma < 1$$

or in $\overline{U}(x_0, R_0)$ if $t^\star < R_0 < R$ and

$$\int_0^1 \left(g_1(t^\star + \theta\, R_0) - g_2(t^\star) \right) d\theta + g_3(t^\star + R_0) + g_0(t^\star) + \beta + \gamma < 1.$$

Theorem 1.2.2. *Suppose that $(\mathcal{C}_1)$ conditions, (1.96) and hypotheses of Lemma 1.2.1 or Lemma 1.2.2 hold. Then, the conclusions of Theorem 1.2.1 hold with $\{s_n\}$ replacing sequence $\{t_n\}$.*

The local convergence analysis of single-step Newton-like method is given in the next section as a special case of the two-point Newton-like method.

We present our results for the semi-local and local convergence of the two-point Newton-like method. As in Ref. [100], suppose there exist $v, w \in \mathcal{X}$ such that $A(v, w)^{-1} \in \mathcal{L}(\mathcal{Y}, \mathcal{X})$ and for each $x, y, z \in \overline{U}(w, r)$, $\theta \in [0, 1]$ the following conditions hold

$$\| A(v, w)^{-1} \left(A(x, y) - A(v, w) \right) \| \leq f_0(\| x - v \|, \| y - w \|) + \beta \tag{1.97}$$

and

$$\begin{aligned}
\| A(v, w)^{-1} &\left((F'(y + \theta\,(z - y)) - A(x, y)) (z - y) + G(x) - G(y) \right) \| \\
&\leq (f_1(\| y - w \| + \theta\, \| z - y \|) - f_2(\| y - w \|) + \\
&\quad f_3(\| z - x \|) + \gamma) \| z - y \|,
\end{aligned} \tag{1.98}$$

where $f_0(r, s)$ is continuous on $[0, R]^2$ and decreasing with respect to the first component, $f_1(r + \bar{r}) - f_2(r)$ $(\bar{r} \geq 0)$, $f_2(r)$, $f_3(r)$ are non-decreasing and continuous functions on $[0, R]$, with $f_0(0, 0) = f_1(0) = f_2(0) = f_3(0) = 0$ and for constants β,

γ which satisfy $\beta \geq 0$, $\gamma \geq 0$ and $\beta + \gamma < 1$. Using (1.97) and (1.98) a semi-local convergence analysis was given by us in [Theorem 2] of Ref. [100]. We now show how to improve these results under the same hypotheses (1.97) and (1.98). Let $r_0 \in [0, R]$, $y_{-1} \in \overline{U}(w, R)$ and $y_0 \in \overline{U}(w, r_0)$. It follows from (1.97) and (1.98) that

$$\| A(v, w)^{-1} \left(A(y_{-1}, y_0) - A(v, w) \right) \| \leq f_0^0(\| v - y_{-1} \|, \| w - y_0 \|) + \beta_0 \qquad (1.99)$$

and

$$\| A(v, w)^{-1} \left((F'(y_0 + \theta (y_1 - y_0)) - A(y_{-1}, y_0)) (y_1 - y_0) + G(y_1) - G(y_0) \right) \|$$
$$\leq \left(f_1^0(\| y_0 - w \| + \theta \| y_1 - y_0 \|) - f_2^0(\| y_0 - w \|) + f_3^0(\| y_1 - y_{-1} \|) + \gamma_0 \right) \| y_1 - y_0 \|,$$
$$(1.100)$$

where f_i^0 $(i = 0, 1, 2, 3)$, β_0, γ_0 are as f_i $(i = 0, 1, 2, 3)$, β, γ, respectively. Clearly

$$f_1^0(r + \overline{r}) - f_2^0(r) \leq f_1(r + \overline{r}) - f_2(r), \qquad (1.101)$$

$$f_0^0(r, t) \leq f_0(r, t), \qquad (1.102)$$

$$f_1^0(r) \leq f_1(r), \qquad (1.103)$$

$$f_2^0(r) \leq f_2(r), \qquad (1.104)$$

$$f_3^0(r) \leq f_3(r), \qquad (1.105)$$

$$\beta_0 \leq \beta \qquad (1.106)$$

and

$$\gamma_0 \leq \gamma. \qquad (1.107)$$

In practice, the computation of the functions f_i and constants β, γ requires that of f_i^0, β_0, γ_0 for $i = 0, 1, 2, 3$. Given y_{-1}, y_0 in $\mathcal{X}$, define c_{-1}, c_0, c_1 by

$$\| y_{-1} - v \| \leq c_{-1}, \quad \| y_{-1} - y_0 \| \leq c_0 \quad \text{and} \quad \| v - w \| \leq c_1. \qquad (1.108)$$

We shall refer to (1.97)–(1.100) and (1.108) as the $(\mathcal{C}_2)$ conditions. The majorizing sequence $\{t_n\}$ for $\{y_n\}$ was given in Ref. [100] by

$$\begin{aligned} t_{-1} &= r_0, \quad t_0 = c_0 + r_0, \quad t_1 = c_0 + r_0 + \eta \text{ for some } \eta \geq 0 \\ t_{n+2} &= t_{n+1} + \delta_n (t_{n+1} - t_n) \quad \text{for each} \quad n = 0, 1, \cdots \end{aligned} \qquad (1.109)$$

where

$$\delta_n = \frac{\int_0^1 (f_1(t_n - t_0 + r_0 + \theta (t_{n+1} - t_n)) - f_2(t_n - t_0 + r_0)) \, d\theta + \gamma + f_3(t_{n+1} - t_{n-1})}{1 - \beta - f_0(t_n - t_{n-1} + c_{-1}, t_{n+1} - t_0 + r_0)}$$

for each $n = 0, 1, \cdots$.

The following estimates hold for each $n = 0, 1, \cdots$

$$\| y_{n+1} - y_n \| \leq t_{n+1} - t_n \qquad (1.110)$$

and

$$\| y_n - x^\star \| \le t^\star - t_n, \tag{1.111}$$

where $t^\star = \lim\limits_{n \longrightarrow \infty} t_n$.

In this section, we use the more precise majorizing sequence $\{s_n\}$ given by

$$s_{-1} = t_{-1}, \quad s_0 = t_0, \quad s_1 = t_1 \quad s_2 = s_1 + \alpha_0 (s_1 - s_0)$$
$$s_{n+2} = s_{n+1} + \alpha_n (s_{n+1} - s_n) \quad \text{for each} \quad n = 1, 2, \cdots, \tag{1.112}$$

where

$$\alpha_0 = \frac{\displaystyle\int_0^1 (f_1^0(s_0 + \theta (s_1 - s_0)) - f_2^0(s_0)) \, d\theta + \gamma_0 + f_3^0(s_1 - s_{-1})}{1 - \beta_0 - f_0^0(s_0 - s_{-1} + c_{-1}, s_1 - s_0 + r_0)}$$

and

$$\alpha_n = \frac{\displaystyle\int_0^1 \Gamma_{\theta,n} \, d\theta + \gamma + f_3(s_{n+1} - s_{n-1})}{1 - \beta - f_0(s_n - s_{n-1} + c_{-1}, s_{n+1} - s_0 + r_0)} \quad \text{for each} \quad n = 1, 2, \cdots,$$

where $\Gamma_{\theta,n} = f_1(s_n - s_0 + r_0 + \theta (s_{n+1} - s_n)) - f_2(s_n - s_0 + r_0)$. In view of (1.101)–(1.107), (1.109) and (1.112), a simple inductive argument shows for each $n = 0, 1, \cdots$ that

$$s_n \le t_n, \quad s_{n+1} - s_n \le t_{n+1} - t_n \quad \text{and} \quad s^* = \lim\limits_{n \longrightarrow \infty} s_n \le t^\star.$$

Hence, $\{s_n\}$ converges to $s^\star$ under the same convergence hypotheses for $\{t_n\}$ given in Theorem 2 of Ref. [100]. However, sequence $\{s_n\}$ is tighter than $\{t_n\}$. Next, we first provide weaker sufficient convergence conditions for sequence $\{s_n\}$. Then, we show that $\{s_n\}$ is indeed a majorizing sequence for $\{y_n\}$.

It is convenient for us to define functions p_n^N, q_n^N, p_∞^N on $[0, 1]$ for each fixed $N = 1, 2, \cdots$ and each $n = N, N + 1, \cdots$ by

$$p_n^N(t) = \int_0^1 \left(f_1(\frac{1 - t^{n-1}}{1 - t} (s_{N+1} - s_N) + s_N - s_0 + r_0 + \theta \, t^{n-1} (s_{N+1} - s_N)) - \right.$$
$$\left. f_2(\frac{1 - t^{n-1}}{1 - t} (s_{N+1} - s_N) + s_N - s_0 + r_0) \right) d\theta + \gamma + f_3((t^{n-1} + t^{n-2}) (s_{N+1} - s_N)) +$$
$$t \, f_0(\frac{1 - t^{n-1}}{1 - t} (s_{N+1} - s_N) + s_N - s_{-1} + c_{-1}, \frac{1 - t^n}{1 - t} (s_{N+1} - s_N) + s_N - s_0 + r_0) -$$
$$t \, (1 - \beta), \tag{1.113}$$

$$q_n^N(t) = p_{n+1}^N(t) - p_n^N(t) \tag{1.114}$$

and

$$p_\infty^N(t) = f_1(\frac{s_{N+1} - s_N}{1 - t} + s_N - s_0 + r_0) - f_2(\frac{s_{N+1} - s_N}{1 - t} + s_N - s_0 + r_0) + f_3(0) +$$
$$\gamma + t \, f_0(\frac{s_{N+1} - s_N}{1 - t} + s_N - s_{-1} + c_{-1}, \frac{s_{N+1} - s_N}{1 - t} + s_N - s_0 + r_0) - t \, (1 - \beta). \tag{1.115}$$

Then, we present the following results on majorizing sequences for two-point Newton-like methods. The proofs are similar to the proofs of Lemmas 1.2.1, 1.2.2 and Theorems 1.2.1, 1.2.2.

Lemma 1.2.6. *Suppose*

$$f_0(s_0 - s_{-1} + c_{-1}, s_1 - s_0 + r_0) < 1 - \beta_0$$

and there exists $\alpha \in (0,1)$ such that

$$0 \leq \alpha_1 \leq \alpha,$$
$$q_n^1(\alpha) \geq 0 \tag{1.116}$$

and

$$p_\infty^1(\alpha) \leq 0. \tag{1.117}$$

Then, sequence $\{s_n\}$ given by (1.112) is well defined, non-decreasing, bounded from above by

$$s_1^{\star\star} = s_1 + \frac{s_2 - s_1}{1 - \alpha}$$

and converges to its unique least upper bound $s^\star \in [0, s_1^{\star\star}]$. Moreover, the following estimates hold

$$0 \leq s_{n+2} - s_{n+1} \leq (s_2 - s_1)\,\alpha^n \quad \text{for each} \quad n = 1, 2, \cdots.$$

Remark 1.2.3. The conclusions of Lemma 1.2.6 hold if (1.116) and (1.117) are replaced by

$$q_n^1(\alpha) \leq 0 \tag{1.118}$$

and

$$p_1^1(\alpha) \leq 0. \tag{1.119}$$

This time we have $p_{k+1}^1(\alpha) \leq p_k^1(\alpha) \leq p_1^1(\alpha) \leq 0$.

We have the following useful and obvious extension of Lemma 1.2.6.

Lemma 1.2.7. *Suppose there exists a minimum natural integer $N > 1$ and $\alpha \in (0,1)$ such that (1.116), (1.117),*

$$s_1 \leq s_2 \leq \cdots \leq s_{N+1},$$
$$0 < \alpha_N \leq \alpha,$$
$$f_0(s_{N+1} - s_{-1} + c_{-1}, s_{N+1} - s_0 + r_0) < 1 - \beta$$

and

$$f_0\Big(\frac{s_{N+1} - s_N}{1 - t}\, s_N - s_{-1} + c_{-1}, \frac{s_{N+1} - s_N}{1 - t}\,(s_N - s_N) + s_N - s_0 + r_0\Big) < 1 - \beta$$

hold. Then, sequence $\{s_n\}$ given by (1.112) is well defined, non-decreasing, bounded from above by

$$s_N^{\star\star} = s_{N-1} + \frac{s_N - s_{N-1}}{1 - \alpha}$$

and converges to its unique least upper bound $s_N^\star$ which satisfies $s_N^\star \in [0, s_N^{\star\star}]$. Moreover, the following estimates hold

$$0 \leq s_{N+n} - s_{N+n-1} \leq (s_{N+1} - s_N)\,\alpha^n \quad \text{for each} \quad n = 1, 2, \cdots.$$

Remark 1.2.4. The upper bound ϱ on the limit point $s^\star$ can be computed as in Lemmas 1.2.3–1.2.5. Simply replace g_i $(i = 0, 1, 2, 3)$ by

$$f_0(t - t_{-1} + c_{-1}, t - t_0 + r_0), \quad f_1(t - t_0 + r_0), \quad f_2(t - t_0 + r_0)$$

and $f_3(t - s)$, respectively.

We shall then call Lemma 1.2.5* the result corresponding to Lemma 1.2.5 with the above changes. Next, we present the semi-local convergence results for two-step Newton-like method. The proofs can be found in Theorem 2 of Ref. [100].

Theorem 1.2.3. *Suppose that the* $(\mathcal{C}_2)$ *conditions and the hypotheses of Lemma 1.2.5* hold. Moreover, suppose that there exist* $y_{-1} \in \overline{U}(w, R)$, $y_0 \in \overline{U}(w, r_0)$ *such that*

$$\| A(y_{-1}, y_0)^{-1} \left(F(y_0) + G(y_0) \right) \| \leq \eta. \tag{1.120}$$

Then, sequence $\{y_n\}$ *generated by two-step Newton-like method is well defined, remains in* $\overline{U}(w, t^\star)$ *for each* $n = 0, 1, \cdots$ *and converges to a solution* $x^\star$ *of equation* $F(x) + G(x) = 0$. *Moreover, the following estimates hold for each* $n = 0, 1, \cdots$

$$\| y_{n+1} - y_n \| \leq t_{n+1} - t_n \leq \varrho - t_n \quad and \quad \| y_n - x^\star \| \leq t^\star - t_n \leq \varrho - t_n.$$

Furthermore, the solution $x^\star$ *is unique in* $\overline{U}(x_0, t^\star)$ *for* $x_{-1} = v$ *if*

$$\int_0^1 (f_1(t^\star + \theta\, t^\star) - f_2(t^\star))\, d\theta + f_3(t^\star) + f_0(t^\star + c_1, t^\star) + \beta + \gamma < 1$$

or in $\overline{U}(x_0, R_0)$ *if* $t^\star \leq R_0 < R$ *and*

$$\int_0^1 (f_1(t^\star + \theta\, R_0) - f_2(t^\star))\, d\theta + f_3(R_0) + f_0(t^\star + c_1, t^\star) + \beta + \gamma < 1.$$

Theorem 1.2.4. *Suppose that the* $(\mathcal{C}_2)$ *conditions hold and there exist* $y_{-1} \in \overline{U}(w, R)$, $y_0 \in \overline{U}(w, r_0)$ *such that (1.120) holds. Moreover, suppose that hypotheses of Lemma 1.2.6 or Lemma 1.2.7 hold. Then, the conclusions of Theorem 1.2.3 hold with* $\{s_n\}$ *replacing* $\{t_n\}$.

In order for us to cover the local convergence case for the two-point Newton-like method, let us suppose as in [Theorem 5] of Ref. [100] that $x^\star$ is a solution of (1.54), $A(x^\star, x^\star)^{-1} \in \mathcal{L}(\mathcal{Y}, \mathcal{X})$ and for any $x, y \in \overline{U}(x^\star, r)$, $\theta \in [0, 1]$

$$\| A(x^\star, x^\star)^{-1} \left(A(x, y) - A(x^\star, x^\star) \right) \| \leq f_4(\| x - x^\star \|, \| y - x^\star \|) + \beta_4 \tag{1.121}$$

and

$$\| A(x^\star, x^\star)^{-1} \left((F'(x^\star + \theta\, (y - x^\star)) - A(x, y)) (y - x^\star) + G(y) - G(x^\star) \right) \|$$
$$\leq (f_5((1 + \theta)\, \| y - x^\star \|) - f_6(\| y - x^\star \|) + f_7(\| x - x^\star \|) + \gamma_4)\, \| y - x^\star \|, \tag{1.122}$$

where f_i $(i = 4, 5, 6, 7)$, β_4, γ_4 are as f_i $(i = 0, 1, 2, 3)$, β_0, γ_0, respectively. These conditions were used in [Theorem 5] of Ref. [100] to provide a local convergence analysis for the two-step Newton-like method. We can improve the error bounds on the distances $\| y_n - x^\star \|$ for each $n = 0, 1, \cdots$. It follows from hypotheses (1.121) and (1.122) that

$$\| A(x^\star, x^\star)^{-1} \left(A(y_{-1}, y_0) - A(x^\star, x^\star) \right) \| \leq f_4^0 (\| y_{-1} - x^\star \|, \| y_{-1} - x^\star \|) + \beta_4^0,$$
$$(1.123)$$

$$\| A(x^\star, x^\star)^{-1} \left(A(y_0, y_1) - A(x^\star, x^\star) \right) \| \leq f_4^1 (\| y_0 - x^\star \|, \| y_1 - x^\star \|) + \beta_4^1, \quad (1.124)$$

$$\| A(x^\star, x^\star)^{-1} \left((F'(x^\star + \theta (y_0 - x^\star)) - A(y_{-1}, y_0)) (y_0 - x^\star) + G(y_0) - G(x^\star) \right) \|$$
$$\leq (f_5^0((1 + \theta) \| y_0 - x^\star \|) - f_6^0(\| y_0 - x^\star \|) + f_7^0(\| y_{-1} - x^\star \|) + \gamma_4^0) \| y_0 - x^\star \|$$
$$(1.125)$$

and

$$\| A(x^\star, x^\star)^{-1} \left((F'(x^\star + \theta (y_1 - x^\star)) - A(y_0, y_1)) (y_1 - x^\star) + G(y_1) - G(x^\star) \right) \|$$
$$\leq (f_5^1((1 + \theta) \| y_1 - x^\star \|) - f_6^1(\| y_1 - x^\star \|) + f_7^1(\| y_0 - x^\star \|) + \gamma_4^1) \| y_1 - x^\star \|,$$
$$(1.126)$$

where f_i^j, β_4^j, γ_4^j ($i = 4, 5, 6, 7$ and $j = 1, 2$) are tighter than f_i, β_4, γ_4 ($i = 4, 5, 6, 7$), respectively. Note that (1.123)–(1.126) require computations only at the initial data. Using the approximation

$$-(y_{n+1} - x^\star) = (A(y_{n-1}, y_n)^{-1} A(y_{-1}, y_0)) A(y_{-1}, y_0)^{-1} \left(\int_0^1 (F'(x^\star + \theta (y_n - x^\star)) - A(y_{n-1}, y_n)) \, d\theta \, (y_n - x^\star) + G(y_n) - G(x^\star) \right),$$

$$(1.127)$$

(1.121)–(1.126), where (1.123)–(1.126) are used for the first two distances $\| y_1 - x^\star \|$, $\| y_2 - x^\star \|$ and (1.121), (1.122) for $\| y_n - x^\star \|$ for each $n = 2, 3, \cdots$ as in [Theorem 5] of Ref. [100] (where only (1.121) and (1.122) were used), we arrive at the following result.

Theorem 1.2.5. *Suppose that there exists a solution of equation $f_8(t) = 0$ in $[0, R]$, where*

$$f_8(t) = \int_0^1 (f_5((1 + \theta) t) - f_6(t)) \, d\theta + f_7(t) + f_4(t, t) + \beta_4 + \gamma_4 - 1.$$

Denote by R_1 the smallest of the solutions in $[0, R]$. Then, sequence $\{y_n\}$ generated by the two-step Newton-like method is well defined, remains in $\overline{U}(x^\star, R_1)$ and converges to $x^\star$ provided that y_{-1}, y_0 in $\overline{U}(x^\star, R_1)$. Moreover, the following error bounds hold for each $n = 0, 1, \cdots$

$$\| y_{n+1} - x^\star \| \leq e_n \, \| y_n - x^\star \|,$$

where

$$e_n = \frac{\int_0^1 (f_5^n((1 + \theta) \| y_n - x^\star \|) - f_6^n(\| y_n - x^\star \|)) \, d\theta + f_7^n(\| y_{n-1} - x^\star \|) + \gamma_4^n}{1 - \beta_4^n - f_4^n(\| y_n - x^\star \|)}$$

for $n = 0, 1$

and

$$e_n = \frac{\int_0^1 (f_5((1+\theta) \parallel y_n - x^\star \parallel) - f_6(\parallel y_n - x^\star \parallel)) \, d\theta + f_7(\parallel y_{n-1} - x^\star \parallel) + \gamma_4}{1 - \beta_4 - f_4(\parallel y_n - x^\star \parallel)}$$

for each $n = 2, 3, \cdots$.

Remark 1.2.5.

(a) We have $f_8(0) = \beta_4 + \gamma_4 - 1 < 0$. Therefore, if $f_8(R) \geq 0$, then it follows from the intermediate value theorem that function f_8 has solution in $[0, R]$. That is $f_8(R) \geq 0$ can replace the hypothesis about the existence of the solution in Theorem 1.2.5.

(b) Let us define $\overline{e_n}$ as e_n but using f_i $(i = 4, 5, 6, 7)$, β_4, γ_4 for each $n = 0, 1, \cdots$. Then, we have $e_n \leq \overline{e_n}$ from which it follows that our new estimates on the distances $\parallel y_n - x^\star \parallel$ are tighter than the ones in Ref. [100]. Note that these advantages are obtained under the same computational cost, since in practice the computation of f_i $(i = 4, 5, 6, 7)$, β_4, γ_4 requires that of f_i^j $(i = 4, 5, 6, 7)$, β_4^j, γ_4^j $(j = 0, 1)$.

(c) The local convergence results for the single-step Newton-like method are obtained immediately from the above if we replace $A(x^\star, x^\star)$, y_{-1}, y_0, y_1, $f_4(t, t)$, $f_4^j(t, t)$ $(j = 0, 1)$, y_n by $A(x^\star)$, x_0, x_0, x_1, $f_4(t)$, $f_4^j(t)$ $(j = 0, 1)$, x_n, respectively.

Let us consider single step methods in the Lipschitz case for Newton's method. We set $A(x) = F'(x)$, $G(x) = 0$ for each $x \in D$, $x_0 = w$, $\beta_0 = \beta = \gamma_0 = \gamma = r_0 = 0$, $g_0(t) = L_0 \, t$, $g_1(t) = g_2(t) = L \, t$, $g_3(t) = 0$, $g_0^0(t) = L_{-2} \, t$, $g_1^0(t) = g_2^0(t) = L_{-1} \, t$. Then the $(\mathcal{C}_1)$ conditions reduce to

$$\parallel F'(x_0)^{-1} (F'(x) - F'(x_0)) \parallel \leq L_0 \parallel x - x_0 \parallel, \tag{1.128}$$

$$\parallel F'(x_0)^{-1} (F'(x) - F'(y)) \parallel \leq L \parallel x - y \parallel, \tag{1.129}$$

$$\parallel F'(x_0)^{-1} (F'(x_0 + \theta (F'(x_0)^{-1} F(x_0))) - F'(x_0)) \parallel \leq L_{-1} \theta \parallel F'(x_0)^{-1} F(x_0) \parallel \tag{1.130}$$

and

$$\parallel F'(x_0)^{-1} (F'(x_0 - F'(x_0)^{-1} F(x_0)) - F'(x_0)) \parallel \leq L_{-2} \parallel F'(x_0)^{-1} F(x_0) \parallel . \tag{1.131}$$

Iteration $\{s_n\}$ becomes

$$s_0 = 0, \quad s_1 = \eta, \quad s_2 = \eta + \frac{L_{-2} \, \eta^2}{2 \, (1 - L_{-1} \eta)}$$
$$s_{n+2} = s_{n+1} + \frac{L \, (s_{n+1} - s_n)^2}{2 \, (1 - L_0 \, s_{n+1})} \quad \text{for each} \quad n = 1, 2, \cdots . \tag{1.132}$$

Note that

$$L_{-2} \leq L_{-1} \leq L_0 \leq L. \tag{1.133}$$

(a) The conditions of Lemma 1.2.2 will hold for $N = 2$ if

$$\eta + \frac{L_{-2}\,\eta^2}{2\,(1 - L_{-1}\,\eta)} < \frac{1}{L_0} \qquad (1.134)$$

and

$$\alpha_1 = \frac{\dfrac{L}{2}\,\dfrac{L_{-2}\,\eta^2}{2\,(1 - L_{-1}\,\eta)}}{1 - L_0\left(\eta + \dfrac{L_{-2}\,\eta^2}{2\,(1 - L_{-1}\,\eta)}\right)} \le \alpha \le \alpha_2 = 1 - \frac{L_0\,\dfrac{L_{-2}\,\eta^2}{2\,(1 - L_{-1}\,\eta)}}{1 - L_0\,\eta}. \qquad (1.135)$$

In order for us to solve the system of inequalites (1.134) and (1.135), it is convenient to define quadratic polynomials p_1 and p_2 by

$$p_1(t) = \beta_1\,t^2 + \beta_2\,t + \beta_3$$

and

$$p_2(t) = \gamma_1\,t^2 + \gamma_2\,t + \gamma_3,$$

where

$$\beta_1 = L\,L_{-2} + 2\,\alpha\,L_0\,(2\,L_{-1} - L_{-2}), \quad \beta_2 = 4\,\alpha\,(L_{-1} - L_0), \quad \beta_3 = -4\,\alpha,$$

$$\gamma_1 = L_0\,(L_{-2} - 2\,(1-\alpha)\,L_{-1}), \quad \gamma_2 = 2\,(1-\alpha)\,(L_0 + L_{-1}) \quad \text{and} \quad \gamma_3 = -2\,(1-\alpha).$$

In view of (1.133), (1.134) and (1.135), polynomial p_1 has a positive root

$$\lambda_1 = \frac{-\beta_2 + \sqrt{\beta_2^2 - 4\,\beta_1\,\beta_3}}{2\,\beta_1}$$

and a negative root.

Inequalities (1.134) and (1.135) are satisfied if

$$p_1(\eta) \le 0 \qquad (1.136)$$

and

$$p_2(\eta) \le 0. \qquad (1.137)$$

Therefore, (1.136) is satisfied if

$$\eta \le \lambda_1. \qquad (1.138)$$

If $\gamma_1 < 0$ and $\Delta_2 := \gamma_2^2 - 4\,\gamma_1\,\gamma_3 \le 0$, (1.137) always holds. If $\gamma_1 < 0$ and $\Delta_2 > 0$, then, p_2 has two positive roots. The smaller is denoted by λ_2 and is given by

$$\lambda_2 = \frac{-\gamma_2 - \sqrt{\Delta_2}}{2\,\gamma_1}. \qquad (1.139)$$

If $\gamma_1 > 0$, polynomial p_2 has a positive root λ_3 given by

$$\lambda_3 = \frac{-\gamma_2 + \sqrt{\Delta_2}}{2\,\gamma_1} \qquad (1.140)$$

and a negative root.

Let us define

$$\frac{L_4^{-1}}{2} = \begin{cases} \lambda_1 & \text{if } \gamma_1 < 0 \quad \text{and} \quad \Delta_2 \leq 0 \\ \min\{\lambda_1, \lambda_2\} & \text{if } \gamma_1 < 0 \quad \text{and} \quad \Delta_2 > 0 \\ \min\{\lambda_1, \lambda_3\} & \text{if } \gamma_1 > 0. \end{cases} \qquad (1.141)$$

Summarizing we conclude that (1.134) and (1.135) are satisfied if

$$h_4 = L_4 \, \eta \leq \frac{1}{2}. \qquad (1.142)$$

In the special case when $L_{-2} = L_{-1} = L_0$, elementary computations show $L_3 = L_4$. That is (1.142) reduces to (1.95), which is weaker than the h conditions given in Ref. [155]. The rest of the h conditions are given as (1.3)–(1.5).

In view of the definition of β_1, the condition h_4 can be improved if L can be replaced by $L_\star$ such that $0 < L_\star < L$. That requires the verification of one condition

$$\| F'(x_0)^{-1}(F'(x_1 + \theta(x_2 - x_1)) - F'(x_1)) \| \leq L_\star \theta \, \| x_2 - x_1 \| \quad \text{for each } \theta \in [0,1],$$

where x_1 and x_2 are given by Newton's iterations. The computation of $L_\star$ is possible and requires computations at the initial data. Note also that iterate s_3 will be given by

$$s_3 = s_2 + \frac{L_\star \, (s_2 - s_1)^2}{2 \, (1 - L_0 \, s_2)} \quad \text{instead of} \quad s_3 = s_2 + \frac{L \, (s_2 - s_1)^2}{2 \, (1 - L_0 \, s_2)}.$$

Moreover, condition h_4 is then replaced by at least as weak $h_5 = L_5 \, \eta \leq 1/2$, where L_5 is defined as L_4 with $L_\star$ replacing L in the definition of β_1.

(b) Another extension of our results is given as follows. Let $N = 1, 2, \cdots$ and $R \in (0, 1/L_0)$. Assume $x_1, x_2, \cdots, x_N$ can be computed by Newton's iterations, $F'(x_N)^{-1} \in \mathcal{L}(\mathcal{Y}, \mathcal{X})$ such that $\| F'(x_N)^{-1} F(x_N) \| \leq R - \| x_N - x_0 \|$. Set $\mathcal{D} = \overline{U}(x_0, R)$ and $\mathcal{D}_N = \overline{U}(x_N, R - \| x_N - x_0 \|)$. Then, for all $x \in \mathcal{D}_N$, we have that

$$\begin{aligned} &\| F'(x_N)^{-1} \, (F'(x) - F'(y)) \| \\ &\leq \| F'(x_N)^{-1} F'(x_0) \| \, \| \, F'(x_0)^{-1} (F'(x) - F'(y)) \| \\ &\leq \frac{L}{1 - L_0 \, \| x_N - x_0 \|} \, \| x - y \| \end{aligned}$$

and

$$\| F'(x_N)^{-1} F(x_N) \| \leq \frac{L}{2} \, \frac{L}{1 - L_0 \, \| x_N - x_0 \|} \, \| x_N - x_{N-1} \|^2 .$$

Set

$$L^N = L_0^N = \frac{L}{1 - L_0 \, \| x_N - x_0 \|} \quad \text{and} \quad \eta^N = \frac{L \, L_N}{2} \, \| x_N - x_{N-1} \|^2 .$$

Then, the Kantorovich hypothesis becomes $h_\star^N = L^N \, \eta^N \leq 1/2$. Clearly, the most interesting case is when $N = 1$. In this case the Kantorovich condition becomes $h_\star^1 = L_6 \, \eta \leq 1/2$, where $L_6 = (L_0 + L \sqrt{L})/2$. Note also that L_6 can be smaller than L.

Table 1.3 Comparison table of the h conditions

ξ	$x^\star$	$h_\star$	h_1	h_2	h_3
.486967	.6978302086	.5174905727	.4736234295	.4584042632	.4368027442
.5245685	.7242710128	.4676444075	.4299718890	.4169293786	.3983631448
.452658	.6727986326	.5646168433	.5146862992	.4973315343	.4727617854
.435247	.6597325216	.5891326340	.5359749755	.5174817371	.4913332192
.425947	.6526461522	.6023932312	.5474704233	.5283539600	.5013421729
.7548589	.8688261621	.2034901726	.1934744795	.1900595014	.1850943832

Let us compare the h conditions on concrete examples.

Example 1.2.1. Let $\mathcal{X} = \mathcal{Y} = \mathbb{R}$, $x_0 = 1$, $\mathcal{D} = [\xi, 2 - \xi]$, $\xi \in [0, .5)$. Define function F as (1.48) (with ξ replaces a) on $\mathcal{D}$.

(a) Consider our h conditions given in this section. Then, we obtain that

$$h_1 = \frac{1}{6}(7 - 3\xi)(1 - \xi) \leq .5 \quad \text{for all} \quad \xi \in [.4648162415, .5),$$

$$h_2 = \frac{1}{12}(8 - 3\xi + (5\xi^2 - 24\xi + 28)^{1/2})(1 - \xi) \leq .5 \quad \text{for all} \quad \xi \in [.450339002, .5)$$

and

$$h_3 = \frac{1}{24}(1 - \xi)(12 - 4\xi + (84 - 58\xi + 10\xi^2)^{1/2} + (12 - 10\xi + 2\xi^2)^{1/2}) \leq .5$$
for all $\quad \xi \in [.4271907643, .5)$.

Next, we pick some values of ξ such that all hypotheses are satisfied, so we can compare the h conditions. Using Maple 13, we can compare our h conditions (see Table 1.3). In Fig. 1.3, we compare the h conditions for $\xi \in (0, .9)$.

(b) Consider the case $\xi = .7548589$ where our h are satisfied. Let

$$L_{-1} = 2 \quad \text{and} \quad L_{-2} = \frac{\xi + 5}{3} = 1.918286300.$$

Hence, we have that

$$\alpha = .5173648648, \quad \alpha_1 = .01192474572 \quad \text{and} \quad \alpha_2 = .7542728992.$$

Then, conditions (1.133)–(1.135) hold. We also have that

$$\beta_1 = 9.613132975, \quad \beta_2 = -.5073095684, \quad \beta_3 = -2.069459459,$$

$$\gamma_1 = -.02751250012, \quad \gamma_2 = 4.097708498, \quad \gamma_3 = -.965270270,$$

$$\lambda_1 = .4911124649, \quad \Delta_2 = 16.68498694 \quad \text{and} \quad \lambda_2 = 148.7039440.$$

Since $\gamma_1 < 0$, $\Delta_2 > 0$ and using (1.141), we get in turn that

$$\frac{1}{2L_4} = \min\{\lambda_1, \lambda_2\} = .4911124649.$$

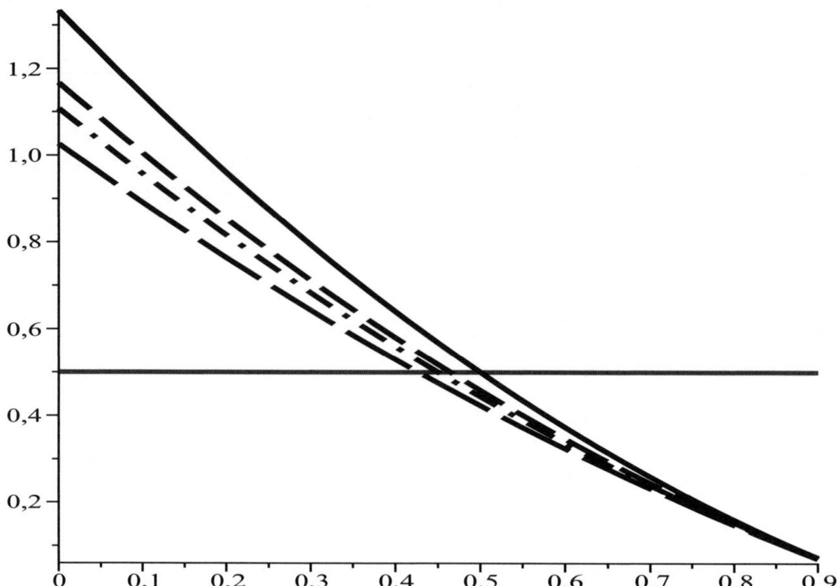

Fig. 1.3   Functions $h_\star$, h_1, h_2 and h_3 (from top to bottom) with respect to ξ in interval $(0, .9)$, respectively. The horizontal red line is of equation $y = .5$.

Table 1.4   Comparison table of conditions h_4 and h_5

ξ	h_4	h_5
.486967	.1767312629	.1377052696
.5245685	.1634740591	.1290677624
.452658	.1888584478	.1454742725
.435247	.1950234005	.1493795529
.425947	.1983192281	.1514558980
.7548589	.08319245167	.07234026489

Then, we deduce that condition (1.142) holds provided that

$$L_4 = 1.018096741 \quad \text{and} \quad h_4 = .08319245167 < .5.$$

If we consider $L_\star = 1.836572600$, we get that

$$h_5 = .07234026489 < h_4.$$

Finally, we pick the same values of ξ as in Table 1.3, so we can compare the h_4 and h_5 conditions (see Table 1.4).

Example 1.2.2. We also consider Example 1.2.1 in one more case. Let $\xi = .74137931$. Then $L = 2.51724138$, $L_0 = 2.25862069$ and $\eta = .0862068966$. Using hypotheses of Lemma 1.2.5 and Remark 1.2.2(a), we have that

$$\varrho = .1218276252, \quad \alpha = .5181703378, \quad 1/L_0 = .4427480915 \quad \text{and} \quad r_3 = .1078366107.$$

Using (1.60), we get that

$$t_2 = .09782207463, \quad t_3 = .09804003491, \quad t_4 = .09804011171$$

and for all $n \geq 5$

$$t_n = t_4 = .09804011171 = t^\star < \varrho^\star = .09839144902 < .4427480915 = 1/L_0.$$

Moreover, if we use the choices in Remark 1.2.2(a) with $N = 1$, we get

$$L_3 = 2.279717419 \quad \text{and} \quad \varrho_3^1 = .09781769768.$$

For $\varrho := \varrho_3^1$, we obtain that

$$\psi(\eta, t) = .1293103450\, t^2 + .7829964326\, t - .08891007985 \quad \text{for} \quad \eta \leq t < \varrho_3^1.$$

Hence, hypotheses of Lemma 1.2.5 are satisfied for $\lambda = \eta$ and $\varrho := \varrho_3^1$. Finally, we have in turn that $r_3 \geq \varrho_3^1$.

For examples in the local case, where $\ell^\star < \ell$ ($\ell^\star$, ℓ are the Lipschitz conditions as in (1.128) and (1.129) where x_0 is replaced by $x^\star$), see section 1.1. We conclude this section by an example where $G \neq 0$ in (1.54).

Example 1.2.3. Let $\mathcal{X}$ and $\mathcal{Y}$ as in Example 1.1.5. Consider the integral equation on $\mathcal{D} = \overline{U}\left(x_0, \dfrac{r}{2}\right)$, $(r \in [0, R])$ given by

$$x(t) = \int_0^1 k(t, s, x(s))ds, \tag{1.143}$$

where the kernel $k(t, s, x(s))$ with $(t, s) \in [0, 1] \times [0, 1]$ is a nondifferentiable operator on $\mathcal{D}$. Define operators F, G on $\mathcal{D}$ by

$$F(x)(t) = I\, x(t) \quad (I \text{ the identity operator on } \mathcal{X}) \tag{1.144}$$

$$G(x)(t) = -\int_0^1 k(t, s, x(s))ds. \tag{1.145}$$

Choose $x_0 = 0$ and assume there exists a constant $k_0 \in [0, 1)$, a real function $k_1(t, s)$ such that

$$\|k(t, s, x) - k(t, s, y)\| \leq k_1(t, s)\|x - y\| \tag{1.146}$$

and

$$\sup_{t \in [0,1]} \int_0^1 k_1(t, s)ds \leq k_0 \tag{1.147}$$

for each $t, s \in [0, 1]$ and $x, y \in \mathcal{D}$. Moreover choose $r_0 = 0$, $y_0 = y_{-1}$, $A(x) = I(x)$, $v_0(r) = r$, $\beta = \gamma = 0$, $v = 0$ and $v_1(r) = k_0$ for each $x, y \in \mathcal{D}$ and $r, s \in [0, 1]$. It can easily be seen that the conditions of Theorem 1.2.1 hold if

$$t^\star = \frac{\eta}{1 - k_0} \leq \frac{r}{2}.$$

1.3 Convergence without Lipschitz Conditions

We provide in this section new local and semi-local convergence results for Newton's method in a Banach space. The sufficient convergence conditions do not include the Lipschitz constant usually associated with Newton's method. Numerical examples demonstrating the expansion of Newton's method are also provided in this section. We are concerned with the problem of approximating a locally unique solution $x^\star$ of equation (1.1) using Newton's method (NM) defined by (1.2). There is a plethora of local as well as semi-local convergence results for (NM) based on the Lipschitz condition:

$$\| F'(x) - F'(y) \| \leq K \| x - y \| \quad \text{for all} \quad x, y \in \mathcal{D}. \tag{1.148}$$

A survey of such results can be found in Refs. [110, 134, 135] (see also Refs. [69, 100, 116, 155, 444] and the references therein). Let

$$\| F'(x_0)^{-1} \| \leq \beta \quad \text{and} \quad \| F'(x_0)^{-1} F(x_0) \| \leq \eta.$$

Then, the famous for its simplicity and clarity Kantorovich hypothesis

$$h_K = \beta K \eta \leq \frac{1}{2} \tag{1.149}$$

is a sufficient condition for semi-local convergence of (NM) (see Refs. [110, 290]). Under (1.149), (NM) converges to a locally unique solution $x^\star$ of equation (1.1). The solution $x^\star$ belongs in a ball centered at x_0 and of a certain radius $r > 0$. Moreover, the order of convergence is quadratic, if $h_K < 1/2$ and linear, if $h_K = 1/2$ (see Ref. [290]).

Semi-local case

Recently, Argyros in Ref. [110] used a combination of (1.148) and the center Lipschitz condition

$$\| F'(x) - F'(x_0) \| \leq K_0 \| x - x_0 \| \quad \text{for all} \quad x \in \mathcal{D}, \tag{1.150}$$

which leads to the weaker sufficient semi-local convergence condition hypothesis

$$h_A = \beta \overline{K} \eta \leq \frac{1}{2}, \tag{1.151}$$

where $\overline{K}$ is given as $\overline{L}$ in section 1.1 with K and K_0 replace L and L_0, respectively.

Local case

Under condition (1.148) the radius of convergence for (NM) was found by Rheinboldt in Ref. [383] and Traub in Refs. [418, 419] to be

$$r = \frac{2}{3\,\beta^\star K}, \tag{1.152}$$

where

$$\| \, F'(x^\star)^{-1} \, \| \leq \beta^\star. \tag{1.153}$$

Argyros in Ref. [110] introduced the center-Lipschitz condition

$$\| \, F'(x) - F'(x^\star) \, \| \leq K_1 \, \| \, x - x^\star \, \| \quad \text{for all} \quad x \in \mathcal{D} \tag{1.154}$$

to obtain the radius

$$r_1 = \frac{2}{\beta^\star \, (2 \, K_1 + K)}. \tag{1.155}$$

Clearly

$$K_1 \leq K \tag{1.156}$$

holds and K/K_1 can be arbitrarily large (see Refs. [98, 155]). By comparing (1.152) and (1.155), we see that

$$r \leq r_1. \tag{1.157}$$

If strict inequality holds in (1.156), then so does in (1.157). The error bounds on distances $\| \, x_n - x^\star \, \|$, $(n \geq 0)$ are also tighter under our approach in Refs. [100, 155].

In this section, we would like to further extend the applicability of (NM). We have noticed that there are many interesting problems in the literature for which Lipschitz condition (1.148) is not satisfied but center-Lipschitz condition (1.150) (or (1.154)) holds. However, in this case, the convergence of (NM) is only linear (see Ref. [110]). So, we are wondering, if it is possible to

(a) preserve the quadratic convergence of (NM) using (1.150) (semi-local case) or (1.154) (local case) but without assuming (1.148);
(b) weaken sufficient convergence condition (1.151);
(c) enlarge radius of convergence (1.155).

We shall refer to (a)–(c) as the ($\mathcal{A}$) advantages. These advantages can be obtained (see Remark 1.3.4).

We need some auxiliary results on majorizing sequences for (NM).

Lemma 1.3.1. *(Ref. [116]) Assume there exist constants $K_0 \geq 0$, $K \geq 0$ with $K_0 \leq K$ and $\eta \geq 0$, such that*

$$q_A = \overline{K} \, \eta \begin{cases} \leq \dfrac{1}{2} \; if \; K_0 \neq 0 \\[2ex] < \dfrac{1}{2} \; if \; K_0 = 0, \end{cases} \tag{1.158}$$

where

$$\overline{K} = \frac{1}{8} \left(K + 4 \, K_0 + \sqrt{K^2 + 8 \, K_0 \, K} \right).$$

Then, sequence $\{v_k\}$ given by

$$v_0 = 0, \ v_1 = \eta, \ v_{n+2} = v_n + \frac{K\,(v_{n+1} - v_n)^2}{2\,(1 - K_0\,v_{n+1})} \quad for\ each \quad n = 0, 1, \cdots \quad (1.159)$$

is well defined, nondecreasing, bounded above by $v^{\star\star}$ and converges to its unique least upper bound $v^\star \in [0, v^{\star\star}]$, where

$$v^{\star\star} = \frac{2\,\eta}{2 - \xi}, \qquad (1.160)$$

$$1 \le \xi = \frac{4\,K}{K + \sqrt{K^2 + 8\,K_0\,K}} < 2 \quad for\ K_0 \ne 0. \qquad (1.161)$$

Moreover, the following estimates hold:

$$K_0\,v^\star \le 1, \qquad (1.162)$$

$$0 \le v_{k+1} - v_k \le \frac{\xi}{2}\,(v_k - v_{k-1}) \le \cdots \le \left(\frac{\xi}{2}\right)^k \eta, \quad (k \ge 1), \qquad (1.163)$$

$$v_{k+1} - v_k \le \left(\frac{\xi}{2}\right)^k (2\,q_A)^{2^k - 1}\,\eta, \quad (k \ge 0) \qquad (1.164)$$

and

$$0 \le v^\star - v_k \le \left(\frac{\xi}{2}\right)^k \frac{(2\,q_A)^{2^k - 1}\,\eta}{1 - (2\,q_A)^{2^k}}, \quad (2\,q_A < 1), \quad (k \ge 0). \qquad (1.165)$$

Lemma 1.3.2. *Let β, ℓ, c, d and η be positive constants. Define*

$$\alpha = \beta\,c\,d, \quad \gamma = \beta\,\ell, \quad a = 1 - \gamma\,\eta, \qquad (1.166)$$

$$\delta = \frac{2\,\alpha}{\alpha - a\,c + \sqrt{(\alpha - a\,c)^2 + 4\,\alpha\,(a\,c + \gamma)}}, \qquad (1.167)$$

$$\mu_0 = \frac{2\,\delta}{\alpha + (c + \gamma)\,\delta + \sqrt{(\alpha + (c + \gamma)\,\delta)^2 - 4\,c\,\gamma\,\delta^2}}, \qquad (1.168)$$

$$\mu_1 = \frac{1}{\alpha + \gamma + (a\,c + \gamma)\,\delta} \qquad (1.169)$$

and

$$\mu = \frac{1}{2}\,\min\{\mu_0, \mu_1\}^{-1}. \qquad (1.170)$$

Suppose that

$$h = \mu\,\eta \le \frac{1}{2}. \qquad (1.171)$$

Then, iteration $\{s_n\}$ given by

$$s_0 = 0, \ s_1 = \eta, \ s_{n+2} = s_{n+1} + \frac{\alpha \left(s_{n+1} - s_n\right)^2}{\left(1 - \gamma \, s_{n+1}\right)\left(1 - c \left(s_{n+1} - s_n\right)\right)} \quad for \ each \quad n = 0, 1, \cdots$$

$$\tag{1.172}$$

is well defined, nondecreasing, bounded from above by

$$s^{\star\star} = \frac{\eta}{1 - \delta} \tag{1.173}$$

and converges to the unique least upper bound $s^\star$ satisfying

$$0 \leq s^\star \leq s^{\star\star}. \tag{1.174}$$

Moreover, the following estimate holds:

$$0 \leq s^\star - s_n \leq \frac{\delta^n \eta}{1 - \delta}. \tag{1.175}$$

Proof. It follows from the definition of the constants and (1.171) that $\alpha > 0$, $\gamma > 0$, $a \in (0, 1)$, $\delta \in (0, 1)$, $\mu_i > 0$ $(i = 0, 1)$, $\mu > 0$ and $h > 0$. We shall show the following estimates using induction:

$$0 \leq s_{k+2} - s_{k+1} \leq \delta \left(s_{k+1} - s_k\right), \tag{1.176}$$

$$\gamma \, s_{k+1} < 1 \tag{1.177}$$

and

$$c \left(s_{k+1} - s_k\right) < 1. \tag{1.178}$$

Estimates (1.177) and (1.178) are true for $k = 0$ by (1.169), (1.171), $a \in (0, 1)$ and $\delta \in (0, 1)$. It follows from (1.172) that estimate (1.176) is true, if

$$\frac{\alpha \left(s_{k+1} - s_k\right)}{\left(1 - \gamma \, s_{k+1}\right)\left(1 - c \left(s_{k+1} - s_k\right)\right)} \leq \delta. \tag{1.179}$$

In particular, (1.179) is true for $k = 0$, if

$$\frac{\alpha \, \delta}{\left(1 - \gamma \, \eta\right)\left(1 - c \, \eta\right)} \leq \delta, \tag{1.180}$$

or

$$c \, \gamma \, \delta \, \eta^2 - \left(\alpha + (c + \gamma) \, \delta\right) \eta + \delta \geq 0, \tag{1.181}$$

or

$$\eta \leq \mu_0. \tag{1.182}$$

Inequality (1.182) is true by (1.168) and (1.171). Let us assume estimates (1.176)–(1.178) are true for all $k \leq n + 1$. Then, using (1.176) we have that

$$s_{k+2} - s_{k+1} \leq \delta^{k+1} \, \eta \tag{1.183}$$

and

$$s_{k+1} \leq \frac{1 - \delta^{k+1}}{1 - \delta} \, \eta. \tag{1.184}$$

In view of the induction hypotheses, the induction for (1.176)–(1.178) shall be completed, if

$$\alpha\left(s_{k+1} - s_k\right) - \left(1 - \gamma\, s_{k+1}\right)\left(1 - c\left(s_{k+1} - s_k\right)\right)\delta \le 0, \tag{1.185}$$

or

$$\alpha\left(s_{k+1} - s_k\right) + \gamma\, s_{k+1}\,\delta + c\,\delta\left(1 - \gamma\, s_{k+1}\right)\left(s_{k+1} - s_k\right) - \delta \le 0, \tag{1.186}$$

or (since, $1 - \gamma\, s_{k+1} \le 1 - \gamma\,\eta$)

$$\alpha\left(s_{k+1} - s_k\right) + \gamma\, s_{k+1}\,\delta + c\,\delta\, a\left(s_{k+1} - s_k\right) - \delta \le 0, \tag{1.187}$$

or

$$\left(\alpha + a\, c\,\delta\right)\left(s_{k+1} - s_k\right) + \gamma\, s_{k+1}\,\delta - \delta \le 0, \tag{1.188}$$

or

$$\left(\alpha + a\, c\,\delta\right)\delta^k\,\eta + \gamma\,\frac{1 - \delta^{k+1}}{1 - \delta}\,\eta\,\delta - \delta \le 0. \tag{1.189}$$

Estimate (1.189) motivates us to set $\delta = s$ and define recurrent polynomials f_n on $[0, 1)$ $(n \ge 1)$ by

$$f_n(s) = \left(\alpha + a\, c\, s\right)s^{n-1}\,\eta + \gamma\,\frac{1 - s^{n+1}}{1 - s}\,\eta - 1. \tag{1.190}$$

We need to find a relationship between two consecutive f_n:

$$\begin{aligned}
f_{n+1}(s) &= \left(\alpha + a\, c\, s\right)s^n\,\eta + \gamma\left(1 + s + \cdots + s^{n+1}\right)\eta - 1\\
&= f_n(s) + \left(\alpha + a\, c\, s\right)s^n\,\eta + \gamma\left(1 + s + \cdots + s^{n+1}\right)\eta - 1 - f_n(s)\\
&= f_n(s) + g(s)\, s^{n-1}\,\eta,
\end{aligned} \tag{1.191}$$

where

$$g(s) = \left(a\, c + \gamma\right)s^2 + \left(\alpha - a\, c\right)s + \alpha. \tag{1.192}$$

Note that δ given by (1.167) is the unique root of polynomial g. In view of (1.190), estimate (1.189) holds, if

$$f_k(\delta) \le 0 \qquad (k \ge 1). \tag{1.193}$$

By (1.167), (1.191) and (1.193), we have

$$f_k(\delta) = f_1(\delta) \qquad (k \ge 1). \tag{1.194}$$

Hence, (1.193) holds if

$$f_1(\delta) \le 0, \tag{1.195}$$

or

$$\frac{\eta}{\mu_1} \le 1, \tag{1.196}$$

which is true by (1.169) and (1.171). Moreover, define

$$f_\infty(s) = \lim_{k\to\infty} f_k(s). \tag{1.197}$$

Then, we have

$$f_\infty(s) = \lim_{k \to \infty} f_k(s) \leq 0. \tag{1.198}$$

The induction is complete. It then follows from (1.176) that sequence $\{s_n\}$ is increasing, bounded above by $s^{\star\star}$ and as such it converges to its unique least upper bound $s^\star$. Finally, we shall show estimate (1.175). Let $m \geq 0$. Using (1.176) we have that

$$0 \leq s_{n+m} - s_n = \sum_{i=n}^{n+m-1} (s_{i+1} - s_i) \leq (\delta^n + \delta^{n+1} + \cdots + \delta^{n+m-1})\,\eta = \frac{1 - \delta^m}{1 - \delta}\,\delta^n\,\eta. \tag{1.199}$$

By letting $m \longrightarrow \infty$ in (1.199) we obtain (1.175). That completes the proof of Lemma 1.3.2. $\qquad\square$

We shall study the order of convergence of iteration $\{s_n\}$.

Lemma 1.3.3. *Let $\lambda > 1$ be a fixed parameter. Define*

$$L = 2\,\alpha\,\lambda, \quad L_0 = \gamma, \quad \mu_2 = \frac{1}{c}\left(1 - \frac{1}{\lambda}\right), \tag{1.200}$$

$$\mu_3 = \frac{1}{8}\left(L + 4\,L_0 + \sqrt{L^2 + 8\,L_0\,L}\right) \tag{1.201}$$

and

$$\mu^\star = \frac{1}{2}\,\min\{\mu_i,\ i = 0, 1, 2, 3\}^{-1}, \tag{1.202}$$

where constants α, γ, μ_0, μ_1 and c are given in Lemma 1.3.2. Suppose that

$$h^\star = \mu^\star\,\eta \leq \frac{1}{2} \tag{1.203}$$

and define iteration $\{t_n\}$ by

$$t_0 = 0, \ t_1 = \eta, \ t_{n+2} = t_{n+1} + \frac{L\,(t_{n+1} - t_n)^2}{2\,(1 - L_0\,t_{n+1})} \quad for\ each \quad n = 0, 1, \cdots. \tag{1.204}$$

Then, iterations $\{t_n\}$, $\{s_n\}$ satisfy the conclusions of Lemmas 1.3.1, 1.3.2, respectively. Moreover, the following estimates hold for each $n = 0, 1, \cdots$

$$s_n \leq t_n, \tag{1.205}$$

$$s_{n+1} - s_n \leq t_{n+1} - t_n \tag{1.206}$$

and

$$s^\star - s_n \leq t^\star - t_n, \tag{1.207}$$

where $t^\star = \lim\limits_{n \longrightarrow \infty} t_n$, $s^\star$ and $\{s_n\}$ are given in Lemma 1.3.2.

Proof. Using Lemma 1.3.1, (1.171) and (1.203) we deduce that iteration $\{t_n\}$, $\{s_n\}$ satisfy the conclusions of Lemmas 1.3.1 and 1.3.2, respectively. By (1.172), (1.203), (1.204), a simple inductive argument and the observation (for $k \geq 0$)

$$\mu \leq \mu_2 \implies t_{k+1} - t_k \leq \mu_2 \implies c\,(t_{k+1} - t_k) \leq 1 - \frac{1}{\lambda} \implies \frac{1}{1 - c\,(t_{k+1} - t_k)} \leq \lambda, \tag{1.208}$$

we conclude that (1.205)–(1.207) are true. That completes the proof of Lemma 1.3.3. $\qquad\square$

Remark 1.3.1. Constants μ and $\mu^\star$ depend on η (since δ and a depend on η). However, in view of (1.166), (1.186) and (1.187), we can certainly set $a = 1$. In which case, both μ and $\mu^\star$ are independent of η. Note also that verification of the Kantorovich-type hypotheses (1.171) or (1.203) requires only simple algebraic computations using the initial data β, ℓ, c, d and η.

We can show the following semi-local convergence result for (NM).

Theorem 1.3.1. *Let $F : \mathcal{D} \subseteq \mathcal{X} \longrightarrow \mathcal{Y}$ be a Fréchet-differentiable operator. Suppose that there exist $x_0 \in \mathcal{D}$ and positive constants β, ℓ, c, d, η, such that for all $x, y \in \mathcal{D}$, the following hold:*

$$F'(x_0)^{-1} \in \mathcal{L}(\mathcal{Y}, \mathcal{X}), \quad \| \, F'(x_0)^{-1} F(x_0) \, \| \leq \eta, \quad \| \, F'(x_0)^{-1} \, \| \leq \beta; \tag{1.209}$$

$$\| \, F'(x_0)^{-1} \left(F'(x) - F'(x_0) \right) \, \| \leq \ell \, \| \, x - x_0 \, \|; \tag{1.210}$$

$$\| \, F(y) - F(x) - F'(x) \, (y - x) \, \| \leq c \, \| \, F(y) - F(x) \, \| \, \| \, y - x \, \|; \tag{1.211}$$

$$\| \, F'(x) \, \| \leq d; \tag{1.212}$$

$$h = \mu \eta \leq \frac{1}{2}$$

and

$$\overline{U}(x_0, s^\star) \subseteq \mathcal{D}, \tag{1.213}$$

where $s^\star$ is given in Lemma 1.3.2.

Then, sequence $\{x_n\}$ generated by (NM) is well defined, remains in $\overline{U}(x_0, s^\star)$ for all $n \geq 0$ and converges to a unique solution $x^\star \in \overline{U}(x_0, s^\star)$ of equation $F(x) = 0$. Moreover the following estimate holds for each $n = 0, 1, \cdots$

$$\| \, x_n - x^\star \, \| \leq s^\star - s_n, \tag{1.214}$$

where iteration $\{s_n\}$ is given by (1.172). Furthermore, the solution $x^\star$ is unique in $\overline{U}(x_0, s_0^\star)$, if

$$s_0^\star \geq s^\star, \tag{1.215}$$

$$U(x_0, s_0^\star) \subseteq \mathcal{D} \tag{1.216}$$

and

$$\beta \ell \, (s^\star + s_0^\star) < 2. \tag{1.217}$$

Proof. We shall show using induction that

$$\| x_{k+1} - x_k \| \le s_{k+1} - s_k \tag{1.218}$$

and

$$\overline{U}(x_{k+1}, s^\star - s_{k+1}) \subseteq \overline{U}(x_k, s^\star - s_k) \tag{1.219}$$

hold for all $k \ge 0$. For every $z \in \overline{U}(x_1, s^\star - s_1)$,

$$\| z - x_0 \| \le \| z - x_1 \| + \| x_1 - x_0 \| \le s^\star - s_1 + s_1 = s^\star - s_0,$$

implies $z \in \overline{U}(x_0, s^\star - s_0)$. Since, also

$$\| x_1 - x_0 \| = \| F'(x_0)^{-1} F(x_0) \| \le \eta = s_1 - s_0,$$

estimates (1.218) and (1.219) hold for $k = 0$. Assume estimates (1.218) and (1.219) hold for all $n \le k$. Then, we have that

$$\| x_{k+1} - x_0 \| \le \sum_{i=1}^{k+1} \| x_i - x_{i-1} \| \le \sum_{i=1}^{k+1} (s_i - s_{i-1}) = s_{k+1}$$

and

$$\| x_k + \theta (x_{k+1} - x_k) - x_0 \| \le s_k + \theta (s_{k+1} - s_k) \le s^\star \quad \text{for all} \quad \theta \in [0, 1].$$

In view of (1.171) and (1.210), we have for $x \in \overline{U}(x_0, s^\star)$ that

$$\| F'(x_0)^{-1} \| \, \| F'(x) - F'(x_0) \| \le \beta \ell \, \| x - x_0 \| = \gamma \, \| x - x_0 \| < 1. \tag{1.220}$$

It follows from (1.220) and the Banach lemma on invertible operators (see Refs. [110, 290]) that $F'(x)^{-1}$ exists and

$$\| F'(x)^{-1} \| \le \frac{\beta}{1 - \gamma \, \| x - x_0 \|}. \tag{1.221}$$

In particular, (1.221) for $x = x_{k+1}$ gives

$$\| F'(x_{k+1})^{-1} \| \le \frac{\beta}{1 - \gamma \, \| x_{k+1} - x_0 \|} \le \frac{\beta}{1 - \gamma \, s_{k+1}}. \tag{1.222}$$

Estimates (1.211), (1.212) and (1.171) lead to (for $y = x_{k+1}$, $x = x_k$)

$$\| F(x_{k+1}) - F(x_k) \| \le \| F'(x_k)(x_{k+1} - x_k) \| + c \, \| F(x_{k+1}) - F(x_k) \| \, \| x_{k+1} - x_k \|$$

or

$$\| F(x_{k+1}) - F(x_k) \| \, (1 - c \, \| x_{k+1} - x_k \|) \le \| F'(x_k)(x_{k+1} - x_k) \| \le d \, \| x_{k+1} - x_k \|$$

or

$$\| F(x_{k+1}) - F(x_k) \| \le \frac{d \, \| x_{k+1} - x_k \|}{1 - c \, \| x_{k+1} - x_k \|} \le \frac{d \, (s_{k+1} - s_k)}{1 - c \, (s_{k+1} - s_k)} \tag{1.223}$$

(since $c \, \| x_{k+1} - x_k \| \le c \, (s_{k+1} - s_k) \le c \, (s_1 - s_0) = c\eta < 1$). Using (NM) we obtain the approximation

$$F(x_{k+1}) = F(x_{k+1}) - F(x_k) - F'(x_k)(x_{k+1} - x_k). \tag{1.224}$$

Then, by (1.211) for $y = x_{k+1}$, $x = x_k$, (1.223) and (1.224) we get that

$$\| F(x_{k+1}) \| \leq \frac{c\,d \,\| x_{k+1} - x_k \|^2}{1- \| x_{k+1} - x_k \|} \leq \frac{c\,d\,(s_{k+1} - s_k)^2}{1 - c\,(s_{k+1} - s_k)}. \tag{1.225}$$

Using (1.166), (1.172), (1.220) and (1.225) we obtain that

$$\begin{aligned}
\| x_{k+2} - x_{k+1} \| &= \| F'(x_{k+1})^{-1} F(x_{k+1}) \| \leq \| F'(x_{k+1})^{-1} \| \, \| F(x_{k+1}) \| \\
&\leq \frac{\beta}{1 - \gamma\,s_{k+1}} \frac{c\,d\,(s_{k+1} - s_k)^2}{1 - c\,(s_{k+1} - s_k)} \\
&= \frac{\alpha\,(s_{k+1} - s_k)^2}{(1 - \gamma\,s_{k+1})\,(1 - c\,(s_{k+1} - s_k))} = s_{k+2} - s_{k+1},
\end{aligned} \tag{1.226}$$

which shows (1.218) for all $k \geq 0$. Moreover, for every $w \in \overline{U}(x_{k+2}, s^\star - s_{k+2})$ we have that

$$\begin{aligned}
\| w - x_{k+1} \| &\leq \| w - x_{k+2} \| + \| x_{k+2} - x_{k+1} \| \\
&\leq s^\star - s_{k+2} + s_{k+2} - s_{k+1} = s^\star - s_{k+1},
\end{aligned}$$

which implies $w \in \overline{U}(x_{k+1}, s^\star - s_{k+1})$. The induction for estimates (1.218) and (1.219) is complete. According to Lemma 1.3.2, sequence $\{s_n\}$ is Cauchy. From (1.218) and (1.219) $\{x_n\}$ $(n \geq 0)$ is also a Cauchy sequence and as such it converges to some $x^\star \in \overline{U}(x_0, s^\star)$ (since $\overline{U}(x_0, s^\star)$ is a closed set). By letting $k \longrightarrow \infty$ in (1.225) and noting that

$$0 < \frac{1}{1 - c\,(s_{k+1} - s_k)} \leq \frac{1}{1 - c\,\eta},$$

we get $F(x^\star) = 0$. That is $x^\star$ is a solution of equation (1.1). Estimate (1.214) follows from (1.218) by using standard majorization techniques in Refs. [110, 134, 135, 290]. Finally to show uniqueness, let $y^\star \in \overline{U}(x_0, s_0^\star)$ be a solution of equation $F(x) = 0$. Let $x_\theta^\star = y^\star + \theta\,(x^\star - y^\star)$, $\theta \in [0,1]$. Using (1.210) and (1.215)–(1.217) we get in turn that

$$\begin{aligned}
\| F'(x_0)^{-1} \| \, \| \int_0^1 (F'(x_\theta^\star) - F'(x_0))\, d\theta \| &\leq \beta\,\ell \int_0^1 \| x_\theta^\star - x_0 \|\, d\theta \\
\leq \gamma \int_0^1 (\theta\, \| x^\star - x_0 \| + (1 - \theta)\, \| y^\star - x_0 \|)\, d\theta &\leq \frac{\gamma}{2}\,(s^\star + s_0^\star) < 1.
\end{aligned} \tag{1.227}$$

It follows from (1.227) and the Banach lemma on invertible operators implies that the linear operator $\mathcal{M} = \int_0^1 F'(x_\theta^\star)\, d\theta$ is invertible. Using the identity $0 = F(x^\star) - F(y^\star) = \mathcal{M}\,(x^\star - y^\star)$, we deduce $x^\star = y^\star$. That completes the proof of Theorem 1.3.1. $\qquad\square$

Remark 1.3.2.

(a) The limit point $s^\star$ can be replaced by $s^{\star\star}$ given in the closed form by (1.173) in hypotheses (1.213)–(1.215) and (1.217).

(b) If (1.171) is replaced by (1.203) then, under the hypotheses of Theorem 1.3.1, conclusions of the theorem hold true with $\{t_n\}$, $t^\star$, $t^{\star\star}$ replacing $\{s_n\}$, $s^\star$, $s^{\star\star}$, respectively.

(c) In view of (1.210), the constant d in hypothesis (1.212) can be replaced by

$$d_0 = \ell\, r^\star + \parallel F'(x_0) \parallel, \qquad (1.228)$$

where $r^\star$ is $s^\star$ or $s^{\star\star}$ and $\mathcal{D} = \overline{U}(x_0, s^\star)$ or $\mathcal{D} = \overline{U}(x_0, s^{\star\star})$, respectively.

(d) Hypotheses (1.211) and (1.212) can be replaced by

$$\parallel F(y) - F(x) - F'(x)\,(y - x) \parallel\, \leq \frac{K}{2} \parallel y - x \parallel^2 \quad \text{for all} \quad x, y \in \mathcal{D}. \qquad (1.229)$$

Set $K_0 = \ell$. In this case, under the hypotheses of Lemma 1.3.1 and the rest of the hypotheses of Theorem 1.3.1, the conclusions hold with $\{v_n\}$, $v^\star$, $v_0^\star$, $v^{\star\star}$ replacing $\{s_n\}$, $s^\star$, $s_0^\star$, $s^{\star\star}$, respectively.

(e) Hypothesis (1.211) can also be replaced by

$$\parallel F(y) - F(x) - F'(x)\,(y - x) \parallel\, \leq c \parallel F'(x)\,(y - x) \parallel \parallel y - x \parallel \quad \text{for all} \quad x, y \in \mathcal{D}. \qquad (1.230)$$

Set $K_0 = \ell$ and $K = 2\,c\,d$. Then, see (d) above.

(f) In the Shen, Li Ref. [405] and the Guo Ref. [246], inexact Newton methods have been studied under Lipschitz condition (1.148). Their semi-local sufficient convergence condition reduces in the special case of Newton's method to the Kantorovich hypothesis (1.149) which is stronger than (1.151) or (1.158) or (1.171). In Refs. [405, 246], the authors assume that $\mathcal{D}$ is a convex subset of $\mathcal{X}$. The convexity of $\mathcal{D}$ is not assumed in the present section. Moreover, hypotheses (1.211), (1.229) or (1.230) do not necessarily imply (1.148) and vice versa. Similarly, in the local case, the convergence radius is given by (1.152) but ours is given by (1.155) (or (1.235)) which can be at least as large (see (1.156), (1.157) and (1.248)). Finally, (1.232) or (1.233) does not necessarily imply (1.148).

Now, we examine the local convergence of (NM).

Theorem 1.3.2. *Let $F : \mathcal{D} \subseteq \mathcal{X} \longrightarrow \mathcal{Y}$ be a Fréchet-differentiable operator. Suppose that there exist $x^\star \in \mathcal{D}$ and positive constants β, ℓ, c, d, $\gamma = \beta\,\ell$, $\alpha = \beta\,c\,d$, such that for all $x \in \mathcal{D}$, the following hold:*

$$F(x^\star) = 0, \quad F'(x^\star)^{-1} \in \mathcal{L}(\mathcal{Y}, \mathcal{X}), \quad \parallel F'(x^\star)^{-1} \parallel\, \leq \beta; \qquad (1.231)$$

$$\parallel F'(x) - F'(x^\star) \parallel\, \leq \ell \parallel x - x^\star \parallel; \qquad (1.232)$$

$$\parallel F(x) - F'(x)\,(x - x^\star) \parallel\, \leq c \parallel F(x) \parallel \parallel x - x^\star \parallel; \qquad (1.233)$$

$$\parallel F'(x) \parallel\, \leq d$$

and

$$\overline{U}(x^\star, R^\star) \subseteq \mathcal{D}, \qquad (1.234)$$

where

$$R^\star = \frac{2}{\alpha + c + \gamma + \sqrt{(\alpha + c + \gamma)^2 - 4\,\gamma\,c}}. \qquad (1.235)$$

Then, sequence $\{x_n\}$ generated by (NM) is well defined, remains in $\overline{U}(x^\star, R^\star)$ for all $n \geq 0$ and converges to $x^\star$ for $x_0 \in \overline{U}(x^\star, R^\star)$. Moreover, the following estimates hold for each $n = 0, 1, \cdots$

$$\| x_{n+1} - x^\star \| \leq \frac{\alpha \| x_n - x^\star \|^2}{(1 - \gamma \| x_n - x^\star \|)(1 - c \| x_n - x^\star \|)}. \tag{1.236}$$

Proof. The point $R^\star$ is well defined, since

$$(\alpha + c + \gamma)^2 - 4\gamma c = \alpha^2 + 2\alpha(c + \gamma) + (c - \gamma)^2 > 0. \tag{1.237}$$

Let $x \in \overline{U}(x^\star, R^\star)$. Then using (1.231), (1.232) and (1.235), we have that

$$\| F'(x^\star)^{-1} \| \| F'(x) - F'(x^\star) \| \leq \beta \ell \| x - x^\star \| = \gamma \| x - x^\star \| \leq \gamma R^\star < 1. \tag{1.238}$$

It follows from (1.238) and the Banach lemma on invertible operators that $F'(x)^{-1}$ exists and

$$\| F'(x)^{-1} \| \leq \frac{\beta}{1 - \gamma \| x - x^\star \|} \leq \frac{\beta}{1 - \gamma R^\star}. \tag{1.239}$$

In particular, (1.239) for $x = x_{k+1}$ gives

$$\| F'(x_k)^{-1} \| \leq \frac{\beta}{1 - \gamma \| x_k - x^\star \|} \leq \frac{\beta}{1 - \gamma R^\star}. \tag{1.240}$$

Using (NM) we obtain the approximation

$$\begin{aligned} x_{k+1} - x^\star &= x_k - F'(x_k)^{-1} F(x_k) - x^\star \\ &= -F'(x_k)^{-1}(F(x_k) - F(x^\star) - F'(x_k)(x_k - x^\star)). \end{aligned} \tag{1.241}$$

Exactly as in the derivation of (1.225) we arrive at:

$$\| F(x_k) - F(x^\star) - F'(x_k)(x_k - x^\star) \| \leq \frac{cd \| x_k - x^\star \|^2}{1 - c \| x_k - x^\star \|}. \tag{1.242}$$

Using (1.240)–(1.242) we get that

$$\begin{aligned} \| x_{k+1} - x^\star \| &\leq \frac{\beta}{1 - \gamma \| x_k - x^\star \|} \frac{cd \| x_k - x^\star \|^2}{1 - c \| x_k - x^\star \|} \\ &= \frac{\alpha \| x_k - x^\star \|^2}{(1 - \gamma \| x_k - x^\star \|)(1 - c \| x_k - x^\star \|)}. \end{aligned} \tag{1.243}$$

By hypotheses $x_0 \in \overline{U}(x^\star, R^\star)$, (1.243) for $k = 0$ and (1.235) we have that

$$\begin{aligned} \| x_1 - x^\star \| &\leq \frac{\alpha \| x_0 - x^\star \|^2}{(1 - \gamma \| x_0 - x^\star \|)(1 - c \| x_0 - x^\star \|)} \\ &< \frac{\alpha R^\star}{(1 - \gamma R^\star)(1 - c R^\star)} \| x_0 - x^\star \| = \| x_0 - x^\star \| < R^\star, \end{aligned} \tag{1.244}$$

which implies $x_1 \in U(x^\star, R^\star)$. Assuming $x_n \in \overline{U}(x^\star, R^\star)$ for all $n \leq k$, we get by (1.243) that

$$\begin{aligned} \| x_{k+1} - x^\star \| &\leq \frac{\alpha \| x_k - x^\star \|^2}{(1 - \gamma \| x_k - x^\star \|)(1 - c \| x_k - x^\star \|)} \\ &< \frac{\alpha R^\star}{(1 - \gamma R^\star)(1 - c R^\star)} \| x_k - x^\star \| = \| x_k - x^\star \| < R^\star, \end{aligned} \tag{1.245}$$

which implies $x_{k+1} \in \overline{U}(x^\star, R^\star)$. The induction is now complete. It also follows from (1.245) that $\lim\limits_{k \longrightarrow \infty} x_k = x^\star$. That completes the proof of Theorem 1.3.2. $\quad\square$

Remark 1.3.3. The results obtained in this section can be given in affine invariant form if $F'(x_0)\,F$ or $F'(x^\star)^{-1}\,F$ replaces operator F in the semi-local and local cases, respectively. We should also set $\beta = 1$ in this case.

Remark 1.3.4. It is possible that any of the inequalities

$$\mu < \overline{K}, \tag{1.246}$$

or

$$\mu^\star < \overline{K} \tag{1.247}$$

and

$$r^\star < R^\star, \tag{1.248}$$

is true. In this case advantages $(\mathcal{A})$ can be obtained. For example if (1.246) is true then, (1.171) (depending on (1.211)) is weaker than (1.158) (depending on (1.148)).

We provide a numerical example to show that center-Lipschitz condition (1.150) holds but not Lipschitz condition (1.148).

Example 1.3.1. Consider the integral equation

$$u(s) = f(s) + \varrho \int_a^b \mathcal{G}(s,t)\,u(t)^{1+\frac{1}{n}}\,dt, \quad n \in \mathbb{N}.$$

Here, f is a given continuous function satisfying $f(s) > 0$, $s \in [a,b]$, ϱ is a real number and the kernel $\mathcal{G}$ is continuous and positive in $[a,b] \times [a,b]$. For example, when $\mathcal{G}(s,t)$ is the Green kernel, the corresponding integral equation is equivalent to the boundary value problem

$$u'' = \varrho\,u^{1+\frac{1}{n}},$$
$$u(a) = f(a), \ u(b) = f(b).$$

These types of problems have been considered in Refs. [110, 134, 135]. Consider operator F as follows

$$F \colon \mathcal{D} \subseteq \mathcal{C}[a,b] \to \mathcal{C}[a,b], \quad \mathcal{D} = \{u \in \mathcal{C}[a,b] \colon u(s) \geq 0,\ s \in [a,b]\}$$

and

$$F(u)(s) = u(s) - f(s) - \varrho \int_a^b \mathcal{G}(s,t)\,u(t)^{1+\frac{1}{n}}\,dt. \tag{1.249}$$

$\mathcal{C}[a,b]$ is equipped with the max-norm. The derivative F' is given by

$$F'(u)\,v(s) = v(s) - \varrho\left(1 + \frac{1}{n}\right)\int_a^b \mathcal{G}(s,t)\,u(t)^{\frac{1}{n}}v(t)\,dt, \quad v \in \mathcal{D}. \tag{1.250}$$

First, note that F' does not satisfy a Lipschitz condition (1.148) in $\mathcal{D}$. Let us consider, for instance, $[a,b] = [0,1]$, $\mathcal{G}(s,t) = 1$ and $y(t) = 0$. Then $F'(y)\,v(s) = v(s)$ and

$$\| F'(x) - F'(y) \| = |\varrho|\left(1 + \frac{1}{n}\right)\int_0^1 x(t)^{\frac{1}{n}}\,dt. \tag{1.251}$$

Suppose that F' is a Lipschitz function, then

$$\| F'(x) - F'(y) \| \le K \| x - y \|.$$

Consequently, we obtain that

$$\int_0^1 x(t)^{\frac{1}{n}} \, dt \le \frac{K}{|\varrho| \left(1 + \frac{1}{n}\right)} \max_{x \in [0,1]} x(s) \tag{1.252}$$

would hold for all $x \in \mathcal{D}$. But this is not true. Consider, for example, the functions

$$x_j(t) = \frac{t}{j}, \quad j \ge 1, \quad t \in [0,1].$$

If these are substituted into (1.252), we obtain that

$$\frac{1}{j^{1/n}\left(1 + \frac{1}{n}\right)} \le \frac{K}{j\,|\varrho|\left(1 + \frac{1}{n}\right)} \iff j^{1-\frac{1}{n}} \le \frac{K}{|\varrho|}, \quad \forall j \ge 1.$$

This inequality is not true when $j \to \infty$. Therefore, condition (1.148) fails in this case. However, condition (1.150) holds. To show this, let $x_0(t) = f(t)$ and $\Xi = \min_{s \in [a,b]} f(s) > 0$. Then, for $v \in \mathcal{D}$,

$$\| (F'(x) - F'(x_0)) v \| = |\varrho| \left(1 + \frac{1}{n}\right) \max_{s \in [a,b]} \left| \int_a^b \mathcal{G}(s,t)\left(x(t)^{\frac{1}{n}} - f(t)^{\frac{1}{n}}\right) v(t) dt \right|$$

$$\le |\varrho| \left(1 + \frac{1}{n}\right) \max_{s \in [a,b]} \int_a^b \frac{\mathcal{G}(s,t)\,|x(t) - f(t)|}{x(t)^{(n-1)/n} + x(t)^{(n-2)/n} f(t)^{1/n} + \cdots + f(t)^{(n-1)/n}} \, dt \, \| v \|.$$

Hence,

$$\| F'(x) - F'(x_0) \| \le \frac{|\varrho|\left(1 + \frac{1}{n}\right)}{\Xi^{(n-1)/n}} \max_{s \in [a,b]} \int_a^b \mathcal{G}(s,t) \, dt \, \| x - x_0 \| = K_0 \, \| x - x_0 \|,$$

where

$$K_0 = \frac{|\varrho|\left(1 + \frac{1}{n}\right)}{\Xi^{(n-1)/n}} \max_{s \in [a,b]} \int_a^b \mathcal{G}(s,t) \, dt$$

and (1.150) is satisfied.

Finally, we provide a numerical example to show how the parameters in Lemma 1.3.3 and Theorem 1.3.1 can be computed.

Example 1.3.2. Let $\mathcal{X} = \mathcal{Y} = \mathbb{R}^2$, $x_0 = (.495, .495)^T$ and $\mathcal{D} = \overline{U}(x_0, \frac{7}{8})$. Define function F on $\mathcal{D}$ by

$$F(x) = (\xi_1^3 - \frac{1}{8}, \xi_2^3 - \frac{1}{8})^T, \qquad x = (\xi_1, \xi_2)^T. \tag{1.253}$$

Using (1.166)–(1.169), (1.200)–(1.203), (1.209)–(1.212) and (1.253), we get that

$$\ell = 3\left(\frac{7}{8} + 2\,|x_0|\right), \quad d = 3\left(|x_0| + \frac{7}{8}\right)^2,$$

$$\beta = 1.360405401, \quad c = 8, \quad \eta = .0050506751,$$

$$\ell = 5.595, \quad d = 5.6307, \quad \gamma = 7.611468219,$$

$$a = .96155694, \quad \alpha = 61.28027753, \quad \delta = .90861451,$$

$$\mu_0 = .12499999 \quad \text{and} \quad \mu_1 = .01207895.$$

For $\lambda = 1.5$ we obtain that

$$L_0 = \gamma = 7.611468219, \quad L = 183.8408325, \quad \mu_2 = .4166666,$$

$$\mu_3 = 53.29988847 \quad \text{and} \quad \mu^\star = \frac{1}{2\,\mu_1} = 41.39432649.$$

Then, (1.203) is satisfied, since

$$h^\star = \mu^\star\,\eta = .20906929 < .5.$$

Consequenlty, condition (1.213) holds for $s^\star$ replaced by $s^{\star\star} = .0092508411$. That is, all hypotheses of Theorem 1.3.1 are satisfied. Hence, (NM) starting at x_0 converges quadratically to $x^\star = (.5, .5)^T$.

1.4   Convex Majorants

We provide a local convergence analysis for Newton's method under a weak majorant condition in a Banach space setting. Our results provide under the same information a larger radius of convergence and tighter error estimates on the distances involved than before Ref. [212]. Special cases and numerical examples are also provided in this section. We are concerned with the problem of approximating a locally unique solution $x^\star$ of equation (1.1). We use Newton's method (NM) given by (1.2). Recently, some results using convex majorants are provided by Argyros in Refs. [109, 115], which improved and extended the corresponding ones given by Ferreira, Gonçalves in Ref. [213], Ferreira, Svaiter in Ref. [217] and Wang in Ref. [427].

We are motivated in this section by a recent paper of Ferreira (see Ref. [212]), which weakened earlier convergence conditions (see Refs. [213, 217, 427]) for the local convergence analysis of (NM) under general majorant condition (see (1.255)). The information used is $I(x^\star, F, f)$, where f is a majorant function (to be precised later in (1.255)). Using $I(x^\star, F, f)$, Ferreira in Ref. [212] provided error estimates on the distances $\| x_n - x^\star \|$ $(n \geq 1)$ as well as what he claimed to be the best possible convergence radius.

In our analysis we are also motivated by optimization considerations and the work in Ref. [212]. Using the same information $I(x^\star, F, f)$, we show that in general the radius of convergence given in Ref. [212] is not as the best possible but it can be enlarged. We also show that the upper bounds on the distances $\| x_n - x^\star \|$ $(n \geq 1)$ can be tighter. These observations are very important in computational mathematics, since they allow a wider choice of initial guesses x_0 and fewer iterations to obtain a desired error tolerance $\epsilon > 0$. Note that similar improvements in both the local and semi-local cases of the works in Refs. [213, 217, 305, 427] have already been obtained by us in Refs. [109, 140, 163] under stronger than (1.255) majorant-type conditions.

We state the main local convergence result for (NM) under the majorant condition.

Theorem 1.4.1. *Let $\mathcal{D}$ be an open and convex set; let $\mathcal{X}$ and $\mathcal{Y}$ be Banach spaces and let $F : \mathcal{D} \subset \mathcal{X} \longrightarrow \mathcal{Y}$ be Fréchet-differentiable. Let $x^\star \in \mathcal{D}$ such that $F(x^\star) = 0$, $R > 0$ and $\kappa := \sup\{t \in [0, R) : U(x^\star, t) \subset \mathcal{D}\}$. Suppose $F'(x^\star)^{-1} \in \mathcal{L}(\mathcal{Y}, \mathcal{X})$ and there exist $f, f_0 : [0, R) \longrightarrow (-\infty, +\infty)$ continuously differentiable such that*

$$\| F'(x^\star)^{-1} \left(F'(x) - F'(x^\star) \right) \| \le f_0'(\| x - x^\star \|) - f_0'(0), \tag{1.254}$$

$$\| F'(x^\star)^{-1} \left(F'(x) - F'(x_\theta) \right) \| \le f'(\| x - x^\star \|) - f'(\theta \| x - x^\star \|), \tag{1.255}$$

for all $x \in U(x^\star, \kappa)$ and $x_\theta = x^\star + \theta (x - x^\star)$, $\theta \in [0, 1]$,

$(\mathcal{H}_1)$ *$f_0(0) = f(0) = 0$ and $f_0'(0) = f'(0) = -1$;*
$(\mathcal{H}_2)$ *f_0', f' are strictly increasing,*

$$f_0(t) \le f(t) \quad and \quad f_0'(t) \le f'(t) \quad t \in [0, R). \tag{1.256}$$

Define: parameter ν_0, function f_1 on $(0, \nu_0)$, parameters ν, ρ_0, r_0 and scalar iteration $\{s_n\}$ by

$$\nu_0 := \sup\{t \in [0, R) : f_0'(t) < 0\},$$

$$f_1(t) := \frac{f'(t)}{f_0'(t)}, \tag{1.257}$$

$$\nu := \sup\{t \in [0, R) : f'(t) < 0\},$$

$$\rho_0 := \sup\{\delta \in [0, \nu) : \left(\frac{f(t)}{f'(t)} - t \right) \frac{f_1(t)}{t} < 1, \ t \in [0, \delta)\}$$

$$r_0 := \min\{\kappa, \rho_0\},$$

$$s_0 = \| x_0 - x^\star \|, \quad s_{n+1} = \left| \left(s_n - \frac{f(s_n)}{f'(s_n)} \right) f_1(s_n) \right| \quad (n \ge 0). \tag{1.258}$$

Then, the following assertions hold:

(a) $\{s_n\}$ is well defined; strictly decreasing; contained in $(0, r_0)$; converges to zero and

$$\lim_{n \longrightarrow 0} \frac{s_{n+1}}{s_n} = 0. \tag{1.259}$$

(b) $\{x_n\}$ generated by (NM), starting from $x_0 \in U(x^\star, r_0) \setminus \{x^\star\}$ is well defined; remains in $U(x^\star, r_0)$ for all $n \geq 0$ and converges to $x^\star$, which is the unique solution of equation (1.1) in $U(x^\star, \sigma_0)$, where

$$\sigma_0 := \sup\{t \in [0, \kappa) : f_0(t) < 0\}$$

and

$$\lim_{n \longrightarrow \infty} \frac{\| x_{n+1} - x_n \|}{\| x_n - x^\star \|} = 0. \tag{1.260}$$

(c) If

$$\left(\frac{f(\rho_0)}{\rho_0 \, f'(\rho_0)} - 1 \right) f_1(\rho_0) = 1 \quad and \quad \rho_0 < \kappa$$

then $r_0 = \rho_0$ is the possible convergence radius.

(d) If scalar sequence $\{t_n\}$ is given by

$$t_0 = \| x_0 - x^\star \|, \quad t_{n+1} = \left| t_n - \frac{f(t_n)}{f'(t_n)} \right| \quad (n \geq 0) \tag{1.261}$$

then

$$s_n \leq t_n \quad (n \geq 0) \tag{1.262}$$

and strict inequality holds for $n > 1$ in (1.262), if $f_0'(t) < f'(t)$, $t \in [0, R)$.

If additionally, given $0 \leq p \leq 1$

$(\mathcal{H}_3)$ The function $t \longrightarrow \left(\dfrac{f(t)}{f'(t)} - t \right) \dfrac{f_1(t)}{t^{p+1}}$ is strictly increasing on $(0, \nu_0)$,

then,

(e) The sequence $\left\{ \dfrac{s_{n+1}}{s_n^{p+1}} \right\}$ is strictly decreasing so that

$$\| x_{n+1} - x^\star \| \leq \frac{s_{n+1}}{s_n^{p+1}} \| x_n - x^\star \|^{p+1} \quad (n \geq 0). \tag{1.263}$$

Furthermore, for $n \geq 0$,

$$\| x_n - x^\star \| \leq \begin{cases} s_0 \left[\dfrac{s_1}{s_0} \right]^n & if \ p = 0 \\[3mm] s_0 \left(\dfrac{s_1}{s_0} \right)^{((p+1)^n - 1)/p} & if \ p \neq 0. \end{cases} \tag{1.264}$$

We shall break down the proof of Theorem 1.4.1 into 10 pieces called lemmas. First we shall show the statements of the theorem involving sequence $\{s_n\}$.

Lemma 1.4.1. *The constants* κ, ν, σ_0 *are positive and* $\left(t - \dfrac{f(t)}{f'(t)} \right) f_1(t) < 0$ *for all* $t \in (0, \nu)$.

Proof. The set $\mathcal{D}$ is open and $x^\star \in \mathcal{D}$, so we deduce that κ is positive. Since f' is continuous in 0 with $f'(0) = -1$, there exists $\delta > 0$ such that $f'(t) < 0$ for all $t \in (0, \delta)$. That is $\nu > 0$. Now, because $f(0) = 0$ and $f'(0) = -1$, there exists $\delta > 0$ such that $f(t) < 0$ for all $t \in (0, \delta)$. Hence, we have $\sigma = \sup\{t \in [0, \kappa) : f(t) < 0\} > 0$ and by $(\mathcal{H}_2)$: $\sigma_0 \geq \sigma > 0$, $f_0(t) < 0$, $t \in (0, \sigma_0)$.

It also follows from $(\mathcal{H}_1)$ and $(\mathcal{H}_2)$ that $0 = f(0) > f(t) - t\, f'(t)$ for $t \in (0, R)$. If $t \in (0, \nu)$ then $f'(t) < 0$, which together with (1.257) complete the proof of Lemma 1.4.1. $\qquad\square$

According to $(\mathcal{H}_2)$, the definition of ν_0 and ν, we have $f_0'(t) < 0$, $f'(t) < 0$ for all $t \in [0, \nu)$, since $\nu \leq \nu_0$. Moreover, function f_1 is well defined on $(0, \nu_0)$. Therefore, Newton's iteration function

$$\eta_{f,f_0} : [0, \nu) \longrightarrow (-\infty, 0]$$
$$t \longrightarrow \left(t - \frac{f(t)}{f'(t)} \right) f_1(t) \tag{1.265}$$

is well defined.

Lemma 1.4.2. *The following assertions hold:*

$$\lim_{t \to 0} \frac{\eta_{f,f_0}(t)}{t} = 0, \tag{1.266}$$

$$\rho_0 > 0 \tag{1.267}$$

and

$$|\eta_{f,f_0}(t)| < t \quad for\ all \quad t \in (0, \rho_0). \tag{1.268}$$

Proof. Using definition (1.265), Lemma 1.4.1, $(\mathcal{H}_1)$ and the definition of ν, a simple algebraic manipulation leads to

$$\begin{aligned}
\frac{|\eta_{f,f_0}(t)|}{t} &= \left(\frac{f(t)}{f'(t)} - t \right) \frac{f_1(t)}{t} \\
&= \left(\frac{1}{f'(t)} \frac{f(t) - f(0)}{t - 0} - 1 \right) f_1(t) \quad \text{for all} \quad t \in (0, \nu),
\end{aligned} \tag{1.269}$$

which leads to (1.266) if we let $t \longrightarrow 0$ in (1.269). It then follows from (1.266) and the first equality in (1.269) that there exists $\delta > 0$ such that

$$0 < \left(\frac{f(t)}{f'(t)} - t \right) \frac{f_1(t)}{t} < 1 \quad \text{for all} \quad t \in (0, \delta). \tag{1.270}$$

Hence, we deduce that $\rho_0 > 0$. Finally, the first equality in (1.269) together with the definition of ρ_0 imply (1.268). That completes the proof of Lemma 1.4.2. $\qquad\square$

In view of (1.265), sequence $\{s_n\}$ can be defined as:

$$s_0 = \| x_0 - x^\star \|, \quad s_{n+1} = |\eta_{f,f_0}(s_n)| \quad (n \geq 0). \tag{1.271}$$

Replace η_f by η_{f,f_0} in the proof of [Corollary 5] in Ref. [212] to obtain:

Lemma 1.4.3. *Sequence $\{s_n\}$ is well defined, strictly decreasing and contained in $(0, \rho_0)$. Moreover, $\{s_n\}$ converges to zero with superlinear rate, i.e., $\displaystyle \lim_{n \longrightarrow \infty} \frac{s_{n+1}}{s_n} = 0$. Furthermore, if $(\mathcal{H}_3)$ holds, then sequence $\left\{ \frac{s_{n+1}}{s_n^{p+1}} \right\}$ is strictly decreasing.*

Proof. It follows from $0 < s_0 = \| x_0 - x^\star \| < r_0 \leq \rho_0$, Lemma 1.4.2 and (1.271) that sequence $\{s_n\}$ is well defined, strictly decreasing and is contained in $(0, \rho_0)$. Hence, it also converges. So, there exists $s^\star$ such that $\lim\limits_{n \to \infty} s_n = s^\star$ with $0 \leq s^\star < \rho_0$, which together with (1.271) imply $0 \leq s^\star = |\eta_{f,f_0}(s^\star)|$. If $s^\star \neq 0$, then Lemma 1.4.2 implies $|\eta_{f,f_0}(s^\star)| < s^\star$. Hence, we deduce $s^\star = 0$. Therefore, the definition of $\{s_n\}$ in (1.271) and Lemma 1.4.2 imply

$$\lim_{n \to \infty} \frac{s_{n+1}}{s_n} = \lim_{n \to \infty} \frac{|\eta_{f,f_0}(s_n)|}{s_n} = 0.$$

Finally, since $\{s_n\}$ is strictly decreasing, the last statement is a consequence of $(\mathcal{H}_3)$. That completes the proof of Lemma 1.4.3. $\square$

Secondly, we need relationships between the majorant function f and nonlinear operator F.

We provide in the following lemma a perturbation result.

Lemma 1.4.4. *If $x \in U(x^\star, t)$, $t \in [0, \min\{\kappa, \nu_0\})$, $\| x - x^\star \| \leq \min\{\kappa, \nu_0\}$, then the following assertions hold*

$$F'(x)^{-1} \in \mathcal{L}(\mathcal{Y}, \mathcal{X})$$

and

$$\| F'(x)^{-1} F'(x^\star) \| \leq -\frac{1}{f_0'(\| x - x^\star \|)} \leq -\frac{1}{f_0'(t)}. \tag{1.272}$$

Proof. Let $x \in U(x^\star, t)$, $t \in [0, \min\{\kappa, \nu_0\})$. Using $f_0'(0) = -1$, (1.254) and the fact that f_0' is strictly increasing, we obtain in turn

$$\begin{aligned}
\| F'(x^\star)^{-1} (F'(x) - F'(x^\star)) \| &\leq f_0'(\| x - x^\star \|) - f_0'(0) \\
&= f_0'(\| x - x^\star \|) + 1 \leq f_0'(t) + 1 < 1.
\end{aligned} \tag{1.273}$$

The last inequality in (1.273) holds by the definitions of κ, ν_0 and the choice of t. It then follows from (1.273) and the Banach Lemma on invertible operators (see Refs. [106, 163, 290]) that $F'(x)^{-1} \in \mathcal{L}(\mathcal{Y}, \mathcal{X})$ so that (1.272) holds. That completes the proof of Lemma 1.4.4. $\square$

Newton's iteration at a point is a zero of the linearization of F at such a point. Hence, we shall study the linearization error at a point in $\mathcal{D}$:

$$E_F(x, y) := F(y) - (F(x) + F'(x)(x - y)) \quad \text{for all} \quad x, y \in \mathcal{D}. \tag{1.274}$$

We shall bound this error by the error in linearization of the majorant function f:

$$e_f(t, u) := f(u) - (f(t) + f'(t)(u - t)) \quad \text{for all} \quad t, u \in [0, R]. \tag{1.275}$$

Lemma 1.4.5. *If $\| x^\star - x \| < \kappa$, then the following assertion holds*

$$\| F'(x^\star)^{-1} E_F(x, x^\star) \| \leq e_f(\| x - x^\star \|, 0).$$

Proof. The proof of Lemma 1.4.5 is given in [Lemma 2.10] of Ref. [210] or in [Lemma 7] of Ref. [212]. $\square$

Lemma 1.4.4 guarantees the invertibility of F' and consequently

$$N_f : U(x^\star, r_0) \longrightarrow \mathcal{Y}$$
$$x \longrightarrow x - F'(x)^{-1} F(x) \tag{1.276}$$

is a well-defined operator.

Lemma 1.4.6. *If $\| x - x^\star \| < r_0$, then the following assertions hold*

$$\| N_F(x) - x^\star \| \leq |\eta_{f,f_0}(\| x - x^\star \|)| \tag{1.277}$$

and

$$N_F(U(x^\star, r_0)) \subset U(x^\star, r_0). \tag{1.278}$$

Proof. It follows from $F(x^\star) = 0$ that the inequality is trivial for $x = x^\star$. If $0 < \| x - x^\star \| < r_0$, Lemma 1.4.4 implies that $F'(x)$ is invertible. Using $F(x^\star) = 0$ and (1.276), we obtain the approximation

$$x^\star - N_F(x) = -F'(x)^{-1} \left(F(x^\star) - F(x) - F'(x) (x^\star - x) \right)$$
$$= -F'(x)^{-1} E_F(x, x^\star). \tag{1.279}$$

Using Lemma 1.4.4, Lemma 1.4.5 and (1.279), we get in turn

$$\| x^\star - N_F(x) \| \leq \| F'(x)^{-1} F'(x^\star) \| \, \| F'(x^\star)^{-1} E_F(x, x^\star) \|$$
$$\leq \frac{e_f(\| x - x^\star \|, 0)}{|f_0'(\| x - x^\star \|)|}. \tag{1.280}$$

By the definition of e_f, η_{f,f_0} and hypothesis $f(0) = 0$, we have

$$\frac{e_f(\| x - x^\star \|, 0)}{|f_0'(\| x - x^\star \|)|} = |\eta_{f,f_0}(\| x - x^\star \|)|, \tag{1.281}$$

which together with (1.280) show (1.277). Let $x \in U(x^\star, r_0)$. It follows from $\| x - x^\star \| < r_0 \leq \rho_0$, (1.277) and Lemma 1.4.2 that

$$\| N_F(x) - x^\star \| \leq |\eta_{f,f_0}(\| x - x^\star \|)| < \| x - x^\star \|,$$

which shows (1.278). That completes the proof of Lemma 1.4.6. $\qquad\square$

Lemma 1.4.7. *If $(\mathcal{H}_3)$ holds and*

$$\| x - x^\star \| \leq t < r_0, \tag{1.282}$$

then the following assertion holds

$$\| N_F(x) - x^\star \| \leq \frac{|\eta_{f,f_0}(t)|}{t^{p+1}} \| x - x^\star \|^{p+1}. \tag{1.283}$$

Proof. Estimate (1.283) is trivial, if $x = x^\star$. Assume $\| x - x^\star \| \leq t$, then $(\mathcal{H}_3)$ and (1.265) imply

$$\frac{|\eta_{f,f_0}(\| x - x^\star \|)|}{\| x - x^\star \|^{p+1}} \leq \frac{|\eta_{f,f_0}(t)|}{t^{p+1}}. \tag{1.284}$$

The result follows from Lemma 1.4.6 and (1.284). That completes the proof of Lemma 1.4.7. $\qquad\square$

Next, we shall establish the uniqueness and optimal convergence radius. The proof of the next two results can be found in the analogous [Lemma 2.13] of Ref. [210] and [Lemma 2.15] of Ref. [210], respectively.

Lemma 1.4.8. *The point $x^\star$ is the unique zero of operator F in $U(x^\star, \sigma_0)$.*

Lemma 1.4.9. *If*

$$\left(\frac{f(\rho_0)}{\rho_0 \, f'(\rho_0)} - 1 \right) f_1(\rho_0) = 1$$

and $\rho_0 < \kappa$, then $r_0 = \rho_0$ is the optimal convergence radius.

Finally, we shall show the statements of Theorem 1.4.1 involving Newton's method sequence $\{x_n\}$. It follows from (1.2) and (1.276) that Newton's method can be written as:

$$x_{n+1} = N_F(x_n) \qquad (n \geq 0). \tag{1.285}$$

Lemma 1.4.10. *Sequence $\{x_n\}$ is well defined, contained in $U(x^\star, r_0)$ and converges to the point $x^\star$, which is the unique zero of F in $U(x^\star, \sigma_0)$. Moreover,*

$$\lim_{n \to \infty} \frac{\| x_{n+1} - x^\star \|}{\| x_n - x^\star \|} = 0. \tag{1.286}$$

Furthermore, if $(\mathcal{H}_3)$ holds, then so do (1.263) and (1.264).

Proof.   Let $x_0 \in U(x^\star, r_0)$ and $r_0 \leq \nu_0$. Using Lemmas 1.4.4, 1.4.6 and (1.285), we deduce that sequence $\{x_n\}$ is well defined and remains in $U(x^\star, r_0)$ for all $n \geq 0$. Using Lemmas 1.4.2, 1.4.6 and (1.285), we obtain in turn

$$\begin{aligned}
\| x_{n+1} - x^\star \| &\leq \| N_F(x_n) - x^\star \| \\
&\leq |\eta_{f,f_0}(\| x_n - x^\star \|)| < \| x_n - x^\star \| \quad (n \geq 0).
\end{aligned} \tag{1.287}$$

Hence $\{\| x_n - x^\star \|\}$ is strictly decreasing and converges to some α. Since $\| x_n - x^\star \|$ is inside $(0, \rho_0)$ and strictly decreasing, we obtain $0 \leq \alpha < \rho_0$. It then follows from (1.287) and the continuity of η_{f,f_0} in $[0, \rho_0)$ that $0 \leq \alpha = |\eta_{f,f_0}(\alpha)|$ and from Lemma 1.4.2, we get $\alpha = 0$. The uniqueness part was shown in Lemma 1.4.8. Next we shall show (1.286) that

$$\frac{\| x_{n+1} - x^\star \|}{\| x_n - x^\star \|} \leq \frac{|\eta_{f,f_0}(\| x_n - x^\star \|)|}{\| x_n - x^\star \|} \quad (n \geq 0) \tag{1.288}$$

since, $\lim\limits_{n \to \infty} \| x_n - x^\star \| = 0$, (1.286) follows from Lemma 1.4.2. We shall show by induction

$$\| x_n - x^\star \| \leq s_n \quad (n \geq 0). \tag{1.289}$$

Since $s_0 = \| x_0 - x^\star \|$, (1.289) holds as equality for $n = 0$. Assume $\| x_k - x^\star \| \leq s_k$. In view of (1.285), Lemma 1.4.7, the induction hypothesis and (1.271), we obtain in turn

$$\begin{aligned}
\| x_{k+1} - x^\star \| &= \| N_F(x_k) - x^\star \| \\
&\leq \frac{|\eta_{f,f_0}(s_k)|}{s_k^{p+1}} \| x_k - x^\star \|^{p+1} \leq |\eta_{f,f_0}(s_k)| = s_{k+1},
\end{aligned} \tag{1.290}$$

which completes the induction for (1.289). Hence (1.263) follows from (1.285), (1.289), Lemma 1.4.7 and (1.271). Finally, to show (1.264) notice that since sequence $\{\frac{s_{k+1}}{s_k^{p+1}}\}$ is strictly decreasing, we have

$$\frac{s_{k+1}}{s_k^{p+1}} \leq \frac{s_1}{s_0^{p+1}} \qquad (k \geq 0). \tag{1.291}$$

It then follows from (1.263) that

$$\| x_{k+1} - x^\star \| \leq \frac{s_1}{s_0^{p+1}} \| x_k - x^\star \|^{p+1} \qquad (k \geq 0). \tag{1.292}$$

The first inequality in (1.264) follows from (1.292) if $p = 0$, whereas the second inequality is also derived from (1.292) if $0 < p \leq 1$. That completes the proof of Lemma 1.4.10. $\qquad\qquad\Box$

Proof.   [of Theorem 1.4.1] The proof of Theorem 1.4.1 now follows the above Lemmas 1.4.1–1.4.10. $\qquad\qquad\Box$

Remark 1.4.1.

(a) If $f_0(t) = f(t)$ $(t \in [0, R))$, then our Theorem 1.4.1 reduces to [Theorem 2, p. 1516] in Ref. [212]. Moreover, in this case, we have

$$s_n = t_n \quad (n \geq 0), \quad \rho_0 = \rho, \quad \sigma_0 = \sigma$$

and

$$r_0 = r,$$

where ρ, r are defined as ρ_0, r_0, respectively by replacing (f_0, f_0') by (f, f'). Otherwise, it constitutes an improvement with advantages as already stated in the introduction of this section (see also (1.262)).

(b) Theorem 1.4.1 uses the same information $(x^\star, F, f)$ as [Theorem 2, p. 1516] in Ref. [212], since f_0 is a special case of f. In practice, the computation of f requires that of f_0. Note also that the existence of function f_0 is implied by (1.255). Hence, (1.254) is not an additional hypothesis. We also have:

$$f_0'(t) \leq f'(t) \qquad t \in [0, R)$$

hold in general and $\dfrac{f'(t)}{f_0'(t)}$. The proof of [Theorem 2, p. 1516] in Ref. [212] uses (1.255) to obtain the estimate

$$\| F'(x)^{-1} F'(x^\star) \| \leq -\frac{1}{f'(\| x - x^\star \|)} \leq -\frac{1}{f'(t)}$$

corresponding to (1.272). However, we note that (1.272) is a tighter estimate than the above.

Now, we provide special cases and numerical examples.

Convergence under Hölder-type condition

Proposition 1.4.1. *Let $\mathcal{D}$ be an open and convex set; let $\mathcal{X}$ and $\mathcal{Y}$ be Banach spaces and let $F : \mathcal{D} \subset \mathcal{X} \longrightarrow \mathcal{Y}$ be a Fréchet-differentiable operator. Let $x^\star \in \mathcal{D}$ such that $F(x^\star) = 0$, $R > 0$ and $\kappa := \sup\{t \in [0, R) : U(x^\star, t) \subset \mathcal{D}\}$. Suppose $F'(x^\star)^{-1} \in \mathcal{L}(\mathcal{Y}, \mathcal{X})$ and there exist $L_0 > 0$, $L > 0$ and $0 < p \le 1$ such that*

$$\| F'(x^\star)^{-1} (F'(x) - F'(x^\star)) \| \le L_0 \| x - x^\star \|^p,$$

$$\| F'(x^\star)^{-1} (F'(x) - F'(x_\theta)) \| \le L (1 - \theta^p) \| x - x^\star \|^p,$$

for all $x \in U(x^\star, \kappa)$ and $x_\theta = x^\star + \theta (x - x^\star)$, $\theta \in [0, 1]$. Let

$$r_0 := \min\{\kappa, \ \left(\frac{p + 1}{L + L_0 (p + 1)} \right)^{1/p} \},$$

$$x_0 \in U(x^\star, r_0) \setminus \{x^\star\}, \quad s_0 = \| x_0 - x^\star \|, \quad s_{n+1} = \frac{L\, p\, s_n^{p+1}}{(p + 1) (1 - L_0\, s_n^p)}.$$

Then, the following assertions hold:

(a) $\{s_n\}$ is well defined; strictly decreasing; contained in $(0, r_0)$; converges to zero and

$$\lim_{n \longrightarrow 0} \frac{s_{n+1}}{s_n} = 0.$$

(b) Sequence $\{x_n\}$ given by (NM), starting from $x_0 \in U(x^\star, r_0) \setminus \{x^\star\}$ is well defined; remains in $U(x^\star, r_0)$ for all $n \ge 0$ and converges to $x^\star$, which is the unique solution of (1.1) in $U(x^\star, (\frac{p + 1}{L_0})^{1/p})$, so that for $n \ge 0$:

$$\| x_{n+1} - x^\star \| \le \frac{L\, p}{(p + 1) (1 - L_0\, s_n^p)} \| x_n - x^\star \|^{p+1}$$

and

$$\| x_n - x^\star \| \le \left(\frac{L\, p \, \| x_0 - x^\star \|^p}{(p + 1) (1 - L_0 \, \| x_0 - x^\star \|^p)} \right)^{((p+1)^n - 1)/p} \| x_0 - x^\star \|.$$

Furthermore, if

$$\varrho_0 = \left(\frac{p + 1}{L\, p + L_0 (p + 1)} \right)^{1/p} < \kappa,$$

then $r = \varrho_0$ is the best possible convergence radius.

Proof. Use Theorem 1.4.1 for functions $f_0, f : [0, \kappa] \longrightarrow \mathbb{R}$ defined by

$$f_0(t) = \frac{L_0\, t^{p+1}}{p + 1} - t \quad \text{and} \quad f(t) = \frac{L\, t^{p+1}}{p + 1} - t.$$

That completes the proof of Proposition 1.4.1. $\qquad\qquad\square$

Remark 1.4.2. If $L = L_0$, our results reduce to the ones in [Theorem 13] of Ref. [212] (see also Refs. [210, 213, 217]). Moreover, if $L_0 < L$, we have

$$\varrho = \left(\frac{p+1}{(2p+1)\,L} \right)^{1/p} < \varrho_0$$

and

$$\parallel x_{n+1} - x^\star \parallel \, \leq \frac{L\,p}{(p+1)\,(1 - L_0\,t_n^p)} \parallel x_n - x^\star \parallel^{p+1} \quad (n \geq 0).$$

That is our results provide a larger convergence radius and tighter error bounds than the ones in Ref. [212]. Note also that we have

$$\frac{\varrho}{\varrho_0} \longrightarrow \left(\frac{p}{2p+1} \right)^{1/p} \quad \text{as} \quad \frac{L_0}{L} \longrightarrow 0.$$

So, our approach provides a radius of convergence at most $\left(\dfrac{p}{2p+1} \right)^{-1/p}$ times larger than the one in Refs. [210, 217].

If the Lipschitz condition

$$\parallel F'(x^\star)^{-1}\,(F'(x) - F'(y)) \parallel \leq L \parallel x - y \parallel,$$

holds for $x, y \in \mathcal{D}$, then $p = 1$ and we have

$$\varrho = \frac{2}{3\,L} \leq \varrho_0 = \frac{2}{2\,L_0 + L}.$$

The radius of convergence ϱ was obtained by Rheinboldt in Ref. [384]. Note however that the results in Refs. [210, 217] were given in non-affine invariant form. The advantages of affine invariant over non-affine invariant results for Newton-type methods have been explained in Refs. [100, 163].

Example 1.4.1. (see Refs. [106, 163]) Let $\mathcal{X} = \mathcal{Y} = \mathbb{R}$. Define function F on $\mathcal{D} = (-1, 1)$, given by

$$F(x) = e^x - 1. \tag{1.293}$$

Then, for $x^\star = 0$, using (1.293), we have $F(x^\star) = 0$ and $F'(x^\star) = e^0 = 1$. Moreover, hypotheses of Proposition 1.4.1 hold for $p = 1$, $L = e > L_0 = e - 1$. Note that

$$\frac{L}{L_0} = \frac{e}{e-1} = 1.581976707$$

and

$$\varrho = \frac{2}{3\,L} = .2452529608 < \varrho_0 = \frac{2}{2\,L_0 + L} = .3249472314.$$

We can also provide the comparison table using the software Maple 13. Using (1.258) and (1.261) for $x_0 = .7158$. Table 1.5 shows that our error bounds (1.258) are tighter than (1.261). Note that hypothesis $(\mathcal{H}_3)$ of Theorem 1.4.1 does not hold, since Γ is not increasing on $(0, \nu_0)$ for all $\nu_0 > 0$ (see Figure 1.4), where

$$f_0(t) = \frac{(e-1)\,t^2}{2} - t, \quad f(t) = \frac{e\,t^2}{2} - t$$

and

$$\Gamma(t) = \left(\frac{f(t)}{f'(t)} - t \right) \frac{f_1(t)}{t^2} = \left(\frac{.5\,e\,t - 1}{.5\,(e-1)\,t - 1} - t \right) \frac{e\,t - 1}{(e-1)\,t^3 - t^2}.$$

Table 1.5 Comparison table

	(1.2)	(1.258)	(1.261)
k	$\parallel x_{k+1} - x_k \parallel$	s_k	t_k
0	.2473838936	.2842	.2842
1	.03614663422	.2145495033	.4826134043
2	.0006692478074	.09909547154	1.015025071
3	2.239999498e-7	.01608560415	.7960154923
4	0	.0003616695761	.7399991923
5	$\sim$	1.778927982e-7	.7357830417
6	$\sim$	4.299999235e-14	.7357588833
7	$\sim$	0	.7357588824
8	$\sim$	$\sim$	.7357588825

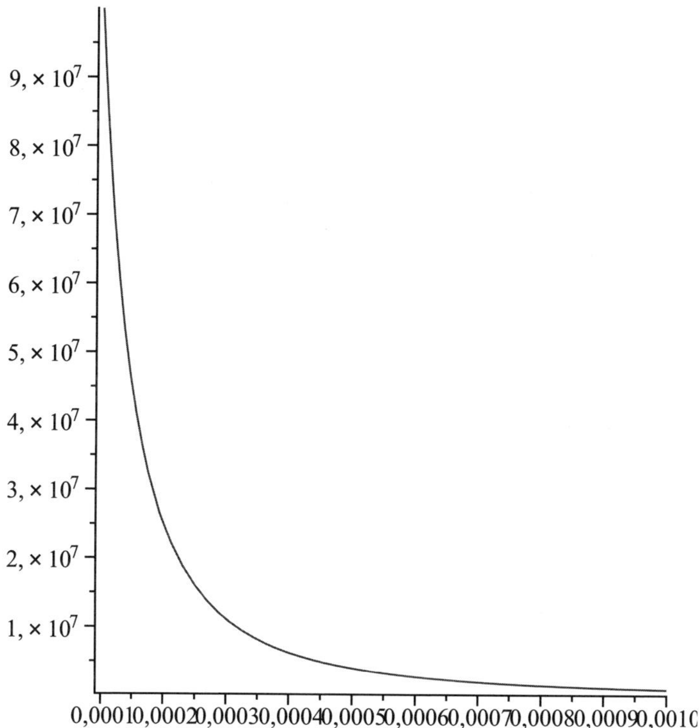

Fig. 1.4 Function Γ on intervall $(.0001, .001)$

Example 1.4.2. (see Refs. [163, 212]) Let $\mathcal{X} = \mathcal{Y} = \mathbb{R}$. Define function F on $\mathcal{D} = (1, 3)$, given by

$$F(x) = \frac{2}{3}\, x^{3/2} - x. \tag{1.294}$$

Then, the zero of F is $x^\star = \dfrac{9}{4} = 2.25$. Using (1.294) and hypotheses of Proposition 1.4.1, $F'(x^\star) = .5$, $L = 2 > L_0 = 1$ and $p = .5$. Moreover, we have

$$\varrho = .1406250000 < \varrho_0 = .1836734694.$$

Example 1.4.3. Consider Example 1.1.7. Under hypotheses of Proposition 1.4.1 for $x^\star(x) = 0$ for all $x \in [0,1]$, we get that $p = 1$, $L = 15$, $L_0 = 7.5$ and

$$\varrho = .04444444444 < \varrho_0 = .06666666667.$$

Convergence under generalized Lipschitz condition

Proposition 1.4.2. *Let $\mathcal{D}$ be an open and convex set; let $\mathcal{X}$ and $\mathcal{Y}$ be Banach spaces and let $F : \mathcal{D} \subset \mathcal{X} \longrightarrow \mathcal{Y}$ be a Fréchet-differentiable operator. Let $x^\star \in \mathcal{D}$ such that $F(x^\star) = 0$, $R > 0$ and $\kappa := \sup\{t \in [0,R) : U(x^\star, t) \subset \mathcal{D}\}$. Suppose $F'(x^\star)^{-1} \in \mathcal{L}(\mathcal{Y}, \mathcal{X})$ and there exist positive integrable functions L_0, $L : [0,R) \longrightarrow \mathbb{R}$ such that*

$$\| F'(x^\star)^{-1} \left(F'(x) - F'(x^\star) \right) \| \le \int_0^{\|x - x^\star\|} L_0(u)\, du,$$

$$\| F'(x^\star)^{-1} \left(F'(x) - F'(x_\theta) \right) \| \le \int_{\theta \, \|x - x^\star\|}^{\|x - x^\star\|} L(u)\, du,$$

for all $x \in U(x^\star, \kappa)$ and $x_\theta = x^\star + \theta\,(x - x^\star)$, $\theta \in [0,1]$. Let $\nu_0 > 0$, $\rho_0 > 0$ and $r_0 > 0$ be the constants defined by

$$\nu_0 = \sup\{t \in [0,R) \;:\; \int_0^t L_0(u)\, du - 1 < 0\},$$

$$\rho_0 = \sup\{t \in [0,\nu_0) \;:\; \frac{\displaystyle\int_0^t L(u)\, du}{t\left(1 - \displaystyle\int_0^t L_0(u)\, du\right)} < 1\},$$

$$r_0 = \min\{\kappa, \rho_0\}.$$

Let

$$x_0 \in U(x^\star, r_0) \setminus \{x^\star\}, \quad s_0 = \| x_0 - x^\star \|, \quad s_{n+1} = \frac{\displaystyle\int_0^{s_n} L(u)\, u\, du}{1 - \displaystyle\int_0^{s_n} L_0(u)\, du}.$$

Then, the following assertions hold:

(a) Sequence $\{s_n\}$ is well defined; strictly decreasing; contained in $(0, r_0)$; converges to zero and

$$\lim_{n \longrightarrow 0} \frac{s_{n+1}}{s_n} = 0.$$

(b) Sequence $\{x_n\}$ given by (NM), starting from $x_0 \in U(x^\star, r_0)\backslash\{x^\star\}$ is well defined; remains in $U(x^\star, r_0)$ for all $n \geq 0$ and converges to $x^\star$, which is the unique solution of (1.1) in $U(x^\star, \sigma_0)$, so that

$$\lim_{n \longrightarrow \infty} \frac{\| x_{n+1} - x^\star \|}{\| x_n - x^\star \|} = 0,$$

where

$$\sigma_0 = \sup\{t \in [0, \kappa) : \int_0^t L_0(u)\,(t - u)\,du - t < 0\}.$$

Moreover, if

$$\varrho_0 = \frac{\displaystyle\int_0^{\rho_0} L(u)\,u\,du}{\rho_0 \left(1 - \displaystyle\int_0^{\rho_0} L_0(u)\,du\right)} = 1$$

and $\rho_0 < \kappa$, then $r_0 = \rho_0$ is the best possible convergence radius. Furthermore, if $(\mathcal{H}_3)$ of Theorem 1.4.1 holds for

$$f(t) = \int_0^t L(u)\,(t - u)\,du - t, \quad f_0(t) = \int_0^t L_0(u)\,(t - u)\,du - t$$

then estimate (1.263) and (1.264) also hold.

Proof. Hypotheses $(\mathcal{H}_1)$–$(\mathcal{H}_3)$ can easily be verified with the above choices of functions f_0 and f. $\qquad\qquad\qquad\qquad\qquad\qquad\qquad\qquad\qquad\qquad\qquad\qquad\qquad\square$

Remark 1.4.3. If $L = L_0$, the results of Proposition 1.4.2 reduce to the ones of [Theorem 14] in Ref. [212]. Otherwise they constitute an improvement with advantages as already stated in Remark 1.4.2. The results obtained in this section can be extended to include equation with a nondifferentiable term, or inexact Newton-type methods along the lines of our relevant works in Refs. [100, 163].

1.5 Nondiscrete Induction

We extend in this section the applicability of Newton's method for approximating a solution of (1.1) using nondiscrete mathematical induction. Under the same computational cost as before, we provide: weaker sufficient convergence conditions; tighter error estimates on the distances involved and more precise information on the location of the solution. Numerical examples are also provided in this section. We are concerned with the problem of approximating a locally unique solution $x^\star$ of equation (1.1) using Newton's method (1.2).

In computer graphics, the intersection of two surfaces is also modeled by nonlinear equation and can be complicated in general, because some closed loops and singularities. This requires finding efficient algorithms for solving this intersection.

We often need to compute and display the intersection $\mathcal{C} = \mathcal{A} \cap \mathcal{B}$ of two surfaces $\mathcal{A}$ and $\mathcal{B}$ in $\mathbb{R}^3$ (see Ref. [312]). If the two surfaces are explicitly given by $\mathcal{A} = \{(u, v, w)^T : w = F_1(u, v)\}$ and $\mathcal{B} = \{(u, v, w)^T : w = F_2(u, v)\}$, then, the solution $x^\star = (u^\star, v^\star, w^\star)^T \in \mathcal{C}$ must satisfy the nonlinear equation $F_1(u^\star, v^\star) = F_2(u^\star, v^\star)$ and $w^\star = F_1(u^\star, v^\star)$. Hence, we must solve an equation in the form of (1.1) with $F := F_1 - F_2$. There is a signifiant litterature addressing the surface intersection problem (see Refs. [116, 296]).

Let $x_0 \in \mathcal{D}$ be such that $F'(x_0)^{-1} \in \mathcal{L}(\mathcal{Y}, \mathcal{X})$ the space of bounded linear operators from $\mathcal{Y}$ into $\mathcal{X}$. We say that $F'(x_0)^{-1} F'(.)$ satisfies a Lipschitz-condition on $\mathcal{D}$ with constant L $(L > 0)$, if

$$\| F'(x_0)^{-1} (F'(x) - F'(y)) \| \leq L \| x - y \| \quad \text{for all} \quad x, y \in \mathcal{D}. \tag{1.295}$$

Set

$$\| F'(x_0)^{-1} F(x_0) \| \leq r_0. \tag{1.296}$$

Then, a sufficient convergence condition for the semi-local convergence of (NM) is the famous Kantorovich hypothesis (KH) given in Refs. [106, 134, 290] by

$$H_K = 2\,L\,r_0 \leq 1. \tag{1.297}$$

In the scalar case (1.297) coincides with the condition given earlier by Ostrowski in Refs. [340, 343]. If strict inequality holds in (1.297), the convergence is quadratic. Otherwise it is only linear. Later Ostrowski in Ref. [343] obtained sharp a priori estimates. Simpler sharp a priori estimates were provided (using different method and proofs) by Gragg and Tapia in Ref. [238] and some works of Pták in Refs. [361, 362, 364, 366]. The celebrated method of nondiscrete induction is first used by Pták in Refs. [363, 365]. Subsequently, Potra and Pták developed in a series of papers and an excellent book (see Refs. [349, 359, 357, 360]) the nondiscrete induction and provided a posteriori estimates which are in general better than those given by Gragg and Tapia in Ref. [238]. Other works on iterative methods and nondiscrete induction can be found in Refs. [354, 356, 359, 358, 367].

Hypothesis (1.297) is not a sufficient condition for the convergence of (NM). In this section we have provided an example where although hypothesis (1.297) is violated but (NM) (1.2) converges to the solution $x^\star$. Therefore, any hypothesis using the same information (F, x_0, L) weaker than (1.297) will expand applicability of (NM).

Let us report on what has been done in this direction. First of all note that in view of (1.295), $F'(x_0)^{-1} F'(.)$ satisfies a center-Lipschitz condition with constant L_0 $(L_0 > 0)$. That is

$$\| F'(x_0)^{-1} (F'(x) - F'(x_0)) \| \leq L_0 \| x - x_0 \| \quad \text{for all} \quad x \in \mathcal{D}. \tag{1.298}$$

Condition (1.298) is not an additional (to (1.295)) hypothesis, since in practice the computation of Lipschitz constant L requires that of center-Lipschitz constant L_0. We can then use (1.298) instead (1.295) to compute upper bounds on the norms $\| F'(x)^{-1} F'(x_0) \|$. This observation has led to the following set of advantages $(\mathcal{A})$ in the discrete case when $L_0 < L$ (see Refs. [100, 163]):

★ A weaker hypothesis than (KH) (1.297);
★ A tighter error bounds on the distances involved;

and

★ An at least as precise information on the location of the solution $x^\star$.

These advantages $(\mathcal{A})$ are obtained under the same information (x_0, F, L). We have provided the following hypothesis (see Refs. [104, 106, 98, 100, 134, 140, 163]) instead of (1.297)

$$H_1 = (5 + 2\sqrt{6})\, L_0\, r_0 \leq 1, \tag{1.299}$$

$$H_2 = (L + L_0)\, r_0 \leq 1, \tag{1.300}$$

$$H_3 = 2\,\overline{L}\, r_0 \leq 1, \tag{1.301}$$

where

$$\overline{L} = \frac{1}{8}\left(L + 4L_0 + (L^2 + 8L_0 L)^{1/2}\right). \tag{1.302}$$

Note that in particular

$$H_K \leq 1 \Longrightarrow H_2 \leq 1 \Longrightarrow H_3 \leq 1, \tag{1.303}$$

but not necessarily vice versa unless $L_0 = L$. We also have

$$\frac{H_3}{H_K} \longrightarrow \frac{1}{4} \quad \text{as} \quad \frac{L_0}{L} \longrightarrow 0, \tag{1.304}$$

$$\frac{H_2}{H_K} \longrightarrow \frac{1}{2} \quad \text{as} \quad \frac{L_0}{L} \longrightarrow 0, \tag{1.305}$$

and

$$\frac{H_3}{H_2} \longrightarrow \frac{1}{2} \quad \text{as} \quad \frac{L_0}{L} \longrightarrow 0, \tag{1.306}$$

which provide a maximum measure on the expandability of (NM) under hypothesis (1.299) or (1.300) or (1.301). By comparing (1.297) to (1.299) we get

$$\frac{L}{L_0} \geq \frac{5 + 2\sqrt{6}}{2} \quad \text{and} \quad H_K \leq 1 \Longrightarrow H_1 \leq 1 \tag{1.307}$$

or

$$\frac{L}{L_0} \leq \frac{5 + 2\sqrt{6}}{2} \quad \text{and} \quad H_1 \leq 1 \Longrightarrow H_K \leq 1. \tag{1.308}$$

Clearly, the first case (1.307) expands the applicability of (NM) when

$$\frac{L}{L_0} > \frac{5 + 2\sqrt{6}}{2}, \quad H_1 \leq 1 \quad \text{and} \quad H_K > 1. \tag{1.309}$$

Hypothesis (1.299) requires the computation of only constant L_0, whereas (1.300) and (1.301) require both constants L_0 and L. In Ref. [104], Argyros further weakened (1.299) in some sense using

$$H_M = 2\, L_0\, r_0 \le 1, \tag{1.310}$$

which is sufficient convergence condition for the modified Newton's method (MNM)

$$y_{n+1} = y_n - F'(y_0)^{-1}\, F(y_n) \quad (n \ge 0), \quad (y_0 = x_0 \in \mathcal{D}). \tag{1.311}$$

But this time a certain number of iterates y_n in (1.311) must be computed until $y_N = x_0$ (N is a finite natural number), for more details, see Ref. [104]. We also note that if (1.299) or (1.310) hold, then the convergence of (NM) is shown only to be linear. Note also that (1.310) is the weakest of the H hypotheses given by (1.297) and (1.299)–(1.301).

In this section we are motivated by optimization considerations and the method of nondiscrete mathematical induction as developed by Potra and Pták in Ref. [359]. We show that the advantages ($\mathcal{A}$) carry out from the discrete to the nondiscrete case using (1.299) or (1.300) or (1.301) instead of (1.297) and smaller rate of convergence ω and corresponding estimate functions s (to be precised after in this section). Note that ω and s are used to measure the error distances involved.

Potra and Pták in Ref. [359] defined functions ω (see Fig. 1.5) and s (see Fig. 1.6) by

$$\omega(r) = \frac{1}{2}\, r^2 (r^2 + a^2)^{-1/2} \tag{1.312}$$

and

$$s(r) = r + (r^2 + a^2)^{1/2} - a, \tag{1.313}$$

where $a \ge 0$. Under hypothesis (1.297), Potra and Pták in Ref. [359] showed that the optimum value for a is given by

$$a = a_P = \left(\frac{1}{L} \left(\frac{1}{L} - 2\, r_0 \right) \right)^{1/2}. \tag{1.314}$$

The error bounds are related with functions w and s by

$$d(x_n, x_{n-1}) \le \omega^{(n)}(r_0), \tag{1.315}$$

and

$$d(x_n, x^\star) \le s(\omega^{(n)}(r_0)), \tag{1.316}$$

where $\omega^{(n)}$ is the n-iterate of the function ω so that

$$\omega^{(0)}(r) = r,\ \omega^{(1)}(r) = \omega(r),\ \omega^{(2)}(r) = \omega(\omega(r)),\ \cdots,\ \omega^{(n)}(r) = \omega(\omega^{(n-1)}(r)).$$

It follows from (1.312)–(1.316) that the larger parameter a is the tighter estimates (1.315) and (1.316) will be. If (1.300) holds, set:

$$a_1 = \frac{1}{L} \left((1 - L_0\, r_0)^2 - L^2\, r_0^2 \right)^{1/2} \ge 0. \tag{1.317}$$

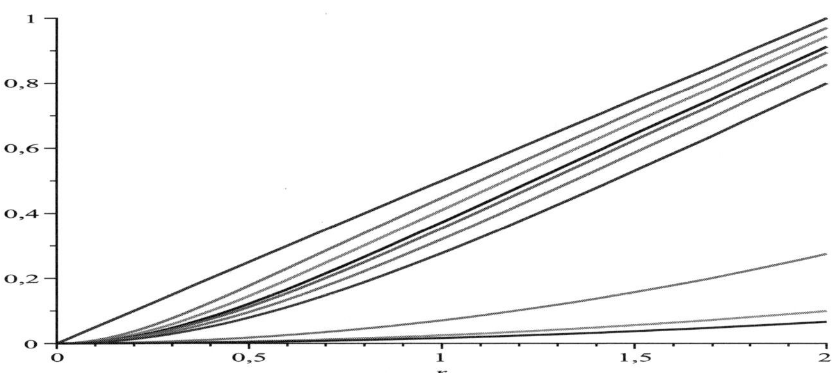

Fig. 1.5 Functions $\omega(r)$ (from top to bottom) on $[0,2]$ for $a = 0, .5, .7, .9, 1, 1.2, 1.5, 7, 20, 30$, respectively.

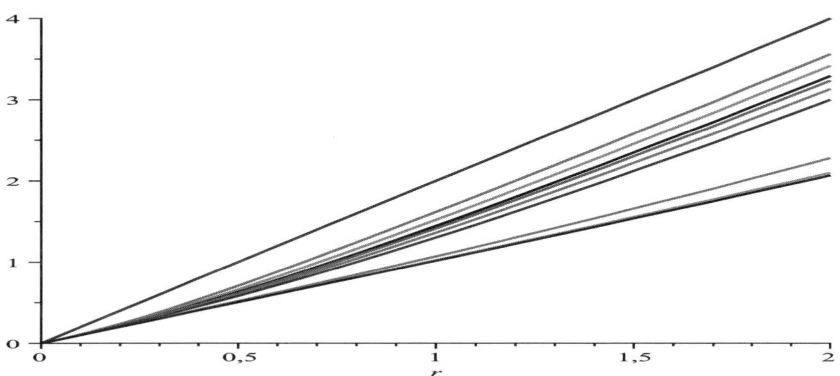

Fig. 1.6 Functions $s(r)$ (from top to bottom) on $[0,2]$ for $a = 0, .5, .7, .9, 1, 1.2, 1.5, 7, 20, 30$, respectively.

Moreover, if (1.310) is satisfied, let

$$a_M = \left(\frac{1}{L_0} \left(\frac{1}{L_0} - 2\,r_0 \right) \right)^{1/2} \geq 0. \tag{1.318}$$

Note that if $L_0 = L$, then $a_M = a_1 = a_P$. Otherwise, we have

$$a_P < a_1 < a_M. \tag{1.319}$$

Other values for parameter a have been given after in this section.

Our introduction of the center-Lipschitz condition in the discrete case has produced advantages $(\mathcal{A})$ for other iterative processes such as Secant method, directional Newton method, Stirling's method, Steffensen's method and Newton-like methods in Refs. [100, 106, 116, 134, 163]. In this section we show that advantages $(\mathcal{A})$ can carry from discrete to nondiscrete case. In particular we provide using the same information (F, x_0, L) a finer convergence analysis than in Refs. [348, 353] for (NM).

Pták inaugurated in his Gatlinburg lecture (see Ref. [363]) the method of Nondiscrete Mathematical Induction (NMI). We refer the reader to the excellent monograph by Potra and Pták in Ref. [360] for more details about the motivation and general principles for (NMI). Let $\mathcal{T}$ be either the positive real axis or an interval of the form

$$\mathcal{T} = \{r \in \mathbb{R} : 0 < r < \alpha\} = (0, \alpha).$$

We need the definition of the rate of convergence.

Definition 1.5.1. A function $\omega : \mathcal{T} \longrightarrow \mathcal{T}$ is called a rate of convergence on $\mathcal{T}$, if the series

$$\sum_{n=0}^{\infty} \omega^{(n)}(r) \tag{1.320}$$

is convergent for each $r \in \mathcal{T}$. The sum (1.320) is denoted by $s(r)$ and is called the corresponding estimate function. Then we write

$$s(r) = \sum_{n=0}^{\infty} \omega^{(n)}(r) \quad \text{for all} \quad r \in \mathcal{T}. \tag{1.321}$$

Functions ω and s satisfy the functional equation

$$s(r) = r + \omega(r). \tag{1.322}$$

It then follows from (1.322) that (with the exception of pathological cases) we have:

$$\omega(r) = s^{-1}(s(r) - r). \tag{1.323}$$

That is given s, function ω can be recovered using functional equation (1.323). The computation of function s is very difficult or impossible in general. We have the following result characterizing rates of convergence.

Proposition 1.5.1. *(see Ref. [360]) Let* $\omega : \mathcal{T} \longrightarrow \mathcal{T}$ *and* $\nu : \mathcal{T} \longrightarrow \mathcal{T}$ *be such that*

$$\nu(r) = r + \nu(\omega(r)) \quad for\ all \quad r \in \mathcal{T}. \tag{1.324}$$

Then the following items hold:

(a) ω is a rate of convergence on $\mathcal{T}$;

and

(b) If the limit $\nu(0) = \lim_{r \searrow 0} \nu(r)$ exists, then

$$s(r) = \sum_{n=0}^{\infty} \omega^{(n)}(r) = \nu(r) - \nu(0) \quad for\ all \quad r \in \mathcal{T}. \tag{1.325}$$

It can easily be seen by verifying (1.324), function ω given by (1.307) is a rate of convergence on $\mathcal{T}$ with corresponding estimate function s given by (1.308).

Another example is given for $\delta \in [0, 1)$ by

$$\omega(r) = \delta\, r \tag{1.326}$$

and

$$s(r) = \frac{r}{1 - \delta}. \tag{1.327}$$

Define $G : \mathcal{D} \longrightarrow \mathcal{Y}$ by

$$G(x) = x - F'(x)^{-1}\, F(x). \tag{1.328}$$

We need the following result relating the (MNI), (1.328) and (NM).

Lemma 1.5.1. *(see Refs.* [359, 360]*)*

(1) Assume that for given pair (G, x_0) there exists a rate of convergence ω on an interval $\mathcal{T}$ and a family of sets $\mathbb{Z}(r) \subseteq \mathcal{X}$ such that inclusion conditions $x_0 \in \mathbb{Z}(r_0)$ for a certain $r_0 \in \mathcal{T}$ and

$$r \in \mathcal{T} \quad and \quad x \in \mathbb{Z}(r) \Longrightarrow G(x) \in U(x, r) \cap \mathbb{Z}(\omega(r)) \tag{1.329}$$

are satisfied. Then, sequence $\{x_n\}$ generated by (NM) is well defined and converges to a point $x^\star$. Moreover, (1.315), (1.316) and the following estimate

$$x_n \in \mathbb{Z}(\omega^{(n)}(r_0)) \tag{1.330}$$

hold.

(2) If, in addition, for a certain $n \geq 1$, we have

$$x_{n-1} \in \mathbb{Z}(d(x_n, x_{n-1})), \tag{1.331}$$

then for this n, the following estimate holds

$$d(x_n, x^\star) \leq f(d(x_n, x_{n-1})) \tag{1.332}$$

for some function $f : [0, \infty) \longrightarrow [0, \infty)$ such that

$$f(r) = s(r) - r. \tag{1.333}$$

Lemma 1.5.1 is essentially a corollary of the induction theorem (see [Proposition 1.7, p. 5] in Ref. [360]). This theorem is related to the graph theorem of functional analysis. The closed graph theorem can be seen as a limit case of the induction theorem for an infinity fast rate of convergence (see, e.g. [Theorem 1.15] in Ref. [360]). We use the following measure of invertibility

$$d(\mathcal{B}) = \inf_{\|x\|=1} \|\,\mathcal{B}(x)\,\| \quad for \quad \mathcal{B} \in \mathcal{L}(\mathcal{X}, \mathcal{Y}). \tag{1.334}$$

If $\mathcal{B}$ is invertible and $\mathcal{B}^{-1} \in \mathcal{L}(\mathcal{Y}, \mathcal{X})$, then

$$d(\mathcal{B}) = \|\,\mathcal{B}^{-1}\,\|^{-1}. \tag{1.335}$$

We also need the following Banach-type result on invertible operators.

Lemma 1.5.2. *If $\mathcal{B}$ and $\mathcal{C}$ belong in $\mathcal{L}(\mathcal{X}, \mathcal{Y})$ such that $\mathcal{B}$ is boundedly invertible and*

$$d(\mathcal{B}) > \| \mathcal{B} - \mathcal{C} \| \tag{1.336}$$

then $\mathcal{C}$ is also boundedly invertible and

$$d(\mathcal{C}) \geq d(\mathcal{B}) - \| \mathcal{B} - \mathcal{C} \| . \tag{1.337}$$

Nondiscrete induction fror iterative processes requires verification of inclusion hypotheses in (1) of Lemma 1.5.1. We shall illustrate how this method works in the case of (NM). The differences between our approach and the one given by Potra and Pták in Refs. [359, 360] will also be given in our description that follows. First, we need to define a suitable non-empty approximate set $\mathbb{Z}$ for some rate of convergence ω. If x is an initial guess, we hope

$$x_+ = x - F'(x)^{-1} F(x) \tag{1.338}$$

to be closer to the solution $x^\star$. Let r be the distance between x and x_+. We must have for $x \in \mathbb{Z}(r)$ that $x_+ \in \mathbb{Z}(\omega(r))$. Potra and Pták in [p. 23] of Ref. [360] have used the approximate set $\mathbb{Z}(r)$ $(r > 0)$ for some rate of convergence ω (first in non-affine invariant form):

$$\mathbb{Z}(r) = \{x \in \mathcal{X} : \| x - x_0 \| \leq g(r), \ F'(x) \text{ is invertible}, \\ \| F'(x)^{-1} F(x) \| \leq r \text{ and } d(F'(x_0)^{-1} F(x)) \geq h(r)\}, \tag{1.339}$$

where g and h are functions to be determined later. This way they produced a plethora of results on (NM) that have improved the error bounds on the distances $d(x_n, x_{n-1})$ and $d(x_n, x^\star)$ of the discrete case but not the sufficient convergence condition (1.297).

Let $x \in \mathbb{Z}(r)$, then for $x_+ \in \mathbb{Z}(\omega(r))$, the following must hold:

$$\| x_+ - x_0 \| \leq g(\omega(r)), \tag{1.340}$$

$$d(F'(x_0)^{-1} F'(x_+)) \geq h(\omega(r)), \tag{1.341}$$

and

$$\| F'(x_+)^{-1} F(x_+) \| \leq \omega(r). \tag{1.342}$$

But we can write

$$\| x_0 - x_+ \| \leq \| x_+ - x \| + \| x - x_0 \| \leq r + g(r). \tag{1.343}$$

We also have

$$\begin{aligned}
d(F'(x_0)^{-1} F'(x_+)) &\geq d(F'(x_0)^{-1} F'(x)) - \| F'(x_0)^{-1} (F'(x_+) - F'(x)) \| \\
&\geq d(F'(x_0)^{-1} F'(x)) - (\| F'(x_0)^{-1} (F'(x_+) - F'(x_0)) \| + \\
&\quad \| F'(x_0)^{-1} (F'(x_0) - F'(x)) \|) \\
&\geq d(F'(x_0)^{-1} F'(x)) - L_0 (\| x_+ - x_0 \| + \| x - x_0 \|) \\
&\geq h(r) - L_0 (r + 2 g(r)).
\end{aligned}$$

$$\tag{1.344}$$

As long as $h(r) - L_0\,(r + 2\,g(r))$ is positive, the Banach lemma on invertible operators and Lemma 1.5.2 guarantee the existence of $F'(x)^{-1}$ and the estimate

$$\| \, F'(x)^{-1}\, F'(x_0) \, \| \le (h(r) - L_0\,(r + 2\,g(r)))^{-1}. \tag{1.345}$$

Using the approximation

$$\begin{aligned} F(x_+) &= F(x_+) - F(x) - F'(x)\,(x_+ - x) \\ &= \int_0^1 (F'(x + t\,(x_+ - x)) - F'(x))\,(x_+ - x)\,dt \end{aligned}$$

and (1.295), we get

$$\| \, F'(x_0)^{-1}\, F(x_+) \, \| \le \frac{1}{2}\,L \, \| \, x_+ - x \, \|^2 . \tag{1.346}$$

Then, we have by (1.345) and (1.346)

$$\begin{aligned} \| \, F'(x_+)^{-1}\, F(x_+) \, \| &\le \| \, F'(x_+)^{-1}\, F'(x_0) \, \| \, \| \, F'(x_0)^{-1}\, F(x_+) \, \| \\ &\le \frac{1}{2}\,L\,(h(r) - L_0\,(r + 2\,g(r)))^{-1}\,r^2. \end{aligned} \tag{1.347}$$

In view of (1.343), (1.345) and (1.347), conditions (1.340)–(1.342) hold if there exist functions h, g and parameter b satisfying the system of inequalities S_{AH}:

$$g(r) + r \le g(\omega(r)), \tag{1.348}$$

$$h(r) - L_0\,(r + 2\,g(r)) \ge h(\omega(r)), \tag{1.349}$$

$$\frac{L}{2}\,r^2\,(h(r) - L_0\,(r + 2\,g(r)))^{-1} \le \omega(r), \tag{1.350}$$

$$g(r) < b \tag{1.351}$$

and

$$0 < h(r) \le 1. \tag{1.352}$$

The system S_{PP} in [p. 25] of Ref. [360] uses inequation

$$h(r) - L\,r \ge h(\omega(r)) \tag{1.353}$$

instead of (1.349). The rest of the inequalities are the same.

We shall see later that replacing (1.353) by (1.349) in a major modification leading to the advantages $(\mathcal{A})$ already stated in the Introduction of this section.

Next, we shall show that system S_{AH} is satisfied in two cases when rate of convergence ω is given by (1.307) or (1.326) and

$$h(r) = L\left(a + \frac{L - L_0}{L}\,(s(r) - r) + \frac{L_0}{L}\,(2\,s(r_0) - s(r))\right) \tag{1.354}$$

$$g(r) = s(r_0) - s(r), \tag{1.355}$$

$$b_0 = s(r_0) < b, \tag{1.356}$$

where s is the estimate function corresponding to rate of convergence ω and $a \geq 0$ to be determined later. In the first case functions ω, s are given by (1.312) and (1.313), respectively.

Proposition 1.5.2. *Let $r_0 \geq 0$ and $L \geq L_0 > 0$. Let also functions ω, s be given by (1.312) and (1.313), respectively. Assume that (1.300) and*

$$b_1 = \frac{1}{L}\left(1 + (L - L_0)\, r_0 - ((1 - L_0\, r_0)^2 - L^2\, r_0^2)^{1/2}\right) < b \tag{1.357}$$

hold. Then, system S_{AH} has a solution (h, g, b_1), where

$$h(r) = L\left(a_1 + \frac{L - L_0}{L}\left((r^2 + a_1^2)^{1/2} - a_1\right) + \frac{L_0}{L}\left(a_1 + 2\,b_1 - r - (r^2 + a_1^2)^{1/2}\right)\right) \tag{1.358}$$

$$g(r) = \frac{1 + (L - L_0)\, r_0}{L} - r - (r^2 + a_1^2)^{1/2} \tag{1.359}$$

and a_1 is given by (1.317). Moreover, we have

$$x_0 \in \mathbb{Z}(r_0). \tag{1.360}$$

Proof. By hypothesis (1.300), $a_1 \geq 0$. Indeed, if $L_0 \neq L$, we have:

$$(L_0^2 - L^2)\,(r_0 - \frac{1}{L_0 - L})\,(r_0 - \frac{1}{L_0 + L}) \geq 0 \Longrightarrow$$
$$(L_0^2 - L^2)\, r_0^2 - 2\,L_0\, r_0 + 1 \geq 0 \Longrightarrow$$
$$(1 - L_0\, r_0)\, r_0^2 - L^2\, r_0^2 \geq 0 \Longrightarrow a_1 \geq 0.$$

If $L_0 = L$, then again we deduce $a_1 \geq 0$, since $2\,L\,r_0 \leq 1$. Moreover it can easily be seen by simple substitution that triplet (h, g, b_1) satisfies system S_{AH}. Note in particular that (1.358) implies (1.352). Finally, inclusion (1.360) follows from (1.338) and (1.352). That completes the proof of Proposition 1.5.2. $\qquad\square$

Remark 1.5.1. If $L_0 = L$, hypothesis (1.300) reduces to (1.297). In this case we have $a = a_1 = a_P$. If $L_0 < L$, Proposition 1.5.2 improves the results in Ref. [360] and $a_1 > a_P$ (see also Example 1.5.1).

In the second case functions ω, s are given by (1.326) and (1.327), respectively.

Proposition 1.5.3. *Let $r_0 \geq 0$ and $L \geq L_0 > 0$. Let also functions ω, s be given by (1.326) and (1.327), respectively for $\delta = \frac{1}{2}$. Assume that (1.300) for $L \geq 3\,L_0$ or $4\,L_0\,r_0 \leq 1$ for $L \leq 3\,L_0$ and*

$$b_2 = 2\,r_0 < b \tag{1.361}$$

hold. Then, system S_{AH} has a solution (h, g, b_2), where

$$h(r) = (L - 3\,L_0)\,r + 4\,L_0\,r_0 \tag{1.362}$$

and

$$g(r) = 2\,(r_0 - r). \tag{1.363}$$

Proof. It is easy to see by substitution that system S_{AH} is satisfied with the above choices of g, h, b_2 and b. That completes the proof of Proposition 1.5.3. $\square$

Remark 1.5.2. It turn out that if the approximate set $\mathbb{Z}$ is defined in a way other than (1.339), then system S_{AH} can be simplified and weaker than before hypotheses are needed (in some cases). This time, we define

$$\mathbb{Z}_0(r) = \{x \in \mathcal{X} : \| x - x_0 \| \leq g(r), \ F'(x) \text{ is invertible},$$
$$\| F'(x)^{-1} F(x) \| \leq r \text{ and } d(F'(x_0)^{-1} F(x)) \geq 1 - L_0 (r + g(r))\}.$$

$$(1.364)$$

The motivation for the introduction of new approximate set $\mathbb{Z}_0$ is due to the estimate

$$d(F'(x_0)^{-1} F'(x_+)) \geq d(F'(x_0)^{-1} F'(x_0)) - \| F'(x_0)^{-1} (F'(x_+) - F'(x_0)) \|$$
$$\geq 1 - L_0 \| x_+ - x_0 \| \geq 1 - L_0 (r + g(r)).$$

$$(1.365)$$

Then, in view of the implications

$$\omega(r) \geq 0 \implies \omega(r) + s(r_0) - s(\omega(r)) \geq r + s(r_0) - s(r)$$
$$\implies 1 - L_0 (r + g(r)) \geq 1 - L_0 (\omega(r) + g(\omega(r))),$$

$$(1.366)$$

inequality (1.349) can be dropped from the system S_{AH}. Denote the resulting system by $S^{\star}_{AH}$ defined by

$$S^{\star}_{AH} \begin{cases} g(r) + r \leq g(\omega(r)) \\ \dfrac{L}{2} r^2 \left(1 - L_0 (r + g(r))\right)^{-1} \leq \omega(r) \\ g(r) < b \\ 0 < L_0 (r + g(r)) \leq 1. \end{cases}$$

Then, we can have results corresponding to Propositions 1.5.2 and 1.5.3, respectively.

Proposition 1.5.4. *Under the hypotheses of Proposition 1.5.2, $S^{\star}_{AH}$ has a solution (g, b_1), where g and b_1 are given in Proposition 1.5.2. Moreover, we have*

$$x_0 \in \mathbb{Z}_0(r_0). \tag{1.367}$$

Proof. It can easily be seen that pair (g, b_1) satisfies system $S^{\star}_{AH}$. In particular for the verification of (1.350), we must show

$$1 - L_0 (r + g(r)) \geq 0 \tag{1.368}$$

and

$$L r^2 \leq 2 \omega(r) (1 - L_0 (r + g(r))). \tag{1.369}$$

We have

$$1 - L_0 (r + g(r)) = 1 - L_0 (r + (r^2 + a_1^2)^{1/2}) \geq 0 \tag{1.370}$$

by the choice of a_1 and $r \in [0, r_0]$. Hence estimate (1.368) holds. We also have that

$$a_1^2 \leq \frac{(1 - L_0 r)^2 - L^2 r^2}{L^2} \implies L^2 a_1^2 \leq (1 - L_0 r)^2 - L^2 r^2$$
$$\implies L^2 (r^2 + a_1^2) \leq (1 - L_0 r)^2 \implies L (r^2 + a_1^2)^{1/2} \leq 1 - L_0 r$$
$$\implies L (r^2 + a_1^2)^{1/2} \leq 1 - L_0 (r + (r^2 + a_1^2)^{1/2}) + L_0 (r^2 + a_1^2)^{1/2}$$
$$\implies L r^2 \leq r^2 (r^2 + a_1^2)^{-1/2} (1 - L_0 (r + (r^2 + a_1^2)^{1/2}) + L_0 (r^2 + a_1^2)^{1/2}) \implies (1.368).$$

That completes the proof of Proposition 1.5.4. $\square$

Proposition 1.5.5. *Let $r_0 \geq 0$ and $L \geq L_0 > 0$. Let also ω, s be given by (1.325) and (1.327), respectively, with*

$$\delta = \frac{2L}{L + (L^2 + 8 L_0 L)^{1/2}}.$$ (1.371)

Suppose that (1.301) and

$$b_3 = \frac{r_0}{1 - \delta} < b$$ (1.372)

hold. Then system $S^\star_{AH}$ has a solution (g, b_3), where

$$g(r) = \frac{1}{1 - \delta}(r_0 - r).$$ (1.373)

Moreover, we have $x_0 \in \mathbb{Z}_0(r_0)$. Furthermore, $\delta \in [1/2, 1)$.

Proof. We shall show how do we arrive at hypothesis (1.301) and the choice of δ. The rest shall follow by substituting (g, b_3) in $S^\star_{AH}$. Indeed, we have that

$$L\, r^2 \leq 2\,\omega(r)\,(1 - L_0\,(r + g(r)))$$

or

$$r\left(L - \frac{2 L_0 \delta^2}{1 - \delta}\right) \leq 0$$

or

$$2 L_0 \delta^2 + L \delta - L \geq 0,$$ (1.374)

which is true as equality by (1.371). We must also show

$$g(r) \geq 0$$ (1.375)

or

$$s(r_0) \geq s(r), \quad \text{or} \quad \frac{r}{1 - \delta} \leq \frac{1}{L_0} \quad \text{or} \quad r_0 L_0 \leq 1 - \delta = 1 - \frac{2L}{L + (L^2 + 8 L_0 L)^{1/2}}$$

or

$$r_0 L_0 \leq \frac{-L + (L^2 + 8 L_0 L)^{1/2}}{L + (L^2 + 8 L_0 L)^{1/2}} \quad \text{or} \quad r_0 L_0 \left(L + (L^2 + 8 L_0 L)^{1/2}\right)^2 \leq 8 L_0 L$$

or

$$r_0 \left(L + 4 L_0 + (L^2 + 8 L_0 L)^{1/2}\right) \leq 4,$$

which is exactly hypothesis (1.301). The proof of Proposition 1.5.5 is complete. $\square$

Remark 1.5.3. If $L = L_0$, then (1.301) reduces to (1.297) and $\delta = 1/2$. If $L_0 < L$, then (1.301) is weaker than (1.300) and (1.297).

The only difference in the proofs of [Sections 1 and 5] in Refs. [360, 359] and ours is that we use different value of parameters a and (1.300) or (1.301) instead of (1.297). Therefore the proofs of semi-local convergence results (corresponding to Propositions 1.5.2, 1.5.3 and 1.5.5) for (NM) are omitted. For brevity, we only provide estimates of the form (1.315), (1.316) and (1.330). Estimates of the form (1.332) can also follow immediately as in Refs. [360, 359] but using different parameters a as in Propositions 1.5.2, 1.5.3 and 1.5.5.

Theorem 1.5.1. *Let $F : \mathcal{D} \subseteq X \longrightarrow \mathcal{Y}$ be a Fréchet-differentiable operator and let $x_0 \in \mathcal{D}$. Assume:*

(i) $F'(x_0)^{-1} \in \mathcal{L}(\mathcal{Y}, X)$;
(ii) $F'(x_0)^{-1} F'$ *satisfies a Lipschitz condition with constant L and a center-Lipschitz condition with constant L_0 on $\mathcal{D}$;*
(iii) $\| F'(x_0)^{-1} F(x_0) \| \leq r_0$;
(iv) Hypotheses of Proposition 1.5.2 hold;

and

(v) $\overline{U}(x_0, b_1) \subseteq \mathcal{D}$.

Then sequence $\{x_n\}$ $(n \geq 0)$ generated by (NM) is well defined, remains in $\overline{U}(x_0, b_1)$ for all $n \geq 0$ and converges to a unique solution $x^\star$ of equation (1.1) in $\overline{U}(x_0, b_1)$. Moreover, the following error estimates hold for all $n \geq 1$:

$$d(x_n, x_{n-1}) \leq \omega^{(n)}(r_0) = \frac{2\,a_1\,\theta_1(r_0)^{2^n}}{1 - \theta_1(r_0)^{2^{n+1}}}, \tag{1.376}$$

$$d(x_n, x^\star) \leq s(\omega^{(n)}(r_0)) = \frac{2\,a_1\,\theta_1(r_0)^{2^n}}{1 - \theta_1(r_0)^{2^n}}, \tag{1.377}$$

and

$$d(x_n, x^\star) \leq (a_1^2 + \| x_n - x_{n-1} \|^2)^{1/2} - a_1, \tag{1.378}$$

where

$$\theta_1(r) = \frac{(r^2 + a_1^2)^{1/2} - a_1}{r}, \tag{1.379}$$

where a_1 is given by (1.317).

Theorem 1.5.2. *Let $F : \mathcal{D} \subseteq X \longrightarrow \mathcal{Y}$ be a Fréchet-differentiable operator and let $x_0 \in \mathcal{D}$.*

(1) Hypotheses (i)–(iii) of Theorem 1.5.1 hold;
(2) Hypotheses of Proposition 1.5.3 hold;

and

(3) $\overline{U}(x_0, b_2) \subseteq \mathcal{D}$.

Then sequence $\{x_n\}$ ($n \geq 0$) generated by (NM) is well defined, remains in $\overline{U}(x_0, b_2)$ for all $n \geq 0$ and converges to a unique solution $x^\star$ of equation (1.1) in $\overline{U}(x_0, b_2)$. Moreover, the following error estimates hold for all $n \geq 1$:

$$d(x_n, x_{n-1}) \leq \omega^{(n)}(r_0) = \left(\frac{1}{2}\right)^n r_0 \tag{1.380}$$

and

$$d(x_n, x^\star) \leq s(\omega^{(n)}(r_0)) = \left(\frac{1}{2}\right)^{n-1} r_0. \tag{1.381}$$

Theorem 1.5.3. *Let $F : \mathcal{D} \subseteq X \longrightarrow \mathcal{Y}$ be a Fréchet-differentiable operator and let $x_0 \in \mathcal{D}$.*

(1) Hypotheses (i)–(iii) of Theorem 1.5.1 hold;
(2) Hypotheses of Proposition 1.5.5 hold;

and

(3) $\overline{U}(x_0, b_3) \subseteq \mathcal{D}$.

Then, the conclusions of Theorem 1.5.2 hold with b_3, $\dfrac{\delta}{2}$ replacing b_2 and $\dfrac{1}{2}$, respectively.

Remark 1.5.4. If $L_0 = L$, the results reduce to the corresponding ones in Refs. [359, 360]. Otherwise they constitute an improvement since (1.300) or (1.301) is weaker than (1.297), error estimates are tighter and the information on the location of the solution $x^\star$ more precise, since our a_1 is larger than a_P. Indeed, under hypotheseis (1.297), the error bounds in Refs. [359, 360] are:

$$d(x_n, x_{n-1}) \leq \frac{2\,a_P\,\theta_P(r_0)^{2^n}}{1 - \theta_P(r_0)^{2^{n+1}}}, \tag{1.382}$$

$$d(x_n, x^\star) \leq \frac{2\,a_P\,\theta_P(r_0)^{2^n}}{1 - \theta_P(r_0)^{2^n}}, \tag{1.383}$$

and

$$d(x_n, x^\star) \leq (a_P^2 + \| x_n - x_{n-1} \|^2)^{1/2} - a_P, \tag{1.384}$$

where

$$\theta_P(r) = \frac{(r^2 + a_P^2)^{1/2} - a_P}{r}, \tag{1.385}$$

where a_P is given by (1.314) and

$$\overline{b}_0 = \frac{1}{L} - \left(\frac{1}{L}\left(\frac{1}{L} - 2\,r_0\right)\right)^{1/2} < b. \tag{1.386}$$

Then, we have by (1.312), (1.314), (1.317), (1.357) and (1.386) that

$$\theta_1(r) < \theta_P(r) \qquad r \in [0, r_0] \quad \text{and} \quad b_1 < \overline{b}_0.$$

Concerning (MNM) defined by (1.311), we have the following semi-local convergence result.

Theorem 1.5.4. *Let $F : \mathcal{D} \subseteq X \longrightarrow \mathcal{Y}$ be a Fréchet-differentiable operator and let $x_0 \in \mathcal{D}$.*

(1) Hypotheses (i)–(iii) of Theorem 1.5.1 and (1.310) hold;

and

(2) $\overline{U}(x_0, b_M) \subseteq \mathcal{D}$, where $b_M = \dfrac{1}{L_0} - \left(\dfrac{1}{L_0} \left(\dfrac{1}{L_0} - 2\,r_0 \right) \right)^{1/2}$.

Then sequence $\{x_n\}$ $(n \geq 0)$ generated by (MNM) given by (1.311) is well defined, remains in $\overline{U}(x_0, b_M)$ for all $n \geq 0$ and converges to a unique solution $x^\star$ of equation (1.1) in $\overline{U}(x_0, b_M)$. Moreover, estimates (1.315), (1.316) and

$$\| x_n - x^\star \| \leq s(\| x_n - x_{n-1} \|) - \| x_n - x_{n-1} \|$$

hold, with

$$\omega(r) = \frac{1}{2}\,L_0\,r^2 + r\left(1 - (L_0^2\,a_M^2 + 2\,L_0\,r)^{1/2} \right),$$

and

$$s(r) = \left(a_M^2 + \frac{2\,r}{L_0} \right)^{1/2} - a_M,$$

where a_M is given by (1.318).

Remark 1.5.5. If $L_0 = L$, Theorem 1.5.4 reduces to the corresponding one in Refs. [359, 360]. Otherwise they constitute an improvement, since $a_P < a_M$, $b_M < \overline{b}_0$ and our ω, s functions are smaller than ones in Refs. [359, 360].

We provide examples where our results apply but earlier one does not. In case all results apply we show that ours provide tighter error results bounds and better information on the location of the solution.

Example 1.5.1. Let $\mathcal{X} = \mathcal{Y} = \mathbb{R}$ be equipped with the max-norm, $x_0 = 1$, $\mathcal{D} = [\varrho, 2 - \varrho]$, $\varrho \in \left[0, \dfrac{1}{2} \right)$, and define function F as (1.48) (with ϱ replace a) on $\mathcal{D}$. Using (1.295), (1.296) and (1.298) we get:

$$r_0 = \frac{1}{3}\,(1 - \varrho), \quad L_0 = 3 - \varrho \ \text{ and } \ L = 2\,(2 - \varrho).$$

Then, we obtain conditions (1.297) and (1.299), respectively, as follows

$$H_K = \frac{4}{3}\,(1 - \varrho)\,(2 - \varrho) > 1$$

and

$$H_1 = \frac{1}{3}\,(5 + 2\sqrt{6})\,(3 - \varrho)\,(1 - \varrho) > 1 \quad \text{for all} \quad \varrho \in \left[0, \frac{1}{2} \right).$$

Hence, there is no guarantee that (NM) converges to $x^\star = \sqrt[3]{\varrho}$, starting at x_0. However, if we consider our conditions (1.310), (1.299) and (1.301), respectively, we get

$$H_M = \frac{2}{3}\left(3 - \varrho\right)\left(1 - \varrho\right) \leq 1 \quad \text{for all} \quad \varrho \in [.418861170, .5),$$

$$H_2 = \frac{1}{3}\left(7 - 3\,\varrho\right)\left(1 - \varrho\right) \leq 1 \quad \text{for all} \quad \varrho \in [.464816242, .5),$$

and

$$H_3 = \frac{1}{6}\left(8 - 3\,\varrho + \left(5\,\varrho^2 - 24\,\varrho + 28\right)^{1/2}\right)\left(1 - \varrho\right) \leq 1 \quad \text{for all} \quad \varrho \in [.450339002, .5).$$

Next we pick three values of ϱ such that all hypotheses are satisfied, so we can compare the values of parameters a and the corresponding error bounds.

Case $\varrho = .49999$
By Maple 13, we have the following results

$$x^\star = .7936952346, \quad H_K = 1.000026667 > 1, \quad H_1 = 4.124673776 > 1,$$

$$H_2 = .9166899999 < 1, \quad H_3 = .8877981560 < 1, \quad H_M = .8333533332 < 1,$$

$$a_1 = .1001396659, \quad a_M = .1632888647,$$

and

$$b_1 = .2609701491, \quad b_2 = .3333400000, \quad b_3 = .3551178419.$$

We cannot compare (1.376) and (1.382) in the case $\varrho \in (.5, 1)$ since (1.297) does not hold and a_P is a complex number in this interval. Note that we have

$$a_P \geq 0 \iff \varrho \in (.5, 2.5).$$

Case $\varrho = .5$
By Maple 13, we have the following results

$$x^\star = .7937005260, \quad H_K = 1, \quad H_2 = .9166666665 < 1,$$

$$a_P = 0, \quad a_1 = .1001542021, \quad b_1 = .5109569091 \quad \text{and} \quad b_2 = .3333333333.$$

Then the convergence is only linear in Refs. [359, 360] (see also estimates (1.382)–(1.384) in Remark 1.5.4) since $a_P = 0$, but our Theorems 1.5.1 and 1.5.2 apply and we can produce the following tables (Tables 1.6 and 1.7) for estimating error bounds (1.376), (1.377) and (1.380), (1.381), respectively.

Table 1.6   Comparison Table

n	x_n	(1.376)	(1.377)
1	.8333333333	.05612119686	.07077314850
2	.8151148834	.01371698569	.01465195163
3	.8059078274	.0009306422249	.0009349659389
4	.8008359800	.000004323620699	.000004323714024
5	.7979271348	9.332457124e-11	9.332457129e-11
6	.7962228487	4.348033039e-20	4.348033039e-20
7	.7952122874	9.438141843e-39	9.438141843e-39
8	.7946089091	4.447068601e-76	4.447068601e-76
9	.7942471777	9.872985216e-151	9.872985216e-151
10	.7940297902	4.866287937e-300	4.866287937e-300

Table 1.7   Comparison Table

n	x_n	(1.380)	(1.381)
1	.8333333333	.08333333335	.1666666666
2	.8151148834	.04166666666	.08333333335
3	.8059078274	.02083333334	.04166666666
4	.8008359800	.01041666666	.02083333334
5	.7979271348	.005208333335	.01041666666
6	.7962228487	.002604166666	.005208333335
7	.7952122874	.001302083334	.002604166666
8	.7946089091	.0006510416665	.001302083334
9	.7942471777	.0003255208334	.0006510416665
10	.7940297902	.0001627604166	.0003255208334

Table 1.8   Comparison Table

n	x_n	(1.376)	(1.382)
1	.8400000000	.07006258936	.07909638766
2	.8229000192	.01700313396	.02822534747
3	.8144944601	.001135124615	.004822385995
4	.8099973466	.000005104594030	.0001494969517
5	.8074963585	1.032319444e-10	1.439489908e-7
6	.8060774320	4.222016307e-20	1.334633444e-13
7	.8052635983	7.062061639e-39	1.147278253e-25
8	.8047939593	1.975853311e-76	8.477782648e-50
9	.8045219996	1.546682206e-151	4.629235881e-98
10	.8043641969	9.477501671e-302	1.380267862e-194

Case $\varrho = .52$

By Maple 13, we have the following results

$$x^\star = .8041451517, \quad H_K = .9472000000 < 1, \quad H_1 = 3.927915060 > 1,$$

$$H_2 = .8703999998 < 1, \quad H_3 = .8438043214 < 1, \quad H_M = .7935999998 < 1,$$

$$a_1 = .1262055091, \quad a_M = .1831905919, \quad a_P = .07762922486$$

and

$$b_1 = .2375782746, \quad b_2 = .3200000000, \quad b_3 = .3402436781.$$

Table 1.9   Comparison Table

n	x_n	(1.377)	(1.383)
1	.8400000000	.08820595264	.1122937620
2	.8229000192	.01814336327	.03319737437
3	.8144944601	.001140229312	.004972026896
4	.8099973466	.000005104697262	.0001496409008
5	.8074963585	1.032319444e-10	1.439491243e-7
6	.8060774320	4.222016307e-20	1.334633444e-13
7	.8052635983	7.062061639e-39	1.147278253e-25
8	.8047939593	1.975853311e-76	8.477782648e-50
9	.8045219996	1.546682206e-151	4.629235881e-98
10	.8043641969	9.477501671e-302	1.380267862e-194

We can now compare our results of Theorem 1.5.1 (see also estimates (1.376)–(1.378) to ones in Refs. [359, 360] (see also estimates (1.382)–(1.384)). Tables 1.8 and 1.9 show that our error bounds (1.376) and (1.377) are finer than (1.382) and (1.383) given in Refs. [359, 360].

Other applications and examples including the solution of nonlinear Chandrasekhar-type integral equations appearing in radiative transfer can also be found in section 1.1.

1.6   Exercises

1.6.1. Define the function $f(x) = x^3$ on the interval $[-2, 2]$ and choose $x_0 = 1$ as the initial point. For $p \in [0, 1]$, define

$$\ell_0 = \sup_{-2 \leq x \leq 2} \frac{|f'(x) - f'(x_0)|}{|x - x_0|^p} \quad \text{and} \quad \ell = \sup_{-2 \leq x \neq y \leq 2} \frac{|f'(x) - f'(y)|}{|x - y|^p}.$$

Show that $\ell_0 \leq \ell$.

1.6.2. Let x^* be a solution of equation (1.1). If the linear operator $F'(x^*)$ has a bounded inverse and $\lim_{\|x-x^*\| \to 0} \|F'(x) - F'(x^*)\| = 0$, then show (1.2) method converges to x^* if x_0 is sufficiently close to x^* and $\|x_n - x^*\| \leq d\varepsilon^n$ $(n \leq 0)$, where ε is any positive number; d is a constant depending on x_0 and ε.

1.6.3. The above result cannot be strengthened, in the sense that for every sequence of positive numbers c_n such that $\lim_{n \to \infty} \frac{c_{n+1}}{c_n} = 0$, there is an equation for which (1.2) converges less rapidly than c_n. Define

$$s_n = \begin{cases} c_{n/2} & \text{if } n \text{ is even} \\ \sqrt{c_{(n-1)/2} c_{(n+1)/2}} & \text{if } n \text{ is odd.} \end{cases}$$

Show: $s_n \to 0$, $\frac{s_{n+1}}{s_n} \to 0$ and $\lim_{n \to \infty} \frac{c_n}{s_{n+k}} = 0$, $(k \geq 1)$.

1.6.4. Assume operator $F'(x)$ satisfies a Hölder condition

$$\|F'(x) - F'(y)\| \leq a \|x - y\|^b,$$

in $U\left(x_0, R\right)$ with $0 < b < 1$. Define $h_0 = b_0 a \eta_0^b \le c_0$, where c_0 is a root of

$$\left(\frac{c}{1+b}\right)^b = (1-c)^{1+b} \quad (0 \le c \le 1)$$

and let $R \ge \dfrac{\eta_0}{1-d_0} = r_0$, where $d_0 = \dfrac{h_0}{(1+b)(1-h_0)}$. Show that (1.2) method converges to a solution x^* of equation $F\left(x\right) = 0$ in $U\left(x_0, r_0\right)$.

1.6.5.

(a) Show that f defined by $f(x, y) = |\sin y| + x$ satisfies a Lipschitz condition with respect to the second variable (on the whole xy-plane).

(b) Does f defined by $f(t, x) = |x|^{1/2}$ satisfy a Lipschitz condition?

Chapter 2

Special Conditions for Newton's Method

In this chapter we present some new results for convergence analysis of Newton's method in a Banach space setting using special conditions.

2.1 $\omega^\star$-Convergence

We provide a new semi-local result for the quadratic convergence of Newton's method under $w^\star$-conditioned second Fréchet-derivative. This way we can handle equations, where the usual Lipschitz-type conditions are non verifiable. An application involving nonlinear integral equations, and two boundary value problems is provided in this section. It turns out that a similar result using ω-conditioned hypotheses can provide usable error estimates indicating only linear convergence for Newton's method. We are concerned with the problem of approximating a locally unique solution $x^\star$ of (1.1) using Newton's method (1.2).

The main condition is given by

$$\| F''(x) \| \le K, \quad K \ge 0 \tag{2.1}$$

for x in some neighborhood of the initial guess x_0. Several authors have used Lipschitz conditions:

$$\| \mathcal{J}\, (F'(x) - F'(y)) \| \le \alpha(\| x - y \|), \quad \text{for} \quad x, y \in \mathcal{D}, \tag{2.2}$$

where $\mathcal{J}$ is the identity operator or $F'(x_0)^{-1}$ and α a continuous non-decreasing function defined on the non-negative axis. Some special cases are given by $\alpha(r) = \ell\, r^p$, $p \in [0, 1]$ and $\alpha(r) = \gamma(r)\, r$, where function γ has the same properties with α. A survey of such results can be found in Ref. [110] and the references therein (see also Refs. [69, 100, 116, 444]). As Ezquerro and Hernández have already noted in Refs. [205, 204], the verification of (2.2) at least for some type of problems is not possible. That is why motivated by (2.1), they introduced the condition

$$\| F''(x) \| \le \omega(\| x \|) \quad \text{for} \quad x \in \mathcal{D}, \tag{2.3}$$

where $\omega\, : \mathcal{I} = [0, +\infty) \longrightarrow \mathcal{I}$ is a continuous function with $\omega(0) \ge 0$, which is either non-decreasing or non-increasing. Ezquerro and Hernández in Refs. [205, 204]

89

call this type of estimate ω-conditioned second derivative. Sufficient semi-local convergence conditions were given but the usable error bounds imply only linear convergence for Newton's method (1.2).

In order for us to rectify this drawback, instead of (2.3), we use the condition

$$\| F'(x_0)^{-1} F''(x) \| \leq w^\star(\| x_0 - x \|) \quad \text{for all} \quad x \in \mathcal{D}, \tag{2.4}$$

where $w^\star : \mathcal{I} = [0, +\infty) \longrightarrow \mathcal{I}$ is a continuous function with $w^\star(0) \geq 0$, which is either non-decreasing or non-increasing. We shall call this type of estimate $w^\star$-conditioned second derivative. Our semi-local convergence analysis is based on an approach using more precise majorizing sequences.

We shall first introduce some preliminary conditions and results. Let $x_0 \in \mathcal{D}$ be such that $F'(x_0)^{-1} \in \mathcal{L}(\mathcal{Y}, \mathcal{X})$. We assume:

$(\mathcal{C}_1)$ $\| F'(x_0)^{-1} F(x_0) \| \leq \eta$,
$(\mathcal{C}_2)$ $\| F'(x_0)^{-1} F''(x) \| \leq w^\star(\| x - x_0 \|)$ for all $x \in \mathcal{D}$, where $w^\star : \mathcal{I} \longrightarrow \mathcal{I}$ is a continuous function with $w^\star(0) \geq 0$, such that

$$w^\star(t\,y) \leq h(t)\,w^\star(y) \quad \text{for all} \quad t \in [0, 1], \ y \in \mathcal{D},$$

and $h : [0, 1] \longrightarrow \mathcal{I}$ a continuous non-decreasing function.

Set

$$H = \int_0^1 h(t)\,dt.$$

$(\mathcal{C}_3)$ Function

$$g(r) = \left(\overline{L}(r) + 4\,\overline{L}_0(r) + \sqrt{\overline{L}^2(r) + 8\,\overline{L}(r)\,\overline{L}_0(r)} \right) \eta - 4$$

has a minimal positive zero, denoted by r_0, where

$$\overline{L}(r) = w^\star(r + \| x_0 \|), \qquad \overline{L}_0(r) = H\,w^\star(r).$$

$(\mathcal{C}_4)$ $\overline{L}_0(r) \leq \overline{L}(r)$ for all $r \in \mathcal{I}_0 = [0, r_0]$.
$(\mathcal{C}_5)$ $\overline{U}(x_0, r_0) \subseteq \mathcal{D}$.
$(\mathcal{C}_6)$ Function

$$q(r) = \int_0^1 \int_0^1 w^\star(s\,(r_0 + t\,(r - r_0)))\,ds\,(r_0 + t\,(r - r_0))\,dt - 1$$

has a minimal positive zero $r^\star \geq r_0$.

The importance of introducing function h, so we can have better error bounds has been explained in Refs. [110, 135, 145].

We need two results on majorizing sequences, before we present the main theorem. The first is Lemma 1.3.1 and the second is as follows.

Lemma 2.1.1. *Assume conditions $(\mathcal{C}_3)$ and $(\mathcal{C}_4)$ hold and set $L_0 = \overline{L}_0(r_0)$ and $L = \overline{L}(r_0)$. Then, the conclusions of Lemma 1.3.1 hold for iteration $\{t_n\}$.*

Proof. Clearly, all hypotheses of Lemma 1.3.1 are satisfied for the above choices of L_0 and L. That completes the proof of Lemma 2.1.1. $\qquad\square$

We can show the main semi-local convergence result for Newton's method under $w^\star$-conditions and $w^\star$ a non-decreasing function.

Theorem 2.1.1. *Let $F : \mathcal{D} \subseteq \mathcal{X} \longrightarrow \mathcal{Y}$ be a twice Fréchet differentiable operator. Let x_0 be such that $F'(x_0)^{-1} \in \mathcal{L}(\mathcal{Y},\mathcal{X})$. Moreover, assume conditions $(\mathcal{C}_1)$–$(\mathcal{C}_6)$ hold, and $t^\star \leq r_0$. Then, sequence $\{x_n\}$ generated by Newton's method (1.2) is well defined, remain in $\overline{U}(x_0, t^\star)$ for all $n \geq 0$ and converges to a solution $x^\star \in \overline{U}(x_0, t^\star)$ of equation $F(x) = 0$. Moreover, the following estimates hold for all $n \geq 0$:*

$$\| x_n - x^\star \| \leq t^\star - t_n, \tag{2.5}$$

where L_0, L and $t^\star$, t_n are given in Lemmas 1.3.1, 2.1.1, respectively. Furthermore, the solution $x^\star$ is unique in $\mathcal{D}_0 = U(x_0, t^\star) \cap \mathcal{D}$.

Proof. We shall show using induction:

$$\| x_{k+1} - x_k \| \leq t_{k+1} - t_k, \tag{2.6}$$

and

$$\overline{U}(x_{k+1}, t^\star - t_{k+1}) \subseteq \overline{U}(x_k, t^\star - t_k) \tag{2.7}$$

hold for all $k \geq 0$. For every $z \in \overline{U}(x_1, t^\star - t_1)$

$$\| z - x_0 \| \leq \| z - x_1 \| + \| x_1 - x_0 \| \leq t^\star - t_1 + t_1 - t_0 = t^\star - t_0,$$

implies $z \in \overline{U}(x_0, t^\star - t_0)$. Since, also

$$\| x_1 - x_0 \| = \| F'(x_0)^{-1} F(x_0) \| \leq \eta = t_1 - t_0,$$

estimates (2.6) and (2.7) hold for $k = 0$.

Given (2.6) and (2.7) hold for all $i \leq k$, we have:

$$\| x_{k+1} - x_0 \| \leq \sum_{i=1}^{i=k+1} \| x_i - x_{i-1} \| \leq \sum_{i=1}^{i=k+1} (t_i - t_{i-1}) = t_{k+1} < t^\star$$

and

$$\| x_k + \rho\,(x_{k+1} - x_k) - x_0 \| \leq t_k + \rho\,(t_{k+1} - t_k) \leq t^\star, \quad \rho \in [0,1].$$

We obtain that $\| x_1 - x_0 \| < t^\star \leq r_0$. That is $x_1 \in \overline{U}(x_0, t^\star)$ and from the above $x_0 + \rho\,(x_1 - x_0) \in \overline{U}(x_0, t^\star)$ for $\rho \in [0,1]$.

Let $x_k \in U(x_0, t^\star)$. We can then get from $(\mathcal{C}_2)$ the estimate

$$\| F'(x_0)^{-1} (F'(x_k) - F'(x_0)) \| = \| \int_0^1 F'(x_0)^{-1} F''(x_0 + t\,(x_k - x_0))\,(x_k - x_0)\,dt \|$$

$$\leq \int_0^1 w^\star(t \| x_k - x_0 \|)\,\| x_k - x_0 \|\,dt \leq \int_0^1 h(t)\,dt\,w^\star(\| x_k - x_0 \|)\,\| x_k - x_0 \|$$

$$\leq H\,w^\star(t_k)\,t_k \leq H\,w^\star(t^\star)\,t_k \leq H\,w^\star(r_0)\,t_k = L_0\,t_k \leq L_0\,t^\star < 1. \tag{2.8}$$

It follows from (2.8) and the Banach lemma on invertible operators that $F'(x_k)^{-1}$ exists and

$$\| F'(x_k)^{-1} F'(x_0) \| \leq (1 - L_0 t_k)^{-1}. \tag{2.9}$$

In view of (1.2) and Taylor's formula:

$$F(x_k) = \int_0^1 F''(x_{k-1} + t(x_k - x_{k-1}))(1 - t)(x_k - x_{k-1})^2 \, dt,$$

we get that

$$\begin{aligned}
\| F'(x_0)^{-1} F(x_k) \| \\
=& \left\| \int_0^1 F'(x_0)^{-1} F''(x_{k-1} + t(x_k - x_{k-1}))(1-t)(x_k - x_{k-1})^2 \, dt \right\| \\
\leq& \frac{1}{2} \int_0^1 w^\star(\| x_{k-1} - x_0 + t(x_k - x_{k-1}) \|) \| x_k - x_{k-1} \|^2 \, dt \\
\leq& \frac{1}{2} \int_0^1 w^\star((1-t)\| x_{k-1} - x_0 \| + t \| x_k - x_0 \| + \| x_0 \|)(t_k - t_{k-1})^2 \, dt \\
\leq& \frac{1}{2} \int_0^1 w^\star(t^\star + \| x_0 \|)(t_k - t_{k-1})^2 \, dt \\
\leq& \frac{1}{2} \int_0^1 w^\star(r_0 + \| x_0 \|)(t_k - t_{k-1})^2 \, dt = \frac{1}{2} L(t_k - t_{k-1})^2.
\end{aligned} \tag{2.10}$$

Using (1.2), Lemma 1.3.1, (2.9) and (2.10), we deduce

$$\| x_{k+1} - x_k \| \leq \| F'(x_k)^{-1} F'(x_0) \| \, \| F'(x_0)^{-1} F(x_k) \| \leq \frac{L(t_k - t_{k-1})^2}{2(1 - L_0 t_k)} = t_{k+1} - t_k,$$

which shows (2.6) for all k. Thus, for every $v \in \overline{U}(x_{k+1}, t^\star - t_{k+1})$, we have

$$\| v - x_k \| \leq \| v - x_{k+1} \| + \| x_{k+1} - x_k \| \leq t^\star - t_{k+1} + t_{k+1} - t_k = t^\star - t_k,$$

which implies $v \in \overline{U}(x_k, t^\star - t_k)$. The induction for (2.6) and (2.7) is complete.

Lemma 2.1.1 implies that $\{t_n\}$ is a Cauchy sequence. By (2.6) and (2.7), sequence $\{x_n\}$ is Cauchy too, in a Banach space $\mathcal{X}$ and as such it converges to some $x^\star \in \overline{U}(x_0, t^\star)$ (since $\overline{U}(x_0, t^\star)$ is a closed set). We get in view of (2.10) by letting $k \longrightarrow \infty$ that $F(x^\star) = 0$. Estimate (2.5) follows from (2.6) by using standard majorization techniques. Finally, to show uniqueness of $x^\star$ in $\mathcal{D}_0$, let us assume $y^\star$ is a solution of equation $F(x)$ in $\mathcal{D}_0$. We need the identity $\mathcal{M}(y^\star - x^\star) = F(y^\star) - F(x^\star)$, where $\mathcal{M} = \int_0^1 F'(x_0)^{-1} F'(x^\star + \theta(y^\star - x^\star)) \, d\theta$. Then, using $(\mathcal{C}_2)$, $(\mathcal{C}_6)$, and (2.10), we obtain in turn that

$$\begin{aligned}
\| F'(x_0)^{-1} (F'(x_0) - \mathcal{M}) \| \\
\leq& \left\| \int_0^1 F'(x_0)^{-1} (F'(x^\star + t(y^\star - x^\star)) - F'(x_0)) \, dt \right\| \\
\leq& \left\| \int_0^1 \int_0^1 F'(x_0)^{-1} F''(x_0 + s((1-t)(x^\star - x_0) + \right. \\
& \left. t(y^\star - x_0))) \, ds \, ((1-t)(x^\star - x_0) + t(y^\star - x_0)) \, dt \right\| \\
<& \int_0^1 \int_0^1 \| F'(x_0)^{-1} F''(s((1-t)(x^\star - x_0) + t(y^\star - x_0))) \| \, ds \, (r_0 + t(r^\star - r_0)) \, dt \\
\leq& \int_0^1 \int_0^1 w^\star(s(r_0 + t(r^\star - r_0))) \, ds \, (r_0 + t(r^\star - r_0)) \, dt = 1.
\end{aligned}$$

$$\tag{2.11}$$

In view of (2.11) and the Banach lemma on invertible operator $\mathcal{M}^{-1}$ exists. It follows that $x^\star = y^\star$. That completes the proof of Theorem 2.1.1. $\qquad\square$

Remark 2.1.1. It follows from $(\mathcal{C}_2)$ and (2.11) that the uniqueness of the solution $x^\star$ is guaranteed in $U(x_0, r^\star)$ if

$$\int_0^1 w^\star(s\, r^\star)\, ds\, r^\star \leq 1 \quad \text{or} \quad H\, w^\star(r^\star)\, r^\star \leq 1.$$

In $(\mathcal{C}_2)$, we assumed that $w^\star$ is a non-decreasing function. Howeover, a result can be given under the condition that $w^\star$ is a non-increasing function. Indeed, let us assume:

$(\mathcal{B}_1) \equiv (\mathcal{C}_1)$
$(\mathcal{B}_2)$ $\| F'(x_0)^{-1} F''(x) \| \leq w^\star(\| x - x_0 \|)$ for all $x \in \mathcal{D}$, where $w^\star : \mathcal{I} \longrightarrow \mathcal{I}$ is a continuous non-increasing function, with $w^\star(0) \geq 0$.
$(\mathcal{B}_3)$ Function g defined in $(\mathcal{C}_3)$ has a minimal positive zero, denoted by r_0 with $r_0 \leq \| x_0 \|$, where $\overline{L}(r) = w^\star(\| x_0 \| - r)$, $\overline{L}_0 = \overline{L}_0(r) = w^\star(0)$, $r \in [0, \| x_0 \|]$.
$(\mathcal{B}_4) \equiv (\mathcal{C}_4)$
$(\mathcal{B}_5)$ $w^\star(0)\, r_0 \leq 1$.

Set now $r^\star = r_0 + 2\,(1 - w^\star(0)\, r_0)$. Then, we can show the following semi-local result for Newton's method under $w^\star$-conditions and $w^\star$ a non-increasing function.

Theorem 2.1.2. *Let* $F : \mathcal{D} \subseteq \mathcal{X} \longrightarrow \mathcal{Y}$ *be a twice Fréchet differentiable operator. Let* x_0 *be such that* $F'(x_0)^{-1} \in \mathcal{L}(\mathcal{Y}, \mathcal{X})$. *Moreover, assume conditions* $(\mathcal{B}_1)$–$(\mathcal{B}_5)$ *hold, and* $t^\star \leq r_0$. *Then, the conclusions of Theorem 2.1.1 hold.*

Proof. We follow the proof of Theorem 2.1.1 until

$$\| F'(x_0)^{-1} (F'(x_k) - F'(x_0)) \| \leq \int_0^1 w^\star(t\, \| x_k - x_0 \|)\, \| x_k - x_0 \|\, dt$$
$$\leq \int_0^1 w^\star(0)\, t_k\, dt = L_0\, t_k \leq L_0\, t^\star < 1.$$

Then, we have instead of (2.10):

$$\| F'(x_0)^{-1} F(x_k) \|$$
$$\leq \frac{1}{2} \int_0^1 w^\star(\| x_{k-1} - x_0 + t\,(x_k - x_{k-1}) \|)\, \| x_k - x_{k-1} \|^2\, dt$$
$$\leq \frac{1}{2} \int_0^1 w^\star(\| x_0 \| - (1 - t)\, \| x_{k-1} - x_0 \| - t\, \| x_k - x_0 \|)\, \| x_k - x_{k-1} \|^2\, dt$$
$$\leq \frac{1}{2} \int_0^1 w^\star(\| x_0 \| - r_0)\, \| x_k - x_{k-1} \|^2\, dt = \frac{1}{2} L\,(t_k - t_{k-1})^2.$$
$$\tag{2.12}$$

Finally, for the uniqueness part, we have in turn

$$\| F'(x_0)^{-1} \left(F'(x_0) - \mathcal{M} \right) \|$$
$$\leq \int_0^1 \int_0^1 w^\star(s \parallel (1-t)\,(x^\star - x_0) \parallel + t \parallel y^\star - x_0 \parallel)\, ds\, (r_0 + t\,(r^\star - r_0))\, dt$$
$$\leq \int_0^1 w^\star(0)\,(r_0 + t\,(r^\star - r_0))\, dt = w^\star(0)\, r_0 + \frac{1}{2}\,(r^\star - r_0) = 1.$$

That completes the proof of Theorem 2.1.2. $\qquad\qquad\qquad\qquad\qquad$ $\square$

Remark 2.1.2. We shall compare our results with earlier ones. Let us introduce the set of conditions used in Refs. [205, 204] but in affine invariant form.

$(\mathcal{A_1}) \equiv (\mathcal{C_1})$

$(\mathcal{A_2})$ $\parallel F'(x_0)^{-1} F''(x) \parallel \leq \omega(\parallel x \parallel)$ for all $x \in \mathcal{D}$, where $\omega : \mathcal{I} \longrightarrow \mathcal{I}$ is a continuous non-decreasing or non-increasing function with $\omega(0) \geq 0$.

$(\mathcal{A_3})$ Equation $3\,\eta\,\varphi(t)\,t - 2\,\eta^2\,\varphi(t) - 2\,t + 2\,\eta = 0$ has a minimal positive solution R, where

$$\varphi(t) = \omega(\parallel x_0 \parallel + t) \quad \text{if } \omega \text{ is increasing,}$$

or

$$\varphi(t) = \omega(\parallel x_0 \parallel - t) \quad \text{if } \omega \text{ is decreasing.}$$

Note that R must be smaller than $\parallel x_0 \parallel$ if ω is non-increasing

$(\mathcal{A_4})$ $\overline{U}(x_0, R) \subseteq \mathcal{D}$.

$(\mathcal{A_5})$ $\alpha_0 = \eta\,\varphi(R) \in (0, \frac{1}{2})$.

$(\mathcal{A_6}) \equiv (\mathcal{C_6})$.

Under $(\mathcal{A_1})$–$(\mathcal{A_6})$, the linear convergence of iteration $\{x_n\}$ to a unique solution of equation $F(x) = 0$ in $U(x_0, R)$ was established.

The advantages of our approach over the corresponding ones in Refs. [205, 204] are already stated in the introduction of this section. Finally, note that verifying set of conditions $(\mathcal{C_1})$–$(\mathcal{C_6})$ (or $(\mathcal{B_1})$–$(\mathcal{B_5})$) requires the same type of difficulties as $(\mathcal{A_1})$–$(\mathcal{A_6})$.

Remark 2.1.3. Estimate $L_0 \leq L$ in Lemma 1.3.1 is only needed to show the quadratic convergence of iteration $\{t_n\}$. However, in Theorem 2.1.2, we have $L_0 \geq L$, which guarantees only linear convergence. But, in view of estimate (2.12), parameters L_0 and L in Theorem 2.1.2 can be defined by $L = L_0 = w^\star(0)$. This way the quadratic convergence of iteration $\{t_n\}$ is recovered according to Theorem 2.1.2. Note that in this case, we replace r_0 by $\parallel x_0 \parallel$ in hypothesis $t^\star \leq r_0$, whereas $(\mathcal{B_3})$ is replaced in Theorem 2.1.2 by $(\mathcal{B_3'})$ $2\,w^\star(0)\,\eta \leq 1$.

We shall apply our results for integral equations of Hammerstein type. Such equations have already been studied in Refs. [110, 205, 204]. The usable error bounds in Refs. [205, 204] indicate only linear convergence for Newton's method (1.2). Here, we shall show that under our sufficient convergence conditions, Newton's method converges quadratically.

Example 2.1.1. We will repeat some terminology-results from Refs. [205, 204]. We consider the nonlinear Hammerstein equation of the second kind:

$$x(s) = y(s) + \int_a^b \mathcal{G}(s,t)\,\mathbb{K}(t,x(t))\,dt, \qquad s \in [a,b] \tag{2.13}$$

where $\mathcal{G}(s,t)$ is the Green's kernel given by:

$$\mathcal{G}(s,t) = \begin{cases} \dfrac{(b-s)\,(t-a)}{b-a} & \text{if } t \le s \\[2mm] \dfrac{(s-a)\,(b-t)}{b-a} & \text{if } s \le t \end{cases} \tag{2.14}$$

$\mathbb{K}(t,u)$ is a continuous function for $t \in [a,b]$, $-\infty < u < +\infty$, $y \in \mathcal{C}[a,b]$, and x is the unknown function sought in $\mathcal{C}[a,b]$. In particular, we shall consider a special case of (2.13) given by

$$x(s) = y(s) + \int_a^b \mathcal{G}(s,t)\,(x(t)^{1+p} + \lambda\,x(t)^2)\,dt, \qquad p \in [0,1], \quad \lambda \in \mathbb{R}, \tag{2.15}$$

where y is a continuous function such that $y(s) > 0$, $s \in [a,b]$. Equation (2.15) is equivalent to the point boundary value problem

$$\begin{aligned} x'' &= -x^{1+p} - \lambda\,x^2 \\ x(a) &= \nu(a), \quad x(b) = \nu(b) \end{aligned} \tag{2.16}$$

where ν is a continuous function. Note that (2.15) is equivalent to solving equation $F(x) = 0$ on

$$\mathcal{D} = \{x \in \mathcal{C}[a,b] \;:\; x(s) > 0, \quad s \in [a,b]\}, \tag{2.17}$$

where $F : \mathcal{C}[a,b] \longrightarrow \mathcal{C}[a,b]$, and

$$F(x)(s) = x(s) - y(s) - \int_a^b \mathcal{G}(s,t)\,(x(t)^{1+p} + \lambda\,x(t)^2)\,dt. \tag{2.18}$$

Equations of type (2.18) are interesting, since the usual studies do not apply, due to the fact that operator F is neither Lipschitz continuous or p-Hölder continuous. We shall compute the quantities and functions needed in Lemma 1.3.1 and Remark 2.1.3, so we can verify the hypotheses of Theorem 2.1.2. Let us consider the max-norm and set

$$N = \max_{s \in [a,b]} \int_a^b |\mathcal{G}(s,t)|\,dt. \tag{2.19}$$

We have that

$$[F'(x)\phi](s) = \phi(s) - \int_a^b \mathcal{G}(s,t)\,((1+p)\,x(t)^p + 2\,\lambda\,x(t))\,\phi(t)\,dt. \tag{2.20}$$

Let $x_0(s)$ be fixed, then we have

$$\| I - F'(x_0) \| \le N\,((1+p)\,\| x_0 \|^p + 2\,|\lambda|\,\| x_0 \|) \tag{2.21}$$

and if

$$N\left((1+p)\parallel x_0 \parallel^p + 2\,|\lambda|\parallel x_0 \parallel\right) < 1, \tag{2.22}$$

the existence of $F'(x_0)^{-1}$ is guaranteed by the Banach lemma on invertible operators and

$$\parallel F'(x_0)^{-1}\parallel\le \beta = \left(1 - N\left((1+p)\parallel x_0 \parallel^p + 2\,|\lambda|\parallel x_0 \parallel\right)\right)^{-1}. \tag{2.23}$$

Moreover, in view of (2.18), we get

$$\parallel F(x_0)\parallel\le\parallel x_0 - y \parallel + N\left(\parallel x_0 \parallel^{1+p} + |\lambda|\parallel x_0 \parallel^p\right).$$

Consequently, we obtain that

$$\parallel F'(x_0)^{-1}\,F(x_0)\parallel\le\eta = \frac{\parallel x_0 - y \parallel + N\left(\parallel x_0 \parallel^{1+p} + |\lambda|\parallel x_0 \parallel^p\right)}{1 - N\left((1+p)\parallel x_0 \parallel^p + 2\,|\lambda|\parallel x_0 \parallel\right)}. \tag{2.24}$$

We also have that

$$[F''(x)\,\phi\,z](s) = -\int_a^b \mathcal{G}(s,t)\left((1+p)\,p\,x(t)^{p-1} + 2\,\lambda\right)z(t)\,\phi(t)\,dt. \tag{2.25}$$

So, for $(\mathcal{B}_2)$ to be satisfied, we define

$$w^\star(r) = \beta\,N\left((1+p)\,p\,(r+\parallel x_0 \parallel)^{p-1} + 2\,|\lambda|\right). \tag{2.26}$$

We consider the special case :

$$x_0(s) = y(s) = 1, \quad [a,b] = [0,1], \quad \lambda = p = \frac{1}{2}.$$

Then, using (2.24), (2.26) and Remark 2.1.3, we get

$$\eta = \frac{3}{11}, \quad L_0 = L = w^\star(0) = \frac{7}{22}, \quad r_0 = 1 \quad \text{and} \quad r^\star = \frac{30}{22}.$$

According to Lemma 1.3.1

$$\delta = 1, \quad t^{\star\star} = 2\,\eta = \frac{6}{11} < 1 = r_0,$$

and $(\mathcal{B}'_3)$ is satisfied, since

$$2\,w^\star(0)\,\eta = 2\,\frac{7}{22}\,\frac{3}{11} = .173553719 < .5.$$

Hence, Theorem 2.1.2 guarantees the existence and uniqueness of a solution $x^\star$ for equation (2.15). Moreover, according to Lemma 1.3.1, Theorem 2.1.2 and Remark 2.1.3 the convergence of iteration $\{x_n\}$ to $x^\star$ is quadratic. Note that on the same equation in Refs. [205, 204], the convergence was shown to be only linear.

Finally, we provide an example with a non-decreasing function $w^\star$.

Example 2.1.2. Let $\mathcal{X} = \mathcal{Y} = \mathbb{R}$. Set $x_0 = .99$, $\gamma \in (0,1]$, and $\mathcal{D} = U(.99, 1-\gamma)$. Define function F on $\mathcal{D}$ by $F(x) = x^3 - \gamma$. Using $(\mathcal{C}_1)$–$(\mathcal{C}_6)$, we get

$$w^\star(r) = \frac{2}{.99^2}\,r = 2.040608101\,r, \quad h(t) = t, \quad H = \frac{1}{2},$$

$$\overline{L}_0(r) = \frac{1}{.99^2}\, r = 1.020304051\, r < \overline{L}(r) = 2.040608101\,(r + .99),$$

$$\eta = \frac{1}{3 \times .99^2}\,(.99^3 - \gamma),$$

$$g(r) = 6.121824303\,\eta\, r + 2.020202020\,\eta +$$
$$.00001\,\eta\,\sqrt{208204071100\, r^2 + 247346436500\, r + 40812162030} - 4,$$

and

$$q(r) = .34010135\,(r - r_0)^2 + 1.02030405\, r_0\,(r - r_0) + 1.02030405\, r_0{}^2,$$

Note that $w^\star$ is a non-decreasing function. For $\gamma = .01$ and using Maple 13, we obtain

$$\eta = .3265989865, \quad r_0 = .7294057602 \quad \text{and} \quad r^\star = 1.229434953.$$

Hence, the conclusions of our Theorem 2.1.1 can apply to solve equation $F(x) = 0$. Moreover, sequence defined by (1.2) starting at x_0 converges to $x^\star = .215443469$.

Remark 2.1.4. In Example 2.1.2, if we set $x_0 = 1$, we show that this choice is not possible for $\eta = \frac{1}{3}\,(1 - \gamma)$. Indeed, by (C_1)–(C_6), we get that

$$w^\star(r) = 2\,r, \quad h(t) = t, \quad H = \frac{1}{2},$$

$$\overline{L}_0(r) = r < \overline{L}(r) = 2\,(r + 1), \quad \eta = \frac{1}{3}\,(1 - \gamma),$$

$$g(r) = 2\left(3\,r + 1 + \sqrt{5\,r^2 + 6\,r + 1}\right)\eta - 4$$

and

$$q(r) = \frac{r^3 - r_0^3}{3\,(r - r_0)} - 1 = r^2 + r\,r_0 + r_0^2 - 1.$$

Using Maple 13, we obtain

$$r_0(\eta) = \frac{3 - \sqrt{5 + 4\,\eta}}{2\,\eta} \quad \text{and} \quad r^\star(r_0) = \frac{\sqrt{4 - 3\,r_0^2} - r_0}{2}.$$

The solutions-set of inequality $r^\star(r_0) \geq r_0$ in $\mathbb{R}^+$ is

$$I_{r_0} = \left[0, \frac{\sqrt{3}}{3}\right] \simeq [0, .5773502692],$$

(see also figure 2.1). Consequently, we can find all $\eta > 0$ such that $r_0(\eta) \in I_{r_0}$, we get $\eta \in (.3840179681, \infty)$. This contradicts the hypothesis $\eta = \frac{1}{3}\,(1 - \gamma) \in [0, 0.333333)$. The choice of starting point $x_0 = 1$ is possible if we replace η by $\eta^\star > .3840179681$.

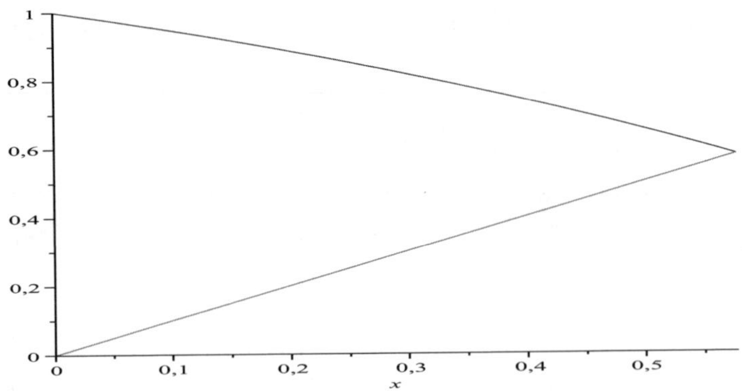

Fig. 2.1   Functions $r^\star(x)$ and x on $[0, .5773502692]$.

2.2   Regular Smoothness

In this section we are concerned with the problem of approximating a locally unique solution $x^\star$ of (1.1) using (1.2). We provide an existence and uniqueness result for nonlinear equations involving regularly smooth operators in a Banach space setting, under weaker hypotheses than before Refs. [223, 320]. Our approach extends the applicability and improves the optimality of Newton-type methods using the same information, under weaker hypotheses. Moreover, the information on the location of the solution is at least as precise as in earlier studies (see Refs. [223, 320]). Numerical example is also provided to show that the results can be applied to solve equations, but earlier ones using the same information cannot.

The optimality of iterative methods for solving (1.1) was inaugurated by Maĭstrovskiĭ in Ref. [320], using Newton's method (NM):

$$x_+ = x - F'(x)^{-1} F(x) \tag{2.27}$$

by considering the finite dimensional case and assuming Lipschitz conditions on the operator F', as well as unlimited values of (F, F'). Maĭstrovskiĭ showed that (NM) is optimal in the sense that no other method using the information: one value of F and one of its derivative per iteration converges faster in any ϵ-neighborhood of the solution $x^\star$.

Galperin introduced in Ref. [223] another concept of optimality. He described this concept as follows: Suppose there exist an $x \in \mathcal{D}$ and a set $\mathcal{S}(x)$, such that equation (1.1) has a solution $x^\star \in \mathcal{S}(x)$ which is unique in $\mathcal{S}(x)$. Then, state a procedure that using information from F at x provides the next approximation x_+. The response is not known when x_+ is being chosen. However, we may know what responses of the operator are possible. This way, we choose x_+ by taking this knowledge into account. A method will then be considered optimal if it minimizes a measure $\mu(x_+)$ of the size of $\mathcal{S}(x_+)$ relative to that of $\mathcal{S}(x)$. This concept of optimality clearly depends on how an operator's responses are predicted. Maĭstrovskiĭ's approach is based on the worst case scenario, if it is assumed that the information at x_+ is the least favorable for the size of $\mathcal{S}(x_+)$. Another approach is the optimality on the average, where we may assume that the information at x_+ is most probable.

An existence and uniqueness result, associating $\mathcal{S}(x)$ with x under ω-regular smoothness of F on $\mathcal{D}$ (to be precised in this section) was provided by Galperin in Ref. [223]. It was then specialized in the case of scalar equations to show that Newton-type method (NTM) resulting from this analysis reduces the entropy of the solution more than (NM) (from .5 of (NM) to $1 - .5\sqrt{2}$ for the new method provided F is Lipschitz smooth). Moreover, it is shown that under optimality on the average, (NM) is optimal.

We introduce the concept of center-ω_0-regular smoothness of F on $\mathcal{D}$, where ω_0 is always at least as tight scalar function as ω (to be also precised in this section). If $\omega_0 = \omega$, our results reduce to the corresponding ones in Ref. [223]. Otherwise, our results are finer with advantages as already stated in the abstract of this section. In order to make the section as self-contained as possible, we reintroduce some definitions and some results on the regular smoothness concept as in Refs. [66, 102, 221, 223]) (see also Refs. [110, 134, 387]).

Definition 2.2.1.

(a) Let $\mathcal{N}$ be the class of increasing concave functions $\omega : \mathbb{R}^+ \longrightarrow \mathbb{R}^+$, with $\omega(0) = 0$. Note that $\mathcal{N}$ contains the functions in the form $\varphi(t) = c\,t^p$, ($c \geq 0$ and $p \in (0, 1]$).

(b) Given an $\omega \in \mathcal{N}$, we say that a mapping $\mathcal{S} : \mathcal{X} \longrightarrow \mathcal{L}(\mathcal{X}, \mathcal{Y})$ is ω-regularly continuous on $\mathcal{D} \subseteq \mathcal{X}$, or, equivalently, that ω is the regularity continuous modulus of $\mathcal{S}$ on $\mathcal{D}$, if the following inequality holds for all $x, y \in \mathcal{D}$

$$
\begin{aligned}
&\omega^{-1}\left(\min\{\| \mathcal{S}(x) \|, \| \mathcal{S}(y) \|\} + \| \mathcal{S}(x) - \mathcal{S}(y) \| \right) \\
&-\omega^{-1}\left(\min\{\| \mathcal{S}(x) \|, \| \mathcal{S}(y) \|\} \right) \leq \| x - y \| .
\end{aligned}
\tag{2.28}
$$

The operator F is ω-regularly smooth on $\mathcal{D}$ if its Fréchet-derivative F' is ω-regularly continuous there. The function ω is called a regular smoothness modulus of F on $\mathcal{D}$. F is regularly smooth on $\mathcal{D}$ if it has a regular smoothness modulus there. We denote by ω^{-1} the function whose closed epigraph $\mathrm{cl}\,\{(s, t) : s \geq 0 \ \text{ and } \ t \geq \omega^{-1}(s)\}$ is symmetrical to the closure of the subgraph of ω with respect to the axis $t = s$. Note that every Lipschitz smooth operator (with a modulus $t \longrightarrow \lambda t$) is, clearly, also regularly smooth with the same modulus.

Let $x_0 \in \mathcal{D}$. In view of (2.28), there exists $\omega_0 \in \mathcal{N}$, such that for all x in $\mathcal{D}$

$$
\begin{aligned}
&\omega_0^{-1}\left(\min\{\| \mathcal{S}(x) \|, \| \mathcal{S}(x_0) \|\} + \| \mathcal{S}(x) - \mathcal{S}(x_0) \| \right) \\
&-\omega_0^{-1}\left(\min\{\| \mathcal{S}(x) \|, \| \mathcal{S}(x_0) \|\} \right) \leq \| x - x_0 \| .
\end{aligned}
\tag{2.29}
$$

Clearly, we have

$$\omega_0(t) \leq \omega(t) \quad \text{for all} \quad t \in [0, \infty) \tag{2.30}$$

holds. If ω_0, ω are linear functions ($\omega(t) = ct$ and $\omega_0(t) = c_0 t$), then (2.28) and (2.29) become Lipschitz and center-Lipschitz continuous conditions, respectively, i.e., the following hold respectively for all $(x, y) \in \mathcal{D}^2$:

$$\| \mathcal{S}(x) - \mathcal{S}(y) \| \leq c \| x - y \| \tag{2.31}$$

and

$$\| \mathcal{S}(x) - \mathcal{S}(x_0) \| \leq c_0 \| x - x_0 \| . \tag{2.32}$$

Estimate (2.30) now gives $c_0 \leq c$. Let $\omega_0 \in \mathcal{N}$. It is convenient for us to define

$$\Omega(t) = \int_0^t \omega_0(\tau)\, d\tau, \tag{2.33}$$

$$\psi(u, t) = \omega_0((u - t)^+ + t) - \omega_0((u - t)^+)$$

$$= \begin{cases} \omega_0(u) - \omega_0(u - t) & \text{if } t \in [0, u] \\ \omega_0(t) & \text{if } t \in [u, \infty), \end{cases} \tag{2.34}$$

$$\Upsilon(u, t) = \int_0^t \psi(u, \tau)\, d\tau,$$

$$= \begin{cases} t\,\omega_0(u) - \Omega(u) + \Omega(u - t) & \text{if } t \in [0, u] \\ u\,\omega_0(u) - 2\,\Omega(u) + \Omega(t) & \text{if } t \in [u, \infty). \end{cases} \tag{2.35}$$

Moreover, if there exists $x_0 \in \mathcal{D}$, such that $F'(x_0) \in \mathcal{L}(\mathcal{Y}, \mathcal{X})$, set

$$F_0 = F'(x_0)^{-1} F, \tag{2.36}$$

$$a = \| F_0(x_0) \|, \quad \kappa = \omega_0^{-1}(1), \quad p(u, t) = t\,\omega_0(u) + \Upsilon(u, t) \tag{2.37}$$

and

$$q(u, t) = t\,\omega_0(t) - \psi(u, t).$$

We shall denote by $\overline{\Omega}$, $\overline{\psi}$, $\overline{\Upsilon}$, $\overline{\kappa}$, $\overline{p}$ and $\overline{q}$ the functions using ω instead ω_0 in the expressions of Ω, ψ, Υ, κ, p and q, respectively. The proofs of the results that follow can be obtained from the corresponding ones in Ref. [223] by simply replacing ω satisfying (2.28) (for $\mathcal{S} = F_0'$) with the actually needed, more precise and easier to verify ω_0 (satisfying (2.29) for $\mathcal{S} = F_0'$). We illustrate this observation by simply providing one of the proofs but we omit the remaining.

We state the following result on existence and uniqueness of solution $x^\star$ of equation (1.1).

Theorem 2.2.1. *Assume that F_0 is ω_0-regularly smooth on $\mathcal{D}$ in the sense of (2.29).*

(a) *If (1.1) has a solution $x^\star \in \mathcal{D}$, then*

$$\| x^\star - x_0 \| \geq p^{-1}(\kappa, a), \tag{2.38}$$

where $p^{-1}(\kappa, a)$ denotes the unique solution (for any $a \geq 0$) of equation

$$p(\kappa, t) = a. \tag{2.39}$$

(b) *If (1.1) has a solution $x^\star \in \mathcal{D}$ and*

$$a \leq q(\kappa, \kappa), \tag{2.40}$$

then

$$p^{-1}(\kappa, a) \leq \| x^\star - x_0 \| \leq q_-^{-1}(\kappa, a), \tag{2.41}$$

where $q_-^{-1}(\kappa, a)$ and $q_+^{-1}(\kappa, a)$ are the solutions (for any $a \geq 0$) of equation

$$q(\kappa, t) = a. \tag{2.42}$$

(c) *If the real a satisfies (2.40) and*

$$\mathcal{D}_0 = \{x \ : \ p^{-1}(\kappa, a) \leq \| x - x_0 \| \leq q_-^{-1}(\kappa, a)\} \subseteq \mathcal{D}, \tag{2.43}$$

then, (1.1) has a solution x_∞ in $\mathcal{D}_0$, which is unique in $U(x_0, q_+^{-1}(\kappa, a))$ and can be obtained as the limit of the modified Newton method

$$x_{n+1} = x_n - F'(x_0)^{-1} F(x_n). \tag{2.44}$$

(d) *The existence radius $q_-^{-1}(\kappa, a)$ is sharp and it is attained for the (ω_0-regularly smooth in $[0, \kappa]$) function $\Gamma \ : \ t \longrightarrow q(\kappa, t) - a$ and $t_0 = 0$.*

Proof.

(a)–(b) In view of the identity

$$0 = F(x^\star) = F_0(x_0) + F_0'(x_0)\,\delta + \int_0^1 (F_0'(x_0 + t\,\delta) - F_0'(x_0))\,\delta\,dt, \tag{2.45}$$

where $\delta = x^\star - x_0$, we get

$$|a - \| \delta \| | \leq \| F_0(x_0) + F_0'(x_0)\,\delta \| \leq \int_0^1 \| F_0'(x_0 + t\,\delta) - F_0'(x_0) \| \, \| \delta \| \, dt. \tag{2.46}$$

Using (2.29) and (2.46), we have

$$
\begin{aligned}
|a - \| \delta \| | &\leq \omega\left(\omega^{-1}(\min\{1, \| F_0'(x_0 + t\,\delta) \|\}) + t \, \| \delta \|\right) \\
&\quad - \min\{1, \| F_0'(x_0 + t\,\delta) \|\} \\
&= \omega\left(\min\{\kappa, \omega^{-1}(\| F_0'(x_0 + t\,\delta) \|)\} + t \, \| \delta \|\right) \\
&\quad - \omega(\min\{\kappa, \omega^{-1}(\| F_0'(x_0 + t\,\delta) \|)\}) \\
&\leq \omega(\min\{(\kappa - t \, \| \delta \|)^+, \kappa\} + t \, \| \delta \|) - \omega(\min\{(\kappa - t \, \| \delta \|)^+, \kappa\}) \\
&= \omega((\kappa - t \, \| \delta \|)^+ + t \, \| \delta \|) - \omega((\kappa - t \, \| \delta \|)^+) \\
&= \psi(\kappa, t \, \| \delta \|).
\end{aligned}
\tag{2.47}
$$

That is we have

$$|a - \parallel \delta \parallel | \leq \int_0^1 \psi(\kappa, t \parallel \delta \parallel) \parallel \delta \parallel \, dt = \int_0^{\parallel \delta \parallel} \psi(\kappa, t) \, dt = \Upsilon(\kappa, \parallel \delta \parallel),$$

or, equivalently,

$$p(\kappa, \parallel \delta \parallel) \geq a \geq q(\kappa, \parallel \delta \parallel). \tag{2.48}$$

The first inequality $p(\kappa, \parallel \delta \parallel) \geq a$ in (2.48) is equivalent to $\parallel \delta \parallel \geq p^{-1}(\kappa, a)$. The second inequality $a \geq q(\kappa, \parallel \delta \parallel)$ in (2.48) is equivalent to

$$q(\kappa, \kappa) < a \tag{2.49}$$

or

$$\begin{cases} q(\kappa, \kappa) \geq a \\ \text{and} \\ \parallel \delta \parallel \leq q_-^{-1}(\kappa, a) \quad \text{or} \quad \parallel \delta \parallel \geq q_+^{-1}(\kappa, a). \end{cases} \tag{2.50}$$

If (2.49) holds, then (1.1) may have no solution in $\mathcal{D}$ as the example of function Γ shows. Then, we have no information on $\parallel \delta \parallel$ in inequality $a \geq q(\kappa, \parallel \delta \parallel)$. Consequently, we are left only with the lower bound (2.41). If (2.50) holds, then $p^{-1}(\kappa, a) \leq \parallel \delta \parallel \leq q_-^{-1}(\kappa, a)$ and $x^\star$ is unique in $U(x_0, q_+^{-1}(\kappa, a))$.

(c) As we noted in Refs. [102, 104] (see also Refs. [110, 134]), the sufficient convergent condition for the modified Newton's method (2.44) is given by (2.40). That is $\lim_{n \to \infty} x_n = x_\infty$ and $F(x_\infty) = 0$. According to (b), we have $x_\infty \in U(x_0, q_+^{-1}(\kappa, a))$.

(d) Let $t \in [0, \kappa]$. Then, we get

$$\Gamma(t) = 0 \iff a \leq \Omega(\kappa) \quad \text{and} \quad t = q_-^{-1}(\kappa, a) = \kappa - \Omega^{-1}(\Omega(\kappa) - a).$$

That completes the proof of Theorem 2.2.1. $\qquad\qquad\square$

Corollary 2.2.1. *Let F_0 be center-Lipschitz continuous on $\mathcal{D}$:*

$$\parallel F_0'(x) - F_0'(x_0) \parallel \leq L_0 \parallel x - x_0 \parallel, \quad for\ all \quad x \in \mathcal{D}. \tag{2.51}$$

Then the following statements hold:

(a) If $x^\star$ is a solution of (1.1), then

$$L_0 \parallel x^\star - x_0 \parallel \geq \sqrt{1 + 2\, a\, L_0} - 1.$$

(b) If $x^\star$ is a solution of (1.1) and $2\, a\, L_0 \leq 1$, then

$$\sqrt{1 + 2\, a\, L_0} - 1 \leq L_0 \parallel x^\star - x_0 \parallel \leq 1 - \sqrt{1 - 2\, a\, L_0}.$$

(c) If $2\, a\, L_0 \leq 1$ and

$$\mathcal{D}_0 = \{x : \sqrt{1 + 2\, a\, L_0} - 1 \leq L_0 \parallel x^\star - x_0 \parallel \leq 1 - \sqrt{1 - 2\, a\, L_0}\} \subseteq \mathcal{D},$$

then, (1.1) has a solution in $\mathcal{D}_0$ and it is unique in $U(x_0, \dfrac{1}{L_0}(1 + \sqrt{1 - 2\, a\, L_0}))$.

(d) The existence radius $\frac{1}{L_0}(1 - \sqrt{1 - 2\,a\,L_0})$ and the uniqueness one $\frac{1}{L_0}(1 + \sqrt{1 - 2\,a\,L_0})$ are sharp, i.e., they attained for the scalar quadratic polynomial $p_1(t) = .5\,L_0\,t^2 - t + a$ and $t_0 = 0$. $\frac{1}{L_0}(1 - \sqrt{1 - 2\,a\,L_0})$ is attained for the polynomial $p_2(t) = .5\,L_0\,t^2 + t - a$ and $t_0 = 0$.

Proof. We adapt the proof of Theorem 2.2.1 to show (a)–(d). Just take

$$\omega_0(t) = L_0\,t, \quad \kappa = \omega_0^{-1}(1) = \frac{1}{L_0}, \quad \Omega(t) = \Upsilon(u, t) = .5\,L_0\,t^2,$$

$$p(u, t) = L_0\,t\,u + .5\,L_0\,t^2, \quad p^{-1}(u, a) = \sqrt{u^2 + \frac{2\,a}{L_0}} - u$$

$$q(u, t) = L_0\,t\,u - .5\,L_0\,t^2, \quad q_-^{-1}(u, a) = u - \sqrt{u^2 - \frac{2\,a}{L_0}}$$

and

$$q_+^{-1}(u, a) = u + \sqrt{u^2 - \frac{2\,a}{L_0}}.$$

In particular

$$p^{-1}(\kappa, a) = \frac{1}{L_0}\left(\sqrt{1 + 2\,a\,L_0} - 1\right), \quad q_-^{-1}(\kappa, a) = \frac{1}{L_0}\left(1 - \sqrt{1 - 2\,a\,L_0}\right)$$

and

$$q_+^{-1}(\kappa, a) = \frac{1}{L_0}\left(1 + \sqrt{1 - 2\,a\,L_0}\right).$$

That completes the proof of Corollary 2.2.1. $\qquad\square$

The scalar version of Theorem 2.2.1 is stated below for $\mathcal{X} = \mathcal{Y} = \mathbb{R}$. As noted in Ref. [110], any center-ω_0-regularly function F defined on $[\alpha, \beta]$ can be extended to $\mathbb{R}$ by considering

$$F(x) = \begin{cases} F(\alpha) + F'(\alpha)\,(x - \alpha) & \text{if } x < \alpha \\ F(\beta) + F'(\beta)\,(x - \beta) & \text{if } x > \beta. \end{cases}$$

Corollary 2.2.2. *Let $F : \mathbb{R} \longrightarrow \mathbb{R}$ be a differentiable function everywhere and satisfies center-ω_0-regularly condition in the sense of (2.29). Denote*

$$a = F(x_0), \quad b = F'(x_0), \quad \kappa = \omega_0^{-1}(|b|).$$

Then the following statements hold:

(a) If $|a| > \Omega(\kappa)$ and $a\,b > 0$, then, F has no zero in the interval

$$(x_0 - p^{-1}(\kappa, |a|), x_0 + q_+^{-1}(\kappa, -|a|)).$$

(b) If $|a| > \Omega(\kappa)$ and $a\,b < 0$, then, F has no zero in the interval

$$(x_0 - q_+^{-1}(\kappa, -|a|), x_0 + p^{-1}(\kappa, |a|)).$$

(c) If $|a| \le \Omega(\kappa)$ and $a\,b > 0$, then, F has a zero $x^\star$ in the interval

$$[x_0 - q_-^{-1}(\kappa, |a|), x_0 - p^{-1}(\kappa, |a|)]$$

and $x^\star$ is unique in $(x_0 - q_+^{-1}(\kappa, |a|), x_0 + q_+^{-1}(\kappa, -|a|))$.
(d) If $|a| \le \Omega(\kappa)$ and $a\,b < 0$, then, F has a zero $x^\star$ in the interval

$$[x_0 + p^{-1}(\kappa, |a|), x_0 + q_-^{-1}(\kappa, |a|)]$$

and no solution in $(x_0 - q_+^{-1}(\kappa, -|a|), x_0 + q_+^{-1}(\kappa, |a|))$.

In order for us to cover the worst-case optimal methods for scalar equations as in Ref. [223] and make the section as self-contained as possible, we adopt the notations (in our setting)

$$a_1 = F(x_1), \quad b_1 = F'(x_1), \quad \kappa_1 = \omega_0(|b_1|),$$

$$v = p^{-1}(\kappa, a), \quad w = q_-^{-1}(\kappa, a), \quad \alpha = \alpha(\delta) = a - p(\kappa, |\delta|),$$

$$\beta = \beta(\delta) = a - q(\kappa, |\delta|), \quad \gamma = \max\{|\alpha|, |\beta|\},$$

$$\xi = \xi(\delta) = w - |\delta|, \quad \sigma = \sigma(\delta) = |\delta| - v,$$

$$A_1 = \{(\kappa, a_1) : |a_1| \le \Omega(\kappa_1), \ |\kappa_1 - \kappa| \le |\delta|, \ \alpha \le a_1 \le \beta, \quad q_-^{-1}(\kappa_1, |a_1|) \le \xi\},$$
$$\tag{2.52}$$

$$A_2 = \{(\kappa, a_1) : |a_1| \le \Omega(\kappa_1), \ |\kappa_1 - \kappa| \le |\delta|, \ \alpha \le a_1 \le \beta, \quad q_-^{-1}(\kappa_1, |a_1|) \le \sigma\}$$
$$\tag{2.53}$$

and

$$M(|\delta|) = \max_{(\kappa_1, a_1) \in A_1 \cup A_2} \{q_-^{-1}(\kappa_1, |a_1|) - p^{-1}(\kappa_1, |a_1|)\} = \max\{M_1(|\delta|), M_2(|\delta|)\},$$

where (for $i = 1, 2$),

$$M_i(|\delta|) = \max_{(\kappa_1, a_1) \in A_i} \{q_-^{-1}(\kappa_1, |a_1|) - p^{-1}(\kappa_1, |a_1|)\}.$$

In view of (2.52) and (2.53), we can get that $M_2(|\delta|)$ is deduced from $M_1(|\delta|)$ by substituting σ for ξ. We denote also the functions g_i ($i = 1, 2, 3, 4$) by

$$g_1(s) = q_-^{-1}(t, s) - p^{-1}(t, s), \quad g_2(t) = q_-^{-1}(t, s) - p^{-1}(t, s),$$

$$g_3(t) = s - p^{-1}(t, q(t, s)) \quad \text{and} \quad g_4(s) = s - p^{-1}(t, q(t, s - c)) \quad (c \in \mathbb{R}).$$

Then, we have the following results.

Lemma 2.2.1. *The following assertions hold:*

(a)

$$|a| \le \Omega(\kappa) \iff w \le \kappa \implies \Omega(\kappa) - |a| = \Omega(\kappa - w).$$

(b)

$$v \le |\delta| \le w \implies \alpha \le 0 \le \beta = \Omega(\kappa - |\delta|) - \Omega(\kappa - w) = q(\kappa - |\delta|, \xi).$$

(c) Functions g_1 and g_4 are increasing on $[0, \Omega(t)]$ and $[c, \infty)$, respectively.
(d) Functions g_2 and g_3 are decreasing on $[\Omega^{-1}(s), \infty)$ and $[s, \infty)$, respectively.
(e)

$$v \le \left| \frac{a}{b} \right| \le \frac{v + w}{2}.$$

Lemma 2.2.2. *(see Ref. [223]) Let $\mathbb{Z} \subseteq \mathbb{R}^2$ and F a function defined on $\mathbb{Z}$. Denote:*

$$Y(x) = \{y : (x, y) \in \mathbb{Z}\} \quad and \quad X = \{x : Y(x) \ne \emptyset\}.$$

Then

$$\sup_{(x,y)\in\mathbb{Z}} F(x, y) = \sup_{x\in X} \sup_{y\in Y(x)} F(x, y).$$

Proposition 2.2.1. *The following assertions hold*

(a) For $v \le |\delta| \le w$, we have

$$M(|\delta|) = \begin{cases} \xi - p^{-1}(\kappa - w + \xi, \Omega(\kappa - w + \xi) - \Omega(\kappa - w)) & if \ \dfrac{w - v}{2} \le \xi \le w - v \\ \sigma - p^{-1}(\kappa - w + \sigma, \Omega(\kappa - w + \sigma) - \Omega(\kappa - w)) & if \ \dfrac{w - v}{2} \le \sigma \le w - v \end{cases}$$

(b) The minimum of $M(|\delta|)$ on $[v, w]$ is attained at $\dfrac{w + v}{2}$, i.e.,

$$\min_{v\le|\delta|\le w} M(|\delta|) = M\left(\frac{w + v}{2}\right) = \frac{w - v}{2} - p^{-1}\left(\kappa - \frac{w + v}{2}, \Omega\left(\kappa - \frac{w + v}{2}\right) - \Omega(\kappa - w)\right).$$

Remark 2.2.1.

(a) The entropy of zero's position within the existence interval $[v, w]$ can be reduced (using one evaluation of F and one of F') by the following

$$r(|a|) = \frac{\dfrac{w - v}{2} - p^{-1}\left(\kappa - \dfrac{w + v}{2}, \Omega\left(\kappa - \dfrac{w + v}{2}\right) - \Omega(\kappa - w)\right)}{w - v}. \tag{2.54}$$

(b) The function r defined by (2.54) is increasing in $[0, \Omega(\kappa)]$ and

$$r([0, \Omega(\kappa)]) = \left[0, .5\left(1 - \frac{1}{\mu} p^{-1}(\mu, \Omega(\mu))\right)\right],$$

where

$$\mu = .5\left(\kappa - p^{-1}(\kappa, \Omega(\kappa))\right) \quad and \quad p^{-1}(\mu, \Omega(\mu)) < \mu.$$

In the particular case where F is center-ω_0-regularly smooth, with $\omega_0(t) = L_0 t$, we have

$$\kappa = \frac{|b|}{L_0}, \quad \Omega(t) = .5 L_0 t^2, \quad p^{-1}(u, a) = \sqrt{u^2 + \frac{2a}{L_0}} - u$$

$$v = \sqrt{\kappa^2 + \frac{2\,a}{L_0}} - \kappa \quad \text{and} \quad w = \frac{|b|}{L} - \sqrt{\kappa^2 - \frac{2\,a}{L_0}}.$$

Then, the function r defined by (2.54) reduces to

$$r(|a|) = \frac{1}{2}\left(1 - \frac{2\,\kappa - \dfrac{w+v}{2} - w}{\sqrt{\left(\kappa - \dfrac{w+v}{2}\right)^2 + \dfrac{1}{2}\,(w-v)\left(2\,\kappa - \dfrac{w+v}{2} - w\right)} + \kappa - \dfrac{w+v}{2}}\right) \tag{2.55}$$

and

$$\max_{t \in [0,\,.5\,L_0^{-1}]} r(t) = 1 - .5\sqrt{2} = .29\cdots$$

(c) Using (NM) by considering $a = -\dfrac{a}{b}$, we have by Lemma 2.2.1 (e): $|\delta| \in [v, \dfrac{w+v}{2}]$. Then, we get $\xi \in [\dfrac{w-v}{2}, w - v]$, $M(|\delta|) = w - |a| - p^{-1}(\kappa - |a|, \Omega(\kappa - |a|) - \Omega(\kappa - w))$ and $r(|a|)$ given by (2.55) is reduced to

$$r^N(|a|) = \frac{M(|\delta|)}{w - v}. \tag{2.56}$$

The function r^N given by (2.56) is increasing in $[0, \Omega(\kappa)]$, with $r^N(0) = 0$ and

$$r^N(\Omega(\kappa)) = \frac{\kappa - \Omega(\kappa) - p^{-1}(\kappa - \Omega(\kappa), \Omega(\kappa - \Omega(\kappa)))}{\kappa - p^{-1}(\kappa, \Omega(\kappa))}. \tag{2.57}$$

Consequently, for solving $F(x) = 0$, where F satisfies the center-ω_0-regularly smoothness condition (2.29), the hybrid method becomes:

$$x_+ = x - \begin{cases} \dfrac{F(x)}{F'(x)} & \text{if } |F(x)| > \Omega(\omega_0^{-1}(|F'(x)|)) \\ \Theta(x) & \text{if } |F(x)| \leq \Omega(\omega_0^{-1}(|F'(x)|)), \end{cases} \tag{2.58}$$

where

$$\Theta(x) = \frac{\mathrm{sign}\,(F(x)\,F'(x))}{2}\,(q_-^{-1}(\omega_0^{-1}(|F'(x)|), |F(x)|) - p^{-1}(\omega_0^{-1}(|F'(x)|), |F(x)|)).$$

Note that (2.58) may use less iterations than (NM) to get into a neighborhood of a zero of (1.1). Finally, (NM) is optimal on the average in Ref. [223]. By Proposition 2.2.1 and using other iterative methods that use the same information per iteration, we can obtain other hybrid methods.

Remark 2.2.2.

(a) If $\omega_0 = \omega$ in (2.30) and (2.51) is replaced by the stronger Lipschitz condition

$$\| F_0'(x) - F_0'(y) \| \leq L \, \| x - y \| \quad \text{for all} \quad x, y \in \mathcal{D}, \tag{2.59}$$

then, all the above results reduce to the corresponding in Ref. [223]. Otherwise (i.e., if $\omega_0(t) < \omega(t)$ for all $t \geq 0$), our results constitute an improvement with advantages as already stated in the abstract of this section. In particular, we have

$$\overline{p}^{-1}(\overline{\kappa}, a) < p^{-1}(\kappa, a), \tag{2.60}$$

$$q_-^{-1}(\kappa, a) < \overline{q}_-^{-1}(\overline{\kappa}, a), \tag{2.61}$$

$$\overline{q}_+^{-1}(\overline{\kappa}, a) < q_+^{-1}(\kappa, a) \tag{2.62}$$

and

$$a \leq \overline{q}(\overline{\kappa}, \overline{\kappa}) \implies a \leq q(\kappa, \kappa) \tag{2.63}$$

but not vice versa (unless if $\omega_0 = \omega$).

Note that condition (2.40) guarantees the convergence of the modified Newton method (2.44) (see Refs. [102, 104]), or simply replace ω in (2.28) by ω_0 given in (2.29) in the proofs of Ref. [221] for the convergence of (2.44), but not necessarily the convergence of (NM). Note that Galperin's condition

$$a \leq \overline{q}(\overline{\kappa}, \overline{\kappa}) \tag{2.64}$$

guarantees the convergence of both methods in Ref. [223]. However, (2.64) is stronger than (2.40) (see also (2.63)).

(b) In the special case, when $\omega(t) = Lt$, condition (2.64) reduces to the famous Kantorovich sufficient convergence condition for solving nonlinear equations using (NM) (see Refs. [110, 134]). However, we have shown in Refs. [100, 110, 134], that Kantorovich convergence condition can always be replaced by (for $\omega_0(t) = L_0 t$)

$$h_A = \overline{L}\, a \leq \frac{1}{2} \tag{2.65}$$

where $\overline{L}$ is given by (1.302).

We provide examples where strict inequality holds in (2.30); and where (2.40) holds but not (2.64). Other such examples where (2.40) (or (2.65)) hold but not Kantorovich convergence condition can be found in Refs. [100, 134]. The following example was used in Ref. [223].

Example 2.2.1.

(a) Let $\mathcal{X} = \mathcal{Y} = \mathbb{R}$, $\mathcal{D} = [-5, 5]$, $x_0 = 2$, $x^\star = 1$ and define function $F(x) = x^3 - 1$ on $\mathcal{D}$. Using (2.28), (2.29), (2.51) and (2.59), we get $\omega(t) = Lt$, $\omega_0(t) = L_0 t$, where $L_0 = 1.75 < L = 2.5$. We also have by (2.37): $\kappa = .57142857$ and $a = .5833333$. Then, in our case (2.38) in Theorem 2.2.1 gives

$$\| x^\star - x_0 \| = 1 > .425216098 = p^{-1}(\kappa, a)$$

whereas the corresponding result in Ref. [223] leads to (for $\overline{\kappa} = .4$)

$$\| x^\star - x_0 \| = 1 > .391622806 = \overline{p}^{-1}(\overline{\kappa}, a).$$

(b) If $x_0 = 1.1$, we have that

$$L_0 = 5.041322314 < L = 8.26446281,$$

$$a = .091184573, \quad p^{-1}(\kappa, a) = .076451668, \quad \overline{p}^{-1}(\overline{\kappa}, a) = .070592449,$$

$$\| \ x^\star - x_0 \ \| = .1, \quad \text{and} \quad 2a = .182369146 > \overline{\kappa} = .121$$

but

$$2a < \kappa = .198360656.$$

That is the results (b), (c), (d) of Theorem 2.2.1 and consequently the rest of the results in Ref. [223] do not hold but ours do. We found that

$$.076451668 < \| \ x^\star - x_0 \ \| = .1 < .142039375 \quad \text{and} \quad U(1.1, .254681937) \subseteq \mathcal{D}.$$

2.3   Smale's α-Theory

We present a tighter convergence analysis than earlier studies such as Refs. [246, 406, 429, 430] of Newton's method using Smale's α-theory by introducing the notion of the center γ_0-condition. Numerical examples and applications are also provided in this section. We are concerned with the problem of approximating a locally unique solution $x^\star$ of (1.1) using (1.2), where F is a twice-Fréchet continuously differentiable operator defined on $\mathcal{D}$ with values in $\mathcal{Y}$.

We expand in this section the applicability of Newton's method under the γ-conditions by introducing the notion of the center γ_0-condition (to be precised in Definition 2.3.1) for some $\gamma_0 \leq \gamma$. This way we obtain tighter upper bounds on the norms of $\| \ F'(x)^{-1} F'(x_0) \ \|$ for each $x \in \mathcal{D}$ (see (2.67), (2.70) and (2.72)) leading to weaker sufficient convergence conditions and a tighter convergence analysis than in earlier studies such as Refs. [246, 406, 409, 429, 430]. The approach of introducing center-Lipschitz condition has already been fruitful for expanding the applicability of Newton's method under the Kantorovich-type theory (see Refs. [100, 116, 155, 163]).

We need the concept of center γ_0-condition and the corresponding Banach lemma.

Definition 2.3.1. Let $\gamma_0 > 0$ and let $0 < r \leq \dfrac{1}{\gamma_0}$ be such that $U(x_0, r) \subseteq \mathcal{D}$. The operator F is said to satisfy the center γ_0-condition at $x_0 \in U(x_0, r)$ if

$$\| \ F'(x_0)^{-1} (F'(x) - F'(x_0)) \ \| \leq \frac{1}{(1 - \gamma_0 \ \| \ x - x_0 \ \|)^2} - 1 \tag{2.66}$$

for each $\quad x \in U(x_0, r)$.

Lemma 2.3.1. *Let $\gamma_0 > 0$ and let $0 < r \leq \dfrac{1}{\gamma_0}$ be such that $U(x_0, r) \subseteq \mathcal{D}$. Suppose that F satisfies the center γ_0-condition (2.66) at x_0 on $U(x_0, r)$. If $\| x - x_0 \| \leq (1 - \dfrac{1}{\sqrt{2}}) \dfrac{1}{\gamma_0}$, then $F'(x)^{-1} \in \mathcal{L}(\mathcal{Y}, \mathcal{X})$ and satisfies*

$$\| F'(x)^{-1} F'(x_0) \| \leq \left(2 - \frac{1}{(1 - \gamma_0 \| x - x_0 \|)^2} \right)^{-1}. \tag{2.67}$$

Proof. We have by (2.66) and the choice of x that

$$\| F'(x_0)^{-1} (F'(x) - F'(x_0)) \| \leq \frac{1}{(1 - \gamma_0 \| x - x_0 \|)^2} - 1 < 1. \tag{2.68}$$

It follows from (2.68) and the Banach lemma on invertible operators that $F'(x)^{-1} \in \mathcal{L}(\mathcal{Y}, \mathcal{X})$, so that (2.67) is satisfied. The proof of Lemma 2.3.1 is complete. $\square$

The following definition and related lemma are taken from Refs. [246, 405].

Definition 2.3.2. *Let $\gamma > 0$ and let $0 < r \leq \dfrac{1}{\gamma}$ be such that $U(x_0, r) \subseteq \mathcal{D}$. The operator F is said to satisfy the γ-condition at x_0 on $U(x_0, r)$ if*

$$\| F'(x_0)^{-1} F''(x) \| \leq \frac{2\gamma}{(1 - \gamma \| x - x_0 \|)^3} \quad \text{for each} \quad x \in U(x_0, r). \tag{2.69}$$

Lemma 2.3.2. *Let $0 < r \leq \dfrac{1}{\gamma}$ be such that $U(x_0, r) \subseteq \mathcal{D}$. Suppose that F satisfies the γ-condition (2.69) at x_0 on $U(x_0, r)$. If $\| x - x_0 \| \leq (1 - \dfrac{1}{\sqrt{2}}) \dfrac{1}{\gamma}$, then $F'(x)^{-1} \in \mathcal{L}(\mathcal{Y}, \mathcal{X})$ and satisfies*

$$\| F'(x)^{-1} F'(x_0) \| \leq \left(2 - \frac{1}{(1 - \gamma \| x - x_0 \|)^2} \right)^{-1}. \tag{2.70}$$

Remark 2.3.1. It follows from (2.69) that

$$\| F'(x_0)^{-1} (F'(x) - F'(x_0)) \| \leq \frac{1}{(1 - \gamma \| x - x_0 \|)^2} - 1 < 1. \tag{2.71}$$

Clearly, we have that

$$\gamma_0 \leq \gamma \tag{2.72}$$

and γ/γ_0 can be arbitrarily large. In case $\gamma_0 < \gamma$, estimate (2.67) is tighter than (2.70). This observation leads to a tighter convergence analysis for Newton's method (see Remark 2.3.2).

We can show the following semi-local convergence theorem for Newton's method.

Theorem 2.3.1. *Suppose that*

(a) There exist $x_0 \in \mathcal{D}$ and $\eta > 0$ such that

$$F'(x_0)^{-1} \in \mathcal{L}(\mathcal{Y}, \mathcal{X}), \qquad \| F'(x_0)^{-1} F(x_0) \| \leq \eta;$$

(b) Operator F satisfies the center γ_0-condition at x_0 on $U(x_0, (1 - \frac{1}{\sqrt{2}}) \frac{1}{\gamma_0})$ and the γ-condition at x_0 on $U(x_0, (1 - \frac{1}{\sqrt{2}}) \frac{1}{\gamma})$ such that

$$U(x_0, (1 - \frac{1}{\sqrt{2}}) \frac{1}{\gamma_0}) \subseteq \mathcal{D};$$

(c) Scalar sequence $\{s_n\}$ defined by

$$s_0 = 0, \;\; s_1 = \eta$$

$$s_{n+2} = s_{n+1} + \frac{\gamma \, (s_{n+1} - s_n)^2}{\left(2 - \dfrac{1}{(1 - \gamma_0 \, s_{n+1})^2}\right)(1 - \gamma \, s_{n+1})(1 - \gamma \, s_n)^2} \tag{2.73}$$

for each $n = 0, 1, \cdots$

satisfies for each $n = 1, 2, \cdots$

$$s_n < \begin{cases} \dfrac{1}{\gamma} & if \;\; \dfrac{\gamma_0}{\gamma} \leq 1 - \dfrac{1}{\sqrt{2}} \\[2ex] (1 - \dfrac{1}{\sqrt{2}}) \dfrac{1}{\gamma_0} & if \;\; \dfrac{\gamma_0}{\gamma} \geq 1 - \dfrac{1}{\sqrt{2}}. \end{cases} \tag{2.74}$$

Then, the following assertions hold

(i) Sequence $\{s_n\}$ is increasing and bounded above by its unique least upper bound $s^\star$ which satisfies $s^\star \in [\eta, 1/\gamma]$.

(ii) Sequence $\{x_n\}$ generated by Newton's method is well defined, remains in $\overline{U}(x_0, s^\star)$ for each $n = 0, 1, \cdots$ and converges to a unique solution $x^\star \in \overline{U}(x_0, s^\star)$ of equation $F(x) = 0$. Moreover, the following estimates hold

$$\| \, x_{n+1} - x_n \, \| \leq s_{n+1} - s_n \tag{2.75}$$

and

$$\| \, x_n - x^\star \, \| \leq s^\star - s_n \quad for \;\; each \;\; n = 0, 1, 2, \cdots. \tag{2.76}$$

Proof.   We use Mathematical Induction to prove that

$$\| \, x_{k+1} - x_k \, \| \leq s_{k+1} - s_k \tag{2.77}$$

and

$$\overline{U}(x_{k+1}, s^\star - s_{k+1}) \subseteq \overline{U}(x_k, s^\star - s_k) \quad for \;\; each \;\; k = 1, 2, \cdots. \tag{2.78}$$

Let $z \in \overline{U}(x_1, s^\star - s_1)$. Then, we obtain that

$$\| \, z - x_0 \, \| \leq \| \, z - x_1 \, \| + \| \, x_1 - x_0 \, \| \leq s^\star - s_1 + s_1 - s_0 = s^\star - s_0,$$

which implies $z \in \overline{U}(x_0, s^\star - s_0)$. Note also that

$$\| \, x_1 - x_0 \, \| = \| \, F'(x_0)^{-1} F(x_0) \, \| \leq \eta = s_1 - s_0.$$

Hence, estimates (2.77) and (2.78) hold for $k = 0$. Suppose these estimates hold for natural integers $n \leq k$. Then, we have that

$$\| x_{k+1} - x_0 \| \leq \sum_{i=1}^{k+1} \| x_i - x_{i-1} \| \leq \sum_{i=1}^{k+1} (s_i - s_{i-1}) = s_{k+1} - s_0 = s_{k+1}$$

and

$$\| x_k + \theta\,(x_{k+1} - x_k) - x_0 \| \leq s_k + \theta\,(s_{k+1} - s_k) \leq s^{\star} \quad \text{for all} \quad \theta \in (0,1).$$

Using (2.66), Lemma 2.3.1 for $x = x_{k+1}$ and the induction hypotheses we get that

$$
\begin{aligned}
\| F'(x_0)^{-1}\,(F'(x_{k+1}) - F'(x_0)) \| &\leq \frac{1}{(1 - \gamma_0\,\| x_{k+1} - x_0 \|)^2} - 1 \\
&\leq \frac{1}{(1 - \gamma_0\,s_{k+1})^2} - 1 < 1.
\end{aligned}
\tag{2.79}
$$

It follows from (2.79) and the Banach lemma on invertible operators (cf. Refs. [106, 163, 290]) that $F'(x_{m+1})^{-1}$ exists and

$$\| F'(x_{k+1})^{-1}\,F'(x_0) \| \leq \left(2 - \frac{1}{(1 - \gamma_0\,s_{k+1})^2} \right)^{-1}.
\tag{2.80}$$

Using (1.2), we obtain the approximation

$$
\begin{aligned}
F(x_{k+1}) &= F(x_{k+1}) - F(x_k) - F'(x_k)\,(x_{k+1} - x_k) \\
&= \int_0^1 \int_0^1 F''(x_k^{\tau\,s})\,\tau\,ds\,d\tau\,(x_{k+1} - x_k)^2,
\end{aligned}
\tag{2.81}
$$

where $x_k^{\tau\,s} = x_k + \tau\,s\,(x_{k+1} - x_k)$ for each $0 \leq \tau, s \leq 1$. Hence, it follows from (2.70) and (2.81) in turn that

$$
\begin{aligned}
\| F'(x_0)^{-1}\,F(x_{k+1}) \| &\leq \int_0^1 \int_0^1 \frac{2\,\gamma\,\tau\,ds\,d\tau\,\| x_{k+1} - x_k \|^2}{(1 - \gamma\,\| x_k^{\tau\,s} - x_0 \|)^3} \\
&\leq \int_0^1 \int_0^1 \frac{2\,\gamma\,\tau\,ds\,d\tau\,\| x_{k+1} - x_k \|^2}{(1 - \gamma\,\| x_k - x_0 \| - \gamma\,\tau\,s\,\| x_{k+1} - x_k \|)^3} \\
&= \frac{\gamma\,\| x_{k+1} - x_k \|^2}{(1 - \gamma\,\| x_k - x_0 \| - \gamma\,\| x_{k+1} - x_k \|)\,(1 - \gamma\,\| x_k - x_0 \|)^2} \\
&\leq \frac{\gamma\,(s_{k+1} - s_k)^2}{(1 - \gamma\,s_{k+1})\,(1 - \gamma\,s_k)^2} \left(\frac{\| x_{k+1} - x_k \|}{s_{k+1} - s_k} \right)^2 \leq \frac{\gamma\,(s_{k+1} - s_k)^2}{(1 - \gamma\,s_{k+1})\,(1 - \gamma\,s_k)^2}.
\end{aligned}
\tag{2.82}
$$

Then, in view of (1.2), (2.73), (2.80) and (2.82) we obtain that

$$
\begin{aligned}
\| x_{k+2} - x_{k+1} \| &= \| (F'(x_{k+1})^{-1}\,F'(x_0))\,(F'(x_0)^{-1}\,F(x_{k+1})) \| \\
&\leq \| F'(x_{k+1})^{-1}\,F'(x_0) \|\,\| F'(x_0)^{-1}\,F(x_{k+1}) \| \\
&\leq \frac{1}{2 - \dfrac{1}{(1 - \gamma_0\,s_{k+1})^2}}\,\frac{\gamma\,(s_{k+1} - s_k)^2}{(1 - \gamma\,s_{k+1})\,(1 - \gamma\,s_k)^2} = s_{k+2} - s_{k+1}.
\end{aligned}
\tag{2.83}
$$

Hence, we showed (2.77) holds for all $k \geq 0$. Furthermore, let $w \in \overline{U}(x_{k+2}, s^{\star} - s_{k+2})$. Then, we have that

$$
\begin{aligned}
\| w - x_{k+1} \| &\leq \| w - x_{k+2} \| + \| x_{k+2} - x_{k+1} \| \\
&\leq s^{\star} - s_{k+2} + s_{k+2} - s_{k+1} = s^{\star} - s_{k+1}.
\end{aligned}
\tag{2.84}
$$

That is $w \in \overline{U}(x_{k+1}, s^\star - s_{k+1})$. The induction for (2.77) and (2.78) is now complete. Lemma 2.3.1 implies that $\{s_n\}$ is a complete sequence. It follows from (2.77) and (2.78) that $\{x_n\}$ is also a complete sequence in a Banach space $\mathcal{X}$ and as such it converges to some $x^\star \in \overline{U}(x_0, s^\star)$ (since $\overline{U}(x_0, s^\star)$ is a closed set). By letting $k \longrightarrow \infty$ in (2.82) we get $F(x^\star) = 0$. Estimate (2.76) is obtained from (2.75) (cf. Refs. [106, 163, 290]) by using standard majorization techniques. Finally, to show the uniqueness part, let $y^\star \in U(x_0, s^\star)$ be a solution of equation (1.1). Using (2.66) for x replaced by $z^\star = x^\star + \tau\,(y^\star - x^\star)$ and $\mathcal{M} = \displaystyle\int_0^1 F'(z^\star)\,d\tau$ we get as in (2.68) that $\|\,F'(x_0)^{-1}\,(\mathcal{M} - F'(x_0))\,\| < 1$. That is $\mathcal{M}^{-1} \in \mathcal{L}(\mathcal{Y}, \mathcal{X})$. Using the identity $0 = F(x^\star) - F(y^\star) = \mathcal{M}\,(x^\star - y^\star)$, we deduce $x^\star = y^\star$. The proof of Theorem 2.3.1 is complete. $\qquad\square$

Remark 2.3.2. According to the proof of Theorem 2.3.1, γ_0 can replace γ in the definition of s_2. Estimate (2.74) can be checked, since scalar sequence is based on the initial data γ_0, γ and η, especially in the case when $s_i = s_{i+n}$ for some finite i. Next, we provide some alternative convergence conditions to (2.74). We have the estimates

$$2 - \frac{1}{(1 - \gamma_0\,s_{k+1})^2} > 2 - \frac{1}{(1 - \gamma_0\,\eta)^2} \quad \text{and} \quad \frac{1}{1 - \gamma\,s_n} \leq \sqrt{2}. \tag{2.85}$$

Estimates (2.85) indicate that less tight than (2.73) sequence $\{t_n\}$ defined by

$$t_0 = 0, \;\; t_1 = \eta$$
$$t_{n+2} = t_{n+1} + \frac{\delta\,(t_{n+1} - t_n)^2}{2\,(1 - \gamma\,t_{n+1})} \quad \text{for each} \quad n = 0, 1, \cdots, \tag{2.86}$$

where

$$\delta = \frac{2\,\gamma}{2 - \dfrac{1}{(1 - \gamma_0\,\eta)^2}} \tag{2.87}$$

can be used as a majorizing sequence of Newton's method (1.2) in Theorem 2.3.1. Several sufficient convergence Kantorovich-type conditions for sequence $\{t_n\}$ have been given by us in Refs. [106, 155, 163]. One such condition is given by

$$\overline{H} = \frac{1}{8}\,(\delta + 4\,\gamma + \sqrt{\delta^2 + 8\,\gamma\,\delta})\,\eta \leq \frac{1}{2}. \tag{2.88}$$

We also have that

$$t_n \leq \frac{\eta}{1 - \mu} \quad \text{for all} \quad n = 0, 1, \cdots,$$

where

$$\mu = \frac{2\,\delta}{\delta + \sqrt{\delta^2 + 8\,\gamma\,\delta}}.$$

Let

$$\Xi = \frac{1}{8}\,(\delta + 4\,\gamma + \sqrt{\delta^2 + 8\,\gamma\,\delta}), \quad \Theta = \frac{1}{2}\left((1 - \frac{1}{\sqrt{2}})\,\frac{1}{\gamma_0}\right)^{-1},$$

$$\Upsilon = \frac{1}{2\,(1-\mu)} \left((1 - \frac{1}{\sqrt{2}})\, \frac{1}{\gamma} \right)^{-1} \quad \text{and} \quad \delta^\star = \max\{\Xi, \Theta, \Upsilon\}.$$

The convergence criterion is given by

$$H = \delta^\star\, \eta \begin{cases} \leq \dfrac{1}{2} & \text{if } \delta^\star = \Xi \text{ or } \delta^\star = \Upsilon \\[2ex] < \dfrac{1}{2} & \text{if } \delta^\star = \Theta. \end{cases} \tag{2.89}$$

Clearly, (2.89) and $\{t_n\}$, respectively, can replace (2.74) and $\{s_n\}$ in Theorem 2.3.1. Using a simple inductive argument we have that for each $n = 1, 2, \cdots$

$$s_n \leq t_n, \quad s_{n+1} - s_n \leq t_{n+1} - t_n \quad \text{and} \quad s^\star \leq t^\star = \lim_{n\to\infty} t_n. \tag{2.90}$$

It follows from the proof of Theorem 2.3.1 that sequence $\{r_n\}$ defined by

$$r_0 = 0, \; r_1 = \eta, \; r_2 = r_1 + \frac{\gamma_0\,(r_1 - r_0)^2}{\left(2 - \dfrac{1}{(1 - \gamma_0\, r_1)^2}\right)(1 - \gamma\, r_1)\,(1 - \gamma\, r_0)^2}$$

$$r_{n+2} = r_{n+1} + \frac{\gamma\,(r_{n+1} - r_n)^2}{\left(2 - \dfrac{1}{(1 - \gamma_0\, r_{n+1})^2}\right)(1 - \gamma\, r_{n+1})\,(1 - \gamma\, r_n)^2} \tag{2.91}$$

for each $n = 1, 2, \cdots$

is a tighter majorizing sequence of $\{x_n\}$ than $\{s_n\}$. In particular, we have that

$$r_n \leq s_n, \quad r_{n+1} - r_n \leq s_{n+1} - s_n \tag{2.92}$$

and

$$r^\star = \lim_{n\to\infty} r_n \leq s^\star. \tag{2.93}$$

If $\gamma_0 < \gamma$, then, strict inequalities hold in (2.92). Sequence $\{r_n\}$ can replace $\{s_n\}$ in Theorem 2.3.1. Moreover, (2.74) (with $\{s_n\}$ replaced by $\{r_n\}$) can be replaced if we consider sequence $\{w_n\}$ defined by

$$w_0 = 0, \; w_1 = \eta, \; w_2 = w_1 + \frac{\delta_0\,(w_1 - w_0)^2}{2\,(1 - \gamma\, w_1)^2}$$

$$w_{n+2} = w_{n+1} + \frac{\delta\,(w_{n+1} - w_n)^2}{2\,(1 - \gamma\, w_{n+1})} \quad \text{for each} \quad n = 1, 2, \cdots, \tag{2.94}$$

where

$$\delta_0 = \frac{2\,\gamma_0}{2 - \dfrac{1}{(1 - \gamma_0\, \eta)^2}}. \tag{2.95}$$

Scalar sequence $\{w_n\}$ was studied by us in Ref. [155]. The sufficient convergence criterion is given by

$$\overline{h} = \frac{1}{8}\,(4\,\gamma_1 + \sqrt{\gamma_1\,\delta + 8\,\gamma_1^2} + \sqrt{\gamma_1\,\delta})\,\eta \leq \frac{1}{2}$$

and

$$w_n \le \lambda = \lambda_1\,\eta = \left(1 + \frac{\delta_0\,\eta}{2\,(1-\mu)\,(1-\gamma_1\,\eta)}\right)\eta \quad \text{for all}\quad n = 0, 1, \cdots,$$

where

$$\gamma_1 = \max\{\gamma, \delta_0\}.$$

Let

$$\Xi_1 = \frac{1}{8}\left(4\,\gamma_1 + \sqrt{\gamma_1\,\delta + 8\,\gamma_1\,\delta} + \sqrt{\gamma_1\,\delta}\right), \quad \Upsilon_1 = \frac{1}{2}\left((1 - \frac{1}{\sqrt{2}})\,\frac{1}{\gamma_0}\right)^{-1}\lambda_1$$

and

$$\delta_1^\star = \max\{\Xi_1, \Theta, \Upsilon_1\}.$$

Then, the convergence criterion is given by

$$h = \delta_1^\star\,\eta \begin{cases} \le 1/2 \text{ if } \delta_1^\star = \Xi_1 \text{ or } \delta_1^\star = \Upsilon_1 \\ < 1/2 \text{ if } \delta_1^\star = \Theta \end{cases} \tag{2.96}$$

which is weaker than (2.89). Other sufficient convergence conditions for the scalar sequences introduced so far have been given in Refs. [155, 157].

Let us state one more result for sequence $\{w_n\}$ (see [Lemma 2.4] in Ref. [155]).

Lemma 2.3.3. *Suppose that there exists a minimum integer $N > 1$ such that iterates w_i $(i = 0, 1, \cdots, N - 1)$ given by (2.94) are well defined,*

$$w_i < w_{i+1} < \frac{1}{\gamma_1} \quad \text{for each}\quad i = 0, 1, \cdots, N - 2 \tag{2.97}$$

and

$$w_N \le \frac{1}{\gamma_1}\,(1 - (1 - \gamma_1\,w_{N-1})\,\mu_0), \tag{2.98}$$

where

$$\mu_0 = \frac{2\,\delta}{\delta + \sqrt{\delta^2 + 8\,\gamma_1\,\delta}}. \tag{2.99}$$

Then, the following assertions hold

$$\gamma_1\,w_N < 1, \tag{2.100}$$

$$w_{N+1} \le \frac{1}{\gamma_1}\,(1 - (1 - \gamma_1\,w_N)\,\mu_0), \tag{2.101}$$

$$\alpha_{N-1} = \frac{\delta\,(w_{N+1} - w_N)}{2\,(1 - \gamma_1\,w_{N+1})} \le \mu_0 \le 1 - \frac{\gamma_1\,(w_{N+1} - w_N)}{1 - \gamma_1\,w_N}, \tag{2.102}$$

sequence $\{w_n\}$ given by (2.94) is well defined, increasing, bounded from above by

$$w_N^{\star\star} = w_{N-1} + \frac{1}{1 - \mu_0}\,(w_N - w_{N-1}) \tag{2.103}$$

and converges to its unique least upper bound $w_N^\star$ which satisfies $0 \le w_N^\star \le w_N^{\star\star}$. Moreover, the following estimates hold

$$0 < w_{N+n} - w_{N+n-1} \le \mu_0^{n-1}\,(w_{N+1} - w_N) \quad \text{for each}\quad n = 1, 2, \cdots. \tag{2.104}$$

Remark 2.3.3. We can compare the above with existing results. The majorizing sequence $\{u_n\}$ essentially used in Ref. [429] is given by

$$u_0 = 0, \ u_1 = \eta$$

$$u_{n+2} = u_{n+1} + \frac{\gamma \left(u_{n+1} - u_n\right)^2}{\left(2 - \dfrac{1}{(1 - \gamma \, u_{n+1})^2}\right)\left(1 - \gamma \, u_{n+1}\right)\left(1 - \gamma \, u_n\right)^2} \tag{2.105}$$

for each $n = 0, 1, \cdots$.

Note that if $\gamma_0 = \gamma$ then $s_n = u_n$ for each $n = 0, 1, \cdots$. A simple inductive argument shows that for each $n = 1, 2, \cdots$

$$s_n \leq u_n, \quad s_{n+1} - s_n \leq u_{n+1} - u_n \tag{2.106}$$

and

$$s^\star \leq u^\star = \lim_{n \to \infty} u_n. \tag{2.107}$$

If $\gamma_0 < \gamma$, strict inequalities hold in (2.106). The sufficient convergence criterion for sequence $\{u_n\}$ is given by

$$q = \gamma \eta \leq 3 - 2\sqrt{2}. \tag{2.108}$$

Clearly, (2.74) is weaker than (2.108) and (2.89), (2.96) (or (2.97), (2.98)) can be weaker than (2.108). However, a direct comparison between $\{u_n\}$ and all sequences introduced above other than $\{s_n\}$ is not practical (see the numerical examples in the next section). Note also that sequence $\{s_n\}$ which is tighter than $\{u_n\}$ converges under criterion (2.108).

Remark 2.3.4. Let us introduce center-Lipschitz condition

$$\| F'(x_0)^{-1} \left(F'(x) - F'(x_0)\right) \| \leq L_0 \, \| x - x_0 \| \quad \text{for each} \ \ x \in \mathcal{D} \tag{2.109}$$

as an alternative to (2.66) or (2.69) for the computation of the upper bounds on $\| F'(x)^{-1} F'(x_0) \|$ for each $x \in \mathcal{D}$. Then, we have instead of (2.67) or (2.70) that

$$\| F'(x)^{-1} F'(x_0) \| \leq \frac{1}{1 - L_0 \, \| x - x_0 \|} \quad \text{for each} \ \ x \in U(x_0, 1/L_0). \tag{2.110}$$

The modification leads to majorizing sequence $\{\varsigma_n\}$ defined by

$$\varsigma_0 = 0, \ \varsigma_1 = \eta$$

$$\varsigma_{n+2} = \varsigma_{n+1} + \frac{\gamma \left(\varsigma_{n+1} - \varsigma_n\right)^2}{\left(1 - L_0 \, \varsigma_{n+1}\right)\left(1 - \gamma \, \varsigma_{n+1}\right)\left(1 - \gamma \, \varsigma_n\right)^2} \tag{2.111}$$

for each $n = 0, 1, \cdots$.

Then, sequence $\{\varsigma_n\}$ can also replace $\{s_n\}$ in Theorem 2.3.1. Noting that $1/(1 - \gamma \, s_k) \leq \sqrt{2}$ for each $k = 1, 2, \cdots$, we arrive at the less tight majorizing sequence $\{\zeta_n\}$ given by

$$\zeta_0 = 0, \ \zeta_1 = \eta$$

$$\zeta_{n+2} = \zeta_{n+1} + \frac{\nu \left(\zeta_{n+1} - \zeta_n\right)^2}{2 \left(1 - \gamma_0 \, \zeta_{n+1}\right)} \quad \text{for each} \ \ n = 0, 1, \cdots, \tag{2.112}$$

where

$$\nu = 4\sqrt{2}\,\gamma. \tag{2.113}$$

Then, the sufficient criterion is given by (see also (2.89))

$$H^\star = \delta_2^\star \eta \le \frac{1}{2}, \tag{2.114}$$

where

$$\delta_2^\star = \frac{1}{2}\,\max\{\frac{1}{4}\,(\nu + 4\,L_0 + \sqrt{\nu^2 + 8\,L_0\,\nu}),\; \frac{1}{2\,(1-\mu_0)}\left((1-\frac{1}{\sqrt{2}})\,\frac{1}{\gamma}\right)^{-1}\}.$$

Similar to Remark 2.3.2 and Lemma 2.3.3, sequences, conditions and comments can follow in this new setting.

Remark 2.3.5. Summing up we conclude that in practice we shall test existing sufficient convergence to see if any is satisfied and also choose the tightest majorizing sequence for the computation of the upper bounds.

Remark 2.3.6. In order for us to cover the local convergence analysis of Newton's method, let us define function $f_\epsilon \,:\, \mathcal{I}_\epsilon = [0, \frac{1}{\epsilon}\,(1 - \frac{1}{\sqrt{2}})] \longrightarrow \mathbb{R}$ by

$$f_\epsilon(t) = (1 - \epsilon\,t)^2\,(2 - t)\,t - (1 - t)^2\,(2\,(1 - \epsilon\,t)^2 - 1). \tag{2.115}$$

Suppose that

$$\epsilon > \frac{1}{2}\,(1 - \frac{1}{\sqrt{2}}). \tag{2.116}$$

Then, we have that

$$f_\epsilon(0) = -1 < 0 \quad\text{and}\quad f_\epsilon(\frac{1}{\epsilon}\,(1 - \frac{1}{\sqrt{2}})) = \frac{1}{\epsilon\sqrt{2}}\,(1 - \frac{1}{\sqrt{2}})\,(2 - \frac{1}{\epsilon}\,(1 - \frac{1}{\sqrt{2}})) > 0.$$

Hence, it follows from the intermediate value theorem that function f_ϵ has a zero in $\overset{\circ}{\mathcal{I}}_\epsilon$. Denote by $\mu_\epsilon^\star$ the minimal such zero. Define function $g_\epsilon \,:\, \mathcal{I}_\epsilon \longrightarrow \mathbb{R}$ by

$$g_\epsilon(t) = \frac{(1 - \epsilon\,t)^2\,(2 - t)\,t}{(1 - t)^2\,(2\,(1 - \epsilon\,t)^2 - 1)}. \tag{2.117}$$

Then, we have that

$$0 \le g_\epsilon(t) < 1 \quad\text{for each}\quad t \in [0, \mu_\epsilon^\star]. \tag{2.118}$$

Set

$$R_\epsilon = \frac{\mu_\epsilon^\star}{\gamma}. \tag{2.119}$$

It follows from the definition of f_ϵ, $\mu_\epsilon^\star$ and g_ϵ that

$$R_1 = \frac{3 - \sqrt{6}}{3\,\gamma} \le R_\epsilon. \tag{2.120}$$

Moreover, strict inequality holds if $\epsilon \neq 1$. Let us assume that F satisfies the center γ_0-condition at $x^\star$ on $U(x^\star, \dfrac{1}{\epsilon\gamma}(1 - \dfrac{1}{\sqrt{2}}))$ with $F(x^\star) = 0$ and the γ-condition at $x^\star$ on $U(x^\star, \dfrac{1}{\gamma})$. Then, for $x_0 \in U(x^\star, R_\epsilon)$, we have the identity

$$
\begin{aligned}
x_{n+1} - x^\star &= F'(x_n)^{-1}\left(F(x^\star) - F(x_n) - F'(x_n)\left(x^\star - x_n\right)\right) \\
&= F'(x_n)^{-1}\, F'(x^\star) \int_0^1 \int_0^1 F'(x^\star)^{-1} F''(x_{n,\star}^{\tau\,s})\,\tau\,ds\,d\tau\,(x^\star - x_n)^2,
\end{aligned}
\tag{2.121}
$$

where

$$
x_{n,\star}^{\tau\,s} = x_n + \tau\,s\left(x^\star - x_n\right) \quad \text{for each} \quad 0 \le t \le 1, \quad 0 \le s \le 1
$$

and the estimates

$$
\| F'(x_n)^{-1}\, F'(x^\star) \| \le \left(2 - \frac{1}{(1 - \epsilon\gamma\,\| x_n - x^\star \|)^2}\right)^{-1}
\tag{2.122}
$$

and

$$
\begin{aligned}
&\left\| \int_0^1 \int_0^1 F'(x^\star)^{-1} F''(x_{n,\star}^{\tau\,s})\,\tau\,ds\,d\tau\,(x^\star - x_n)^2 \right\| \le \\
&\int_0^1 \int_0^1 \frac{2\gamma\,\| x_{n,\star}^{\tau\,s} - x^\star \|\,ds\,d\tau}{(1 - \gamma s\,\| x_{n,\star}^{\tau\,s} - x^\star \|)^3}\,\| x_{n,\star}^{\tau\,s} - x^\star \| \le \\
&\left(\frac{1}{(1 - \gamma\,\| x_n - x^\star \|)^2} - 1\right)\,\| x_n - x^\star \| .
\end{aligned}
\tag{2.123}
$$

That is we have that

$$
\begin{aligned}
\| x_{n+1} - x^\star \| &\le \frac{(1 - \epsilon\gamma\,\| x_n - x^\star \|)^2\,(1 - (1 - \gamma\,\| x_n - x^\star \|)^2)}{(2\,(1 - \epsilon\gamma\,\| x_n - x^\star \|)^2 - 1)\,(1 - \gamma\,\| x_n - x^\star \|)^2}\,\| x_n - x^\star \| \\
&< g_\epsilon(\mu_\epsilon^\star)\,\| x_n - x^\star \| = \| x_n - x^\star \|,
\end{aligned}
\tag{2.124}
$$

which shows that $x_{n+1} \in U(x^\star, R_\epsilon)$ and $\displaystyle\lim_{n\to\infty} = x^\star$. Hence we arrived at the following result of the local convergence for Newton's method.

Theorem 2.3.2. *Suppose that*

(a) There exists $x^\star \in \mathcal{D}$ such that $F(x^\star) = 0$ and $F'(x^\star)^{-1} \in \mathcal{L}(\mathcal{Y}, \mathcal{X})$.

(b) Operator F satisfies the center γ_0-condition at $x^\star$ on $U(x^\star, \dfrac{1}{\epsilon\gamma}(1 - \dfrac{1}{\sqrt{2}}))$ for ϵ satisfying (2.116) and the γ-condition at $x^\star$ on $U(x^\star, \dfrac{1}{\gamma})$.

Then, if $x_0 \in U(x^\star, R_\epsilon)$, sequence $\{x_n\}$ generated by Newton's method is well defined, remains in $U(x^\star, R_\epsilon)$ for each $n = 0, 1, \cdots$ and converges to $x^\star$. Moreover, the following estimate holds

$$
\| x_{n+1} - x_n \| \le \frac{\gamma\,(2 - \gamma\,\| x_n - x^\star \|)\,(1 - \epsilon\gamma\,\| x_n - x^\star \|)^2}{(1 - \gamma\,\| x_n - x^\star \|)^2\,(2\,(1 - \epsilon\,\| x_n - x^\star \|)^2 - 1)}\,\| x_n - x^\star \|^2 .
\tag{2.125}
$$

Remark 2.3.7. If $\epsilon = 1$ (i.e. $\gamma_0 = \gamma$), our results reduce to the ones given by Wang in Ref. [429]. Otherwise, if

$$\frac{1}{2}\left(1 - \frac{1}{\sqrt{2}}\right) \leq \frac{\gamma_0}{\gamma} < 1, \tag{2.126}$$

then, according to (2.120) our convergence radius is larger.

Remark 2.3.8. Let us define function $f_{\epsilon_1} : \mathcal{I}_{\epsilon_1} = [0, \frac{1}{\epsilon_1}] \longrightarrow \mathbb{R}$ for $\epsilon_1 > 0$ by

$$f_{\epsilon_1}(t) = (2 - t)\,t - (1 - t)^2\,(1 - \epsilon_1\,t). \tag{2.127}$$

Suppose that

$$\epsilon_1 > \frac{1}{2}. \tag{2.128}$$

Then, we have

$$f_{\epsilon_1}(0) = -1 < 0 \quad \text{and} \quad f_{\epsilon_1}\left(\frac{1}{\epsilon_1}\right) > 0.$$

Denote by $\mu^{\star}_{\epsilon_1}$ the minimal zero of f_{ϵ_1} on $\overset{\circ}{\mathcal{I}}_{\epsilon_1}$. Define function $g_{\epsilon_1} : \mathcal{I}_{\epsilon_1} \longrightarrow \mathbb{R}$ by

$$g_{\epsilon_1}(t) = \frac{(2 - t)\,t}{(1 - t)^2\,(1 - \epsilon_1\,t)}. \tag{2.129}$$

Then, we have that

$$0 \leq g_{\epsilon_1}(t) < 1 \quad \text{for each} \quad t \in [0, \mu^{\star}_{\epsilon_1}].$$

Set

$$L_0 = \epsilon_1\,\gamma \quad \text{and} \quad R_{\epsilon_1} = \frac{\mu^{\star}_{\epsilon_1}}{\gamma}. \tag{2.130}$$

Hence, we arrived at the following result.

Theorem 2.3.3. *Suppose that*

(a) There exists $x^{\star} \in D$ such that $F(x^{\star}) = 0$ and $F'(x^{\star})^{-1} \in \mathcal{L}(\mathcal{Y}, \mathcal{X})$.

(b) Operator F satisfies the center L_0-condition at $x^{\star}$ on $U(x^{\star}, \frac{1}{\epsilon_1})$ for c_1 satisfying

(2.128) and the γ-condition at $x^{\star}$ on $U(x^{\star}, \frac{1}{\gamma})$.

Then, if $x_0 \in U(x^{\star}, R_{\epsilon_1})$, sequence $\{x_n\}$ generated by Newton's method is well defined, remains in $U(x^{\star}, R_{\epsilon_1})$ for each $n = 0, 1, \cdots$ and converges to $x^{\star}$. Moreover, the following estimate holds

$$\| x_{n+1} - x_n \| \leq \frac{\gamma\,(2 - \gamma\,\| x_n - x^{\star} \|)\,\| x_n - x^{\star} \|^2}{(1 - \gamma\,\| x_n - x^{\star} \|)^2\,(1 - \epsilon_1\,\| x_n - x^{\star} \|)}. \tag{2.131}$$

We provide now some numerical examples.

Table 2.1 Comparison Table

n	s_n	u_n	$s_{n+1} - s_n$	$u_{n+1} - u_n$
1	.1	.1	.0058103058	.0059006211
2	.1058103058	.1059006211	.0000219435	.0000230132
3	.1058322493	.1059236343	3e-10	4e-10
4	.1058322496	.1059236347	0	0
5	$\sim$	$\sim$	$\sim$	$\sim$

Example 2.3.1.

(a) Consider $\gamma = 1.8$, $\gamma_0 = .44$ and $\eta = .1$. Using (2.73) and (2.74), we get that

$$\frac{\gamma_0}{\gamma} = .2444444444 \leq 1 - \frac{1}{\sqrt{2}} = .2928932190, \quad \frac{1}{\gamma} = .5555555556,$$

$$s_2 = .1242332266, \quad s_3 = .1265315555, \quad s_4 = .1265548041$$

and

$$s_n = s_5 = .1265548065 \quad \text{for each} \quad n = 6, 7, \cdots.$$

That is $s_n < 1/\gamma$ for each $n = 1, 2, \cdots$ and condition (2.74) holds. Hence, our Theorem 2.3.1 is applicable. By Remark 2.3.3 (see condition (2.108)), we have that

$$q = .18 > 3 - 2\sqrt{2} = .171572876.$$

Hence the older convergence criteria in Ref. [429] do not hold.

(b) Consider now $\gamma = .5$, $\gamma_0 = .44$ and $\eta = .1$. Using (2.73) and (2.74), we get that

$$\frac{\gamma_0}{\gamma} = .88 > 1 - \frac{1}{\sqrt{2}} = .2928932190, \quad (1 - \frac{1}{\sqrt{2}})\frac{1}{\gamma_0} = .665666408,$$

$$s_2 = .1058103058, \quad s_3 = .1058322493, \quad s_4 = .1058322496$$

and

$$s_n = s_4 = .1058322496 \quad \text{for each} \quad n = 5, 6, \cdots.$$

That is $s_n < (1 - (1/\sqrt{2}))/\gamma_0$ for each $n = 1, 2, \cdots$ and condition (2.74) holds. Hence, our Theorem 2.3.1 is applicable. We also have that

$$q = .05 \leq .171572876.$$

Hence the convergence criteria in Ref. [429] is also applicable. We can now compare our results of Theorem 2.3.1 (see also sequence $\{s_n\}$ given by (2.73)) to ones in Ref. [429] (see also sequence $\{u_n\}$ given by (2.105)). Table 2.1 shows that our error bounds using sequence $\{s_n\}$ are tighter than those given in Ref. [429]. Finally, we can validate in Comparison Table 2.2 that sequences $\{r_n\}$ (see (2.91)) is tighter than $\{s_n\}$.

Table 2.2   Comparison Table

n	s_n	r_n	$s_{n+1} - s_n$	$r_{n+1} - r_n$
1	.1	.1	.0058103058	.0051300426
2	.1058103058	.1051130691	.0000219435	.0000170868
3	.1058322493	.1051471294	3e-10	2e-10
4	.1058322496	.1051471296	0	0
5	$\sim$	$\sim$	$\sim$	$\sim$

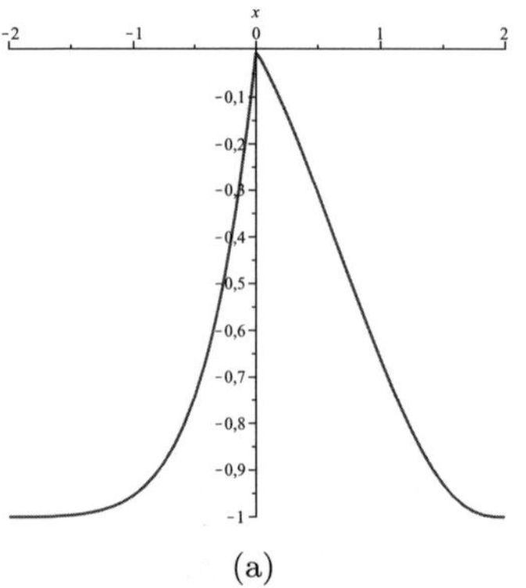

(a)

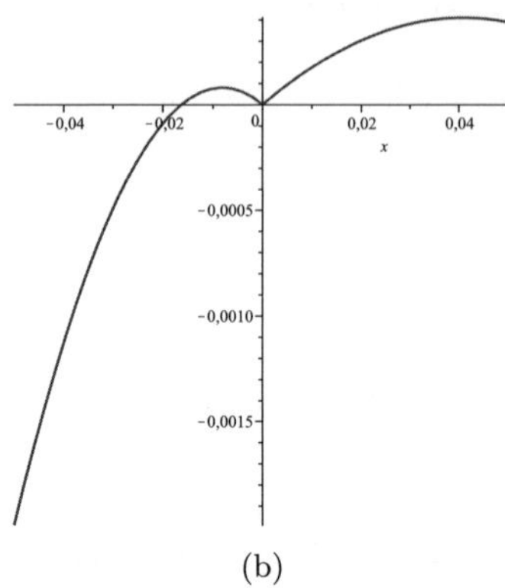

(b)

Fig. 2.2   (a) Function φ_γ on interval $[-2, 2]$ for $\gamma = .5$. (b) Function φ_{γ_0} on interval $[-.05, .05]$ for $\gamma_0 = .49$.

Example 2.3.2. Let $\mathcal{X} = \mathbb{R}$, $\mathcal{D} = U(0, 1)$. Define function $F(x) = e^x - 1$ on $\mathcal{D}$. It can easily be seen that the hypotheses of Theorem 2.3.2 hold for $\gamma = \gamma_0 = .5$. Indeed, using Definitions 2.3.1, 2.3.2 (see also inequalities (2.66) and (2.69)), we can consider functions

$$\varphi_\gamma(x) = e^x \left(1 - \gamma \left|x\right|\right)^3 - 2\gamma$$

and

$$\varphi_{\gamma_0}(x) = \left|e^x - 1\right| \left(1 - \gamma_0 \left|x\right|\right)^2 - 1 - \left(1 - \gamma_0 \left|x\right|\right)^2.$$

Then, (2.66) and (2.69) are equivalent to solving $\varphi_\gamma(x) \leq 0$ and $\varphi_{\gamma_0}(x) \leq 0$, respectively, in a neighborhood of $x^\star = 0$. We deduce that $\gamma \geq .5$ and $\gamma_0 \geq .5$. Note that for $\gamma_0 < .5$, (2.66) does not hold (see Figure 2.2). Using Remark 2.3.6, we can give in Table 2.3 the values of R_ϵ for various values of parameters ϵ satisfying condition (2.116).   By (2.120), we also get that $R_1 = .367006838$. Hence, the radius of convergence R_1 is smaller than ours. Inequalities (2.66) and (2.69) also hold in particular case for $\gamma_0 = (e - 1)/2 = .8591409140 < \gamma = e/2 = 1.359140914$ (see Figure 2.4). In this case we have that

$$\epsilon = .632120588 \quad \text{and} \quad R_1 = .2135894310 < R_\epsilon = .2574184161.$$

Table 2.3   Values of R_ϵ

ϵ	$\mu_\epsilon^\star$	R_ϵ
.156458	.2759901233	.5519802466
.243155	.2661390930	.5322781860
.545149	.2310139217	.4620278434
.745243	.2088038065	.4176076130
.78	.2051342696	.4102685392
.98568954	.1848226548	.3696453096

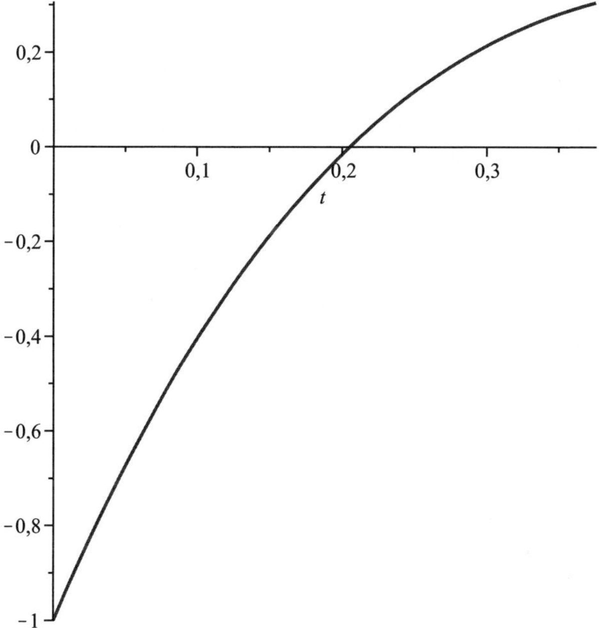

Fig. 2.3   Function f_ϵ given by (2.115) for $\epsilon = .78$ on interval $[0, .3755041264)$.

2.4   Exercises

2.4.1.

(a) Let $F\colon D \subseteq X \to X$ be an analytic operator. Assume:

- there exists $\alpha \in [0, 1)$ such that

$$\|F'(x)\| \le \alpha \quad (x \in D); \tag{2.132}$$

-

$$\gamma = \sup_{\substack{k>1 \\ x \in D}} \left\| \frac{1}{k!} F^{(k)}(x) \right\|^{\frac{1}{k-1}} \qquad \text{is finite;}$$

- there exists $x_0 \in D$ such that

$$\|x_0 - F(x_0)\| \le \eta \le \frac{3 - \alpha - 2\sqrt{2 - \alpha}}{\gamma}, \quad \gamma \ne 0;$$

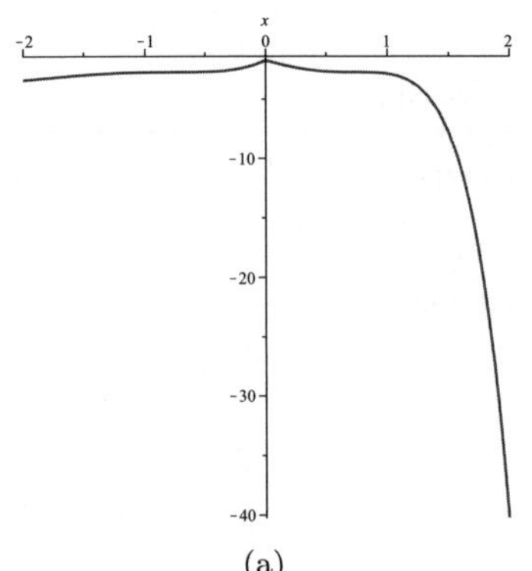 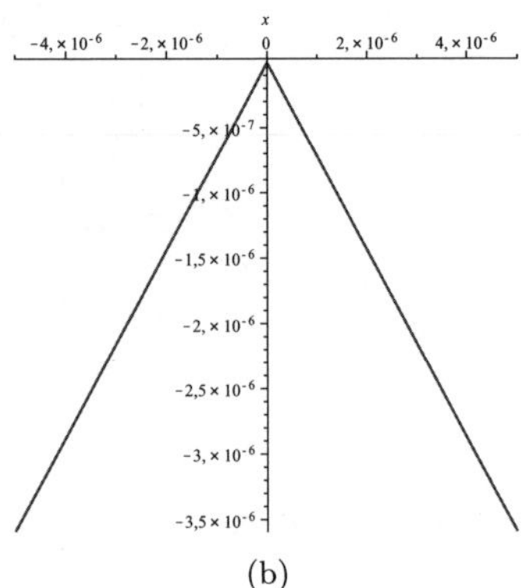

$$\text{(a)} \qquad\qquad\qquad\qquad\qquad\qquad \text{(b)}$$

Fig. 2.4   (a) Function φ_γ on interval $[-2, 2]$ for $\gamma = 1.359140914$. (b) Function φ_{γ_0} on interval $[-.000005, .000005]$ for $\gamma_0 = .8591409140$.

- $\bar{U}(x_0, r_1) \subseteq D$, where r_1, r_2 with $0 \le r_1 \le r_2$ are the two zeros of function f given by $f(r) = \gamma(2 - \alpha)r^2 - (1 + \eta\gamma - \alpha)r + \eta$.

Show: method of successive substitutions is well defined, remains in $\bar{U}(x_0, r_1)$ for all $n \ge 0$ and converges to a fixed point $x^* \in \bar{U}(x_0, r_1)$ of operator F. Moreover, x^* is the unique fixed point of F in $\bar{U}(x_0, r_2)$. Furthermore, the following estimates hold for all $n \ge 0$:

$$\|x_{n+2} - x_{n+1}\| \le \beta\|x_{n+1} - x_n\| \quad \text{and} \quad \|x_n - x^*\| \le \frac{\beta^n}{1 - \beta}\eta,$$

where

$$\beta = \frac{\gamma\eta}{1 - \gamma\eta} + \alpha.$$

The above result is based on the assumption that the sequence

$$\gamma_k = \left\| \frac{1}{k!} F^{(k)}(x) \right\|^{\frac{1}{k-1}} \quad (x \in D), \quad (k > 1)$$

is bounded above by γ. This kind of assumption does not always hold. Let us then not assume sequence $\{\gamma_k\}$ $(k > 1)$ is bounded and define "function" f_1 by

$$f_1(r) = \eta - (1 - \alpha)r + \sum_{k=2}^{\infty} \gamma_k^{k-1} r^k.$$

(b) Let $F \colon D \subseteq X \to X$ be an analytic operator. Assume (2.132) holds and for $x_0 \in D$ function f_1 has a minimum positive zero r_3 such that

$$\bar{U}(x_0, r_3) \subseteq D.$$

Show: method of successive substitutions is well defined, remains in $\bar{U}(x_0, r_3)$ for all $n \ge 0$ and converges to a unique fixed point $x^* \in \bar{U}(x_0, r_3)$ of operator F. Moreover the following estimates hold for all $n \ge 0$

$$\|x_{n+2} - x_{n+1}\| \le \beta_1 \|x_{n+1} - x_n\|$$

and

$$\|x_n - x^*\| \le \frac{\beta_1^n}{1 - \beta_1}\eta \quad \text{where} \quad \beta_1 = \sum_{k=2}^{\infty} \gamma_k^{k-1} \eta^{k-1} + \alpha.$$

2.4.2.

(a) It is convenient to define:

$$\gamma = \sup_{k>1} \left\| \frac{1}{k!} F^{(k)}(x^*) \right\|^{\frac{1}{k-1}}$$

with $\gamma = \infty$, if the supremum does not exist. Let $F \colon D \subseteq X \to X$ be an analytic operator and $x^* \in D$ be a fixed point of F. Moreover, assume that there exists α such that

$$\|F'(x^*)\| \leq \alpha \quad \text{and} \quad \bar{U}(x^*, r^*) \subseteq D, \tag{2.133}$$

where

$$r^* = \begin{cases} \infty & \text{if } \gamma = 0 \\ \frac{1}{\gamma} \frac{1-\alpha}{2-\alpha} & \text{if } \gamma \neq 0. \end{cases}$$

Then, if

$$\beta = \alpha + \frac{\gamma r^*}{1 - \gamma r^*} < 1,$$

show: the method of successive substitutions remains in $\bar{U}(x^*, r^*)$ for all $n \geq 0$ and converges to x^* for any $x_0 \in U(x^*, r^*)$. Moreover, the following estimates hold for all $n \geq 0$:

$$\|x_{n+1} - x^*\| \leq \beta_n \|x_n - x^*\| \leq \beta \|x_n - x^*\|,$$

where

$$\beta_0 = 1, \quad \beta_{n+1} = \alpha + \frac{\gamma r^* \beta_n}{1 - \gamma r^* \beta_n} \quad (n \geq 0).$$

The above result was based on the assumption that the sequence

$$\gamma_k = \left\| \frac{1}{k!} F^{(k)}(x^*) \right\|^{\frac{1}{k-1}} \quad (k \geq 2)$$

is bounded γ. In the case where the assumption of boundedness does not necessarily hold, we have the following local alternative.

(b) Let $F \colon D \subseteq X \to X$ be an analytic operator and $x^* \in D$ be a fixed point of F. Moreover, assume: $\max_{r>0} \sum_{k=2}^{\infty} (\gamma_k r)^{k-1}$ exists and is attained at some $r_0 > 0$. Set

$$p = \sum_{k=2}^{\infty} (\gamma_k r_0)^{k-1};$$

there exist α, δ with $\alpha \in [0, 1)$, $\delta \in (\alpha, 1)$ such that (2.133) holds,

$$p + \alpha - \delta \leq 0 \quad \text{and} \quad \bar{U}(x^*, r_0) \subseteq D.$$

Show: the method of successive substitutions $\{x_n\}$ $(n \geq 0)$ remains in $\bar{U}(x^*, r_0)$ for all $n \geq 0$ and converges to x^* for any $x_0 \in \bar{U}(x^*, r_0)$. Moreover the following error bounds hold for all $n \geq 0$:

$$\|x_{n+1} - x^*\| \leq \alpha \|x_n - x^*\| + \sum_{k=2}^{\infty} \gamma_k^{k-1} \|x_n - x^*\|^k \leq \delta \|x_n - x^*\|.$$

Chapter 3

Newton's Method on Special Spaces

The study of numerical algorithms on special spaces for solving eigenvalue or optimization problems is very important in computational mathematics. In this chapter we use Newton's method to approximate a zero of a mapping on special spaces as for examples Lie groups, Hilbert space and Banach space with convergence structure.

3.1 Lie Groups

We use in this section Newton's method to approximate a zero of a mapping from a Lie group into its Lie algebra. Under the same computational cost as before, we show convergence with the following advantages over earlier works: weaker sufficient convergence conditions, tighter error bounds on the distances involved and at least as precise information on the location of the solution (in the semi-local case); larger radius of convergence and tighter error bounds (in the local case). Numerical examples are also given for solving equations in cases not covered before.

We are concerned with the problem of approximating a zero $x^\star$ of C^1-mapping $F : G \longrightarrow Q$, where G is a Lie group and Q the Lie algebra of G that is the tangent space $T_e G$ of G at e, equipped with the Lie bracket $[.,.] : Q \times Q \longrightarrow Q$ (see Refs. [110, 196, 249, 258, 421]). Numerical algorithms on manifolds (see Refs. [2, 139, 303, 310, 424, 428]) is very important in Computational Mathematics. Newton-type methods are the most popular iterative procedures used to solve equations, when these equations contain differentiable operators. A local as well as a semi-local convergence of Newton-type methods has been given by several authors under various conditions (for example see Refs. [3, 443]). Newton's method (NM) with initial point $x_0 \in G$ was first introduced by Owren and Welfert in Ref. [344] in the form

$$x_{n+1} = x_n \cdot \exp\left(-dF_{x_n}^{-1} F(x_n)\right) \quad (n \geq 0). \tag{3.1}$$

Let $\eta =\parallel dF_{x_0}^{-1} F(x_0) \parallel$ and $L > 0$ be a Lipschitz constant (precised in (3.24)). Wang and Li in Ref. [425] recently used the Kantorovich hypothesis

$$h_0 = L\eta \leq \frac{1}{2} \tag{3.2}$$

125

together with majorizing sequence

$$\overline{t_0} = 0, \quad \overline{t_1} = \eta, \quad \overline{t_{n+1}} = \overline{t_n} - h'(\overline{t_n})^{-1} h(\overline{t_n}) = \overline{t_n} + \frac{L\left(\overline{t_n} - \overline{t_{n-1}}\right)^2}{2\left(1 - L\,\overline{t_n}\right)}, \qquad (3.3)$$

where

$$h(t) = \frac{L}{2} t^2 - t + \eta \qquad (3.4)$$

and

$$\lim_{n \longrightarrow \infty} \overline{t_n} = \varrho_0 = \frac{1 - \sqrt{1 - 2\,h_0}}{L} \qquad (3.5)$$

to provide a convergence analysis for (NM) that improved earlier works in Ref. [307, 310].

Hypothesis (3.2) is famous for its simplicity and clarify sufficient convergence condition for solving nonlinear equations due to Kantorovich in Ref. [290]. However, condition (3.2) can be violated. Based on the same information (F, x_0, L), we provided weaker sufficient convergence conditions (see Refs. [105, 110, 113, 139])

$$h_1 = \ell_1\,\eta \le \frac{1}{2}, \qquad (3.6)$$

or

$$h_2 = \ell_2\,\eta \le \frac{1}{2}, \qquad (3.7)$$

using the majorizing sequence

$$s_0 = 0, \quad s_1 = \eta, \quad s_{n+2} = s_{n+1} + \frac{L\left(s_{n+1} - s_n\right)^2}{2\left(1 - L_0\,s_{n+1}\right)} \quad (n \ge 0), \qquad (3.8)$$

where $L_0 \in [0, L]$ is a center-Lipschitz constant (precised in (3.25), which always exists if L does),

$$\ell_1 = \frac{L + L_0}{2}, \quad \ell_2 = \frac{1}{8}\left(L + 4\,L_0 + \sqrt{L^2 + 8\,L_0\,L}\right),$$

$$s^\star = \lim_{n \longrightarrow \infty} s_n \le \varrho_1 = \frac{\eta}{1 - \alpha} \quad \text{and} \quad \alpha = \frac{2\,L}{L + \sqrt{L^2 + 8\,L_0\,L}}. \qquad (3.9)$$

These advantages are obtained since we use the needed, less expensive to compute and more precise (3.25) (i.e., L_0) instead of (3.24) (i.e., L) in the computation for the upper bounds on $\| dF_x^{-1} dF_{x_0} \|$ $(x \in G)$. Indeed, we get (see (3.26))

$$\| dF_x^{-1} dF_{x_0} \| \le \left(1 - L_0 \sum_{i=0}^{k} \| z_i \|\right)^{-1} \qquad (3.10)$$

instead of the less tight and more expensive to compute estimate (for $L_0 < L$)

$$\| dF_x^{-1} dF_{x_0} \| \le \left(1 - L \sum_{i=0}^{k} \| z_i \|\right)^{-1} \qquad (3.11)$$

used in Ref. [425]. This modification leads to more precise majorizing sequence $\{s_n\}$ (for $\{x_n\}$) than $\{\overline{t_n}\}$, which makes our approach different and finer from the one in Ref. [425] under the same computational cost and information, since in practice the computation of L requires that of L_0.

In this section, we are motivated by the work in Ref. [425] and optimization considerations. Using again a combination of Lipschitz and center-type Lipschitz conditions, but even more precise majorizing sequences, we provide a convergence analysis with some advantages (under the same computational cost) over the work in Ref. [425]. We are wondering if in particular the weakest hypothesis (3.7) can be improved. It turns out that in the case of nonlinear equations when F is an operator between two Banach spaces, the sufficient convergence condition (see Ref. [155]) is

$$h_3 = \ell_3\,\eta \leq \frac{1}{2}, \tag{3.12}$$

where

$$\ell_3 = \frac{1}{8}\left(\sqrt{L_0\,L} + 4\,L_0 + \sqrt{L_0\,(L + 8\,L_0)}\right)$$

and the majorizing sequence given by

$$t_0 = 0, \quad t_1 = \eta, \quad t_2 = \eta + \frac{L_0\,\eta^2}{2\,(1 - L_0\,\eta)}, \quad t_{n+2} = t_{n+1} + \frac{L\,(t_{n+1} - t_n)^2}{2\,(1 - L_0\,t_{n+1})} \quad (n \geq 1), \tag{3.13}$$

where

$$t^\star = \lim_{n\longrightarrow\infty} t_n \leq \varrho_3 = \eta + \frac{1}{1-\alpha}\frac{L_0\,\eta^2}{2\,(1 - L_0\,\eta)}. \tag{3.14}$$

The applicability of (NM) for solving nonlinear equations is extended even further. The convergence of majorizing sequences is quadratic unless if equality holds in the corresponding h conditions, in which case it is only linear.

We introduce standard concepts and notations (see Refs. [110, 196, 249, 258]). A Lie group $(G,\cdot)$ is a Hausdorff topological group with countable bases which also has the structure of a smooth manifold such that the group product and the inversion are smooth operations in the differentiable structure given on the manifold. The dimension of a Lie group is that of the underlying manifold and we shall always assume that it is finite. The symbol e designates the identity element of G. Let Q be the Lie algebra of the Lie group G which is the tangent space T_eG of G at e, equipped with Lie bracket $[\cdot,\cdot] : Q \times Q \to Q$. We will make use of the left translation of the Lie group G. We define for each $y \in G$

$$\begin{aligned} L_y : G &\longrightarrow G \\ z &\longrightarrow y \cdot z \end{aligned} \tag{3.15}$$

the left multiplication in the group. The differential of L_y at e denoted by $(dL_y)_e$ determines an isomorphism of $Q = T_eG$ with the tangent space T_yG via the relation

$$(dL_y)_e(Q) = T_yG,$$

or, equivalently,

$$Q = (dL_y)_e^{-1}(T_yG) = (dL_{y-1})_y(T_yG).$$

The exponential map is noted by *exp* and defined by

$$\exp : Q \longrightarrow G$$
$$u \longrightarrow \exp(u),$$

which is certainly the most important construction associated to G and Q. Given $u \in Q$, the left invariant vector field $X_u : y \longrightarrow (dL_y)_e(u)$ determines a one-parameter subgroup of G: $\sigma_u : \mathbb{R} \longrightarrow G$ such that

$$\sigma_u(0) = e \quad \text{and} \quad \sigma_u'(t) = X_u(\sigma_u(t)) = (dL_{\sigma_u(t)})_e(u) \quad \forall\, t \in \mathbb{R}. \tag{3.16}$$

Consequently, the exponential map is defined by the relation

$$\exp(u) = \sigma_u(1).$$

Note that the exponential map is not surjective in general. However, the exponential map is a diffeomorphism on an open neighborhood $\mathcal{N}(0)$ of $0 \in Q$. Let

$$N(e) = \exp(\mathcal{N}(0)).$$

Then for each $y \in N(e)$, there exists $v \in \mathcal{N}(0)$ such that $y = \exp(v)$. Furthermore, if

$$\exp(u) = \exp(v) \in N(e)$$

for some $u, v \in \mathcal{N}(0)$, then $u = v$. If G is Abelian, *exp* is also a homomorphism from Q to G, i.e.,

$$\exp(u + v) = \exp(u) \cdot \exp(v) \quad \text{for all} \quad u, v \in Q. \tag{3.17}$$

In the non-abelian case, exp is not a homomorphism and (3.17) must be replaced by

$$\exp(\omega) = \exp(u) \cdot \exp(v), \tag{3.18}$$

where ω is given by the Bake–Campbell–Hausdorff (BCH) formula

$$\omega = u + v + \frac{1}{2}[u, v] + \frac{1}{12}([u[u, v]] + [v[v, u]]) + \cdots, \tag{3.19}$$

for all u, v in an open neighborhood of $0 \in Q$.

Consider $f : G \longrightarrow Q = T_eG$ be a C^1-mapping. The differential of f at a point $x \in G$ is a linear map $f_x' : T_xG \longrightarrow Q$ defined by

$$f_x'(\triangle_x) = \frac{d}{dt} f(x \cdot \exp(t((dL_{x-1})_x)(\triangle_x)))\,|_{t=0} \quad \text{for any} \quad \triangle_x \in T_xG. \tag{3.20}$$

The differential f_x' can be expressed via a function $df_x : Q \longrightarrow Q$ given by

$$df_x = (f \circ L_x)_e' = f_x' \circ (dL_x)_e.$$

Thus, by (3.20), it follows that

$$df_x(u) = f_x'((dL_x)_e(u)) = \frac{d}{dt} f(x \cdot \exp(tu))\,|_{t=0} \quad \text{for any} \quad u \in Q.$$

Therefore the following lemma is clear.

Lemma 3.1.1. *Let $x \in G$, $u \in Q$ and $t \in \mathbb{R}$. Then*

$$\frac{d}{dt} f(x \cdot \exp(-tu)) = -df_{x \cdot \exp(-tu)}(u) \tag{3.21}$$

and

$$f(x \cdot \exp(tu)) - f(x) = \int_0^t df_{x \cdot \exp(su)}(u)\, ds. \tag{3.22}$$

We shall study the semi-local/local convergence of (NM). We assume $\langle \cdot, \cdot \rangle$ the inner product and $\| \cdot \|$ on Q. As in Ref. [425] we define a distance on G for $x, y \in G$ as follows:

$$m(x, y) = \inf \Big\{ \sum_{i=1}^{k} \| z_i \| : \text{there exist } k \geq 1 \text{ and } z_1, \cdots, z_k \in Q \tag{3.23}$$
$$\text{such that } y = x \cdot \exp z_1 \cdots \exp z_k \Big\}.$$

By convention $\inf \emptyset = +\infty$. It is easy to see that $m(\cdot, \cdot)$ is a distance on G and the topology induced is equivalent to the original one on G. Let $w \in G$ and $r > 0$, we denote by

$$U(w, r) = \{ y \in G : m(w, y) < r \}$$

the open ball centered at w and of radius r. Moreover, we denote the closure of $U(w, r)$ by $\overline{U}(w, r)$. Let also $L(Q)$ denotes the set of all linear operators on Q.

We need the definition of Lipschitz continuity for a mapping.

Definition 3.1.1. Let $M : G \longrightarrow L(Q)$, $x_0 \in G$ and $r > 0$. We say that M satisfies the L-Lipschitz condition on $U(x_0, r)$ if

$$\| M(x \cdot \exp u) - M(x) \| \leq L \| u \| \tag{3.24}$$

holds for any $u \in Q$ and $x \in U(x_0, r)$ such that $\| u \| + m(x, x_0) < r$.

It follows from (3.24) that there exists L_0 such that

$$\| M(x \cdot \exp u) - M(x_0) \| \leq L_0 \left(\| u \| + m(x, x_0) \right) \tag{3.25}$$

holds for any $u \in Q$ and $x \in U(x_0, r)$ such that $\| u \| + m(x, x_0) < r$.

We then say M satisfies the L_0-center Lipschitz condition at $x_0 \in G$ on $U(x_0, r)$. Note that if M satisfies the L-Lipschitz condition on $U(x_0, r)$, then it also satisfies the L_0-center Lipschitz condition at $x_0 \in G$ on $U(x_0, r)$. Let us show that (3.24) indeed implies (3.25). If (3.24) holds, then for $x = x_0$ there exists $K \in (0, L]$ such that

$$\| M(x_0 \cdot \exp u) - M(x_0) \| \leq K \| u \| \quad \text{for any} \quad u \in Q \quad \text{such that} \quad \| u \| < r.$$

There exist $k \geq 1$ and $z_0, z_1, \cdots, z_k \in Q$ such that $x = x_0 \cdot exp\, z_0 \cdots exp\, z_k$. Let $y_0 = x_0$, $y_{i+1} = y_i \cdot exp\, u_i$, $i = 0, 1, \cdots, k$. Then, we have that $x = y_{k+1}$. We can write the identity

$$M(x \cdot exp\, u) - M(x_0) = (M(x \cdot exp\, u) - M(x)) + (M(x) - M(x_0))$$
$$= (M(x \cdot exp\, u) - M(x)) + (M(y_k \cdot exp\, u_k) - M(y_k)) + (M(y_k) - M(x_0))$$
$$= (M(x \cdot exp\, u) - M(x)) + (M(y_k \cdot exp\, u_k) - M(y_k)) + (M(y_{k-1} \cdot exp\, u_{k-1}) - M(x_0))$$
$$= (M(x \cdot exp\, u) - M(x)) + (M(y_k \cdot exp\, u_k) - M(y_k)) + \cdots +$$
$$(M(y_1 \cdot exp\, u_1) - M(y_1)) + (M(x_0 \cdot exp\, u_0) - M(x_0)).$$

Using (3.24) and the triangle inequality, we obtain in turn that

$$\| M(x \cdot exp\, u) - M(x_0) \|$$
$$\leq \| M(x \cdot exp\, u) - M(x) \| + \| M(y_k \cdot exp\, u_k) - M(y_k) \| + \cdots$$
$$+ \| M(y_1 \cdot exp\, u_1) - M(y_1) \| + \| M(x_0 \cdot exp\, u_0) - M(x_0) \|$$
$$\leq L \left(\| u \| + \| u_k \| + \cdots + \| u_1 \|\right) + K \| u_0 \|$$
$$= L \left(\| u \| + \| u_k \| + \cdots + \| u_0 \|\right) + (K - L) \| u_0 \|$$
$$\leq L \left(\| u \| + \| u_k \| + \cdots + \| u_0 \|\right),$$

which implies (3.25).

We need a Banach type lemma on invertible mappings. The conclusion of the next result is weaker and more precise than in [425] .

Lemma 3.1.2. *Let $x_0 \in G$ such that $dF_{x_0}^{-1}$ exists and let $0 < r \leq \dfrac{1}{L_0}$. Suppose $dF_{x_0}^{-1} dF$ satisfies the L_0-center Lipschitz condition at $x_0 \in G$ on $U(x_0, r)$; for $x \in U(x_0, r)$, there exist $k \geq 1$ and $z_0, z_1, \cdots, z_k \in Q$, such that $x = x_0 \cdot exp\, z_0 \cdots exp\, z_k$ and $\sum_{i=0}^{k} \| z_i \| < r$. Then, linear mapping dF_x^{-1} exists and*

$$\| dF_x^{-1} dF_{x_0} \| \leq \left(1 - L_0 \sum_{i=0}^{k} \| z_i \| \right)^{-1}. \tag{3.26}$$

Proof. Let $y_k = x_0 \cdot exp\, z_0 \cdot exp\, z_1 \cdots exp\, z_{k-1}$. Then, we deduce $y_k \in U(x_0, r)$, since $\sum_{i=0}^{k-1} \| z_i \| < r$. Note that $x = y_k \cdot exp\, z_k$. Using (3.25) for $M = dF_{x_0}^{-1} dF$, we get in turn

$$\| dF_{x_0}^{-1} (dF_x - dF_{x_0}) \| \leq L_0 (\| z_k \| + m(y_k, x_0)) \leq L_0 \sum_{i=0}^{k} \| z_i \| < L_0\, r \leq 1. \tag{3.27}$$

It follows from (3.27) and the Banach lemma on invertible operators that dF_x^{-1} exists and (3.26) holds. That completes the proof of Lemma 3.1.2. $\qquad\square$

Remark 3.1.1. The corresponding proof in [Lemma 2.2] of Ref. [425] uses (3.24) (see also (3.11)) instead of the less expensive and tighter (for $L_0 < L$) (3.25). We decided to include this proof to stretch the fact that our approach differs from Ref. [425] after this point, since (3.26) leads to tighter majorizing sequence and estimates.

We can show the main semi-local convergence result for (NM).

Theorem 3.1.1. *Let $F : G \longrightarrow Q$ be a C^1-mapping. Let $x_0 \in G$ be such that $dF_{x_0}^{-1}$ exists and let $r > 0$. Set*

$$\| dF_{x_0}^{-1} F(x_0) \| = \eta. \tag{3.28}$$

Suppose $dF_{x_0}^{-1} dF$ satisfies the L-Lipschitz condition on $U(x_0, r)$; (3.12) holds and

$$\overline{U}(x_0, \varrho_3) \subset U(x_0, r), \tag{3.29}$$

where ϱ_3 is given in (3.14). Then, sequence $\{x_n\}$ $(n \geq 0)$ generated by (NM) is well defined, remains in $U(x_0, \varrho_3)$ for all $n \geq 0$ and converges to a zero $x^\star \in \overline{U}(x_0, \varrho_3)$ of mapping F. Moreover, the following estimates hold for all $n \geq 1$:

$$m(x_{n+1}, x_n) \leq \| dF_{x_n}^{-1} F(x_n) \| \leq t_{n+1} - t_n \tag{3.30}$$

and

$$m(x^\star, x_n) \leq t^\star - t_n, \tag{3.31}$$

where $\{t_n\}$ $(n \geq 0)$ and $t^\star$ are given in (3.13) and (3.14), respectively.

Proof. We shall first show using induction that (3.30) holds for all $n \geq 0$. Set $z_n = dF_{x_n}^{-1} F(x_n)$. We have by (3.28) that z_0 is well defined and $x_1 = x_0 \cdot \exp z_0$. We have $m(x_1, x_0) \leq \| z_0 \|$ and $\| z_0 \| = \| dF_{x_0}^{-1} F(x_0) \| = \eta = t_1 - t_0$, which shows (3.30) for $n = 0$.

Assume z_n is well defined and (3.30) holds for all $n \leq k - 1$. Then, we have that

$$\sum_{i=0}^{k-1} \| z_i \| \leq t_k - t_0 = t_k \leq \varrho_3 \quad \text{and} \quad x_k = x_0 \cdot \exp z_0 \cdots \exp z_{k-1}. \tag{3.32}$$

It follows from Lemma 3.1.2 and (3.29) that $dF_{x_k}^{-1}$ exists and

$$\| dF_{x_k}^{-1} dF_{x_0} \| \leq (1 - L_0 t_k)^{-1}. \tag{3.33}$$

We also have that

$$F(x_1) = F(x_1) - F(x_0) - dF_{x_0} z_0 = \int_0^1 (dF_{x_0 \cdot \exp(t z_0)} - dF_{x_0}) z_0 \, dt \tag{3.34}$$

and

$$\begin{aligned}
\| dF_{x_0}^{-1} F(x_1) \| &\leq \int_0^1 \| dF_{x_0}^{-1} (dF_{x_0 \cdot \exp(t z_0)} - dF_{x_0}) \| \| z_0 \| \, dt \\
&\leq \int_0^1 L_0 \| t z_0 \| \| z_0 \| \, dt \leq \frac{L_0}{2} (t_1 - t_0)^2.
\end{aligned} \tag{3.35}$$

Then, we obtain that

$$\begin{aligned}
\| z_1 \| = \| -dF_{x_1}^{-1} F(x_1) \| &\leq \| dF_{x_1}^{-1} dF_{x_0} \| \| dF_{x_0}^{-1} F(x_1) \| \\
&\leq \frac{L_0 (t_1 - t_0)^2}{2 (1 - L_0 t_1)} = t_2 - t_1,
\end{aligned} \tag{3.36}$$

which shows (3.30) for $n = 1$. Similarly for $k > 1$, we have that

$$F(x_k) = F(x_k) - F(x_{k-1}) - dF_{x_{k-1}} z_{k-1}$$
$$= \int_0^1 (dF_{x_{k-1} \cdot exp\,(t\,z_{k-1})} - dF_{x_{k-1}}) \, z_{k-1} \, dt \tag{3.37}$$

$$\| \, dF_{x_0}^{-1} \, F(x_k) \, \| \leq \int_0^1 \| \, dF_{x_0}^{-1} \, (dF_{x_{k-1} \cdot exp\,(t\,z_{k-1})} - dF_{x_{k-1}}) \, \| \, \| \, z_{k-1} \, \| \, dt$$
$$\leq \int_0^1 L \, \| \, t \, z_{k-1} \, \| \, \| \, z_{k-1} \, \| \, dt \leq \frac{L}{2} \, (t_k - t_{k-1})^2. \tag{3.38}$$

Hence, we get that

$$\| \, z_k \, \| = \| \, -dF_{x_k}^{-1} \, F(x_k) \, \| \leq \| \, dF_{x_k}^{-1} \, dF_{x_0} \, \| \, \| \, dF_{x_0}^{-1} \, F(x_k) \, \|$$
$$\leq \frac{L \, (t_k - t_{k-1})^2}{2 \, (1 - L_0 \, t_k)} = t_{k+1} - t_k. \tag{3.39}$$

We also have $m(x_{k+1}, x_k) \leq \| \, z_k \, \|$, since $x_{k+1} = x_k \cdot exp \, z_k$, which completes the induction for (3.30). It follows from (3.30) that $\{x_k\}$ is a Cauchy sequence in a Banach space and as such it converges to some $x^\star \in \overline{U}(x_3, \varrho_3)$ (since $\overline{U}(x_3, \varrho_3)$ is a closed set). By letting $k \longrightarrow \infty$ in (3.38), we get that $x^\star$ is a zero of F. Estimate (3.31) follows from (3.30) by using standard majorization techniques. That completes the proof of Theorem 3.1.1. $\qquad\qquad \square$

Remark 3.1.2. If $L = L_0$, Theorem 3.1.1 reduces to [Theorem 3.1, p. 328] in Ref. [425]. Otherwise (i.e., if $L_0 < L$), it constitutes an improvement under the same information as already stated in the introduction of this section. Clearly, the conclusions of Theorem 3.1.1 hold true if (h_1, ϱ_1) or (h_2, ϱ_1) replace (h_3, ϱ_3).

Next, we study the convergence domain of (NM) around a zero $x^\star$ of mapping F. First from Lemma 3.1.3 until Corollary 3.1.4, we present the local result for (NM) when G is an Abelian group. Then, the corresponding local results follow when G is not necessarily an Abelian group.

Lemma 3.1.3. *Let G be an Abelian group and $0 < r \leq \dfrac{1}{L_\star}$. Let $F : G \longrightarrow Q$ be a C^1-mapping. Let $x^\star \in G$ be a zero of mapping F. Suppose $dF_{x^\star}^{-1}$ exists and let $L_\star > 0$; there exist $j \geq 1$ and $z_1, z_2, \cdots, z_j \in Q$ such that*

$$x_0 = x^\star \cdot exp \, z_1 \cdots exp \, z_j \quad for \quad \sum_{i=1}^{j} \| \, z_i \, \| < r; \tag{3.40}$$

$dF_{x^\star}^{-1} \, dF$ satisfies the $L_\star$-center Lipschitz condition at $x^\star$ on $U(x^\star, r)$. Then, linear mapping $dF_{x_0}^{-1}$ exists and

$$\| \, dF_{x_0}^{-1} \, F(x_0) \, \| \leq \frac{(2 + L_\star \sum\limits_{i=1}^{j} \| \, z_i \, \|) \sum\limits_{i=1}^{j} \| \, z_i \, \|}{2 \, (1 - L_\star \sum\limits_{i=1}^{j} \| \, z_i \, \|)}. \tag{3.41}$$

Proof. Using hypothesis $dF_{x^\star}^{-1} dF$ satisfies the $L_\star$-center Lipschitz condition at $x^\star$ on $U(x^\star, \frac{1}{L_\star})$, we have that

$$\| dF_{x^\star}^{-1} (dF_{x \cdot exp\, u} - dF_{x^\star}) \| \leq L_\star (\| u \| + m(x, x^\star)) \tag{3.42}$$

for all $u \in Q$, $x \in U(x^\star, r)$ such that $\| u \| + m(x, x^\star) < r$. Let $z = z_1 + z_2 + \cdots + z_j$. Then, since G is an Abelian group, $exp\, z_1 \cdot exp\, z_2 \cdots exp\, z_j = exp\,(z_1 + z_2 + \cdots + z_j) = exp\, z$, so, we can write $x_0 = x^\star \cdot exp\, z$. It then follows from Lemma 3.1.2 that $dF_{x_0}^{-1}$ exists and

$$\| dF_{x_0}^{-1} dF_{x^\star}) \| \leq (1 - L_\star \sum_{i=1}^{j} \| z_i \|)^{-1}. \tag{3.43}$$

We also have the following identity

$$\begin{aligned}
F(x_0) &= F(x^\star \cdot exp\, z) - F(x^\star) \\
&= \int_0^1 dF_{x^\star \cdot exp\,(tz)}\, z\, dt = \int_0^1 (dF_{x^\star \cdot exp\,(tz)} - dF_{x^\star})\, z\, dt + dF_{x^\star}\, z.
\end{aligned} \tag{3.44}$$

In view of (3.42) and (3.44), we get that

$$\begin{aligned}
\| dF_{x^\star}^{-1} F(x_0) \| &\leq \int_0^1 \| dF_{x^\star}^{-1} (dF_{x^\star \cdot exp\,(t z)} - dF_{x^\star}) \| \| z \| \, dt + \| z \| \\
&\leq \int_0^1 \| L_\star (t \| z \|) \| z \| \, dt + \| z \| \\
&\leq \left(\frac{L_\star}{2} \sum_{i=1}^{j} \| z_i \| + 1 \right) \sum_{i=1}^{j} \| z_i \| .
\end{aligned} \tag{3.45}$$

Moreover, by (3.43) and (3.45), we obtain that

$$\| dF_{x_0}^{-1} F(x_0) \| \leq \| dF_{x_0}^{-1} dF_{x^\star} \| \, \| dF_{x^\star}^{-1} F(x_0) \| \leq \frac{(2 + L_\star \sum_{i=1}^{j} \| z_i \|) \sum_{i=1}^{j} \| z_i \|}{2 (1 - L_\star \sum_{i=1}^{j} \| z_i \|)}.$$

That completes the proof of Lemma 3.1.3. $\qquad\square$

Remark 3.1.3. The proof of Lemma 3.1.3 reduces to [Lemma 3.2] in Ref. [425] if $L_\star = L$. Otherwise (i.e., if $L_\star < L$) it constitutes an improvement. We have also included the proof although similar to the corresponding one in Ref. [425] because it is not straightforward to see that $L_\star$ can replace L in the derivation of the crucial estimate (3.41). Note also that (3.42) can hold for some $L_\star^1 \in (0, L_\star]$. If $L_\star^1 < L_\star$, then according to the proof of Lemma 3.1.3, $L_\star^1$ can replace $L_\star$ in (3.41).

Let us define parameter α for $\beta = \dfrac{L_\star}{L}$ by

$$\alpha = \begin{cases} 4 \dfrac{\sqrt{1 + 3\beta} - (1 + \beta)}{\beta\,(1 - \beta)} & \text{if } L_\star \neq L \\ 1 & \text{if } L_\star = L. \end{cases} \tag{3.46}$$

Then, it can easily be seen that $\alpha \geq 1$. We have the local convergence result for (NM).

Theorem 3.1.2. *Let G be an Abelian group. Let $0 < r \leq \dfrac{\alpha}{4\,L}$, where α in given in (3.46). Let $F : G \longrightarrow Q$ be a C^1-mapping. Suppose there exists $x^\star \in G$ such that $F(x^\star) = 0$, $dF_{x^\star}^{-1}$ exists and $dF_{x^\star}^{-1}\, dF$ satisfies the L-Lipschitz condition on $U(x^\star, 3\,r/(1 - L_\star\, r))$. Then, sequence $\{x_n\}$ ($n \geq 0$) generated by (NM) starting at $x_0 \in U(x^\star, r)$ is well defined, remains in $U(x^\star, 3\,r/(1 - L_\star\, r))$ for all $n \geq 0$ and converges to a zero $y^\star$ of mapping F such that $m(x^\star, y^\star) < 3\,r/(1 - L_\star\, r)$.*

Proof. It follows from hypothesis $x_0 \in U(x^\star, r)$ that there exist $j \geq 1$ and $z_1, z_2, \cdots, z_j \in Q$ such that (3.40) holds. If $dF_{x^\star}^{-1}\, dF$ is L-Lipschitz on $U(x^\star, 3\,r/(1 - L_\star\, r))$, then it is $L_\star$-center Lipschitz at $x^\star$ on $U(x^\star, 3\,r/(1 - L_\star\, r))$. Note also that $\dfrac{\alpha}{4\,L} \leq \dfrac{1}{L_\star}$ by the choice of α. It follows from Lemma 3.1.3 that linear mapping $dF_{x_0}^{-1}$ exists and

$$\eta = \| dF_{x_0}^{-1}\, F(x_0) \| \leq \frac{\left(2 + L_\star \displaystyle\sum_{i=1}^{j} \| z_i \|\right) \displaystyle\sum_{i=1}^{j} \| z_i \|}{2\left(1 - L_\star \displaystyle\sum_{i=1}^{j} \| z_i \|\right)}. \tag{3.47}$$

Set

$$\overline{L} = \frac{L}{1 - L_\star \displaystyle\sum_{i=1}^{j} \| z_i \|} \quad \text{and} \quad \overline{r} = \frac{(2 + L_\star\, r)\, r}{1 - L_\star\, r}.$$

We shall show mapping $dF_{x_0}^{-1}\, dF$ satisfies the $\overline{L}$-Lipschitz condition on $U(x_0, \overline{r})$. Indeed, let $x \in U(x_0, \overline{r})$ and $z \in Q$ be such that $\| z \| + m(x_0, x) < \overline{r}$. Then, we get that

$$\| z \| + m(x, x^\star) < \| z \| + m(x, x_0) + m(x_0, x^\star) < \overline{r} + r \leq \frac{3\,r}{1 - L_\star\, r}.$$

Using Lemma 3.1.2, we have that

$$\| dF_{x_0}^{-1}\, (dF_{x \cdot \exp z} - dF_x) \| \leq \| dF_{x_0}^{-1}\, dF_{x^\star} \| \, \| dF_{x^\star}^{-1}\, (dF_{x \cdot \exp z} - dF_x) \|$$
$$\leq \frac{L\, \| z \|}{1 - L_\star \displaystyle\sum_{i=1}^{j} \| z_i \|} = \overline{L}\, \| z \|.$$

Set

$$\overline{h} = \overline{L}\,\eta \quad \text{and} \quad \overline{r_1} = \frac{2\,\eta}{1 + \sqrt{1 - 2\,\overline{h}}} \leq 2\,\eta. \tag{3.48}$$

Then, by (3.47), we obtain that

$$\overline{r_1} \leq \frac{(2 + L_\star \sum_{i=1}^{j} \| z_i \|) \sum_{i=1}^{j} \| z_i \|}{1 - L_\star \sum_{i=1}^{j} \| z_i \|} \leq \frac{(2 + L_\star r)\, r}{1 - L_\star r} = \overline{r} \tag{3.49}$$

and

$$\overline{h} \leq \frac{(2 + L_\star r)\, L\, r}{2\,(1 - L_\star r)^2} \leq \frac{1}{2}, \tag{3.50}$$

by the choice of r and α. According to Theorem 3.1.1 and Remark 3.1.2, $\{x_k\}$ converges to some zero $y^\star$ of mapping F and $m(x_0, y^\star) \leq \overline{r_1}$. Furthermore, we obtain that

$$m(x^\star, y^\star) \leq m(x^\star, x_0) + m(x_0, y^\star) \leq r + \overline{r_1} \leq r + \overline{r} \leq \frac{3\,r}{1 - L_\star r}.$$

That completes the proof of Theorem 3.1.2. $\qquad\square$

The proofs of remaining results are omitted, since they can be obtained from the corresponding ones in Ref. [425].

Corollary 3.1.1. *Let G be an Abelian group. Let $F : G \longrightarrow Q$ be a $\mathcal{C}^1$-mapping. Let $x^\star \in G$ be a zero of mapping F. Suppose $dF_{x^\star}^{-1}$ exists and $dF_{x^\star}^{-1}\, dF$ satisfies the L-Lipschitz condition on $U(x^\star, \frac{1}{L_\star})$. Let also $x_0 \in U(x^\star, \frac{\alpha}{4\,L})$. Then, sequence $\{x_n\}$ $(n \geq 0)$ generated by (NM) starting at x_0 is well defined, remains in $U(x^\star, \frac{\alpha}{4\,L})$ for all $n \geq 0$ and converges to a zero $y^\star$ of mapping F such that $m(x^\star, y^\star) < \frac{1}{L_\star}$.*

We denote in the following Corollary $B(0, r) = \{z \in Q : \| z \| < r\}$.

Corollary 3.1.2. *Let G be an Abelian group. Let $F : G \longrightarrow Q$ be a $\mathcal{C}^1$-mapping. Let $x^\star \in G$ be a zero of mapping F. Suppose $dF_{x^\star}^{-1}$ exists and $dF_{x^\star}^{-1}\, dF$ satisfies the L-Lipschitz condition on $U(x^\star, \frac{1}{L_\star})$. Let also $x_0 \in U(x^\star, \frac{1}{L_\star})$. Let σ be the maximal number such that $U(e, \sigma) \subseteq exp\,(B(0, \frac{1}{L_\star}))$. Set $r = \min \left\{ \frac{\sigma}{3 + L_\star \sigma}, \frac{\alpha}{4\,L} \right\}$ and $N(x^\star, r) = x^\star\, exp(B(0, r))$. Then, sequence $\{x_n\}$ $(n \geq 0)$ generated by (NM) starting at $x_0 \in N(x^\star, r)$ converges to $x^\star$.*

Corollary 3.1.3. *Let G be an Abelian group. Let $F : G \longrightarrow Q$ be a $\mathcal{C}^1$-mapping, where G is a compact connected Lie group equipped with a bi-invariant Riemannian metric. Let $x^\star \in G$ be a zero of mapping F and $0 < r < \frac{\alpha}{4\,L}$. Suppose $dF_{x^\star}^{-1}$ exists and $dF_{x^\star}^{-1}\, dF$ satisfies the L-Lipschitz condition on $U(x^\star, \frac{3\,r}{1 - L_\star r})$. Let also $x_0 \in U(x^\star, r)$. Then, sequence $\{x_n\}$ $(n \geq 0)$ generated by (NM) starting at x_0, remains in $\overline{U}(x^\star, \frac{3\,r}{1 - L_\star r})$ for all $n \geq 0$ and converges to $x^\star$.*

Corollary 3.1.4. *Let G be an Abelian group. Let $F : G \longrightarrow Q$ be a C^1-mapping, where G is a compact connected Lie group equipped with a bi-invariant Riemannian metric. Let $x^\star \in G$ be a zero of mapping F. Suppose $dF_{x^\star}^{-1}$ exists and $dF_{x^\star}^{-1} dF$ satisfies the L-Lipschitz condition on $U(x^\star, \frac{1}{L_\star})$. Let also $x_0 \in U(x^\star, \frac{\alpha}{4L})$. Then, sequence $\{x_n\}$ $(n \geq 0)$ generated by (NM) starting at x_0, remains in $\overline{U}(x^\star, \frac{\alpha}{4L})$ for all $n \geq 0$ and converges to $x^\star$.*

Lemma 3.1.4. *Let $0 < r \leq \dfrac{1}{L}$. Let $F : G \longrightarrow Q$ be a C^1-mapping. Let $x^\star \in G$ be a zero of mapping F. Suppose $dF_{x^\star}^{-1}$ exists; there exist $j \geq 1$ and $z_1, z_2, \cdots, z_j \in Q$ such that $x_0 = x^\star \cdot exp\, z_1 \cdots exp\, z_j$ for $\displaystyle\sum_{i=1}^{j} \| z_i \| < r$ and $dF_{x^\star}^{-1} dF$ satisfies the L-Lipschitz condition at $x^\star$ on $U(x^\star, r)$. Then, the following assertions hold*

(i) $\| dF_{x^\star}^{-1} (dF_{x^\star \cdot exp\, z} - dF_{x^\star}) \| \leq K_\star \| z \|$ *for each $z \in Q$ with $\| z \| < r$ and some $K_\star \in (0, L]$.*

(ii) $\| dF_{x^\star}^{-1} (F'(x_0) - F'(x^\star)) \| \leq L_\star \displaystyle\sum_{i=1}^{j} \| z_i \|$ *for some $L_\star \in (0, L]$.*

(iii) Linear mapping $dF_{x_0}^{-1}$ exists and

$$\| dF_{x_0}^{-1} F(x_0) \| \leq \frac{(2 + L \displaystyle\sum_{i=1}^{j} \| z_i \|) \displaystyle\sum_{i=1}^{j} \| z_i \|}{2 (1 - L_\star \displaystyle\sum_{i=1}^{j} \| z_i \|)}.$$

Proof.

(i) This assertion follows from the hypothesis that $dF_{x^\star}^{-1} dF$ satisfies the L-Lipschitz condition at $x^\star$ on $U(x^\star, r)$.

(ii) Let $w_0 = x^\star$, $w_{i+1} = w_i \cdot exp\, z_{i+1}$, $i = 1, 2, \cdots, j - 1$. Then, we have $w_j = w_{j-1} \cdot exp\, z_j = x_0$. Using the L-Lipschitz condition, we get in turn that

$$\| dF_{x^\star}^{-1} (dF_{w_j} - dF_{x^\star}) \|$$
$$\leq \| dF_{x^\star}^{-1} (dF_{w_{j-1} \cdot exp\, z_j} - dF_{w_{j-1}}) \| + \cdots + \| dF_{x^\star}^{-1} (dF_{w_1 \cdot exp\, z_2} - dF_{w_1}) \| +$$
$$\| dF_{x^\star}^{-1} (dF_{x^\star \cdot exp\, z_1} - dF_{x^\star}) \| \leq L(\| z_j \| + \cdots + \| z_2 \|) + K_\star \| z_1 \|$$
$$= L (\| z_j \| + \cdots + \| z_1 \|) + (K_\star - L) \| z_1 \| \leq L (\| z_j \| + \cdots + \| z_1 \|)$$

which shows (ii).

(iii) We have by (ii) that

$$\| dF_{x^\star}^{-1} (dF_{x_0} - dF_{x^\star}) \| \leq L_\star (\| z_j \| + \cdots + \| z_1 \|) \leq L\, r < 1.$$

Hence, $dF_{x_0}^{-1}$ exists and

$$\| dF_{x_0}^{-1} dF_{x^\star} \| \leq (1 - L_\star \sum_{i=1}^{j} \| z_i \|)^{-1}. \tag{3.51}$$

We have that

$$dF_{x^\star}^{-1}\left(F(w_i) - F(w_{i-1})\right) = dF_{x^\star}^{-1} \int_0^1 dF_{w_{i-1}\cdot exp\,(t\,z_i)}\, z_i\, dt$$

$$= \int_0^1 dF_{x^\star}^{-1}\left(dF_{w_{i-1}\cdot exp\,(t\,z_i)} - dF_{x^\star}\right) z_i\, dt + z_i.$$

Hence, we get that

$$\| dF_{x^\star}^{-1}\left(dF_{w_{k-1}\cdot exp\,(z_k)} - dF_{w_{k-1}}\right) \| \le L \parallel z_k \parallel \quad \text{for each} \quad k = 1, 2, \cdots, j.$$

Therefore, we obtain that

$$\| dF_{x^\star}^{-1}\left(dF_{w_{i-1}\cdot exp\,(t\,z_i)} - dF_{x^\star}\right) \| \le \sum_{k=1}^{i-1} \| dF_{x^\star}^{-1}\left(dF_{w_{k-1}\cdot exp\,(z_k)} - dF_{w_{k-1}}\right) \|$$

$$+ \| dF_{x^\star}^{-1}\left(dF_{w_{i-1}\cdot exp\,(t\,w_i)} - dF_{w_{i-1}}\right) \| \le L \sum_{k=1}^{i-1} \| z_k \| + L\,t \parallel z_i \parallel .$$

We have that $F(x_0) = \sum_{i=1}^{j}(F(w_i) - F(w_{i-1}))$. That is, we can get

$$\| dF_{x^\star}^{-1} F(x_0) \|$$

$$\le \sum_{i=1}^{j}\left(\int_0^1 \| dF_{x^\star}^{-1}\left(dF_{w_{i-1}\cdot exp\,(t\,z_i)} - dF_{x^\star}\right) \| \| z_i \| \, dt + \| z_i \| \right)$$

$$\le \sum_{i=1}^{j}\left(\int_0^1 L\left(\sum_{k=1}^{i-1} \| z_k \| + t \parallel z_i \parallel \right) \| z_i \| \, dt + \| z_i \| \right) \tag{3.52}$$

$$\le \left(\frac{L}{2} \sum_{i=1}^{j} \| z_i \| + 1 \right) \sum_{i=1}^{j} \| z_i \| .$$

The result now follows from (3.51), (3.52) and

$$\| dF_{x_0}^{-1} F(x_0) \| \le \| dF_{x_0}^{-1} dF_{x^\star} \| \| dF_{x^\star}^{-1} F(x_0) \| \le \frac{(2 + L \sum_{i=1}^{j} \| z_i \|) \sum_{i=1}^{j} \| z_i \|}{2\left(1 - L_\star \sum_{i=1}^{j} \| z_i \|\right)}.$$

The proof of Lemma 3.1.4 is complete. $\qquad\square$

Let us define parameter δ by

$$\delta = \begin{cases} 4\, \dfrac{\sqrt{2\,(1+\beta)} - (1+\beta)}{1 - \beta^2} & \text{if } L_\star \ne L \\[2mm] 1 & \text{if } L_\star = L. \end{cases} \tag{3.53}$$

We have that $\delta \ge 1$.

Theorem 3.1.3. *Let $0 < r \le \dfrac{\delta}{4\,L}$, where δ is given in (3.53). Let $F : G \longrightarrow Q$ be a C^1-mapping. Suppose there exists $x^\star \in G$ such that $F(x^\star) = 0$, $dF_{x^\star}^{-1}$ exists and*

$dF_{x^\star}^{-1}\, dF$ satisfies the L-Lipschitz condition on $U(x^\star, R = \dfrac{3 + (L - L_\star)\, r}{1 - L_\star\, r}\, r)$. Then, sequence $\{x_n\}$ $(n \geq 0)$ generated by (NM) starting at $x_0 \in U(x^\star, r)$ is well defined, remains in $U(x^\star, R)$ for all $n \geq 0$ and converges to a zero $y^\star$ of mapping F such that $m(x^\star, y^\star) < R$.

Proof.   The proof is similar to the proof of Theorem 3.1.2. Note that $\bar{r}$ in the proof of Theorem 3.1.2 is replaced by $\bar{\bar{r}} = ((2 + L\, r)\, r)/(1 - L_\star\, r)$ and η satisfies the new condition

$$\eta \leq \frac{\left(2 + L \displaystyle\sum_{i=1}^{j} \|\, z_i\, \|\right) \displaystyle\sum_{i=1}^{j} \|\, z_i\, \|}{2\left(1 - L_\star \displaystyle\sum_{i=1}^{j} \|\, z_i\, \|\right)}.$$

The proof of Theorem 3.1.3 is complete. $\square$

Corollary 3.1.5. *Let* $F : G \longrightarrow Q$ *be a* C^1-*mapping. Let* $x^\star \in G$ *be a zero of mapping* F. *Suppose* $dF_{x^\star}^{-1}$ *exists and* $dF_{x^\star}^{-1}\, dF$ *satisfies the* L-*Lipschitz condition on* $U(x^\star, \dfrac{1}{L_\star})$. *Let also* $x_0 \in U(x^\star, \dfrac{\delta}{4\, L})$. *Then, the conclusions of Corollary 3.1.1 hold.*

Corollary 3.1.6. *Let* $F : G \longrightarrow Q$ *be a* C^1-*mapping. Let* $x^\star \in G$ *be a zero of mapping* F. *Suppose* $dF_{x^\star}^{-1}$ *exists and* $dF_{x^\star}^{-1}\, dF$ *satisfies the* L-*Lipschitz condition on* $U(x^\star, \dfrac{1}{L_\star})$. *Let also* $x_0 \in U(x^\star, \dfrac{1}{L_\star})$. *Let* σ *and* $N(x^\star, r)$ *as in Corollary 3.1.2.*

Set $r = \min\left\{ r_0, r_1, \dfrac{\delta}{4\, L} \right\}$, *where*

$$r_0 = \begin{cases} \dfrac{-(L_\star + 3\, L) + \sqrt{(L_\star + 3\, L)^2 + 4\, L\, (L - L_\star)}}{2\, L\, (L - L_\star)} & \textit{if } L_\star \neq L \\[2ex] \dfrac{1}{4\, L} & \textit{if } L_\star = L \end{cases}$$

and

$$r_1 = \begin{cases} \dfrac{-(3 + \sigma\, L_\star) + \sqrt{(3 + \sigma\, L_\star)^2 + 4\, \sigma\, (L - L_\star)}}{2\, (L - L_\star)} & \textit{if } L_\star \neq L \\[2ex] \dfrac{\sigma}{3 + \sigma\, L} & \textit{if } L_\star = L. \end{cases}$$

Then, sequence $\{x_n\}$ $(n \geq 0)$ *generated by (NM) starting at* $x_0 \in N(x^\star, r)$ *converges to* $x^\star$.

Corollary 3.1.7. *Let* $F : G \longrightarrow Q$ *be a* C^1-*mapping, where* G *is a compact connected Lie group equipped with a bi-invariant Riemannian metric. Let* $x^\star \in G$ *be a zero of mapping* F *and* $0 < r < \dfrac{\delta}{4\, L}$. *Suppose* $dF_{x^\star}^{-1}$ *exists and* $dF_{x^\star}^{-1}\, dF$ *satisfies the* L-*Lipschitz condition on* $U(x^\star, R)$, *where* R *is defined in Theorem 3.1.3. Let also* $x_0 \in U(x^\star, r)$. *Then, sequence* $\{x_n\}$ $(n \geq 0)$ *generated by (NM) starting at* x_0, *remains in* $\overline{U}(x^\star, R)$ *for all* $n \geq 0$ *and converges to* $x^\star$.

Corollary 3.1.8. *Let $F : G \longrightarrow Q$ be a $\mathcal{C}^1$-mapping, where G is a compact connected Lie group equipped with a bi-invariant Riemannian metric. Let $x^\star \in G$ be a zero of mapping F. Suppose $dF_{x^\star}^{-1}$ exists and $dF_{x^\star}^{-1} \, dF$ satisfies the L-Lipschitz condition on $U(x^\star, \frac{1}{L_\star})$. Let also $x_0 \in U(x^\star, \frac{\delta}{4L})$. Then, sequence $\{x_n\}$ $(n \geq 0)$ generated by (NM) starting at x_0, remains in $\overline{U}(x^\star, \frac{\delta}{4L})$ for all $n \geq 0$ and converges to $x^\star$.*

Remark 3.1.4.

(a) The local results reduce to the corresponding ones in Ref. [425] if $L = L_\star$. Otherwise (i.e., if $L_\star < L$), they constitute an improvement under the same computational cost with advantages as already stated in the introduction of this section. Note also that $\alpha > 1$, $\delta > 1$ if $L_\star < L$ and $\alpha \longrightarrow \infty$, $\delta \longrightarrow \infty$ if $\frac{L}{L_\star} \longrightarrow \infty$.

(b) The local results if G is an Abelian group are weaker and tighter than the ones when G is not necessarily an Abelian group. We have for example that $\bar{r} < \bar{\bar{r}}$, $\delta < \alpha$, $\frac{3r}{1 - L_\star r} < R$ and the upper bound on η is smaller if $L_\star < L$.

(c) It is obvious that finer results can be immediately obtained if conditions (3.6) or (3.7) or (3.12) are used instead of (3.2) for $L_0 < L$ or $L_\star < L$. However, we decided to leave this part of analysis to the motivated reader. We refer the reader to Ref. [155] for such results involving nonlinear equations in a Banach space setting.

Following Mahony in Refs. [316, 317] and Wang and Li in Ref. [425], we provide some applications to optimization problems. Let $f : G \longrightarrow \mathbb{R}$ be a $\mathcal{C}^2$-mapping. Consider the optimization problem:

$$\min_{x \in G} f(x). \tag{3.54}$$

Let $X \in Q$. Denote by $\tilde{X}$ the left-invariant vector field associated with X and given by

$$\tilde{X}(x) = (L'_x)_e \, X \quad \text{for} \quad x \in G, \tag{3.55}$$

where $\tilde{X} f$ is the Lie derivative of f with respect to the left-invariant vector field $\tilde{X}$. That is, we have for each $x \in G$

$$(\tilde{X} f)(x) = \frac{d}{dt}\bigg|_{t=0} f(x \cdot exp \, t \, X). \tag{3.56}$$

Let $\{X_1, \cdots, X_n\}$ be an orthonormal basis of Q. Then $grad \, f$ (see Ref. [421]) is a vector field on G given by

$$
\begin{aligned}
grad \, f(x) &= (\tilde{X}_1, \cdots, \tilde{X}_n)(\tilde{X}_1 \, f(x), \cdots, \tilde{X}_n \, f(x))^T \\
&= \sum_{j=1}^{n} \tilde{X}_j \, f(x) \, \tilde{X}_j \quad \text{for} \quad x \in G.
\end{aligned} \tag{3.57}
$$

Then, (NM) according to Mahony in Ref. [317] is given

Algorithm 3.1.1. Find $X^k \in G$ such that

$$\tilde{X}^k = (L'_x)_e \, X^k \quad \text{and} \quad grad \, f(x_k) + grad \, (\tilde{X}^k \, f)(x_k) = 0.$$

Set $x_{k+1} = x_k \cdot exp \, X^k$.
Set $k \longleftarrow k + 1$ and repeat.

Let $F : G \longrightarrow Q$ be a mapping defined by

$$F(x) = (L'_x)_e^{-1} \, grad \, f(x) \qquad \text{for} \quad x \in G. \tag{3.58}$$

Define linear operator $H_x \, f : Q \longrightarrow Q$ by

$$(H_x \, f)X = (L'_x)_e^{-1} \, grad(\tilde{X} \, f)(x) \quad \text{for} \quad X \in Q. \tag{3.59}$$

Then, $H(\cdot)$ defines a mapping from G to $L(Q)$ such that

$$d \, F_x = H_x \, f. \tag{3.60}$$

It follows from (3.60) that sequence generated by Algorithm 3.1.1 for f coincides with the one generated by (NM) for F defined by (3.58). Let $x_0 \in G$ be such that $(H_{x_0} \, f)^{-1}$ exists and set

$$\eta_f = \| \, (H_{x_0} \, f)^{-1} \, (L'_{x_0})_e^{-1} \, grad \, f(x_0) \, \| \, .$$

With the above notation the previous main results can be rewritten as follows:

Theorem 3.1.4. *Suppose $h_3 = \ell_3 \, \eta_f \leq 1/2$; $(H_{x_0} \, f)^{-1} \, (H_{(\cdot)} \, f)$ satisfies L-Lipschitz condition and L_0-center Lipschitz condition at $x_0 \in G$ on $U(x_0, \varrho_3)$. Then, sequence generated by Algorithm 3.1.1, starting at x_0 is well defined and converges to a critical point $x^\star$ of f with $grad \, f(x^\star) = 0$. Moreover, if $H_{x_0} \, f$ is positive definite,*

$$\| \, (H_{x_0} \, f)^{-1} \, \| \, \| \, H_{x \cdot exp \, z} \, f - H_x \, f \, \| \leq L \, \| \, z \, \|$$

and

$$\| \, (H_{x_0} \, f)^{-1} \, \| \, \| \, H_{x_0 \cdot exp \, z} \, f - H_{x_0} \, f \, \| \leq L_0 \, \| \, z \, \|$$

for all $x \in G$ and $z \in Q$ with $m(x_0, x) + \| \, z \, \| < \varrho_3$, then $x^\star$ is a local solution of problem (3.54).

Corollary 3.1.9. *Let G be an Abelian group. Let $x^\star \in G$ be a local optimal solution of problem (3.54) such that $(H_{x^\star} \, f)^{-1}$ exists. Suppose $(H_{x^\star} \, f)^{-1} \, (H_{(\cdot)} \, f)$ satisfies L-Lipschitz condition and $L_\star$-center Lipschitz condition at $x^\star$ on $U(x^\star, 1/L_\star)$. Let also $x_0 \in U(x^\star, \alpha/4 \, L)$. Then, the sequence generated by Algorithm 3.1.1 starting at x_0 is well defined, remains in $\overline{U}(x^\star, \alpha/4 \, L)$ for all $n \geq 0$ and converges to some critical point $y^\star$ of f such $m(x^\star, y^\star) < 1/L_\star$.*

Remark 3.1.5. Similarly Corollaries 3.1.2–3.1.4, 3.1.6–3.1.8 can also be phrased in this setting.

Next, we provide numerical example, where tighter analysis than Ref. [425] is given.

Example 3.1.1. Let $d\nu_x : Q^2 \longrightarrow Q$ for a $\mathcal{C}^2$-map $\nu : G \longrightarrow Q$. Consider the second differential equation (see Ref. [425])

$$d^2\,\nu_x\,u_1\,u_2 = \left(\frac{\partial^2}{\partial t_2 \partial t_2}\nu(x \cdot exp\,t_2\,u_2 \cdot exp\,t_1\,u_1)\right)_{t_2=t_1=0}. \tag{3.61}$$

We can write that

$$d\nu_{x \cdot exp\,t\,u} - d\nu_x = \int_0^T d^2\nu_{x \cdot exp\,s\,u}\,u\,ds \quad \text{for all} \quad u \in Q \text{ and } t \in \mathbb{R}. \tag{3.62}$$

Let $N \in \mathbb{N}^\star$ and let $e = \mathcal{I}_N$ the identity matrix of order N. Consider the linear group $G = SL(N, \mathbb{R}) = \{x \in \mathbb{R}^{N \times N} : det\,x = 1\}$ endowed with standard matrix multiplication. Then G is a connected Lie group and Q is defined by $Q = T_e\,G = sl(N, \mathbb{R}) = \{x \in \mathbb{R}^{N \times N} : tr\,x = 0\}$. Q is endowed with the inner product $< x, y > = tr\,(x^T\,y)$ for each x and y in Q. The corresponding norm is the Frobenius norm. $\mathbb{I}_Q$ denotes the identity on Q. The exponential map $exp : Q \longrightarrow G$ is given by $exp\,(u) = \sum_{n \in \mathbb{N}} u^n/n!$ for each $u \in Q$ and its inverse is $exp^{-1}\,(x) = \sum_{n \in \mathbb{N}^\star} (-1)^{n-1}\,(x - e)^n/n$ for each $x \in G$ with $\| x - e \| < 1$. Let $g : G \times \mathbb{R} \longrightarrow Q$ be a differentiable map and let x_0 be a random initial guess. Consider the initial value problem on G as follows (see Ref. [344])

$$\begin{cases} x' = x \cdot g(x) \\ x(0) = x_0 \end{cases} \tag{3.63}$$

where $g(x) = (\sin x)\,(2x - 5\,x^2) - ((\sin x)\,(2\,x - 5\,x^2))^T$ for each $x \in G$. The function sin is defined as

$$\sin x = \sum_{k \in \mathbb{N}^\star} (-1)^{k-1}\,\frac{x^{2\,k-1}}{(2\,k-1)!} \quad \text{for each} \quad x \in G.$$

Then, we can write g as

$$g(x) = 2\sum_{k \in \mathbb{N}^\star} (-1)^{k-1}\frac{x^{2k} - (x^{2\,k})^T}{(2k-1)!} - 5\sum_{k \in \mathbb{N}^\star} (-1)^{k-1}\frac{x^{2k+1} - (x^{2k+1})^T}{(2k-1)!} \quad \text{for all } x \in G. \tag{3.64}$$

Let Δ be the size of the Euler discretization method of (3.63). Then, (3.63) leads to the fixed-point problem

$$x = x_0 \cdot exp\,(\Delta\,g(x)). \tag{3.65}$$

Let $F : G \longrightarrow Q$ defined by

$$F(x) = exp^{-1}\,(x_0^{-1} \cdot x) - \Delta\,g(x) \quad \text{for each} \quad x \in G.$$

Then the resolution of (3.65) is equivalent to finding a zero of F. Consider the special case $\Delta = 1$ and $x_0 = exp\,v_0$ with $v_0 \in Q$ satisfying $\| v_0 \| \leq 1/11$. Let

$\varphi(\cdot) = exp^{-1}(x_0^{-1} \cdot (\cdot))$ and $x = exp\, v_1\, exp\, v_2 \cdots exp\, v_k$ for some $v_i \in Q$ $(1 \le i \le k)$ with $\sum_{i=0}^{k} \| v_i \| < 1/4$. By Refs. [344, 425], we have that

$$\| x_0^{-1} x - e \| \le \frac{5}{4} \sum_{i=0}^{k} \| v_i \| < \frac{5}{16} < 1. \tag{3.66}$$

By the definition of exp^{-1} and (3.66), we have that

$$d\varphi_x(u) = \sum_{j \in \mathbb{N}} (-1)^{j-1} \frac{\sum_{i=0}^{j-1} (x_0^{-1} x - e)^i \, x_0^{-1} \, x\, u (x_0^{-1} x - e)^{j-1-i}}{j} \qquad \text{for each} \quad u \in Q \tag{3.67}$$

and

$$\| d\varphi_x - \mathbb{I}_Q \| \le \frac{2 \| x_0^{-1} x - e \|}{1 - \| x_0^{-1} x - e \|} \le \frac{10 \sum_{i=0}^{k} \| v_i \|}{4 - 5 \sum_{i=0}^{k} \| v_i \|}. \tag{3.68}$$

We also have that

$$\| d^2\varphi_x \| \le \frac{3 + \| x_0^{-1} x - e \|}{(1 - \| x_0^{-1} x - e \|)^2} \le \frac{3 + \frac{5}{4} \sum_{i=0}^{k} \| v_i \|}{(1 - \frac{5}{4} \sum_{i=0}^{k} \| v_i \|)^2}. \tag{3.69}$$

Let $v \in Q$ and $x \in U(x_0, 1/50)$. Then there exist $k \ge 1$, $v_1, \cdots, v_k \in Q$ such that $x = exp\, v_1\, exp\, v_2 \cdots exp\, v_k$ and $\| v \| + \sum_{i=1}^{k} \| v_i \| < 1/50$. Let $s \in [0,1]$ and $y = x \cdot exp\, s\, v$. By (3.68) and since $\| v_0 \| < 1/10$, we have that

$$\| d\varphi_{x_0} - \mathbb{I}_Q \| \le \frac{10 \| v_0 \|}{4 - 5 \| v_0 \|} \le \frac{2}{7}. \tag{3.70}$$

By (3.69) and since $\sum_{i=0}^{k} \| v_i \| + \| v \| < (1/10) + \sum_{i=1}^{k} \| v_i \| + \| v \| < 3/25$, we have that

$$\| d^2\varphi_y \| \le \frac{3 + \frac{3}{20}}{(1 - \frac{5}{4}(\frac{1}{10} + \sum_{i=1}^{k} \| v_i \| + s \| v \|))^2} \le \frac{144}{35\,(1 - \frac{10}{7}(\sum_{i=1}^{k} \| v_i \| + s \| v \|))^2}. \tag{3.71}$$

We also have that $dg_{x_0} = (-6\cos 1 - 16\sin 1)\,\mathbb{I}_Q$. Hence, we get that

$$dg_{x_0}^{-1} = \frac{1}{-6\cos 1 - 16\sin 1}\,\mathbb{I}_Q \quad \text{and} \quad (\mathbb{I}_Q - dg_{x_0})^{-1} = \frac{1}{1 + 6\cos 1 + 16\sin 1}\,\mathbb{I}_Q. \tag{3.72}$$

By (3.70), we obtain that

$$\| (\mathbb{I}_Q - dg_{x_0})^{-1} \| \, \| \, dF_{x_0} - (\mathbb{I}_Q - dg_{x_0}) \| \leq \frac{2}{7}\,\frac{1}{1 + 6\cos 1 + 16\sin 1} < 1.$$

By the Banach lemma on invertible operators, we have that

$$\| dF_{x_0}^{-1} \| \leq \frac{\dfrac{1}{1 + 6\cos 1 + 16\sin 1}}{1 - \left(\dfrac{1}{1 + 6\cos 1 + 16\sin 1}\right)\dfrac{2}{7}} \leq \frac{1}{10}. \tag{3.73}$$

By (3.64), we have that

$$\| d^2 g_y \| \leq 2 \sum_{k \in \mathbb{N}^\star} 2\,\frac{(2k)^2\,\| y^{2k} \|}{(2k-1)!} + 5 \sum_{k \in \mathbb{N}^\star} 2\,\frac{(2k+1)^2\,\| y^{2k+1} \|}{(2k-1)!}. \tag{3.74}$$

We also have that $\| y^i \| \leq 5\,i/4\,\left(\sum_{m=1}^{k} \| v_m \| + s\,\| v \|\right) + 1$ for each $i \geq 1$. By (3.74), we deduce that

$$\| d^2 g_y \| \leq \frac{5}{4}\left(\sum_{i=1}^{k} \| v_i \| + s\,\| v \|\right) \sum_{j \in \mathbb{N}^\star} 2\,\frac{4(2j)^3 + 10(2j+1)^3}{(2j-1)!} + \\ \sum_{j \in \mathbb{N}^\star} 2\,\frac{4(2j)^2 + 10(2j+1)^2}{(2j-1)!}. \tag{3.75}$$

Using elementary analysis and (3.75), we obtain that

$$\| d^2 g_y \| \leq 1320\,\mathbf{e}\left(\sum_{i=1}^{k} \| v_i \| + s\,\| v \|\right) + 136\,\mathbf{e} \tag{3.76}$$

where $\mathbf{e}$ is the exponential number. Combining (3.71) and (3.73), we obtain the following estimate

$$\| dF_{x_0}^{-1} d^2 F_y \| \leq \| dF_{x_0}^{-1} \| \, \| d^2 \varphi_y \| + \| dF_{x_0}^{-1} d^2 g_y \|$$
$$\leq \frac{72}{175\left(1 - \dfrac{10}{7}\left(\sum_{i=1}^{k} \| v_i \| + s\,\| v \|\right)\right)^2} + \frac{\mathbf{e}}{10}\left(1320\left(\sum_{i=1}^{k} \| v_i \| + s\,\| v \|\right) + 136\right). \tag{3.77}$$

Since $y = x \cdot \exp s\,v$ and by (3.62), we can write

$$\| dF_{x_0}^{-1}(dF_{x \cdot \exp v} - dF_x) \| \leq \int_0^1 \| dF_{x_0}^{-1} d^2 F_{x \cdot \exp s\,v} \| \, \| v \| \, ds \leq$$

$$\int_0^1 \left(\frac{72}{175\left(1 - \dfrac{10}{7}\left(\sum_{i=1}^{k} \| v_i \| + s\,\| v \|\right)\right)^2} \right. \\ \left. + \frac{\mathbf{e}}{10}\left(1320\left(\sum_{i=1}^{k} \| v_i \| + s\,\| v \|\right) + 136\right) \right) \| v \| \, ds \leq 44.58088305\,\| v \| \leq 51\,\| v \|. \tag{3.78}$$

We conclude that $dF_{x_0}^{-1} \, dF$ is L-Lipschitz at x_0 on $U(x_0, 1/50)$ with $L = 51$. Since $r_0 < 1/L < 1/50$, then, $dF_{x_0}^{-1} \, dF$ is L-Lipschitz at x_0 on $U(x_0, r_0)$. Note that $F(x_0) = -v_0$. Using (3.73), we have that

$$h_0 = L\eta = L \parallel dF_{x_0}^{-1} F(x_0) \parallel \leq L \parallel dF_{x_0}^{-1} \parallel \parallel v \parallel \leq \frac{51}{10} \parallel v_0 \parallel < \frac{1}{2}.$$

We conclude by Theorem 3.1.1 that (NM) starting at x_0 converges to a zero $x^\star$ of F. However, if $\parallel v_0 \parallel < 100/1015$, then, all the computations remain the same. But we can only deduce that

$$h_0 = L\eta \leq \frac{51}{10} \frac{100}{1015} = .5024630542 > \frac{1}{2}.$$

Hence, since the Kantorovich hypothesis (3.2) is not satisfied, Theorem 3.1 in [p. 328] of Ref. [308] cannot guarantee the convergence of (NM) to $x^\star$. If we choose v, v_1 such that $\parallel v \parallel + \parallel v_1 \parallel \leq 1/51$ and $\parallel v \parallel + \sum_{i=1}^{k} \parallel v_i \parallel < 1/50$, then we get that

$$\parallel dF_{x_0}^{-1} (dF_{x_0 \cdot exp \, v} - dF_{x_0}) \parallel \leq \int_0^1 \parallel dF_{x_0}^{-1} \, d^2 F_{x_0 \cdot exp \, s \, v} \parallel \parallel v \parallel \, ds$$

$$\int_0^1 \left(\frac{72}{175 \left(1 - \frac{10}{7}(\parallel v_1 \parallel + s \parallel v \parallel)\right)^2} + \frac{e}{10} \left(1320 \left(\parallel v_1 \parallel + s \parallel v \parallel\right) + 136\right) \right) \parallel v \parallel \, ds$$

$$\leq 44.43966955 \parallel v \parallel .$$

Hence, we can choose $K = 44.43966955$. Moreover, according to the discussion above Lemma 3.1.2, we can certainly choose $L_0 = 50 < L = 51$. Our strongest condition (3.6) is now satisfied, since

$$h_1 = \ell_1 \eta \leq \frac{50 + 51}{2} \frac{1}{10} \frac{100}{1015} = .497536946 < \frac{1}{2}.$$

Hence, our Theorem 3.1.1 guarantees the convergence of (NM) starting at x_0 to a zero $x^\star$ of F in $U(x_0, s^\star)$, where $s^\star$ is given in (3.9). Finally, note that if $\parallel v_0 \parallel < 1/11$, then clearly our Theorem 3.1.1 provides tighter error estimates and a more precise information on the location of the solution.

3.2   Hilbert Space

We provide semi-local convergence results for continuous modified Newton-type methods to solve nonlinear operator equations in a real Hilbert space setting. Using a combination of Lipschitz and center Lipschitz continuous conditions, we provide a finer convergence analysis than before under weaker conditions and computational cost (see Refs. [4, 7, 373, 375–377, 371]). This way we expand the applicability of Newton-type continuous methods under the same computational cost as before. We are concerned with the problem of approximating a solution of equation (1.1), where F is defined on a closed subset D of a real Hilbert space X with values in a real Hilbert space Y.

Newton-type methods (NTM) are undoubtedly the most popular methods for generating a sequence approximating a solution of equation (1.1). Recent results on local as well as semi-local convergence results for (NTM) on a Banach space setting can be found in Refs. [98, 100, 110] and the references therein. We first consider the continuous analog of a modified Newton's method (MNM)

$$\dot{x}(t) = -[F'(x_0)]^{-1}F(x(t)), \quad x(0) = x_0 \in X \tag{3.79}$$

along the lines of the elegant works by Airapetyan in Ref. [4], Ram, Smirnora and Favini in Refs. [5, 7, 373, 375–377]. We associate (3.79) with the autonomous dynamical system

$$\dot{x}(t) = \Phi(x(t)), \quad 0 \le t < +\infty, \quad x(0) = x_0, \tag{3.80}$$

where x_0 is an initial approximation to $x^\star$ and $x(t)$ is the trajectory.

Then, we improve the results of Ref. [377], by simply showing that we can replace the Lipschitz condition

$$\|F'(x) - F'(y)\| \le M\,\|x - y\|, \quad \forall x, y \in U(x_0, r) \subseteq D \tag{3.81}$$

by the actually needed center-Lipschitz condition

$$\|F'(x) - F'(x_0)\| \le M_0\|x - x_0\|, \forall x \in U(x_0, r) \subseteq D, \tag{3.82}$$

where r is a known radius. The results of Ref. [377] for the local case are improved following an analogous way. Note that if condition (3.81) holds, then M_0 exists,

$$M_0 \le M. \tag{3.83}$$

The importance of (NTM) for solving well and ill posed problems and a short history of their continuous analogs can be found in Ref. [377].

We need a slightly modified existence and uniqueness result for the semi-local convergence of (MNM) (see [p. 40] in Ref. [377]).

Lemma 3.2.1. *Let X, Y be real Hilbert spaces, D be a closed subset of X, $F : D \to Y$ and $\Phi : D \to Y$. Assume there exist $c_1 > 0$ and $c_2 > 0$ such that for*

$$r = \frac{c_2\|F(x_0)\|}{c_1}, \tag{3.84}$$

operators F and Φ are Fréchet differentiable on $U(x_0, r)$; For all $y \in U(x_0, r)$, the following conditions hold:

$$(F'(y)\Phi(y), F(y)) \le -c_1\|F(y)\|^2, \tag{3.85}$$

$$\|\Phi(y)\| \le c_2\|F(y)\| \tag{3.86}$$

and

$$U(x_0, r) \subseteq D. \tag{3.87}$$

Then, there exists a global solution $x = x(t)$ to system (3.80) in $U(x_0, r)$, such that

$$\lim_{t \to +\infty} x(t) = x^\star, \tag{3.88}$$

where $x^\star$ is a solution of equation $F(x) = 0$ in $U(x_0, r)$. Moreover, the following estimates hold:

$$\|x(t) - x^\star\| \le re^{-c_1 t}, \tag{3.89}$$

and

$$\|F(x(t))\| \le \|F(x_0)\|e^{-c_1 t}. \tag{3.90}$$

Note that if $D = X = Y$, Lemma 3.2.1 reduces to the corresponding one in [p. 40] of Ref. [377].

We shall show the following semi-local convergence theorem for (MNM) which relates the asymptotic behavior of a solution $x(t)$ to (3.79) and solutions to (1.1).

Theorem 3.2.1. *Let X, Y be real Hilbert spaces, D be a closed subset of X, $F : D \to Y$. Assume Operator F is Fréchet-differentiable, its Fréchet derivative satisfies (3.82) on $U(x_0, r)$,*

$$r \in \{r_1, r_2\}, \tag{3.91}$$

where provided that

$$h_A = 4M_0 m_1^2 \|F(x_0)\| \le 1, \quad m_1 = \|F'(x_0)^{-1}\|, \tag{3.92}$$

$$r_1 = \frac{1 - \sqrt{1 - 4M_0 m_1^2 \|F(x_0)\|}}{2M_0 m_1}, \tag{3.93}$$

$$r_2 = \frac{1 + \sqrt{1 - 4M_0 m_1^2 \|F(x_0)\|}}{2M_0 m_1}, \tag{3.94}$$

(3.87) holds and $h_A \ne 0$, when $r = r_2$. Then, there exists a unique solution $x = x(t)$, $t \in [0, +\infty)$, to system (3.80) in $U(x_0, r)$ and

$$\lim_{t \to +\infty} x(t) = x^\star,$$

where $x^\star$ is a solution of equation $F(x) = 0$. Moreover, the following estimates hold:

$$\|x(t) - x^\star\| \le r e^{-c_1(r)\, t}, \quad c_1(r) = 1 - M_0\, m_1\, r \tag{3.95}$$

and

$$\|F(x(t))\| \le \|F(x_0)\| e^{-c_1(r)t}. \tag{3.96}$$

Proof.   As in Ref. [377], set

$$\Phi(x(t)) = -F'(x_0)^{-1} F(x(t)). \tag{3.97}$$

Using (3.82) we can obtain in turn for all $y \in U(x_0, r)$:

$$\begin{aligned}
(F'(y)\Phi(y), F(y)) &= -(F'(y)[F'(x_0)]^{-1} F(y), F(y)) \\
&= -\|F(y)\|^2 + (\{I - F'(y)[F'(x_0)]^{-1}\} F(y), F(y)) \\
&= -\|F(y)\|^2 + (\{F'(x_0) - F'(y)\} F'(x_0)^{-1} F(y), F(y)) \\
&\le -\|F(y)\|^2 + M_0 m_1 r \|F(y)\|^2 \\
&= -(1 - M_0 m_1 r)\|F(y)\|^2 = -c_1(r)\|F(y)\|^2,
\end{aligned} \tag{3.98}$$

and

$$\|\Phi(y)\| \le m_1 \|F(y)\|. \tag{3.99}$$

Choose

$$c_1 = c_1(r) \quad \text{and} \quad c_2 = m_1. \tag{3.100}$$

We have that

$$\frac{c_2 \|F(x_0)\|}{c_1} = \frac{m_1 \|F(x_0)\|}{c_1(r)} = r, \tag{3.101}$$

which holds true by (3.92) and the choices of r_1 and r_2. The results now follow by Lemma 3.2.1. That completes the proof of Theorem 3.2.1. $\square$

Remark 3.2.1.

(a) If $D = X = Y$ and $M_0 = M$, then Theorem 3.2.1 reduces to Theorem 2.3 in Ref. [377].

(b) The conditions used in Ref. [377] under (3.81) and $D = X = Y$ are

$$h_R = 4Mm_1^2\|F(x_0)\| \leq 1, \tag{3.102}$$

whereas the corresponding error estimates are

$$\|x(t) - x^\star\| \leq 2m_1\|F(x_0)\|e^{-t/2} \tag{3.103}$$

and

$$\|F(x(t))\| \leq \|F(x_0)\|e^{-t/2}. \tag{3.104}$$

By comparing (3.92) to (3.102), we see that

$$h_R \leq 1 \quad \Rightarrow \quad h_A \leq 1, \tag{3.105}$$

but not necessarily vice versa unless if $M_0 = M$. In view of (3.105), the applicability of (MNM) has been extended. Moreover, in view of (3.95), (3.96), (3.103) and (3.104), our error estimates are tighter and the information on the location of the solution at least as precise.

(c) In order for us to obtain a result for the continuous Newton method

$$\dot{x}(t) = -[F'(x(t))]^{-1}F(x(t)), \quad x(0) = x_0 \in X, \tag{3.106}$$

set $\Phi(y) = -[F'(y)]^{-1}F(y)$,

$$c_2 = \frac{m_1}{1 - m_1 r M_0} \quad \text{and} \quad c_1 = 1. \tag{3.107}$$

Indeed, in view of (3.82), we have

$$\|F'(x_0)^{-1}\|\|F'(x_0) - F'(x)\| \leq m_1 M_0\|x - x_0\| \leq m_1 M_0 r < 1. \tag{3.108}$$

It follows from (3.108) and the Banach lemma on invertible operators that $F'(x)^{-1}$ exists and

$$\|F'(x)^{-1}\| \leq \frac{m_1}{1 - m_1 M_0 r}, \tag{3.109}$$

from which the definition of c_2 follows.

(d) A simple iteration method is obtained for $\Phi(y) = -F(y)$. Simply, choose $F' \geq c_1 > 0$ and $c_2 = 1$.

(e) The gradient method is obtained for $\Phi(y) = -[F'(y)]^\star F(y)$. Choose, $c_2 = M_1 = \sup_{x \in U(x_0, r)} \|F'(x)\|$ and $c_1 = \mu_1^{-2}$. Here

$$\mu_1 = \sup_{x \in U(x_0, r)} \|F'(x)^{-1}\|.$$

(f) The continuous Gauss-Newton's method is obtained for

$$\Phi(y) = -[F'^\star(y)F'(y)]^{-1}F'^\star(y)F(y),$$

set $c_1 = 1$ and $c_2 = \mu_1^2 M_1$, where μ_1 and M_1 are the same as in (e) above.

(g) In order for us to make the comparison easier with the results in Ref. [377], we also provided the results in non-affine invariant form. However, we note that the results can be immediately obtained in affine invariant form if operator F is replaced by $F'(x_0)^{-1}F$. In this case one should set $m_1 = 1$.

The advantages of presenting results in affine instead of non-affine invariant form are well-known in the literatures (see e.g. Ref. [110]).

A problem is ill-posed when the Fréchet derivative is not boundedly invertible. It was then suggested in Ref. [377] the regularized version of (MNM):

$$\dot{x}(t) = -[F'(x_0) + \varepsilon(t)I]^{-1}(F(x(t)) + \varepsilon(t)(x(t) - x_0)), \quad x(0) = x_0 \in H, \quad 0 < \varepsilon(t), \tag{3.110}$$

where x_0 is chosen so that $(F'(x_0)y, y) \geq 0$ for all $y \in D$.

By simply exchanging hypothesis (3.82) in Theorem 3.1 of Ref. [377] by the analog of (3.81) for the local case, we arrive at:

Theorem 3.2.2. *Let F, D, X, Y be as in Theorem 3.2.1. Assume $x^\star$ is a solution of equation $F(x) = 0$; there exists a positive function $\varepsilon(t) \in C^1[0, +\infty)$ converging monotonically to zero as $t \to +\infty$, such that $\dfrac{\dot{\varepsilon}(t)}{\varepsilon(t)}$ is nondecreasing, and $\varepsilon(0) > |\dot{\varepsilon}(0)|$; operator F is Fréchet-differentiable and M_0-center-Lipschitz:*

$$\|F'(x) - F'(x^\star)\| \leq M_0\|x - x^\star\|,$$

with

$$F'(x) = F'(x_0)G(x_0, x), \quad G(x_0, x) \in L(X),$$

$$\|G(x_0, x) - I\| \leq C(G)\|x - x_0\|$$

for all $x, x_0 \in U(x^\star, \rho)$, $C(G) > 0$,

$$\rho = \frac{\varepsilon(0) - |\dot{\varepsilon}(0)|}{M_0 + C(G)\varepsilon(0)};$$

$F'(x_0)$ is non-negative definite:

$$(F'(x_0)y, y) \geq 0 \quad for \ all \ y \in X;$$

there exists $v \in X$ such that $x^\star - x_0 = F'(x_0)v$,

$$\varepsilon(0) - |\dot{\varepsilon}(0)| \geq [M_0 + C(G)\varepsilon(0)]\varepsilon(0)\sqrt{\frac{2\|v\|}{M_0}},$$

and

$$U(x_0, \rho) \subseteq D.$$

Then, there exists a unique solution $x = x(t)$ of (3.110) for all $t \in [0, +\infty)$ such that

$$\|x(t) - x^\star\| \leq \frac{\varepsilon(0) - |\dot{\varepsilon}(0)|}{\varepsilon(0)[M_0 + C(G)\varepsilon(0)]}\varepsilon(t).$$

Corollary 3.2.1. *Let the operator F in (1.1) have the form:*

$$F(x) = \Psi(x) - z.$$

Assume that Ψ is given and instead of z, we have a δ-approximation z_δ:

$$\|z - z_\delta\| \le \delta.$$

Then, the following estimates hold:

$$\frac{1}{2}\frac{d}{dt}\|x(t) - x^\star\|^2 \le -(1 - C(G)\rho)\|x(t) - x^\star\|^2 + (\varepsilon(t)\|v\| + \frac{\delta}{\varepsilon(t)})\|x(t) - x^\star\|$$
$$+ \frac{M_0}{2\varepsilon(t)}\|x(t) - x^\star\|^3,$$

$$\dot{f}(t) \le -(1 - C(G)\rho - \frac{|\dot{\varepsilon}(0)|}{\varepsilon(0)})f(t) + 2\|v\| + \frac{M_0}{2}f^2(t), \quad f(0) = \frac{\|x_0 - x^\star\|}{\varepsilon(0)}$$

and

$$\|x(\tau_\delta) - x^\star\| \le \frac{\rho}{\varepsilon(0)\|v\|^{\frac{1}{2}}}\delta^{\frac{1}{2}},$$

provided that hypotheses of Theorem 3.2.2 and

$$\varepsilon(0) - |\dot{\varepsilon}(0)| \ge 2[M_0 + C(G)\varepsilon(0)]\varepsilon(0)\sqrt{\frac{\|v\|}{M_0}}$$

hold.

Remark 3.2.2. If $D = X = Y$ and $M_0 = M$, then Theorem 3.2.2 and Corollary 3.2.1 reduce to the corresponding ones in Ref. [377]. Otherwise these results constitute an improvement with advantages as stated in the local case of the introduction of this section.

We provide two examples in a Hilbert space setting where Lipschitz condition (3.81) does not hold but center-Lipschitz condition (3.82) does.

Example 3.2.1. Let $X = Y = \mathbb{R}$. $\mathbb{R}$ is a Hilbert space with inner product defined by $< x, y >= xy$. In fact, we obtain $\|x\| =< x, x >^{\frac{1}{2}} = (x^2)^{\frac{1}{2}} = |x|$. The Euclidean metric is defined by $\|x - y\| =< x - y, x - y >^{\frac{1}{2}} = |x - y|$. Let $D = [0, +\infty)$, $x_0 = 1$ and define function F on D by

$$F(x) = \frac{x^{1+\frac{1}{i}}}{1 + \frac{1}{i}} + c_1 x + c_2, \tag{3.111}$$

where c_1, c_2 are given real numbers and $i > 2$ an integer. Then, using (3.111), we get $F'(x) = x^{\frac{1}{i}} + c_1$, which is not Lipschitz in any neighborhood of 0. That is operator F' is not Lipschitz on D, i.e. (3.81) is not satisfied. However, center-Lipschitz condition (3.82) holds for $M_0 = 1$. Indeed, we have in turn that

$$\|F'(x) - F'(x_0)\| = |x^{\frac{1}{i}} - x_0^{\frac{1}{i}}|$$
$$= \frac{|x - x_0|}{x_0^{\frac{i-1}{i}} + x_0^{\frac{i-2}{i}} x^{\frac{1}{i}} + \ldots + x_0^{\frac{1}{i}} x^{\frac{i-2}{i}} + x^{\frac{i-1}{i}}} \le \|x - x_0\|, \tag{3.112}$$

which shows (3.82).

Example 3.2.2. Let $X = H^1[0,1]$ and $Y = L^2(0,1)$. Consider an operator F defined on X by

$$F(x)(s) = \int_0^1 K(s,t)P(t,x(t))dt - v, \quad s \in [0,1],$$

where $K(s,t) \in L^\infty((0,1)^2)$ and $P(t,u)$ continuously differentiable with respect to u on $0 \le t,s \le 1$, $-\infty < u < +\infty$ and $v \in X$ is given. Then, we get

$$(F'(x)h)(s) = \int_0^1 K(s,t)P_x(t,x(t))h(t)dt. \tag{3.113}$$

Moreover, assume

$$\|P_x(t,x(t)) - P_y(t,y(t))\|_{L^2} \le \overline{M}\|x - y\|_{L^\infty} \quad \text{for all } x,y \in X. \tag{3.114}$$

Let $x_0 \in X$ be fixed. Then, it follows from (3.114) that there exists $\overline{M_0}$ such that $\overline{M_0} \in [0, \overline{M}]$ and

$$\|P_x(t,x(t)) - P_{x_0}(t,x_0(t))\|_{L^2} \le \overline{M_0}\|x - x_0\|_{L^\infty} \quad \text{for all } x \in X. \tag{3.115}$$

Using (3.113) and (3.114) we obtain in turn:

$$\|(F'(x) - F'(y))h\|_{L^2} = \{\int_0^1 [\int_0^1 K(t,\theta)P_x(t,x(\theta) - P_y(t,y(\theta))h(t)d\theta)]^2 dt\}^{\frac{1}{2}}$$
$$\le \overline{M}\|K\|_{L^\infty}\|x - y\|_{L^\infty}\|h\|_{L^2}.$$

Set $M = \overline{M}\|K\|_{L^\infty}$. Similarly, using (3.113) and (3.115) we get

$$\|(F'(x) - F'(x_0))h\|_{L^2} \le \overline{M_0}\|K\|_{L^\infty}\|x - x_0\|_{L^\infty}\|h\|_{L^2}.$$

Set $M_0 = \overline{M_0}\|K\|_{L^\infty}$. Note that the $L^\infty(0,1)$-norms of $x - y$, $x - x_0$ and h can be estimated by their X-norms times some constants, due to Sobolev's embedding theorems.

Conditions (3.81), (3.82) are now satisfied with the above choices of M and M_0. Clearly, one can now choose K, P, x_0 so that: (3.83) holds as a strict inequality, our condition (3.92) holds but the condition (3.102) given in Ref. [377] is violated.

These advantages extend in the case of the example given in Remark 3.3 of Ref. [377] related to problem (3.110). Indeed, on top of the above choices of M_0 and M, consider operator $T : H^1[0,1] \to H^1[0,1]$ and solve equation $T(x) = 0$. Clearly, $T'(x_0)$ is non-negative definite. Under the additional assumptions $|P_x(t,x_0)| \ge \gamma > 0$ for any $t \in (0,1)$ and $y(t,u) \in C^3((0,1) \times (-\infty,+\infty))$, we can set as in [377]:

$$(G(x_0,x)h)(t) = \frac{P_x(t,x(t))}{P_x(t,x_0(t))}h(t)$$

to obtain

$$(T'(x)h)(t) = (T'(x_0)G(x_0,x)h)(t).$$

Then, for any $h \in H$ we get in turn:

$$\|(G(x_0,x) - I)h\|_{L^2} \le \frac{\|P_{xx}\|_{L^\infty}}{\gamma}\|x - x_0\|_{L^\infty}\|h\|_{L^2},$$

and (see Refs. [286, 377])

$$\left\|\frac{d}{dt}(G(x_0,x) - I)h\right\|_{L^2} \le \frac{\|P_{xx}\|_{L^\infty}}{\gamma}[\|x' - x_0'\|_{L^2}\|h\|_{L^\infty} + \|x - x_0\|_{L^\infty}\|h'\|_{L^2}].$$

3.3 Convergence Structure

We present in this section a semi-local convergence theorem for Newton's method (NM) on spaces with a convergence structure. Using our new idea of recurrent functions, we provide a tighter analysis, with weaker hypotheses than Refs. [22, 106, 325, 327]. We are motivated by the elegant works by Meyer in Ref. [327], Ezquerro-Hernández in Ref. [205], Proinov in Refs. [368, 369], and optimization considerations (see, also Refs. [22, 106, 325, 326]). Meyer provided in Ref. [327] a general structure for the convergence analysis of (NM), which enables us to consider under the same framework, Kantorovich-type semi-local convergence results and theorems based on monotonicity considerations. Here, we use our new idea of recurrent functions to provide a tighter semi-local analysis of (NM) with weaker conditions and under the same computational cost as before in Refs. [22, 106, 205, 325, 327, 368, 369]. Other approaches on the Kantorovich-type semi-local convergence of (NM) can also be found in Refs. [100, 368, 369].

We introduce concepts that can be found in Refs. [22, 106, 325, 327]:

Definition 3.3.1. A Banach space $\mathcal{X}$ with convergence structure is a triple $(\mathcal{X}, \mathcal{V}, \mathcal{W})$ satisfying:

(1) $(\mathcal{X}, \| \cdot \|)$ is a real Banach space.
(2) $(\mathcal{V}, \mathbb{C}, \| \cdot \|_{\mathcal{V}})$ is a real Banach space which is partially ordered by the closed convex cone $\mathbb{C}$; the norm $\| \cdot \|_{\mathcal{V}}$ is assumed to be monotone on $\mathbb{C}$.
(3) $\mathcal{W}$ is a closed convex cone in $\mathcal{X} \times \mathcal{V}$ satisfying $\{0\} \times \mathbb{C} \subset \mathcal{W} \subset \mathcal{X} \times \mathbb{C}$.
(4) The following map $/./ : \mathcal{D} \longrightarrow \mathbb{C}$ is well defined:

$$/x/ := \inf \{p \in \mathbb{C} : (x, p) \in \mathcal{W}\} \tag{3.116}$$

where

$$\mathcal{D} := \{x \in \mathcal{X} : \exists\, p \in \mathbb{C},\, (x, p) \in \mathcal{W}\}.$$

(5) For every $x \in \mathcal{D}$, we have

$$\| x \| \leq \| /x/ \|_{\mathcal{V}} . \tag{3.117}$$

We use standard properties of partial orderings on $\mathcal{V}$ and $\mathcal{X} \times \mathcal{V}$ (see Ref. [338]). The set $\mathcal{D}$ satisfies $\mathcal{D} + \mathcal{D} \subset \mathcal{D}$ and for $s > 0$: $s\mathcal{D} \subset \mathcal{D}$, whereas:

$$U(a) := \{x \in \mathcal{X} : (x, a) \in \mathcal{W}\}$$

defines a generalized neighborhood of zero. Definition 3.3.1 is motivated by the following examples (see Refs. [22, 106, 327]):

Example 3.3.1. Let $\mathcal{X} = \mathbb{R}^n$ equipped with the maximum-norm:

(a) $\mathcal{V} = \mathbb{R}$, $\mathcal{W} = \{(x, p) \in \mathbb{R}^{n+1} : \| x \|_\infty \leq p\}$. This case concerns classical convergence analysis in a Banach space.

(b) $\mathcal{V} = \mathbb{R}^n$, $\mathcal{W} = \{(x,p) \in \mathbb{R}^{2n} : |x| \leq p\}$, i.e., if $x = (x_i)_{i=1}^n$ and $p = (p_i)_{i=1}^n$, then $|x| \leq p \Longleftrightarrow x_i \leq p_i$, for all $1 \leq i \leq n$. This case concerns componentwise analysis and error estimates.

(c) $\mathcal{V} = \mathbb{R}^n$, $\mathcal{W} = \{(x,p) \in \mathbb{R}^{2n} : 0 \leq x \leq p\}$. This case is used in monotone convergence analysis.

Remark 3.3.1. The convergence analysis is based on monotonocity considerations in $\mathcal{X} \times \mathcal{V}$. Let $(x_n, p_n) \in \mathcal{W}^N$ be an increasing sequence, then:

$$(x_n, p_n) \leq (x_{n+m}, p_{n+m}) \Longrightarrow 0 \leq (x_{n+m} - x_m, p_{n+m} - p_m).$$

If $p_n \longrightarrow p$, we obtain $0 \leq (x_{m+n} - x_n, p - p_n)$. Using (3.117) of Definition 3.3.1, we have

$$\| x_{n+m} - x_n \| \leq \| p - p_n \|_\mathcal{V} \xrightarrow[n \to \infty]{} 0.$$

Then $\{x_n\}$ is a Cauchy sequence. When deriving error estimates we shall use sequence $p_n = a_0 - a_n$ with a decreasing sequence $\{x_n\} \in \mathbb{C}$ to obtain estimate:

$$0 \leq (x_{m+n} - x_n, a_n - a_{n+m}) \leq (x_{m+n} - x_n, a_n).$$

If $x_n \longrightarrow x$ this implies the estimate $/x - x_n/ \leq a_n$.

Definition 3.3.2. We denote the space of multilinear, symmetric, bounded operators $A : \mathcal{X}^n \longrightarrow \mathcal{X}$ on a Banach space $\mathcal{X}$ by $\mathcal{L}(\mathcal{X}^n)$ and for an ordered Banach space $\mathcal{V}$:

$$\mathcal{L}_+(\mathcal{V}^n) = \{L \in \mathcal{L}(\mathcal{V}^n) : 0 \leq x_i \, (1 \leq i \leq n) \Longrightarrow 0 \leq L(x_1, \cdots, x_n)\}.$$

A map $L \in \mathbb{C}^1(\mathcal{V}_L \longrightarrow V)$ on a open subset V_L of an ordered Banach space $\mathcal{V}$ is defined to be order convex on an interval $[a, b] \subset V_L$, if

$$c, d \in [a, b], \ c \leq d \Longrightarrow L'(d) - L'(c) \in \mathcal{L}_+(\mathcal{V}).$$

Definition 3.3.3. As the set of bounds for an operator $A \in \mathcal{L}(\mathcal{X}^n)$, we define:

$$B(A) = \{L \in \mathcal{L}_+(\mathcal{V}^n) : (x_i, p_i) \in \mathcal{W} \Longrightarrow (A(x_1, \cdots, x_n), L(p_1, \cdots, p_n)) \in \mathcal{W}\}.$$

Lemma 3.3.1. *(see Ref. [325]) Let $A : [0, 1] \longrightarrow \mathcal{L}(\mathcal{X}^n)$ and $L : [0, 1] \longrightarrow \mathcal{L}_+(\mathcal{V}^n)$ be continuous maps, then:*

$$\forall t \in [0, 1] : L(t) \in B(A(t)) \Longrightarrow \int_0^1 L(t) \, dt \in B\left(\int_0^1 A(t) \, dt\right)$$

which will be used for the remainder of Taylor's formula.

Finally the following conventions are needed: Let $T : \mathcal{Y} \longrightarrow \mathcal{Y}$ be a map on a subset $\mathcal{Y}$ of a normed space. Then $T^n(x)$ denotes the result of n-fold application of T and in case of convergence:

$$T^\infty(x) = \lim_{n \longrightarrow \infty} T^n(x).$$

In particular, we define a right inverse through:

Definition 3.3.4. Let $A \in \mathcal{L}(\mathcal{X})$ and $y \in \mathcal{X}$ be given. Then we can write

$$A^{\star} y := z \iff z \in T^{\infty}(0), \ T(x) := (\mathcal{I} - A)\,x + y \iff z = \sum_{j=0}^{\infty} (\mathcal{I} - A)^j \, y$$

provided this limit exists.

We provide sufficient semi-local convergence conditions for (NM) to determine a zero $x^{\star}$ of operator G on Banach space. Our results are stated for the operator

$$F(x) = A\,G(x_0 + x), \tag{3.118}$$

where x_0 is the initial guess for (NM) and A is an approximation of $G'(x_0)^{-1}$. That is the following result is affine invariant in the sense of Ref. [195].

Theorem 3.3.1. *Let $\mathcal{X}$ be a Banach space with a convergence structure $(\mathcal{X}, \mathcal{V}, \mathcal{W})$ with $\mathcal{V} = (V, \mathbb{C}, \| \, . \, \|_V)$, an operator $F \in \mathbb{C}^1(\mathcal{X}_F \longrightarrow \mathcal{X})$ with $\mathcal{X}_F \subseteq \mathcal{X}$, an operator $L \in \mathbb{C}^1(\mathcal{V}_L \longrightarrow \mathcal{V})$ with $\mathcal{V}_L \subseteq \mathcal{V}$, an operator $L_0 \in \mathbb{C}^1(\mathcal{V}_{L_0} \longrightarrow \mathcal{V})$ with $\mathcal{V}_L \subseteq \mathcal{V}_{L_0}$ and a point $a \in \mathbb{C}$ such that the following hypotheses are satisfied:*

$$U(a) \subseteq \mathcal{X}_F \quad and \quad [0, a] \subseteq \mathcal{V}_L; \tag{3.119}$$

L is order-convex on $[0, a]$, satisfying

$$L'(|x| + |y|) - L'(|x|) \in B(F'(x) - F'(x + y)) \tag{3.120}$$

for all x, $y \in U(a)$ with $|x| + |y| \le a$; L_0 is order-convex on $[0, a]$, satisfying $L_0' \le L'$ on $\mathcal{V}_L$ and

$$L_0'(|x|) - L_0'(0) \in B(F'(0) - F'(x)) \tag{3.121}$$

for all $x \in U(a)$ with $|x| \le a$;

$$L_0'(0) \in B(I - F'(0)) \quad and \quad (-F(0), L(0)) \in \mathcal{W}; \tag{3.122}$$

$$L(a) \le a \tag{3.123}$$

and

$$L'(a)^n \, a \longrightarrow 0 \quad as \quad n \longrightarrow \infty. \tag{3.124}$$

Then, sequence $\{x_n\}$ generated by (NM):

$$x_0 = 0, \quad x_{n+1} = x_n + F'(x_n)^{\star} \, (-F(x_n)) \tag{3.125}$$

is well defined, remains in $U(a)$ for all $n \ge 0$, and converges to the unique zero $x^{\star}$ of operator F in $U(a)$. Moreover, the following estimates hold true for all $n \ge 0$:

$$/x_{n+1} - x_n / \le d_{n+1} - d_n, \tag{3.126}$$

$$/x_{n+1} - x^{\star}/ \le b - d_n, \tag{3.127}$$

where

$$b = L^{\infty}(0) \tag{3.128}$$

is the minimal fixed point of operator L in $[0, a]$ and sequence $\{d_n\}$ is given by

$$d_0 = 0, \quad d_{n+1} = L(d_n) + L_0'(|x_n|)\, c_n, \quad c_n = /x_{n+1} - x_n/. \tag{3.129}$$

Furthermore, sequence $(x_n, d_n) \in (\mathcal{X} \times \mathcal{V})^N$ is well defined, remains in $\mathcal{W}^N$ and is monotone.

Remark 3.3.2.

(a) If $L_0' = L'$ on $\mathcal{V}_L$, then hypotheses of Theorem 3.3.1 reduce to the ones in [Theorem 5] in Ref. [327]. However, note that in general

$$L_0' \leq L' \tag{3.130}$$

holds. Condition (3.121) is not an additional hypothesis. In practice, computing operator L' also requires determining L_0'. Moreover, operator L_0' always exists (if L' exists). The benefits out of introducing condition (3.121) are given in Remark 3.3.3.

(b) Assume conditions (3.119)–(3.121) of Theorem 3.3.1 hold and

$$\exists t \in (0,1) \ : \ L(a) \leq t\, a.$$

Then, there exists $a' \in [0, t\, a]$ satisfying all conditions of Theorem 3.3.1. The zero $z \in U(a')$ is unique in $U(a)$ (see Ref. [327]).

Let $L \in \mathbb{C}^1(\mathcal{V}_L \to \mathcal{V})$ be a map satisfying the conditions of Theorem 3.3.1. Then L is monotone. Therefore, sequences $b_n = L^n(0)$ and $c_n = L^n(a)$ are monotone with

$$0 \leq b_n \leq b_{n+1} \leq c_{n+1} \leq c_n \leq a.$$

In view of (3.124), we conclude that the sequence $\{c_n - b_n\}$ converges to zero (see Ref. [325]). That is, we obtain $b = L^\infty(0)$ is well defined and is the smallest solution of $L(p) \leq p$ in $[0, a]$.

In case of Remark 3.3.2(b), we obtain that

$$0 \leq c_n - b_n \leq t^n\, a.$$

That is $b = L^\infty(0)$ is well defined and the following inequalities hold:

$$L'(b)\,(a - b) \leq L(a) - L(b) \leq t\,(a - b)$$

$$b = L(b) \leq L(a) \leq t\, a \implies b \leq \frac{t}{1-t}\,(a-b),$$

so,

$$L'(b)^n\, b \leq \frac{t}{1-t}\, L'(b)^n\,(a-b) \leq \frac{t^{n+1}}{1-t}\,(a-b) \longrightarrow 0.$$

Therefore, $a' = b$ satisfies the additional hypothesis of Remark 3.3.2(b).

As a special case, we obtain the following result for affine maps:

Corollary 3.3.1. *(see Ref. [327]) Let $L \in L_+(\mathcal{V})$ and $a, p \in \mathbb{C}$ be given such that:*

$$L\, p + a \leq p \quad and \quad L^n\, p \longrightarrow 0.$$

Then, the map

$$(I - L)^\star \ : \ [0, a] \longrightarrow [0, a]$$

is well defined and continuous.

As a substitute for the Banach Lemma we use:

Lemma 3.3.2. *(see Ref. [327]) Let $A \in L(\mathcal{X})$, $L \in B(A)$, $y \in \mathcal{D}$ and $p \in \mathbb{C}$ be given as in Theorem 3.3.1 such that:*

$$L\,p + /y/ \leq p \quad and \quad L^n p \longrightarrow 0.$$

Then, $x = (I - A)^\star y$ is well defined, $x \in \mathcal{D}$ and

$$/x/ \leq (I - L)^\star /y/ \leq p.$$

The proof of Lemma 3.3.2 is easy. Simply note that the sequence $\{b_n\}$ defined by

$$b_0 = 0, \quad b_{n+1} = L\,b_n + /y/ \leq p$$

is well defined and converges to

$$b = (I - L)^\star /y/ \leq p.$$

If we consider $x_{n+1} = A\,x_n + y$, $x_0 = 0$, then the sequence (x_n, b_n) is monotone in $\mathcal{X} \times \mathcal{V}$. Hence, the statement follows from the general principles seen earlier in this section.

Proof. [of Theorem 3.3.1] Conditions of Theorem 3.3.1 are satisfied for b replacing a. We shall show that for each n, there exists x_{n+1} solving

$$p = (I - F'(x_n))\,p + (-F(x_n)). \tag{3.131}$$

Let $n = 1$ and $p = b$. Using (3.120)–(3.123) we get:

$$\begin{aligned} |I - F'(0)|\,b + |-F(0)| &\leq L_0'(0)\,b + |-F(0)| \leq L'(0)\,b + |-F(0)| \\ &\leq L(b) - L(0) + L(0) = L(b) = b. \end{aligned} \tag{3.132}$$

Hence, x_1 is well defined and $(x_1, b) \in \mathcal{W}$. We also have that

$$x_1 = (I - F'(0))\,x_1 + (-F(0)). \tag{3.133}$$

So,

$$|x_1| \leq L_0'(0)\,|x_1| + L(0) = d_1 \tag{3.134}$$

and consequently,

$$d_1 = L_0'(0)\,|x_1| + L(0) \leq L'(0)\,b + L(0) \leq L(b) - L(0) + L(0) = L(b) = b. \tag{3.135}$$

Let us assume that (x_k, d_k) is well defined and monotone for all $k \leq n$, with

$$0 \leq (x_{k-1}, d_{k-1}) \leq (x_k, d_k), \quad d_k \leq b. \tag{3.136}$$

Using (3.121) and (3.122), we have that

$$\begin{aligned} |I - F'(x_k)| &\leq |(I - F'(0)) + (F'(0) - F'(x_k))| \leq |I - F'(0)| + |F'(0) - F'(x_k)| \\ &\leq L_0'(0) + L_0'(|x_k|) - L_0'(0) = L_0'(|x_k|), \end{aligned} \tag{3.137}$$

so,

$$L_0'(|x_k|) \in B(I - F'(x_k)). \tag{3.138}$$

In view of Lemma 3.3.2, we must solve for p:

$$L_0'(|x_k|)\, p + |-F(x_k)| \leq p. \tag{3.139}$$

We need an estimate on $|-F(x_k)|$. By Taylor's theorem, (3.120) and Lemma 3.3.2, we obtain in turn that

$$\begin{aligned}
|-F(x_k)| &= |-F(x_k) + F(x_{k-1}) + F'(x_{k-1})(x_k - x_{k-1})| \\
&\leq \int_0^1 [L'(|x_{k-1}| + t\, c_{k-1}) - L'(|x_{k-1}|)]\, c_{k-1}\ dt \\
&= L(|x_{k-1}| + c_{k-1}) - L(|x_{k-1}|) - L'(|x_{k-1}|)\ c_{k-1} \\
&\leq L(d_{k-1} + d_k - d_{k-1}) - L(d_{k-1}) - L'(|x_{k-1}|)\ c_{k-1} \\
&\leq L(d_k) - L(d_{k-1}) - L_0'(|x_{k-1}|)\ c_{k-1} = L(d_k) - d_k.
\end{aligned} \tag{3.140}$$

Let $p = b - d_k$. Then, we have by (3.137), and (3.140):

$$\begin{aligned}
L_0'(|x_k|)\, p + |-F(x_k)| + d_k &\leq L'(d_k)\,(b - d_k) + L(d_k) \\
&\leq L(b) - L(d_k) + L(d_k) = L(b) = b.
\end{aligned} \tag{3.141}$$

Hence, x_{k+1} is well defined by Lemma 3.3.2 and $c_k \leq b - d_k$. Therefore, d_{k+1} is well defined too and we obtain that

$$\begin{aligned}
d_{k+1} &\leq L(d_k) + L_0'(d_k)\,(b - d_k) \leq L(d_k) + L'(d_k)\,(b - d_k) \\
&\leq L(d_k) + L(b) - L(d_k) = L(b) = b.
\end{aligned} \tag{3.142}$$

In view of the estimate

$$c_k + d_k \leq L_0'(|x_k|)\, c_k + |-F(x_k)| + d_k \leq L_0'(|x_k|)\, c_k + L(d_k) = d_{k+1}, \tag{3.143}$$

we deduce the monotonicity

$$(x_k, d_k) \leq (x_{k+1}, d_{k+1}), \tag{3.144}$$

which also implies (3.126). It follows inductively from (3.129) that:

$$L^k(0) \leq d_k \leq b, \tag{3.145}$$

which together with $L^k(0) \longrightarrow b$, imply $d_k \longrightarrow b$ as $k \longrightarrow \infty$. Sequence $\{x_n\}$ converges to some $x^\star \in U(b)$. By setting $k \longrightarrow \infty$ in (3.140), we deduce that $x^\star$ is a zero of operator F. Estimate (3.127) now follows from (3.126) by using standard majorization techniques.

The uniqueness statement is given in Ref. [325], where the modified Newton method $x_{n+1} = x_n - F(x_n)$ is considered. Clearly, sequence $(x_n, L^n(0))$ is monotone in $\mathcal{X} \times \mathcal{V}$. Moreover, if there exists a zero $y^\star \in U(a)$ of F, then, it was shown in Ref. [325] that

$$|y^\star - x_n| \leq L^n(a) - L^n(0) \longrightarrow 0 \quad \text{as} \quad n \longrightarrow \infty, \tag{3.146}$$

which implies $\lim\limits_{n \to \infty} x_n = y^\star$. But, we showed $\lim\limits_{n \to \infty} x_n = x^\star$. Hence, we deduce $x^\star = y^\star$. That completes the proof of Theorem 3.3.1. $\qquad\square$

Remark 3.3.3. If equality holds in (3.130), then our Theorem 3.3.1 reduces to [Theorem 5] in Ref. [327]. Otherwise (i.e., if $L_0' < L'$), the former Theorem improves the latter. Indeed, the majorizing sequence $\{q_n\}$ used in Ref. [327] is given by

$$q_0 = 0, \quad q_{n+1} = L(q_n) + L'(|x_n|)\, c_n. \tag{3.147}$$

In view of (3.129) and (3.147), a simple inductive argument shows

$$d_n < q_n \quad \text{and} \quad d_{n+1} - d_n < q_{n+1} - q_n, \quad (n \geq 1).$$

Hence, sequence $\{d_n\}$ is a tighter majorizing sequence for $\{x_n\}$ than $\{q_n\}$. As already noted in Remark 3.3.2, these advantages are obtained under the same hypotheses and computational cost as in Ref. [327].

By simply replacing L' by L_0' in the definitions of the operators involved, we can also improve the a posteriori estimates given in Ref. [327] under the hypotheses of Theorem 3.3.1. More, precisely in order for us to obtain a posteriori estimates, we define:

$$R_n\,(p) = (I - L_0'\,(|x_n|))^\star\, S_n\,(p) + c_n \tag{3.148}$$

where

$$S_n\,(p) = L\,(|x_n| + p) - L\,(|x_n|) - L_0'\,(|x_n|)\, p. \tag{3.149}$$

Operator S_n is monotone on the interval $I_n = [0, a - |x_n|]$; Moreover, if there exists $p_n \in \mathbb{C}$ such that $|x_n| + p_n \leq a$, and

$$S_n\,(p_n) + L_0'\,(|x_n|)\,(p_n - c_n) \leq p_n - c_n, \tag{3.150}$$

then operator $R_n : [0, p_n] \longrightarrow [0, p_n]$ is well defined by Corollary 3.3.1 and monotone.

We then have that

$$\begin{aligned} d_n + c_n \leq d_{n+1} &\Rightarrow L\,(a) - L\,(d_n) - L_0'\,(|x_n|)\; c_n \leq a - d_n - c_n \\ &\Rightarrow S_n\,(a - d_n) + L_0'\,(|x_n|)\,(a - d_n - c_n) \leq a - d_n - c_n, \end{aligned} \tag{3.151}$$

which implies $a - d_n$ is a suitable choice for p_n.

Other ways for choosing suitable p_n are given by the following:

Proposition 3.3.1. *Assume:*

$$R_n(p) \leq p \quad for\ some \quad p \in I_n. \tag{3.152}$$

Then we have that

$$c_n \leq R_n(p) = \overline{p} \leq p \tag{3.153}$$

and

$$R_{n+1}(\overline{p} - c_n) \leq \overline{p} - c_n. \tag{3.154}$$

Proof.   Using (3.140), the order convexity of L_0, L and the estimate

$$S_n\,(p) + L_0'\,(|x_n|)\,(\overline{p} - c_n) = \overline{p} - c_n, \tag{3.155}$$

we get that

$$S_{n+1}\,(\overline{p} - c_n) + |-F(x_{n+1})| + L_0'\,(|x_{n+1}|)\,(\overline{p} - c_n) \leq \overline{p} - c_n. \tag{3.156}$$

That completes the proof of Proposition 3.3.1. $\qquad\square$

Proposition 3.3.2. *Assume that conditions of Theorem 3.3.1 hold and there exists a solution $p_n \in I_n$ satisfying*

$$R_n(p) \leq p. \tag{3.157}$$

Define sequence

$$a_n = p_n, \quad a_{m+1} = R_m\left(a_m\right) - c_m \qquad (m \geq n). \tag{3.158}$$

Then, the following a posteriori estimate holds

$$/x^\star - x_m/ \leq a_m. \tag{3.159}$$

Proof. An induction argument shows $R_m\left(a_m\right) \leq a_m$, so, $a_{m+1} + c_m \leq a_m$ and consequently, we deduce the monotonicity of $(x_m, a_n - a_m)$ in $\mathcal{X} \times \mathcal{V}$. That completes the proof of Proposition 3.3.2. $\qquad\qquad\square$

The properties of R_n imply the existence of $R_n^\infty(0)$, which is a suitable choice for p_n in Proposition 3.3.2. Hence we arrive at:

Corollary 3.3.2. *Assume that conditions of Theorem 3.3.1 hold and there exists $p \in I_n$ satisfying $R_n(p) \leq p$. Then the following a posterioiri estimates holds*

$$|x^* - x_n| \leq R_n^\infty(0) \leq p. \tag{3.160}$$

As already noted in Ref. [327], in view of the estimate

$$S_n(p) + / - F(x_n)/ + L_0'(|x_n|)\, p \leq p \Longrightarrow R_n(p) \leq p, \tag{3.161}$$

one may consider further majorization:

$$Q_n(p) = L(|x_{n-1}| + c_{n-1} + p) - L(|x_{n-1}|) - L_0'(|x_{n-1}|)\, c_{n-1}. \tag{3.162}$$

Note that an application to a two-point boundary value problem was given in Ref. [327] in the case $L_0' = L_0$. Further discussion on applications and practical aspects can be found in Refs. [22, 106, 325, 326, 422].

Remark 3.3.4. If $L_0' = L'$, then our a posteriori estimates reduce to the ones in Ref. [327]. However, if $L_0' < L'$, then our a posteriori estimates are tighter. So far, we showed how to improve on the error estimates given in Ref. [327]. We are now wondering if we can also weaken the sufficient convergence conditions (3.123) and (3.124). It turns out that this can be done, using our new idea of recurrent functions (see Ref. [106]).

We need to define some operator sequences.

Definition 3.3.5. Let $\eta \in \mathbb{C}$. Define operators:

$$f_n, h_n, \beta_n : [0,1) \longrightarrow \mathcal{X}$$

and

$$\delta : I_\delta = \left[1, \frac{1}{1-\gamma}\right] \times [0,1)^4 \longrightarrow \mathcal{X}, \quad \gamma \in [0,1)$$

by

$$f_n(\gamma) = \left\{ \int_0^1 \left(L'\left(\left(\frac{1-\gamma^{n-1}}{1-\gamma} + t\,\gamma^{n-1}\right)\eta\right) - L'\left(\frac{1-\gamma^{n-1}}{1-\gamma}\,\eta\right)\right) dt + \gamma\, L_0'\left(\frac{1-\gamma^n}{1-\gamma}\,\eta\right)\right\} - \gamma,$$

$$(3.163)$$

$$h_n(\gamma) = \int_0^1 \left(L'\left(\left(\frac{1-\gamma^n}{1-\gamma} + t\,\gamma^n\right)\eta\right) - L'\left(\left(\frac{1-\gamma^{n-1}}{1-\gamma} + t\,\gamma^{n-1}\right)\eta\right)\right) dt +$$
$$\left(L'\left(\frac{1-\gamma^{n-1}}{1-\gamma}\,\eta\right) - L'\left(\frac{1-\gamma^n}{1-\gamma}\,\eta\right)\right) +$$
$$\gamma\left(L_0'\left(\frac{1-\gamma^{n+1}}{1-\gamma}\,\eta\right) - L_0'\left(\frac{1-\gamma^n}{1-\gamma}\,\eta\right)\right),$$

$$(3.164)$$

$$\overline{\beta_n}(\gamma) = \int_0^1 \left(L'\left(\left(\frac{1-\gamma^{n+1}}{1-\gamma} + t\,\gamma^{n+1}\right)\eta\right) + L'\left(\left(\frac{1-\gamma^{n-1}}{1-\gamma} + t\,\gamma^{n-1}\right)\eta\right) - 2\,L'\left(\left(\frac{1-\gamma^n}{1-\gamma} + t\,\gamma^n\right)\eta\right)\right) dt +$$
$$\left(2\,L'\left(\frac{1-\gamma^n}{1-\gamma}\,\eta\right) - L'\left(\frac{1-\gamma^{n-1}}{1-\gamma}\,\eta\right) - L'\left(\frac{1-\gamma^{n+1}}{1-\gamma}\,\eta\right)\right) +$$
$$\gamma\left(L_0'\left(\frac{1-\gamma^{n+2}}{1-\gamma}\,\eta\right) + L_0'\left(\frac{1-\gamma^n}{1-\gamma}\,\eta\right) - 2\,L_0'\left(\frac{1-\gamma^{n+1}}{1-\gamma}\,\eta\right)\right),$$

$$\beta_n(\gamma) = \overline{\beta_n}(\gamma), \qquad (3.165)$$

$$\overline{\delta}(v_1, v_2, v_3, v_4, \gamma) = \int_0^1 \left(L'((v_1 + v_2 + v_3 + t\,v_4)\,\eta) + L'((v_1 + t\,v_2)\,\eta) - 2\,L'((v_1 + v_2 + t\,v_3)\,\eta)\right) dt +$$
$$\left(2\,L'((v_1 + v_2)\,\eta) - L'(v_1\,\eta) - L'((v_1 + v_2 + v_3)\,\eta)\right) +$$
$$\gamma\left(L_0'((v_1 + v_2 + v_3 + v_4)\,\eta) + L_0'((v_1 + v_2)\,\eta) - 2\,L_0'((v_1 + v_2 + v_3)\,\eta)\right),$$

$$\delta(v_1, v_2, v_3, v_4, \gamma) = \overline{\delta}(v_1, v_2, v_3, v_4, \gamma). \qquad (3.166)$$

Moreover, define function $f_\infty : [0,1) \longrightarrow \mathcal{X}$ by

$$f_\infty(\gamma) = \lim_{n \longrightarrow \infty} f_n(\gamma). \qquad (3.167)$$

It then follows from (3.163), and (3.167) that

$$f_\infty(\gamma) = b\left(L'\left(\frac{\eta}{1-\gamma}\right) + \gamma\, L_0'\left(\frac{\eta}{1-\gamma}\right)\right) - \gamma. \qquad (3.168)$$

It can also easily be seen from (3.163)–(3.166) that the following identities hold:

$$f_{n+1}(\gamma) = f_n(\gamma) + h_n(\gamma), \tag{3.169}$$

$$h_{n+1}(\gamma) = h_n(\gamma) + \beta_n(\gamma), \tag{3.170}$$

and for

$$v_1 = \sum_{i=0}^{n-2} \gamma^i, \quad v_2 = \gamma^{n-1}, \quad v_3 = \gamma^n, \quad v_4 = \gamma^{n+1}, \tag{3.171}$$

we have that

$$\delta(v_1, v_2, v_3, v_4, \gamma) = \beta_n(\gamma). \tag{3.172}$$

Finally, let us define sequence $\{t_n\}$ by

$$t_0 = 0, \quad t_1 = L_0(0) + L_0'(0)\,\eta,$$

$$t_{n+1} = t_n + L_0'(t_n)(t_n - t_{n-1}) + \int_0^1 (L'(t_{n-1} + t(t_n - t_{n-1})) \tag{3.173}$$
$$-L'(t_{n-1}))dt(t_n - t_{n-1}).$$

We need the following result on majorizing sequences for (NM).

Lemma 3.3.3. *Assume that there exist* η, $a \in \mathbb{C}$ *and* $\alpha \in (0,1)$ *such that*

$$\frac{\eta}{1-\alpha} \in [0,a]; \tag{3.174}$$

$$0 \le L_0'(t_1) + \int_0^1 (L'(t\,t_1) - L'(0))\,dt \le \alpha\,I; \tag{3.175}$$

$$\delta(v_1, v_2, v_3, v_4, \gamma) \ge 0 \quad \text{on} \quad I_\delta, \tag{3.176}$$

$$h_1(\alpha) \ge 0 \tag{3.177}$$

and

$$f_\infty(\alpha) \le 0, \tag{3.178}$$

where 0, I *are the zero endomorphism and the identity opertor on* $\mathcal{X}$, *respectively.*
Then iteration $\{t_n\}$ $(n \ge 0)$ *given by (3.173) is non-decreasing, bounded from above*
by

$$t^{\star\star} = \frac{\eta}{1-\alpha} \tag{3.179}$$

and converges to its unique least upper bound $t^\star$ *satisfying*

$$t^\star \in \langle 0, t^{\star\star}\rangle. \tag{3.180}$$

Moreover the following error bounds hold for all $n \ge 0$:

$$0 \le t_{n+1} - t_n \le \alpha\,(t_n - t_{n-1}) \le \alpha^n\,\eta \tag{3.181}$$

and

$$t^\star - t_n \le \frac{\eta}{1-\alpha}\,\alpha^n. \tag{3.182}$$

Proof. Estimate (3.181) is true, if

$$0 \leq L_0'(t_n) + \int_0^1 \left(L'(t_{n-1} + t\,(t_n - t_{n-1})) - L(t_{n-1}) \right) dt \leq \alpha\, I \qquad (3.183)$$

holds for all $n \geq 1$. In view of (3.175) and (3.179), estimate (3.183) holds for $n = 1$. We also have by (3.173) and (3.183) that

$$0 \leq t_2 - t_1 \leq \alpha\,(t_1 - t_0).$$

Let us assume that (3.181) and (3.183) hold for all $k \leq n$. Then, we have

$$t_n \leq \frac{1 - \alpha^n}{1 - \alpha}\,\eta. \qquad (3.184)$$

Moreover, (3.181) and (3.183) shall hold if

$$\left\{ \int_0^1 \left(L'\left(\left(\frac{1 - \alpha^{n-1}}{1 - \alpha} + t\,\alpha^{n-1}\right) \eta \right) - L'\left(\frac{1 - \alpha^{n-1}}{1 - \alpha}\,\eta\right) \right) dt + \right.$$
$$\left. \alpha\, L_0'\left(\frac{1 - \alpha^n}{1 - \alpha}\,\eta\right) \right\} - \alpha \leq 0. \qquad (3.185)$$

Estimate (3.185) motivates us to define functions f_n (for $\gamma = \alpha$) and show instead

$$f_n(\alpha) \leq 0. \qquad (3.186)$$

We have by (3.169)–(3.172), (3.176) and (3.177) that

$$f_{n+1}(\alpha) \geq f_n(\alpha). \qquad (3.187)$$

In view of (3.167) and (3.187), estimate (3.186) shall hold, if (3.178) is true. The induction is complete.

It follows that iteration $\{t_n\}$ is non-decreasing, bounded from above by $t^{\star\star}$ (given by (3.179)) and as such it converges to $t^\star$ satisfying (3.180). Finally, estimate (3.182) follows from (3.181) by using standard majorizing techniques. That completes the proof of Lemma 3.3.3. $\qquad\square$

We can show the following semi-local convergence result for (NM).

Theorem 3.3.2. *Let $\mathcal{X}$ be a Banach space with a convergence structure $(\mathcal{X}, \mathcal{V}, \mathcal{W})$ with $\mathcal{V} = (V, \mathbb{C}, \|\,.\,\|_V)$, an operator $F \in \mathbb{C}^1(\mathcal{X}_F \longrightarrow \mathcal{X})$ with $\mathcal{X}_F \subseteq \mathcal{X}$, an operator $L \in \mathbb{C}^1(\mathcal{V}_L \longrightarrow \mathcal{V})$ with $\mathcal{V}_L \subseteq \mathcal{V}$, an operator $L_0 \in \mathbb{C}^1(\mathcal{V}_{L_0} \longrightarrow \mathcal{V})$ with $\mathcal{V}_L \subseteq \mathcal{V}_{L_0}$ and a point $a \in \mathbb{C}$ such that the following hypotheses are satisfied: (3.119)–(3.122);*

$$L(\eta) \leq \eta, \qquad \eta = L_0^\infty(0), \quad \eta \leq a; \qquad (3.188)$$

$$L'(\eta)^n\,\eta \longrightarrow 0 \quad \text{as} \quad n \longrightarrow \infty \qquad (3.189)$$

and hypotheses of Lemma 3.3.3 hold for $|x_1| \leq \eta$, where x_1 solves

$$p = (I - F'(0))\,p + (-F(0)). \qquad (3.190)$$

Then, sequence $\{x_n\}$ generated by (NM) is well defined, remains in $U(t^\star)$ for all $n \geq 0$, and converges to a zero $x^\star$ of operator F in $U(t^\star)$. Moreover, the following estimates hold true for all $n \geq 0$:

$$/x_{n+1} - x_n/ \leq t_{n+1} - t_n, \qquad (3.191)$$

$$/x_{n+1} - x^\star/ \leq t^\star - t_n. \qquad (3.192)$$

Furthermore, sequence $(x_n, t_n) \in (\mathcal{X} \times \mathcal{V})^N$ is well defined, remains in $\mathcal{W}^N$ and is monotone.

Proof.   As in Theorem 3.3.1, we have b_0 is the smallest fixed point of operator L_0 in $[0, a]$ guaranteed to exist by (3.122), (3.188), (3.189), and Lemma 3.3.2 since

$$L_0'(0)\,\eta + |-F(0)| \le L_0(\eta) - L_0(0) + L_0(0) = L_0(\eta) = \eta. \tag{3.193}$$

Equation (3.190) is satisfied for $p = \eta$. Therefore, x_1 is well defined and $(x_1, \eta) \in \mathcal{W}$. We also have that

$$|x_1| \le L_0'(0)\,|x_1| + L(0) = L_0'(0)\,\eta + L_0(0) = t_1$$

and

$$t_1 \le L_0(\eta) - L_0(0) + L_0(0) = L_0(\eta) \le \eta \le t^\star.$$

We also get that

$$
\begin{aligned}
& L_0'(|x_1|)\,(t_1 - t_0) + |-F(x_1)| \\
& \le L_0'(t_1)\,(t_1 - t_0) + \int_0^1 (L'(|x_0| + t\,c_0) - L'(|x_0|))\,c_0\,dt \\
& = L_0'(t_1)\,(t_1 - t_0) + \int_0^1 (L'(t\,c_0) - L'(0))\,c_0\,dt = t_2 - t_1 \le \alpha\,(t_1 - t_0),
\end{aligned}
$$

which together with Lemma 3.3.2 implies x_2 is well defined and (3.191) holds for $n = 1$.

Let us assume (x_k, t_k) is well defined and monotone for all $k \le n$, i.e.,

$$0 \le (x_{k-1}, t_{k-1}) \le (x_k, t_k), \quad \text{and} \quad t_k \le t^\star, \quad k = 1, \cdots, n.$$

In view of (3.121), (3.123), (3.137), (3.140), the definition of sequence $\{t_n\}$, we have in turn for $p = t_k - t_{k-1}$:

$$
\begin{aligned}
& L_0'(|x_k|)\,(t_k - t_{k-1}) + |-F(x_k)| \\
& \le L_0'(t_k)\,(t_k - t_{k-1}) + \int_0^1 [L'(|x_{k-1}| + t\,c_{k-1}) - L'(|x_{k-1}|)]\,c_{k-1}\,dt \\
& \le L_0'(t_k)\,(t_k - t_{k-1}) + \int_0^1 [L'(|t_{k-1}| + t\,(t_k - t_{k-1})) - L'(|t_{k-1}|)]\,(t_k - t_{k-1})\,dt \\
& = t_{k+1} - t_k \le \alpha\,(t_k - t_{k-1}).
\end{aligned}
$$

$$\tag{3.194}$$

It follows from (3.194) and Lemma 3.3.2 that x_{k+1} is well defined, and (3.191) holds for all n. Sequence t_{k+1} is well defined too and bounded above by $t^\star$. According to what we saw earlier in this section, $\{x_n\}$ converges to some $x^\star \in U(t^\star)$. By setting $k \longrightarrow \infty$ in the upper bound of $|-F(x_k)|$ given in (3.194), we obtain that $x^\star$ is a zero of operator F. Estimate (3.192) follows from (3.191) as in Theorem 3.3.1. That completes the proof of Theorem 3.3.2. $\qquad\square$

Remark 3.3.5.

(a) Note that $t^{\star\star}$, given in closed form by (3.179), can replace $t^\star$ in Theorem 3.3.1.

(b) In view of the proof of Theorem 3.3.2, it follows that sequence $\{s_n\}$ given by

$$s_0 = 0, \quad s_1 = L_0(0) + L_0'(0)\,|x_1|,$$

$$s_{n+1} = s_n + L_0'(|x_n|)\,c_n + \int_0^1 \left(L'(|x_{n-1}| + t\,c_{n-1}) - L'(|x_{n-1}|)\right)c_{n-1}\,dt$$

is also a finer majorizing sequence for $\{x_n\}$ than t_n.

(c) **The monotone case.** This is a particular case of Theorem 3.3.1 (or Theorem 3.3.2) but is omitted here, since it follows along the lines of Theorem 13 in Ref. [327], where $\mathcal{X}$ is itself partially ordered and satisfies the conditions for $\mathcal{V}$ in Definition 3.3.1. We set $\mathcal{X} = \mathcal{V}$, $\mathcal{D} = \mathbb{C}^2$ and $/./ = I$ (see Case 3 in Example 3.3.1). The (NM) is given by

$$u_0 = u, \quad u_{n+1} = u_n + (A\,G'(u_n))^\star\,(-A\,G(u_n)),$$

$$G \in \mathbb{C}^1(\mathcal{V}_G \longrightarrow \mathcal{Y}), \quad A \in L(\mathcal{Y} \longrightarrow \mathcal{X}),$$

$$u, v \in \mathcal{V}, \quad L_0(p) = p - F(p), \quad [u, v] \subseteq \mathcal{V}_G, \quad \text{and} \quad a = v - u.$$

Application 3.3.1. Let $\mathcal{X}$ be a Banach space with real norm $\| \cdot \|$. We shall check the conditions of Theorem 3.3.1, and [Theorem 5] in Ref. [327]. Let us assume for simplicity that $F'(0) = I$ and that there exists a monotone operator $E : [0, a] \longrightarrow \mathbb{R}$, such that

$$\| F'(x_0)^{-1}\,(F'(x) - F'(y))\,\| \leq E(\| x - y \|)\ \| x - y \| \tag{3.195}$$

for all $x, y \in U(a)$. Define L by

$$L(p) = \eta + \int_0^p ds \int_0^s dt\,E(t), \quad \eta \geq \| F'(x_0)^{-1}\,F(x_0)\,\|\,. \tag{3.196}$$

We have to solve (3.123) for $E(t) \leq E(a) = \ell$, i.e.,

$$\eta + \frac{1}{2}\,\ell\,a^2 \leq a, \tag{3.197}$$

which is possible if

$$h_K = \ell\,\eta \leq \frac{1}{2}. \tag{3.198}$$

Condition (3.198) is famous for its simplicity, and clarify Kantorovich sufficient convergence hypothesis for (NM) (see Refs. [22, 134, 338]). In view of (3.195), there exists a monotone operator $E_0 : [0, a] \longrightarrow \mathbb{R}$, such that

$$\| F'(x_0)^{-1}\,(F'(x) - F'(x_0))\,\| \leq E_0(\| x - x_0 \|)\ \| x - x_0 \| \tag{3.199}$$

for all $x \in U(a)$. Define operator L_0 by

$$L_0(p) = \eta + \int_0^p ds \int_0^s dt\,E_0(t). \tag{3.200}$$

Set $E_0(a) = \ell_0$. Then it can easily be seen that hypotheses of Lemma 3.3.3 and Theorem 3.3.2 hold, if

$$h_{AH} = \bar{\ell}\,\eta \leq \frac{1}{2}, \tag{3.201}$$

where

$$\bar{\ell} = \frac{1}{8}\left(\ell + 4\,\ell_0 + \sqrt{\ell^2 + 8\,\ell_0\,\ell}\right) \tag{3.202}$$

and

$$\alpha = \frac{2\,\ell}{\ell + \sqrt{\ell^2 + 8\,\ell_0\,\ell}}. \tag{3.203}$$

Hence, in this special case, our Theorem 3.3.2 is weaker than [Theorem 5] in Ref. [327].

3.4   Riemannian Manifolds

We present in this section a semi-local convergence analysis of a bilinear operator free third order method on Riemannian manifolds. Using a combination of generalized Lipschitz and center-Lipschitz conditions, we provide a convergence analysis which expands the applicability of the method even in the setting of nonlinear equations considered in earlier studies such as Refs. [11, 57, 294].

In this section we present a semi-local convergence of a bilinear operator free third order iterative method. In 2006, Kou, Li and Wang in Ref. [294] presented a new modification of the Newton method cubically convergent for solving nonlinear equations $f(x) = 0$, where $f : \Omega \subseteq \mathbb{R} \longrightarrow \mathbb{R}$ with Ω an open interval (see Ref. [294]). The method is defined by

$$\begin{aligned}
y_n &= x_n + \frac{f(x_n)}{f'(x_n)} \\
x_{n+1} &= y_n - \frac{f(x_n)}{f'(x_n)} \quad \text{for each} \quad n = 0, 1, \cdots,
\end{aligned} \tag{3.204}$$

where $x_0 \in \Omega$ is an initial point. This method is preferred if the computational costs of the first derivative are greater than or equal to those of the function itself. Note also that (3.204) doesn't involve the second derivative. Argyros studied high order method on a Banach space setting where bilinear operator is replaced by suitable divided difference of order one (see, e.g. Ref. [57] and the references therein). In 2010, Amat, Bermúdez, Busquier and Plaza in Ref. [11] presented a generalization of (3.204) to Banach spaces (see Ref. [11]) defined by

$$\begin{aligned}
y_n &= x_n + DF(x_n)^{-1} F(x_n) \\
x_{n+1} &= y_n - DF(x_n)^{-1} F(x_n) \quad \text{for each} \quad n = 0, 1, \cdots,
\end{aligned}$$

where F is nonlinear operator defined on an open convex subset Ω of a Banach space E with values in a Banach space E_1. The semi-local convergence of this method was given in Ref. [11] under Kantorovich-type conditions. They studied actually the equivalent method defined by

$$x_{n+1} = x_n - DF\left(x_n\right)^{-1}\left[y_n, x_n; F\right] DF\left(x_n\right)^{-1} F\left(x_n\right) \quad \text{for each} \quad n = 0, 1, \cdots,$$

where the operator $[y_n, x_n; F]$ denotes a divided difference of first order and defined by

$$[y, x; F]\left(x - y\right) = F\left(x\right) - F\left(y\right) \quad \text{for each} \quad x, y \in \Omega \quad \text{with} \quad x \neq y.$$

Note that if F is Fréchet differentiable at $x \in \Omega$, then $DF\left(x\right) = [x, x; F]$.

In the last years, attention has been paid in studying iterative procedure on manifolds, since there are many numerical problems posed on manifolds that arise naturally in many contexts. Some examples include eigenvalue problems, minimization problems with orthogonality constraints, optimization problems with equality constraints, invariant subspace computations, see for instance (see Refs. [8, 105, 139, 196, 292, 309]). For these problems, one has to compute solutions of equations or to find zeros of a vector field on Riemannian manifolds. That is why the objective in this section is to extend this method in the context of Riemannian manifolds.

First, we define derivatives and divided difference on Riemannian manifolds. We refer the reader to (see Refs. [1, 8, 106, 169, 191, 196, 216, 298, 309, 410, 411]) and the references therein for more details.

Definition 3.4.1. A differentiable manifold of dimension m is a set M and a family of injective mappings $x_\alpha : U_\alpha \subset \mathbb{R}^m \longrightarrow M$ of open sets U_α of $\mathbb{R}^m$ into M such that

(i) $\cup_\alpha x_\alpha\left(U_\alpha\right) = M$.
(ii) for any α, β with $x_\alpha\left(U_\alpha\right) \cap x_\beta\left(U_\beta\right) = W \neq \phi$, $x_\alpha^{-1}(W)$ and $x_\beta^{-1}(W)$ are open sets in $\mathbb{R}^m$ and the mappings $x_\beta^{-1} \circ x_\alpha$ are differentiable.
(iii) The family $\{(U_\alpha, x_\alpha)\}$ is maximal relative to the conditions (i) and (ii).

The pair (U_α, x_α) (or the mapping x_α) with $p \in x_\alpha\left(U_\alpha\right)$ is called a parametrization (or system of coordinates) of M at p; $x_\alpha\left(U_\alpha\right)$ is then called neighborhood at p and $\left(x_\alpha\left(U_\alpha\right), x_\alpha^{-1}\right)$ is called a coordinate chart. A family $\{(U_\alpha, x_\alpha)\}$ satisfying (i) and (ii) is called a differentiable structure on M.

Let M denote a differentiable manifold, given $p \in M$ and T_pM denotes the tangent space at p to M, let $x : U \subset \mathbb{R}^m \longrightarrow M$ be a system of coordinates around p whit $x\left(x_1, x_2, ..., x_m\right) = p$. Let basis $\left\{ \left.\dfrac{\partial}{\partial x_1}\right|_p, \left.\dfrac{\partial}{\partial x_2}\right|_p, ..., \left.\dfrac{\partial}{\partial x_m}\right|_p \right\}$ be the associated basis of T_pM. The *tangent bundle* TM is defined as

$$TM = \{(p, v); \ p \in M \text{ and } v \in T_pM\} = \bigcup_{p \in M} T_pM$$

the set TM admits a differentiable structure of dimension $2m$ and the functions $X \in C^k\left(M, T_{(.)}M\right)$ are called *vector fields of class C^k* (see Ref. [196]). Next, we define the concept of Riemannian metric

Definition 3.4.2. A Riemannian metric on a differentiable manifold M is a correspondence which associates to each point p of M an inner product $\langle ., . \rangle_p$ (that is, a symmetric, bilinear, positive-definite form) on the tangent space T_pM, which varies differentiably in the following sense: $x : U \subset \mathbb{R}^m \longrightarrow M$ is a system of coordinates around p whit $x\left(x_1, x_2, \cdots, x_m\right) = p$, then

$$g_{ij}\left(x_1, x_2, \cdots, x_m\right) := \left\langle \left.\frac{\partial}{\partial x_i}\right|_p, \left.\frac{\partial}{\partial x_j}\right|_p \right\rangle_p = \left\langle dx^{-1}\left(\left.\frac{\partial}{\partial x_i}\right|_p\right), dx^{-1}\left(\left.\frac{\partial}{\partial x_j}\right|_p\right) \right\rangle,$$

in which dx^{-1} is the tangent map of x^{-1}, is a differentiable function on U for each $i, j = 1, 2, \cdots, n$. The functions g_{ij} are called *the local representatives of the Riemannian metric.*

This definition does not depend on the choice of a coordinate system (see Ref. [196]).

In the rest of the section we will always assume that M is equipped with a Riemannian metric g. The inner product $\langle ., . \rangle_p$ induces in a natural way the norm $||.||_p$. The subscript p is usually deleted if there is no confusion.

Let $\gamma : (-\epsilon, \epsilon) \longrightarrow M$ be a piecewise smooth curve. If we choose a parametrization $x : U \subset \mathbb{R}^m \longrightarrow M$, we can express the curve γ in this parametrization by

$$x^{-1} \circ \gamma\left(t\right) = \left(x_1\left(t\right), x_2\left(t\right), \cdots, x_n\left(t\right)\right).$$

Thus, the vector $\gamma'\left(t\right)$ can be expressed in the parametrization x by

$$\gamma'\left(t\right) = \sum_{i=1}^{n} x_i'\left(t\right) \left.\frac{\partial}{\partial x_i}\right|_{x^{-1}\circ\gamma(t)}.$$

If p and q are two points of a manifold M, let $\gamma : [0, 1] \longrightarrow M$ be a piecewise smooth curve connecting p and q. The *arc length of γ* is defined by

$$l\left(\gamma\right) = \int_0^1 ||\gamma'\left(t\right)|| \, dt = \int_0^1 \left\langle \frac{d\gamma}{dt}, \frac{d\gamma}{dt} \right\rangle^{1/2} dt \tag{3.205}$$

and the *Riemannian distance* from p to q (see Ref. [196, 411]) by

$$d\left(p, q\right) = \inf_{\gamma} l\left(\gamma\right). \tag{3.206}$$

Definition 3.4.3. Let $\chi\left(M\right)$ be the set of all vector fields of class C^∞ on M and $\mathcal{D}\left(M\right)$ the ring of real-valued functions of class C^∞ defined on M, that is:

$$\chi\left(M\right) = C^\infty\left(M, T_{(.)}M\right) \quad \text{and} \quad \mathcal{D}\left(M\right) = C^\infty\left(M, \mathbb{R}\right).$$

An *affine connection* ∇ on M is a mapping

$$\begin{aligned} \nabla : \chi\left(M\right) \times \chi\left(M\right) &\longrightarrow \chi\left(M\right) \\ \left(X, Y\right) &\longmapsto \nabla_X Y \end{aligned} \tag{3.207}$$

that satisfies the following properties

(i) $\nabla_{fX+gY} Z = f\nabla_X Z + g\nabla_Y Z$;

(ii) $\nabla_X (Y + Z) = \nabla_X Y + \nabla_X Z$;

(iii) $\nabla_X (fY) = f\nabla_X Y + X(f) Y$,

where $X, Y, Z \in \chi(M)$ and $f, g \in \mathcal{D}(M)$.

Definition 3.4.4. If X is a C^1 vector field on M, the *covariant derivative* of X determined by the connection ∇ defines on each $p \in M$ a linear application of $T_p M$ in itself

$$\begin{aligned}
\mathcal{D}X(p) : T_p M &\longrightarrow T_p M \\
v &\longmapsto \mathcal{D}X(p)(v) = \nabla_Y X(p)
\end{aligned} \tag{3.208}$$

where Y is a vector field satisfying $Y(p) = v$. The value $\mathcal{D}X(p)(v)$ depends only on the tangent vector $v = Y(p)$ since ∇ is linear in Y. In this way we can write

$$\mathcal{D}X(p)(v) = \nabla_v X(p).$$

Let us consider the curve $\gamma : [a, b] \longrightarrow M$ and vector field X along γ, i.e. $X(p) \in T_{\gamma(t)} M$, where $\gamma(t) = p$ for all t. We say that a vector field X is parallel along of γ (with respect to ∇) if $\mathcal{D}X(p)(\gamma'(t)) = 0$. The affine connection is *compatible* with the metric $\langle .,. \rangle$, when for any smooth curve γ and any pair of parallel vector fields P and P' along γ, we have that $\langle P, P' \rangle$ is constant; equivalently,

$$\frac{d}{dt} \langle X, Y \rangle = \langle \nabla_{\gamma'(t)} X, Y \rangle + \langle X, \nabla_{\gamma'(t)} Y \rangle,$$

where X and Y are vector fields along the differentiable curve $\gamma : I \longrightarrow M$ (see Ref. [196, 411]). We say that ∇ is *symmetric* if

$$\nabla_X Y - \nabla_Y X = [X, Y] \quad \text{for all } X, Y \in \chi(M).$$

Levi-Civita Theorem establishes that there exists a unique affine connection ∇ on M compatible with the metric and symmetric (see Ref. [411]). This connection is called the *Levi-Civita connection.*

Definition 3.4.5. A parametrized curve $\gamma : I \longrightarrow M$ is a geodesic at $t_0 \in I$ if $\nabla_{\gamma'(t)} \gamma'(t) = 0$ at the point t_0. If γ is a geodesic at t, for all $t \in I$, we say that γ is a geodesic. If $[a, b] \subseteq I$, the restriction of γ to $[a, b]$ is called a geodesic segment joining $\gamma(a)$ to $\gamma(b)$.

In some cases, by abuse of language, we refer to the image $\gamma(I)$, of a geodesic γ, as a geodesic. A basic property of a geodesic is that $\gamma'(t)$ is parallel along of $\gamma(t)$; this implies that $\|\gamma'(t)\|$ is constant. Let $B(p, r)$ and $B[p, r]$ denote respectively the *open geodesic* and the *closed geodesic* balls with center p and radius r, that is $B(p, r) = \{q \in M : d(p, q) < r\}$ and $B[p, r] = \{q \in M : d(p, q) \leq \}$, respectively. The Hopf-Rinnov theorem (see Ref. [196]), establishes that if M is a complete metric space then for every $p, q \in M$ there exists a geodesic γ, called *minimizing geodesic*, joining p to q with $l(\gamma) = d(p, q)$. Moreover, if $v \in T_p M$, there exists a unique locally minimizing geodesic γ such that $\gamma(0) = p$ and $\gamma'(0) = v$. The point $\gamma(1)$ is called the image of v by the *exponential map at p*, that is, there exists a well-defined map

$$\exp_p : T_p M \longrightarrow M \tag{3.209}$$

such that $\exp_p(v) = \gamma(1)$ and for any $t \in [0,1]$, $\gamma(t) = \exp_p(tv)$. The exponential map has many important properties (see Refs. [196, 410]). When the exponential map is defined for each value of the parameter $t \in \mathbb{R}$, we will say that the Riemannian manifold M is geodesically complete or, simply, complete. The Hopf-Rinnov theorem, also establishes that the property of the Riemannian manifold of being geodesically complete is equivalent to being complete as a metric space.

Definition 3.4.6. Let γ be a piecewise smooth curve. For any $a, b \in \mathbb{R}$, we define *the parallel transport along γ* which is denoted by P_γ as

$$\begin{aligned} P_{\gamma,a,b} : T_{\gamma(a)} M &\longrightarrow T_{\gamma(b)} M \\ v &\longmapsto V(\gamma(b)), \end{aligned} \tag{3.210}$$

where V is the unique vector field along γ such that $\nabla_{\gamma'(t)} V = 0$ and $V(\gamma(a)) = v$.

It is easy to show that $P_{\gamma,a,b}$ is linear and one-one, so that $P_{\gamma,b,a}$ is an isomorphism between the tangent spaces $T_{\gamma(a)} M$ and $T_{\gamma(b)} M$. Its inverse is the parallel transport along the reversed portion of γ from $V(\gamma(b))$ to $V(\gamma(a))$. Thus $P_{\gamma,a,b}$ is an isometry between $T_{\gamma(a)} M$ and $T_{\gamma(b)} M$.

Moreover, for a positive integer i and for all $(v_1, v_2, \cdots, v_i) \in T_{\gamma(b)} M \times T_{\gamma(b)} M \times \cdots \times T_{\gamma(b)} M$, we define P_γ^i as

$$P_{\gamma,a,b}^i : \underbrace{T_{\gamma(a)} M \times \cdots \times T_{\gamma(a)} M}_{i\text{-times}} \longrightarrow \underbrace{T_{\gamma(b)} M \times \cdots \times T_{\gamma(b)} M}_{i\text{-times}},$$

where

$$P_{\gamma,a,b}^i(v_1, v_2, \cdots, v_i) = (P_{\gamma,a,b}(v_1), P_{\gamma,a,b}(v_2), \cdots, P_{\gamma,a,b}(v_i)).$$

The parallel transport has the following important properties

$$P_{\gamma,a,b} \circ P_{\gamma,b,d} = P_{\gamma,a,d} \quad \text{and} \quad P_{\gamma,b,a}^{-1} = P_{\gamma,a,b}. \tag{3.211}$$

Next, we extend the concept of covariant derivative to higher order

$$\begin{aligned} \mathcal{D}X : C^k(TM) &\longrightarrow C^{k-1}(TM) \\ (v,.) &\longmapsto \mathcal{D}X(Y) = \nabla_Y X, \end{aligned} \tag{3.212}$$

where TM is the tangent bundle. Similar to the higher order Fréchet-derivative. We define the *higher order covariant derivatives* (see Refs. [298, 388]), as the multilinear map or j-tensor:

$$\mathcal{D}^j X : \underbrace{C^k(TM) \times C^k(TM) \times \cdots \times C^k(TM)}_{j\text{-times}} \longrightarrow C^{k-j}(TM)$$

given by

$$\mathcal{D}^j X \left(Y_1, Y_2, \cdots, Y_{j-1}, Y\right) = \nabla_Y \mathcal{D}^{j-1} \left(X \left(Y_1, Y_2, \cdots, Y_{j-1}\right)\right) \tag{3.213}$$

$$- \sum_{i=1}^{j-1} \mathcal{D}^{j-1} X \left(Y_1, Y_2, \cdots, \nabla_Y Y_i, \cdots, Y_{j-1}\right)$$

for each $Y_1, Y_2, \cdots, Y_{j-1} \in C^k \left(TM\right)$. In the case $j = 2$, we have

$$\mathcal{D}^2 X : C^k \left(TM\right) \times C^k \left(TM\right) \longrightarrow C^{k-2} \left(TM\right)$$

and

$$\mathcal{D}^2 X \left(Y_1, Y\right) = \nabla_Y \mathcal{D} X \left(Y_1\right) - \mathcal{D} X \left(\nabla_Y Y_1\right) = \nabla_Y \left(\nabla_{Y_1} X\right) - \nabla_{\nabla_Y Y_1} X. \tag{3.214}$$

The multilinearity refers to the structure of $C^k \left(M\right)$-module, where the value of $\mathcal{D}^j X \left(Y_1, Y_2, \cdots, Y_{j-1}, Y\right)$ at $p \in M$ only depends on the j-tuple of tangent vectors

$$\left(v_1, v_2, \cdots, v_j\right) = \left(Y_1 \left(p\right), Y_2 \left(p\right), \cdots, Y_{j-1} \left(p\right), Y \left(p\right)\right) \in \left(T_p M\right)^j.$$

Therefore, for any $p \in M$, we define the map

$$\mathcal{D}^j X \left(p\right) : \left(T_p M\right)^j \longrightarrow T_p M$$

by

$$\mathcal{D}^j X \left(p\right) \left(v_1, v_2, \cdots, v_j\right) = \mathcal{D}^j X \left(Y_1, Y_2, \cdots, Y_{j-1}, Y\right) \left(p\right). \tag{3.215}$$

Definition 3.4.7. Let M be a Riemannian manifold, $\Omega \subseteq M$ an open convex set and $X \in \chi \left(M\right)$. The covariant derivative $\mathcal{D} X = \nabla_{(.)} X$ is Lipschitz with constant $L > 0$, if for any geodesic γ and $a, b \in \mathbb{R}$ so that $\gamma \left[a, b\right] \subseteq \Omega$, it holds that:

$$\left\| P_{\gamma, b, a} \mathcal{D} X \left(\gamma \left(b\right)\right) P_{\gamma, a, b} - \mathcal{D} X \left(\gamma \left(a\right)\right) \right\| \leq L \int_a^b \left\| \gamma' \left(t\right) \right\| dt. \tag{3.216}$$

We will write $\mathcal{D} X \in Lip_L \left(\Omega\right)$, (see Refs. [8, 216]).

Note that $P_{\gamma, b, a} \mathcal{D} X \left(\gamma \left(b\right)\right) P_{\gamma, b, a}$ and $\mathcal{D} X \left(\gamma \left(a\right)\right)$ are both operators defined in the same tangent plane $T_{\gamma(a)} M$. If M is the Euclidean space, the above definition coincides with the usual Lipschitz definition for the operator $DF : M \longrightarrow M$.

Proposition 3.4.1. *Let γ be a curve in M and X a C^1 vector field on M, then the covariant derivative of X in the direction of $\gamma' \left(s\right)$ is*

$$\mathcal{D} X \left(\gamma \left(s\right)\right) \gamma' \left(s\right) = \nabla_{\gamma'(s)} X_{\gamma(s)} = \lim_{h \to 0} \frac{1}{h} \left(P_{\gamma, s+h, s} X \left(\gamma \left(s + h\right)\right) - X \left(\gamma \left(s\right)\right)\right). \tag{3.217}$$

Note that if $M = \mathbb{R}^n$, the previous proposition agrees with the definition of classic directional derivative in $\mathbb{R}^n$ (see Ref. [411]). Let us recall that if $A : T_p M \longrightarrow T_p M$ is linear, then $\|A\| = \sup \left\{\|Av\| : v \in T_p M, \|v\| = 1\right\}$. The following is an important lemma, that allows one to know when an operator is invertible and also allows to give an estimate for its inverse.

Lemma 3.4.1. *(Banach's lemma) Let A be an invertible bounded linear operator in a Banach space E and B a bounded linear operator B in E. If $\left\|A^{-1}B - I\right\| < 1$, then B^{-1} exists and the following estimates hold*

$$\left\|B^{-1}\right\| \le \frac{\left\|A^{-1}\right\|}{1 - \left\|A^{-1}B - I\right\|} \le \frac{\left\|A^{-1}\right\|}{1 - \left\|A^{-1}\right\|\left\|B - A\right\|}$$

and

$$\left\|B^{-1}A\right\| \le \frac{1}{1 - \left\|A^{-1}B - I\right\|} \le \frac{1}{1 - \left\|A^{-1}\right\|\left\|B - A\right\|}.$$

Next, we define divided difference in this new context.

Definition 3.4.8. Let M be a Riemannian manifold, $\Omega \subseteq M$ an open convex set. Asume that γ is a curve in M , $[a,b] \subset dom\,(\gamma)$ and $X : M \longrightarrow TM$ a C^0 vector field on M. We define the divided difference of first order for the vector field X on the points $\gamma\,(s)$, $\gamma\,(s+h)$ in the direction $\gamma'\,(s)$ by

$$[\gamma\,(s+h), \gamma\,(s)\,; X]\,\gamma'\,(s) = \frac{1}{h}\,(P_{\gamma,s+h,s}X\,(\gamma\,(s+h)) - X\,(\gamma\,(s)))\,. \tag{3.218}$$

Note that in the case that M be a Banach space, if γ is the geodesic joining x and y, such that $\gamma\,(s) = x + s\,(y - x)$, $t \in \mathbb{R}$, then (3.218) implies $[y, x; X]\,(y - x) = X\,(y) - X\,(x)$. This is the classic definition of divided difference of first order in Banach spaces (see Ref. [11]). Also if there exist $\mathcal{D}X\,(p)$ then $\mathcal{D}X\,(p) = [p, p; X]$.

Let us suppose that we have the method

$$\begin{aligned}
q_n &= \exp_{p_n}\left(\mathcal{D}X\,(p_n)^{-1}\,X\,(p_n)\right), \\
p_{n+1} &= \exp_{p_n}\left(\mathcal{D}X\,(p_n)^{-1}\,(X\,(p_n) - P_{\gamma,1,0}X\,(q_n))\right).
\end{aligned} \tag{3.219}$$

If

$$\gamma\,(s) = \exp_{p_n}\left(s\mathcal{D}X\,(p_n)^{-1}\,X\,(p_n)\right),$$

then, using the definition of divided difference of first order. It is clear that

$$[\gamma\,(1), \gamma\,(0)\,; X]\,\gamma'\,(0) = P_{\gamma,1,0}X\,(\gamma\,(1)) - X\,(\gamma\,(0))\,.$$

Thus,

$$[q_n, p_n; X]\,\mathcal{D}X\,(p_n)^{-1}\,X\,(p_n) = P_{\gamma,1,0}X\,(q_n) - X\,(p_n)\,.$$

Therefore, we get that

$$\mathcal{D}X\,(p_n)^{-1}\,(X\,(p_n) - P_{\gamma,1,0}X\,(q_n)) = -\mathcal{D}X\,(p_n)^{-1}\,[q_n, p_n; X]\,\mathcal{D}X\,(p_n)^{-1}\,X\,(p_n)\,.$$

Then (3.219) is transformed into

$$\begin{aligned}
q_n &= \exp_{p_n}\left(\mathcal{D}X\,(p_n)^{-1}\,X\,(p_n)\right) \\
p_{n+1} &= \exp_{p_n}\left(-\mathcal{D}X\,(p_n)^{-1}\,[q_n, p_n; X]\,\mathcal{D}X\,(p_n)^{-1}\,X\,(p_n)\right).
\end{aligned} \tag{3.220}$$

We will also use the following notations

$$\Gamma_n = \mathcal{D}X\left(p_n\right) \quad \text{and} \quad \Phi_n = \mathcal{D}X\left(p_n\right)\left[q_n, p_n; X\right]^{-1} \mathcal{D}X\left(p_n\right). \tag{3.221}$$

We present a semi-local convergence analysis of method (3.220) using the concept of generalized ω-condition.

Definition 3.4.9. We say that the divided difference of first order satisfies the ω-condition, if

$$\left\|\left[p_1, q_1; X\right] - P_{\gamma,1,0}\left[p_2, q_2; X\right]\right\| \leq \omega\left(d\left(p_1, p_2\right), d\left(q_1, q_2\right)\right); \quad p_1, p_2, q_1, q_2 \in \Omega, \tag{3.222}$$

where γ is a geodesic joining the points $\gamma\left(0\right) = q_1$, $\gamma\left(1\right) = q_2$ and $\omega : \mathbb{R}_+ \times \mathbb{R}_+ \to \mathbb{R}_+$ is a nondecreasing, continuous in both variables.

We suppose throughout this section that

$$\omega\left(0, t\right) = \omega\left(t, 0\right) = \frac{1}{2}\omega\left(t, t\right). \tag{3.223}$$

Theorem 3.4.1. *Let M be a complete Riemannian manifold, $\Omega \subseteq M$ be an open convex set and X a C^1 vector field on M. Suppose that the divided difference of first order satisfies the ω-condition and let $p_0 \in \Omega$. Moreover, suppose that*

(1) $\left\|\Gamma_0^{-1}\right\| \leq \beta.$
(2) $\max\left\{\left\|\Gamma_0^{-1}X\left(p_0\right)\right\|, \left\|\Phi_0^{-1}X\left(p_0\right)\right\|\right\} \leq \eta.$
(3) The function

$$f(t) = t\left(1 - \frac{m}{1 - 2\beta\omega\left(t, t\right)}\right) - \eta \tag{3.224}$$

has a minimal positive zero R, where $m = \beta\omega\left(\eta, \eta\right).$
(4) $\overline{B(p_0, R)} \subset \Omega$ *and*

$$\beta\omega\left(R, R\right) < \frac{1}{3}. \tag{3.225}$$

Then,

$$N = \frac{m}{1 - 2\beta\omega\left(R, R\right)} \in \left(0, 1\right),$$

method (3.220) is well defined, $\{p_n\}$ remains in $B(p_0, R)$ for each $n = 0, 1, \cdots$ and converges to the unique solution of $X\left(p\right) = 0$ in $B\left[p_0, R\right]$. Moreover, the following estimates hold for each $n = 1, 2, \cdots$

$$\tau_{n+1} - \tau_n \leq N\left(\tau_n - \tau_{n-1}\right), \tag{3.226}$$

$$\tau^\star = \lim_{n \to \infty} \tau_n \leq R \tag{3.227}$$

and

$$\tau^\star - \tau_n \leq \frac{N^n \eta}{1 - N}, \tag{3.228}$$

where

$$\tau_0 = 0, \ \tau_1 = \eta, \ \tau_{n+1} = \tau_n + N\left(\tau_n - \tau_{n-1}\right) \quad \text{for each} \quad n = 1, 2, \cdots. \tag{3.229}$$

Proof. First, we note that

$$1 - 2\beta\omega\left(R, R\right) \geq 1 - \frac{2}{3} = \frac{1}{3} \geq 0.$$

Thus, we get that

$$\frac{m}{1 - 2\beta\omega\left(R, R\right)} \geq 0. \tag{3.230}$$

On the other hand, as R is a zero of (3.224), we obtain that

$$\left(1 - \frac{m}{1 - 2\beta\omega\left(R, R\right)}\right) = \frac{\eta}{R}$$

and by (3.230)

$$\left(1 - \frac{m}{1 - 2\beta\omega\left(R, R\right)}\right) \leq 1.$$

Therefore, we have that $\eta/R \leq 1$ and consequently $\eta \leq R$. The previous inequality also implies $\omega\left(\eta, \eta\right) \leq \omega\left(R, R\right).$ Therefore, we get that

$$\frac{m}{1 - 2\beta\omega\left(R, R\right)} = \frac{\beta\omega\left(\eta, \eta\right)}{1 - 2\beta\omega\left(R, R\right)} \leq \frac{\beta\omega\left(R, R\right)}{1 - 2\beta\omega\left(R, R\right)} \leq \frac{\dfrac{1}{3}}{1 - 2\dfrac{1}{3}} = 1.$$

Hence, we conclude by (3.230) that

$$0 \leq \frac{m}{1 - 2\beta\omega\left(R, R\right)} \leq 1. \tag{3.231}$$

Now, let us begin by considering the families of geodesic

$$\gamma_n\left(t\right) = \exp_{p_n}\left(tu_n\right),$$
$$\sigma_n\left(t\right) = \exp_{p_n}\left(tv_n\right) \quad \text{for each} \quad n = 0, 1, \cdots,$$

where

$$u_n = DX\left(p_n\right)^{-1} X\left(p_n\right) \quad \text{and} \quad v_n = -DX\left(p_n\right)^{-1}\left[q_n, p_n; X\right] DX\left(p_n\right)^{-1} X\left(p_n\right).$$

Thus, we have that $\gamma_n\left(0\right) = p_n$, $\gamma_n\left(1\right) = q_n$, $\sigma_n\left(0\right) = p_n$ and $\sigma_n\left(1\right) = p_{n+1}$. In this way, the functionals $\left[q_n, p_n; X\right]$ and $\left[p_{n+1}, p_n; X\right]$ are well defined through the geodesics γ_n and σ_n. Then, our objective is to bound $d\left(p_2, p_1\right)$ using notation (3.221). To achieve this, it is necessary to find bounds for

$$\left\|\Gamma_1^{-1} P_{\sigma_0, 0, 1}\left(\left[p_1, p_0; X\right] - \Phi_0\right)\right\| \quad \text{and} \quad \left\|\Phi_1^{-1}\Gamma_1\right\| .$$

(a) For $\left\|\Gamma_1^{-1} P_{\sigma_0, 0, 1}\left(\left[p_1, p_0; X\right] - \Phi_0\right)\right\|$. Let us prove that Γ_1 is invertible. Indeed, we have in turn that

$$\left\|\Gamma_0^{-1}\right\| \left\|\Gamma_0 - P_{\sigma_0, 1, 0}\Gamma_1\right\| = \left\|\Gamma_0^{-1}\right\| \left\|DX\left(p_0\right) - P_{\sigma_0, 1, 0}DX\left(p_1\right)\right\|$$
$$= \left\|\Gamma_0^{-1}\right\| \left\|\left[p_0, p_0; X\right] - P_{\sigma_0, 1, 0}\left[p_1, p_1; X\right]\right\|$$
$$\leq \left\|\Gamma_0^{-1}\right\| \omega\left(d\left(p_0, p_1\right), d\left(p_0, p_1\right)\right)$$
$$\leq \beta\omega\left(\eta, \eta\right) \leq \beta\omega\left(R, R\right) < 1.$$

Thus, by Banach's lemma on invertible operators, Γ_1 is invertible and

$$\left\|\Gamma_1^{-1}\right\| \leq \frac{\left\|\Gamma_0^{-1}\right\|}{1 - \left\|\Gamma_0^{-1}\right\| \left\|\Gamma_0 - P_{\sigma_0,1,0}\Gamma_1\right\|} \leq \frac{\beta}{1 - \beta\omega(\eta,\eta)}. \tag{3.232}$$

In particular, we note that Φ_1^{-1} and p_2 are well defined. Then, because

$$\begin{aligned}
\left\|\Gamma_0^{-1}\right\| \left\|\Gamma_0 - [q_0, p_0; X]\right\| &\leq \beta \left\|[p_0, p_0; X] - [q_0, p_0; X]\right\| \\
&\leq \beta\omega\left(d(p_0, q_0), d(p_0, p_0)\right) \\
&\leq \beta\omega\left(\left\|\Gamma_0^{-1} X(p_0)\right\|, 0\right) \leq \beta\omega(\eta, 0) < 1,
\end{aligned}$$

we conclude, by Banach's lemma on invertible operators, that $[q_0, p_0; X]$ is invertible and

$$\left\|[q_0, p_0; X]^{-1}\right\| \leq \frac{\beta}{1 - \beta\omega(\eta, 0)}. \tag{3.233}$$

It is clear that Φ_0 is well defined and

$$\begin{aligned}
\left\|I_{T_{p_0}M} - [q_0, p_0; X]^{-1}\Gamma_0\right\| &= \left\|I_{T_{p_0}M} - [q_0, p_0; X]^{-1}\Gamma_0\right\| \\
&\leq \left\|[q_0, p_0; X]^{-1}\right\| \left\|[q_0, p_0; X] - \Gamma_0\right\| \\
&\leq \left(\frac{\beta}{1 - \beta\omega(\eta, 0)}\right)\omega(\eta, 0) \\
&= \frac{\frac{1}{2}\beta\omega(\eta, \eta)}{1 - \frac{1}{2}\beta\omega(\eta, \eta)} \leq \frac{\frac{1}{2}}{1 - \frac{1}{2}} = 1.
\end{aligned}$$

Therefore, we obtain in turn that

$$\begin{aligned}
&\left\|\Gamma_1^{-1} P_{\sigma_0,0,1}\left([p_1, p_0; X] - \Phi_0\right)\right\| \\
&= \left\|\Gamma_1^{-1} P_{\sigma_0,0,1}\left([p_1, p_0; X] - \Phi_0 + \Gamma_0 - \Gamma_0\right)\right\| \\
&\leq \left\|\Gamma_1^{-1} P_{\sigma_0,0,1}\left([p_1, p_0; X] - \Gamma_0\right)\right\| + \left\|\Gamma_1^{-1} P_{\sigma_0,0,1}\left(\Gamma_0 - \Phi_0\right)\right\| \\
&\leq \left\|\Gamma_1^{-1}\right\| \left\|[p_1, p_0; X] - \Gamma_0\right\| + \left\|\Gamma_1^{-1} P_{\sigma_0,0,1}\Gamma_0\right\| \left\|\left(I_{T_{p_0}M} - \Gamma_0^{-1}\Phi_0\right)\right\| \\
&\leq \left\|\Gamma_1^{-1}\right\| \left\|[p_1, p_0; X] - \Gamma_0\right\| + \left\|\Gamma_1^{-1} P_{\sigma_0,0,1}\Gamma_0\right\| \left\|\left(I_{T_{p_0}M} - \Gamma_0^{-1}\Phi_0\right)\right\| \\
&\leq \left\|\Gamma_1^{-1}\right\| \left\|[p_1, p_0; X] - [p_0, p_0; X]\right\| + \left\|\Gamma_1^{-1} P_{\sigma_0,0,1}\Gamma_0\right\| \left\|\left(I_{T_{p_0}M} - \Gamma_0^{-1}\Phi_0\right)\right\| \\
&\leq \left\|\Gamma_1^{-1}\right\| \left\|[p_1, p_0; X] - [p_0, p_0; X]\right\| + \left\|\Gamma_1^{-1} P_{\sigma_0,0,1}\Gamma_0\right\| \left\|\left(I_{T_{p_0}M} - [q_0, p_0; X]^{-1}\Gamma_0\right)\right\| \\
&\leq \left\|\Gamma_1^{-1}\right\| \left\|[p_1, p_0; X] - [p_0, p_0; X]\right\| + \left\|\Gamma_1^{-1} P_{\sigma_0,0,1}\Gamma_0\right\| \left\|\left(I_{T_{p_0}M} - [q_0, p_0; X]^{-1}\Gamma_0\right)\right\| \\
&\leq \frac{\beta}{1 - \beta\omega(\eta, \eta)}\omega\left(d(p_1, p_0), 0\right) + \frac{1}{1 - \beta\omega(\eta, \eta)}\frac{\beta\omega(\eta, 0)}{1 - \beta\omega(\eta, 0)} \\
&\leq \frac{\beta}{1 - \beta\omega(\eta, \eta)}\omega(\eta, 0) + \frac{1}{1 - \beta\omega(\eta, \eta)}\frac{\beta\omega(\eta, 0)}{1 - \beta\omega(\eta, 0)} \\
&= \frac{\beta\omega(\eta, \eta)}{2 - 2\beta\omega(\eta, \eta)}\frac{4 - \beta\omega(\eta, \eta)}{2 - \beta\omega(\eta, \eta)}.
\end{aligned}$$
$$\tag{3.234}$$

Note that

$$\frac{\beta\omega(\eta, \eta)}{2 - 2\beta\omega(\eta, \eta)}\frac{4 - \beta\omega(\eta, \eta)}{2 - \beta\omega(\eta, \eta)} \leq 1.$$

In view of $\beta\omega\left(\eta,\eta\right) \leq \beta\omega\left(R,R\right) \leq \dfrac{1}{3}$ and $\beta\omega\left(\eta,\eta\right) \leq \dfrac{4}{3}$, we obtain that

$$\frac{\beta\omega\left(\eta,\eta\right)}{2 - 2\beta\omega\left(\eta,\eta\right)} \leq \frac{1}{4} \quad\text{and}\quad \frac{4 - \beta\omega\left(\eta,\eta\right)}{2 - \beta\omega\left(\eta,\eta\right)} \leq 4.$$

(b) For $\left\|\Phi_1^{-1}\Gamma_1\right\|$. First, let us note that

$$\Phi_0\left(v_0\right) =$$
$$\mathcal{D}X\left(p_0\right)\left[q_0, p_0; X\right]^{-1}\mathcal{D}X\left(p_0\right)\left(-\mathcal{D}X\left(p_0\right)^{-1}\left[q_0, p_0; X\right]\mathcal{D}X\left(p_0\right)^{-1}X\left(p_0\right)\right)$$
$$= -X\left(p_0\right).$$

Then, we get that $X\left(p_0\right) = -\Phi_0\left(v_0\right)$. According to Definition 3.4.9, we obtain that

$$\left[\sigma_0\left(1\right), \sigma_0\left(0\right); X\right]\sigma_0'\left(0\right) = P_{\sigma_0,1,0}X\left(\gamma\left(1\right)\right) - X\left(\gamma\left(0\right)\right).$$

That is

$$\left[p_1, p_0; X\right]v_0 = P_{\sigma_0,1,0}X\left(p_1\right) - X\left(p_0\right).$$

Therefore, we get that

$$P_{\sigma_0,1,0}X\left(p_1\right) = \left[p_1, p_0; X\right]v_0 - \Phi_0\left(v_0\right) = \left(\left[p_1, p_0; X\right] - \Phi_0\right)v_0 \qquad (3.235)$$

and

$$\begin{aligned}
d\left(q_1, p_1\right) = \left\|u_1\right\| &= \left\|\Gamma^{-1}X\left(p_1\right)\right\| \\
&= \left\|\Gamma^{-1}P_{\sigma_0,0,1}\left(\left[p_1, p_0; X\right] - \Phi_0\right)v_0\right\| \\
&\leq \left\|\Gamma^{-1}P_{\sigma_0,0,1}\left(\left[p_1, p_0; X\right] - \Phi_0\right)\right\|\left\|v_0\right\| \leq \left\|v_0\right\| \leq \eta.
\end{aligned}$$

Then, we also have that

$$\begin{aligned}
\left\|I_{T_{p_1}M} - \Gamma_1^{-1}\left[q_1, p_1; X\right]\right\| &\leq \left\|\Gamma_1^{-1}\right\|\left\|\Gamma_1 - \left[q_1, p_1; X\right]\right\| \\
&\leq \left\|\Gamma_1^{-1}\right\|\left\|\left[p_1, p_1; X\right] - \left[q_1, p_1; X\right]\right\| \\
&\leq \left\|\Gamma_1^{-1}\right\|\omega\left(d\left(p_1, q_1\right), 0\right) \leq \frac{\beta\omega\left(\eta, 0\right)}{1 - \beta\omega\left(\eta, \eta\right)} < 1,
\end{aligned}$$

since

$$\begin{aligned}
\left\|\left[q_1, p_1; X\right]^{-1}\Gamma_1\right\| &= \left\|\Gamma_1^{-1}\left[q_1, p_1; X\right]\right\|^{-1} \\
&\leq \frac{1}{1 - \left\|I_{T_{p_1}M} - \Gamma_1^{-1}\left[q_1, p_1; X\right]\right\|} \\
&\leq \frac{1}{1 - \dfrac{\beta\omega\left(\eta, 0\right)}{1 - \beta\omega\left(\eta, \eta\right)}} = \frac{2 - 2\beta\omega\left(\eta, \eta\right)}{2 - 3\beta\omega\left(\eta, \eta\right)},
\end{aligned}$$

given that

$$\left\|\left[q_1, p_1; X\right]^{-1}\right\| \leq \left(\frac{\beta}{1 - \beta\omega\left(\eta, \eta\right)}\right)\left(\frac{2 - 2\beta\omega\left(\eta, \eta\right)}{2 - 3\beta\omega\left(\eta, \eta\right)}\right) = \frac{2\beta}{2 - 3\beta\omega\left(\eta, \eta\right)}.$$

This way, we can have that

$$\left\| I_{T_{p_1} M} - \Gamma_1^{-1}\Phi_1 \right\| = \left\| I_{T_{p_1} M} - [q_1, p_1; X]^{-1}\Gamma_1 \right\|$$

$$\leq \left\| [q_1, p_1; X]^{-1} \right\| \left\| [q_1, p_1; X] - \Gamma_1 \right\|$$

$$\leq \left\| [q_1, p_1; X]^{-1} \right\| \left\| [q_1, p_1; X] - \Gamma_1 \right\|$$

$$\leq \frac{2\beta}{2 - 3\beta\omega(\eta, \eta)} \left\| [q_1, p_1; X] - [p_1, p_1; X] \right\|$$

$$\leq \frac{2\beta}{2 - 3\beta\omega(\eta, \eta)}\omega\left(d\left(q_1, p_1\right), 0\right) < \frac{\beta\omega(\eta, \eta)}{2 - 3\beta\omega(\eta, \eta)}.$$

Finally, we get that

$$\left\| \Phi_1^{-1}\Gamma_1 \right\| \leq \frac{1}{1 - \dfrac{\beta\omega(\eta, \eta)}{2 - 3\beta\omega(\eta, \eta)}} = \frac{2 - 3\beta\omega(\eta, \eta)}{2 - 4\beta\omega(\eta, \eta)}. \tag{3.236}$$

We can now estimate $d\left(p_2, p_1\right)$. Using (3.234) and (3.236), we obtain that

$$d\left(p_2, p_1\right) = \left\| v_1 \right\| = \left\| \Phi_1^{-1} X\left(p_1\right) \right\|$$

$$= \left\| \Phi_1^{-1}\Gamma_1\Gamma_1^{-1}P_{\sigma_0, 0, 1}\left([p_1, p_0; X] - \Phi_0\right)v_0 \right\|$$

$$\leq \left\| \Phi_1^{-1}\Gamma_1 \right\| \left\| \Gamma_1^{-1}P_{\sigma_0, 0, 1}\left([p_1, p_0; X] - \Phi_0\right) \right\| \left\| v_0 \right\|$$

$$\leq \frac{2 - 3\beta\omega(\eta, \eta)}{2 - 4\beta\omega(\eta, \eta)}\frac{\beta\omega(\eta, \eta)}{2 - 2\beta\omega(\eta, \eta)}\frac{4 - \beta\omega(\eta, \eta)}{2 - \beta\omega(\eta, \eta)}\eta.$$

Note that we can have the following estimate

$$\frac{2 - 3\beta\omega(\eta, \eta)}{2 - 4\beta\omega(\eta, \eta)}\frac{\beta\omega(\eta, \eta)}{2 - 2\beta\omega(\eta, \eta)}\frac{4 - \beta\omega(\eta, \eta)}{2 - \beta\omega(\eta, \eta)} \leq \frac{\beta\omega(\eta, \eta)}{1 - 2\beta\omega(\eta, \eta)}. \tag{3.237}$$

Because (3.237) is equivalent to

$$-2\beta\omega(\eta, \eta) - \left(\beta\omega(\eta, \eta)\right)^2 \leq 0,$$

then

$$d\left(p_2, p_1\right) \leq \frac{\beta\omega(\eta, \eta)}{1 - 2\beta\omega(\eta, \eta)}\eta \leq N\eta.$$

This way, we obtain that

$$d\left(p_2, p_0\right) \leq d\left(p_2, p_1\right) + d\left(p_1, p_0\right) \leq N\eta + \eta = (N + 1)\eta$$

and as R is root of

$$t\left(1 - \frac{m}{1 - 2\beta\omega(t, t)}\right) - \eta = 0.$$

Then, we can write that $R\left(1 - N\right) - \eta = 0$. Thus, we also have that

$$(N + 1)\eta = \frac{\eta}{R}\left(2R - \eta\right) \leq R.$$

Therefore, we conclude that $d\left(p_2, p_0\right) \leq R$ and consequently $p_2 \in B\left(p_0, R\right)$. Using similar arguments, in an inductive way, we can prove that

$$d\left(p_{n+1}, p_n\right) \leq N^n d\left(p_1, p_0\right) \leq N^n \eta \qquad (3.238)$$

and

$$d\left(q_n, p_n\right) \leq N^n d\left(q_0, p_0\right) \leq N^n \eta. \qquad (3.239)$$

Then, from the triangular inequality

$$d\left(p_{n+1}, p_0\right) \leq \sum_{k+0}^{n} N^n \eta.$$

Moreover, given that

$$\sum_{k+0}^{n} N^n \eta = \left(\frac{1 - N^{n+1}}{1 - N}\right) \eta = \left(1 - \left(1 - \frac{\eta}{R}\right)^{n+1}\right) R \leq R,$$

we get that $d\left(p_{n+1}, p_0\right) \leq R$, and consequently $p_{n+1} \in B\left(p_0, R\right)$. Then (3.238) shows that sequence $\{p_n\}_{n \in \mathbb{N}}$ is Cauchy in M, which is complete and as such it converges to some $p_* \in B\left[p_0, R\right]$. Furthermore, we shall prove that p_* is a singularity of X. Indeed, we have that

$$\|X\left(p_n\right)\| = \left\|\Gamma_n \Gamma_n^{-1} X\left(p_n\right)\right\| \leq \|\Gamma_n\| d\left(p_n, q_n\right)$$

and

$$\|\Gamma_n\| \leq \|\Gamma_0\| + \omega\left(R, R\right).$$

By passing to limit when $n \longrightarrow \infty$ we obtain that $X\left(p_*\right) = 0$. Finally, to show the uniqueness part, let q_* be a singularity of X in $B\left[p_0, R\right]$, where σ is the minimizing geodesic joining the points p_0 and p_* such that $\sigma\left(0\right) = p_0$ and $\sigma\left(1\right) = p_*$. Then, we get that

$$\left\|\Gamma_0^{-1}\right\| \|\Gamma_0 - P_{\sigma,1,0}\left[q_*, p_*; X\right]\| \leq \omega\left(d\left(p_0, p_*\right), d\left(p_0, q_*\right)\right) \leq \omega\left(R, R\right) < 1.$$

It follows that the operator $\left[q_*, p_*; X\right]$ is invertible. If γ is the minimizing geodesic joining the points p_*, q_* such that $\gamma\left(0\right) = p_*$, $\gamma\left(1\right) = q_*$ and since

$$\left[\gamma\left(1\right), \gamma\left(0\right); X\right] \gamma'\left(0\right) = P_{\gamma,1,0} X\left(\gamma\left(1\right)\right) - X\left(\gamma\left(0\right)\right),$$

we have that $\left[q_*, p_*; X\right] \gamma'\left(0\right) = 0$. Thus, we get that $\gamma'\left(0\right) = 0$. Hence, we conclude that $d\left(p_*, q_*\right) = \|\gamma'\left(0\right)\| = 0$. That is $p_* = q_*$. The proof of Theorem 3.4.1 is complete. $\qquad \square$

We present a tighter semi-local convergence analysis for method (3.220) using the concepts of generalized ω and center-ω_0 conditions. First, we need to introduce some functions and provide sufficient conditions for the existence of small positive zeros for such functions. Secondly, we shall study the convergence of a scalar sequence $\{t_n\}$ which shall be shown to be majorizing for $\{x_n\}$.

Let $\beta > 0$ and $\eta > 0$. Let also $\omega_0 : \mathbb{R}_+ \times \mathbb{R}_+ \longrightarrow \mathbb{R}_+$ be continuous and nondecreasing in both variables. Define function g_1 on $\mathbb{R}_+$ by

$$g_1(t) = \beta\left(2\,\omega(0, \eta) + \omega_0(t, t)\right) - 1. \qquad (3.240)$$

Suppose that

$(\mathcal{C}_1)$ Function g_1 has a minimal zero. Denote such zero by R_1.

Then, function g given by

$$g(t) = \frac{\beta\left(\omega(t,0)\left(1 - \beta\,\omega_0(t,t)\right) + \omega(0,\eta)\right)\left(1 - \omega(0,\eta) - \beta\,\omega_0(t,t)\right)}{(1 - \beta\,\omega_0(t,t))^2\left(1 - 2\,\beta\,\omega(0,\eta) - \beta\,\omega_0(t,t)\right)} \tag{3.241}$$

is well defined on $[0, R_1)$.

Moreover, suppose that

$(\mathcal{C}_2)$ Function g_2 given by

$$g_2(t) = g(t) - 1 \tag{3.242}$$

has a minimal zero. Denote such a zero by R_2.

Note that we have that

$$0 < g(t) < 1, \tag{3.243}$$

$$\frac{\beta\left(\omega(t,0)\left(1 - \beta\,\omega_0(t,t)\right) + \omega(0,\eta)\right)}{(1 - \beta\,\omega_0(t,t))^2} < 1 \tag{3.244}$$

and

$$\frac{1 - \omega(0,\eta) - \beta\,\omega_0(t,t)}{1 - 2\,\beta\,\omega(0,\eta) - \beta\,\omega_0(t,t)} \geq 0 \tag{3.245}$$

for each $t \in [0, R_2)$.

Furthermore, suppose that

$(\mathcal{C}_3)$ Function h given by

$$h(t) = t\left(1 - g(t)\right) - \eta \tag{3.246}$$

has a minimal root $R^\star \in (0, R_2)$.

Then, we have that

$$N_0 = g(R^\star) \in (0, 1). \tag{3.247}$$

There are many possibilities under conditions $(\mathcal{C}_1)$–$(\mathcal{C}_3)$. Let us provide some for each condition:

$(\mathcal{C}_1^\star)$ $\beta\left(2\,\omega(0,\eta) + \omega_0(0,0)\right) < 1$.

We have that $g_1(0) = \beta\left(2\,\omega(0,\eta) + \omega_0(0,0)\right) - 1 < 0$ and $g_1(t) > 0$ for sufficiently large $t > 0$. It follows from the intermediate value theorem that function g_1 has zeros in $(0, t)$ for sufficiently large t.

Define function $g_i : \mathbb{R}_+ \longrightarrow \mathbb{R}_+$ for $i = 3, 4, 5$ by

$$g_3(t) = \beta\left(\omega(t,0)\left(1 - \beta\,\omega_0(t,t)\right) + \omega(0,\eta)\right)\left(1 - \beta\,\omega(0,\eta) - \beta\,\omega_0(t,t)\right),$$

$$g_4(t) = (1 - \beta\,\omega_0(t,t))^2\left(1 - 2\,\beta\,\omega(0,\eta) - \beta\,\omega_0(t,t)\right)$$

and $g_5(t) = g_3(t) - g_4(t)$. Consider the following assumption

$(\mathcal{C}_2^\star)$ $g_5(0) < 0$ and $g_5(R_1) > 0$.

It follows that function g_5 has zeros in $(0, R_1)$. Then, $(\mathcal{C}_2)$ and (3.243)–(3.245) hold. Hence, we can consider the following assumption

$(\mathcal{C}_3^\star)$ Assumption $(\mathcal{C}_2)$ or $(\mathcal{C}_2^\star)$ hold.

Moreover, suppose that $h(R_2) > 0$. In this case we also have $h(0) = -\eta < 0$. Hence, the existence of R follows. Let us suppose that $w(0,0) = w_0(0,0) = 0$. In this case we have

$(\mathcal{C}_1^{\star\star})$ $\beta\, w(0,\eta) < 1$.
$(\mathcal{C}_2^{\star\star})$ $\beta\, w(0,\eta) < (3 - \sqrt{5})/2 = .381966011\cdots$ and

$$(w(0, R_1) + w(0, \eta))\,(w(0, R_1)\,(w(R_1, 0) - w(0, R_1)) + w(0, \eta)^2) + w(0, R_1)\,w(0, \eta) \geq 0.$$

Note that the condition $(\mathcal{C}_2^\star)$, respectively, reduce to the ones above.

From now on we shall refer to $(\mathcal{C}_1)$–$(\mathcal{C}_3)$, $(\mathcal{C}_1^\star)$–$(\mathcal{C}_3^\star)$, $(\mathcal{C}_1^{\star\star})$–$(\mathcal{C}_3^{\star\star})$ as the $(\mathcal{C})$ conditions. In the special case when $w(0,t) = w(t,0) = w(t,t)/2$ and $w_0(t,t) = w(t,t)$, the $(\mathcal{C})$ conditions are weaker than conditions 3 and 4 in Theorem 3.4.1 (see (3.237), (3.241) and (3.242)).

Next, we present a convergence result for a scalar sequence. This sequence will later be shown to be majorizing for $\{x_n\}$.

Lemma 3.4.2. *Suppose that the $(\mathcal{C})$ conditions hold. Then, scalar sequence $\{t_n\}$ defined by*

$$t_0 = 0, \; t_1 = \eta, \; t_2 = t_1 + e_1\,(t_1 - t_0)$$
$$t_{n+2} = t_{n+1} + e_{n+1}\,(t_{n+1} - t_n) \quad for \; each \quad n = 0, 1, \cdots, \tag{3.248}$$

where

$$e_n = e_n^1\, e_n^2 \quad for \; each \quad n = 0, 1, \cdots \tag{3.249}$$

$$e_1^1 = \frac{\beta\,(w_0(d(\eta, 0), 0)\,(1 - \beta\, w_0(0, d(\eta, 0))) + w_0(\eta, 0))}{(1 - \beta\, w_0(d(\eta, 0), d(\eta, 0)))\,(1 - \beta\, w_0(0, d(\eta, 0)))},$$

$$e_1^2 = \frac{1 - \beta\, w(0, d(\eta, 0)) - \beta\, w_0(d(\eta, 0), d(\eta, 0))}{1 - 2\,\beta\, w(0, d(\eta, 0)) - \beta\, w_0(d(\eta, 0), d(\eta, 0))},$$

$$e_{n+1}^1 = \frac{\beta\,(w(d(t_{n+1}, 0), 0)\,(1 - \beta\, w_0(d(t_n, 0), d(t_n, 0))) + w(0, d(t_{n+1}, t_n)))}{(1 - \beta\, w_0(d(t_{n+1}, 0), d(t_{n+1}, 0)))\,(1 - \beta\, w_0(d(t_n, 0), d(t_n, 0)))},$$

$$e_{n+1}^2 = \frac{1 - \beta\, w(d(t_{n+1}, t_n), 0) - \beta\, w_0(d(t_{n+1}, 0), d(t_{n+1}, 0))}{1 - 2\,\beta\, w(d(t_{n+1}, t_n), 0) - \beta\, w_0(d(t_{n+1}, 0), d(t_{n+1}, 0))}$$

is well defined, nondecreasing, bounded from above by $R^\star$ given in $(\mathcal{C}_3)$ and converges to its unique least upper bounds $t^\star$ which satisfies

$$\eta \leq t^\star \leq R^\star. \tag{3.250}$$

Moreover, the following estimates hold for each $n = 0, 1, \cdots$

$$t_{n+2} - t_{n+1} \leq N_0 \left(t_{n+1} - t_n\right) \tag{3.251}$$

and

$$t^\star - t_n \leq \frac{N_0^n \, \eta}{1 - N_0}, \tag{3.252}$$

where N_0 is given by (3.247).

Proof. Simply notice that by (3.241)–(3.244) and (3.247), sequence $\{t_n\}$ is well defined, nondecreasing and

$$e_n \leq N_0 < 1 \quad \text{for each} \quad n = 0, 1, \cdots, \tag{3.253}$$

which shows (3.251). Estimate (3.252) follows from (3.251) by using standard majorization techniques. Finally, if we let $n = 0$ in (3.252) we deduce (3.250) by (3.246) and (3.248). That completes the proof of Lemma 3.4.2. $\qquad \square$

We present the following useful generalization of Lemma 3.4.2.

Lemma 3.4.3. *Suppose that there exists an integer $k \geq 1$ such that iterates $\{t_i\}$ are well defined for each $i = 0, 1, \cdots, k$;*

$$t_{i-1} \leq t_i \quad for \ each \quad i = 1, 2, \cdots;$$

the $(\mathcal{C})$ conditions are satisfied for $t_k - t_{k-1}$ replacing η. Then, the conclusions of Lemma 3.4.2 for $\{t_n\}$ hold and estimates (3.251) and (3.252) hold for $n \geq k$.

Lemma 3.4.4. *Suppose that sequence $\{t_n\}$ is well defined and*

$$\beta \left(\omega_0(d(t_{n+1}, 0), d(t_{n+1}, 0)) + 2\,\omega(d(t_{n+1}, t_n), 0)\right) < 1 \quad for \ each \quad i = 0, 1, \cdots. \tag{3.254}$$

Then, $\{t_n\}$ is nondecreasingly convergent to its unique least upper bound $t^\star$.

Proof. It follows from (3.248) and (3.254) that sequence $\{t_n\}$ is nondecreasing and bounded above. Hence, it converges to $t^\star$. The proof of Lemma 3.4.4 is complete. $\qquad \square$

Remark 3.4.1. It is worth noting that hypotheses of Lemmas 3.4.2 and 3.4.3 imply (3.254). Condition (3.254) seems to be the weakest possible convergence condition for sequence $\{t_n\}$. The verification of (3.254) is difficult in general but it is possible, especially if iterates are equal to each other after a finite number of steps.

We have the following relationship between sequences $\{\tau_n\}$ and $\{t_n\}$.

Corollary 3.4.1. *Suppose that (3.223), (3.224), (3.225) hold and $\omega_0(s, t) = \omega(s, t)$ for each $(s, t) \in \mathbb{R}_+ \times \mathbb{R}_+$. Then, the following assertions hold for each $n = 0, 1, \cdots$*

$$t_n \leq \tau_n, \tag{3.255}$$

$$t_{n+1} - t_n \leq \tau_{n+1} - \tau_n, \tag{3.256}$$

$$N_0 \leq N, \tag{3.257}$$

$$R^\star \leq R \tag{3.258}$$

and

$$t^\star \leq \tau^\star. \tag{3.259}$$

Proof. Estimates (3.255) and (3.256) follow using a simple inductive argument, the definitions of $\{\tau_n\}$, $\{t_n\}$ and (3.237). Moreover, estimates (3.257) and (3.258) follow from the definitions of N_0, N, $R^\star$ and R. Furthermore, (3.259) follows from (3.255) by letting $n \longrightarrow \infty$. The proof of Corollary 3.4.1 is complete. $\qquad\square$

Corollary 3.4.2. *Suppose hypotheses of Theorem 3.4.1 hold. Then, the conclusions of Theorem 3.4.1 hold with N_0, $R^\star$, $\{t_n\}$ replacing respectively N, R and $\{\tau_n\}$.*

Next, we shall show that sequence $\{t_n\}$ is majorizing for $\{x_n\}$. But first we need the definition of center ω_0-condition.

Definition 3.4.10. We say that the divided difference of first order satisfies the ω_0-condition at $p_0 \in \Omega$, if

$$\|[p, q; X] - DF(p_0)\| \leq \omega_0 \left(d\left(p, p_0\right), d\left(q, p_0\right)\right); \quad p, q \in \Omega, \tag{3.260}$$

where $\omega_0 : \mathbb{R}_+ \times \mathbb{R}_+ \to \mathbb{R}_+$ is a nondecreasing, continuous in both variables.

Remark 3.4.2. It follows from (3.222) that (3.260) always holds. Hence, (3.260) is not an addition to (3.222) hypothesis. In practice, the computation of function ω requires the computation of function ω_0. Moreover, we have that

$$\omega_0(s, t) \leq \omega(s, t) \quad \text{for each} \quad (s, t) \in \mathbb{R}_+ \times \mathbb{R}_+ \tag{3.261}$$

and ω/ω_0 can be arbitrarily large.

In Theorem 3.4.1 the upper bounds on the norms $\| DF(p)^{-1} \|$, $\| [p, q; X]^{-1} \|$ were computed using (3.222) and we obtained, respectively, the following estimates

$$\| DF(p_1)^{-1} \| \leq \frac{\beta}{1 - \beta\,\omega(d(p_1, p_0), d(p_1, p_0))} \tag{3.262}$$

and

$$\| [p, q; X]^{-1} \| \leq \frac{\beta}{1 - \beta\,\omega(d(p_1, p_0), 0)}. \tag{3.263}$$

However, if we use the less expensive and the at least as precise (3.260), we obtain respectively, instead (3.262) and (3.263)

$$\| DF(p_1)^{-1} \| \leq \frac{\beta}{1 - \beta\,\omega_0(d(p_1, p_0), d(p_1, p_0))} \tag{3.264}$$

$$\| [p, q; X]^{-1} \| \leq \frac{\beta}{1 - \beta\,\omega_0(d(p_1, p_0), 0)}. \tag{3.265}$$

Hence, if we use (3.260) instead of (3.222) for the computation of inverses and simply follow the proof of Theorem 3.4.1, we arrive at the following semi-local convergence result for $\{x_n\}$.

Theorem 3.4.2. *Let M be a complete Riemannian manifold, $\Omega \subseteq M$ be an open convex set and X a C^1 vector field on M. Suppose that the divided difference of first order satisfies the ω-condition and the center ω_0-condition at $p_0 \in \Omega$. Moreover, suppose that $\left\|\Gamma_0^{-1}\right\| \leq \beta$; $\max\left\{\left\|\Gamma_0^{-1}X(p_0)\right\|, \left\|\Phi_0^{-1}X(p_0)\right\|\right\} \leq \eta$, $(\mathcal{C})$ hypotheses hold and $\overline{B(p_0, R^\star)} \subset \Omega$. Then, method (3.220) is well defined, $\{p_n\}$ remains in $B(p_0, R^\star)$ for each $n = 0, 1, \cdots$ and converges to the unique solution of $X(p) = 0$ in $B[p_0, R^\star]$. Moreover, the following estimates hold for each $n = 1, 2, \cdots$*

$$\overline{e}_n^1 \leq e_n^1, \quad \overline{e}_n^2 \leq e_n^2, \quad \overline{e}_n \leq e_n, \quad \| p_{n+1} - p_n \| \leq \overline{e}_n \| p_n - p_{n-1} \|,$$

$$\| p_{n+1} - p_n \| \leq t_{n+1} - t_n \quad and \quad \| p_n - p_* \| \leq t^\star - t_n,$$

where

$$\overline{e}_1^1 = \frac{\beta\left(\omega_0(d(p_1, p_0), 0)\left(1 - \beta\,\omega_0(0, d(p_1, 0))\right) + \omega_0(p_1, 0)\right)}{\left(1 - \beta\,\omega_0(d(p_1, 0), d(p_1, 0))\right)\left(1 - \beta\,\omega_0(0, d(p_1, 0))\right)},$$

$$\overline{e}_1^2 = \frac{1 - \beta\,\omega(0, d(p_1, 0)) - \beta\,\omega_0(d(p_1, 0), d(p_1, 0))}{1 - 2\,\beta\,\omega(0, d(p_1, 0)) - \beta\,\omega_0(d(p_1, 0), d(p_1, 0))}$$

and for each $n = 1, 2, \cdots$

$$\overline{e}_{n+1}^1 = \frac{\beta\left(\omega(d(p_{n+1}, 0), 0)\left(1 - \beta\,\omega_0(d(p_n, 0), d(p_n, 0))\right) + \omega(0, d(p_{n+1}, p_n))\right)}{\left(1 - \beta\,\omega_0(d(p_{n+1}, 0), d(p_{n+1}, 0))\right)\left(1 - \beta\,\omega_0(d(p_n, 0), d(p_n, 0))\right)},$$

$$\overline{e}_{n+1}^2 = \frac{1 - \beta\,\omega(d(p_{n+1}, p_n), 0) - \beta\,\omega_0(d(p_{n+1}, 0), d(p_{n+1}, 0))}{1 - 2\,\beta\,\omega(d(p_{n+1}, p_n), 0) - \beta\,\omega_0(d(p_{n+1}, 0), d(p_{n+1}, 0))}.$$

Furthermore, if there exists $R^{\star\star} \geq R^\star$ such that $\omega_0(R^{\star\star}, R^\star) < 1$ and $B[p_0, p_*] \subseteq \Omega$, then, the solution p_* is unique in $B[p_0, R^{\star\star}]$.

Proof. According to the proof of Theorem 3.4.1, (3.263) and (3.264) we only need to show the uniqueness part. As in Theorem 3.4.1, if $q_* \in B[p_0, R^\star]$ and $X(q_*) = 0$. Then, we have by (3.260) that

$$\left\|\Gamma_0^{-1}\right\|\left\|\Gamma_0 - P_{\sigma,1,0}[q_*, p_*; X]\right\| \leq \omega_0\left(d(p_0, p_*), d(p_0, q_*)\right) \leq \omega\left(R^\star, R^{\star\star}\right) < 1.$$

Hence, again we deduce $p_* = q_*$. The proof of Theorem 3.4.2 is complete. $\qquad\square$

Remark 3.4.3.

(a) Clearly, hypotheses of Lemma 3.4.3 or Lemma 3.4.4 can replace the $(\mathcal{C})$ conditions in Theorem 3.4.2.

(b) Theorem 3.4.2 improves Theorem 3.4.1 and the corresponding result in Ref. [11] even if $\omega_0 = \omega$ (see also (3.237), Corollary 3.4.1 and uniqueness of the solution domains).

(c) There are cases when (3.222) is not satisfied but (3.260) is satisfied (see Ref. [106]). In order for us to cover this case we note that for each $p, q, \overline{p}, \overline{q} \in \Omega$ if (3.222) holds then we have that

$$\| [p, q; X] - [\overline{p}, \overline{q}; X] \| \leq \omega(d(p, \overline{p}), d(q, \overline{q}))$$

whereas if (3.260) only holds we have that

$$\| [p, q; X] - [\overline{p}, \overline{q}; X] \|$$
$$\leq \| [p, q; X] - [p_0, p_0; X] \| + \| [p_0, p_0; X] - [\overline{p}, \overline{q}; X] \|$$
$$\leq \omega_0(d(p, p_0), d(q, p_0)) + \omega_0(d(\overline{p}, p_0), d(\overline{q}, p_0)).$$

These estimates suggest that all the results obtained here can be rewritten by dropping (3.222), using instead (3.260) and by making the changes suggested by the following diagrams

$$\omega(0, \| y_n - x_n \|)$$
$$\longrightarrow \omega_0(\| x_n - x_0 \|, \| x_n - x_0 \|) + \omega_0(\| x_n - x_0 \|, \| y_n - x_0 \|) \longrightarrow 2\,\omega_0(t_n, t_n),$$

$$w(t, 0) \longrightarrow 2\,\omega_0(t, t) \quad \text{and} \quad \omega(0, t) \longrightarrow 2\,\omega_0(t, t).$$

The results in this setting improve the corresponding ones in Ref. [11].

3.5   Newton-type Method on Riemannian Manifolds

In the last years, attention has been paid in studying Newton's method on manifolds. In this section, we generalize this study considering some Newton-type iterative methods. A characterization of the convergence under Kantorovich type conditions and optimal estimates of the error are found. Using normal coordinates the order of convergence is derived. The sufficient semi-local convergence criteria are weaker and the majorizing sequences are tighter for the special cases of simplified Newton and Newton methods than in earlier studies such as Refs. [100, 104, 110, 290].

Let us suppose that F is an operator defined on an open convex subset Ω of a Banach space E. Let us denote by $DF(x_n)$ the first Fréchet derivatives of F at x_n. Given an integer m and an initial point $x_0 \in E$, we move from x_n to x_{n+1} through an intermediate sequence $\{y_n^i\}_{i=0}^m$, $y_n^0 = x_n$, which is a generalization of Newton ($m = 1$) and simplified Newton ($m = \infty$) methods

$$\begin{cases} y_n^1 = y_n^0 - DF\left(y_n^0\right)^{-1} F\left(y_n^0\right) \\ y_n^2 = y_n^1 - DF\left(y_n^0\right)^{-1} F\left(y_n^1\right) \\ \quad \vdots \\ y_n^m = x_{n+1} = y_n^{m-1} - DF\left(y_n^0\right)^{-1} F\left(y_n^{(m-1)}\right). \end{cases}$$

This family of methods was introduced by E. Shamanskii in Ref. [404]. Under appropriate conditions, these iterative methods converge to a root x_* of the equation $F(x) = 0$. Moreover, if x_0 is sufficiently near x_* the method has order of convergence at least $m + 1$. See Refs. [292, 404, 418].

It is also possible to obtain a version of the fundamental theorem of calculus for manifolds:

Theorem 3.5.1. *Let c be a geodesic in M and X be a C^1 vector field on M, then*

$$P_{c,t,0}X\left(c\left(t\right)\right) = X\left(c\left(0\right)\right) + \int_0^t P_{c,s,0}\left(\mathcal{D}X\left(c\left(s\right)\right)c'\left(s\right)\right)ds. \tag{3.266}$$

Proof. Let us consider the curve

$$f\left(s\right) = P_{c,s,0}X\left(c\left(s\right)\right)$$

in $T_{c(0)}M$, therefore

$$\begin{aligned}
f'\left(s\right) &= \lim_{h\to 0}\frac{f\left(s+h\right)-f\left(s\right)}{h}\\
&= \lim_{h\to 0}\frac{1}{h}\left(P_{c,s+h,0}X\left(c\left(s+h\right)\right)-P_{c,s,0}X\left(c\left(s\right)\right)\right)\\
&= \lim_{h\to 0}\frac{1}{h}\left(P_{c,s,0}\circ P_{c,s+h,s}X\left(c\left(s+h\right)\right)-P_{c,s,0}X\left(c\left(s\right)\right)\right),
\end{aligned}$$

since $P_{c,s,0}$ is linear and continuous, we obtain

$$\begin{aligned}
f'\left(s\right) &= P_{c,s,0}\left(\lim_{h\to 0}\frac{1}{h}\left(P_{c,s+h,s}X\left(c\left(s+h\right)\right)-X\left(c\left(s\right)\right)\right)\right)\\
&\quad P_{c,s,0}\left(\lim_{h\to 0}\frac{1}{h}\left(P_{c,s+h,s}X\left(c\left(s+h\right)\right)-X\left(c\left(s\right)\right)\right)\right).
\end{aligned}$$

Using (3.217) we have

$$\begin{aligned}
f'\left(s\right) &= P_{c,s,0}\left(\nabla_{c'(s)}X\left(c\left(s\right)\right)\right) \tag{3.267}\\
&= P_{c,s,0}\left(\mathcal{D}X\left(c\left(s\right)\right)c'\left(s\right)\right)
\end{aligned}$$

and from

$$\int_0^t f'\left(s\right)ds = f\left(t\right)-f\left(0\right)$$

we conclude

$$P_{c,t,0}X\left(c\left(t\right)\right) = X\left(c\left(0\right)\right) + \int_0^t P_{c,s,0}\left(\mathcal{D}X\left(c\left(s\right)\right)c'\left(s\right)\right)ds.$$

$\square$

It is not hard to prove (using induction) that

$$f^{(n)}\left(s\right) = P_{c,s,0}\mathcal{D}^{(n)}X\left(c\left(s\right)\right)\underbrace{\left(c'\left(s\right),c'\left(s\right),\cdots,c'\left(s\right)\right)}_{n-times} \tag{3.268}$$

Theorem 3.5.2. *Let c be a geodesic in M and X be a C^2 vector field on M, then*

$$P_{c,t,0}\mathcal{D}X\left(c\left(t\right)\right)c'\left(t\right) = \mathcal{D}X\left(c\left(0\right)\right)c'\left(0\right) + \int_0^t P_{c,s,0}\left(\mathcal{D}^2X\left(c\left(s\right)\right)\left(c'\left(s\right),c'\left(s\right)\right)\right)ds.$$

$$\tag{3.269}$$

Proof.   Let us consider the vector field along the geodesic $c\,(s)$

$$Y\left(c\left(s\right)\right) = \mathcal{D}X\left(c\left(s\right)\right)c'\left(s\right).$$

By the previous theorem

$$P_{c,t,0}Y\left(c\left(t\right)\right) = Y\left(c\left(0\right)\right) + \int_0^t P_{c,s,0}\left(\mathcal{D}Y\left(c\left(s\right)\right)c'\left(s\right)\right)ds.$$

Hence

$$P_{c,t,0}\mathcal{D}X\left(c\left(t\right)\right)c'\left(t\right) = \mathcal{D}X\left(c\left(0\right)\right)c'\left(0\right) + \int_0^t P_{c,s,0}\left(\mathcal{D}\left(\mathcal{D}X\left(c\left(s\right)\right)c'\left(s\right)\right)c'\left(s\right)\right)ds$$

by (3.214)

$$\mathcal{D}^2 X\left(c\left(s\right)\right)\left(c'\left(s\right),c'\left(s\right)\right) = \nabla_{c'(s)}\mathcal{D}\left(X\left(c\left(s\right)\right)\left(c'\left(s\right)\right)\right) - \mathcal{D}X\left(c\left(s\right)\right)\left(\nabla_{c'(s)}c'\left(s\right)\right)$$
$$\mathcal{D}\left(\mathcal{D}X\left(c\left(s\right)\right)c'\left(s\right)\right)c'\left(s\right) - \mathcal{D}X\left(c\left(s\right)\right)\left(\nabla_{c'(s)}c'\left(s\right)\right),$$

since $c\,(s)$ is a geodesic, we have $\nabla_{c'(s)}c'\left(s\right) = 0$, hence

$$\mathcal{D}^2 X\left(c\left(s\right)\right)\left(c'\left(s\right),c'\left(s\right)\right) = \mathcal{D}\left(\mathcal{D}X\left(c\left(s\right)\right)c'\left(s\right)\right)c'\left(s\right).$$

Therefore

$$P_{c,t,0}\mathcal{D}X\left(c\left(t\right)\right)c'\left(t\right) = \mathcal{D}X\left(c\left(0\right)\right)c'\left(0\right) + \int_0^t \mathcal{D}^2 X\left(c\left(s\right)\right)\left(c'\left(s\right),c'\left(s\right)\right)ds.$$

$\square$

In a similar way, using an induction strategy, we can prove that

$$P_{c,t,0}\mathcal{D}^n X\left(c\left(s\right)\right)P_{c,0,t}^n - \mathcal{D}^n X\left(c\left(0\right)\right)$$
$$= \int_0^s P_{c,t,0}\left(\mathcal{D}^n X\left(c\left(t\right)\right)P_{c,0,t}^n\left(c'\left(0\right),\ldots,c'\left(0\right)\right)\right)dt. \tag{3.270}$$

Theorem 3.5.3. *Let c be a geodesic in M, $[0,1] \subseteq Dom\,(c)$ and X be a C^2 vector field on M, then*

$$P_{c,1,0}X\left(c\left(1\right)\right) =$$
$$X\left(c\left(0\right)\right) + \mathcal{D}X\left(c\left(0\right)\right)c'\left(0\right) + \int_0^1 \left(1-t\right)P_{c,s,0}\mathcal{D}^2 X\left(c\left(t\right)\right)\left(c'\left(t\right),c'\left(t\right)\right)dt. \tag{3.271}$$

Proof.   Consider the curve $f\,(s) = P_{c,s,0}X\left(c\left(s\right)\right)$ in $T_{c(0)}M$. By (3.268), we have that $f''\,(s) = P_{c,s,0}\mathcal{D}^2 X\left(c\left(s\right)\right)\left(c'\left(s\right),c'\left(s\right)\right)$. From Taylor's theorem, we get that

$$f\left(1\right) = f\left(0\right) + f'\left(0\right)\left(1-0\right) + \int_0^1 \left(1-t\right)f''\left(t\right)dt.$$

Therefore

$$P_{c,1,0}X\left(c\left(1\right)\right) = X\left(c\left(0\right)\right) + \mathcal{D}X\left(c\left(0\right)\right)c'\left(0\right) +$$
$$\int_0^1 \left(1-t\right)P_{c,t,0}\mathcal{D}^2 X\left(c\left(t\right)\right)\left(c'\left(t\right),c'\left(t\right)\right)dt.$$

$\square$

When conditions are imposed on x_0 and on F, as in the Kantorovich Theorem Ref. [290], to assure the convergence of the sequence $\{x_n\}_{n\in\mathbb{N}}$ to a solution of $F(x) = 0$ we say that *conditions of Kantorovich type* are imposed. Let us recall Kantorovich's Theorem in Banach spaces, see Ref. [290].

Theorem 3.5.4. *(Kantorovich) Let E be a Banach space, $\Omega \subseteq E$ be an open convex set, $F : \Omega \longrightarrow \Omega$ be a continuous operator, such that, $F \in C^1$ and DF is Lipschitz in Ω*

$$\|DF(x) - DF(y)\| \leq l\,\|x - y\|, \text{ for all } x, y \in \Omega, l > 0.$$

Suppose that for some $x_0 \in \Omega$, $DF(x_0)$ is invertible and that for some $a > 0$ and $b \geq 0$:

$$\left\|DF(x_0)^{-1}\right\| \leq a, \quad \left\|DF(x_0)^{-1} F(x_0)\right\| \leq b,$$

$$h = a\,b\,l \leq \frac{1}{2} \tag{3.272}$$

and

$$B(x_0, t_*) \subseteq \Omega \text{ where } t_* = \frac{1}{a\,l}\left(1 - \sqrt{1 - 2\,h}\right).$$

If

$$v_k = -DF(x_k)^{-1} F(x_k),$$
$$x_{k+1} = x_k + v_k.$$

Then $\{x_k\}_{k\in\mathbb{N}} \subseteq B(x_0, t_*)$ and $x_k \longrightarrow p_*$, which is the unique zero of F in $B[x_0, t_*]$. Furthermore, if $h < \dfrac{1}{2}$ and $B(x_0, r) \subseteq \Omega$ with

$$t_* < r \leq t_{**} = \frac{1}{a\,l}\left(1 + \sqrt{1 - 2\,h}\right),$$

then p_* is also the unique zero of F in $B(x_0, r)$. Also, the error bound is:

$$\|x_k - x_*\| \leq (2\,h)^{2^k} \frac{b}{h}; \; k = 1, 2, \ldots.$$

Although the concepts will be defined later on, to extend the method to Riemannian manifolds, preliminarily we will say that the derivative of F at x_n is replaced by the covariant derivative of X at p_n :

$$\nabla_{(.)} X(p_n) : T_{p_n} M \longrightarrow T_{p_n} M$$
$$v \longrightarrow \nabla_Y X,$$

where Y is a vector field satisfying $Y(p) = v$. We adopt the notation $DX(p)\,v = \nabla_y X(p)$. Hence $DX(p)$ is a linear mapping of $T_p M$ into $T_p M$. So, in this new context $-F'(x_n)^{-1} F(x_n)$ is written as $-DX(p_n)^{-1} X(p_n)$ or $-\left(\nabla_{X(p_n)} X\right)^{-1} (p_n)$.

Now we can write Kantorovich's theorem in the new context. We will say that a singularity of a vector field X, is a point $p \in M$ for which $X(p) = 0$.

Theorem 3.5.5. *(Kantorovich's theorem in Riemannian manifold) Let M be a Riemannian manifold, $\Omega \subseteq M$ be an open convex set, $X \in \chi(M)$ and $\mathcal{D}X \in Lip_l(\Omega)$. Suppose that for some $p_0 \in \Omega$, $\mathcal{D}X(p_0)$ is invertible and that for some $a > 0$ and $b \geq 0$:*

$$(1) \left\| \mathcal{D}X(p_0)^{-1} \right\| \leq a \qquad\qquad \left(\left\| (\nabla_{(.)}X(p_o))^{-1} \right\| \leq a \right)$$

$$(2) \left\| \mathcal{D}X(p_0)^{-1} X(p_0) \right\| \leq b \qquad \left(\left\| (\nabla_{X(p_o)}X(p_o))^{-1} \right\| \leq b \right)$$

$$(3) \ h = a\,b\,l \leq \frac{1}{2}$$

$$(4) \ B(p_0, t_*) \subseteq \Omega \text{ where } t_* = \frac{1}{a\,l}\left(1 - \sqrt{1 - 2\,h}\right).$$

If

$$v_k = -\mathcal{D}X(p_k)^{-1} X(p_k),$$
$$p_{k+1} = \exp_{p_k}(v_k),$$

then $\{p_k\}_{k \in \mathbb{N}} \subseteq B(p_0, t_*)$ and $p_k \longrightarrow p_*$ which is the unique singularity of X in $B[p_0, t_*]$. Furthermore, if $h < \frac{1}{2}$ and $B(p_0, r) \subseteq \Omega$ with

$$t_* < r \leq t_{**} = \frac{1}{a\,l}\left(1 + \sqrt{1 - 2\,h}\right),$$

then p_* is also the unique singularity of F in $B(p_0, r)$. The error bound is:

$$d(p_k, p_*) \leq \frac{b}{h}(2\,h)^{2^k} \ ; \ k = 1, 2, \ldots. \tag{3.273}$$

Next we will prove the semi-local convergence of the simplified Newton's method in Riemannian manifolds (fixing $\mathcal{D}X(p_0)^{-1}$ in each iteration). Our main result is:

Theorem 3.5.6. *Let M be a Riemannian manifold, $\Omega \subseteq M$ be an open convex set, $X \in \chi(M)$ and $\mathcal{D}X \in Lip_l(\Omega)$. Suppose that for some $p_0 \in \Omega$, $\mathcal{D}X(p_0)$ is invertible and that for some $a > 0$ and $b \geq 0$:*

$$(1) \left\| \mathcal{D}X(p_0)^{-1} \right\| \leq a,$$

$$(2) \ \left\| \mathcal{D}X(p_0)^{-1} X(p_0) \right\| \leq b,$$

$$(3) \ h = a\,b\,l \leq \frac{1}{2},$$

$$(4) \ B(p_0, t_*) \subseteq \Omega \quad \text{where} \quad t_* = \frac{1}{a\,l}\left(1 - \sqrt{1 - 2\,h}\right).$$

If

$$v_k = -P_{\sigma_k, 0, 1}\mathcal{D}X(p_0)^{-1} P_{\sigma_k, 1, 0}X(p_k),$$
$$p_{k+1} = \exp_{p_k}(v_k), \tag{3.274}$$

where $\{\sigma_k : [0, 1] \longrightarrow M\}_{k \in \mathbb{N}}$ is the minimizing geodesic family connecting p_0, p_k, then $\{p_k\}_{k \in \mathbb{N}} \subseteq B(p_0, t_*)$ and $p_k \longrightarrow p_*$ which is the only one singularity of X in $B[p_0, t_*]$. Furthermore, if $h < \frac{1}{2}$ and $B(p_0, r) \subseteq \Omega$ with

$$t_* < r \leq t_{**} = \frac{1}{a\,l}\left(1 + \sqrt{1 - 2\,h}\right),$$

then p_* is also the only singularity of F in $B(p_0, r)$. The error bound is:

$$d(p_k, p_*) \leq \frac{b}{h}\left(1 - \sqrt{1 - 2h}\right)^{k+1}; \quad k = 1, 2, \ldots. \tag{3.275}$$

First, we establish some results that are of primary relevance in this proof.

Lemma 3.5.1. *Let M be a Riemannian manifold, $\Omega \subseteq M$ an open convex set, $X \in \chi(M)$ and $\mathcal{D}X \in Lip_l(\Omega)$. Take $p \in B(p_0, r) \subseteq \Omega$, $v \in T_pM$, $\sigma : [0,1] \longrightarrow M$ be a minimizing geodesic connecting p_0, p and*

$$\gamma(t) = \exp_p(tv).$$

Then

$$P_{\gamma,t,0}X(\gamma(t)) = X(p) + P_{\sigma,0,1}t\mathcal{D}X(p_0)P_{\sigma,1,0}v + R(t)$$

with

$$\|R(t)\| \leq l\left(\frac{t}{2}\|v\| + d(p_0, p)\right)t\|v\|.$$

Proof. From Theorem 3.5.1, it follows that

$$P_{\gamma,t,0}X(\gamma(t)) - X(\gamma(0)) = \int_0^t P_{\gamma,s,0}(\mathcal{D}X(\gamma(s))\gamma'(s))\,ds,$$

since γ is a minimizing geodesic, then $\gamma'(t)$ is parallel and $\gamma'(s) = P_{\gamma,0,s}\gamma'(0)$. Moreover $\gamma'(0) = v$ then

$$P_{\gamma,t,0}X(\gamma(t)) - X(p) = \int_0^t P_{\gamma,s,0}(\mathcal{D}X(\gamma(s))P_{\gamma,0,s}v)\,ds.$$

Thus

$$P_{\gamma,t,0}X(\gamma(t)) - X(p) - P_{\sigma,0,1}t\mathcal{D}X(p_0)P_{\sigma,1,0}v$$

$$= \int_0^t P_{\gamma,s,0}(\mathcal{D}X(\gamma(s))P_{\gamma,0,s}v)\,ds - P_{\sigma,0,1}\mathcal{D}X(p_0)P_{\sigma,1,0}v$$

$$= \int_0^t (P_{\gamma,s,0}\mathcal{D}X(\gamma(s))P_{\gamma,0,s}v - P_{\sigma,0,1}\mathcal{D}X(p_0)P_{\sigma,1,0}v)\,ds,$$

letting

$$R(t) = \int_0^t (P_{\gamma,s,o}\mathcal{D}X(\gamma(s))P_{\gamma,o,s}v - P_{\sigma,0,1}\mathcal{D}X(p_0)P_{\sigma,1,0}v)\,ds,$$

and since $\mathcal{D}X \in Lip_l(\Omega)$, we obtain that

$\|R(t)\|$

$$\leq \int_0^t \|(P_{\gamma,s,o}\mathcal{D}X(\gamma(s))P_{\gamma,o,s} - \mathcal{D}X(p) + \mathcal{D}X(p) - P_{\sigma,0,1}\mathcal{D}X(p_0)P_{\sigma,1,0})\|\,\|v\|\,ds$$

$$\leq \int_0^t (\|P_{\gamma,s,o}\mathcal{D}X(\gamma(s))P_{\gamma,o,s} - \mathcal{D}X(p)\| + \|\mathcal{D}X(p) - P_{\sigma,0,1}\mathcal{D}X(p_0)P_{\sigma,1,0}\|)\,\|v\|\,ds$$

$$= \int_0^1 (\| P_{\gamma,s,o}\mathcal{D}X(\gamma(s))P_{\gamma,o,s} - \mathcal{D}X(\gamma(0)) \| +$$

$$\| \mathcal{D}X(\sigma(1)) - P_{\sigma,0,1}\mathcal{D}X(\sigma(0))P_{\sigma,1,0} \|)\,\| v \|\,ds$$

$$\leq l \int_0^t \left(\int_0^s \|\gamma'(\tau)\|\,d\tau + d(p_0, p)\right)\|v\|\,ds = l\int_0^t \left(\int_0^s \|\gamma'(0)\|\,d\tau + d(p_0, p)\right)\|v\|\,ds$$

$$= l\int_0^t (s\|\gamma'(0)\| + d(p_0, p))\|v\|\,ds = l\left(\frac{t^2}{2}\|v\| + td(p_0, p)\right)\|v\|.$$

Therefore,

$$\|R(t)\| \leq l\left(\frac{t}{2}\|v\| + d(p_0, p)\right) t\,\|v\|.$$

$\square$

Corollary 3.5.1. *Let M be a Riemannian manifold, $\Omega \subseteq M$ be an open convex set, $X \in \chi(M)$, $\mathcal{D}X \in Lip_l(\Omega)$. Take $p \in \Omega$, $v \in T_pM$ and let be*

$$\gamma(t) = \exp_p(tv).$$

If $\gamma[0, t) \subseteq \Omega$ and $P_{\sigma,0,1}\mathcal{D}X(p_0)\,P_{\sigma,1,0}v = -X(p)$, then

$$\|P_{\gamma,1,0}X(\gamma(1))\| \leq l\left(\frac{1}{2}\|v\| + d(p_0, p)\right)\|v\|. \tag{3.276}$$

Now we can prove the simplified Kantorovich theorem on Riemannian manifolds. The proof of this theorem will be divided into two parts. First, we will prove that simplified Kantorovich method is well defined, i.e. $\{p_k\}_{k\in\mathbb{N}} \subseteq B(p_0, t_*)$; we will also prove the convergence of the method. In the second part, we will establish uniqueness.

We consider the auxiliary real function $f : \mathbb{R} \longrightarrow \mathbb{R}$, defined by

$$f(t) = \frac{l}{2}t^2 - \frac{1}{a}t + \frac{b}{a}. \tag{3.277}$$

Its discriminant is

$$\triangle = \frac{1}{a^2}(1 - 2\,l\,b\,a),$$

which is positive, because $a\,b\,l \leq \dfrac{1}{2}$. Thus f has at least one real root (unique when $h = \dfrac{1}{2}$). If t_* is the smallest root, a direct calculation show that $f'(t) < 0$ for $0 \leq t < t_*$, so f is strictly decreasing in $[0, t_*]$. Therefore the (scalar) Newton's method can be applied to f, in other words:

If $t_0 \in [0, t_*)$, for $k = 0, 1, 2, \ldots$, we define

$$t_{k+1} = t_k - \frac{f(t_k)}{f'(0)}.$$

Then $\{t_k\}_{k\in\mathbb{N}}$ is well defined, strictly increasing and converges to t_*. Furthermore, if $h = a\,b\,l < \dfrac{1}{2}$, then (see Ref. [290])

$$t_* - t_k \leq \frac{b}{h}\left(1 - \sqrt{1 - 2\,h}\right)^{k+1}, \quad k = 1, 2, \ldots. \tag{3.278}$$

Let us take as starting point $t_0 = 0$. We want to show that Newton's iterations are well defined for any $q \in B(p_0, t_*) \subseteq \Omega$.

We define

$$\mathbf{K}(t) = \left\{q \in \overline{B}(p_0, t) : \left\|P_{\sigma,0,1}\mathcal{D}X(p_0)^{-1}P_{\sigma,1,0}X(q)\right\| \leq \frac{f(t)}{|f'(0)|} = af(t),\ 0 \leq t < t_*\right\} \tag{3.279}$$

where $\sigma : [0, 1] \longrightarrow M$ is the minimizing geodesic connecting p_0 and q. Note that $\mathbf{K}(t) \neq \emptyset$ since $p_0 \in \mathbf{K}(t)$.

Proposition 3.5.1. *Under the hypotheses of either the Kantorovich or the simplified Kantorovich method, if $q \in B(p_0, t_*)$, then $\mathcal{D}X(q)$ is nonsingular and*

$$\left\| \mathcal{D}X(q)^{-1} \right\| \leq \frac{1}{|f'(\lambda)|} \quad where \quad \lambda = d(p_0, q) < t_*.$$

Proof. Let $\lambda = d(p_0, q)$ and $\alpha : [0, 1] \longrightarrow M$ be a geodesic with $\alpha(0) = p_0$, $\alpha(1) = q$ and $\|\alpha'(0)\| = \lambda$. Define $\phi : T_q M \longrightarrow T_q M$ by letting

$$\phi = P_{\alpha,1,0} \mathcal{D}X(p_0) P_{\alpha,0,1}. \tag{3.280}$$

Since $P_{\alpha,1,0}$ and $P_{\alpha,0,1}$ are linear, isometric and $\mathcal{D}X(p_0)$ is nonsingular, we have that ϕ is linear, nonsingular and

$$\left\| \phi^{-1} \right\| = \left\| \mathcal{D}X(p_0)^{-1} \right\| \leq a = \frac{1}{|f'(0)|},$$

with $\alpha([0, 1]) \subseteq B(p_0, t_*)$. Since $d(p_0, q) < t_*$, $\mathcal{D}X \in Lip_L(\Omega)$ and $\|\alpha'(0)\| = \lambda$. Therefore

$$\|\mathcal{D}X(q) - \phi\| \leq l\lambda. \tag{3.281}$$

By (3.280) and (3.281), we have that

$$\left\| \phi^{-1} \right\| \|\mathcal{D}X(q) - \phi\| \leq al\lambda \leq alt_* = al\frac{1}{al}\left(1 - \sqrt{1 - 2\,a\,b\,l}\right) \leq 1.$$

Using Banach's lemma, we conclude that $\mathcal{D}X(q)$ is nonsingular and

$$\left\| \mathcal{D}X(q)^{-1} \right\| \leq \frac{\left\| \phi^{-1} \right\|}{1 - \left\| \phi^{-1} \right\| \|\mathcal{D}X(q) - \phi\|} \leq \frac{a}{1 - al\lambda} \leq \frac{1}{|f'(\lambda)|}. \qquad \square$$

Therefore, for any $q \in B(p_0, t_*)$, we can apply the Kantorovich methods.

Lemma 3.5.2. *Let $q \in \mathbf{K}(t)$, define*

$$t_+ = t - \frac{f(t)}{|f'(0)|}$$

$$q_+ = \exp_q\left(-P_{\sigma,0,1}\mathcal{D}X(p_0)^{-1} P_{\sigma,1,0}X(q)\right).$$

Then $t < t_+ < t_*$ and $q_+ \in \mathbf{K}(t_+)$.

Proof. Consider the geodesic $\gamma : [0, 1] \longrightarrow M$ defined by

$$\gamma(\theta) = \exp_q\left(-\theta P_{\sigma,0,1}\mathcal{D}X(p_0)^{-1} P_{\sigma,1,0}X(q)\right),$$

we have that

$$d(p_0, \gamma(\theta)) \leq d(p_0, q) + d(q, \gamma(\theta))$$

$$\leq t + \left\| \theta P_{\sigma,0,1}\mathcal{D}X(p_0)^{-1} P_{\sigma,1,0}X(q) \right\| \leq t + \theta\frac{f(t)}{|f'(0)|}.$$

Since

$$\gamma\left(1\right) = \exp_q\left(-P_{\sigma,0,1}\mathcal{D}X\left(p_0\right)^{-1}P_{\sigma,1,0}X\left(q\right)\right) = q_+,$$

this implies that

$$d\left(p_0, q_+\right) = d\left(p_0, \gamma\left(1\right)\right) \le t + \frac{f\left(t\right)}{\left|f'\left(0\right)\right|} = t_+,$$

therefore

$$q_+ \in B\left(p_0, t_+\right) \subset B\left(p_0, t_*\right).$$

Moreover, if $\sigma_+\left[0,1\right] \longrightarrow M$ is the minimizing geodesic connecting p_0 and q_+, then

$$\left\|-P_{\sigma_+,0,1}\mathcal{D}X\left(p_0\right)^{-1}P_{\sigma_+,1,0}X\left(q_+\right)\right\| \le \left\|\mathcal{D}X\left(p_0\right)^{-1}\right\|\left\|X\left(q_+\right)\right\|.$$

Furthermore, if $v = -P_{\sigma,0,1}\mathcal{D}X\left(p_0\right)^{-1}P_{\sigma,1,0}X\left(q\right),$ then

$$P_{\sigma,0,1}\mathcal{D}X\left(p_0\right)P_{\sigma,1,0}v = P_{\sigma,0,1}\mathcal{D}X\left(p_0\right)P_{\sigma,1,0}\left(-P_{\sigma,0,1}\mathcal{D}X\left(p_0\right)^{-1}P_{\sigma,1,0}X\left(q\right)\right)$$

$$= -P_{\sigma,0,1}\mathcal{D}X\left(p_0\right)\mathcal{D}X\left(p_0\right)^{-1}P_{\sigma,1,0}X\left(q\right) = -X\left(q\right).$$

By Theorem 3.5.2,

$$\left\|X\left(q_+\right)\right\| = \left\|X\left(\gamma\left(1\right)\right)\right\| \le l\left(\frac{1}{2}\left\|v\right\| + d\left(p_0, p\right)\right)\left\|v\right\|$$

$$\le l\left(\frac{1}{2}\left\|-P_{\sigma_+,0,1}\mathcal{D}X\left(p_0\right)^{-1}P_{\sigma_+,1,0}X\left(q\right)\right\| + t\right)\left\|-P_{\sigma_+,0,1}\mathcal{D}X\left(p_0\right)^{-1}P_{\sigma_+,1,0}X\left(q\right)\right\|$$

$$\le l\left(\frac{1}{2}\left(\frac{f\left(t\right)}{\left|f'\left(0\right)\right|}\right) + t\right)\left(\frac{f\left(t\right)}{\left|f'\left(0\right)\right|}\right).$$

Thus, by (3.279), after some calculations

$$\left\|-P_{\sigma_+,0,1}\mathcal{D}X\left(p_0\right)^{-1}P_{\sigma_+,1,0}\right\|\left\|X\left(q_+\right)\right\|$$

$$\le \left(\frac{1}{\left|f'\left(0\right)\right|}\right)l\left(\frac{1}{2}\left(\frac{f\left(t\right)}{\left|f'\left(0\right)\right|}\right) + t\right)\left(\frac{f\left(t\right)}{\left|f'\left(0\right)\right|}\right)$$

$$= \frac{1}{8}l\left(2b - 2t + alt^2\right)\left(2b + 2t + alt^2\right) = \frac{f\left(t_+\right)}{\left|f'\left(0\right)\right|},$$

we thus conclude $\left\|X\left(q_+\right)\right\| \le f\left(t_+\right)/\left|f'\left(0\right)\right|$ and therefore $q_+ \in \mathbf{K}\left(t_+\right).$ $\square$

Now we are going to prove that starting from any point of $\mathbf{K}\left(t\right)$ the simplified Newton method converges.

Corollary 3.5.2. *Take $0 \le t < t_*$ and $q \in \mathbf{K}\left(t\right),$ and define*

$$\tau_0 = t$$

$$\tau_{k+1} = \tau_k - \frac{f\left(\tau_k\right)}{\left|f'\left(0\right)\right|} \quad for\ each \quad k = 0, 1, \cdots.$$

Then the sequence generated by Newton's method starting with the point $q_0 = q$ is well defined for any k and

$$q_k \in \mathbf{K}\left(\tau_k\right). \tag{3.282}$$

Moreover $\left\{q_k\right\}_{k\in\mathbb{N}}$ converges to some $q_* \in B\left(p_0, t_*\right),$ $X\left(q_*\right) = 0$ and

$$d\left(q_k, q_*\right) \le t_* - \tau_k \quad for\ each \quad k = 0, 1, \cdots.$$

Proof. It is clear that the sequence $\{\tau_k\}_{k\in\mathbb{N}}$ is the sequence generated by Newton's method for solving $f(t)=0$. Therefore, $\{\tau_k\}_{k\in\mathbb{N}}$ is well defined, strictly increasing and it converges to the root t_* (see the definition of f). By hypothesis, $q_0 \in \mathbf{K}(\tau_0)$; suppose that the points $q_0, q_1, \ldots, q_k$ are well defined. Then, using Banach's Lemma, we conclude that q_{k+1} is well defined. Furthermore,

$$d(q_{k+1}, q_k) \leq \left\| -P_{\sigma_k,0,1}\mathcal{D}X(p_0)^{-1} P_{\sigma_k,1,0}X(q_k) \right\|.$$

Since

$$q_{k+1} = \exp_{q_k}\left(-P_{\sigma_k,0,1}\mathcal{D}X(p_0)^{-1} P_{\sigma_k,1,0}X(q_k) \right)$$

and $\sigma_k : [0,1] \longrightarrow M$ is the minimizing geodesic connecting p_0, q_k, from Lemma 3.5.2 and using (3.282) we obtain that

$$d(q_{k+1}, q_k) \leq \frac{f(\tau_k)}{|f'(0)|} = \tau_{k+1} - \tau_k. \tag{3.283}$$

Hence, for $k \geq s$, $s \in \mathbb{N}$,

$$d(q_k, q_s) \leq \tau_s - \tau_k. \tag{3.284}$$

It follows that $\{q_k\}_{k\in\mathbb{N}}$ is a Cauchy sequence. Since M is complete, it converges to some $q_* \in M$. Moreover, $q_k \in \mathbf{K}(\tau_k) \subseteq B[p_0, t_*]$, therefore $q_* \in B[p_0, t_*]$. Next, we prove that $X(q_*) = 0$. We have that

$$\|X(q_k)\| = \left\| P_{\sigma_k,0,1}\mathcal{D}X(p_0) P_{\sigma_k,1,0}P_{\sigma_k,0,1}\mathcal{D}X(p_0)^{-1} P_{\sigma_k,1,0}X(q_k) \right\|$$

$$\leq \|\mathcal{D}X(p_0)\| \left\| \mathcal{D}X(p_0)^{-1} X(q_k) \right\|$$

$$\leq (\|\mathcal{D}X(p_0)\|) \frac{f(\tau_k)}{|f'(0)|} = (\|\mathcal{D}X(p_0)\|)(\tau_{k+1} - \tau_k).$$

Passing to the limit in k, we conclude $X(q_*) = 0$. Finally, letting $s \longrightarrow \infty$ in (3.284), we get that $d(q_*, q_k) \leq t_* - \tau_k$. $\qquad\square$

Finally, from (3.278)

$$d(q_*, q_k) \leq \frac{b}{h}\left(1 - \sqrt{1 - 2h} \right)^{k+1}, \quad k = 1, 2, \ldots.$$

By hypothesis, $p_0 \in \mathbf{K}(0)$, thus by Lemma 3.5.2, the sequence $\{p_k\}_{k\in\mathbb{N}}$ generated by (3.274) is well defined, contained in $B(p_0, t_*)$ and converges to some p_*, which is a singular point of X in $B[p_0, t_*]$. Moreover, if $h < 1/2$, then

$$d(p_k, p_*) \leq \frac{b}{h}\left(1 - \sqrt{1 - 2h} \right)^{k+1}.$$

This proof will be made in an indirect way, by contradiction. But before we are going to establish some results.

Lemma 3.5.3. *Take $0 \leq t < t_*$ and $q \in \mathbf{K}(t)$, let*

$$A^{-1} = -P_{\sigma,0,1}\mathcal{D}X(p_0)^{-1} P_{\sigma,1,0}$$
$$v = A^{-1}X(q),$$

where $\sigma : [0,1] \longrightarrow M$ is a minimizing geodesic connecting p_0, p_k. Define for $\theta \in \mathbb{R}$

$$\tau(\theta) = t + \theta a f(t),$$
$$\gamma(\theta) = \exp_q(\theta v).$$

Then, for $\theta \in [0,1]$, we have $t < \tau(\theta) < t_*$ and $\gamma(\theta) \in \mathbf{K}(\tau(\theta))$.

Proof. Because γ is a minimizing geodesic, for all $\theta \in [0,1]$ we have

$$d(p_0, \gamma(\theta)) \le d(p_0, q) + d(q, \gamma(\theta))$$
$$\le t + \theta \|v\| \le t + \theta a f(t) = \tau(\theta).$$

This implies that

$$t \le \tau(\theta) \le \tau(1) \le t_* \text{ and } \gamma([0, \theta]) \subset B(p_0, t_*). \tag{3.285}$$

Using Lemma 3.5.1, we obtain

$$X(\gamma(\theta)) = P_{\gamma, 0, \theta}\left(X(p) + P_{\sigma, 0, 1}\theta DX(p_0) P_{\sigma, 1, 0}v + R(\theta)\right),$$

with

$$R(\theta) = \int_0^\theta \left(P_{\gamma, s, 0}DX(\gamma(s)) P_{\gamma, 0, s}v - P_{\sigma, 0, 1}DX(p_0) P_{\sigma, 1, 0}v\right) ds,$$

and

$$\|R(\theta)\| \le L\left(\frac{\theta}{2}\|v\| + d(p_0, q)\right)\theta\|v\|.$$

After some calculations, this yields

$$\left\|A^{-1}X(\gamma(\theta))\right\| = \left\|A^{-1}P_{\gamma, 0, \theta}\left(X(q) - \int_0^\theta P_{\gamma, s, 0}DX(\gamma(s)) P_{\gamma, 0, s}v\right) ds\right\|$$
$$= \left\|A^{-1}P_{\gamma, 0, \theta}\left((1 - \theta) X(q) - \int_0^\theta \left(P_{\gamma, s, 0}DX(\gamma(s)) P_{\gamma, 0, s} - DX(q)\right)v\right) ds\right\|$$
$$\le \left\|A^{-1}P_{\gamma, 0, \theta}(1 - \theta) X(q)\right\| + \left\|A^{-1}P_{\gamma, 0, \theta}\int_0^\theta \left(P_{\gamma, s, 0}DX(\gamma(s)) P_{\gamma, 0, s} - DX(q)\right)vds\right\|$$
$$\le (1 - \theta) a f(t) + a\|R(\theta)\| \le (1 - \theta) a f(t) + aL\left(\frac{\theta}{2}\|v\| + d(p_0, q)\right)\theta\|v\|$$
$$\le (1 - \theta) a f(t) + aL\left(\frac{\theta}{2}a f(t) + t\right)\theta a f(t)$$
$$= \frac{1}{8}\left(2b - 2t + aLt^2\right)\left(-4\theta + a^2L^2\theta^2 t^2 + 4aL\theta t + 2abL\theta^2 - 2aL\theta^2 t + 4\right)$$
$$= a f(\tau(\theta)).$$

Therefore $\gamma(\theta) \in \mathbf{K}(\tau(\theta))$ and the Lemma is proved. $\qquad\square$

Lemma 3.5.4. *Let $0 \le t < t_*$ and $q \in \mathbf{K}(t)$. Suppose that $q_* \in B[p_0, t_*]$ is a singularity of the vector field X and $t + d(q, q_*) = t_*$. Then $d(p_0, q) = t$. Moreover, letting*

$$t_+ = t + a f(t),$$
$$q_+ = \exp_q\left(A^{-1}X(q)\right),$$

then $t < t_+ < t_*$, $q_+ \in \mathbf{K}(t_+)$ and $t_+ + d(q_+, q_*) = t_*$.

Proof. Consider the minimizing geodesic $\alpha : [0,1] \longrightarrow M$ joining q to q_*. Since $q \in \mathbf{K}(t)$, we have that

$$d(p_0, \alpha(\theta)) \leq d(p_0, q) + d(q, \alpha(\theta)) \leq t + \theta d(q, q_*) \leq t + d(q, q_*) = t_*.$$

It follows that $\alpha([0,1]) \subset B(p_0, t_*)$. Taking $u = \alpha'(0)$, by Lemma 3.5.1 we have

$$P_{\alpha,1,0}X(\alpha(1)) = X(q) + P_{\sigma,0,1}DX(p_0) P_{\sigma,1,0}u + R(1),$$

with

$$\|R(1)\| \leq L\left(\frac{1}{2}\|u\| + d(p_0, q)\right)\|u\|.$$

Therefore

$$\|R(1)\| \leq L\left(\frac{1}{2}d(q, q_*) + d(p_0, q)\right)d(q, q_*) \tag{3.286}$$

$$= L\left(\frac{1}{2}(t_* - t) + d(p_0, q)\right)(t_* - t)$$

$$\leq L\left(\frac{1}{2}(t_* - t) + t\right)(t_* - t) = L\frac{1}{2}(t_* + t)(t_* - t).$$

On the other hand, since $|f(t)|$ is strictly decreasing in $[0, t_*]$ and $0 \leq d(p_0, q) \leq t < t_*$,

$$\|R(1)\| = \|X(q) + Au\| \geq \frac{1}{\left\|DX(p_0)^{-1}\right\|}\left\|A^{-1}X(q) + u\right\| \tag{3.287}$$

$$\geq |f'(0)|\left\|A^{-1}X(q) + u\right\| \geq |f'(0)|\left(\|u\| - \left\|A^{-1}X(q)\right\|\right)$$

$$\geq |f'(0)|\left(\|u\| - af(t)\right) = -f'(0)(t_* - t) - f(t) > 0.$$

Because $f''(t) = L$, $0 = f'(t_*) = f(t) + f'(t)(t_* - t) + \frac{1}{2}f''(t)(t_* - t)^2$ and $f'(t) = f'(0) + \int_0^t f''(t)\,dt$. Therefore, we have that

$$0 = f(t) + (f'(0) + tL)(t_* - t) + \frac{1}{2}L(t_* - t)^2.$$

Hence, we deduce that

$$\frac{1}{2}L(t_* + t)(t_* - t) = -f'(0)(t_* - t) - f(t).$$

Thus, the last term in (3.286) is equal to the last term in inequality (3.287). In particular

$$\frac{1}{\left\|DX(p_0)^{-1}\right\|} = |f'(0)| = a,$$

$$\|u\| - \left\|A^{-1}X(q)\right\| = \left\|A^{-1}X(q) + u\right\| > 0, \tag{3.288}$$
$$\left\|A^{-1}X(q)\right\| = af(t),$$
$$L\left(\frac{1}{2}(t_* - t) + d(p_0, q)\right)(t_* - t) = L\left(\frac{1}{2}(t_* - t) + t\right)(t_* - t).$$

From the last equation in (3.288), we obtain $d\left(p_0, q\right) = t$. The second equation in (3.288) implies that u and $A^{-1}X\left(q\right)$ are linearly dependent vectors in $T_q M$, so that there exists $r \in \mathbb{R}$ such that $A^{-1}X\left(q\right) = -ru$. Thus, the second equation implies $1 - |r| = |1 - r|$ and because $r \neq 0$ and $r \neq 1$, we have $0 < r < 1$, thus $q_+ = \exp_q\left(ru\right) = \alpha\left(r\right)$. Moreover, given that α is a minimizing geodesic joining q to q_*, we have that q, $\alpha\left(r\right)$ and q_* are in the same geodesic line, thus $d\left(q, \alpha\left(r\right)\right) + d\left(\alpha\left(r\right), q_*\right) = d\left(q, q_*\right)$. Therefore, we have that $d\left(q, q_+\right) + d\left(q_+, q_*\right) = d\left(q, q_*\right)$. Moreover, we get that

$$d\left(q, q_+\right) = \|ru\| = \left\|A^{-1}X\left(q\right)\right\| = af\left(t\right) = t_+ - t.$$

Hence, we deduce that

$$d\left(q_+, q_*\right) = d\left(q, q_*\right) - d\left(q, q_+\right) = \left(t_* - t\right) - \left(t_+ - t\right) = t_* - t_+.$$

That is, $d\left(q_+, q_*\right) + t_+ = t_*$. $\qquad\square$

Corollary 3.5.3. *Suppose that $q_* \in B\left[p_0, t_*\right]$ is a zero of the vector field X. If there exist $\tilde{t}$ and $\tilde{q}$ such that $0 \leq \tilde{t} < t_*, \tilde{q} \in \mathbf{K}\left(\tilde{t}\right)$ and $\tilde{t} + d\left(\tilde{q}, q_*\right) = t_*$, then $d\left(p_0, q_*\right) = t_*$.*

Proof. Changing τ_0 by $\tilde{t}$ and q_0 by $\tilde{q}$ in Corollary 3.5.2, we obtain $q_k \in \mathbf{K}\left(\tau_k\right)$ for all $k \in \mathbb{N}$. We have also $\{\tau_k\}_{k\in\mathbb{N}}$ converges to t_*, $\{q_k\}_{k\in\mathbb{N}}$ converges to some $\tilde{q}_* \in B\left(p_0, t_*\right)$ and $X\left(q_*\right) = 0$. Moreover, by Lemma 3.5.4 and applying induction, it is easy to show that for all k, $d\left(p_0, q_k\right) = \tau_k$ and $d\left(q_k, q_*\right) = t_* - \tau_k$. Passing to the limit, we obtain $d\left(p_0, \tilde{q}_*\right) = t_*$ and $d\left(\tilde{q}_*, q_*\right) = 0$. Therefore $\tilde{q}_* = q_*$ and $d\left(p_0, q_*\right) = t_*$. $\qquad\square$

The following two Lemmas complete the proof of the uniqueness.

Lemma 3.5.5. *The limit p_* of $\{p_k\}_{k\in\mathbb{N}}$ is the unique singularity of X in $B\left[p_0, t_*\right]$.*

Proof. Let $q_* \in B\left[p_0, t_*\right]$ be a singularity of the vector field X. Using induction, we will show that $d\left(p_k, q_*\right) + t_k \leq t_*$. We need to consider two cases:

Case 1. $\left(d\left(p_0, q_*\right) < t_*\right)$. First we show by induction that for all $k \in \mathbb{N}$,

$$d\left(p_k, q_*\right) + t_k < t_*. \tag{3.289}$$

Indeed, for $k = 0$ (3.289) is immediately true, because $t_0 = 0$. Now, suppose the property is true for some k. Let us take the geodesic $\gamma_k\left(\theta\right) = \exp_{p_k}\left(-\theta v_k\right)$, where v_k is defined in (3.274). From Lemma 3.5.3, for all $\theta \in [0, 1]$,

$$\gamma_k\left(\theta\right) \in \mathbf{K}\left(t_k + \theta\left(t_{k+1} - t_k\right)\right). \tag{3.290}$$

Define $\phi : [0, 1] \longrightarrow M$ by

$$\phi\left(\theta\right) = d\left(\gamma_k\left(\theta\right), q_*\right) + t_k + \theta\left(t_{k+1} - t_k\right). \tag{3.291}$$

We know that $\phi\left(0\right) = d\left(p_k, q_*\right) + t_k < t_*$. We next show, by contradiction, that $\phi\left(\theta\right) \neq t_*$ for all $\theta \in [0, 1]$. Suppose that there exists a $\tilde{\theta} \in [0, 1]$ such that $\phi\left(\tilde{\theta}\right) =

t_*. Let $\tilde{q} = \gamma_k\left(\tilde{\theta}\right)$ and $\tilde{t} = t_k + \tilde{\theta}\left(t_{k+1} - t_k\right)$. By (3.290) and (3.291), we have $\tilde{q} \in \mathbf{K}\left(\tilde{t}\right)$ and $d\left(\tilde{q}, q_*\right) + \tilde{t} = t_*$. Applying Corollary 3.5.3, we conclude that $d\left(p_0, q_*\right) = t_*$, which contradicts our assumption. Thus $\phi\left(\theta\right) \neq t_*$ for all $\theta \in [0,1]$. Since $\phi\left(0\right) < t_*$ and ϕ is continuous, we have that $\phi\left(\theta\right) < t_*$ for all $\theta \in [0,1]$. In particular, by (3.291), we have $d\left(\gamma_k\left(1\right), q_*\right) + t_{k+1} = \phi\left(1\right) < t_*$. Thus, we deduce $d\left(p_{k+1}, q_*\right) + t_{k+1} < t_*$, in this way (3.289) is true for all $k \in \mathbb{N}$.

Case 2. $\left(d\left(p_0, q_*\right) = t_*\right)$. Using induction, let us prove that for all $k \in \mathbb{N}$,

$$d\left(p_k, q_*\right) + t_k = t_*. \tag{3.292}$$

Indeed, for $k = 0$, this is immediately true, because $t_0 = 0$. Now, suppose that (3.292) is true for some k. Since $p_k \in \mathbf{K}\left(t_k\right)$, by Lemma 3.5.4 we conclude that $d\left(p_{k+1}, q_*\right) + t_{k+1} = t_*$. Finally, by (3.289) and (3.292) we conclude that for all $k \in \mathbb{N}$, that $d\left(p_k, q_*\right) + t_k \leq t_*$. Passing to the limit $k \longrightarrow \infty$, we obtain $d\left(p_*, q_*\right) = 0$ and $p_* = q_*$. $\qquad\square$

Lemma 3.5.6. *If* $h = abL < \dfrac{1}{2}$ *and* $B\left(p_0, r\right) \subseteq \Omega$, *with*

$$t_* < r \leq t_{**} = \frac{1}{aL}\left(1 + \sqrt{1 - 2h}\right),$$

then the limit p_* *of the sequence* $\{p_k\}_{k \in \mathbb{N}}$ *is the unique singularity of the vector field* X *in* $B\left(p_0, r\right)$.

Proof. Let $q_* \in B\left(p_0, r\right)$ be a singularity of the vector field X in $B\left(p_0, r\right)$. Let us consider the minimizing geodesic $\alpha : [0,1] \longrightarrow M$ joining p_0 to q_*. By Lemma 3.5.1,

$$P_{\alpha,1,0}X\left(\alpha\left(1\right)\right) = X\left(p_0\right) + P_{\sigma,0,1}\mathcal{D}X\left(p_0\right)P_{\sigma,1,0}u + R\left(1\right),$$

where

$$\left\|R\left(1\right)\right\| \leq L\left(\frac{1}{2}\left\|u\right\| + d\left(p_0, p_0\right)\right)\left\|u\right\| = \frac{L}{2}d\left(p_0, q_*\right)^2 \text{ and } \left\|u\right\| = d\left(p_0, q_*\right).$$

$$\tag{3.293}$$

In a similar way to the inequality (3.287), it is easy to prove that

$$\left\|R\left(1\right)\right\| \geq \frac{1}{a}\left(\left\|u\right\| - \left\|\mathcal{D}X\left(p_0\right)^{-1}X\left(p_0\right)\right\|\right) \geq \frac{1}{a}d\left(p_0, q_*\right) - \frac{b}{a}.$$

Therefore

$$\frac{L}{2}d\left(p_0, q_*\right)^2 \geq \frac{1}{a}d\left(p_0, q_*\right) - \frac{b}{a}.$$

Hence, we deduce $f\left(d\left(p_0, q_*\right)\right) \geq 0$, since $d\left(p_0, q_*\right) \leq r \leq t_{**}$. We conclude that $d\left(p_0, q_*\right) \leq t_*$. Finally, from Lemma 3.5.5, we have $p_* = q_*$. $\qquad\square$

The analysis of the order of convergence is performed in a local way, that is, in a neighborhood of the zero of the vector field. Then, we can define the order of convergence in Riemannian manifolds in the following way:

Definition 3.5.1. Let M be a manifold and let $\{p_k\}_{k\in\mathbb{N}}$ be a sequence on M converging to p_*. If there exists a system of coordinates (U, x) of M with $p_* \in U_\alpha$, constants $p > 0, c \geq 0$ and $K \geq 0$ such that, for all $k \geq K$, $\{p_k\}_{k=K}^{\infty} \subseteq U_\alpha$ the following inequality holds:

$$\left\| x^{-1}\left(p_{k+1}\right) - x^{-1}\left(p_*\right) \right\| \leq c \left\| x^{-1}\left(p_k\right) - x^{-1}\left(p_*\right) \right\|^p , \tag{3.294}$$

then we said that $\{p_k\}_{k\in\mathbb{N}}$ converges to p_* with order at least p.

It can be shown that the definition above does not depend on the choice of the coordinate system and the multiplicative constant c depends on the chart, but for any chart, there exists such a constant (see Ref. [1]). Notice that in normal coordinates of 0_{p_k}, we have $\left\| \exp_{p_k}^{-1}(p) - \exp_{p_k}^{-1}(q) \right\| = d(p, q)$. Thus, in normal coordinates, (3.294) is transformed into $d(p_{k+1}, p_*) \leq cd(p_k, p_*)^p$.

Lemma 3.5.7. *Let M be a Riemannian manifold, $\Omega \subseteq M$ be an open set, $X \in \chi_x(M)$ and $\mathcal{D}X \in Lip_l(\Omega)$. Let us take $p \in \Omega$, $v \in T_pM$ and $\gamma(t) = \exp_p(tv)$. If $\gamma[0, t) \subseteq \Omega$, then $P_{\gamma,t,0}X(\gamma(t)) = X(p) + t\mathcal{D}X(p)v + R(t)$, with $\|R(t)\| \leq (l/2)t^2 \|v\|^2$.*

Proof. From Theorem 3.5.1, it follows that

$$P_{\gamma,t,0}X(\gamma(t)) - X(\gamma(0)) = \int_0^t P_{\gamma,s,0}\left(\mathcal{D}X(\gamma(s))\gamma'(s)\right)ds.$$

Given that γ is a geodesic, we have that $\gamma'(t)$ is parallel and $\gamma'(s) = P_{\gamma,0,s}\gamma'(0)$. Moreover, since $\gamma'(0) = v$ then

$$P_{\gamma,t,0}X(\gamma(t)) - X(p) = \int_0^t P_{\gamma,s,0}\left(\mathcal{D}X(\gamma(s))P_{\gamma,0,s}v\right)ds.$$

Therefore, we have that

$$P_{\gamma,t,0}X(\gamma(t)) - X(p) - t\mathcal{D}X(p)v = \int_0^t P_{\gamma,s,0}\left(\mathcal{D}X(\gamma(s))P_{\gamma,0,s}v\right)ds - t\mathcal{D}X(p)v$$

$$= \int_0^t \left(P_{\gamma,s,0}\left(\mathcal{D}X(\gamma(s))P_{\gamma,0,s}v\right) - \mathcal{D}X(p)v\right)ds.$$

Let

$$R(t) = \int_0^t \left(P_{\gamma,s,0}\mathcal{D}X(\gamma(s))P_{\gamma,0,s}v - \mathcal{D}X(p)v\right)ds.$$

By hypothesis, $\mathcal{D}X \in Lip_L(\Omega)$, hence

$$\|R(t)\| \leq \int_0^t \left\|\left(P_{\gamma,s,0}\mathcal{D}X(\gamma(s))P_{\gamma,0,s}v - \mathcal{D}X(p)v\right)\right\|ds$$

$$\leq \int_0^t \left\|\left(P_{\gamma,s,0}\mathcal{D}X(\gamma(s))P_{\gamma,0,s} - \mathcal{D}X(p)\right)\right\| \|v\|\,ds$$

$$\leq \int_0^t \left(L\int_0^s \|\gamma'(\tau)\|\,d\tau\right)\|v\|\,ds.$$

Since γ is a geodesic, $\|\gamma'(\tau)\|$ is constant. Therefore, we have $\|\gamma'(\tau)\| = \|\gamma'(0)\| = \|v\|$. We deduce in turn that

$$\|R(t)\| \leq \int_0^t \left(L\int_0^s \|v\|\,d\tau\right)\|v\|\,ds = \int_0^t L\|v\|\,s\,\|v\|\,ds = \frac{L}{2}t^2 \|v\|^2. \qquad \square$$

Lemma 3.5.8. *(Order of convergence)*

i) *The convergence order of the Newton method in Riemannian manifold is two (quadratic convergence).*

ii) *The convergence order of the simplified Newton method in Riemannian manifold is one (linear convergence).*

Proof. Let k be sufficiently large in such a way that p_*, p_k, $p_{k+1}, \ldots$, belong to a normal neighborhood U of p_k. Let us consider the geodesic γ_k joining p_k to p_* defined by $\gamma_k(t) = \exp_{p_k}(tu_k)$, where $u_k \in T_{p_k}M$ and $d(p_k, p_*) = \|u_k\|$. We know that if p, q are in one normal neighborhood U of p_k, then $\left\|\exp_{p_k}^{-1}(p) - \exp_{p_k}^{-1}(q)\right\| = d(p, q)$. We prove now i). By Lemma 3.5.1, we have that

$$P_{\gamma, t, o}X(p_*) = X(p_k) + \mathcal{D}X(p_k)u_k + R(1),$$

with

$$\|R(1)\| \le \frac{L}{2}\|u_k\|^2 \text{ and } \|u_k\| = d(p_k, p_*).$$

Hence,

$$0 = \mathcal{D}X(p_k)^{-1}X(p_k) + u_k + \mathcal{D}X(p_k)^{-1}R(1).$$

Since

$$-\mathcal{D}X(p_k)^{-1}X(p_k) = \exp_{p_k}^{-1}(p_{k+1}) \text{ and } u_k = \exp_{p_k}^{-1}(p_*),$$

we have that

$$\exp_{p_k}^{-1}(p_{k+1}) - \exp_{p_k}^{-1}(p_*) = \mathcal{D}X(p_k)^{-1}R(1).$$

Thus, we get in turn that

$$d(p_{k+1}, p_*) \le \left\|\mathcal{D}X(p_k)^{-1}\right\|\frac{L}{2}\|u_k\|^2.$$

Moreover, by Banach's Lemma, we get that

$$\left\|\mathcal{D}X(p_k)^{-1}\right\| \le \frac{a}{1 - ald(p_k, p_0)} \le \frac{a}{1 - al\tau_k} \le \frac{a}{1 - alt_*} = \frac{a}{\sqrt{1 - 2abl}}.$$

Therefore, we have that $d(p_{k+1}, p_*) \le Cd(p_k, p_*)^2$, with $C = (La)/(2\sqrt{1 - 2abL})$. We prove now ii). Let p_0 be sufficiently near to p_* in such a way that p_0 is in the normal neighborhood U of 0_{p_k}, By Lemma 3.5.1, if $\sigma_k : [0, 1] \longrightarrow M$ is the minimizing geodesic connecting p_0, p_k, then, we have that

$$P_{\gamma, 1, 0}X(p_*) = X(p_k) + P_{\sigma_k, 0, 1}\mathcal{D}X(p_0)P_{\sigma_k, 1, 0}u_k + R(1),$$

with

$$\|R(1)\| \le L\left(\frac{1}{2}\|u_k\| + d(p_0, p_k)\right)\|u_k\|.$$

Therefore, we deduce that

$$0 = P_{\sigma_k, 0, 1}\mathcal{D}X(p_0)^{-1}P_{\sigma_k, 1, 0}X(p_k) + u_k + P_{\sigma_k, 0, 1}\mathcal{D}X(p_0)^{-1}P_{\sigma_k, 1, 0}R(1).$$

Since

$$-P_{\sigma_k,0,1}DX\left(p_0\right)^{-1}P_{\sigma_k,1,0}X\left(p_k\right) = \exp_{p_k}^{-1}\left(p_{k+1}\right) \text{ and } u_k = \exp_{p_k}^{-1}\left(p_*\right),$$

we have that

$$\exp_{p_k}^{-1}\left(p_{k+1}\right) - \exp_{p_k}^{-1}\left(p_*\right) = P_{\sigma_k,0,1}DX\left(p_0\right)^{-1}P_{\sigma_k,1,0}R\left(1\right).$$

We conclude that

$$d\left(p_{k+1},p_*\right) = \left\|\exp_{p_k}^{-1}\left(p_{k+1}\right) - \exp_{p_k}^{-1}\left(p_*\right)\right\| = \left\|P_{\sigma_k,0,1}DX\left(p_0\right)^{-1}P_{\sigma_k,1,0}R\left(1\right)\right\|$$

$$\leq \left\|DX\left(p_0\right)^{-1}\right\|\left\|R\left(1\right)\right\| \leq al\left(\frac{1}{2}\left\|u_k\right\| + d\left(p_0,p_k\right)\right)\left\|u_k\right\|$$

$$= al\left(\frac{1}{2}d\left(p_k,p_*\right) + d\left(p_0,p_k\right)\right)d\left(p_k,p_*\right)$$

$$= al\left(\frac{1}{2}\frac{d\left(p_k,p_*\right)}{d\left(p_0,p_k\right)} + 1\right)d\left(p_0,p_k\right)d\left(p_k,p_*\right).$$

If k is sufficiently large, then $d\left(p_k,p_*\right) \leq d\left(p_0,p_k\right)$. Therefore, we deduce that

$$\left(\frac{1}{2}\frac{d\left(p_k,p_*\right)}{d\left(p_0,p_k\right)} + 1\right) \leq \frac{3}{2}.$$

For p_0 sufficiently close to p_*, we get $d\left(p_{k+1},p_*\right) \leq K_0 d\left(p_0,p_k\right)d\left(p_k,p_*\right)$, with $K_0 \leq 3aL/2$. $\qquad\qquad\square$

Remark 3.5.1. Note that if instead of putting in the Kantorovich method the point p_0, we fix p_j sufficiently close to p_*, we will obtain a new convergent method. Indeed the calculations made in the previous lemma become

$$d\left(p_{k+1},p_*\right) \leq K_j d\left(p_j,p_*\right)d\left(p_k,p_*\right),$$

with $K_j \leq 3aL/2$. Thus, we have that

$$d\left(p_{k+1},p_*\right) \leq K d\left(p_j,p_*\right)d\left(p_k,p_*\right), \tag{3.295}$$

with $K \leq 3aL/2$.

We recall the Shamanskii family of iterative methods. Given an integer m and an initial point x_0 in a Banach Space, we move from x_n to x_{n+1} through an intermediate sequence $\{y_n^i\}_{i=0}^m$, $y_n^0 = x_n$, which is a generalization of Newton ($m = 1$) and simplified Newton ($m = \infty$) methods

$$\begin{cases} y_n^1 = y_n^0 - DF\left(y_n^0\right)^{-1}F\left(y_n^0\right) \\ y_n^2 = y_n^1 - DF\left(y_n^0\right)^{-1}F\left(y_n^1\right) \\ \;\;\vdots \\ y_n^m = x_{n+1} = y_n^{m-1} - DF\left(y_n^0\right)^{-1}F\left(y_n^{(m-1)}\right). \end{cases}$$

For a problem on Riemannian manifolds, let us consider the family

$$
\begin{cases}
q_n^1 = \exp_{p_n}\left(-\mathcal{D}X\left(p_n\right)^{-1}X\left(p_n\right)\right) \\[2mm]
q_n^2 = \exp_{q_n^1}\left(-P_{\sigma_1,0,1}\mathcal{D}X\left(p_n\right)^{-1}P_{\sigma_1,1,0}X\left(q_n^1\right)\right) \\[2mm]
\ \vdots \\[2mm]
q_n^m = p_{n+1} = \exp_{q_n^{m-1}}\left(-P_{\sigma_{m-1},0,1}\mathcal{D}X\left(p_n\right)^{-1}P_{\sigma_{m-1},1,0}X\left(q_n^{(m-1)}\right)\right),
\end{cases}
\tag{3.296}
$$

where $\sigma_k : [0,1] \longrightarrow M$ be the minimizing geodesic joining the points p_n and q_n^k; $k = 1, 2, \ldots, (m-1)$, thus:

$$
\sigma_k\left(0\right) = p_n \text{ and } \sigma_k\left(1\right) = q_n^k.
$$

Theorem 3.5.7. *Under the hypotheses of Kantorovich's theorem, the method described in (3.296) converges with order of convergence $m+1$.*

Proof. Let us observe that

$$
d\left(p_{n+1},p_n\right) \leq d\left(p_{n+1},q_n^{(m-1)}\right) + d\left(q_n^{(m-1)},q_n^{(m-2)}\right) + \cdots + d\left(q_n^2,q_n^1\right) + d\left(q_n^1,p_n\right).
$$

Now, if we define $p_{n+1} = q_n^m$, $p_n = q_n^1$, looking at each step as a different method according to (3.296), then by Kantorovich theorem in the first step and by the simplified Kantorovich theorem for the following steps, each of the sequences $\{q_n^m\}_{m\in\mathbb{N}}$ for fixed n, is convergent to the same point $p_* \in M$. Therefore, $\{p_n\}_{p\in\mathbb{N}}$ is convergent to p_*. Moreover, for Lemma 3.5.8 i) and (3.295),

$$
d\left(p_{n+1},p_*\right) \leq K d\left(p_n,p_*\right) d\left(q_n^{(m-1)},p_*\right) \leq K d\left(p_n,p_*\right) K d\left(p_n,p_*\right) d\left(q_n^{(m-2)},p_*\right)
$$

$$
\leq \cdots \leq K^{m-1}d\left(p_n,p_*\right)^{m-1}d\left(q_n^1,p_*\right) \leq K^{m-1}d\left(p_n,p_*\right)^{m-1}Cd\left(p_n,p_*\right)^2.
$$

Therefore,

$$
d\left(p_{n+1},p_*\right) \leq CK^{m-1}d\left(p_n,p_*\right)^{m+1}.
$$
$\square$

We have used Lipschitz condition (3.216) and the famous Kantorovich sufficient convergence criterion (3.272) in connection to majorizing function f for the semi-local convergence of both simplified Newton and Newton methods. According to the proof of Lemma 3.5.1, corresponding majorizing sequences for these methods are given by (see Ref. [216])

$$
t_0 = 0, \ t_1 = b
$$
$$
t_{k+1} = t_k + \frac{f(t_k)}{|f'(0)|} = t_k + \frac{a\,l}{2}\left(t_k - t_{k-1}\right)^2 \quad \text{for each} \quad k = 1, 2, \cdots
$$

for the simplified Newton method and

$$
u_0 = 0, \ u_1 = b
$$
$$
u_{k+1} = u_k + \frac{f(u_k)}{|f'(u_k)|} = u_k + \frac{a\,l\left(u_k - u_{k-1}\right)^2}{2\left(1 - a\,l\,u_k\right)} \quad \text{for each} \quad k = 1, 2, \cdots
$$

for the Newton method. Kantorovich criterion (3.272) may not be satisfied on a particular problem but Newton methods may still converge to p_* (see Ref. [216]). Next, we shall show that condition (3.272) can be weakened by introducing the center Lipschitz condition and relying on tighter majorizing sequences instead of majorizing function f.

Definition 3.5.2. Let E be a Banach space, $\Omega \subseteq E$ be an open convex set, $F : \Omega \longrightarrow \Omega$ be a continuous operator, x_0 be a point in Ω, such that $F \in C^1$ and DF is center-Lipschitz in Ω at x_0

$$\| DF(x) - DF(x_0) \| \le l_0 \| x - x_0 \| \quad \text{for each} \quad x \in \Omega \quad \text{and some} \quad l_0 > 0.$$

As in the case of Definition 3.4.7 we will write $DF \in Lip_{l_0}(\Omega)$ at $x_0 \in \Omega$.

We present the semi-local convergence of the simplified Newton method using only the center-Lipschitz condition.

Theorem 3.5.8. *Let M be a Riemannian manifold, $\Omega \subseteq M$ be an open convex set, $X \in \chi(M)$. Suppose that for some $p_0 \in \Omega$, $DX \in Lip_{l_0}(\Omega)$ at p_0, $DX(p_0)$ is invertible and that for some $a > 0$ and $b \ge 0$, the following hold*

$$\left\| DX\left(p_0\right)^{-1} \right\| \le a, \quad \left\| DX\left(p_0\right)^{-1} X\left(x_0\right) \right\| \le b, \quad h_0 = a\,b\,l_0 \le \frac{1}{2}$$

and

$$B\left(p_0, t_*^0\right) \subseteq \Omega \quad where \quad t_*^0 = \frac{1}{a\,l_0}\left(1 - \sqrt{1 - 2\,h_0}\right).$$

Then, sequence $\{p_k\}$ generated by (3.274) is such that $\{p_k\} \subseteq B(p_0, t_*^0)$ and $p_k \to p_*$, which the only singularity of X in $B(p_0, t_*^0)$. Moreover, if $h_0 < 1/2$ and $B(p_0, r) \subseteq \Omega$ with

$$t_*^0 < r \le t_{**}^0 = \frac{1}{a\,l_0}\left(1 + \sqrt{1 - 2\,h_0}\right)$$

and p_* is also the only singularity of F in $B(p_0, r)$. Furthermore, the following error bounds are satisfied for each $k = 1, 2, \cdots$

$$d(p_k, p_{k-1}) \le t_k^0 - t_{k-1}^0, \quad d(p_k, p_*) \le t_*^0 - t_k^0$$

and

$$d(p_k, p_*) \le \frac{b}{h_0}\left(1 - \sqrt{1 - 2\,h_0}\right)^{k+1},$$

where sequence $\{t_k^0\}$ is defined by

$$t_0^0 = 0, \ t_1^0 = b$$
$$t_{k+1}^0 = t_k^0 + \frac{a\,l_0}{2}\left(t_k^0 - t_{k-1}^0\right)^2 \quad for \ each \quad k = 1, 2, \cdots.$$

Proof. Simply notice that l_0, h_0, $\{t_k^0\}$, t_*^0, t_{**}^0 can replace l, h, $\{t_k\}$, t_*, t_{**}, respectively, in the proof of Theorem 3.5.6. $\qquad\qquad\square$

Remark 3.5.2. Under Kantorovich criterion (3.272) a simple inductive argument shows that

$$t_k^0 \leq t_k \quad \text{and} \quad t_{k+1}^0 - t_k^0 \leq t_{k+1} - t_k \quad \text{for each} \quad k = 0, 1, \cdots .$$

Moreover, we have that

$$t_*^0 \leq t_*, \quad t_{**} \leq t_{**}^0, \quad h \leq \frac{1}{2} \Longrightarrow h_0 \leq \frac{1}{2}$$

and

$$\frac{h_0}{h} \longrightarrow 0 \quad \text{as} \quad \frac{l_0}{l} \longrightarrow 0.$$

Furthermore, strict inequality holds in these estimates (for $k > 1$) if $l_0 < l$.

The convergence order of simplified method is only linear, whereas the convergence order of Newton method is quadratic if $h < 1/2$. If criterion $h \leq 1/2$ is not satisfied but weaker $h_0 \leq 1/2$ is satisfied, we can start with the simplified method until a certain iterate x_N (N a finite natural integer) at which criterion $h \leq 1/2$ is satisfied. Such an integer N exists. Since the simplified Newton method converges (see Refs. [104, 110, 120]). This approach was not possible before since $h \leq 1/2$ was used at the convergence criterion for both methods.

Remark 3.5.3. Under the hypotheses of Theorem 3.5.5, we see in the proof of this Theorem, sequences $\{r_k\}$, $\{s_k\}$ defined by

$$
\begin{aligned}
&r_0 = 0, \ r_1 = b, \ r_2 = r_1 + \frac{a\, l_0\, (r_1 - r_0)^2}{2\, (1 - a\, l_0\, r_1)^2} \\
&r_{k+1} = r_k + \frac{a\, l\, (r_k - r_{k-1})^2}{2\, (1 - a\, l_0\, r_k)} \quad \text{for each} \quad k = 2, 3, \cdots
\end{aligned}
\tag{3.297}
$$

$$
\begin{aligned}
&s_0 = 0, \ s_1 = b \\
&s_{k+1} = s_k + \frac{a\, l\, (s_k - s_{k-1})^2}{2\, (1 - a\, l_0\, s_k)} \quad \text{for each} \quad k = 1, 2, \cdots
\end{aligned}
\tag{3.298}
$$

are also majorizing sequences for $\{p_k\}$ such that

$$r_k \leq s_k \leq u_k, \quad d(p_k, p_{k-1}) \leq r_k - r_{k-1} \leq s_k - s_{k-1} \leq u_k - u_{k-1}$$

and

$$r_* = \lim_{k \to \infty} r_k \leq s_* = \lim_{k \to \infty} s_k \leq t_* = \lim_{k \to \infty} u_k.$$

Simply notice that for the computation of the upper bound on $\| DX(q)^{-1} \|$ (see (3.281)), we can use the center-Lipschitz condition

$$\| DX(q)^{-1} \| \leq \frac{\| \phi^{-1} \|}{1 - \| \phi^{-1} \| \, \| DX(q) - \phi \|} \leq \frac{a}{1 - a\, l_0\, \lambda}$$

instead of the less tight (if $l_0 < l$) and more expensive to compute estimate

$$\| DX(q)^{-1} \| \leq \frac{a}{1 - a\, l\, \lambda}$$

obtained in the proofs of Theorems 3.5.5 and 3.5.6 using the Lipschitz condition. Hence, the results of Theorem 3.5.5 involving sequence $\{u_k\}$ can be rewritten using tighter sequences $\{r_k\}$ or $\{s_k\}$. Note that the introduction of the center-Lipschitz condition is not an additional hypothesis to Lipschitz condition since in practice, the computation of l requires the computation of l_0. So far we showed that under Kantorovich criterion (3.272) the estimates of the distances $d(p_k, p_{k-1})$, $d(p_k, p_*)$ are improved (if $l_0 < l$) using tighter sequences $\{r_k\}$, $\{s_k\}$ for the computation on the upper bounds of these distances. Moreover, the information on the location of the solution is at least as precise.

Note that Kantorovich criterion (3.272) can be weakened if one directly (and not through majorizing function f) studies the convergence of sequences $\{r_k\}$ and $\{s_k\}$.

3.6   Traub-type Method on Riemannian Manifolds

In this section we present semi-local convergence results for Traub-type high convergence order iterative procedures for approximating zeros of a vector field on Riemannian manifolds. A characterization of the convergence under Kantorovich-type conditions and error estimates are also given in this section. We use all fundamental properties and notation of Riemannian manifolds as sections 3.4 and 3.5. Note that (3.269) is equivalent to

$$P_{\gamma,t,0}DX\left(\gamma\left(t\right)\right)P_{\gamma,0,t} - DX\left(\gamma\left(0\right)\right) = \int_0^s P_{\gamma,s,0}D^2X\left(\gamma\left(s\right)\right)P_{\gamma,0,s}\left(\gamma'\left(0\right),.\right)ds.$$

We can prove that

$$P_{\gamma,t,0}D^nX\left(\gamma\left(t\right)\right)P_{\gamma,0,t}^n - D^nX\left(\gamma\left(0\right)\right) = \int_0^s P_{\gamma,s,0}\left(D^nX\left(\gamma\left(s\right)\right)P_{\gamma,0,s}^n\left(\gamma'\left(0\right),...,.\right)\right)ds.$$

$$(3.299)$$

Theorem 3.6.1. *Let γ be a geodesic in M, $[0,1] \subseteq Dom\left(\gamma\right)$ and let X be a C^2 vector field on M. Then*

$$P_{\gamma,s,0}X\left(\gamma\left(s\right)\right) = X\left(\gamma\left(0\right)\right) + sDX\left(\gamma\left(0\right)\right)\gamma'\left(0\right) + \frac{1}{2}s^2D^2X\left(\gamma\left(0\right)\right)\left(\gamma'\left(0\right),\gamma'\left(0\right)\right)$$

$$+ \frac{1}{2}\int_0^s \left(s-t\right)^2 P_{\gamma,t,0}D^3X\left(\gamma\left(t\right)\right)\left(\gamma'\left(t\right),\gamma'\left(t\right),\gamma'\left(t\right)\right)dt.$$

Proof. Consider the curve $f\left(s\right) = P_{\gamma,s,0}X\left(\gamma\left(s\right)\right)$ in $T_{\gamma(0)}M$. By (3.268) and Taylor's theorem, we obtain that

$$f'''\left(s\right) = P_{\gamma,s,0}D^3X\left(\gamma\left(s\right)\right)\left(\gamma'\left(s\right),\gamma'\left(s\right),\gamma'\left(s\right)\right)$$

and

$$f\left(s\right) = f\left(0\right) + f'\left(0\right)\left(s-0\right) + \frac{1}{2}f''\left(0\right)\left(s-0\right)^2 + \frac{1}{2}\int_0^s \left(s-t\right)^2 f'''\left(t\right)dt.$$

We conclude that

$$P_{\gamma,s,0}X\left(\gamma\left(s\right)\right) = X\left(\gamma\left(0\right)\right) + s\mathcal{D}X\left(\gamma\left(0\right)\right)\gamma'\left(0\right) + \frac{1}{2}s^2\mathcal{D}^2X\left(\gamma\left(0\right)\right)\left(\gamma'\left(0\right),\gamma'\left(0\right)\right)$$
$$+ \frac{1}{2}\int_0^s\left(s-t\right)^2 P_{\gamma,t,0}\mathcal{D}^3X\left(\gamma\left(t\right)\right)\left(\gamma'\left(t\right),\gamma'\left(t\right),\gamma'\left(t\right)\right)dt.$$

$$\square$$

Theorem 3.6.2. *Let M be a Riemannian manifold, $\Omega \subseteq M$ an open convex set, $X \in C^3\left(M, T_{(.)}M\right)$. Suppose that there exist a real $c > 0$ such that for any geodesic γ and $\tau_1, \tau_2 \in \mathbb{R}$ with $\gamma\left[\tau_1,\tau_2\right] \subseteq \Omega$ it holds that*

$$\left\|P_{\gamma,\tau_2,\tau_1}\mathcal{D}^2\left(\gamma\left(\tau_2\right)\right)P^2_{\gamma,\tau_1,\tau_2} - \mathcal{D}^2\left(\gamma\left(\tau_1\right)\right)\right\| \le c\int_{\tau_1}^{\tau_2}\left\|\gamma'\left(t\right)\right\|dt,$$

provided that $[0,1] \subseteq [\tau_1,\tau_2]$. Then, the following hold

$$\left\|\left(P_{\gamma,s,0}X\left(\gamma\left(s\right)\right) - X\left(\gamma\left(0\right)\right) - s\mathcal{D}X\left(\gamma\left(0\right)\right)\gamma'\left(0\right)\right.\right.$$
$$\left.\left. - \tfrac{1}{2}s^2\mathcal{D}^2X\left(\gamma\left(0\right)\right)\left(\gamma'\left(0\right),\gamma'\left(0\right)\right)\right)\right\| \le \frac{1}{6}s^{"}\left\|\gamma'\left(0\right)\right\|^3 \tag{3.300}$$

and

$$\left\|\left(P_{\gamma,s,0}\mathcal{D}X\left(\gamma\left(s\right)\right)P_{\gamma,0,s} - \mathcal{D}X\left(\gamma\left(0\right)\right)\right)\right\| \le \left(\left\|\mathcal{D}^2X\left(\gamma\left(0\right)\right)\right\| + \frac{1}{2}sc\left\|\gamma'\left(0\right)\right\|\right)s\left\|\gamma'\left(0\right)\right\|.$$
$$\tag{3.301}$$

Proof. We start with the following Taylor's expansion:

$$P_{\gamma,s,0}X\left(\gamma\left(s\right)\right) = X\left(\gamma\left(0\right)\right) + s\mathcal{D}X\left(\gamma\left(0\right)\right)\gamma'\left(0\right) + \frac{1}{2}s^2\mathcal{D}^2X\left(\gamma\left(0\right)\right)\left(\gamma'\left(0\right),\gamma'\left(0\right)\right)$$
$$+ \frac{1}{2}\int_0^s\left(s-t\right)^2 P_{\gamma,t,0}\mathcal{D}^3X\left(\gamma\left(t\right)\right)\left(\gamma'\left(t\right),\gamma'\left(t\right),\gamma'\left(t\right)\right)dt.$$

Then, we have that

$$\left\|P_{\gamma,s,0}X\left(\gamma\left(s\right)\right) - X\left(\gamma\left(0\right)\right) - s\mathcal{D}X\left(\gamma\left(0\right)\right)\gamma'\left(0\right) - \frac{1}{2}s^2\mathcal{D}^2X\left(\gamma\left(0\right)\right)\left(\gamma'\left(0\right),\gamma'\left(0\right)\right)\right\|$$
$$\le \frac{1}{2}\int_0^s\left(s-t\right)^2\left\|\mathcal{D}^3X\left(\gamma\left(t\right)\right)\right\|\left\|\gamma'\left(0\right)\right\|^3 dt \le \frac{1}{6}s^3\left\|\gamma'\left(0\right)\right\|^3.$$

Using (3.269) and (3.299), we obtain that

$$P_{\gamma,t,0}\mathcal{D}^2X\left(\gamma\left(s\right)\right)P^2_{\gamma,0,t} - \mathcal{D}^2X\left(\gamma\left(0\right)\right) = \int_0^s P_{\gamma,t,o}\left(\mathcal{D}^3X\left(\gamma\left(t\right)\right)P^3_{\gamma,0,t}\left(\gamma'\left(0\right),.,.\right)\right)dt$$

and

$$P_{\gamma,s,0}\mathcal{D}X\left(\gamma\left(s\right)\right)P_{\gamma,0,s} - \mathcal{D}X\left(\gamma\left(0\right)\right) = \int_0^s P_{\gamma,t,0}\left(\mathcal{D}^2X\left(\gamma\left(t\right)\right)P^2_{\gamma,0,t}\left(\gamma'\left(0\right),.\right)\right)dt.$$

Therefore, we get in turn that

$$
\|P_{\gamma,s,0}\mathcal{D}X\left(\gamma\left(s\right)\right)P_{\gamma,0,s} - \mathcal{D}X\left(\gamma\left(0\right)\right)\| = \left\|\int_0^s P_{\gamma,t,0}\left(\mathcal{D}^2 X\left(\gamma\left(t\right)\right)P^2_{\gamma,0,t}\left(\gamma'\left(0\right),.\right)\right)dt\right\|
$$

$$
= \left\|\int_0^s\left(\mathcal{D}^2 X\left(\gamma\left(0\right)\right)\left(\gamma'\left(0\right)\right) + \int_0^t P_{\gamma,\tau,o}\left(\mathcal{D}^3 X\left(\gamma\left(\tau\right)\right)P^3_{\gamma,0,\tau}\left(\gamma'\left(0\right),\gamma'\left(0\right),.\right)\right)d\tau\right)dt\right\|
$$

$$
\leq \left\|\int_0^s \mathcal{D}^2 X\left(\gamma\left(0\right)\right)\left(\gamma'\left(0\right)\right)dt\right\| + \left\|\int_0^s\int_0^t P_{\gamma,\tau,o}\left(\mathcal{D}^3 X\left(\gamma\left(\tau\right)\right)P^3_{\gamma,0,\tau}\left(\gamma'\left(0\right),\gamma'\left(0\right),.\right)\right)d\tau dt\right\|
$$

$$
\leq s\left\|\mathcal{D}^2 X\left(\gamma\left(0\right)\right)\right\|\left\|\gamma'\left(0\right)\right\| + \int_0^s\int_0^t \left\|\mathcal{D}^3 X\left(\gamma\left(\tau\right)\right)\right\|\left\|\gamma'\left(0\right)\right\|^2 d\tau dt
$$

$$
\leq s\left\|\mathcal{D}^2 X\left(\gamma\left(0\right)\right)\right\|\left\|\gamma'\left(0\right)\right\| + \int_0^s\int_0^t c\left\|\gamma'\left(0\right)\right\|^2 d\tau dt
$$

$$
\leq s\left\|\mathcal{D}^2 X\left(\gamma\left(0\right)\right)\right\|\left\|\gamma'\left(0\right)\right\| + \frac{s^2 c}{2}\left\|\gamma'\left(0\right)\right\|^2 = \left(\left\|\mathcal{D}^2 X\left(\gamma\left(0\right)\right)\right\| + \frac{sc}{2}\left\|\gamma'\left(0\right)\right\|\right)s\left\|\gamma'\left(0\right)\right\|.\qquad\square
$$

Recently, Amat et al. in Ref. [9] proposed a new family of higher order procedures, which generalizes classical third order methods. Under Kantorovich conditions, they proved the convergence of these procedures in Banach spaces. This family of methods is given by

$$
\begin{aligned}
y_n &= x_n - DF\left(x_n\right)^{-1} F\left(x_n\right)\\
x_{n+1} &= y_n - \mathcal{T}_F(x_n)\, DF\left(x_n\right)^{-1} F\left(y_n\right) \quad \text{for each} \quad n = 0,1,2,\cdots
\end{aligned}
\tag{3.302}
$$

where F is a nonlinear operator defined on an open convex subset Ω of a Banach space E into itself, I is the identity operator in E, $L_F\left(x_n\right)$, $\mathcal{T}_F(x_n)$ are defined by

$$
L_F\left(x_n\right) = DF\left(x_n\right)^{-1} D^2 F\left(x_n\right) DF\left(x_n\right)^{-1} F\left(x_n\right)
$$

and

$$
\mathcal{T}_F(x_n) = I + L_F(x_n) + L_F^2(x_n)\, G(x_n).
$$

In Ref. [9], the authors assume that $DF\left(x_n\right)^{-1}$ exists and $G\left(x_n\right) : \Omega \subseteq E \longrightarrow E$ is a given nonlinear operator. Some special cases of these methods are:

★ Two-step method of order three defined by

$$
\begin{aligned}
y_n &= x_n - DF(x_n)^{-1} F(x_n)\\
x_{n+1} &= y_n - DF(x_n)^{-1} F(x_n) \quad \text{for each} \quad n = 0,1,2,\cdots.
\end{aligned}
$$

★ Two-step method of order four defined by

$$
\begin{aligned}
y_n &= x_n - DF(x_n)^{-1} F(x_n)\\
x_{n+1} &= y_n - DF(y_n)^{-1} F(y_n) \quad \text{for each} \quad n = 0,1,2,\cdots.
\end{aligned}
$$

★ Two-step method of order four defined by

$$
\begin{aligned}
y_n &= x_n - DF(x_n)^{-1} F'(x_n)\\
x_{n+1} &= y_n - DF(x_n)^{-1}\left(I + L_F(x_n)\right) F(y_n) \quad \text{for each} \quad n = 0,1,2,\cdots.
\end{aligned}
$$

★ Two-step method of order five defined by

$$y_n = x_n - DF(x_n)^{-1} F(x_n)$$
$$x_{n+1} = y_n - \mathcal{S}(x_n) F(y_n) \quad \text{for each} \quad n = 0, 1, 2, \cdots,$$

where

$$\mathcal{S}(x_n) = I + L_F(x_n) + L_F^2(x_n) \frac{1}{2}(\frac{5}{2} I - L_{F'}(x_n)).$$

Many other choices of operator $\mathcal{T}_F$ lead to other popular high order iterative procedures such that Halley-type or Chebyshev-type procedures (see Refs. [110, 120]). Concerning the convergence order of such methods for the case when $E = \mathbb{R}$, we refer the reader to a Theorem of Traub in Ref. [418] which states that for sufficiently smooth $G_F(x)$ these methods have convergence order four.

Our objective is to extend this method to the context of Riemannian manifolds. We shall also establish a convergence and uniqueness theorem. We suppose that M is a complete Riemannian manifold, $\Omega \subseteq M$ an open convex set, $X \in C^2$ and we want to find an approximation to the singularity of the vector field $X : M \longrightarrow TM$. Let us consider the following family of high order iterative methods:

$$\begin{aligned}
u_n &= -\mathcal{D}X\left(p_n\right)^{-1} X\left(p_n\right) \\
v_n &= \mathcal{D}X\left(p_n\right)^{-1} P_{\sigma_n,1,0} X\left(q_n\right) P_{\sigma_n,0,1} \\
q_n &= \exp_{p_n}\left(u_n\right) \\
L_X\left(p_n\right) &= \mathcal{D}X\left(p_n\right)^{-1} \left(\mathcal{D}^2 X\left(p_n\right)\left(u_n, .\right)\right) \\
w_n &= -\left(I_{T_{p_n} M} + L_X\left(p_n\right) + L_X\left(p_n\right)^2 G\left(p_n\right)\right) v_n \\
p_{n+1} &= \exp_{p_n}\left(u_n + w_n\right) \quad \text{for each} \quad n = 0, 1, 2, \cdots,
\end{aligned} \qquad (3.303)$$

where $G : M \longrightarrow TM$ is a vector field (usually depending on the vector field X and its covariants derivatives) and $\{\sigma_n\}_{n\in\mathbb{N}}$ is a geodesic family defined by

$$\sigma_n\left(t\right) = \exp_{p_n}\left(tu_n\right).$$

Let $p_0 \in \Omega$. Assume that the conditions of Kantorovich hold:

(C1) $\Gamma_0 = \mathcal{D}X\left(p_0\right)^{-1}$ exists and $\|\Gamma_0\| \le \beta$,

(C2) $\|\Gamma_0 X\left(p_0\right)\| \le \eta$,

(C3) $\left\|\mathcal{D}^2 X\left(p\right)\right\| \le M$ for all $p \in \Omega$,

(C4) $\left\|P_{\gamma,c,d}\mathcal{D}^2 X\left(\gamma\left(c\right), \cdot\right) P_{\gamma,d,c}^2 - \mathcal{D}^2 X\left(\gamma\left(d\right), \cdot\right)\right\| \le K \int_d^c \|\gamma'\left(t\right)\| \, dt$, where γ is a geodesic and $\gamma\left[d, c\right] \subseteq \Omega$.

Under these hypotheses, it is possible to find a cubic polynomial f in an interval $[a, b]$, $a \ge 0$ such that:

$$f\left(b\right) < 0 < f\left(a\right), \ f'\left(t\right) < 0 \ f''\left(t\right) > 0 \text{ and } f'''\left(t\right) > 0,$$

in $[a, t^*]$ with t^* the single simple solution of $f\left(t\right) = 0$ and verifying
For $t_0 \in [a, b]$ and $f\left(t_0\right) > 0$:

(H1) $\|\Gamma_0\| \leq -\dfrac{1}{f'(t_0)}$,

(H2) $\|\Gamma_0 X(p_0)\| \leq -\dfrac{f(t_0)}{f'(t_0)}$,

(H3) $\|\mathcal{D}^2 X(p,\cdot)\| \leq f''(t)$ for all $p \in \Omega$, such that $d(p,p_0) \leq t - t_0 \leq t^* - t_0$,

(H4) $\left\| P_{\gamma,c,d} \mathcal{D}^2 X(\gamma(c),\cdot) P^2_{\gamma,d,c} - \mathcal{D}^2 X(\gamma(d),\cdot) \right\| \leq |f''(u) - f''(v)|$,   with

$$\int_d^c \|\gamma'(t)\|\, dt \leq |u - v|, \ \gamma[d,c] \subseteq \Omega \text{ and } u,v \in [a,t^*].$$

This construction is in Ref. [388], some properties of the polynomial $f(t)$ are:

(1) $f(t)$ is decreasing in the interval $[a,t^*)$,

(2) $f(t) > 0$ in $[a,t^*]$,

(3) $f'(t)$ is increasing and $f(t)$ is convex in $[a,t^*]$,

(4) $f''(t)$ is increasing in $[a,t^*]$,

(5) $N_f(t) = t - \dfrac{f(t)}{f'(t)}$ is increasing in $[a,t^*)$, $N_f(t^*) = t^*$ and $N_f'(t^*) = 0$,

(6) $L_f(t) = \dfrac{f(t)\, f''(t)}{f'(t)^2} > 0$ in $[a,t^*)$.

For the validity of the previous statements, it suffices to consider the polynomial

$$f(t) := \frac{K}{6}t^3 + \frac{M}{2}t^2 - \frac{1}{\beta}t + \frac{\eta}{\beta},$$

with

$$\eta \leq \frac{4K + M^2\beta - M\beta\sqrt{M^2 + 2K\beta}}{3\beta K\left(M + \sqrt{M^2 + 2K\beta}\right)}. \tag{3.304}$$

This polynomial f has two positive real roots t^* and t^{**} (see Ref. [388] for more details).

Let us suppose additionally that there exists a function g_f associated to the vector field $L_X(p_n)$ defined in (3.303) satisfying:

(C5) $\left\| L_X(p)^2\, G(p) \right\| \leq L_f(t)^2\, g(t)$ for $d(p,p_0) \leq t - t_0 \leq t^* - t_0$,

(C6) $1 + L_f(t) + L_f(t)^2\, g(t) \geq 0$ in $[a,t^*]$,

(C7) $m'(t) > 0$ in $[a,t^*]$, where

$$m(t) = t - \frac{f(t)}{f'(t)} - \left(1 + L_f(t) + L_f(t)^2\, g(t)\right)\frac{f\left(t - \frac{f(t)}{f'(t)}\right)}{f'(t)}.$$

Proposition 3.6.1. *(see Refs. [9, 388]) If (C6) and (C7) are true, then the sequence*

$$
\begin{aligned}
s_n &= t_n - \frac{f(t_n)}{f'(t_n)} \\
t_{n+1} &= s_n - \left(1 + L_f(t_n) + L_f(t_n)^2\, g(t_n)\right)\frac{f(s_n)}{f'(t_n)},
\end{aligned}
\tag{3.305}
$$

starting from the above t_0 converges monotonically to t^*. This is the smaller real, simple root of $f(t) = 0$ in $[a, b]$.

Denote by $(\mathcal{C})$ conditions (C1)–(C7). Next, we can prove the semi-local convergence of (3.303). Using the notation of (3.303), we have the following result.

Theorem 3.6.3. *Assume $p_0 \in \Omega$ and $t_0 \in [a, t^*]$. Suppose that the $(\mathcal{C})$ conditions and (3.304) are true. If $B(p_0, t^*) \subseteq \Omega$, then the sequence (3.303) is well defined, remains in $B[p_0, t^*]$ for each $n = 0, 1, 2, \cdots$ and it converges to the root p_*, which is the solution of $X(p) = 0$ in $B[p_0, t^*]$. Moreover, the following hold*

(i) For each $n = 0, 1, 2, \cdots$, we have $d(p_*, p_n) \leq t^* - t_n$, where $\{t_n\}$ is defined in (3.305).

(ii) If the number t^* also satisfies the condition

$$13K\beta t^* + 6M\beta \leq \sqrt{6}\sqrt{6M^2\beta^2 + 13K\beta},$$

then, the root p_* is unique in $B[p_0, t^*]$.

Proof. We use mathematical induction. It suffices to show for all $k \in \mathbb{N}$.

  i. $\left\| \mathcal{D}X(p_k)^{-1} \right\| \leq -\dfrac{1}{f'(t_k)}$,

  ii. $\left\| \mathcal{D}^2 X(p_k)^{-1} \right\| \leq f''(t_k)$,

  iii. $\|X(p_k)\| \leq f(t_k)$,

  iv. $\|L_X(p_k)\| \leq L_f(t_k)$,

  v. $d(p_{n+1}, p_n) \leq t_{n+1} - t_n$.

The case $k = 0$ follows from the initial conditions on p_0 and t_0. We assume that all the conditions are valid for $k = n$, and we check them for $k = n + 1$. Let us consider the family of geodesics

$$\gamma_n(t) = \exp_{p_n}(t(u_n + w_n)) \qquad \text{for each} \quad n = 0, 1, 2, \cdots.$$

  i. By (3.301), we have that

$$\|(P_{\gamma_n,1,0}\mathcal{D}X(\gamma_{n+1}(1))P_{\gamma_n,0,1} - \mathcal{D}X(p_n))\|$$
$$\leq \left(\left\| \mathcal{D}^2 X(\gamma_n(0)) \right\| + \tfrac{1}{2}K \|\gamma_n'(0)\| \right) \|\gamma_n'(0)\|.$$

Hence, we get in turn that

$$\left\| \mathcal{D}X(p_n)^{-1}(P_{\gamma_n,1,0}\mathcal{D}X(p_{n+1})P_{\gamma_n,0,1} - \mathcal{D}X(p_n)) \right\|$$

$$\leq \left\| \mathcal{D}X(p_n)^{-1} \right\| \left(\left\| \mathcal{D}^2 X(p_n) \right\| + \frac{1}{2}K\|u_n + w_n\| \right) \|u_n + w_n\|$$

$$\leq -\frac{1}{f'(t_n)} \left(f''(t_n) + \frac{1}{2}Kd(p_{n+1}, p_n) \right) d(p_{n+1}, p_n)$$

$$\leq -\frac{1}{f'(t_n)} \left(f''(t_n) + \frac{1}{2}K(t_{n+1} - t_n) \right) (t_{n+1} - t_n)$$

$$= -\frac{1}{f'(t_n)} (f'(t_{n+1}) - f'(t_n)) = 1 - \frac{f'(t_{n+1})}{f'(t_n)} \leq 1,$$

where the last inequality is due to that $f'(t)$ is increasing. Therefore, using Banach's lemma, we obtain that

$$\left\| \mathcal{D}X\left(p_{n+1}\right)^{-1} \right\| = \frac{-\dfrac{1}{f'(t_n)}}{1 - \left(1 - \dfrac{f'(t_{n+1})}{f'(t_n)}\right)} = -\frac{1}{f'(t_{n+1})}.$$

ii. It is clear that

$$\begin{aligned}
\left\| \mathcal{D}^2 X\left(p_{n+1}\right) \right\| &= \left\| P_{\gamma_n,1,0}\mathcal{D}^2 X\left(p_{n+1}\right) P^2_{\gamma_n,0,1} \right\| \\
&\leq \left\| P_{\gamma_n,1,0}\mathcal{D}^2 X\left(p_{n+1}\right) P^2_{\gamma_n,0,1} - \mathcal{D}^2 X\left(p_n\right) \right\| + \left\| \mathcal{D}^2 X\left(p_n\right) \right\| \\
&\leq K d\left(p_{n+1},p_n\right) + \left\| \mathcal{D}^2 X\left(p_n\right) \right\| \leq K d\left(t_{n+1} - t_n\right) + f''\left(t_n\right) = f''\left(t_{n+1}\right).
\end{aligned}$$

iii. First, using the Taylor expansion (3.271), we observe that

$$P_{\sigma_n,1,0}X\left(q_n\right) = X\left(p_n\right) + \mathcal{D}X\left(p_n\right)\sigma'_n\left(0\right) + \frac{1}{2}\mathcal{D}^2 X\left(p_n\right)\left(\sigma'_n\left(0\right),\sigma'_n\left(0\right)\right) + R_{\sigma n},$$

where

$$R_{\sigma n} = \int_0^1 \left(1-t\right)\left(P_{\sigma_n,t,0}\mathcal{D}^2 X\left(\sigma_n\left(t\right)\right) P^2_{\sigma_n,0,t} - \mathcal{D}^2 X\left(p_n\right)\right)\left(\sigma'_n\left(0\right),\sigma'_n\left(0\right)\right)dt.$$

Consequently, we have that

$$\begin{aligned}
\left\| X\left(q_n\right) \right\| &\leq \left\| X\left(p_n\right) + \mathcal{D}X\left(p_n\right)u_n + \tfrac{1}{2}\mathcal{D}^2 X\left(p_n\right)\left(u_n,u_n\right) \right\| + \left\| R_{\sigma n} \right\| \\
&\leq \left\| X\left(p_n\right) + \mathcal{D}X\left(p_n\right)\left(-\mathcal{D}X\left(p_n\right)^{-1}X\left(p_n\right)\right) + \tfrac{1}{2}\mathcal{D}^2 X\left(p_n\right)\left(u_n,u_n\right) \right\| + \left\| R_{\sigma n} \right\| \\
&\leq \frac{1}{2}\left\| \mathcal{D}^2 X\left(p_n\right)\left(u_n,u_n\right) \right\| + \left\| R_{\sigma n} \right\|.
\end{aligned}$$

Since

$$\begin{aligned}
\left\| R_{\sigma n} \right\| &\leq \int_0^1 \left(1-t\right)\left\| P_{\sigma_n,t,0}\mathcal{D}^2 X\left(\sigma_n\left(t\right)\right) P^2_{\sigma_n,0,t} - \mathcal{D}^2 X\left(p_n\right) \right\| \left\| u_n \right\|^2 dt \\
&\leq K \int_0^1 \left(1-t\right)t\left\| u_n \right\|^3 dt = \frac{K}{6}\left(d\left(q_n,p_n\right)\right)^3,
\end{aligned}$$

we have that

$$\begin{aligned}
\left\| X\left(q_n\right) \right\| &\leq \frac{1}{2}f''\left(t_n\right)\left(s_n - t_n\right)^2 + \frac{K}{6}\left(s_n - t_n\right)^3 \\
&= f\left(t_n\right) + f'\left(t_n\right)\left(-\frac{f\left(t_n\right)}{f'\left(t_n\right)}\right) + \frac{1}{2}f''\left(t_n\right)\left(s_n - t_n\right)^2 + \frac{K}{6}\left(s_n - t_n\right)^3 \\
&= f\left(t_n\right) + f'\left(t_n\right)\left(s_n - t_n\right) + \frac{1}{2}f''\left(t_n\right)\left(s_n - t_n\right)^2 + \frac{K}{6}\left(s_n - t_n\right)^3 = f\left(s_n\right).
\end{aligned}$$

Again, using the Taylor expansion (3.271), we get in turn

$$P_{\gamma_n,1,0}X\left(p_{n+1}\right) = X\left(p_n\right) + \mathcal{D}X\left(p_n\right)\gamma'_n\left(0\right) + \frac{1}{2}\mathcal{D}^2 X\left(p_n\right)\left(\gamma'_n\left(0\right),\gamma'_n\left(0\right)\right) + R_n,$$

where

$$R_n = \int_0^1 \left(1-t\right)\left(P_{\gamma_n,t,0}\mathcal{D}^2 X\left(\gamma_n\left(t\right)\right) P^2_{\gamma_n,0,t} - \mathcal{D}^2 X\left(p_n\right)\right)\left(\gamma'_n\left(0\right),\gamma'_n\left(0\right)\right)dt.$$

Therefore,

$$P_{\gamma_n,1,0}X(p_{n+1}) = X(p_n) + DX(p_n)(u_n + w_n) + \tfrac{1}{2}\mathcal{D}^2 X(p_n)(\gamma_n'(0),\gamma_n'(0)) + R_n$$
$$= X(p_n) + DX(p_n)\left(-DX(p_n)^{-1}X(p_n) + w_n\right) + \tfrac{1}{2}\mathcal{D}^2 X(p_n)(\gamma_n'(0),\gamma_n'(0)) + R_n$$
$$= DX(p_n)w_n + \tfrac{1}{2}\mathcal{D}^2 X(p_n)(\gamma_n'(0),\gamma_n'(0)) + R_n.$$

Using the following estimates

$$\left\| DX(p_n)w_n + \tfrac{1}{2}\mathcal{D}^2 X(p_n)(\gamma_n'(0),\gamma_n'(0)) \right\|$$
$$\leq \left\| DX(p_n)\left(I_{T_{p_n}M} + L_X(p_n) + L_X(p_n)^2 G(p_n)\right)v_n \right\|$$
$$+ \tfrac{1}{2}\left\| \mathcal{D}^2 X(p_n)(\gamma_n'(0),\gamma_n'(0)) \right\|$$
$$\leq -f'(t_n)\left(1 + L_f(t_n) + L_f(t_n)^2 g(t_n)\right)\frac{f(s_n)}{f'(t_n)} + \tfrac{1}{2}f'(t_n)(t_{n+1} - t_n)^2$$
$$= f'(t_n)(t_{n+1} - s_n) + \tfrac{1}{2}f'(t_n)(t_{n+1} - t_n)^2.$$

Then, we also have that

$$\|R_n\| \leq \int_0^1 (1-t)\left\| P_{\gamma_n,t,0}\mathcal{D}^2 X(\gamma_n(t))P_{\gamma_n,0,t}^2 - \mathcal{D}^2 X(p_n)\right\| \|\gamma_n'(0)\|^2 \, dt$$
$$\leq \int_0^1 (1-t)K\int_0^t \|\gamma_n'(\tau)\|\,d\tau dt\,\|\gamma_n'(0)\|^2 = K\int_0^1 (1-t)t\,dt\,\|\gamma_n'(0)\|^3$$
$$= \tfrac{K}{6}(t_{n+1} - t_n)^3.$$

We deduce in turn that

$$\|X(p_{n+1})\| \leq f'(t_n)(t_{n+1} - s_n) + \tfrac{1}{2}f'(t_n)(t_{n+1} - t_n)^2 + \tfrac{K}{6}(t_{n+1} - t_n)^3$$
$$\leq f'(t_n)\left(t_{n+1} + \frac{f(t_n)}{f'(t_n)} - t_n\right) + \tfrac{1}{2}f'(t_n)(t_{n+1} - t_n)^2 + \tfrac{K}{6}(t_{n+1} - t_n)^3$$
$$= f(t_n) + f(t_n)(t_{n+1} - t_n) + \tfrac{1}{2}f'(t_n)(t_{n+1} - t_n)^2 + \tfrac{K}{6}(t_{n+1} - t_n)^3 = f(t_{n+1}).$$

iv. It is clear that

$$\|L_X(p_{n+1})\| \leq \left\| DX(p_{n+1})^{-1}\right\| \left\| \mathcal{D}^2 X(p_{n+1})\right\| \|u_{n+1}\| \leq L_f(t_{n+1}).$$

v. Finally, if we let the geodesic γ_{k+1} be defined by

$$\gamma_{n+1}(t) = \exp_{p_{n+1}}\left(t(u_{n+1} + w_{n+1})\right).$$

Then, we have that

$$d(p_{n+1},p_n) = \|u_{n+1} + w_{n+1}\|$$
$$\leq \left\| DX(p_{n+1})^{-1}X(p_{n+1})\right\| + \left\| I_{T_{p_{n+1}}M} + L_X(p_{n+1}) + L_X(p_{n+1})^2 G(p_{n+1})v_{n+1}\right\|$$
$$\leq -\frac{f(t_{n+1})}{f'(t_{n+1})} - \left(1 + L_f(t_{n+1}) + L_f(t_{n+1})^2 g(t_{n+1})\right)\frac{f(s_{n+1})}{f'(t_{n+1})}$$
$$= s_n - t_n + t_{n+1} - s_n = t_{n+1} - t_n.$$

Using item **v**, we have for each $k \geq n$, $n \in \mathbb{N}$ that

$$d(p_k,p_n) \leq t_k - t_n.$$

It follows that $\{t_k\}_{k\in\mathbb{N}}$ is Cauchy sequence. Since M is complete, it converges to the same $p_* \in M$. Moreover, passing to the limit $k \longrightarrow \infty$, for each $n \in \mathbb{N}$, we get in turn that

$$d(p_*,p_n) \leq t^* - t_n,$$

passing to the limit in **iii**, we obtain that $\|X(p_*)\| \le f(t^*) = 0$. Then, we have $X(p_*) = 0$.

To show uniqueness, show that the operator

$$A = \int_0^1 P_{\gamma,t,0} \mathcal{D}X(\gamma(t)) P_{\gamma,0,t} dt$$

is invertible. Let us assume that there exists a singularity $q^* \in B(p_0, t_0)$ and let $\gamma : [0,1] \longrightarrow M$ be a minimizing geodesic joining q^* and p^*, where $\gamma(0) = p^*$ and $\gamma(1) = q^*$. By (3.301), we have that

$$\left\|\mathcal{D}X(p^*)^{-1}\right\| \|A - \mathcal{D}X(p^*)\| \le \left\|\mathcal{D}X(p^*)^{-1}\right\| \int_0^1 \|P_{\gamma,t,0}\mathcal{D}X(\gamma(t))P_{\gamma,0,t} - \mathcal{D}X(p^*)\| \, dt$$

$$= \left\|\mathcal{D}X(\gamma(0))^{-1}\right\| \int_0^1 \|P_{\gamma,t,0}\mathcal{D}X(\gamma(t))P_{\gamma,0,t} - \mathcal{D}X(\gamma(0))\| \, dt$$

$$\le -\frac{1}{f'(t^*)} \int_0^1 \left(\left\|\mathcal{D}^2X(\gamma(0))\right\| + \frac{1}{2}Kt\left\|\gamma'(0)\right\|\right) t\left\|\gamma'(0)\right\| \, dt$$

$$\le -\frac{1}{f'(t^*)} \int_0^1 \left(\left\|\mathcal{D}^2X(\gamma(0))\right\| t + \frac{1}{2}Kt^2\left\|\gamma'(0)\right\|\right) \left\|\gamma'(0)\right\| \, dt$$

$$= -\frac{1}{f'(t^*)} \left(\frac{1}{2}\left\|\mathcal{D}^2X(p^*)\right\| + \frac{1}{6}Kd(p^*,q^*)\right) d(p^*,q^*)$$

$$\le -\frac{1}{f'(0)} \left(\frac{1}{2}f''(t^*) + \frac{1}{6}K\left(d(p^*,p_0) + d(q^*,p_0)\right)\right) \left(d(p^*,p_0) + d(q^*,p_0)\right)$$

$$\le -\frac{2}{f'(t^*)} \left(\frac{1}{2}f''(t^*) + \frac{1}{3}K(t^* - t_0)\right) (t^* - t_0)$$

$$= -\frac{2}{f'(t^*)} \left(\frac{1}{2}(M + Kt^*) + \frac{1}{3}Kt^*\right) t^* \le -\frac{\left(M + \dfrac{5}{3}Kt^*\right)t^*}{f'(t^*)}.$$

Notice that

$$-\frac{\left(M + \dfrac{5}{3}Kt^*\right)t^*}{f'(t^*)} \le 1 \text{ if and only if } 13K\beta(t^*)^2 + 12M\beta t^* - 6 \le 0.$$

The roots of $h(t) = 13K\beta t^2 + 12M\beta t - 6$ are

$$\frac{1}{13K\beta}\left(-6M\beta + \sqrt{6}\sqrt{6M^2\beta^2 + 13K\beta}\right) \text{ and } -\frac{1}{13K\beta}\left(6M\beta + \sqrt{6}\sqrt{6M^2\beta^2 + 13K\beta}\right).$$

Given that $h(0) = -6$, we have for $t \ge 0$ that

$$13K\beta t^2 + 12M\beta t - 6 \le 0 \text{ if and only if } t \le \frac{1}{13K\beta}\left(-6M\beta + \sqrt{6}\sqrt{6M^2\beta^2 + 13K\beta}\right).$$

Using hypothesis

$$0 < t^* \le \frac{1}{13K\beta}\left(-6M\beta + \sqrt{6}\sqrt{6M^2\beta^2 + 13K\beta}\right),$$

we obtain that

$$-\frac{\left(M + \dfrac{5}{3}Kt^*\right)t^*}{f'(t^*)} \le 1.$$

Therefore, A is invertible.

Finally, we have that

$$0 = X\left(q^*\right) - X\left(p^*\right) = \int_0^1 P_{\gamma,t,0}\mathcal{D}X\left(\gamma\left(t\right)\right)P_{\gamma,0,t}\left(\gamma'\left(0\right)\right)dt.$$

Therefore, $\gamma'\left(0\right) = 0$. Thus, we deduce that $0 = \|\gamma'\left(0\right)\| = d\left(p^*,q^*\right)$ and $p^* = q^*$. $\Box$

Theorem 3.6.4. *Without lost of generality, let us assume $t_0 = 0$. Suppose that X has a single singularity p_* in $B\left[p_0, t^*\right]$. If $B\left(p_0, t^{**}\right) \subseteq \Omega$, then p_* is the unique singularity of X in $B\left[p_0, r\right]$, where $t^* < r \leq t^{**}$.*

Proof.   Let $q_* \in B\left[p_0, r\right]$ be a singularity of a vector field X. Let us consider the minimizing geodesic $\gamma : [0, 1] \longrightarrow M$ joining p_0 to q_*. By Theorem 3.6.1, we have that

$$P_{\gamma,s,0}X\left(\gamma\left(1\right)\right) = X\left(\gamma\left(0\right)\right) + \mathcal{D}X\left(\gamma\left(0\right)\right)\gamma'\left(0\right) + \frac{1}{2}\mathcal{D}^2X\left(\gamma\left(0\right)\right)\left(\gamma'\left(0\right),\gamma'\left(0\right)\right) + R,$$

where

$$R = \frac{1}{2}\int_0^1 \left(1 - t\right)^2 P_{\gamma,t,0}\mathcal{D}^3X\left(\gamma\left(t\right)\right)P_{\gamma,t,0}^3\left(\gamma'\left(0\right),\gamma'\left(0\right),\gamma'\left(0\right)\right)dt.$$

We also have that

$$\gamma'\left(0\right) = -\mathcal{D}X\left(p_0\right)^{-1}\left(X\left(p_0\right) + \frac{1}{2}\mathcal{D}^2X\left(p_0\right)\left(\gamma'\left(0\right),\gamma'\left(0\right)\right) + R\right).$$

Hence, we obtain the following estimate

$$d\left(p_0, q_*\right) = \|\gamma'\left(0\right)\|$$
$$\leq \left\|\mathcal{D}X\left(p_0\right)^{-1}\right\|\left(\|X\left(p_0\right)\| + \frac{1}{2}\left\|\mathcal{D}^2X\left(p_0\right)\right\|\|\gamma'\left(0\right)\|^2 + \|R\|\right)$$
$$\leq \eta + \frac{1}{2}\beta M d\left(p_0, q_*\right)^2 + \frac{1}{6}\beta K d\left(p_0, q_*\right)^3.$$

Therefore,

$$\frac{1}{6}K d\left(p_0, q_*\right)^3 + \frac{1}{2}M d\left(p_0, q_*\right)^2 - \frac{1}{\beta}d\left(p_0, q_*\right) + \frac{\eta}{\beta} \geq 0.$$

That is, we get $f\left(d\left(p_0, q_*\right)\right) \geq 0$. Since $d\left(p_0, q_*\right) \leq r \leq t^{**}$, we have that $d\left(p_0, q_*\right) \leq t^*$. Finally, by Theorem 3.6.3, we deduce that $p_* = q_*$. $\Box$

The semi-local convergence analysis of method (3.302) obtained before in this section was based on conditions (C1)–(C4) and the possibility of constructing cubic polynomial f whose coefficient satisfy sufficient conditions (3.304). However, simple numerical examples can be used to show that for example (C4) or (3.304) is not satisfied (see Ref. [120]). Hence, Theorem 3.6.3 cannot guarantee that (3.302) converges in these cases. However, (3.302) may converge even if (C4) or (3.304) is not satisfied. That is why in this section, we present a new semi-local convergence analysis that avoids (C4)–(C7) and (3.304).

It is convenient for our analysis to use instead of (3.302) the similar method defined by

$$
\begin{aligned}
y_n &= x_n - DF\left(x_n\right)^{-1} F\left(x_n\right) \\
x_{n+1} &= y_n - DF\left(x_n\right)^{-1} \mathcal{T}_F(x_n)\, F\left(y_n\right) \quad \text{for each} \quad n = 0, 1, 2, \cdots .
\end{aligned}
\tag{3.306}
$$

In Ref. [121], we used an Ostrowski-type representation of $F(x_n)$ and $F(y_n)$ for each $n = 0, 1, 2, \cdots$ to study the semi-local convergence of certain Newton-type methods in a Banach space setting. In the case of (3.306), these representations become

$$
F(y_n) = \int_0^1 \left(DF(x_n + \theta\left(y_n - x_n\right)) - DF(x_n)\right)\left(y_n - x_n\right) d\theta
\tag{3.307}
$$

and

$$
\begin{aligned}
F(x_{n+1}) = {} & \int_0^1 \left(DF(y_n + \theta\left(x_{n+1} - y_n\right)) - DF(y_n)\right)\left(x_{n+1} - y_n\right) d\theta \\
& + (DF(y_n) - DF(y_n))\left(x_{n+1} - y_n\right) \\
& + (I - \mathcal{T}_F(x_n)) \int_0^1 \left(DF(x_n + \theta\left(y_n - y_n\right)) - DF(x_n)\right)\left(y_n - x_n\right) d\theta .
\end{aligned}
\tag{3.308}
$$

Since, we are using (3.306) instead of (3.302), we consider a slightly different family of high order iterative procedures than (3.303) defined by

$$
\begin{aligned}
u_n &= -\mathcal{D}X\left(p_n\right)^{-1} X\left(p_n\right) \\
v_n &= P_{\sigma_n,1,0} X\left(q_n\right) P_{\sigma_n,0,1} \\
q_n &= \exp_{p_n}\left(u_n\right) \\
L_X\left(p_n\right) &= \mathcal{D}X\left(p_n\right)^{-1}\left(\mathcal{D}^2 X\left(p_n\right)\left(u_n, .\right)\right) \\
w_n &= -\mathcal{D}X\left(p_n\right)^{-1} \left(I_{T_{p_n} M} + L_X\left(p_n\right) + L_X\left(p_n\right)^2 G\left(p_n\right)\right) v_n \\
p_{n+1} &= \exp_{p_n}\left(u_n + w_n\right) \quad \text{for each} \quad n = 0, 1, 2, \cdots .
\end{aligned}
\tag{3.309}
$$

We need the defintions of Lipschitz and center Lipschitz conditions on a Riemannian manifold.

Definition 3.6.1. Consider Definition 3.4.7. Let $p_0 \in \Omega$ be fixed. If condition (3.216) holds in particular at $p_0 \in \Omega$, we will say that the covariant derivative is center Lipschitz with constant L_0. We write $\mathcal{D}X \in Lip_{L_0}\left(\Omega\right)$ at $p_0 \in \Omega$. Note that $P_{\gamma,b,a}\mathcal{D}X\left(\gamma\left(b\right)\right) P_{\gamma,a,b}$ and $\mathcal{D}X\left(\gamma\left(a\right)\right)$ are both operators defined in the same tangent plane $T_{\gamma(a)} M$. If M is the Euclidean space, then these definitions coincide with the usual Lipschitz and center Lipschitz operator $DF : M \longrightarrow M$ given, respectively by

$$
\| DX(p) - DX(q) \| \le L \| p - q \| \quad \text{for each} \quad p, q \in \Omega
$$

and

$$
\| DX(p) - DX(p_0) \| \le L_0 \| p - p_0 \| \quad \text{for each} \quad p \in \Omega .
$$

Then, identities (3.307) and (3.308) justify the introduction of the conditions

(C8) $\mathcal{D}X \in Lip_{L/\beta}(\Omega)$;
(C9) $\mathcal{D}X \in Lip_{L_0/\beta}(\Omega)$ at $p_0 \in \Omega$;
(C10) $\| T_X(p) \| \leq b$ for all $p \in \Omega$;
(C11) $\| I - T_X(p) \| \leq c$ for all $p \in \Omega$.

Note that (C8) $\Longrightarrow$ (C9), $L_0 \leq L$ holds in general and L/L_0 can be arbitrarily large. If $\mathcal{D}^2 X$ exists on Ω, then we can set $L = M$. (C9) is not an additional condition to (C8), since in practice the evaluation of L requires the evaluation of L_0.

Next, we find sufficient conditions for the convergence of scalar sequences that will be shown to be majorizing for (3.309). Let $L_0 > 0$, $L > 0$, $r \geq 0$, $s \geq 0$ and $\varrho > 0$. It is convenient for us to define functions δ, α, h_i $(i = 1, 2, 3)$ by

$$\delta(t) = \frac{r \, L \, t}{2}, \quad \delta := \delta(\varrho), \tag{3.310}$$

$$\alpha(t) = \frac{\varphi(t) \, t}{1 - L_0 \, (1 + \delta(t)) \, t}, \quad \alpha := \alpha(\varrho), \tag{3.311}$$

$$h_1(t) = (\varphi(t) + L_0 \, (1 + \delta(t))) \, t - 1, \tag{3.312}$$

$$h_2(t) = \frac{r \, L}{2} \, \alpha(t) \, t + L_0 \, \delta(t) \, (1 + \delta(t)) \, t - \delta(t) \tag{3.313}$$

and

$$h_3(t) = \varphi(t) \, t + L_0 \, (1 + \delta(t)) \, (1 + \alpha(t)) \, t - 1, \tag{3.314}$$

where

$$\varphi(t) = \frac{L}{2} \, \delta^2(t) + L \, \delta(t) + \frac{s \, L}{2}, \quad \varphi := \varphi(\varrho).$$

Denote by ϱ_1, ϱ_2, ϱ_3 the minimal positive zeros of functions h_1, h_2, h_3, respectively. Note that $\alpha(t)$ is well defined on $(0, \varrho_1)$ and $\alpha \in (0, 1)$ by the choice of ϱ_1. Set

$$\varrho_0 = \min \{\varrho_i, \; i = 1, 2, 3\}. \tag{3.315}$$

Then, we have that for all $t \in (0, \varrho_0)$

$$\alpha(t) \in (0, 1), \quad h_1(t) < 0, \quad h_2(t) \leq 0 \quad \text{and} \quad h_3(t) \leq 0.$$

We can show the following result on the convergence of majorizing sequences for the iterative procedure (3.309).

Lemma 3.6.1. *Let* $L_0 > 0$, $L > 0$, $r \geq 0$, $s \geq 0$, $M \geq 0$ *and* $\varrho > 0$. *Suppose that*

$$\varrho \begin{cases} \leq \varrho_0 \; if \; \varrho_0 \leq \varrho_1 \\ < \varrho_0 \; if \; \varrho_0 = \varrho_1. \end{cases} \tag{3.316}$$

Then, scalar sequence $\{t_n\}$ *generated by*

$$t_0 = 0, \; s_0 = \varrho, \; t_{n+1} = s_n + \frac{r \, L \, (s_n - t_n)^2}{2 \, (1 - L_0 \, t_n)}$$

$$s_{n+1} = t_{n+1} + \frac{\frac{L}{2} (t_{n+1} - s_n)^2 + L \, (s_n - t_n) \, (t_{n+1} - s_n) + \frac{c \, L}{2} (s_n - t_n)^2}{1 - L_0 \, t_{n+1}} \tag{3.317}$$

is increasing, bounded from above by

$$t^{\star\star} = \frac{1+\delta}{1-\alpha}\,\varrho \tag{3.318}$$

and converges to its unique least upper bound $t^\star$ which satisfies

$$\varrho \leq t^\star \leq t^{\star\star}. \tag{3.319}$$

Moreover, the following estimates hold for each $n = 0, 1, 2, \cdots$

$$0 < t_{n+1} - s_n \leq \delta\,(s_n - t_n) \leq \delta\,\alpha^n\,\varrho \tag{3.320}$$

and

$$0 < s_{n+1} - t_{n+1} \leq \alpha\,(s_n - t_n) \leq \alpha^{n+1}\,\varrho. \tag{3.321}$$

Proof. We use mathematical induction to prove (3.320) and (3.321). Estimates (3.320) and (3.321) hold for $n = 0$ by (3.310), (3.311) and (3.317), since

$$t_1 - s_0 = \frac{r\,L}{2}\,(s_0 - t_0)\,(s_0 - t_0) = \delta\,(s_0 - t_0)$$

and

$$s_1 - t_1 \leq \frac{\dfrac{L}{2}\,\delta^2(s_0 - t_0)^2 + L\,\delta\,(s_0 - t_0)^2 + \dfrac{c\,L}{2}\,(s_0 - t_0)^2}{1 - L_0\,(1+\delta)\,\varrho}$$

$$\leq \frac{\varphi(s_0 - t_0)}{1 - L_0\,(1+\delta)\,\varrho}\,(s_0 - t_0) = \alpha\,(s_0 - t_0).$$

Let us assume (3.320) and (3.321) hold for all $k \leq n$. Then, we have that

$$t_{k+1} - s_k \leq \delta\,(s_k - t_n) \leq \delta\,\alpha^k\,\varrho,$$

$$s_{k+1} - t_{k+1} \leq \alpha\,(s_k - t_n) \leq \alpha^k\,\varrho$$

and

$$
\begin{aligned}
t_{k+1} &\leq s_k + \delta\,\alpha^k\,\varrho \leq t_k + \alpha^k\,\varrho + \delta\,\alpha^k\,\varrho \\
&\leq t_{k-1} + \alpha^{k-1}\,\varrho + \alpha^k\,\varrho + \delta\,\alpha^{k-1}\,\varrho + \delta\,\alpha^k\,\varrho \\
&\leq t_2 + (\alpha^2\,\varrho + \alpha^3\,\varrho + \cdots + \alpha^k\,\varrho) + (\delta\,\alpha^2\,\varrho + \delta\,\alpha^3\,\varrho + \cdots + \delta\,\alpha^k\,\varrho) \\
&\leq s_1 + \delta\,\alpha\,\varrho + (\alpha^2\,\varrho + \alpha^3\,\varrho + \cdots + \alpha^k\,\varrho) + (\delta\,\alpha^2\,\varrho + \delta\,\alpha^3\,\varrho + \cdots + \delta\,\alpha^k\,\varrho) \\
&\leq t_1 + \alpha\,\varrho + \delta\,\alpha\,\varrho + (\alpha^2\,\varrho + \alpha^3\,\varrho + \cdots + \alpha^k\,\varrho) + (\delta\,\alpha^2\,\varrho + \delta\,\alpha^3\,\varrho + \cdots + \delta\,\alpha^k\,\varrho) \\
&\leq \varrho + \delta\,\varrho + \alpha\,\varrho + \delta\,\alpha\,\varrho + (\alpha^2\,\varrho + \alpha^3\,\varrho + \cdots + \alpha^k\,\varrho) + (\delta\,\alpha^2\,\varrho + \delta\,\alpha^3\,\varrho + \cdots + \delta\,\alpha^k\,\varrho) \\
&= \frac{1 - \alpha^{k+1}}{1 - \alpha}\,(1+\delta)\,\varrho < \frac{1+\delta}{1-\alpha}\,\varrho = t^{\star\star}.
\end{aligned}
$$

Evidently, estimates (3.320) and (3.321) are true provided that

$$\frac{r\,L\,(s_k - t_k)}{2\,(1 - L_0\,t_k)} \leq \delta \tag{3.322}$$

and

$$\frac{\varphi(s_k - t_k)}{1 - L_0\,t_k} \leq \alpha. \tag{3.323}$$

Estimate (3.322) can be written as

$$\frac{r\,L}{2}\,\alpha^k\,\varrho + \delta\,L_0\,(1+\delta)\,\frac{1-\alpha^k}{1-\alpha}\,\varrho - \delta \le 0. \tag{3.324}$$

Inequality (3.324) motivates us to define recurrent functions f_k on $[0,1)$ for each $k = 1, 2, \cdots$ by

$$f_k(t) = \frac{r\,L}{2}\,t^k\,\varrho + \delta\,L_0\,(1+\delta)\,\frac{1-t^k}{1-\alpha}\,\varrho - \delta. \tag{3.325}$$

We need a relationship between two consecutive functions f_k. We have by (3.325) that

$$\begin{aligned}
f_{k+1}(t) &= \frac{r\,L}{2}\,t^{k+1}\,\varrho + \delta\,L_0\,(1+\delta)\,\frac{1-t^{k+1}}{1-\alpha}\,\varrho - \delta \\
&= f_k(t) + (t-1)\,(\frac{r\,L}{2}\,t + \delta\,L_0\,(1+\delta))\,t^{k-1}\,\varrho.
\end{aligned} \tag{3.326}$$

It follows from (3.326) that

$$f_{k+1}(t) \le f_k(t) \le \cdots \le f_1(t). \tag{3.327}$$

In view of (3.324) and (3.327) it suffices to show

$$f_1(\alpha) \le 0, \tag{3.328}$$

which is true by the choice of ϱ_2, (3.313) and (3.316). Similarly, estimate (3.323) can be written

$$\varphi\,\alpha^{k-1}\,\varrho + L_0\,(1+\delta)\,\frac{1-\alpha^{k+1}}{1-\alpha}\,\varrho - 1 \le 0. \tag{3.329}$$

Define recurrent functions g_k on $[0,1)$ for each $k = 1, 2, \cdots$ by

$$g_k(t) = \varphi\,t^{k-1}\,\varrho + L_0\,(1+\delta)\,\frac{1-\alpha^{k+1}}{1-\alpha}\,\varrho - 1. \tag{3.330}$$

Then using (3.330), we get that

$$g_{k+1}(t) = g_k(t) + (t-1)\,(\varphi + L_0\,(1+\delta))\,(1+t))\,t^{k-1}\,\varrho. \tag{3.331}$$

It follows from (3.331) that

$$g_{k+1}(t) \le g_k(t) \le \cdots \le g_1(t). \tag{3.332}$$

We can show instead of (3.329) that

$$g_1(\alpha) \le 0,$$

which is true by the choice of ϱ_3, (3.314) and (3.316). The induction for (3.320) and (3.321) is complete. Hence, sequence $\{t_n\}$ is increasing, bounded from above by $t^{\star\star}$ given by (3.318) and converges to its unique least upper bound $t^\star$. The proof of Lemma 3.6.1 is complete. $\qquad\square$

We have the following useful and obvious extension of Lemma 3.6.1.

Lemma 3.6.2. *Suppose there exists $N \in \mathbb{N}$ such that*

$$t_0 < s_0 < t_1 < s_1 < \cdots < t_N < s_N < t_{N+1} < \frac{1}{L_0} \tag{3.333}$$

and

$$s_N - t_N \begin{cases} \leq \varrho_0 \ \ if \ \varrho_0 \leq \varrho_1 \\ < \varrho_0 \ \ if \ \varrho_0 = \varrho_1. \end{cases} \tag{3.334}$$

Then, the conclusion of Lemma 3.6.1 hold for sequence $\{t_n\}$. Moreover, the following estimates hold for each $n = 0, 1, 2, \cdots$

$$0 < t_{N+n+1} - s_{N+n} \leq \delta_N \left(s_{N+n} - t_{N+n} \right) \tag{3.335}$$

and

$$0 < s_{N+n+1} - t_{N+n+1} \leq \alpha_N \left(s_{N+n} - t_{N+n} \right), \tag{3.336}$$

where

$$\delta_N = \delta(s_N - t_N), \quad \alpha_N = \alpha(s_N - t_N) \quad and \quad t_N^{\star\star} = \frac{1 + \delta_N}{1 - \alpha_N} \left(s_N - t_N \right).$$

Remark 3.6.1. Note that for $N = 0$, Lemma 3.6.2 reduces to Lemma 3.6.1 with $\alpha_0 = \alpha$ and $\delta_0 = \delta$. From now on we denote by $(\mathcal{C}^\star)$ conditions (C1), (C2), (C8)–(C11) and the conditions of Lemma 3.6.1 or Lemma 3.6.2. Simply uses (C9) instead of (C3) for the computation of the upper bounds on $\mathcal{D}X(p_{n+1})^{-1}$ to obtain that

$$\| \mathcal{D}X(p_{n+1})^{-1} \| \leq \frac{\beta}{1 - L_0 \, d(p_{n+1}, p_n)}. \tag{3.337}$$

Moreover, using the Ostrowski-type identities (3.307) and (3.308) in combination with (C6)–(C8), (3.337) instead of (C4) and the Taylor expansion, we get as in the proof of Theorem 3.6.3 that

$$d(p_{n+1}, p_n) \leq t_{n+1} - t_n. \tag{3.338}$$

Hence, we arrived at the following semi-local convergence result for method (3.309) under the $(\mathcal{C}^\land)$ conditions.

Theorem 3.6.5. *Assume $p_0 \in \Omega$. Suppose that the $(\mathcal{C}^\star)$ conditions are true. If $B\left(p_0, t^\star\right) \subseteq \Omega$, then the sequence (3.309) is well defined, remains in $B[p_0, t^\star]$ for each $n = 0, 1, 2, \cdots$ and it converges to the root $p_\ast$, which is the solution of $X(p) = 0$ in $B\left[p_0, t^\star\right]$. Moreover, the following hold*

(i) For each $n = 0, 1, 2, \cdots$

$$d\left(p_\ast, p_n\right) \leq t^\star - t_n, \tag{3.339}$$

where $\{t_n\}$ is defined in (3.317).

(ii) If

$$(L + L_0)\, t^{\star} < 1, \tag{3.340}$$

then, the root p_* is unique in $B\,[p_0, t^*]$.

Proof. It follows from (3.338) and Lemma 3.6.1 or Lemma 3.6.2 that sequence $\{t_n\}$ is complete. Hence, $\{p_n\}$ is complete in Ω and as such it converges to some p_*, which is a root of $X(p) = 0$, since $\| X(p_{n+1}) \| \longrightarrow 0$ as $n \to \infty$. Estimate (3.339) follows from (3.338) by using standard majoration techniques. For the uniqueness part, we use (C9) and (3.340) instead of (C4) and (C5) used in the proof of Theorem 3.6.3. We get in turn that

$$\left\| \mathcal{D}X\,(p^*)^{-1} \right\| \| A - \mathcal{D}X\,(p^*) \| \leq \frac{\beta}{1 - L_0\, t^{\star}} \, \| A - \mathcal{D}X(\gamma(0)) \|$$

$$\leq \frac{L}{2\,(1 - L_0\, t^{\star})} \, (d(p_*, q_*))$$

$$\leq \frac{L}{2\,(1 - L_0\, t^{\star})} \, (d(p_*, p_0) + d(q_*, p_0)) \leq \frac{L\,(t^{\star} - t_0)}{1 - L_0\, t^{\star}} < 1.$$

It follows that A is invertible. The proof of Theorem 3.6.5 is complete. $\square$

Remark 3.6.2.

(a) The limit point $t^{\star}$ can be replaced by $t^{\star\star}$ (given in closed form by (3.318)) in Theorem 3.6.3.

(b) Special case **I**: two-point Newton method. Let $\mathcal{T}_F(x) = I$. Then, we can choose $r = 1$ and $s = 0$. In this case, method (3.309) reduces to the second two-step Newton method.

3.7 Exercises

3.7.1 Let $\mathcal{M}$ be a real complete n-dimensional Riemannian manifold. Let $p', p \in \mathcal{M}$ and let $c : [0, 1] \longrightarrow \mathcal{M}$ be a piecewise smooth curve connecting p' to p. The distance from p' to p by $d(p', p) = \inf_{c} \, l(c)$, where $l(c) = \int_0^1 \| c'(t) \| \, dt$.

Let $p^{\star}$ be a point in $\mathcal{M}$ such that $\mathcal{D}\mathcal{X}(p^{\star})^{-1}$ exists. $\mathcal{D}\mathcal{X}(p^{\star})^{-1}\mathcal{D}\mathcal{X}$ is said to satisfy the radius Lipschitz condition with $\mathcal{S}$ average in $U(p^{\star}, r)$, $r \in [0, R_1]$, if, for each point p in $U(p^{\star}, r)$ and any minimizing geodesic $c : [0, 1] \longrightarrow \mathcal{M}$ connecting $p^{\star}$ and p,

$$\| \mathcal{D}\mathcal{X}(p^{\star})^{-1}\mathcal{P}_{c,1,0}(\mathcal{D}\mathcal{X}(p)) - \mathcal{P}_{c,\tau,1}\,\mathcal{X}(c(\tau))\,\mathcal{P}_{c,1,\tau} \|$$
$$\leq \int_{\tau\, d(p^{\star},p)}^{d(p^{\star},p)} \mathcal{S}(u)\, du \quad \text{for all } \tau \in [0, 1]. \tag{3.341}$$

Assume there exist $R > 0$ and a positive integrable function $\mathcal{S}$ on $[0, R]$ such that

$$\int_0^R \mathcal{S}(u)\, du > 1. \tag{3.342}$$

Show:

(a) There exists $R_1 < R$ such that

$$\int_0^{R_1} \mathcal{S}(u)\, du = 1. \tag{3.343}$$

(b) There exist $r_0 \le R_1$ and a positive integrable function $\mathcal{S}_0$ on $[0, r_0]$, such that

$$\int_0^{r_0} \mathcal{S}_0(u)\, du < 1. \tag{3.344}$$

(c) For each point p in $U(p^\star, r)$, $r \in [0, r_0]$, any minimizing geodesic $c : [0,1] \longrightarrow \mathcal{M}$ connecting $p^\star$ to p, the center Lipschitz condition with $\mathcal{S}_0$ average in the ball $U(p^\star, r)$:

$$\| \, \mathcal{D}\mathcal{X}(p^\star)^{-1} \left(\mathcal{P}_{c,1,0}\, \mathcal{D}\mathcal{X}(p)\, \mathcal{P}_{c,0,1} - \mathcal{D}\mathcal{X}(p^\star) \right) \, \| \le \int_0^{d(p^\star, p)} \mathcal{S}_0(u)\, du \tag{3.345}$$

is satisfied.

(d) Suppose that (3.344) is satisfied and $\mathcal{D}\mathcal{X}(p^\star)^{-1}\mathcal{D}\mathcal{X}$ satisfies the center Lipschitz condition with $\mathcal{S}_0$ average in $U(p^\star, r)$, $r \in [0, r_0]$. Show that $\mathcal{D}\mathcal{X}(p)^{-1}$ exists on $U(p^\star, r)$ and

$$\| \, \mathcal{D}\mathcal{X}(p)^{-1}\, \mathcal{P}_{c,0,1}\, \mathcal{D}\mathcal{X}(p^\star) \, \| \le b_0 = \left(1 - \int_0^{d(p, p^\star)} \mathcal{S}_0(u)\, du \right)^{-1},$$

where $c : [0,1] \longrightarrow \mathcal{M}$ is a minimizing geodesic connecting $p^\star$ to p.

Chapter 4

Secant Method

The Secant method uses a consistent approximation operator of the Fréchet derivative and is an alternative method of Newton's method. The Secant method is also known under the name of Regula Falsi or the method of Chords. The convergence analysis of Secant method using new Lipschitz and Hölder type conditions are presented in this chapter.

4.1 Semi-local Convergence

We extend the applicability of the method of chords in some cases in order to approximate a locally unique solution of a nonlinear equation in a Banach space setting. The error bounds are tighter and the information on the location of the solution at least as precise under the same information as before in Refs. [101, 163, 192, 23, 297, 352, 397, 435]. Application and examples are also provided in this section. We are concerned with the problem of approximating a locally unique solution $x^\star$ of (1.1) using the method of chords (MOC) in the form

$$x_{n+1} = x_n - \delta F(x_{n-1}, x_n)^{-1} F(x_n) \quad (n \geq 0), \quad (x_{-1}, x_0 \in \mathcal{D}) \qquad (4.1)$$

where $\delta F(x, y) \in \mathcal{L}(\mathcal{X}, \mathcal{Y})$ $(x, y \in \mathcal{D})$ is a consistent approximation of the Fréchet-derivative of F. (MOC) is an alternative method of Newton's method (NM) defined by (1.2).

Bosarge and Falb in Ref. [178], Dennis in Ref. [192], Hernández et al. in Ref. [268], Argyros in Ref. [110], Potra in Refs. [352, 356] and others (see Refs. [267, 338, 442]) have provided sufficient convergence conditions for (MOC) based on Lipschitz-type conditions on δF (for more details, see also relevant works in Refs. [134, 120, 297, 368, 369, 397, 435]). In the previous mentioned references, the conditions usually associated with the semi-local convergence of (MOC) (4.1) are:

$(\mathcal{H}_1)$ F is a nonlinear operator defined on a convex subset $\mathcal{D}$ of a Banach space $\mathcal{X}$ with values in a Banach space $\mathcal{Y}$;

219

$(\mathcal{H}_2)$ x_{-1} and x_0 are two points belonging to the interior $\mathcal{D}^0$ of $\mathcal{D}$ and satisfying the inequality $0 < \| x_0 - x_{-1} \| \le c$;

$(\mathcal{H}_3)$ F is Fréchet-differentiable on $\mathcal{D}^0$ and there exists an operator $\delta F : \mathcal{D}^0 \times \mathcal{D}^0 \to \mathcal{L}(\mathcal{X}, \mathcal{Y})$, such that the linear operator $A = \delta F(x_{-1}, x_0)$ is invertible, its inverse A^{-1} is bounded; $0 < \| A^{-1} F(x_0) \| \le \eta$;

$$\| A^{-1} \left(\delta F(x, y) - F'(z) \right) \| \le L \left(\| x - z \| + \| y - z \| \right),$$
$$\text{for all} \quad x, y, z \in \mathcal{D} \quad \text{and some} \quad L > 0;$$

$$\overline{U}(x_0, \nu) \subseteq \mathcal{D}^0,$$

for some $\nu > 0$ depending on L, c, η; and the popular condition

$$L\,c + 2\,\sqrt{L\,\eta} \le 1 \tag{4.2}$$

holds.

The majorizing sequence used is given by

$$v_{-1} = 0, \ v_0 = c, \ v_1 = c + \eta, \ v_{n+2} = v_{n+1} + \frac{L \left(v_{n+1} - v_{n-1} \right) \left(v_{n+1} - v_n \right)}{1 - L \left(v_{n+1} + v_n - v_0 \right)} \tag{4.3}$$

to obtain convergence of sequence $\{x_n\}$. The error bounds are

$$\| x_{n+1} - x_n \| \le v_{n+1} - v_n, \quad \| x_{n+1} - x^\star \| \le v^\star - v_n \quad \text{and} \quad v^\star = \lim_{n \to \infty} v_n.$$

The sufficient convergence condition (4.2) is easily violated. Indeed, let $L = 1$, $\eta = .18$ and $c = .185$. Then, (4.2) does not hold, since

$$L\,c + 2\,\sqrt{L\,\eta} = 1.033528137 > 1.$$

Hence, there is no guarantee that equation (1.1) under the information (L, c, η) has a solution that can be found using (MOC). In view of $(\mathcal{H}_3)$ there exists L_0 such that

$(\mathcal{H}_3^\star)$ $\| A^{-1} \left(\delta F(x, y) - F'(x_0) \right) \| \le L_0 \left(\| x - x_0 \| + \| y - x_0 \| \right)$ for all $x, y \in \mathcal{D}$ and some $L_0 > 0$.

$(\mathcal{H}_3^\star)$ is not an additional hypothesis, since in practice the computational of L requires that of L_0.

Using the more precise (for $L_0 < L$) majorizing sequence for $\{x_n\}$:

$$w_{-1} = 0, \ w_0 = c, \ w_1 = c + \eta, \ w_{n+2} = w_{n+1} + \frac{L \left(w_{n+1} - w_{n-1} \right) \left(w_{n+1} - w_n \right)}{1 - L_0 \left(w_{n+1} + w_n - w_0 \right)}, \tag{4.4}$$

we provided sufficient convergence conditions that can in some cases be weaker than (4.3). Moreover we showed

$$w_n \le v_n, \tag{4.5}$$

$$w_{n+1} - w_n \le v_{n+1} - v_n \tag{4.6}$$

and

$$w^\star \leq v^\star, \quad w^\star = \lim_{n \longrightarrow \infty} w_n. \tag{4.7}$$

Furthermore (4.5) and (4.6) hold as strict inequalities for $n \geq 2$, if $L_0 < L$ (see Refs. [101, 163]).

We expand the applicability of (MOC) by introducing the majorizing sequence $\{t_n\}$ given by

$$t_{-1} = -c, \ t_0 = 0, \ t_1 = \eta, \ t_{n+2} = t_{n+1} + \frac{L\,(t_{n+1} - t_{n-1})\,(t_{n+1} - t_n)}{1 - L_0\,(t_{n+1} + t_n + c)}, \tag{4.8}$$

which converges, if any of the following are satisfied:

$$\begin{cases} \left(L_0 + L + \sqrt{L^2 + 4\,L_0\,L}\right) c \leq 1 \\ \text{and} \\ \left(L + 2\,L_0 + \sqrt{L^2 + 4\,L_0\,L}\right)\eta + L_0\,c \leq 1 \end{cases} \tag{4.9}$$

or

$$\begin{cases} \left(L_0 + L + \sqrt{L^2 + 4\,L_0\,L}\right) c \geq 1 \\ \text{and} \\ \left(\eta + c\right)\left(L + 4\,L_0 + \sqrt{L^2 + 4\,L_0\,L}\right) \leq 2 \end{cases} \tag{4.10}$$

or

$$\begin{cases} \left(L_0 + L + \sqrt{L^2 + 4\,L_0\,L}\right) c \geq 1, \\ \left(\eta + c\right)\left(L + 2\,L_0 + \sqrt{L^2 + 4\,L_0\,L}\right) \leq 2 \\ \text{and} \\ \left(L + 2\,L_0 + \sqrt{L^2 + 4\,L_0\,L}\right)\eta + L_0\,c \leq 1. \end{cases} \tag{4.11}$$

These conditions are different from (4.2) and others in the literature (see for example Refs. [101, 163, 163]). A simple induction argument shows that $t_n \leq v_n$, $t_{n+1} - t_n \leq v_{n+1} - v_n$ and $t^\star \leq v^\star$, $t^\star = \lim_{n \longrightarrow \infty} t_n$. Moreover, we show that the applicability of (MOC) is expanded in some cases under the same information as before; the error bounds on $\| x_{n+1} - x_n \|$, $\| x_n - x^\star \|$ $(n \geq 0)$ are tighter and the information on the location of the solution more precise. Note in particular that using the above values and $L_0 = .8$, we get for (4.9)

$$\left(L_0 + L + \sqrt{L^2 + 4\,L_0\,L}\right) c = .7121371783 < 1$$

and

$$\left(L + 2\,L_0 + \sqrt{L^2 + 4\,L_0\,L}\right)\eta + L_0\,c = .9848902275 < 1.$$

That is sequence $\{t_n\}$ converges.

We need the following result on majorizing sequences for (MOC).

Lemma 4.1.1. *Let $L_0 > 0$, $L > 0$, $c \geq 0$ and $\eta \geq 0$ be given constants. Define parameters α and γ by*

$$\alpha = \max\{\kappa_0, \kappa\}, \qquad \gamma = \frac{1 - L_0\,(2\,\eta + c)}{1 - L_0\,c},$$

where

$$\kappa_0 = \frac{L\,(\eta + c)}{1 - L_0\,(c + \eta)} \quad and \quad \kappa = \frac{-L + \sqrt{L^2 + 4\,L_0\,L}}{2\,L_0}.$$

Assume that (4.9) holds. Then, scalar sequence $\{t_n\}$ $(n \geq -1)$ given by (4.8) is non-decreasing, bounded from above by

$$t^{\star\star} = \frac{\eta}{1 - \alpha} + (\gamma - \alpha)\,\eta \tag{4.12}$$

and converges to its unique least upper bound $t^\star$ such that

$$0 \leq t^\star \leq t^{\star\star}. \tag{4.13}$$

Moreover, the following estimates hold for all $n \geq 0$:

$$0 \leq t_{n+2} - t_{n+1} \leq \alpha\,(t_{n+1} - t_n) \leq \alpha^{n+1}\,\eta \tag{4.14}$$

and

$$t^\star - t_n \leq \frac{\alpha^n\,\eta}{1 - \alpha}. \tag{4.15}$$

Proof.   We shall first show

$$0 \leq \alpha \leq \gamma. \tag{4.16}$$

Case $\kappa_0 \leq \kappa$. We must have

$$\frac{L\,(\eta + c)}{1 - L_0\,(\eta + c)} \leq \kappa,$$

or

$$\eta \leq \frac{\kappa\,(1 - L_0\,c) - L\,c}{L + \kappa\,L_0} \tag{4.17}$$

and

$$\kappa \leq \frac{1 - L_0(2\,\eta + c)}{1 - L_0\,c}, \tag{4.18}$$

or

$$\eta \leq \frac{(1 - \kappa)\,(1 - L_0\,c)}{2\,L_0}. \tag{4.19}$$

For

$$\frac{(1 - \kappa)\,(1 - L_0\,c)}{2\,L_0} \leq \frac{\kappa\,(1 - L_0\,c) - L\,c}{L + \kappa\,L_0} \tag{4.20}$$

we must have

$$(1 - L_0\,c)\,(-L_0\,\kappa^2 - L\,\kappa + L) - L_0\,\kappa\,(1 - L_0\,c) + 2\,L_0\,c \leq 0. \tag{4.21}$$

By the definition of κ, $L_0\,\kappa^2 + L\,\kappa - L = 0$, so for (4.21) to be satisfied if

$$2\,L\,c \leq \kappa\,(1 - L_0\,c) \tag{4.22}$$

or the first inequality of (4.9) must hold. Then, in this case (4.19) is equivalent to the second inequality in (4.9). If

$$\frac{(1 - \kappa)\,(1 - L_0\,c)}{2\,L_0} \geq \frac{\kappa\,(1 - L_0\,c) - L\,c}{L + \kappa\,L_0}, \tag{4.23}$$

then, we must have

$$2\,L\,c \geq \kappa\,(1 - L_0) \tag{4.24}$$

or the first inequality of (4.10) holds. Then, in this case (4.17) is equivalent to the second inequality in (4.10).

Case $\kappa_0 \geq \kappa$. We must have (4.19) and

$$\eta \geq \frac{\kappa\,(1 - L_0\,c) - L\,c}{L + \kappa\,L_0}. \tag{4.25}$$

Then, for consistency

$$\frac{(1 - \kappa)\,(1 - L_0\,c)}{2\,L_0} \geq \eta \geq \frac{\kappa\,(1 - L_0\,c) - L\,c}{L + \kappa\,L_0} \tag{4.26}$$

leading to the second inequality of (4.9), the first inequality of (4.10) and the second inequality of (4.11).

In view of (4.9), we have $\kappa \in (0,1)$ and $\alpha \in (0,1)$. We shall show using mathematical induction on k

$$0 \leq t_{k+2} - t_{k+1} \leq \alpha\,(t_{k+1} - t_k). \tag{4.27}$$

By (4.8) and (4.27) for $k = 0$, we must show

$$0 \leq \frac{L\,(t_1 - t_{-1})}{1 - L_0\,(t_1 + c)} \leq \alpha \quad \text{or} \quad 0 \leq \frac{L\,(c + \eta)}{1 - L_0\,(c + \eta)} \leq \alpha,$$

which is true from (4.9) and the choice of α. Let assume that (4.27) holds for $k \leq n + 1$. The induction hypothesis gives

$$\begin{aligned}
t_{k+2} &\leq t_{k+1} + \alpha\,(t_{k+1} - t_k) \\
&\leq t_k + \alpha\,(t_k - t_{k-1}) + \alpha\,(t_{k+1} - t_k) \\
&\leq t_1 + \alpha\,(t_1 - t_0) + \cdots + \alpha\,(t_{k+1} - t_k) \\
&\leq \eta + \alpha\,\eta + \cdots + \alpha^{k+1}\,\eta + (\gamma - \alpha)\,\eta \\
&= \frac{1 - \alpha^{k+2}}{1 - \alpha}\,\eta + (\gamma - \alpha)\,\eta < \frac{\eta}{1 - \alpha} + (\gamma - \alpha)\,\eta = t^{\star\star}.
\end{aligned} \tag{4.28}$$

Moreover, we have that

$$\begin{aligned}
&L\,(t_{k+2} - t_k) + \alpha\,L_0\,(t_{k+2} - t_0 + t_{k+1}) \\
&\leq L\left((t_{k+2} - t_{k+1}) + (t_{k+1} - t_k)\right) + \kappa\,L_0\left(\frac{1 - \alpha^{k+2}}{1 - \alpha} + \frac{1 - \alpha^{k+1}}{1 - \alpha}\right) \\
&\leq L\,(\alpha^k + \alpha^{k+1})\,\eta + \frac{\alpha\,L_0}{1 - \alpha}\,(2 - \alpha^{k+1} - \alpha^{k+2})\,\eta + \alpha\,L_0\,c.
\end{aligned} \tag{4.29}$$

We now prove (4.27). By (4.14) and (4.29), we must have estimate

$$L\,(\alpha^k + \alpha^{k+1})\,\eta + \frac{\alpha\,L_0}{1 - \alpha}\,(2 - \alpha^{k+1} - \alpha^{k+2})\,\eta + \alpha\,L_0\,c \leq \alpha \tag{4.30}$$

or

$$L\,(\alpha^{k-1} + \alpha^k)\,\eta + L_0\left((1 + \alpha + \cdots + \alpha^k) + (1 + \alpha + \cdots + \alpha^{k+1})\right)\eta + L_0\,c - 1 \leq 0. \tag{4.31}$$

In view of (4.31), we are motivated to define (for $\alpha = s$) the functions for $k \geq 1$

$$f_k(s) = L\,(s^{k-1} + s^k)\,\eta + L_0\left(2\,(1 + s + \cdots + s^k) + s^{k+1}\right)\eta + L_0\,c - 1. \tag{4.32}$$

We need the relationship between two consecutive functions f_k. Using (4.32), we obtain

$$f_{k+1}(s) = L\left(s^k + s^{k+1}\right)\eta + L_0\left(2\left(1 + s + \cdots + s^{k+1}\right) + s^{k+2}\right)\eta + L_0\,c - 1$$
$$= f_k(s) + L\left(s^{k+1} - s^{k-1}\right)\eta + L_0\left(s^{k+1} + s^{k+2}\right)\eta$$
$$= g(s)\,s^{k-1}\,\eta + f_k(s).$$

$$(4.33)$$

Note that κ is the unique positive root of polynomial g given by

$$g(s) = L_0\,s^3 + (L_0 + L)\,s^2 - L. \tag{4.34}$$

Instead of (4.31), we shall show $f_k(\alpha) \leq 0 \qquad (k \geq 0)$. Using (4.33) and (4.9), we get in turn

$$f_{k+1}(\alpha) \geq f_k(\alpha) \tag{4.35}$$

since $g(\alpha) \geq 0$. Define function f_∞ on $[0, 1)$ by

$$f_\infty(s) = \lim_{k \to \infty} f_k(s). \tag{4.36}$$

We obtain by (4.35) and (4.36) that $f_\infty(\alpha) \geq f_k(\alpha)$ for all $k \geq 0$. That is instead of (4.31) we can show

$$f_\infty(\alpha) = \frac{2\,L_0\,\eta}{1 - \alpha} + L_0\,c - 1, \leq 0$$

which is true by (4.9). Hence, we showed sequence $\{t_n\}$ $(n \geq -1)$ is non-decreasing and bounded from above by $t^{\star\star}$, so that (4.13) holds. It follows that there exists $t^\star \in [0, t^{\star\star}]$, so that $\lim_{n \to \infty} t_n = t^\star$. Estimate (4.15) follows from (4.14) by using standard majorization techniques. That completes the proof of Lemma 4.1.1. $\square$

We can also show the following result about the convergence order $(1 + \sqrt{5})/2$ of majorizing sequence $\{t_n\}$.

Lemma 4.1.2. *Under the hypotheses of Lemma 4.1.1, further assume*

$$q = \beta\,c < 1, \tag{4.37}$$

where

$$\beta = \frac{L\,(1 + \kappa)}{1 - L_0\,c\left(\dfrac{1 + \kappa}{1 - \kappa}\right)}. \tag{4.38}$$

Then, the following estimates hold:

$$t_n - t_{n-1} \leq q^{\theta_{n-1}-1}\,\eta \qquad (n \geq 2) \tag{4.39}$$

and

$$t^\star - t_n \leq e_n\,\eta \qquad (n \geq 1), \tag{4.40}$$

where

$$p = \frac{1 + \sqrt{5}}{2}, \quad e_n = \frac{q^{p^n/\sqrt{5}}}{q\left(1 - q^{p^n\,(p-1)/\sqrt{5}}\right)}$$

and $\{\theta_n\}$ is the Fibonacci's sequence

$$\theta_0 = \theta_1 = 1, \qquad \theta_{n+1} = \theta_n + \theta_{n-1}, \quad (n \geq 1). \tag{4.41}$$

Proof. It follows from (4.12), (4.13), (4.30) and (4.38) that

$$t_{n+2} - t_{n+1} \leq \beta \left(t_n - t_{n-1}\right)\left(t_{n+1} - t_n\right) \quad (n \geq 1). \tag{4.42}$$

Estimate (4.39) holds for $n = 0$ from the initial conditions and (4.41). Assume (4.39) holds for all $k \leq n - 1$. Then, using the induction hypotheses and (4.42), we have

$$t_{k+1} - t_k \leq q\, q^{\theta_{k-2}-1}\left(t_k - t_{k-1}\right) = q^{\theta_{k-1}}\left(t_k - t_{k-1}\right) \leq q^{\theta_{k-2}}\, q^{\theta_{k-1}}\, \eta = q^{\theta_k}\,\eta, \tag{4.43}$$

which shows (4.39) holds for all n. We also have that

$$t_{k+1} - t_1 \leq \sum_{i=1}^{k}(t_{i+1} - t_i) \leq \frac{\eta}{q}\left(q^{\theta_1} + q^{\theta_2} + \cdots + q^{\theta_k}\right) < t_0^\star = \frac{\eta}{q}\sum_{i=1}^{\infty} q^{\theta_i}. \tag{4.44}$$

Note that

$$\theta_k = \frac{1}{\sqrt{5}}\left(\left(\frac{1+\sqrt{5}}{2}\right)^{k+1} - \left(\frac{1-\sqrt{5}}{2}\right)^{k+1}\right) \geq \frac{1}{\sqrt{5}}\left(\frac{1+\sqrt{5}}{2}\right)^{k} = \frac{p^k}{\sqrt{5}}. \tag{4.45}$$

So, for any $k \geq 0$, $m \geq 0$, we get in turn:

$$\begin{aligned}
t_{k+m} - t_k &\leq (t_{k+1} - t_k) + (t_{k+2} - t_{k+1}) + \cdots + (t_{k+m} - t_{k+m-1})\\
&\leq \frac{\eta}{q}\left(q^{\theta_k} + q^{\theta_{k+1}} + \cdots + q^{\theta_{k+m-1}}\right)\\
&\leq \frac{\eta}{q}\left(q^{p^k/\sqrt{5}} + q^{p^{k+1}/\sqrt{5}} + \cdots + q^{p^{k+m-1}/\sqrt{5}}\right).
\end{aligned} \tag{4.46}$$

Using Bernoulli's inequality, we get by (4.46):

$$\begin{aligned}
t_{k+m} &- t_k\\
&\leq \frac{\eta}{q}\, q^{p^k/\sqrt{5}}\left(1 + q^{(p^{k+1}-p^k)/\sqrt{5}} + q^{(p^{k+2}-p^k)/\sqrt{5}} + \cdots + q^{(p^{k+m-1}-p^k)/\sqrt{5}}\right)\\
&= \frac{\eta}{q}\, q^{p^k/\sqrt{5}}\left(1 + q^{p^k(p-1)/\sqrt{5}} + q^{p^k(p^2-1)/\sqrt{5}} + \cdots + q^{p^k(p^{m-1}-1)/\sqrt{5}}\right)\\
&\leq \frac{\eta}{q}\, q^{p^k/\sqrt{5}}\left(1 + q^{p^k(p-1)/\sqrt{5}} + q^{p^k(1+2(p-1)-1)/\sqrt{5}} + \cdots + q^{p^k(p^{m-1}-1)/\sqrt{5}}\right)\\
&= \frac{\eta}{q}\, q^{p^k/\sqrt{5}}\left(1 + q^{p^k(p-1)/\sqrt{5}} + \left(q^{p^k(p-1)/\sqrt{5}}\right)^2 + \cdots + \left(q^{p^k(p-1)/\sqrt{5}}\right)^{m-1}\right)\\
&= \frac{\eta}{q}\, q^{p^k/\sqrt{5}}\,\frac{1 - q^{p^k(p-1)m/\sqrt{5}}}{1 - q^{p^k(p-1)/\sqrt{5}}},
\end{aligned}$$

$$\tag{4.47}$$

which shows (4.40) if $m \longrightarrow \infty$. That completes the proof of Lemma 4.1.2. $\qquad\square$

We shall study (MOC) for triplets (F, x_{-1}, x_0) belonging to the class $\mathcal{C}(L, L_0, \eta, c, \alpha)$ defined as follows:

Definition 4.1.1. Let L, L_0, η, c and α be non-negative constants satisfying the hypotheses of Lemma 4.1.1. A triplet (F, x_{-1}, x_0) belongs to the class $\mathcal{C}(L, L_0, \eta, c, \alpha)$ if:

$(\mathcal{A}_1)$ F is a nonlinear operator defined on a convex subset $\mathcal{D}$ of a Banach space $\mathcal{X}$ with values in a Banach space $\mathcal{Y}$;

$(\mathcal{A}_2)$ x_{-1} and x_0 are two points belonging to the interior $\mathcal{D}^0$ of $\mathcal{D}$ and satisfying the inequality $\| x_0 - x_{-1} \| \leq c$;

$(\mathcal{A}_3)$ F is Fréchet-differentiable on $\mathcal{D}^0$ and there exists an operator $\delta F : \mathcal{D}^0 \times \mathcal{D}^0 \to \mathcal{L}(\mathcal{X}, \mathcal{Y})$, such that, $A^{-1} = \delta F(x_{-1}, x_0)^{-1} \in \mathcal{L}(\mathcal{Y}, \mathcal{X})$, $\| A^{-1} F(x_0) \| \leq \eta$ and for all $x, y, z \in \mathcal{D}$, the following hold

$$\| A^{-1} \left(\delta F(x, y) - F'(z) \right) \| \leq L \left(\| x - z \| + \| y - z \| \right)$$

and

$$\| A^{-1} \left(\delta F(x, y) - F'(x_0) \right) \| \leq L_0 \left(\| x - x_0 \| + \| y - x_0 \| \right);$$

$(\mathcal{A}_4)$ $\overline{U}(x_0, t^\star) \subseteq \mathcal{D}_c = \{ x \in \mathcal{D} : F \text{ is continuous at } x \} \subseteq \mathcal{D}$, where $t^\star$ is given in Lemma 4.1.1.

The semi-local convergence theorem for (MOC) is as follows.

Theorem 4.1.1. *If* $(F, x_{-1}, x_0) \in \mathcal{C}(L, L_0, \eta, c, \alpha)$, *then, the sequence* $\{x_n\}$ $(n \geq -1)$ *generated by (MOC) is well defined, remains in* $\overline{U}(x_0, t^\star)$ *for all* $n \geq 0$ *and converges to a unique solution* $x^\star \in \overline{U}(x_0, t^\star)$ *of (1.1). Moreover the following estimates hold for all* $n \geq 0$

$$\| x_n - x_{n-1} \| \leq t_n - t_{n-1} \tag{4.48}$$

and

$$\| x_n - x^\star \| \leq t^\star - t_n, \tag{4.49}$$

where $\{t_n\}$, $(n \geq 0)$ *is given by (4.8). Furthermore, if there exists* R, *such that*

$$\overline{U}(x_0, R) \subseteq \mathcal{D}, \quad R \geq t^\star \quad \text{and} \quad L_0 \left(t^\star + R \right) + \| A^{-1} \left(F'(x_0) - A \right) \| < 1, \tag{4.50}$$

then, the solution $x^\star$ *is unique in* $\overline{U}(x_0, R)$.

Proof. First, we show that $L = \delta F(x_k, x_{k+1})$ is invertible for $x_k, x_{k+1} \in \overline{U}(x_0, t^\star)$. By $(\mathcal{A}_2)$ and $(\mathcal{A}_3)$, we have that

$$\begin{aligned}
\| I - A^{-1} L \| &= \| A^{-1} \left(L - A \right) \| \\
&\leq \| A^{-1}(L - F'(x_0)) \| + \| A^{-1}(F'(x_0) - A) \| \\
&\leq L_0 \left(\| x_k - x_0 \| + \| x_{k+1} - x_0 \| + \| x_0 - x_{-1} \| \right) \\
&\leq L_0 \left(t_k - t_0 + t_{k+1} - t_0 + c \right) < 1,
\end{aligned} \tag{4.51}$$

by (4.30). Using the Banach Lemma on invertible operators and (4.51), L is invertible and

$$\| L^{-1} A \| \leq \left(1 - L_0 \left(t_{k+1} + t_k + c \right) \right)^{-1}. \tag{4.52}$$

By $(\mathcal{A}_3)$, we have

$$\| A^{-1} \left(F'(u) - F'(v) \right) \| \leq 2 L \| u - v \|, \quad u, v \in \mathcal{D}^0. \tag{4.53}$$

We can write the identity

$$F(x) - F(y) = \int_0^1 F'(y + t(x - y))\, dt\, (x - y) \tag{4.54}$$

then, for all $x, y, u, v \in \mathcal{D}^0$, we obtain

$$\| A^{-1} (F(x) - F(y) - F'(u)(x - y)) \|$$
$$\leq L (\| x - u \| + \| y - u \|) \, \| x - y \| \tag{4.55}$$

and

$$\| A^{-1} (F(x) - F(y) - \delta F(u, v) (x - y)) \|$$
$$\leq L (\| x - v \| + \| y - v \| + \| u - v \|) \, \| x - y \| . \tag{4.56}$$

By a continuity argument (4.53)–(4.56) remain valid if x and/or y belong to $\mathcal{D}_c$. Now we show (4.48). If (4.36) holds for all $n \leq k$ and if $\{x_n\}$ $(n \geq 0)$ is well defined for $n = 0, 1, 2, \cdots, k$, then

$$\| x_n - x_0 \| \leq t_n - t_0 < t^\star - t_0, \quad n \leq k. \tag{4.57}$$

That is (4.1) is well defined for $n = k + 1$. For $n = -1$ and $n = 0$, (4.48) reduces to $\| x_{-1} - x_0 \| \leq c$ and $\| x_0 - x_1 \| \leq \eta$. Suppose (4.48) holds for $n = -1, 0, 1, \cdots, k$ $(k \geq 0)$. By (4.52), (4.56) and

$$F(x_{k+1}) = F(x_{k+1}) - F(x_k) - \delta F(x_{k-1}, x_k) (x_{k+1} - x_k) \tag{4.58}$$

and

$$\| A^{-1} F(x_{k+1}) \| \leq L (\| x_{k+1} - x_k \| + \| x_k - x_{k-1} \|) \, \| x_{k+1} - x_k \|$$
$$\leq L (t_{k+1} - t_k + t_k - t_{k-1}) (t_{k+1} - t_k) \tag{4.59}$$
$$= L (t_{k+1} - t_{k-1}) (t_{k+1} - t_k).$$

We obtain the following estimate

$$\| x_{k+2} - x_{k+1} \| = \| \delta F(x_k, x_{k+1})^{-1} F(x_{k+1}) \|$$
$$\leq \| \delta F(x_k, x_{k+1})^{-1} A \| \, \| A^{-1} F(x_{k+1}) \|$$
$$\leq \frac{L (t_{k+1} - t_{k-1}) (t_{k+1} - t_k)}{1 - L_0 (\| x_{k+1} - x_0 \| + \| x_k - x_0 \| + c)}$$
$$\leq \frac{L (t_{k+1} - t_k + t_k - t_{k-1})}{1 - L_0 (t_{k+1} - t_0 + t_k - t_0 + t_0 - t_{-1})} (t_{k+1} - t_k)$$
$$= t_{k+2} - t_{k+1}.$$

The induction for (4.48) is complete. It follows from (4.48) and Lemma 4.1.1 that $\{x_n\}$ $(n \geq -1)$ is Cauchy in a Banach space $\mathcal{X}$ and as such it converges to some $x^\star \in \overline{U}(x_0, t^\star)$ (since $\overline{U}(x_0, t^\star)$ is a closed set). By letting $k \to \infty$ in (4.59), we obtain $F(x^\star) = 0$. Estimate (4.49) follows from (4.48) by using standard majoration techniques.

Finally, for showing the uniqueness in $\overline{U}(x_0, R)$, let $y^\star \in \overline{U}(x_0, R)$ be a solution (1.1). Set $\mathcal{M} = \int_0^1 F'(y^\star + t (y^\star - x^\star))\, dt$. It then follows by $(\mathcal{A}_3)$ and (4.50):

$$\| A^{-1} (A - \mathcal{M}) \| = L_0 (\| y^\star - x_0 \| + \| x^\star - x_0 \|) + \| A^{-1} (F'(x_0) - A) \|$$
$$\leq L_0 \left(t^\star + R \right) + \| A^{-1} (F'(x_0) - A) \| < 1. \tag{4.60}$$

It follows from (4.60) and the Banach lemma on invertible operators that $\mathcal{M}^{-1}$ exists on $U(x_0, t^\star)$. Using the identity $F(x^\star) - F(y^\star) = \mathcal{M} (x^\star - y^\star)$, we deduce $x^\star = y^\star$. That completes the proof of Theorem 4.1.1. $\qquad\square$

Remark 4.1.1.

(a) The point $t^{\star\star}$ given in closed form by (4.13) can replace $t^\star$ in the hypotheses of Theorem 4.1.1.

(b) In the uniqueness part (see (4.50)), we can replace $\overline{L_0} = \| A^{-1} (F'(x_0) - A) \|$ by the less tight $L_0 c$, since by $(\mathcal{A}_3)$,

$$\overline{L_0} = \| A^{-1} (F'(x_0) - A) \| \leq L_0 \| x_0 - x_{-1} \| \leq L_0 c. \qquad (4.61)$$

In fact according to (4.51), majorizing sequence $\{t_n\}$ can be replaced by the at least as tight

$$t_{-1} = -c, \; t_0 = 0, \; t_1 = \eta, \; t_{n+2} = t_{n+1} + \frac{L (t_{n+1} - t_{n-1}) (t_{n+1} - t_n)}{1 - (\overline{L_0} + L_0 (t_{n+1} - t_n))}, \qquad (4.62)$$

which is also a majorizing sequence for $\{x_n\}$ (converging under the same hypotheses to some $\overline{t^\star} \leq t^\star$).

(c) A more popular hypothesis used instead of the frst inequality in $(\mathcal{A}_3)$ is

$$\| A^{-1} (\delta F(x, y) - \delta F(u, v)) \| \leq \overline{L} (\| x - u \| + \| y - v \|), \qquad (4.63)$$

for all $x, y, u, v \in \mathcal{D}$. Note that (4.63) implies both hypotheses in $(\mathcal{A}_3)$ but not necessarily vice versa. Note also that $L_0 \leq \overline{L}, \; L \leq \overline{L}$ and $\dfrac{\overline{L}}{L_0}, \; \dfrac{\overline{L}}{L}$ can be arbitrarily large.

(d) Our sufficient convergence conditions differ from the ones in Refs. [101, 120, 163].

Remark 4.1.2. It turns out from the proof of Theorem 4.1.1 that $\{s_n\}$ given by

$$s_{-1} = -c, \; s_0 = 0, \; s_1 = \eta, \; s_{n+2} = s_{n+1} + \frac{L_1 (s_{n+1} - s_{n-1}) (s_{n+1} - s_n)}{1 - L_0 (s_{n+1} + c + s_n)}, \qquad (4.64)$$

where

$$L_1 = \begin{cases} L_0 & \text{if } n = 0 \\ L & \text{if } n > 0 \end{cases}$$

is a finer majorizing sequence for $\{x_n\}$ than $\{t_n\}$, if $L_0 < L$. Moreover, we have

$$s_n < t_n, \quad s_{n+1} - s_n < t_{n+1} - t_n, \qquad (4.65)$$

and

$$0 \leq s^\star - s_n \leq t^\star - t_n, \quad s^\star = \lim_{n \to \infty} s_n. \qquad (4.66)$$

Returning back to the example given in the Introduction of this section but choosing $L_0 = .92$, we have again that (4.2) is violated. However, (4.16) is also violated, since

$$\kappa_0 = .549533273, \quad \kappa = .632244981 \quad \text{and} \quad \gamma = .600867679.$$

At this point we are wondering if sequence $\{s_n\}$ converges under weaker than (4.9) condition. Let us define recurrent functions h_n (corresponding to f_n of Lemma 4.1.1) by

$$\begin{aligned} h_n(s) = {} & L (s^n + s^{n-1}) (s_2 - \eta) + s L_0 ((1 + s + \cdots + s^n) + \\ & (1 + s + \cdots + s^{n-1})) (s_2 - \eta) + s (2 L_0 \eta + L_0 c - 1), \end{aligned} \qquad (4.67)$$

then

$$h_{n+1}(s) = h_n(s) + g(s)\,(s_2 - \eta)\,s^{n-1} \tag{4.68}$$

and

$$h_\infty(s) = s\left(\frac{2\,L_0}{1-s}\,(s_2 - \eta) + 2\,L_0\,\eta + L_0\,c - 1\right). \tag{4.69}$$

Then, with the above changes but following the proof of Lemma 4.1.1, we arrive at the corresponding result for majorizing sequence $\{s_n\}$:

Lemma 4.1.3. *Let $L_0 > 0$, $L > 0$, $c \geq 0$ and $\eta \geq 0$ be given constants. Assume:*

$$0 \leq \beta = \max\{\frac{L\,s_2}{1 - L_0\,(s_2 + \eta + c)}, \kappa\} \leq \frac{1 - L_0\,(c + 2\,s_2)}{1 - L_0\,(c + 2\,\eta)}, \tag{4.70}$$

where s_2 is given by (4.64) and κ is defined in Lemma 4.1.1. Then, scalar sequence $\{s_n\}$ $(n \geq -1)$ given by (4.64) is non-decreasing, bounded from above by

$$s^{\star\star} = \eta + \frac{1}{1-\beta}\,(s_2 - \eta) \tag{4.71}$$

and converges to its unique least upper bound $s^\star$ such that

$$0 \leq s^\star \leq s^{\star\star}. \tag{4.72}$$

Moreover, the following estimates hold for all $n \geq 1$:

$$s_{n+2} - s_{n+1} \leq \beta^n\,(s_2 - \eta). \tag{4.73}$$

Remark 4.1.3.

(a) According to the proof of Lemma 4.1.1, hypothesis (4.70) and sequence $\{s_n\}$ can replace (4.9) and $\{t_n\}$ in Theorem 4.1.1.

(b) Using the numerical values of Remark 4.1.2, we get

$$s_2 = .27100271, \quad \frac{L\,s_2}{1 - L_0\,(s_2 + \eta + c)} = .653211383, \quad \beta = .653211383$$

and

$$\frac{1 - L_0\,(2\,s_2 + c)}{1 - L_0\,(2\,\eta + c)} = .664169703.$$

That is hypotheses of Lemma 4.1.3 are satisfied.

(c) Hypothesis (4.70) is slightly more difficult to verify than (4.9) (because of s_2). However the verification of these conditions requires only computations at the original data L_0, L, c and η.

(d) It turns out that Lemma 4.1.3 admits a useful extension as follows: Let $N \geq 1$ be a fixed integer. Let us define recurrent functions h_n^N (corresponding to h_n of Lemma 4.1.3) by

$$\begin{aligned}
h_n^N(s) = {} & L\,(s^n + s^{n-1})\,(s_{N+1} - s_N) + s\,L_0\,((1 + s + \cdots + s^n) + \\
& (1 + s + \cdots + s^{n-1}))\,(s_{N+1} - s_N) + s\,(2\,L_0\,s_{N+1} + \\
& L_0\,s_N + L_0\,c - 1),
\end{aligned} \tag{4.74}$$

then

$$h_{n+1}^N(s) = h_n^N(s) + g(s)(s_{N+1} - s_N)s^{n-1} \tag{4.75}$$

and

$$h_\infty^N(s) = s\left(\frac{2L_0}{1-s}(s_{N+1} - s_N) + 2L_0 s_N + L_0 s_{N-1} + L_0 c - 1\right). \tag{4.76}$$

Note that if $N = 1$, we obtain the corresponding h functions given by (4.67)–(4.69), respectively.

Hence, we arrive at the following result.

Lemma 4.1.4. *Let $L_0 > 0$, $L > 0$, $c \geq 0$ and $\eta \geq 0$ be given constants. Let also $N \geq 1$ be a fixed integer. Assume that iterates s_i, $i = 0, 1, \cdots, N+1$ given by (4.64) are well defined;*

$$s_{i-1} \leq s_i, \quad i = 0, 1, \cdots, N+1; \tag{4.77}$$

$$0 \leq \beta_N = \max\{\frac{L(s_{N+1} - s_{N-1})}{1 - L_0(s_{N+1} + s_N + c)}, \kappa\} \leq \frac{1 - L_0(2s_{N+1} + s_{N-1} + c)}{1 - L_0(2s_N + s_{N-1} + c)}, \tag{4.78}$$

where κ is defined in Lemma 4.1.1. Then, scalar sequence $\{s_n\}$ $(n \geq -1)$ given by (4.64) is non-decreasing, bounded from above by

$$s_N^{\star\star} = \eta + \frac{1}{1 - \beta}(s_{N+1} - s_N) \tag{4.79}$$

and converges to its unique least upper bound $s^\star$ such that

$$0 \leq s_N^\star \leq s_N^{\star\star}. \tag{4.80}$$

Moreover, the following estimates hold for all $n \geq 1$:

$$0 \leq s_{N+n} - s_{N+n-1} \leq \beta^{n-1}(s_{N+1} - s_N). \tag{4.81}$$

Remark 4.1.4.

(a) If $N = 1$, Lemma 4.1.4 reduces to Lemma 4.1.3.

(b) The verification of (4.78) is more difficult than (4.70) but nevertheless it can be done. Let $L_0 = .95$. Then conditions (4.2), (4.16) and (4.70) are violated. In particular, we have

$$s_2 = .2755453502, \quad s_3 = .3427951408, \quad \kappa = \beta = .6267843315$$

and

$$\frac{1 - L_0(c + 2s_2)}{1 - L_0(c + 2\eta)} = .6235641982.$$

However, (4.78) is not satisfied for $N = 2$, since

$$\beta_2 = \max\{\frac{L(s_3 - s_1)}{1 - L_0(s_3 + s_1 + c)}, \kappa\} = \kappa = .6267843315$$

and

$$\frac{1 - L_0(2s_3 + s_1 + c)}{1 - L_0(2s_2 + s_1 + c)} = .01495007920.$$

Table 4.1   Comparison Table of bound errors

	(4.8)	(4.3)	(4.4)	(4.64)
n	$t_{n+1} - t_n$	$v_{n+1} - v_n$	$w_{n+1} - w_n$	$s_{n+1} - s_n$
1	.0927966102	.1034645669	.0927966102	.0742372881
2	.0516874807	.0834298221	.0516874807	.0374029061
3	.0199586020	.0947064307	.0199586020	.0100546296
4	.0045129145	-1.250121362	.0045129145	.0012645839
5	.0003714927	1.264895298	.0003714927	.0000388678
6	.0000061850	.0165794721	.0000061850	1.380e-7
7	8.0e-9	-.1376680823	8.0e-9	0
8	0	-.5014996853	0	$\sim$
9	$\sim$	.5290112930	$\sim$	$\sim$
10	$\sim$	.0251617461	$\sim$	$\sim$

(c) Hypotheses (4.77) and (4.78) can replace (4.9) in Theorem 4.1.1.

We present two numerical examples for validating the theoretical results of this section.

Example 4.1.1. Consider the example given in the introduction of this section but choosing $L_0 = .8$. In Table 4.1, we compare the diffferent sequences cited in this section (i.e., $\{t_n\}$, $\{s_n\}$, $\{v_n\}$, $\{w_n\}$ given by (4.8), (4.64), (4.3) and (4.4), respectively). We validate by Table 4.1 that our majorizing sequences $\{t_n\}$ and $\{v_n\}$ given by (4.8) and (4.3), respectively are majorizing sequences for (MOC) under condition (4.9). We also show by the table that sequence (4.64) is finer than (4.8) (see (4.65) and (4.66) in Remark 4.1.2). Note that using Table 4.1, sequence $\{v_n\}$ is not necessarily a majorizing sequence for (MOC) since condition (4.2) does not hold.

Example 4.1.2. Define the scalar function F by $F(x) = c_0 x + c_1 + c_2 \sin e^{c_3 x}$, $x_0 = 0$, where c_i, $i = 0, 1, 2, 3$ are given parameters. Define linear operator $\delta F(x, y)$ by

$$\delta F(x, y) = \int_0^1 F'(y + t(x - y))\, dt = c_0 + c_2 \frac{\sin e^{c_3\, x} - \sin e^{c_3\, y}}{x - y}.$$

Then it can easily be seen that for c_3 large, c_2 sufficiently small and L/L_0 can be arbitrarily large. That is (4.8) may be satisfied but not (4.2).

4.2   Secant-type Method and Nondiscrete Induction

The celebrated nondiscrete mathematical induction has been used to improve error bounds of distances involved in the discrete case but not the sufficient convergence conditions for Secant-type methods. We are concerned in this section with the problem of approximating a locally unique solution $x^\star$ of (1.1) (with f replacing F) using the Secant method (SM) defined by (4.1).

Using the information provided by the triplet (f, x_{-1}, x_0) and under some hypotheses, sequence $\{x_n\}$ converges to $x^\star$. The R-order of convergence is 1.618033988. The Simplified Secant Method

$$x_{n+1} = x_n - \delta f(x_{-1}, x_0)^{-1} f(x_n) \quad (n \geq 0), \quad (x_{-1}, x_0 \in \mathcal{D}) \tag{4.82}$$

uses just one inversion but the convergence is only linear. An intermediate algorithm consists of choosing a natural number m and keeping the same linear operator for portions of the procedure of m steps each. More precisely, given two points $x_0^m := x_0$ and $x_0^{m-1} := x_{-1}$, define sequences $\{x_n^j\}$ $(n \geq 1)$, $1 \leq j \leq m$ by the following Algorithm

$$\begin{aligned} x_n^0 &= x_{n-1}^m \\ x_n^{j+1} &= x_n^j - \delta f(x_{n-1}^{m-1}, x_{n-1}^m)^{-1} f(x_n^j), \quad j = 0, 1, \cdots, m-1, \quad n \geq 1. \end{aligned} \tag{4.83}$$

Schmidt and Schwetlick in Ref. [396] studied the convergence of Algorithm (4.83), where of $\delta f(x, y)$ is a divided difference of f on x and y $(x, y \in \mathcal{D}_f)$. The order of convergence was found to be $\dfrac{1}{2}(m + (m^2 + 4)^{1/2})$. Note that for $m = 1$, we obtain (SM). The parameter m is chosen according to the dimension of space in order to maximize the numerical efficiency.

In the nondiscrete case, Nondiscrete Mathematical Induction (NMI) has been used to study methods (4.1) and (4.83) under the same conditions in Refs. [359, 360]. (NMI) has improved the error estimates on the distances involved but not the sufficient convergence conditions of the discrete case (see e.g. (4.103)).

In this section, we are motivated by the works of Potra, Pták in Refs. [359, 360] and Schmidt, Schwetlick in Ref. [396]. Using (NMI) and the same information as before, we expand the applicability of Algorithm (4.83) by providing a new semi-local convergence analysis with advantages $\mathcal{A}$:

$(\mathcal{A}_1)$ weaker sufficient convergence conditions;
$(\mathcal{A}_2)$ tighter error bounds on the distances involved;

and

$(\mathcal{A}_3)$ an at least as precise information on the location of the solution $x^\star$.

The results specialize to (SM) for $m = 1$, to Newton's method (NM)

$$x_{n+1} = x_n - f'(x_n)^{-1} f(x_n), \quad (n \geq 0), \quad (x_0 \in \mathcal{D}_f) \tag{4.84}$$

for $m = 1$, $\delta f = \nabla f$ and to Traub's method (see Ref. [418])

$$\begin{aligned} x_n^0 &= x_{n-1}^m \\ x_n^{j+1} &= x_n^j - f'(x_n^0)^{-1} f(x_n^j), \quad j = 0, 1, \cdots, m-1, \quad n \geq 1. \end{aligned} \tag{4.85}$$

Note that the R-convergence orders of (NM) and Traub's method are 2 and $m + 1$, respectively. Clearly Traub's method (4.85) is a generalization of (NM).

We give in this section the general concepts of (NMI), rates of convergence and their relationship to nonlinear equations. The semi-local convergence of Algorithm

(4.83) and numerical examples are also given in this section. First, we reproduce material that can be found in the excellent book of Potra and Pták (see [Algorithms of type (p, m), Section 3] in Ref. [360]). Let p be a natural number. Consider an algorithm given by an operator $G : \mathcal{X}^p \longrightarrow \mathcal{X}$, the recursive formula

$$x_{n+1} = G(x_{n-p+1}, x_{n-p+2}, \cdots, x_n), \qquad n \geq 0$$

and p initial guesses denoted by

$$u_0 = (x_{-p+1}, x_{-p+2}, \cdots, x_0) \in \mathcal{X}^p.$$

If $p = 2$, (SM) is an example of such an algorithm. The formula can be simplified if we introduce a mapping $\tilde{G} : \mathcal{X}^p \longrightarrow \mathcal{X}^p$ given by

$$\tilde{G}(x_1, x_2, \cdots, x_p) = (x_2, x_3, \cdots, x_p, G(x_1, x_2, \cdots, x_p)).$$

Then, we can write

$$u_{n+1} = \tilde{G}(u_n) \qquad (n \geq 0). \tag{4.86}$$

If $\mathcal{P}(l)$ stands for the projection onto the last component, then $x_n = \mathcal{P}(l)\, u_n\ (n \geq 1)$. Let m be a fixed natural number. Let also $y \in \mathcal{X}^{p+m}$ be denoted by

$$y = (y_{-p+1}, y_{-p+2}, \cdots, y_1, \cdots, y_m)$$

and introduce projection mappings $\mathcal{P}_j : \mathcal{X}^{p+m} \longrightarrow \mathcal{X}$ by

$$\mathcal{P}_j\, y = y_j \qquad -p+1 \leq j \leq m.$$

We associate a mapping $F : \mathcal{X}^p \longrightarrow \mathcal{X}^{m+p}$ to $G : \mathcal{X}^p \longrightarrow \mathcal{X}^m$ by

$$F(u) = (u, G(u)). \tag{4.87}$$

Let us define Algorithm

$$\begin{aligned} z_1 &= F(u_0) \\ z_{n+1} &= F\mathcal{P}(z_n) \quad (n \geq 1). \end{aligned} \tag{4.88}$$

Consider $z_0 \in \mathcal{X}^{m+p}$ such that $u_0 = \mathcal{P}(z_0)$. If $\{u_n\}$ is a sequence generated by (4.86), then

$$\begin{aligned} u_n &= \mathcal{P}(z_n) \quad (n \geq 0) \\ u_n &= (\mathcal{P}_1(z_n), \mathcal{P}_2(z_n), \cdots, \mathcal{P}_m(z_n)). \end{aligned} \tag{4.89}$$

In general G is defined only on subset $\mathcal{D}$ of $\mathcal{X}^p$. If $u_0 \in \mathcal{D}$, we may get $z_1 = F(u_0)$. If $u_1 = \mathcal{P}(z_1) \in \mathcal{D}$, we may have z_2, etc. We want $\mathcal{P}(z_n) \in \mathcal{D}\ (n \geq 1)$. We need the definition of a set of admissible systems of initial guesses for Algorithm (4.88).

Definition 4.2.1. Let $G : \mathcal{X}^p \longrightarrow \mathcal{X}^m$ and $F : \mathcal{D} \subset \mathcal{X}^p \longrightarrow \mathcal{X}^{p+m}$ be mappings connected with (4.87). Let also $\mathcal{P} : \mathcal{X}^{p+m} \longrightarrow \mathcal{X}^p$ denote the projection on the last p components defined recursively by

$$\mathcal{D}_0 = \mathcal{D}, \quad \mathcal{D}_{n+1} = \{u \in \mathcal{D}_n : \mathcal{P}\, F(u) \in \mathcal{D}_n\} \quad (n \geq 0).$$

Let $\mathcal{D}_\infty = \cap_{n \geq 0} \mathcal{D}_n$. $\mathcal{D}_\infty$ is called the set of admissible systems of starting points for Algorithm (4.88). If u_0 belongs to $\mathcal{D}_\infty$, then Algorithm (4.88) is well defined. Consequently, Algorithm (4.88), where F is given in (4.87) will be called an iterative Algorithm of type (p, m).

Rates of convergence of type (p, m)

Let $\mathcal{T}$ be the set of all positive real numbers or an interval $\{t : 0 < t \le t_0\}$. Let $w : \mathcal{T}^p \longrightarrow \mathcal{T}^m$ and let $w_j : \mathcal{T}^p \longrightarrow \mathcal{T}$ $(j = 1, 2, \cdots, m)$ be its components. If $t \in \mathcal{T}^p$, we write

$$w(t) = (w_1(t), w_2(t), \cdots, w_m(t)),$$

or simply

$$w = (w_1, w_2, \cdots, w_m).$$

We denote by w_k, $1 - p \le k \le 0$ the projection which associates $t = (t_1, t_2, \cdots, t_p) \in \mathcal{T}^p$ its $(p + k)$-th component. That is

$$w_k(t) = t_{p+k}, \quad 1 - p \le k \le 0.$$

Define iterates $w_j^{(r)}$ by

$$w_{m+k}^{(0)} = w_k, \quad 1 - p \le k \le 0$$

$$w_j^{(n+1)} = w_j \circ (w_{m-p+1}^{(n)}, w_{m-p+2}^{(n)}, \cdots, w_m^{(n)}), \quad 1 - p \le j \le m, \ n \ge 0.$$

Let us consider the function $\tilde{w} : \mathcal{T}^p \longrightarrow \mathcal{T}^p$ by

$$\tilde{w} = (w_{m-p+1}, w_{m-p+2}, \cdots, w_m) \tag{4.90}$$

and denote by $\tilde{w}^{(n)}$ its iterates given by

$$\tilde{w}^{(0)}(t) = t, \quad \tilde{w}^{(n+1)}(t) = \tilde{w} \circ \tilde{w}^{(n)}(t), \quad n \ge 0.$$

If we define the iterates $w^{(n)}$ of mapping w by

$$w^{(n+1)}(t) = w \circ \tilde{w}^{(n)}(t), \quad n \ge 0,$$

then, we have that

$$w^{(n)} = (w_1^{(n)}, \cdots, (w_m^{(n)})).$$

Definition 4.2.2. A mapping $w : \mathcal{T}^p \longrightarrow \mathcal{T}^m$ is called a rate of convergence of type (p, m) on $\mathcal{T}$ if the series

$$s(t) = \sum_{n=1}^{\infty} \sum_{j=0}^{m-1} w_j^{(n)}(t) \tag{4.91}$$

is convergent for all $t \in \mathcal{T}^p$. In view of (4.91), we have

$$s(t) = t_p + \sum_{n=1}^{\infty} \sum_{j=1}^{m} w_j^{(n)}(t) \tag{4.92}$$

and

$$s(t) = s(\tilde{w}(t)) + \sum_{j=0}^{m-1} w_j(t). \tag{4.93}$$

The function $s : \mathcal{T}^p \longrightarrow \mathbb{R}^+$ is called the estimate function corresponding to w. We also define for $j = 1, 2, \cdots, m$, the function $s_j : \mathcal{T}^p \longrightarrow \mathbb{R}^+$ by

$$s_0 = s, \quad s_j = s - (w_0 + \cdots + w_{j-1}). \tag{4.94}$$

Convergence analysis for Algorithms of type (p, m)

We shall use the following result relating the method of (NMI) and Algorithm of type (p, m).

Proposition 4.2.1. *Let* $G : \mathcal{D} \longrightarrow \mathcal{X}^m$ *and* $F : \mathcal{D} \longrightarrow \mathcal{X}^{m+p}$ *be mappings connected by (4.87). Set* $F_k = \mathcal{P}_k F$ $(-p+1 \leq k \leq m)$. *Let* $\mathbb{Z}$ *be a mapping which associates with each* $t \in \mathcal{T}^p$ *a subset* $\mathbb{Z}(t) \subseteq \mathcal{D}$. *Let* w *be a rate of convergence of type* (p, m) *on* $\mathcal{T}$. *Let* u_0 *be a given element of* $\mathcal{D}$ *and* t_0 *be a given element of* $\mathcal{T}^p$. *Assume:*

$$u_0 \in \mathbb{Z}(t_0), \quad \mathcal{P} F \mathbb{Z}(t) \subset \mathbb{Z}(\tilde{w}(t)), \quad d(F_k(u), F_{k+1}(u)) \leq w_k(t) \qquad (4.95)$$

are satisfied for all $t \in \mathcal{T}^p$, $u \in \mathbb{Z}(t)$ *and* $k = 0, 1, \cdots, m - 1$. *Then,*

(i) u_0 is an admissible system of starting points for Algorithm (4.88);

(ii) there exists $x^\star \in \mathcal{X}$ such that each of the $(m + p)$-sequences $\mathcal{P}_k(z_n)$ $(n \geq 0,$ $-p + 1 \leq k \leq m)$ converges to $x^\star$;

(iii) the following estimates hold for all $n \geq 1$

$$\mathcal{P}(z_n) \in \mathbb{Z}(\tilde{w}^{(n)}(t_0)), \qquad (4.96)$$

$$d(\mathcal{P}_k(z_n), \mathcal{P}_{k+1}(z_n)) \leq w_k^{(n)}(t_0), \quad 0 \leq k \leq m - 1, \qquad (4.97)$$

$$d(\mathcal{P}_k(z_n), x^\star) \leq s_k(\tilde{w}_k^{(n-1)}(t_0)), \quad 0 \leq k \leq m; \qquad (4.98)$$

(iv) if for some $n \in \mathbb{N}$, $d_n \in \mathcal{T}^p$,

$$\mathcal{P}(z_{n-1}) \in \mathbb{Z}(d_n), \qquad (4.99)$$

then

$$d(\mathcal{P}_k(z_n), x^\star) \leq s_k(d_n), \quad 0 \leq k \leq m. \qquad (4.100)$$

Algorithms of type (p, m) and nonlinear equations

We need the relationship between Algorithm of type (p, m) and equation (1.1).

Definition 4.2.3. Let $\mathcal{C}$ be a class of pairs (f, s_0), $s_0 = (x_{-q+1}, \cdots, x_0) \in \mathcal{D}_f^q$ (q a natural number). We define an Algorithm of type (p, m) for the class $\mathcal{C}$ as an application which associates with each pair (f, s_0) from $\mathcal{C}$ a mapping $G : \mathcal{D} \subseteq \mathcal{D}_f^p \longrightarrow \mathcal{X}^m$ such that

(i) $u_0 = (x_{-p+1}, \cdots, x_0)$ is an admissible system of starting points for (4.88);

(ii) each of the m sequences $\{\mathcal{P}_k(z_n)\}$ $(n \geq 1)$, $1 \leq k \leq m$ generated by (4.88) converges to the same limit point $x^\star \in \mathcal{X}$, which is the solution of $f(x) = 0$.

We shall study Algorithm (4.83) for triplets (f, x_0, x_{-1}) in the class $\mathcal{C}(L, q_0, r_0)$ defined as follows:

Definition 4.2.4. Let $L > 0$, $q_0 \geq 0$ and $r_0 \geq 0$ be given. A triplet (f, x_0, x_{-1}) belongs to the class $\mathcal{C}(L, q_0, r_0)$ if:

$(\mathcal{H}_1)$ f is a nonlinear operator defined on a convex subset $\mathcal{D}_f$ of a Banach space $\mathcal{X}$ with values in a Banach space $\mathcal{Y}$;

$(\mathcal{H}_2)$ x_{-1} and x_0 belong in $\mathcal{D}_f$ $(x_{-1} \neq x_0)$ and $\| x_0 - x_{-1} \| \leq q_0 < \mu$;

$(\mathcal{H}_3)$ f is Fréchet-differentiable in the open ball $U = U(x_0, \mu)$ and continuous on $\overline{U}$;

$(\mathcal{H}_4)$ There exists a mapping $\delta f : U \times U \to \mathcal{L}(\mathcal{X}, \mathcal{Y})$ such that the linear operator $S_0 = \delta F(x_{-1}, x_0)$ is invertible, its inverse $T_0 = S_0^{-1} \in \mathcal{L}(\mathcal{Y}, \mathcal{X})$ and

$$\| T_0 \left(\delta f(x, y) - f'(z) \right) \| \leq L \left(\| x - z \| + \| y - z \| \right), \tag{4.101}$$
$$\text{for all} \quad x, y, z \in U.$$

Note that in view of $(\mathcal{H}_4)$, there always exists $L_0 > 0$ such that

$$\| T_0 \left(\delta f(x, y) - f'(x_0) \right) \| \leq L_0 \left(\| x - x_0 \| + \| y - x_0 \| \right), \tag{4.102}$$
$$\text{for all} \quad x, y \in U.$$

$(\mathcal{H}_5)$ The following inequalities are satisfied $\| T_0 \, f(x_0) \| \leq r_0$,

$$L_0 \, q_0 + (L_0 + L) \, r_0 + 2 \, L \left(r_0 \left(r_0 + q_0 \right) + a^2 \right)^{1/2} \leq 1 \quad \text{for all} \quad a \geq 0 \tag{4.103}$$

and

$$\mu_0 = r_0 - a + \left(r_0 \left(r_0 + q_0 \right) + a^2 \right)^{1/2} \leq \mu. \tag{4.104}$$

Note that (4.102) is not an addition to (4.101) hypothesis. In practice, the computation of constant L requires that of L_0. Using (4.102) instead of less precise (4.101) for the computation on upper bounds $\| \delta f(x_{n-1}, x_n)^{-1} S_0 \|$ $(n \geq 0)$ generates tighter error estimates. This observation leads to advantages $\mathcal{A}$. Iteration (4.83) is an Algorithm of type $(2, m)$. Given (f, x_0, x_{-1}) in $\mathcal{C}(L, q_0, r_0)$ and $u = (y, x) \in U^2$, define

$$F_{-1}(u) = y, \quad F_0(u) = x$$
$$F_{j+1}(u) = F_j(u) - \delta f(y, x)^{-1} f(F_j(u)), \quad j = 0, 1, \cdots, m - 1. \tag{4.105}$$

Denote by $\mathcal{D}$ the set of $u \in \mathcal{D}^2$ such that $\delta f(y, x)$ is invertible and $F_j(u) \in U$, $j = 0, 1, \cdots, m - 1$. Then, mapping $F : \mathcal{D} \longrightarrow \mathcal{X}^{m+2}$ given by

$$F(u) = (F_{-1}(u), F_0(u), \cdots, F_m(u))$$

is of the form (4.87) and

$$x_n^j = \mathcal{P}_j(z_n), \quad j = 0, 1, \cdots, m \quad \text{and} \quad n \geq 0. \tag{4.106}$$

There are differences between the cases $m = 1$ and $m \geq 2$, but nevertheless they can be treated the same way if we adopt: if an Algorithm requires the computation of some quantity depending on $k = 0, 1, \cdots, p$ and p happens to be negative, ignore this instruction and move to the next one. Similarly, if sums of the form $a_0 + a_1 + \cdots + a_p$ are taken to be equal to 0 if p is negative.

     We shall study the semi-local convergence of Algorithm (4.83) using Proposition 4.2.1. But first, we need to introduce convergence of type $(2, m)$ and the corresponding estimate function.

Lemma 4.2.1. *(see Ref. [360]) Let $t = (q, r) \in \mathcal{T}^2$ and define functions*

$$g(t) = r + (r(q+r) + a^2)^{1/2} \tag{4.107}$$

$$w_{-1}(t) = q \tag{4.108}$$

and

$$w_0(t) = r. \tag{4.109}$$

Moreover, define sequence of functions $\{w_k\}$ by

$$w_{k+1}(t) = \frac{w_k(t)\,(w_{-1}(t) + 2\,(w_0(t) + \cdots + w_{k-1}(t)))}{2\,g(t) + w_{-1}(t)}, \tag{4.110}$$

$$w_m(t) = \frac{w_{m-1}(t)\,(w_{-1}(t) + w_{m-1}(t) + 2\,(w_0(t) + \cdots + w_{m-2}(t)))}{2\,g(t) - 2\,(w_0(t) + \cdots + w_{m-2}(t)) - w_{m-1}(t)}. \tag{4.111}$$

Then, the mapping $w(t) = (w_1(t), \cdots, w_m(t))$ is a rate of convergence of type $(2, m)$ on $\mathcal{T}$ and the corresponding estimate function is given by

$$s(t) = g(t) - a. \tag{4.112}$$

We from now associate w given by (4.111) with (4.89) and (4.94). The proof of the following semi-local convergence theorem is similar to [Theorem 4.13, p. 73] in Ref. [360], but there are crucial differences where (4.102) is used instead of (4.101) for the computation of the upper bounds on norms of the inverses involved.

Theorem 4.2.1. *Assume*

$$(f, x_0, x_{-1}) \in \mathcal{C}(L, q_0, r_0), \tag{4.113}$$

where

$$a = a_{AH}(L, L_0) = \frac{1}{2L}\left((1 - L_0\,q_0 - (L_0 + L)\,r_0)^2 - 4\,L^2\,r_0\,(r_0 + q_0)\right)^{1/2}. \tag{4.114}$$

Then, sequence generated by Algorithm (4.83) with $x_0^m = x_0$, $x_0^{m-1} = x_{-1}$ is well defined, remains in $\overline{U}$, produces $(m+1)$-sequences $\{x_n^j\}$, $(n \geq 0)$, $0 \leq j \leq m$, which converge to a unique solution $x^\star$ of equation $f(x) = 0$ in $\overline{U}(x_0, \mu_0)$. Moreover, the following estimates hold for all $j = 0, 1, \cdots, m$, $n \geq 1$:

$$\| x_n^j - x^\star \| \leq s_j(\tilde{w}^{(n-1)}(q_0, r_0)) \tag{4.115}$$

and

$$\| x_n^j - x^\star \| \leq s_j(\| x_{n-1}^{m-1} - x_{n-1}^m \|, \| x_n^0 - x_n^1 \|). \tag{4.116}$$

Proof.   Let $t = (q, r) \in \mathcal{T}^2$, $t_0 = (q_0, r_0)$ and define

$$\mathbb{Z}(t) = \Big\{ u = (y, x) \in \mathcal{X}^2 \ : \ y \in U, \ \| x - y \| \leq q, \ \| x - x_0 \| \leq s(t_0) - s(t),$$
$$S = \delta f(y, x) \text{ is invertible}, \ \| S^{-1} f(x) \| \leq r$$
$$\text{and} \ \| (T_0 S)^{-1} \| \leq \frac{1}{L(2g(t) + q)} \Big\}.$$

$$(4.117)$$

We shall show hypotheses (4.95), (4.99) of Proposition 4.2.1 are satisfied and $\mathbb{Z}(t) \subset \mathcal{D}$. We have $(x_{-1}, x_0) = u_0 \in \mathbb{Z}(t_0)$. We shall show $u \in \mathbb{Z}(t)$ implies

$$F_k(u) \in U, \quad -1 \leq k \leq m \tag{4.118}$$

and

$$\| F_k(u) - F_{k+1}(u) \| \leq w_{k+1}(t), \quad -1 \leq k \leq m - 1. \tag{4.119}$$

If $k = -1$, (4.118) and (4.119) are satisfied since $y \in U$ and $\| y - x \| \leq q$. For $k = 0$, they also hold since $x \in U$ and $\| S^{-1} f(x) \| \leq r$.

Suppose (4.118) and (4.119) hold for $k = -1, 0, \cdots, i$ for some fixed i with $0 \leq i \leq m - 1$. We have in turn

$$\| x_0 - F_{i+1}(u) \| \leq \| x_0 - x \| + \| F_{i+1}(u) - x \|$$
$$\leq \| x_0 - x \| + \sum_{j=0}^{i} \| F_{i+1}(u) - F_i(u) \| \tag{4.120}$$
$$\leq s(t_0) - s(t) + \sum_{j=0}^{i} w_j(t) = s(t_0) - s_{i+1}(t).$$

That is $F_{i+1}(u) \in U$, which implies (4.118). It also follows from (4.117) and (4.118) that $\mathbb{Z}(t) \subseteq \mathcal{D}$. Let $S_m = \delta f(F_{m-1}(u), F_m(u))$, $f_j = f(F_j(u))$. Using (4.101) and (4.105), we get

$$\| F_{i+1}(u) - F_{i+2}(u) \| = \| S^{-1} f_{i+1} \|$$
$$= \| (T_0 S)^{-1} (f_{i+1} - f_i - S(F_{i+1}(u) - F_i(u))) \|$$
$$\leq \| (T_0 S)^{-1} \| \, L \, (\| F_{i+1}(u) - x \| + \| F_i(u) - x \| +$$
$$\| y - x \|) \, \| F_{i+1}(u) - F_i(u) \|$$
$$\leq \frac{w_i(t)}{2g(t) + q} \, (w_i(t) + 2(w_0(t) + \cdots + w_{-1}(t)) + q)$$
$$= w_{i+1}(t),$$

$$(4.121)$$

which shows (4.119). We must show $(F_{m-1}(u), F_m(u)) \in \mathbb{Z}(\tilde{w}(t))$ for $u \in \mathbb{Z}(t)$. This implication is certainly true, if

$$\| F_{m-1}(u) - F_m(u) \| \leq w_{m-1}(t), \tag{4.122}$$

$$\| F_m(u) - x_0 \| \leq s(t_0) - s_m(t), \tag{4.123}$$

$$\| s_m^{-1} f_m \| \leq w_m(t) \tag{4.124}$$

and

$$\| (T_0 \, S_m)^{-1} \| \le L^{-1} \, (2 \, (g(\tilde{w}(t)) + w_{m-1}(t))^{-1}. \tag{4.125}$$

Inequality (4.122) is obtained from (4.119) and (4.123) follows from (4.121) for $i = m - 1$. Using (4.102) (instead of (3.1) used in Ref. [360]), (4.114) and (4.121), we get in turn

$$\begin{aligned}
\| T_0 \, (S_m - S_0) \| &\le \| T_0 \, (S_m - f'(x_0)) \| + \| T_0 \, (f'(x_0) - S_0) \| \\
&\le L_0 \, (\| x_0 - x_{-1} \| + \| F_m(u) - x_0 \| + \| F_{m-1}(u) - x_0 \|) \\
&\le L_0 \left(2 \, (s(t_0) - s_m(t)) + q_0 - w_{m-1}(t) \right) \\
&\le 1 - L \, (2 \, g(\tilde{w}(t)) + w_{m-1}(t)).
\end{aligned} \tag{4.126}$$

In view of (4.126) and the Banach lemma on invertible operators that $S_m^{-1} \in \mathcal{L}(\mathcal{Y}, \mathcal{X})$ and (4.125) holds. The rest of the proof is identical to the corresponding one in Ref. [360] and is thus omitted. That completes the proof of Theorem 4.2.1. $\qquad\square$

Remark 4.2.1.

(a) If $L_0 = L$, then Theorem 4.2.1 reduces to [Theorem 4.13] in Ref. [360]. Otherwise (i.e., if $L_0 < L$) it constitutes an improvement. Indeed, let us denote by a_P, μ_P, w_P, s_P corresponding to $a(L_0, L)$, μ_0, w and s, obtained by setting $L_0 = L$ in (4.114), (4.111), (4.112), respectively. Then, we have that

$$a_P = \frac{1}{2 \, L} \left((1 - L \, q_0)^2 - 4 \, L_0 \, r_0 \right)^{1/2} < a_{AH}(L_0, L), \tag{4.127}$$

$$\mu_0 < \mu_P, \tag{4.128}$$

$$w(t) < w_P(t) \tag{4.129}$$

and

$$s(t) < s_P(t). \tag{4.130}$$

Note also that the condition corresponding to (4.103) is

$$4 \, L_0 \, r_0 \le (1 - L \, q_0)^2. \tag{4.131}$$

Note that

$$(4.131) \implies (4.103), \tag{4.132}$$

but not necessarily vice versa unless $L_0 = L$.

(b) The result for the Secant method (4.1) are obtained from above if $m = 1$. Set $\delta f(y, x) = F'(x)$ and $q_0 = 0$ to get the results for Traub's method (4.85). Furthermore, if $m = 1$, we obtain results for Newton's method (4.84). Items (4.127)–(4.132) justify advantages $\mathcal{A}$ (as already stated in the introduction of this section) of using our approach over the one in Ref. [360].

We provide now numerical example. For simplicity, we only consider the case $m = 1$. That is the case of (SM) given by (4.1).

Example 4.2.1. Consider equation (1.48) with parameter a replaced by ϱ. Define linear operator $\delta f(x, y)$ by

$$\delta f(x, y) = \int_0^1 f'(y + t(x - y)) \, dt = x^2 + y^2 + xy. \qquad (4.133)$$

Using (4.133), we can write for all $(x, y, z) \in \mathcal{D}^3$:

$$\delta f(x, y) - f'(z) = (x - y + 3z)(x - z) + (y + 2x)(y - z).$$

Then, using Definition 4.2.4 and Maple 13, we get that

$$r_0 = \frac{1}{2.9701}(1 - \varrho), \quad L_0 = \frac{1}{2.9701}(6 - 3\varrho) \quad \text{and} \quad L = \frac{1}{2.9701}(8 - 5\varrho).$$

Inequality (4.103) holds if

$$\varrho \in (.5185273827, 1).$$

The (SM) case for $\varrho = .52$. By Definition 4.2.4 and using Maple 13, we get:

$$r_0 = .1616107202, \quad L_0 = 1.494899162, \quad L = 1.818120602$$

and

$$r_P = .1325500827.$$

Condition (4.131) is violated since

$$r_0 = .1616107202 > r_P = .1325500827.$$

Hence, there is no guarantee that (SM) converges to $x^\star = .8041451517$, starting at $(x_{-1}, x_0) = (.99, 1)$. Moreover, all conditions of Theorem 4.2.1 hold and we can apply the (SM) for solving equation (1.48).

The (SM) case for $\varrho = .57$. Using Definition 4.2.4 and Maple 13, we obtain:

$$r_0 = .1447762702, \quad L_0 = 1.444395812, \quad L = 1.733948352,$$

$$r_P = .1392229604$$

and condition (4.103) holds for $a = .01$ since

$$L_0 q_0 + (L_0 + L) r_0 + 2L (r_0 (r_0 + q_0) + .01^2)^{1/2} = .9948689500 < 1.$$

Condition (4.131) is violated since

$$r_0 = .1447762702 > r_P = .1392229604.$$

Hence, there is no guarantee that (SM) converges to $x^\star = .8291344342$, starting at $(x_{-1}, x_0) = (.99, 1)$. Moreover, all condition of Theorem 4.2.1 hold and we can apply the (SM) for solving equation (1.48).

4.3 Efficient Secant-type Method

A local convergence analysis for an efficient Secant-type method (STM) for solving nonlinear equations is given using both the Lipschitz continuous and center-Lipshchitz continuous divided differences of order one. An estimate of the radius of the convergence ball of (STM) is provided, the error estimate matching its convergence order is established. We are concerned with the problem of approximating a locally unique solution $x^\star$ of (1.1) by (STM).

Another famous iterative process for solving nonlinear equations is Chebyshev's method (see Refs. [121, 208]):

$$\begin{cases} y_n = x_n - F'(x_n)^{-1} \, F(x_n), \\ x_{n+1} = y_n - \dfrac{1}{2} \, F'(x_n)^{-1} F''(x_n)(y_n - x_n)^2, \quad n \geq 0, \quad x_0 \in \Omega, \end{cases} \tag{4.134}$$

which has cubic convergence. This one-point iterative method depends explicitly on the first and second derivatives of F. Ezquerro and Hernández proposed in Ref. [208] some modifications of Chebyshev's method that avoid the computation of the second derivative of F. In this section we recall this method as the Chebyshev-Newton-type method (CNTM), which has the form:

$$\begin{cases} y_n = x_n - F'(x_n)^{-1} \, F(x_n), \\ z_n = x_n + a \, (y_n - x_n), \\ x_{n+1} = x_n - \dfrac{1}{a^2} \, F'(x_n)^{-1} \, ((a^2 + a - 1) \, F(x_n) + F(z_n)), \quad n \geq 0, \quad x_0 \in \Omega, \end{cases}$$
$$\tag{4.135}$$

where $a \in (0, 1]$ is a parameter.

Recently, Argyros et al. introduced in Ref. [121] a new family of iterative methods free from derivatives by replacing the first derivative F' with a divided difference of order one for the operator F. We call this new family as the Chebyshev-Secant-type method (CSTM):

$$\begin{cases} y_n = x_n - A_n^{-1} \, F(x_n), \quad A_n = [x_{n-1}, x_n; F], \\ z_n = x_n + a \, (y_n - x_n), \\ x_{n+1} = x_n - A_n^{-1} \, (b \, F(x_n) + c \, F(z_n)), \quad n \geq 0, \quad x_{-1}, x_0 \in \Omega, \end{cases} \tag{4.136}$$

where $a \in [0, 1]$, $b \geq 0$ and $c \geq 0$ are parameters. A semi-local convergence theorem for (CSTM) is established in Ref. [121] under relation

$$(1 - a) \, c \, = \, 1 - b \tag{4.137}$$

and some other mild conditions. Moreover, several numerical examples were also provided in Ref. [121] to show the efficiency of (CSTM). However, the order of convergence of $\{x_n\}$ to $x^\star$ generated by (CSTM) in the second example in Ref. [121] has only 1.625, which is near 1.618 of the Secant method.

In this section we study the local convergence of (STM) not examined in Ref. [121]. The semi-local convergence of (4.136) and consequently of (4.139) when A_n is replaced by a general $B_n \in L(X, Y)$ has been given in Ref. [138]. In the present section, a more precise local convergence analysis for the (CSTM) is given. In fact, we show that in order to achieve higher order, apart from (4.137), the following relation

$$ac = 1 \tag{4.138}$$

is also needed. Under relations (4.137) and (4.138), we can rewrite (4.137) as follows

$$\begin{cases} y_n = x_n - A_n^{-1} \, F(x_n), \quad A_n = [x_{n-1}, x_n; F], \\ z_n = x_n + a \, (y_n - x_n), \\ x_{n+1} = x_n - A_n^{-1} \, ((2 - \dfrac{1}{a}) \, F(x_n) + \dfrac{1}{a} \, F(z_n)), \quad n \geq 0, \quad x_{-1}, x_0 \in \Omega, \end{cases} \tag{4.139}$$

where $a \in (0, 1]$ is a parameter. We shall call (4.139) Secant-type method (STM). Note that (CSTM) and (STM) are useful alternatives of (CNTM), especially in cases where the computation of $F''(x_n)$ or $F'(x_n)^{-1}$ are expensive or impossible to compute or their analytic representation is unavailable. Our results show that (4.139) is a family of methods with order two for any choice of $a \in (0, 1]$. Moreover, a unified estimate of radius of convergence ball of (STM) as well as a unified error analysis matching its convergence order are given in this section.

We shall show a lemma needed for our main theorem.

Lemma 4.3.1. *Assume $a \in [0, 1]$, $c \geq 0$ are parameters, M and M_0 are positive constants with $M_0 \leq M$. Let q be a function defined on $[0, +\infty)$ by*

$$\begin{aligned} q(r) = {} & a^2 c (Mr)^3 + 4ac(Mr)^2(1 - M_0 r) \\ & + 4(|1 - ac| + (1 - a)ac)Mr(1 - M_0 r)^2 - 8(1 - M_0 r)^3. \end{aligned} \tag{4.140}$$

Then, q has a least positive zero in $(0, R_0]$ with R_0 given by

$$R_0 = \frac{2}{M + 2M_0}. \tag{4.141}$$

Proof. For the case of $ac = 0$, q can be written as

$$q(r) = 4Mr(1 - M_0 r)^2 - 8(1 - M_0 r)^3 = 4(1 - M_0 r)^2(Mr - 2(1 - M_0 r)), \tag{4.142}$$

which has the least positive zero R_0 given by (4.141) since

$$R_0 < \frac{1}{M_0}. \tag{4.143}$$

Next, we assume $ac > 0$. It follows from the definition of R_0 that $1 - M_0 R_0 = \dfrac{1}{2} M R_0$ and thus

$$\begin{aligned} q(R_0) &= a^2 c (M R_0)^3 + 2ac(M R_0)^3 + (|1 - ac| + (1 - a)ac)(M R_0)^3 - (M R_0)^3 \\ &= (3ac + |1 - ac| - 1)(M R_0)^3 \geq 2 \min\{ac, 1\}(M R_0)^3 > 0. \end{aligned} \tag{4.144}$$

In view of $q(0) = -8 < 0$, by the intermediate value theorem, there is a zero of q in $(0, R_0)$ and the least positive zero R of q must satisfy $0 < R < R_0$. That completes the proof of the lemma. $\qquad\square$

We can show the following local convergence theorem for (CSTM).

Theorem 4.3.1. *Assume*

(a) $F : \Omega \subseteq X \to Y$ *is a Fréchet-differentiable operator and there exists divided difference $[x,y;F]$ satisfying*

$$[x, y; F](x - y) = F(x) - F(y) \quad for \ all \ x, y \in \Omega; \qquad (4.145)$$

(b) $x^\star$ *is a solution of equation (1.1), $F'(x_\star)^{-1}$ exists and there is a constant $M > 0$ such that*

$$\|F'(x_\star)^{-1}([x, y; F] - [u, v; F])\| \leq \frac{M}{2}(\|x - u\| + \|y - v\|) \quad for \ all \ x, y, u, v \in \Omega; \qquad (4.146)$$

(c) *There is a constant $M_0 > 0$ such that*

$$\|F'(x_\star)^{-1}([x, y; F] - F'(x^\star))\| \leq \frac{M_0}{2}(\|x - x^\star\| + \|y - x^\star\|) \quad for \ all \ x, y \in \Omega; \qquad (4.147)$$

(d) *The relation (4.137) is true;*

(e) $U(x^\star, R) \subseteq \Omega$, *where R is the least positive zero of function q defined by (4.140).*

Then, sequence $\{x_n\}$ generated by (CSTM) from any initial points x_{-1} and x_0 in $U(x^\star, R)$ is well defined, remains in $U(x^\star, R)$, converges to $x^\star$ with order of $1.618 \cdots$ at least. Moreover, the following error estimates are satisfied

$$\|e_{n+1}\| \leq \frac{1}{R}\|e_n\|\|e_{n-1}\|, \quad n \geq 1, \qquad (4.148)$$

where $e_n = x_n - x^\star$ ($n \geq -1$). Furthermore, if relation (4.138) is also true, the least positive zero of function q can be expressed explicitly by

$$R = \frac{1}{\dfrac{\sqrt{4 + a^2} + 2 - a}{4}M + M_0}, \qquad (4.149)$$

$\{x_n\}$ *converges to $x^\star$ with order at least two and the following estimates are satisfied*

$$\|e_{n+1}\| \leq \frac{1}{R}\left(\frac{\gamma}{R}\|e_n\|\|e_{n-1}\|^2 + (1 - \gamma)\|e_n\|^2\right), \quad n \geq 1, \qquad (4.150)$$

where

$$\gamma = \left(\frac{\sqrt{4 + a^2} + a - 2}{2a}\right)^2. \qquad (4.151)$$

Proof. At first, let us assume conditions (a)-(e) are satisfied. We will prove all the results before (4.149) of the theorem by induction. Denote by $u_n = y_n - x^\star$, $v_n = z_n - x^\star$ for $n \geq 0$. Using the basic identify $F'(x^\star) = [x^\star, x^\star; F]$, condition (4.147), $x_{-1}, x_0 \in U(x^\star, R)$ and Lemma 4.3.1, we have that

$$\|I - F'(x^\star)^{-1}A_0\| = \|F'(x^\star)^{-1}([x_{-1}, x_0; F] - [x^\star, x^\star; F])\|$$
$$\leq \frac{M_0}{2}(\|e_{-1}\| + \|e_0\|) < M_0 R \leq M_0 R_0 < 1. \qquad (4.152)$$

By the Banach lemma, A_0 is invertible and

$$\|A_0^{-1}F'(x^\star)\| \le \frac{1}{1 - \dfrac{M_0}{2}(\|e_{-1}\| + \|e_0\|)} < \frac{1}{1 - M_0 R}. \tag{4.153}$$

Thus, y_0 is well defined. By the definition of divided difference, we have that

$$\begin{aligned}
F(x_0) &= F(x_0) - F(x^\star) = -(F(x^\star) - F(x_0)) \\
&= -[x^\star, x_0; F](x^\star - x_0) = [x^\star, x_0; F]e_0.
\end{aligned} \tag{4.154}$$

Using (4.136) and (4.154), we have that

$$\begin{aligned}
u_0 &= y_0 - x^\star = x_0 - x^\star - A_0^{-1}F(x_0) \\
&= A_0^{-1}F'(x^\star)\,F'(x^\star)^{-1}([x_{-1}, x_0; F] - [x^\star, x_0; F])e_0.
\end{aligned} \tag{4.155}$$

Combining (4.153), condition (4.146) and Lemma 4.3.1 yields

$$\begin{aligned}
\|u_0\| &\le \frac{\dfrac{M}{2}\|e_{-1}\|\|e_0\|}{1 - \dfrac{M_0}{2}(\|e_{-1}\| + \|e_0\|)} \le \frac{M\|e_{-1}\|\|e_0\|}{2(1 - M_0 R)} \le \frac{MR\|e_0\|}{2(1 - M_0 R)} \\
&\le \frac{MR_0\|e_0\|}{2(1 - M_0 R_0)} = \|e_0\| < R.
\end{aligned} \tag{4.156}$$

That is, $y_0 \in U(x^\star, R)$. Since z_0 is a convex combination of x_0 and y_0 and $U(x^\star, R) \subseteq \Omega$ is a convex set, we can deduce that $z_0 \in U(x^\star, R)$. In fact, we have

$$v_0 = au_0 + (1 - a)e_0 \tag{4.157}$$

and thus

$$\|v_0\| \le a\|u_0\| + (1 - a)\|e_0\| \le \|e_0\| < R. \tag{4.158}$$

Using the same technique as in (4.154), we have that

$$F(z_0) = [x^\star, z_0; F]v_0. \tag{4.159}$$

Using (4.136), (4.137), (4.157) and (4.159), we get that

$$\begin{aligned}
e_1 &= e_0 - A_0^{-1}(bF(x_0) + cF(z_0)) \\
&= A_0^{-1}\big([x_{-1}, x_0; F]e_0 - (b[x^\star, x_0; F]e_0 + c[x^\star, z_0; F]v_0)\big) \\
&= A_0^{-1}\big([x_{-1}, x_0; F]e_0 - b[x^\star, x_0; F]e_0 - c[x^\star, z_0; F](au_0 + (1 - a)e_0)\big) \\
&= A_0^{-1}\big([x_{-1}, x_0; F]e_0 - b[x^\star, x_0; F]e_0 - (1 - b)[x^\star, z_0; F]e_0 - ac[x^\star, z_0; F]u_0\big) \\
&= A_0^{-1}\big(([x_{-1}, x_0; F] - [x^\star, x_0; F])e_0 + (1 - b)([x^\star, x_0; F] - [x^\star, z_0; F])e_0\big) \\
&\quad + acA_0^{-1}([x^\star, x_0; F] - [x^\star, z_0; F] + [x_{-1}, x_0; F] - [x^\star, x_0; F])u_0 - acu_0.
\end{aligned} \tag{4.160}$$

Define

$$\begin{aligned}
D_0 &= A_0^{-1}([x_{-1}, x_0; F] - [x^\star, x_0; F]), \\
E_0 &= A_0^{-1}([x^\star, x_0; F] - [x^\star, z_0; F]).
\end{aligned} \tag{4.161}$$

Then, we have from (4.155) that $u_0 = D_0 e_0$. Moreover, we can rewrite (4.160) as

$$e_1 = D_0 e_0 + (1-b)E_0 e_0 + acE_0 u_0 + acD_0 u_0 - acu_0 \tag{4.162}$$
$$= (1-ac)D_0 e_0 + (1-b)E_0 e_0 + acE_0 D_0 e_0 + acD_0^2 e_0.$$

On the other hand, by condition (4.146), we have from (4.153) and (4.161) that

$$\|D_0\| = \|A_0^{-1} F'(x^\star)\, F'(x^\star)^{-1}([x_{-1}, x_0; F] - [x^\star, x_0; F])\|$$
$$\leq \frac{\dfrac{M}{2}\|e_{-1}\|}{1 - \dfrac{M_0}{2}(\|e_{-1}\| + \|e_0\|)}, \tag{4.163}$$

$$\|E_0\| = \|A_0^{-1} F'(x^\star)\, F'(x^\star)^{-1}([x^\star, x_0; F] - [x^\star, z_0; F])\|$$
$$\leq \frac{\dfrac{M}{2}\|z_0 - x_0\|}{1 - \dfrac{M_0}{2}(\|e_{-1}\| + \|e_0\|)} = \frac{\dfrac{aM}{2}\|y_0 - x_0\|}{1 - \dfrac{M_0}{2}(\|e_{-1}\| + \|e_0\|)}$$
$$\leq \frac{\dfrac{aM}{2}(\|u_0\| + \|e_0\|)}{1 - \dfrac{M_0}{2}(\|e_{-1}\| + \|e_0\|)} = \frac{\dfrac{aM}{2}(\|D_0 e_0\| + \|e_0\|)}{1 - \dfrac{M_0}{2}(\|e_{-1}\| + \|e_0\|)} \tag{4.164}$$
$$\leq \frac{\dfrac{aM}{2}(\|D_0\| + 1)\|e_0\|}{1 - \dfrac{M_0}{2}(\|e_{-1}\| + \|e_0\|)} \leq \frac{\dfrac{aM}{2}\left(\dfrac{\dfrac{M}{2}\|e_{-1}\|}{1 - \dfrac{M_0}{2}(\|e_{-1}\| + \|e_0\|)} + 1\right)\|e_0\|}{1 - \dfrac{M_0}{2}(\|e_{-1}\| + \|e_0\|)}$$
$$= \frac{aM^2\|e_{-1}\|\|e_0\|}{4\left(1 - \dfrac{M_0}{2}(\|e_{-1}\| + \|e_0\|)\right)^2} + \frac{aM\|e_0\|}{2\left(1 - \dfrac{M_0}{2}(\|e_{-1}\| + \|e_0\|)\right)}.$$

Therefore, we have from (4.162)–(4.164) that

$$\|e_1\| \leq |1-ac|\|D_0\|\|e_0\| + (1-b)\|E_0\|\|e_0\| + ac\|E_0\|\|D_0\|\|e_0\| + ac\|D_0\|^2\|e_0\|$$
$$\leq \frac{|1-ac|M\|e_{-1}\|\|e_0\|}{2\left(1 - \dfrac{M_0}{2}(\|e_{-1}\| + \|e_0\|)\right)} + \frac{a(1-b)M^2\|e_{-1}\|\|e_0\|^2}{4\left(1 - \dfrac{M_0}{2}(\|e_{-1}\| + \|e_0\|)\right)^2} +$$
$$\frac{a(1-b)M\|e_0\|^2}{2\left(1 - \dfrac{M_0}{2}(\|e_{-1}\| + \|e_0\|)\right)} + \frac{a^2cM^3\|e_{-1}\|^2\|e_0\|^2}{8\left(1 - \dfrac{M_0}{2}(\|e_{-1}\| + \|e_0\|)\right)^3} +$$
$$\frac{a^2cM^2\|e_{-1}\|\|e_0\|^2}{4\left(1 - \dfrac{M_0}{2}(\|e_{-1}\| + \|e_0\|)\right)^2} + \frac{acM^2\|e_{-1}\|^2\|e_0\|}{4\left(1 - \dfrac{M_0}{2}(\|e_{-1}\| + \|e_0\|)\right)^2}. \tag{4.165}$$

Define

$$h_0 = \frac{M}{2\left(1 - \dfrac{M_0}{2}(\|e_{-1}\| + \|e_0\|)\right)} \quad \text{and} \quad H = \frac{M}{2(1 - M_0 R)}.$$

Then, $h_0 \leq H$ and we have from (4.165) that

$$
\begin{aligned}
\|e_1\| &\leq |1 - ac|h_0\|e_{-1}\|\|e_0\| + a(1-b)h_0^2\|e_{-1}\|\|e_0\|^2 + a(1-b)h_0\|e_0\|^2 \\
&\quad + a^2 ch_0^3\|e_{-1}\|^2\|e_0\|^2 + a^2 ch_0^2\|e_{-1}\|\|e_0\|^2 + ach_0^2\|e_{-1}\|^2\|e_0\| \\
&\leq |1 - ac|H\|e_{-1}\|\|e_0\| + a(1-b)H^2\|e_{-1}\|\|e_0\|^2 + a(1-b)H\|e_0\|^2 \\
&\quad + a^2 cH^3\|e_{-1}\|^2\|e_0\|^2 + a^2 cH^2\|e_{-1}\|\|e_0\|^2 + acH^2\|e_{-1}\|^2\|e_0\| \\
&\leq \left([|1-ac| + a(1-b)]HR + [a(1-b) + a^2 c + ac](HR)^2 + a^2 c(HR)^3\right)\|e_0\| \\
&= \frac{q(R) + 8(1 - M_0 R)^3}{8(1 - M_0 R)^3}\|e_0\| = \|e_0\| < R.
\end{aligned}
$$

$$(4.166)$$

This means $x_1 \in U(x^\star, R)$.

Now, we suppose $\{x_n\}(0 \leq n \leq k)$ is well defined and $x_n \in U(x_\star, R)(0 \leq n \leq k)$, where k is a fixed integer and $k \geq 1$. Using a similar argumentation about x_{-1} and x_0, we can deduce the following results:

(i) A_k is invertible and

$$
\|A_k^{-1} F'(x^\star)\| \leq \frac{1}{1 - \dfrac{M_0}{2}(\|e_{k-1}\| + \|e_k\|)} < \frac{1}{1 - M_0 R}; \tag{4.167}
$$

(ii) y_k is well defined, $y_k \in U(x^\star, R)$ and

$$
\|u_k\| \leq \frac{\dfrac{M}{2}\|e_{k-1}\|\|e_k\|}{1 - \dfrac{M_0}{2}(\|e_{k-1}\| + \|e_k\|)} \leq \frac{M\|e_{k-1}\|\|e_k\|}{2(1 - M_0 R)} \leq \frac{MR\|e_k\|}{2(1 - M_0 R)} \leq \|e_k\| < R; \tag{4.168}
$$

(iii) z_k is well defined, $z_k \in U(x^\star, R)$;
(iv) x_{k+1} is well defined and

$$
\begin{aligned}
\|e_{k+1}\| &\leq |1 - ac|H\|e_{k-1}\|\|e_k\| + a(1-b)H^2\|e_{k-1}\|\|e_k\|^2 + a(1-b)H\|e_k\|^2 \\
&\quad + a^2 cH^3\|e_{k-1}\|^2\|e_k\|^2 + a^2 cH^2\|e_{k-1}\|\|e_k\|^2 + acH^2\|e_{k-1}\|^2\|e_k\| \\
&\leq \frac{\left([|1-ac| + a(1-b)]HR + [a(1-b) + a^2 c + ac](HR)^2 + a^2 c(HR)^3\right)\|e_k\|\|e_{k-1}\|}{R} \\
&= \frac{\dfrac{q(R) + 8(1 - M_0 R)^3}{8(1 - M_0 R)^3}\|e_k\|\|e_{k-1}\|}{R} = \frac{\|e_k\|\|e_{k-1}\|}{R} \leq \|e_k\| < R.
\end{aligned}
$$

$$(4.169)$$

Therefore, (4.148) holds for $n = k$. By induction, sequence $\{x_n\}$ generated by (CSTM) from any initial points x_{-1} and x_0 in $U(x^\star, R)$ is well defined, remains in $U(x^\star, R)$ and the error estimates (4.148) are satisfied. Using the classical technique as the Secant method, we can deduce that sequence $\{x_n\}$ converges to $x^\star$ with order of $1.618...$ at least. In what follows, we assume apart from satisfying conditions (a)–(e), relation (4.138) is also true. By (4.140), we can rewrite function q as

$$\begin{aligned}
q(r) &= a(Mr)^3 + 4(Mr)^2(1 - M_0 r) + 4(1 - a)Mr(1 - M_0 r)^2 - 8(1 - M_0 r)^3 \\
&= 8(1 - M_0 r)^3\Big(a\big(\frac{Mr}{2(1 - M_0 r)}\big)^3 + 2\big(\frac{Mr}{2(1 - M_0 r)}\big)^2 + (1 - a)\big(\frac{Mr}{2(1 - M_0 r)}\big) - 1\Big) \\
&= 8(1 - M_0 r)^3\big(\frac{Mr}{2(1 - M_0 r)} + 1\big)\Big(a\big(\frac{Mr}{2(1 - M_0 r)}\big)^2 + (2 - a)\big(\frac{Mr}{2(1 - M_0 r)}\big) - 1\Big) \\
&= 8a(1 - M_0 r)^3\big(\frac{Mr}{2(1 - M_0 r)} + 1\big)\big(\frac{Mr}{2(1 - M_0 r)} - \frac{\sqrt{4 + a^2} + a - 2}{2a}\big) \\
&\quad \times\big(\frac{Mr}{2(1 - M_0 r)} + \frac{\sqrt{4 + a^2} + a - 2}{2a}\big).
\end{aligned}$$

$$(4.170)$$

That is to say, the least positive zero R can be solved by

$$\frac{MR}{2(1 - M_0 R)} - \frac{\sqrt{4 + a^2} + a - 2}{2a} = 0 \tag{4.171}$$

and (4.149) is obtained at once.

Returning to (4.169), we have from relations (4.137) and (4.138) that for $n \geq 1$

$$\begin{aligned}
\|e_{n+1}\| &\leq (1 - a)H^2\|e_{n-1}\|\|e_n\|^2 + (1 - a)H\|e_n\|^2 \\
&\quad + aH^3\|e_{n-1}\|^2\|e_n\|^2 + aH^2\|e_{n-1}\|\|e_n\|^2 + H^2\|e_{n-1}\|^2\|e_n\| \\
&\leq \big((1 - a)H^2 R + (1 - a)H + aH^3 R^2 + aH^2 R\big)\|e_n\|^2 + H^2\|e_n\|\|e_{n-1}\|^2 \\
&\leq H^2\|e_n\|\|e_{n-1}\|^2 + \big(H^2 R + (1 - a)H + aH^3 R^2\big)\|e_n\|^2.
\end{aligned}$$

$$(4.172)$$

Using (4.171) and the definitions of H and γ, we get

$$H^2 = \frac{\gamma}{R^2}. \tag{4.173}$$

From the derivation of (4.170), we can deduce that

$$a(HR)^3 + 2(HR)^2 + (1 - a)HR - 1 = 0. \tag{4.174}$$

In view of (4.173) and (4.174), we have

$$H^2 R + (1 - a)H + aH^3 R^2 = \frac{1 - \gamma}{R}. \tag{4.175}$$

This means (4.151) is true.

At last, we shall show $\{x_n\}$ converges to $x^\star$ with order at least two. In fact, to discover the order of convergence of (CSTM), we postulate the following asymptotic relationship:

$$\|e_{n+1}\| \sim C\|e_n\|^p \tag{4.176}$$

where C is a positive constant. This means that the radio $\dfrac{|e_{n+1}|}{C|e_n|^p}$ tends to 1 as $n \to \infty$ and implies p-order convergence. Hence,

$$\|e_n\| \sim C\|e_{n-1}\|^p \quad and \quad \|e_{n-1}\| \sim (C^{-1}\|e_n\|)^{1/p}. \tag{4.177}$$

Clearly, we have $p \geq 1.618...$ as proved above. Suppose $p < 2$. We can rewrite (4.150) as

$$\frac{\|e_{n+1}\|}{\|e_n\|^p} \leq \frac{1}{R}\left(\frac{\gamma}{R}\|e_n\|^{1-p}\|e_{n-1}\|^2 + (1-\gamma)\|e_n\|^{2-p}\right), \quad n \geq 1. \tag{4.178}$$

Then, we have

$$\frac{\|e_{n+1}\|}{\|e_n\|^p} \to C > 0, \quad \text{as} \quad n \to \infty \tag{4.179}$$

and

$$\frac{1}{R}\left(\frac{\gamma}{R}\|e_n\|^{1-p}\|e_{n-1}\|^2 + (1-\gamma)\|e_n\|^{2-p}\right) \sim \frac{1}{R}\left(\frac{\gamma}{RC^2}\|e_n\|^{\frac{(2-p)(1+p)}{p}} + (1-\gamma)\|e_n\|^{2-p}\right)$$
$$\to 0, \quad \text{as} \quad n \to \infty. \tag{4.180}$$

In view of (4.179) and (4.180), we will deduce that $C \leq 0$ by letting $n \to \infty$ in both sides of (4.178), this is a contradiction since C is a positive constant. Therefore, $p \geq 2$ is obtained. That completes the proof of the theorem. $\qquad\square$

Remark 4.3.1.

(a) In view of (4.146), condition (4.147) always holds. That is (4.147) is not an additional to (4.146) hypothesis.

(b) Estimate (4.141) is the radius of convergence for Newton's method obtained by Argyros in Ref. [101], Argyros and Hilout in Ref. [138], that has enlarged the corresponding one given by Traub and Wozniakowski in Ref. [420]

$$R_1 = \frac{2}{3M} \tag{4.181}$$

provided that $M_0 < M$. That is we have $R_1 < R_0$ if $M_0 < M$.

(c) Note that (CSTM) is reduced to the Secant method (SM) if $a = 0$, and any choices of b and c satisfying (4.137). By Theorem 4.3.1 and the proof process of Lemma 4.3.1, we can see that estimate (4.141) is also the radius of convergence for the Secant method.

(d) Using Theorem 4.3.1, (STM) is a family of methods with quadratic convergence for any parameter $a \in (0, 1]$. If we set $a = 1$ and $a = \frac{1}{2}$ in (4.139) in turn, we can obtain two useful formulas

$$\begin{cases} y_n = x_n - A_n^{-1} F(x_n), & A_n = [x_{n-1}, x_n; F], \\ x_{n+1} = y_n - A_n^{-1} F(y_n), & n \geq 0, \ x_{-1}, x_0 \in \Omega \end{cases} \tag{4.182}$$

and

$$\begin{cases} y_n = x_n - A_n^{-1} F(x_n), & A_n = [x_{n-1}, x_n; F], \\ x_{n+1} = x_n - 2A_n^{-1} F\left(\frac{x_n + y_n}{2}\right), & n \geq 0, \ x_{-1}, x_0 \in \Omega, \end{cases} \tag{4.183}$$

respectively. Although the number of the evaluations of the function value increases by one at each step, the convergence order of (4.182) or (4.183) is raised from 1.618... to 2.

(e) Using estimate (4.149), we can deduce that among the members of family (4.139), method (4.182) has the biggest radius given by

$$R = \frac{1}{\dfrac{\sqrt{5}+1}{4}M + M_0}, \tag{4.184}$$

which shows method (4.182) allows a wider choice for the initial points than the other one in family (4.139).

We provide some examples to show the applications of our theoretical results.

Example 4.3.1. Consider Example 1.293. Then, $x^\star = 0$ is a solution of (1.1) and $F'(x^\star) = 1$. Note that for any $x, y, u, v \in \Omega$, we have

$$|F'(x^\star)^{-1}([x, y; F] - [u, v; F])| = |\int_0^1 (F'(tx + (1-t)y) - F'(tu + (1-t)v))dt|$$

$$= |\int_0^1 \int_0^1 (F''(\theta(tx + (1-t)y) + (1-\theta)$$
$$(tu + (1-t)v))(tx + (1-t)y - (tu + (1-t)v))d\theta dt|$$

$$= |\int_0^1 \int_0^1 (e^{\theta(tx+(1-t)y)+(1-\theta)(tu+(1-t)v)}(tx + (1-t)y - (tu + (1-t)v))d\theta dt|$$

$$\leq \int_0^1 e|t(x - u) + (1-t)(y - v)|dt \leq \frac{e}{2}(|x - u| + |y - v|)$$

$$\tag{4.185}$$

and

$$|F'(x^\star)^{-1}([x, y; F] - [x^\star, x^\star; F])| = |\int_0^1 F'(tx + (1-t)y)dt - F'(x^\star)|$$

$$= |\int_0^1 (e^{tx+(1-t)y} - 1)dt|$$

$$= |\int_0^1 (tx + (1-t)y)(1 + \frac{tx + (1-t)y}{2!} + \frac{(tx + (1-t)y)^2}{3!} + \cdots)dt| \tag{4.186}$$

$$\leq |\int_0^1 (tx + (1-t)y)(1 + \frac{1}{2!} + \frac{1}{3!} + \cdots)dt| \leq \frac{e-1}{2}(|x - x^\star| + |y - x^\star|).$$

That is to say, the Lipschitz condition (4.146) and the center-Lipschitz condition (4.147) are true for $M = e$ and $M_0 = e - 1$, respectively. Using Theorem 4.3.1, we can deduce that the radius of convergence for method (4.182) is given by

$$R = \frac{1}{\dfrac{\sqrt{5}+1}{4}M + M_0} = \frac{1}{\dfrac{\sqrt{5}+1}{4}e + e - 1} = 0.255270179, \tag{4.187}$$

which is bigger than the corresponding radius

$$R' = \frac{1}{\dfrac{\sqrt{5}+1}{4}M + M} = \frac{1}{\dfrac{\sqrt{5}+1}{4}e + e} = 0.203358754 \tag{4.188}$$

obtained by only using the Lipschitz condition (4.146).

Table 4.2 Comparison results of error estimates for Example 4.3.1

n	the right-side of (4.150) by using both M and M_0	the right-side of (4.150) by using only M
0	0.142303821	0.193506148
1	0.097600776	0.122757871
2	2.52305E-05	3.16716E-05
3	2.25904E-09	2.83570E-09
4	1.55876E-20	1.95667E-20

Let us choose $x_{-1} = 0.199$ and $x_0 = 0.2$. Suppose sequence $\{x_n\}$ is generated by method (4.182). Table 4.2 gives a comparison result of error estimates for Example 4.3.1, which shows that tighter error estimates can be obtained from (4.150) by using both the Lipschitz condition (4.146) and the center-Lipschitz condition (4.147) instead of by using only the Lipschitz condition (4.146).

Example 4.3.2. Consider Example 1.1.7. Then, the divided difference of F is defined by $[x, y; F] = \int_0^1 F'(tx + (1-t)y)dt$ for $x, y \in X$. Using Theorem 4.3.1, we can deduce that the radius of convergence for method (4.182) is given by

$$R = \frac{1}{\dfrac{\sqrt{5}+1}{4}M + M_0} = \frac{1}{\dfrac{15(\sqrt{5}+1)}{4} + 7.5} = 0.050928802, \tag{4.189}$$

which is bigger than the corresponding radius

$$R' = \frac{1}{\dfrac{\sqrt{5}+1}{4}M + M} = \frac{1}{\dfrac{15(\sqrt{5}+1)}{4} + 15} = 0.036852427 \tag{4.190}$$

obtained by only using the Lipschitz condition (4.146).

Example 4.3.3. Let $X = Y = \mathbb{R}$, $\Omega = (-\infty, \infty)$ and define function F on D by $F(x) = x^2 - 1$. Then, $x^\star = 1$ is a solution of $F(x) = 0$, F is Fréchet-differentiable, $F'(x^\star) = 2$ and the divided difference $[x, y; F]$ can be written as $[x, y; F] = x + y$, for all $x, y \in \Omega$. Hence, for any $x, y, u, v \in \Omega$, we have

$$|F'(x^\star)^{-1}([x, y; F] - [u, v; F])| = \frac{1}{2}|x + y - (u + v)| \leq \frac{1}{2}(|x - u| + |y - v|), \tag{4.191}$$

which shows the Lipschitz condition (4.146) and the center-Lipschitz condition (4.147) are true for $M = M_0 = \dfrac{1}{2}$. Therefore, we can see that all conditions of Theorem 4.3.1 are satisfied.

Let $x_{-1} = 1.5$ and $x_0 = 1.4$. Assume sequence $\{x_n\}$ is generated by (CSTM) (or (STM)) with different choices of parameters a, b and c. Table 4.3 gives the comparison result for $|x_n - x^\star|$ for this example, which shows that (STM) (or (CSTM) satisfying both conditions (4.137) and (4.138)) is faster than (CSTM) satisfying only

Table 4.3 Comparison results of $|x_n - x^\star|$ for Example 4.3.3 using various methods

n	SM	CSTM $(a = b = 1, c = 0.5)$	CSTM $(a = b = 0.5, c = 1)$	STM $(a = 1)$	STM $(a = 0.5)$
1	6.90e-02	4.44e-02	5.38e-02	1.98e-02	3.87e-02
2	1.12e-02	4.28e-03	5.39e-03	5.62e-04	1.34e-03
3	3.70e-04	4.74e-05	7.58e-05	5.53e-08	9.19e-07
4	2.06e-06	5.07e-08	1.03e-07	4.37e-15	6.22e-13
5	3.81e-10	6.02e-13	1.95e-12	3.34e-30	2.28e-25
6	3.92e-16	7.63e-21	5.00e-20	1.60e-59	3.51e-50
7	7.47e-26	1.15e-33	2.44e-32	4.47e-119	7.65e-100
8	1.46e-41	2.19e-54	3.05e-52	2.85e-237	3.81e-199
9	5.47e-67	6.28e-88	1.86e-84	1.42e-474	9.21e-398
ρ	1.6188076	1.6188739	1.6186348	2.0077130	2.0001332

condition (4.137). The computational order of convergence (COC) is also shown in Table 4.3. Here, COC is defined in Ref. [433] by

$$\rho \approx \frac{\ln |(x_{n+1} - x^\star)/(x_n - x^\star)|}{\ln |(x_n - x^\star)/(x_{n-1} - x^\star)|}. \tag{4.192}$$

The last line of Table 4.3 shows that the computational order of convergence for (CSTM) satisfying only condition (4.137) is near to 1.618, meanwhile (STM) (or (CSTM) satisfying both conditions (4.137) and (4.138)) has quadratic convergence.

In the last example we are not interested in checking if the hypotheses of Theorem 4.3.1 are satisfied or not, but comparing the numerical behavior of (STM) with earlier methods.

Example 4.3.4. In this example we present an application of the previous analysis to the Chandrasekhar equation given by

$$x(s) = 1 + \frac{s}{4}\, x(s) \int_0^1 \frac{x(t)}{s+t}\, dt, \quad s \in [0,1]. \tag{4.193}$$

We determine where a solution is located, along with its region of uniqueness. Later, the solution is approximated by an iterative method of (STM). Note that solving (4.193) is equivalent to solving $F(x) = 0$, where $F : C[0,1] \to C[0,1]$ and

$$[F(x)](s) = x(s) - 1 - \frac{s}{4}\, x(s) \int_0^1 \frac{x(t)}{s+t}\, dt, \quad s \in [0,1]. \tag{4.194}$$

To obtain a numerical solution of (4.193), we first discretize the problem and approach the integral by a Gauss-Legendre numerical quadrature with eight nodes,

$$\int_0^1 f(t)\, dt \approx \sum_{j=1}^8 w_j f(t_j),$$

where if we denote $x_i = x(t_i)$, $i = 1, 2, \dots, 8$, equation (4.193) is transformed into

Table 4.4   Iterates t_n and w_n for $n = 1, \cdots, 8$

$t_1 = 0.019855072, \quad t_2 = 0.101666761, \quad t_3 = 0.237233795, \quad t_4 = 0.408282679,$
$t_5 = 0.591717321, \quad t_6 = 0.762766205, \quad t_7 = 0.898333239, \quad t_8 = 0.980144928,$
$w_1 = 0.050614268, \quad w_2 = 0.111190517, \quad w_3 = 0.156853323, \quad w_4 = 0.181341892,$
$w_5 = 0.181341892, \quad w_6 = 0.156853323, \quad w_7 = 0.111190517, \quad w_8 = 0.050614268.$

Table 4.5   Comparison results of $\|\bar{x}_{n+1} - \bar{x}_n\|$ for Example 4.3.4 using various methods

n	SM	CSTM $(a = b = 1, c = 0.5)$	CSTM $(a = b = 0.5, c = 1)$	STM $(a = 1)$	STM $(a = 0.5)$
1	2.40e-01	2.44e-01	2.42e-01	2.49e-01	2.45e-01
2	9.29e-03	5.02e-03	7.12e-03	6.14e-04	4.87e-03
3	3.13e-04	8.58e-05	1.23e-04	5.76e-07	6.18e-06
4	3.24e-07	2.34e-08	4.81e-08	1.91e-15	3.28e-12
5	9.51e-12	9.38e-14	2.76e-13	4.34e-30	1.33e-24
6	2.54e-19	9.03e-23	5.46e-22	8.04e-62	1.40e-49
7	1.81e-31	3.15e-37	5.62e-36	6.45e-123	1.69e-99
8	3.14e-51	9.72e-61	1.05e-58	1.47e-247	2.03e-199

the following nonlinear system:

$$x_i = 1 + \frac{x_i}{4} \sum_{j=1}^{8} a_{ij} x_j, \quad i = 1, 2, \ldots, 8, \quad \text{where} \quad a_{ij} = \frac{t_i w_j}{t_i + t_j}.$$

Denote now $\bar{x} = (x_1, x_2, \ldots, x_8)^T$, $\bar{1} = (1, 1, \ldots, 1)^T$, $A = (a_{ij})$ and write the last nonlinear system in the matrix form:

$$\bar{x} = \bar{1} + \frac{1}{4} \bar{x} \odot (A\bar{x}), \tag{4.195}$$

where $\odot$ represents the inner product. If we choose $\bar{x}_0 = (1, 1, \ldots, 1)^T$ and $\bar{x}_{-1} = (.99, .99, \ldots, .99)^T$. Assume sequence $\{\bar{x}_n\}$ is generated by (CSTM) (or (STM)) with different choices of parameters a, b and c. Table 4.5 gives the comparison result for $\|\bar{x}_{n+1} - \bar{x}_n\|$ equipped with the max-norm for this example, which shows that (STM) (or (CSTM) satisfying both conditions (4.137) and (4.138)) is faster than (CSTM) satisfying only condition (4.137).

4.4   Secant-like Method and Recurrent Functions

We use the method of recurrent functions to provide a new semi-local convergence analysis for secant-like methods in order to approximate a locally unique solution of a nonlinear equation in a Banach space setting. Our convergence criteria and sufficient convergence conditions are tighter than in earlier studies such as Refs. [209, 266–268, 356, 360]. Numerical examples are given to illustrate the advantages of the new approach. We are concerned with the problem of approximating a locally unique solution $x^\star$ of (1.1). A very important aspect of iterative procedures is the choice of good starting guesses. In Ref. [104], Argyros used the simplified

Newton method defined by

$$x_0 \text{ is initial point}$$
$$x_{n+1} = x_n - F'(x_0)^{-1} F(x_n) \qquad \text{for each} \quad n = 0, 1, 2, \cdots \qquad (4.196)$$

as a predictor and then Newton's method defined (1.2) as a corrector method to generate a sequence approximating $x^\star$. Moreover, in Ref. [209] the simplified Secant method defined by

$$x_{-1}, x_0 \text{ are initial points}$$
$$x_{n+1} = x_n - A_0^{-1} F(x_n), \quad A_0 = [x_{-1}, x_0; F] \quad \text{for each} \quad n = 0, 1, 2, \cdots \qquad (4.197)$$

was used as a predictor and as a corrector the Secant-like method defined by

$$x_{-1}, x_0 \text{ are initial points}$$
$$y_n = \lambda x_n + (1 - \lambda) x_{n-1}, \ \lambda \in [0, 1]$$
$$x_{n+1} = x_n - B_n^{-1} F(x_n), \quad B_n = [y_n, x_n; F] \qquad \text{for each} \quad n = 0, 1, 2, \cdots . \qquad (4.198)$$

The family of Secant-like methods reduces to the Secant method if $\lambda = 0$ and to Newton's method if $\lambda = 1$. It was shown in Ref. [360] (see also Refs. [101, 110, 267, 268] and the references therein) that the R-order of convergence is at least $(1 + \sqrt{5})/2$ if $\lambda \in [0, 1)$, the same as that of the Secant method. In the real case the closer x_n and y_n are, the higher the speed of convergence. Moreover in Ref. [266], it was shown that as λ approaches 1 the speed of convergence is close to that of Newton's method. Moreover, the advantages of using Secant-like method instead of Newton's method is that the former method avoids the computation of $F'(x_n)^{-1}$ at each step. The hypotheses used for the semi-local convergence of simplified Secant-like method and Secant-like method are (see Refs. [110, 209, 266–268]):

$(\mathcal{C}_1)$ There exists a divided difference of order one denoted by $[x, y; F] \in \mathcal{L}(\mathcal{X}, \mathcal{Y})$ satisfying $[x, y; F](x - y) = F(x) - F(y)$ for all $x, y \in \mathcal{D}$;

$(\mathcal{C}_2)$ There exist x_{-1}, x_0 in $\mathcal{D}$ and $c > 0$ such that $\| x_0 - x_{-1} \| \leq c$;

$(\mathcal{C}_3)$ There exist $x_{-1}, x_0 \in \mathcal{D}$ and $M > 0$ such that $A_0^{-1} \in \mathcal{L}(\mathcal{Y}, \mathcal{X})$ and

$$\| A_0^{-1} ([x, y; F] - [u, v; F]) \| \leq M (\| x - u \| + \| y - v \|) \quad \text{for all} \quad x, y, u, v \in \mathcal{D};$$

$(\mathcal{C}_3^\star)$ There exist $x_{-1}, x_0 \in \mathcal{D}$ and $L > 0$ such that $A_0^{-1} \in \mathcal{L}(\mathcal{Y}, \mathcal{X})$ and

$$\| A_0^{-1} ([x, y; F] - [v, y; F]) \| \leq L \| x - v \| \quad \text{for all} \quad x, y, v \in \mathcal{D};$$

$(\mathcal{C}_3^{\star\star})$ There exist $x_{-1}, x_0 \in \mathcal{D}$ and $K > 0$ such that $F(x_0)^{-1} \in \mathcal{L}(\mathcal{Y}, \mathcal{X})$ and

$$\| F'(x_0)^{-1} ([x, y; F] - [v, y; F]) \| \leq K \| x - v \| \quad \text{for all} \quad x, y, v \in \mathcal{D};$$

$(\mathcal{C}_4)$ There exists $\eta > 0$ such that $\| A_0^{-1} F(x_0) \| \leq \eta$;

$(\mathcal{C}_4^\star)$ There exists $\eta > 0$ for each $\lambda \in [0, 1]$ such that $\| B_0^{-1} F(x_0) \| \leq \eta$.

We shall refer to $(\mathcal{C}_1)$–$(\mathcal{C}_4)$ as the $(\mathcal{C})$ conditions. From analyzing the semi-local convergence of simplified Secant method, it was shown in Ref. [209] that the convergence criteria are milder than those of Secant-like method given in Ref. [267].

Consequently, the decreasing and accessibility regions of (4.198) can be improved if (4.197) is used at the beginning of the approximations to the solution $x^\star$. Moreover, the semi-local convergence of (4.198) is guaranteed whenever that of method (4.197) does.

In the present section we show: an even larger convergence domain can be obtained under the same or weaker sufficient convergence criteria for both methods (4.197) and (4.198). In view of $(\mathcal{C}_3)$, we have that

$(\mathcal{C}_5)$ There exists $M_0 > 0$ such that

$$\| A_0^{-1}([x,y;F]-[x_{-1},x_0;F]) \| \le M_0 (\| x-x_{-1} \| + \| y-x_0 \|) \quad \text{for all} \quad x,y \in \mathcal{D}.$$

We shall also use the conditions

$(\mathcal{C}_6)$ There exist $x_0 \in \mathcal{D}$ and $M_1 > 0$ such that $F'(x_0)^{-1} \in \mathcal{L}(\mathcal{Y},\mathcal{X})$ and

$$\| F'(x_0)^{-1}([x,y;F]-F'(x_0)) \| \le M_1 (\| x-x_0 \| + \| y-x_0 \|) \quad \text{for all} \quad x,y \in \mathcal{D};$$

$(\mathcal{C}_7)$ There exist $x_0 \in \mathcal{D}$ and $M_2 > 0$ such that $F'(x_0)^{-1} \in \mathcal{L}(\mathcal{Y},\mathcal{X})$ and

$$\| F'(x_0)^{-1}(F'(x)-F'(x_0)) \| \le M_2 (\| x-x_0 \| + \| y-x_0 \|) \quad \text{for all} \quad x,y \in \mathcal{D}.$$

Note that $M_0 \le M$, $M_2 \le M_1$, $L \le M$ hold in general and M/M_0, M_1/M_2, M/L can be arbitrarily large. We shall refer to $(\mathcal{C}_1)$, $(\mathcal{C}_2)$, $(\mathcal{C}_4)$, $(\mathcal{C}_5)$ as the $(\mathcal{C}^\star)$ conditions, $(\mathcal{C}_1)$, $(\mathcal{C}_2)$, $(\mathcal{C}_3^{\star\star})$, $(\mathcal{C}_4^\star)$, $(\mathcal{C}_6)$ as the $(\mathcal{C}^{\star\star})$ conditions and $(\mathcal{C}_1)$, $(\mathcal{C}_2)$, $(\mathcal{C}_3^\star)$, $(\mathcal{C}_4^\star)$, $(\mathcal{C}_5)$ as the $(\mathcal{C}^{\star\star\star})$ conditions. $(\mathcal{C}_5)$ is not additional hypothesis to $(\mathcal{C}_3)$, since in practice the computation of constant M requires that of M_0. Note that if $(\mathcal{C}_6)$ holds, then we can set $M_2 = 2M_1$ in $(\mathcal{C}_7)$. We present the semi-local convergence of the simplified Secant method under the $(\mathcal{C}^\star)$ conditions. However, first we state the corresponding result from Ref. [209] in affine invariant form.

Theorem 4.4.1. *Suppose that the $(\mathcal{C})$ conditions,*

$$\overline{U}(x_0, R) \subseteq \mathcal{D}, \tag{4.199}$$

$$M c < 1 \quad \text{and} \quad M\eta < \frac{\sqrt{2(1+M^2 c^2)} - (1+Mc)}{2} \tag{4.200}$$

are satisfied, where

$$R = \frac{1 - Mc + 2M\eta - \sqrt{(1-Mc)^2 - 4(1+Mc)M\eta - 4M^2\eta^2}}{4M}. \tag{4.201}$$

Then, sequence $\{x_n\}$ generated by the simplified Secant method is well defined, remains in $\overline{U}(x_0, R)$ for each $n = -1, 0, 1, \cdots$ and converges to a unique solution $x^\star \in U(x_0, R_1) \cap \mathcal{D}$ of equation $F(x) = 0$, where

$$R_1 = \frac{1}{M} - R - c. \tag{4.202}$$

Moreover, the following estimates are satisfied

$$\| x_{n+1} - x_n \| \le M (c+\eta)(M(2R+c))^{n-1} \tag{4.203}$$

and

$$\| A_0^{-1} F(x_n) \| \le M(c+2R) \| x_n - x_{n-1} \| \quad for\ each \quad n = 0, 1, \cdots . \tag{4.204}$$

Next, we present the semi-local convergence result for simplified Secant method corresponding to Theorem 4.4.1.

Theorem 4.4.2. *Suppose that the $(\mathcal{C}^\star)$ conditions,*

$$\overline{U}(x_0, R_0) \subseteq \mathcal{D}, \tag{4.205}$$

$$M_0\, c < 1 \quad and \quad M_0\, \eta < \frac{\sqrt{2\left(1 + M_0^2\, c^2\right)} - (1 + M_0\, c)}{2} \tag{4.206}$$

are satisfied, where

$$R_0 = \frac{1 - M_0\, c + 2\, M_0\, \eta - \sqrt{(1 - M_0\, c)^2 - 4\,(1 + M_0\, c)\, M_0\, \eta - 4\, M_0^2\, \eta^2}}{4\, M_0}. \tag{4.207}$$

Then, sequence $\{x_n\}$ generated by the simplified Secant method is well defined, remains in $\overline{U}(x_0, R_0)$ for each $n = -1, 0, 1, \cdots$ and converges to a solution $x^\star \in U(x_0, R_0)$ of equation $F(x) = 0$. The solution $x^\star$ is unique in $U(x_0, R_2) \cap \mathcal{D}$, where

$$R_2 = \frac{1}{M_0} - R_0 - c. \tag{4.208}$$

Moreover, the following estimates are satisfied

$$\| x_{n+1} - x_n \| \le M_0\,(c + \eta)\,(M_0\,(2\, R_0 + c))^{n-1} \tag{4.209}$$

and

$$\| A_0^{-1}\, F(x_n) \| \le M_0\,(c + 2\, R_0)\, \| x_n - x_{n-1} \| \quad for\ each \quad n = 0, 1, \cdots. \tag{4.210}$$

Furthermore, if $(\mathcal{C}_7)$ holds, then the solution $x^\star$ is unique in $U(x_0, R_3) \cap \mathcal{D}$, where

$$R_3 = \frac{2}{M_2} - R_0. \tag{4.211}$$

Proof. The proof until the uniqueness part follows as in Theorem 4.4.1 by simply replacing the stronger $(\mathcal{C}_3)$ by the weaker and really needed for the proof $(\mathcal{C}_5)$ to show the uniqueness part, let first consider the linear operator $\mathcal{T} = \int_0^1 F'(x^\star + \theta\,(y^\star - x^\star))\, d\theta$, where $F(y^\star) = 0$ and $y^\star \in U(x_0, R_2) \cap \mathcal{D}$. Then, we have by $(\mathcal{C}_5)$ and (4.210)

$$
\begin{aligned}
\| A_0^{-1}\,(\mathcal{T} - A_0) \| &= \left\| A_0^{-1} \int_0^1 \left(F'(x^\star + \theta\,(y^\star - x^\star)) - [x_{-1}, x_0; F]\right) d\theta \right\| \\
&\le M_0 \int_0^1 \left(\| x^\star + \theta\,(y^\star - x^\star) - x_0 \| + \| x^\star + \theta\,(y^\star - x^\star) - x_{-1} \|\right) d\theta \\
&\le M_0 \int_0^1 \left(2\, \| x^\star + \theta\,(y^\star - x^\star) - x_0 \| + \| x_0 - x_{-1} \|\right) d\theta \\
&\le M_0 \int_0^1 \left(2\, \| (1 - \theta)\,(x^\star - x_0) + \theta\,(y^\star - x_0) \| + \| x_0 - x_{-1} \|\right) d\theta \\
&\le M_0\,\left(2\,\frac{1}{2}\,(\| x^\star - x_0 \| + \| y^\star - x_0 \|) + c\right) < M_0\,(R_0 + R_2 + c) = 1.
\end{aligned}
\tag{4.212}
$$

It follows from (4.212) and Banach lemma on invertible operators that $\mathcal{T}^{-1} \in \mathcal{L}(\mathcal{Y}, \mathcal{X})$. Then, using the identity $0 = F(y^\star) - F(x^\star) = \mathcal{T}\,(y^\star - x^\star)$, we deduce that $x^\star = y^\star$. If $y^\star \in U(x_0, R_3) \cap \mathcal{D}$, then using $(\mathcal{C}_7)$, (4.211), respectively, instead of $(\mathcal{C}_5)$, (4.208), we get that

$$\| A_0^{-1} (\mathcal{T} - A_0) \| \leq M_2 \int_0^1 \| x^\star + \theta (y^\star - x^\star) - x_0 \| \, d\theta < \frac{M_2}{2} (R_0 + R_3) = 1.$$

Hence, again we deduce that $x^\star = y^\star$. The proof of Theorem 4.4.2 is complete. $\square$

Remark 4.4.1. If $M_0 = M$, Theorem 4.4.2 reduces to Theorem 4.4.1. Otherwise, i.e. if $M_0 < M$ Theorem 4.4.2 constitutes an improvement over Theorem 4.4.1, since

(i) the computational cost of M_0 is smaller than that of M;
(ii) the convergence criteria (4.206) are weaker than (4.200);
(iii) error estimates (4.209) and (4.210) are tighter than (4.203) and (4.204), respectively;
(iv) the information on the location of the solution $x^\star$ more precise, since $R_0 < R$ and the uniqueness ball is larger, since $R_1 < R_2$.
(v) Radius R_3 can be larger than R_1 or R_2.

We present now the semi-local convergence of Secant-like method. First, we need some results on majorizing sequences for Secant-like method.

Lemma 4.4.1. *Let $c \geq 0$, $\eta > 0$, $M_1 > 0$, $K > 0$ and $\lambda \in [0, 1]$. Set $t_{-1} = 0$, $t_0 = c$ and $t_1 = c + \eta$. Define scalar sequences $\{q_n\}$, $\{t_n\}$, $\{\alpha_n\}$ for each $n = 0, 1, \cdots$ by*

$$q_n = (1 - \lambda) (t_n - t_0) + (1 + \lambda) (t_{n+1} - t_0),$$

$$t_{n+2} = t_{n+1} + \frac{K (t_{n+1} - t_n + (1 - \lambda) (t_n - t_{n-1}))}{1 - M_1 \, q_n} (t_{n+1} - t_n), \qquad (4.213)$$

$$\alpha_n = \frac{K (t_{n+1} - t_n + (1 - \lambda) (t_n - t_{n-1}))}{1 - M_1 \, q_n}, \qquad (4.214)$$

function $\{f_n\}$ for each $n = 1, 2, \cdots$ by

$$f_n(t) = K \eta \, t^n + K (1 - \lambda) \eta \, t^{n-1} + M_1 \, \eta \, ((1 - \lambda) (1 + t + \cdots + t^n) + \newline (1 + \lambda) (1 + t + \cdots + t^{n+1})) - 1 \qquad (4.215)$$

and polynomial p by

$$p(t) = M_1 (1 + \lambda) t^3 + (M_1 (1 - \lambda) + K) t^2 - K \lambda t - K (1 - \lambda). \qquad (4.216)$$

Denote by α the smallest root of polynomial p in $(0, 1)$. Suppose that

$$0 < \alpha \leq 1 - 2 M_1 \, \eta. \qquad (4.217)$$

Then, sequence $\{t_n\}$ is non-decreasing, bounded from above by $t^{\star\star}$ defined by

$$t^{\star\star} = \frac{\eta}{1 - \alpha} + c \qquad (4.218)$$

and converges to its unique least upper bound $t^\star$ which satisfies

$$c + \eta \leq t^\star \leq t^{\star\star}. \tag{4.219}$$

Moreover, the following estimates are satisfied for each $n = 0, 1, \cdots$

$$0 \leq t_{n+1} - t_n \leq \alpha^n \eta \tag{4.220}$$

and

$$t^\star - t_n \leq \frac{\alpha^n \eta}{1 - \alpha}. \tag{4.221}$$

Proof. We shall first prove that polynomial p has roots in $(0, 1)$. If $\lambda \neq 1$, $p(0) = -(1 - \lambda) K < 0$ and $p(1) = 2 M_1 > 0$. If $\lambda = 1$, $p(t) = t \overline{p}(t)$, $\overline{p}(0) = -K < 0$ and $\overline{p}(1) = 2 M_1 > 0$. In either case it follows from the intermediate value theorem that there exist roots in $(0, 1)$. Denote by α the minimal root of p in $(0, 1)$. Note that, in particular for Newton's method (i.e. for $\lambda = 1$) and for Secant method (i.e. for $\lambda = 0$), we have, respectively by (4.216) that

$$\alpha = \frac{2 K}{K + \sqrt{K^2 + 4 M_1 K}} \tag{4.222}$$

and

$$\alpha = \frac{2 K}{K + \sqrt{K^2 + 8 M_1 K}}. \tag{4.223}$$

It follows from (4.213) and (4.214) that estimate (4.220) is satisfied if

$$0 \leq \alpha_n \leq \alpha. \tag{4.224}$$

Estimate (4.224) is true by (4.217) for $n = 0$. Then, we have by (4.213) that

$$t_2 - t_1 \leq \alpha \, (t_1 - t_0) \implies t_2 \leq t_1 + \alpha \, (t_1 - t_0)$$
$$\implies t_2 \leq \eta + t_0 + \alpha \, \eta = c + (1 + \alpha) \, \eta = c + \frac{1 - \alpha^2}{1 - \alpha \eta} < t^{\star\star}.$$

Suppose that

$$t_{k+1} - t_k \leq \alpha^k \eta \quad \text{and} \quad t_{k+1} \leq c + \frac{1 - \alpha^{k+1}}{1 - \alpha} \, \eta. \tag{4.225}$$

Estimate (4.224) shall be true for $k + 1$ replacing n if

$$0 \leq \alpha_{k+1} \leq \alpha \tag{4.226}$$

or

$$f_k(\alpha) \leq 0. \tag{4.227}$$

We need a relationship between two consecutive recurrent functions f_k for each $k = 1, 2, \cdots$. It follows from (4.215) and (4.216) that

$$f_{k+1}(\alpha) = f_k(\alpha) + p(\alpha) \alpha^{k-1} \eta = f_k(\alpha), \tag{4.228}$$

since $p(\alpha) = 0$. Define function f_∞ on $(0, 1)$ by

$$f_\infty(t) = \lim_{n \to \infty} f_n(t). \tag{4.229}$$

Then, we get from (4.215) and (4.229) that

$$
\begin{aligned}
f_\infty(\alpha) &= \lim_{n \to \infty} f_n(\alpha) \\
&= K\,\eta \lim_{n \to \infty} \alpha^n + K\,(1 - \lambda)\,\eta \lim_{n \to \infty} \alpha^{n-1} + \\
&\quad M_1\,\eta \left((1 - \lambda) \lim_{n \to \infty} (1 + \alpha + \cdots + \alpha^n) + \right. \\
&\quad \left. (1 + \lambda) \lim_{n \to \infty} (1 + \alpha + \cdots + \alpha^{n+1}) \right) - 1 \\
&= M_1\,\eta \left(\frac{1 - \lambda}{1 - \alpha} + \frac{1 + \lambda}{1 - \alpha} \right) - 1 = \frac{2\,M_1\,\eta}{1 - \alpha} - 1,
\end{aligned}
\tag{4.230}
$$

since $\alpha \in (0, 1)$. In view of (4.227), (4.228) and (4.230) we can show instead of (4.227) that $f_\infty(\alpha) \leq 0$, which is true by (4.217). The induction for (4.220) is complete. It follows that sequence $\{t_n\}$ is non-decreasing, bounded from above by $t^{\star\star}$ given by (4.218) and as such it converges to $t^\star$ which satisfies (4.219). Estimate (4.221) follows from (4.220) by using standard majorization techniques. The proof of Lemma 4.4.1 is complete. $\qquad\square$

Lemma 4.4.2. *Let $c \geq 0$, $\eta > 0$, $M_1 > 0$, $K > 0$ and $\lambda \in [0, 1]$. Set $r_{-1} = 0$, $r_0 = c$ and $r_1 = c + \eta$. Define scalar sequences $\{r_n\}$ for each $n = 1, \cdots$ by*

$$
\begin{aligned}
r_2 &= r_1 + \beta_1\,(r_1 - r_0) \\
r_{n+2} &= r_{n+1} + \beta_n\,(r_{n+1} - r_n),
\end{aligned}
\tag{4.231}
$$

where

$$\beta_1 = \frac{M_1\,(r_1 - r_0 + (1 - \lambda)\,(r_0 - r_{-1}))}{1 - M_1\,q_1},$$

$$\beta_n = \frac{K\,(r_{n+1} - r_n + (1 - \lambda)\,(r_n - r_{n-1}))}{1 - M_1\,q_n} \quad \text{for each} \quad n = 2, 3, \cdots$$

and function $\{g_n\}$ on $[0, 1)$ for each $n = 1, 2, \cdots$ by

$$
\begin{aligned}
g_n(t) = {}& K\,(t + (1 - \lambda))\,t^{n-1}\,(r_2 - r_1) + \\
& M_1\,t \left((1 - \lambda)\,\frac{1 - t^{n+1}}{1 - t} + (1 + \lambda)\,\frac{1 - t^{n+2}}{1 - t} \right)(r_2 - r_1) + (2\,M_1\,\eta - 1)\,t.
\end{aligned}
\tag{4.232}
$$

Suppose that

$$0 \leq \beta_1 \leq \alpha \leq 1 - \frac{2\,M_1\,(r_2 - r_1)}{1 - 2\,M_1\,\eta}, \tag{4.233}$$

where α is defined in Lemma 4.4.1. Then, sequence $\{r_n\}$ is non-decreasing, bounded from above by $r^{\star\star}$ defined by

$$r^{\star\star} = c + \eta + \frac{r_2 - r_1}{1 - \alpha} \tag{4.234}$$

and converges to its unique least upper bound $r^\star$ which satisfies

$$c + \eta \leq r^\star \leq r^{\star\star}. \tag{4.235}$$

Moreover, the following estimates are satisfied for each $n = 1, \cdots$

$$0 \leq r_{n+2} - r_{n+1} \leq \alpha^n\,(r_2 - r_1). \tag{4.236}$$

Proof. We shall use mathematical induction to show that

$$0 \le \beta_n \le \alpha. \tag{4.237}$$

Estimate (4.237) is true for $n = 0$ by (4.233). Then, we have by (4.231) that

$$0 \le r_3 - r_2 \le \alpha \left(r_2 - r_1 \right) \implies r_3 \le r_2 + \alpha \left(r_2 - r_1 \right)$$
$$\implies r_3 \le r_2 + \left(1 + \alpha \right) \left(r_2 - r_1 \right) - \left(r_2 - r_1 \right)$$
$$\implies r_3 \le r_1 + \frac{1 - \alpha^2}{1 - \alpha} \left(r_2 - r_1 \right) \le r^{\star\star}.$$

Suppose (4.237) holds for each $n \le k$, then, using (4.231), we obtain that

$$0 \le r_{k+2} - r_{k+1} \le \alpha^k \left(r_2 - r_1 \right) \quad \text{and} \quad r_{k+2} \le r_1 + \frac{1 - \alpha^{k+1}}{1 - \alpha} \left(r_2 - r_1 \right). \tag{4.238}$$

Estimate (4.237) is certainly satisfied, if

$$g_k(\alpha) \le 0, \tag{4.239}$$

where g_k is defined by (4.232). Using (4.232), we obtain the following relationship between two consecutive recurrent functions g_k for each $k = 1, 2, \cdots$

$$g_{k+1}(\alpha) = g_k(\alpha) + p(\alpha)\, \alpha^{k-1} \left(r_2 - r_1 \right) = g_k(\alpha), \tag{4.240}$$

since $p(\alpha) = 0$. Define function g_∞ on $[0, 1)$ by

$$g_\infty(t) = \lim_{k \to \infty} g_k(t). \tag{4.241}$$

Then, we get from (4.232) and (4.241) that

$$g_\infty(\alpha) = \alpha \left(\frac{2\, M_1 \left(r_2 - r_1 \right)}{1 - \alpha} + 2\, M_1\, \eta - 1 \right). \tag{4.242}$$

In view of (4.239)–(4.242) to show (4.239), it suffices to have $g_\infty(\alpha) \le 0$, which holds by the right hand hypothesis in (4.233). The induction for (4.237) (i.e. for (4.236)) is complete. The rest of the proof is omitted (as it is identical to the proof of Lemma 4.4.1). The proof of Lemma 4.4.2 is complete. $\qquad\square$

We have the following useful and obvious extensions of Lemma 4.4.1 and Lemma 4.4.2, respectively.

Lemma 4.4.3. *Let $N = 0, 1, 2, \cdots$ be fixed. Suppose that*

$$t_1 \le t_2 \le \cdots \le t_N \le t_{N+1}, \tag{4.243}$$

$$\frac{1}{M_1} > (1 - \lambda)\left(t_N - t_0 \right) + (1 + \lambda)\left(t_{N+1} - t_0 \right) \tag{4.244}$$

and

$$0 \le \alpha_N \le \alpha \le 1 - 2\, M_1 \left(t_{N+1} - t_N \right). \tag{4.245}$$

Then, sequence $\{t_n\}$ generated by (4.213) is nondecreasing, bounded from above by $t^{\star\star}$ and converges to $t^\star$ which satisfies $t^\star \in [t_{N+1}, t^{\star\star}]$. Moreover, the following estimates are satisfied for each $n = 0, 1, \cdots$

$$0 \le t_{N+n+1} - t_{N+n} \le \alpha^n \left(t_{N+1} - t_N \right) \tag{4.246}$$

and

$$t^\star - t_{N+n} \le \frac{\alpha^n}{1 - \alpha} \left(t_{N+1} - t_N \right). \tag{4.247}$$

Lemma 4.4.4. *Let $N = 1, 2, \cdots$ be fixed. Suppose that*

$$r_1 \leq r_2 \leq \cdots \leq r_N \leq r_{N+1}, \tag{4.248}$$

$$\frac{1}{M_1} > (1 - \lambda)(r_N - r_0) + (1 + \lambda)(r_{N+1} - r_0) \tag{4.249}$$

and

$$0 \leq \beta_N \leq \alpha \leq 1 - \frac{2\,M_1\,(r_{N+1} - r_N)}{1 - 2\,M_1\,(r_N - r_{N-1})}. \tag{4.250}$$

Then, sequence $\{r_n\}$ generated by (4.231) is nondecreasing, bounded from above by $r^{\star\star}$ and converges to $r^\star$ which satisfies $r^\star \in [r_{N+1}, r^{\star\star}]$. Moreover, the following estimates are satisfied for each $n = 0, 1, \cdots$

$$0 \leq r_{N+n+1} - r_{N+n} \leq \alpha^n (r_{N+1} - r_N) \tag{4.251}$$

and

$$r^\star - r_{N+n} \leq \frac{\alpha^n}{1 - \alpha}(r_{N+1} - r_N). \tag{4.252}$$

Next, we present the following semi-local convergence result for Secant-like method under the $(C^{\star\star})$ conditions.

Theorem 4.4.3. *Suppose that the $(C^{\star\star})$, Lemma 4.4.1 (or Lemma 4.4.3) conditions and*

$$\overline{U}(x_0, t^\star) \subseteq \mathcal{D} \tag{4.253}$$

hold. Then, sequence $\{x_n\}$ generated by the Secant-like method is well defined, remains in $\overline{U}(x_0, t^\star)$ for each $n = -1, 0, 1, \cdots$ and converges to a solution $x^\star \in \overline{U}(x_0, t^\star - c)$ of equation $F(x) = 0$. Moreover, the following estimates are satisfied for each $n = 0, 1, \cdots$

$$\| x_{n+1} - x_n \| \leq t_{n+1} - t_n \tag{4.254}$$

and

$$\| x_n - x^\star \| \leq t^\star - t_n. \tag{4.255}$$

Furthermore, if there exists $r \geq t^\star$ such that

$$\overline{U}(x_0, r) \subseteq \mathcal{D} \tag{4.256}$$

and

$$r + t^\star < \frac{1}{M_1} \quad or \quad r + t^\star < \frac{2}{M_2}, \tag{4.257}$$

then, the solution $x^\star$ is unique in $\overline{U}(x_0, r)$.

Proof. We use mathematical induction to prove that

$$\parallel x_{k+1} - x_k \parallel \le t_{k+1} - t_k \tag{4.258}$$

and

$$\overline{U}(x_{k+1}, t^\star - t_{k+1}) \subseteq \overline{U}(x_k, t^\star - t_k) \tag{4.259}$$

for each $k = -1, 0, 1, \cdots$. Let $z \in \overline{U}(x_0, t^\star - t_0)$. Then, we obtain that

$$\parallel z - x_{-1} \parallel \le \parallel z - x_0 \parallel + \parallel x_0 - x_{-1} \parallel \le t^\star - t_0 + c = t^\star = t^\star - t_{-1},$$

which implies $z \in \overline{U}(x_{-1}, t^\star - t_{-1})$. Let also $w \in \overline{U}(x_0, t^\star - t_1)$. We get that

$$\parallel w - x_0 \parallel \le \parallel w - x_1 \parallel + \parallel x_1 - x_0 \parallel \le t^\star - t_1 + t_1 - t_0 = t^\star = t^\star - t_0.$$

That is $w \in \overline{U}(x_0, t^\star - t_0)$. Note that

$$\parallel x_{-1} - x_0 \parallel \le c = t_0 - t_{-1} \quad \text{and} \quad \parallel x_1 - x_0 \parallel = \parallel B_0^{-1} F(x_0) \parallel \le \eta = t_1 - t_0 < t^\star,$$

which implies $x_1 \in \overline{U}(x_0, t^\star) \subseteq \mathcal{D}$. Hence, estimates (4.254) and (4.255) hold for $k = -1$ and $k = 0$. Suppose (4.254) and (4.255) hold for all $n \le k$. Then, we obtain that

$$\parallel x_{k+1} - x_0 \parallel \le \sum_{i=1}^{k+1} \parallel x_i - x_{i-1} \parallel \le \sum_{i=1}^{k+1} (t_i - t_{i-1}) = t_{k+1} - t_0 \le t^\star$$

and

$$\parallel y_k - x_0 \parallel \le \lambda \parallel x_k - x_0 \parallel + (1 - \lambda) \parallel x_{k-1} - x_0 \parallel \le \lambda t^\star + (1 - \lambda) t^\star = t^\star.$$

Hence, $x_{k+1}, y_k \in \overline{U}(x_0, t^\star)$. Let $E_k := [x_{k+1}, x_k; F]$ for each $k = 0, 1, \cdots$. Using (4.198), Lemma 4.4.1 and the induction hypotheses, we get that

$$\begin{aligned}
\parallel F'(x_0)^{-1} (B_{k+1} - F'(x_0)) \parallel &\le M_1 (\parallel y_{k+1} - x_0 \parallel + \parallel x_{k+1} - x_0 \parallel) \\
&\le M_1 ((1 - \lambda) \parallel x_k - x_0 \parallel + \lambda \parallel x_{k+1} - x_0 \parallel + \parallel x_{k+1} - x_0 \parallel) \\
&\le M_1 ((1 - \lambda)(t_k - t_0) + (1 + \lambda)(t_{k+1} - t_0)) < 1,
\end{aligned} \tag{4.260}$$

since, $y_{k+1} - x_0 = \lambda (x_{k+1} - x_0) + (1 - \lambda)(x_k - x_0)$ and

$$\begin{aligned}
\parallel y_{k+1} - x_0 \parallel &= \parallel \lambda (x_{k+1} - x_0) + (1 - \lambda)(x_k - x_0) \parallel \\
&\le \lambda \parallel x_{k+1} - x_0 \parallel + (1 - \lambda) \parallel x_k - x_0 \parallel.
\end{aligned}$$

It follows from (4.260) and the Banach lemma on invertible operators that B_{k+1}^{-1} exists and

$$\parallel B_{k+1}^{-1} F'(x_0) \parallel \le \frac{1}{1 - \Theta_k} \le \frac{1}{1 - M_1 \, q_{k+1}}, \tag{4.261}$$

where $\Theta_k = M_1 ((1 - \lambda) \parallel x_k - x_0 \parallel + (1 + \lambda) \parallel x_{k+1} - x_0 \parallel)$. In view of (4.198), we obtain the identity

$$F(x_{k+1}) = F(x_{k+1}) - F(x_k) - B_k (x_{k+1} - x_k) = (E_k - B_k)(x_{k+1} - x_k). \tag{4.262}$$

Then, using the induction hypotheses, the $(C^{\star\star})$ condition and (4.262), we get in turn that

$$
\begin{aligned}
\| F'(x_0)^{-1} F(x_{k+1}) \| &= \| F'(x_0)^{-1} (E_k - B_k)(x_{k+1} - x_k) \| \\
&\leq K \, \| x_{k+1} - y_k \| \, \| x_{k+1} - x_k \| \\
&\leq K \, (\| x_{k+1} - x_k \| + (1 - \lambda) \, \| x_k - x_{k-1} \|) \, \| x_{k+1} - x_k \| \\
&\leq K \, (t_{k+1} - t_k + (1 - \lambda)(t_k - t_{k-1}))(t_{k+1} - t_k),
\end{aligned}
$$
$$(4.263)$$

since, $x_{k+1} - y_k = x_{k+1} - x_k + (1 - \lambda)(x_k - x_{k-1})$ and

$$
\| x_{k+1} - y_k \| \leq \| x_{k+1} - x_k \| + (1 - \lambda) \, \| x_k - x_{k-1} \| \leq t_{k+1} - t_k + (1 - \lambda)(t_k - t_{k-1}).
$$

It now follows from (4.198), (4.213), (4.261)–(4.263) that

$$
\begin{aligned}
\| x_{k+2} - x_{k+1} \| &\leq \| B_{k+1}^{-1} F'(x_0) \| \, \| F'(x_0)^{-1} F(x_{k+1}) \| \\
&\leq \frac{K \, (t_{k+1} - t_k + (1 - \lambda)(t_{k+1} - x_k))(t_{k+1} - t_k)}{1 - M_1 \, q_{k+1}} = t_{k+2} - t_{k+1},
\end{aligned}
$$

which completes the induction for (4.258). Furthermore, let $v \in \overline{U}(x_{k+2}, t^\star - t_{k+2})$. Then, we have that

$$
\begin{aligned}
\| v - x_{k+1} \| &\leq \| v - x_{k+2} \| + \| x_{k+2} - x_{k+1} \| \\
&\leq t^\star - t_{k+2} + t_{k+2} - t_{k+1} = t^\star - t_{k+1},
\end{aligned}
$$

which implies $v \in \overline{U}(x_{k+1}, t^\star - t_{k+1})$. The induction for (4.258) and (4.259) is complete. Lemma 4.4.1 implies that $\{t_k\}$ is a complete sequence. It follows from (4.258) and (4.259) that $\{x_k\}$ is a complete sequence in a Banach space $\mathcal{X}$ and as such it converges to some $x^\star \in \overline{U}(x_0, t^\star)$ (since $\overline{U}(x_0, t^\star)$ is a closed set). By letting $k \longrightarrow \infty$ in (4.263), we get that $F(x^\star) = 0$. Moreover, estimate (4.255) follows from (4.254) by using standard majorization techniques. To show the uniqueness part, let $y^\star \in \overline{U}(x_0, r)$ be such $F(y^\star) = 0$, where r satisfies (4.256) and (4.257). We have that

$$
\begin{aligned}
\| F'(x_0)^{-1} ([y^\star, x^\star; F] - F'(x_0)) \| &\leq M_1 \, (\| y^\star - x_0 \| + \| x^\star - x_0 \|) \\
&\leq M_1 \, (t^\star + r) < 1.
\end{aligned}
$$
$$(4.264)$$

It follows by (4.264) and the Banach lemma on invertible operators that linear operator $[y^\star, x^\star; F]^{-1}$ exists. Then, using the identity $0 = F(y^\star) - F(x^\star) = [y^\star, x^\star; F](y^\star - x^\star)$, we deduce that $x^\star = y^\star$. The proof of Theorem 4.4.3 is complete. $\qquad\square$

In order for us to present the semi-local result for Secant-like method under the $(C^{\star\star\star})$ conditions, we first need a result on a majorizing sequence. The proof is given in Lemma 4.4.1.

Remark 4.4.2. Clearly, (4.233) (or (4.250)), $\{r_n\}$ can replace (4.217) (or (4.245)), $\{t_n\}$, respectively in Theorem 4.4.3.

Lemma 4.4.5. *Let $c \geq 0$, $\eta > 0$, $L > 0$, $M_0 > 0$ with $M_0 c < 1$ and $\lambda \in [0,1]$. Set*

$$s_{-1} = 0, \ s_0 = c, \ s_1 = c + \eta, \ \tilde{K} = \frac{L}{1 - M_0 c} \quad and \quad \tilde{M}_1 = \frac{M_0}{1 - M_0 c}.$$

Define scalar sequences $\{\tilde{q}_n\}$, $\{s_n\}$, $\{\tilde{\alpha}_n\}$ for each $n = 0, 1, \cdots$ by

$$\tilde{q}_n = (1 - \lambda)(s_n - s_0) + (1 + \lambda)(s_{n+1} - s_0),$$

$$s_{n+2} = s_{n+1} + \frac{\tilde{K}(s_{n+1} - s_n + (1 - \lambda)(s_n - s_{n-1}))}{1 - \tilde{M}_1 \tilde{q}_n}(s_{n+1} - s_n),$$

$$\tilde{\alpha}_n = \frac{\tilde{K}(s_{n+1} - s_n + (1 - \lambda)(s_n - s_{n-1}))}{1 - \tilde{M}_1 \tilde{q}_n},$$

function $\{\tilde{f}_n\}$ for each $n = 1, 2, \cdots$ by

$$\tilde{f}_n(t) = \tilde{K}\eta t^n + \tilde{K}(1 - \lambda)\eta t^{n-1} + \tilde{M}_1 \eta((1 - \lambda)(1 + t + \cdots + t^n) + (1 + \lambda)(1 + t + \cdots + t^{n+1})) - 1$$

and polynomial $\tilde{p}$ by

$$\tilde{p}(t) = \tilde{M}_1(1 + \lambda)t^3 + (\tilde{M}_1(1 - \lambda) + \tilde{K})t^2 - \tilde{K}\lambda t - \tilde{K}(1 - \lambda).$$

Denote by $\tilde{\alpha}$ the smallest root of polynomial $\tilde{p}$ in $(0, 1)$. Suppose that

$$0 \leq \tilde{\alpha}_0 \leq \tilde{\alpha} \leq 1 - 2\tilde{M}_1 \eta. \tag{4.265}$$

Then, sequence $\{s_n\}$ is non-decreasing, bounded from above by $s^{\star\star}$ defined by

$$s^{\star\star} = \frac{\eta}{1 - \tilde{\alpha}} + c$$

and converges to its unique least upper bound $s^\star$ which satisfies $c + \eta \leq s^\star \leq s^{\star\star}$. Moreover, the following estimates are satisfied for each $n = 0, 1, \cdots$

$$0 \leq s_{n+1} - s_n \leq \tilde{\alpha}^n \eta \quad and \quad s^\star - s_n \leq \frac{\tilde{\alpha}^n \eta}{1 - \tilde{\alpha}}.$$

Next, we present the semi-local convergence result for Secant-like method under the $(\mathcal{C}^{\star\star\star})$ conditions.

Theorem 4.4.4. *Suppose that the $(\mathcal{C}^{\star\star\star})$ conditions, (4.265) (or Lemma 4.4.2 conditions with $\tilde{\alpha}_n$, $\tilde{\alpha}$, $\tilde{M}_1$ replacing, respectively, α_n, α, M_1) and $\overline{U}(x_0, s^\star) \subseteq \mathcal{D}$ hold. Then, sequence $\{x_n\}$ generated by the Secant-like method is well defined, remains in $\overline{U}(x_0, s^\star)$ for each $n = -1, 0, 1, \cdots$ and converges to a solution $x^\star \in \overline{U}(x_0, s^\star)$ of equation $F(x) = 0$. Moreover, the following estimates are satisfied for each $n = 0, 1, \cdots$*

$$\| x_{n+1} - x_n \| \leq s_{n+1} - s_n \quad and \quad \| x_n - x^\star \| \leq s^\star - s_n.$$

Furthermore, if there exists $r \geq s^\star$ such that $\overline{U}(x_0, r) \subseteq \mathcal{D}$ and $r + s^\star + c < 1/M_0$, then, the solution $x^\star$ is unique in $\overline{U}(x_0, r)$.

Proof. The proof is analogous to Theorem 4.4.3. Simply notice that in view of $(\mathcal{C}_5)$, we obtain instead of (4.260) that

$$\| A_0^{-1} (B_{k+1} - A_0) \| \leq M_0 (\| y_{k+1} - x_{-1} \| + \| x_{k+1} - x_0 \|)$$
$$\leq M_0 ((1 - \lambda) \| x_k - x_0 \| + \lambda \| x_{k+1} - x_0 \| + \| x_0 - x_{-1} \| + \| x_{k+1} - x_0 \|)$$
$$\leq M_0 ((1 - \lambda) (s_k - s_0) + (1 + \lambda) (s_{k+1} - s_0) + c) < 1,$$

leading to B_{k+1}^{-1} exists and

$$\| B_{k+1}^{-1} A_0 \| \leq \frac{1}{1 - \Xi_k},$$

where $\Xi_k = M_0 ((1 - \lambda) (s_k - s_0) + (1 + \lambda) (s_{k+1} - s_0) + c)$. Moreover, using $(\mathcal{C}_3^\star)$ instead of $(\mathcal{C}_3^{\star\star})$, we get that

$$\| A_0^{-1} F(x_{k+1}) \| \leq L (s_{k+1} - s_k + (1 - \lambda) (s_k - s_{k-1})) (s_{k+1} - s_k).$$

Hence, we have that

$$\| x_{k+2} - x_{k+1} \| \leq \| B_{k+1}^{-1} A_0 \| \| A_0^{-1} F(x_{k+1}) \|$$
$$\leq \frac{L (s_{k+1} - s_k + (1 - \lambda) (s_k - s_{k-1})) (s_{k+1} - s_k)}{1 - M_0 ((1 + \lambda) (s_{k+1} - s_0) + (1 - \lambda) (s_k - s_0) + c)}$$
$$\leq \frac{\tilde{K} (s_{k+1} - s_k + (1 - \lambda) (s_k - s_{k-1})) (s_{k+1} - s_k)}{1 - \tilde{M}_1 ((1 + \lambda) (s_{k+1} - s_0) + (1 - \lambda) (s_k - s_0))} = s_{k+2} - s_{k+1}.$$

The uniqueness part is given in Theorem 4.4.2 with r, $s^\star$ replacing R_2 and R_0, respectively. The proof of Theorem 4.4.4 is complete. $\qquad\square$

Remark 4.4.3.

(a) Condition (4.253) can be replaced by $\overline{U}(x_0, t^{\star\star}) \subseteq \mathcal{D}$, where $t^{\star\star}$ is given in the closed form by (4.258).

(b) The majorizing sequence $\{u_n\}$ essentially used in Ref. [209] is defined by

$$u_{-1} = 0, \ u_0 = c, \ u_1 = c + \eta$$
$$u_{n+2} = u_{n+1} + \frac{M (u_{n+1} - u_n + (1 - \lambda) (u_n - u_{n-1}))}{1 - M q_n^\star} (u_{n+1} - u_n), \qquad (4.266)$$

where

$$q_n^\star = (1 - \lambda) (u_n - u_0) + (1 + \lambda) (u_{n+1} - u_0).$$

Then, if $K < M$ or $M_1 < M$, a simple inductive argument shows that for each $n = 2, 3, \cdots$

$$t_n < u_n, \quad t_{n+1} - t_n < u_{n+1} - u_n \quad \text{and} \quad t^\star \leq u^\star = \lim_{n \to \infty} u_n. \qquad (4.267)$$

Clearly $\{t_n\}$ converges under the $(\mathcal{C})$ conditions, (4.199) and (4.200). Similarly if $L \leq M$, $\{s_n\}$ is a tighter sequence than $\{u_n\}$. In general, we shall test the convergence criteria and use the tightest sequence to estimate the error bounds.

(c) Clearly the conclusions of Theorem 4.4.4 hold if $\{s_n\}$, (4.265) are replaced by $\{\tilde{r}_n\}$, (4.233), where $\{\tilde{r}_n\}$ is defined as $\{r_n\}$ with M_0 replacing M_1 in the definition of β_1 (only at the numerator) and the tilda letters replacing the non-tilda letters in (4.233).

Let us consider the equation

$$F(x) + G(x) = 0 \tag{4.268}$$

where F is as before and $G : \mathcal{D} \longrightarrow \mathcal{Y}$ is only continuous. The corresponding Secant-like method is defined by

$$x_{-1},\, x_0 \text{ are initial points}$$
$$y_n = \lambda\, x_n + (1 - \lambda)\, x_{n-1},\ \lambda \in [0, 1] \tag{4.269}$$
$$x_{n+1} = x_n - B_n^{-1}\left(F(x_n) + G(x_n)\right) \quad \text{for each} \quad n = 0, 1, 2, \cdots,$$

with B_n as defined in (4.198). Suppose there exist $H > 0$, $H_1 > 0$ such that

$(\mathcal{C}_8)$ $\| F'(x_0)^{-1}\left(G(x) - G(y)\right) \| \leq H \| x - y \|$ for all $x, y \in \mathcal{D}$;

$(\mathcal{C}_9)$ $\| F'(x_0)^{-1}\left(G(x_1) - G(x_0)\right) \| \leq H_1 \| x_1 - x_0 \|$.

Clearly $H_1 \leq H$ holds and H/H_1 can be arbitrarily large. Note that $(\mathcal{C}_9)$ always follows from $(\mathcal{C}_8)$. We shall denote by $(\mathcal{Q})$ the $(\mathcal{C}^{\star\star})$ together with $(\mathcal{C}_8)$ and $(\mathcal{C}_9)$ conditions. Then, we can present all the corresponding results for the nondifferentiable case along the same lines as the differentiable case already presented above. However for brevity, we shall only present the results corresponding respectively, to Lemma 4.4.2 and Theorem 4.4.3.

Lemma 4.4.6. *Let $c \geq 0$, $\eta > 0$, $M_1 > 0$, $K > 0$, $H_1 > 0$, $H > 0$ and $\lambda \in [0, 1]$. Set $r_{-1} = 0$, $r_0 = c$ and $r_1 = c + \eta$. Define scalar sequences $\{r_n\}$ for each $n = 1, \cdots$ by*

$$r_2 = r_1 + \beta_1\left(r_1 - r_0\right)$$
$$r_{n+2} = r_{n+1} + \beta_n\left(r_{n+1} - r_n\right),$$

where

$$\beta_1 = \frac{M_1\left(r_1 - r_0 + (1 - \lambda)\left(r_0 - r_{-1}\right)\right) + H_1}{1 - M_1\, q_1},$$

$$\beta_n = \frac{K\left(r_{n+1} - r_n + (1 - \lambda)\left(r_n - r_{n-1}\right)\right) + H}{1 - M_1\, q_n} \quad \text{for each} \quad n = 2, 3, \cdots$$

and function $\{g_n\}$ on $[0, 1)$ for each $n = 1, 2, \cdots$ by

$$g_n(t) = K\left(t + (1 - \lambda)\right) t^{n-1}\left(r_2 - r_1\right) +$$
$$M_1\, t\left((1 - \lambda)\frac{1 - t^{n+1}}{1 - t} + (1 + \lambda)\frac{1 - t^{n+2}}{1 - t}\right)\left(r_2 - r_1\right) + \left(2\, M_1\, \eta - 1\right) t + H.$$

Suppose that function φ defined on $[0, 1)$ by

$$\varphi(t) = \frac{2\, M_1\, t}{1 - t}\left(r_2 - r_1\right) + \left(2\, M_1\, \eta - 1\right) t + H$$

has a minimal zero in $(0, 1)$ denoted by $\alpha^\star$ such that $0 \leq \beta_1 \leq \alpha \leq \alpha^\star$, where α is defined in Lemma 4.4.1. Then, sequence $\{r_n\}$ is non-decreasing, bounded from above by $r^{\star\star}$ defined by

$$r^{\star\star} = c + \eta + \frac{r_2 - r_1}{1 - \alpha}$$

and converges to its unique least upper bound $r^\star$ which satisfies $c + \eta \leq r^\star \leq r^{\star\star}$. Moreover, the following estimates are satisfied for each $n = 1, \cdots$

$$0 \leq r_{n+2} - r_{n+1} \leq \alpha^n\left(r_2 - r_1\right).$$

Proof. Simply use the new $\{\beta_n\}$, $\{r_n\}$, $\{g_n\}$ in the proof of Lemma 4.4.2. $\square$

Theorem 4.4.5. *Suppose that the (Q), Lemma 4.4.2 conditions and $\overline{U}(x_0, r^\star) \subseteq \mathcal{D}$ hold for $\| B_0^{-1}(F(x_0) + G(x_0)) \| \leq \eta$. Then, sequence $\{x_n\}$ generated by (4.269) is well defined, remains in $\overline{U}(x_0, r^\star)$ for each $n = -1, 0, 1, \cdots$ and converges to a solution $x^\star \in \overline{U}(x_0, r^\star - c)$ of equation $F(x) + G(x) = 0$. Moreover, the following estimates are satisfied for each $n = 0, 1, \cdots$*

$$\| x_{n+1} - x_n \| \leq r_{n+1} - r_n \quad and \quad \| x_n - x^\star \| \leq r^\star - r_n.$$

Furthermore, if there exists $T \geq r^\star - c$ such that $\overline{U}(x_0, T) \subseteq \mathcal{D}$ and

$$0 < \xi := \frac{K\left((1-\lambda)\,c + T\right) + H}{1 - 2\,M_1\,(r^\star - c)} < 1 \quad for \ some \quad \xi \in (0, 1), \tag{4.270}$$

then, the solution $x^\star$ is unique in $\overline{U}(x_0, T)$.

Proof. The proof until the uniqueness part follows as in Theorem 4.4.3, but using the identity

$$F(x_{k+1}) + G(x_{k+1}) = ([x_{k+1}, x_k; F] - B_k)(x_{k+1} - x_k) + (G(x_{k+1}) - G(x_k))$$

instead of (4.262). Finally for the uniqueness part, let $y^\star \in \overline{U}(x_0, T)$ be such that $F(y^\star) + G(y^\star) = 0$. Then, we get from (4.269) the identity

$$\begin{aligned}
x_{n+1} - y^\star &= x_n - B_n^{-1}(F(x_n) + G(x_n)) - x^\star \\
&= -B_n(F(x_n) - F(x^\star) - B_n(x_n - y^\star) + (G(x_n) - G(y^\star))) \\
&= -B_n([x_n, y^\star; F] - [y_n, x_n; F])(x_n - y^\star) + (G(x_n) - G(y^\star)).
\end{aligned} \tag{4.271}$$

We have that

$$x_n - y_n = (1 - \lambda)(x_n - x_{n-1}) \tag{4.272}$$

and

$$y_n - y^\star = \lambda(x_n - y^\star) + (1 - \lambda)(x_{n-1} - y^\star). \tag{4.273}$$

Using (4.261), (4.263), (C_8) and (4.270)–(4.273), we obtain in turn that

$$\begin{aligned}
\| x_{n+1} &- y^\star \| \\
&\leq \frac{K\left((1-\lambda)\, \| x_n - x_{n-1} \| + \lambda\, \| x_n - y^\star \| + (1-\lambda)\, \| x_{n-1} - y^\star \|\right) + H}{1 - M_1\left((1-\lambda)\,r_n + (1+\lambda)\,r_{n+1}\right)} \, \| x_n - y^\star \| \\
&\leq \frac{K\left((1-\lambda)\,c + T\right) + H}{1 - 2\,M_1\,(r^\star - c)} \, \| x_n - y^\star \| = \xi\, \| x_n - y^\star \| \leq \xi^{n+1}\, \| x_0 - y^\star \| < \xi^{n+1}\, T,
\end{aligned}$$

which shows $\lim_{n \to \infty} x_n = y^\star$. But we have that $\lim_{n \to \infty} x_n = x^\star$. Hence, we deduce $x^\star = y^\star$. The proof of Theorem 4.4.5 is complete. $\square$

4.5 Directional Secant-type Method

A semi-local convergence analysis for directional Secant-type methods in n-variables is provided in this Section. Using weaker hypotheses than in the related work by An and Bai in Ref. [20], and motivated by optimization considerations, we provide under the same computational cost a semi-local convergence analysis for directional Secant method. A numerical example where our results apply to solve an equation but not the ones in Ref. [20] is also provided. In a second example, we show how to implement the method. We are concerned with the problem of approximating a solution $x^\star$ of equation (1.1), where F is a differentiable mapping defined on a convex subset $\mathcal{D}$ of $\mathbb{R}^n$ (n a natural number) with values in $\mathbb{R}$.

An and Bai in Ref. [20] used the directional Secant method (DSM)

$$x_{k+1} = x_k + h_k$$
$$h_k = -A_k\, F(x_k), \quad A_k = \frac{\theta_k}{F(v_k) - F(x_k)} \cdot d_k, \quad v_k = x_k + \theta_k\, d_k$$
$$(k \geq 0, \quad x_0 \in \mathbb{R}^n, \quad d_k \in \mathbb{R}^n, \quad \| d_k \| = 1, \quad \theta_k \geq 0)$$

to generate a sequence $\{x_k\}$ converging to $x^\star$. (DSM) is a useful alternative to (DNM) (see Refs. [116, 176, 300]):

$$x_{k+1} = x_k - \frac{F(x_k)}{\nabla F(x_k) \cdot d_k}\, d_k \quad (k \geq 0),$$

where

$$\nabla F(x_k) = \left(\frac{\partial F(x_k)}{\partial x_1}, \frac{\partial F(x_k)}{\partial x_2}, \cdots, \frac{\partial F(x_k)}{\partial x_n} \right),$$

is the gradient of F and d_k is a direction at x_k. (DNM) converges quadratically to $x^\star$, if x_0 is close enough to $x^\star$ (see Ref. [300]). However, as already noted in Ref. [300], the computation of the gradient $\nabla F(x_k)$ may be too expensive as it is the case when the number n of unknowns is large. In some applications, the mapping F may not be differentiable, or the gradient is impossible to compute. The (DSM) avoids these obstacles. Note that if $n = 1$, (DSM) reduces to the classical Secant method (see Refs. [100, 353, 356]). We consider two choices of the directions d_k for locating $x^\star$. The first is called the directional near-gradient Secant method (DNGSM) and the second is called the directional maximal-component-modulus Secant method (DMCMSM). The quadratic convergence of (DSM) was established for directions d_k sufficiently close to the gradients $\nabla F(x_k)$ and under standard Newton–Kantorovich-type hypotheses (see Ref. [20]).

We are motivated by the works in Refs. [20, 116, 300], and optimization considerations. By introducing the center-Lipschitz condition and using it, in combination with the Lipschitz condition (along the lines of our works in Refs. [98, 134]), we provide a semi-local convergence analysis with the following advantages over the work in Ref. [20]:

(1) Weaker hypotheses;

(2) Larger convergence domain for (DSM);

(3) Finer error bounds on the distances $\| x_{k+1} - x_k \|$, $\| x_k - x^\star \|$ $(k \geq 0)$;

(4) An at least as precise information on the location of the solution $x^\star$.

We use the Euclidean norms for both vector and matrix. The unit direction d_k is chosen such that

$$d_k \approx \frac{\nabla F(x_k)}{\| \nabla F(x_k) \|}.$$

As in Ref. [20], we need two lemmas to establish semi-local convergence of (DNGSM).

Lemma 4.5.1. *(see [3.2.2] in Ref. [338]) Let $C \subseteq \mathbb{R}^n$ be a convex set and $F :$ $C \longrightarrow \mathbb{R}$ a differentiable mapping. Then, for any $x, y \in C$, there exists a vector $\mu \in (x, y)$, such that:*

$$F(y) - F(x) = \nabla F(\mu)\,(y - x), \tag{4.274}$$

where

$$(x, y) = \{z \,:\, z = x + \theta\, y, \quad 0 < \theta < 1\}, \tag{4.275}$$

represents the open straight line between the points x and y.

Lemma 4.5.2. *(see [3.3.10] in Ref. [338]) Assume that $F : C \longrightarrow \mathbb{R}$ is a twice differentiable mapping.*

    Then, for any $x, y \in C$, there exists a vector $\lambda \in (x, y)$, such that:

$$F(y) - F(x) - \nabla F(x)\,(y - x) = \frac{1}{2}\,(y - x)^T\, F''(\lambda)\,(y - x). \tag{4.276}$$

We need also the definition of Lipschitz and center-Lipschitz continuity.

Definition 4.5.1. Let $F : U_0 \subseteq \mathbb{R}^n \longrightarrow \mathbb{R}$ be a differentiable mapping. If there exists a constant $\overline{M} \geq 0$, such that:

$$\| \nabla F(x) - \nabla F(y) \| \leq \overline{M}\, \| x - y \| \quad \text{for all} \quad x, y \in U_0 \tag{4.277}$$

then, we say ∇F is Lipschitz continuous on U_0 with constant $\overline{M}$.

    Note that in view of (4.277), there exists $M_0 \geq 0$ such that

$$\| \nabla F(x) - \nabla F(x_0) \| \leq M_0\, \| x - x_0 \| \quad \text{for all} \quad x \in U_0. \tag{4.278}$$

Then, we say that ∇F is center-Lipschitz continuous on U_0 with constant M_0.

We can state the main semi-local convergence result for (DNGSM).

Theorem 4.5.1. *Let $\mathcal{D}_0 \subseteq \mathbb{R}^n$ be a convex set, $F : \mathcal{D}_0 \subseteq \mathbb{R}^n \longrightarrow \mathbb{R}$ be a differentiable mapping. Let $x_0 \in \mathcal{D}_0$, satisfy*

$$F(x_0) \neq 0 \quad and \quad \nabla F(x_0) \neq 0, \tag{4.279}$$

$d_0 \in \mathbb{R}^n$ *be a given unit vector,* θ_0 *be a given positive parameter,* h_0 *and* x_1 *be defined as*

$$h_0 = \frac{\theta_0\, F(x_0)}{F(x_0 + \theta_0\, d_0) - F(x_0)}\, d_0, \tag{4.280}$$

$$x_1 = x_0 + h_0. \tag{4.281}$$

We suppose that for $U_0 \subseteq \mathcal{D}_0$, ∇F *is center-Lipschitz on* U_0 *with constant* M_0 *and* $F \in \mathcal{C}^2[U_0]$*. Let*

$$M = \sup_{x \in U_0}\ \parallel F''(x) \parallel . \tag{4.282}$$

Moreover, let the sequence $\{x_k\}$ *be generated by (DNGSM). Further, assume that the unit direction* d_k *and positive parameters* α, β, θ_k, *satisfy for all* $k \geq 0$:

$$\frac{|\nabla F(x_k) \cdot d_k|}{\parallel \nabla F(x_k) \parallel} \geq \frac{|\nabla F(x_0) \cdot d_0|}{\parallel \nabla F(x_0) \parallel}, \tag{4.283}$$

$$\theta_k \leq \alpha \parallel h_k \parallel, \tag{4.284}$$

$$|\nabla F(x_k) \cdot d_k| \leq \beta\, |\nabla F(x) \cdot d_k|, \qquad x \in (x_k, x_k + \theta_k\, d_k) \tag{4.285}$$

and

$$p = \overline{L} \parallel h_0 \parallel \leq \frac{1}{2}, \tag{4.286}$$

where

$$L_0 = M_0 \parallel \nabla F(x_0) \parallel^{-1}, \quad L = \beta\,(1+\alpha)\, M\, |\nabla F(x_0) \cdot d_0|^{-1}, \tag{4.287}$$

$$U_0 = \{x \in \mathbb{R}^n : \parallel x - x_1 \parallel \leq t^\star\} \subseteq \mathcal{D}_0, \quad t_1 = \parallel h_0 \parallel$$

and δ, $\overline{L}$ *are given in Lemma 1.3.1.*

Then, sequence $\{x_k\}$ *generated by (DNGSM) is well defined, remains in* U_0 *for all* $k \geq 0$ *and converges to a solution* $x^\star \in U_0$ *of equation* $F(x) = 0$*. Moreover, the following hold*

$$\begin{aligned} \nabla F(x) &\neq 0 \quad for\ all \quad x \in U_0 \quad and \\ \nabla F(x^\star) &\neq 0 \quad unless \quad \parallel x^* - x_0 \parallel = t^\star. \end{aligned} \tag{4.288}$$

Furthermore, the following estimates hold for all $k \geq 0$:

$$\parallel x_{k+1} - x_k \parallel \leq t_{k+1} - t_k \leq \left(\frac{\delta}{2}\right)^k (2\,p)^{2^k - 1} \parallel h_0 \parallel, \tag{4.289}$$

$$\parallel x_k - x^\star \parallel \leq t^\star - t_k \leq \left(\frac{\delta}{2}\right)^k \frac{(2\,p)^{2^k - 1} \parallel h_0 \parallel}{1 - (2\,p)^{2^k}}, \quad (2\,p < 1), \tag{4.290}$$

where $t^\star$, $\{t_k\}$ *are given in Lemma 1.3.1 (with* K_0, K, $\overline{K}$, v_n, $v^\star$, $v^{\star\star}$, q_A *replaced by* L_0, L, $\overline{L}$ t_n, $t^\star$, $t^{\star\star}$, q*, respectively),* $\eta = \parallel h_0 \parallel$ *and* $q = p$*.*

Proof. We shall show using mathematical induction on $k \geq 0$:

$$\| x_{k+1} - x_k \| \leq t_{k+1} - t_k \tag{4.291}$$

and

$$\overline{U}(x_{k+1}, t^\star - t_{k+1}) \subseteq \overline{U}(x_k, t^\star - t_k). \tag{4.292}$$

For every $z \in \overline{U}(x_1, t^\star - t_1)$

$$\| z - x_0 \| \leq \| z - x_1 \| + \| x_1 - x_0 \| \leq t^\star - t_1 + t_1 - t_0 = t^\star - t_0,$$

shows $z \in \overline{U}(x_0, t^\star - t_0)$. Since, also

$$\| x_1 - x_0 \| = \| h_0 \| \leq \eta = t_1 - t_0,$$

estimates (4.291) and (4.292) hold for $k = 0$. Assume (4.291) and (4.292) hold for all $i \leq k$. Then we have:

$$\| x_{i+1} - x_0 \| \leq \| x_{i+1} - x_i \| + \| x_i - x_{i-1} \| + \cdots + \| x_1 - x_0 \|$$
$$\leq (t_{i+1} - t_i) + (t_i - t_{i-1}) + \cdots + (t_1 - t_0) = t_{i+1}$$

and

$$\| x_i + t\,(x_{i+1} - x_i) - x_0 \| \leq t_i + t\,(t_{i+1} - t_i) \leq t^\star, \quad t \in [0,1].$$

Using conditions (4.278) and (4.287) for $x = x_i$, we get in turn:

$$\| \nabla F(x_i) \| \geq \| \nabla F(x_0) \| - \| \nabla F(x_i) - \nabla F(x_0) \| \geq \| \nabla F(x_0) \| - M_0 \, \| x_i - x_0 \|$$
$$\geq \| \nabla F(x_0) \| - M_0 \, (t_i - t_0)$$
$$\geq \| \nabla F(x_0) \| \, (1 - L_0 \, t_i) > 0 \quad \text{(by Lemma 1.3.1 and (4.279)).}$$
$$\tag{4.293}$$

We need the approximation

$$F(x_i) = \int_{x_{i-1}}^{x_i} (x_i - x)^T \, F''(x) \, dx + r_{i-1}, \tag{4.294}$$

where

$$r_i = F(x_i) + \nabla F(x_i)\, h_i \qquad (i \geq 0). \tag{4.295}$$

Using (4.282), (4.284) and the induction hypotheses, we have that

$$\left| \int_{x_{i-1}}^{x_i} (x_i - x)^T \, F''(x) \, dx \right| = \left| \int_0^1 (1 - t)\,(h_{i-1})^T \, F''(x_{i-1} + t\, h_{i-1})\, h_{i-1} \, dt \right|$$
$$\leq \frac{1}{2}\, M \, \| h_{i-1} \|^2$$
$$\tag{4.296}$$

and

$$|r_{i-1}| = \left| F(x_{i-1}) - \frac{\theta_{i-1}\, F(x_{i-1})\, \nabla F(x_{i-1}) \cdot d_{i-1}}{F(x_{i-1} + \theta_{i-1}\, d_{i-1}) - F(x_{i-1})} \right|$$

$$= \left| \frac{F(x_{i-1})}{2} \, \frac{(\theta_{i-1}\, d_{i-1})^T \, F''(\lambda)\,(\theta_{i-1}\, d_{i-1})}{F(x_{i-1} + \theta_{i-1}\, d_{i-1}) - F(x_{i-1})} \right|, \quad \lambda \in (x_{i-1}, x_{i-1} + \theta_{i-1}\, d_{i-1})$$

$$\leq \frac{1}{2}\, M\, \theta_{i-1} \left| \frac{\theta_{i-1}\, F(x_{i-1})}{F(x_{i-1} + \theta_{i-1}\, d_{i-1}) - F(x_{i-1})} \right| \leq \frac{1}{2}\, M\, \alpha \, \| h_{i-1} \|^2.$$
$$\tag{4.297}$$

In view of (4.285) and (4.293)–(4.297), we get

$$|F(x_i)| \leq \frac{1}{2} \, M \, \parallel h_{i-1} \parallel^2 + \frac{1}{2} \, \alpha \, M \, \parallel h_{i-1} \parallel^2 = \frac{1}{2} \, (1 + \alpha) \, M \, \parallel h_{i-1} \parallel^2, \quad (4.298)$$

so,

$$
\begin{aligned}
\parallel h_i \parallel &\leq \left| \frac{\theta_i \, F(x_i)}{F(x_i + \theta_i \, d_i) - F(x_i)} \right| \\
&\leq \frac{1}{2} \, (1 + \alpha) \, M \, \parallel h_{i-1} \parallel^2 \, \left| \frac{\theta_i}{F(x_i + \theta_i \, d_i) - F(x_i)} \right| \\
&\leq \frac{1}{2} \, (1 + \alpha) \, M \, \parallel h_{i-1} \parallel^2 \, \frac{1}{|\nabla F(\mu_i) \cdot d_i|}, \quad (\mu_i \in (x_i, x_i + \theta_i \, d_i)) \\
&\leq \frac{\beta}{2} \, \frac{(1 + \alpha) \, M \, \parallel h_{i-1} \parallel^2}{|\nabla F(x_i) \cdot d_i|} \leq \frac{\beta}{2} \, \frac{(1 + \alpha) \, M \, \parallel h_{i-1} \parallel^2}{|\nabla F(x_0) \cdot d_0|} \, \frac{\parallel \nabla F(x_0) \parallel}{\parallel \nabla F(x_i) \parallel} \\
&\leq \frac{\beta \, (1 + \alpha) \, M \, \parallel h_{i-1} \parallel^2}{2 \, |\nabla F(x_0) \cdot d_0|} \, \frac{\parallel \nabla F(x_0) \parallel}{\parallel \nabla F(x_0) \parallel \, (1 - L_0 \, t_i)} \\
&\leq \frac{L}{2} \, \frac{\parallel h_{i-1} \parallel^2}{1 - L_0 \, t_i} \leq \frac{L}{2} \, \frac{(t_i - t_{i-1})^2}{1 - L_0 \, t_i} = t_{i+1} - t_i,
\end{aligned}
$$

$$(4.299)$$

which shows (4.291) for all i. Moreover, for every $w \in \overline{U}(x_{i+2}, t^\star - t_{i+2})$, we obtain:

$$
\begin{aligned}
\parallel w - x_{i+1} \parallel &\leq \parallel w - x_{i+2} \parallel + \parallel x_{i+2} - x_{i+1} \parallel \\
&\leq t^\star - t_{i+2} + t_{i+2} - t_{i+1} = t^\star - t_{i+1},
\end{aligned}
$$

showing (4.292) for all $k \geq 0$. Lemma 1.3.1 implies that $\{t_n\}$ is a Cauchy sequence. It then follows from (4.291), (4.292), that $\{x_n\}$ is a Cauchy sequence too and as such it converges to some $x^\star \in U_0$ (since U_0 is a closed set). The vector $x^\star$ is a zero of F, since by (4.298):

$$0 \leq |F(x_i)| \leq \frac{1}{2} \, (1 + \alpha) \, M \, (t_i - t_{i-1})^2 \longrightarrow 0 \quad \text{as} \quad i \to \infty. \quad (4.300)$$

Furthermore, we shall show (4.288). Using (4.278) for $x \in U_0$, (1.162) and the definition of L_0, we get

$$\parallel \nabla F(x) - \nabla F(x_0) \parallel \, \leq M_0 \, \parallel x - x_0 \parallel \, \leq M_0 \, t^\star \, \leq \, \parallel \nabla F(x_0) \parallel .$$

If $\parallel x - x_0 \parallel \, < t^\star$, then we have:

$$\parallel \nabla F(x) - \nabla F(x_0) \parallel \, \leq M_0 \, \parallel x - x_0 \parallel \, < M_0 \, t^\star \, \leq \, \parallel \nabla F(x_0) \parallel,$$

or

$$\parallel \nabla F(x_0) \parallel \, > \, \parallel \nabla F(x) - \nabla F(x_0) \parallel,$$

which shows $\nabla F(x) \neq 0$. Finally, the left-hand side inequality in (4.290) follows from (4.289) by using standard majorization techniques. That completes the proof of Theorem 4.5.1. $\qquad \square$

We state Theorems 3.1 and 3.2 in a condensed form (see p. 294 in Ref. [20]), so we can compare them with our Theorem 4.5.1.

Theorem 4.5.2. *Let $\mathcal{D}_0 \subseteq \mathbb{R}^n$ be a convex set, $F : \mathcal{D}_0 \subseteq \mathbb{R}^n \longrightarrow \mathbb{R}$ be a differentiable mapping. Let $x_0 \in \mathcal{D}_0$, satisfy $F(x_0) \neq 0$ and $\nabla F(x_0) \neq 0$; $d_0 \in \mathbb{R}^n$ be a given unit vector, θ_0 be a given positive parameter, h_0 and x_1 be defined by (4.280) and (4.281) respectively. Let $U_0 \subseteq \mathcal{D}_0$, $F \in \mathcal{C}^2[U_0]$ and $M = \sup_{x \in U_0} \| F''(x) \|$ be such that*

$$\varsigma_0 = \ell_0 \parallel h_0 \parallel \leq \frac{1}{2}, \tag{4.301}$$

where

$$\ell_0 = 2\, M\, |\nabla F(x_0) \cdot d_0|^{-1} \tag{4.302}$$

and

$$U_0 = \{x \in \mathbb{R}^n : \parallel x - x_1 \parallel \leq \parallel h_0 \parallel\}.$$

In addition, let the sequence $\{x_k\}$ be generated by (DNGSM), and assume that the unit direction d_k and positive parameter θ_k, satisfy for all $k \geq 0$:

$$\frac{|\nabla F(x_{k+1}) \cdot d_{k+1}|}{\parallel \nabla F(x_{k+1}) \parallel} \geq \frac{|\nabla F(x_k) \cdot d_k|}{\parallel \nabla F(x_k) \parallel}, \quad \theta_k \leq \frac{1}{2} \parallel h_k \parallel \tag{4.303}$$

and

$$|\nabla F(x_k) \cdot d_k| \leq 2\,|\nabla F(x) \cdot d_k|, \qquad x \in (x_k, x_k + \theta_k\, d_k).$$

Then, the following hold:

(a) $\{x_k\} \subseteq U_0$ and there exists $x^\star \in U_0$, such that $x_k \longrightarrow x^\star$ and $F(x^\star) = 0$;
(b) $\nabla F(x) \neq 0 \qquad (x \in U_0)$;
(c) For $k \geq 0$,

$$\nabla F(x_k) \cdot d_k \neq 0, \quad \parallel h_k \parallel \leq \frac{1}{2} \parallel h_{k-1} \parallel, \tag{4.304}$$

$$\begin{aligned}
\parallel x_{k+1} - x_k \parallel &\leq \frac{3\, M}{2\,|\nabla F(x_k) \cdot d_k|} \parallel x_k - x_{k-1} \parallel^2 \\
&\leq \frac{3\, M \parallel \nabla F(x_0) \parallel}{2\, a\, |\nabla F(x_0) \cdot d_0|} \parallel x_k - x_{k-1} \parallel^2
\end{aligned} \tag{4.305}$$

and

$$\begin{aligned}
\parallel x_{k+1} - x^\star \parallel &\leq \frac{3\, M}{2\,|\nabla F(x_k) \cdot d_k|} \parallel x_k - x_{k-1} \parallel^2 \\
&\leq \frac{3\, M \parallel \nabla F(x_0) \parallel}{2\, a\, |\nabla F(x_0) \cdot d_0|} \parallel x_k - x_{k-1} \parallel^2,
\end{aligned} \tag{4.306}$$

where

$$a = \min_{x \in U_0} \parallel \nabla F(x) \parallel > 0; \tag{4.307}$$

(d) If for

$$\gamma = \frac{\| \nabla F(x_0) \| \; |\nabla F(x_0) \cdot d_0|^{-1}}{a}, \tag{4.308}$$

$$\ell = \frac{3 \gamma M}{4}, \tag{4.309}$$

$$r = \ell \; \| h_0 \| < \frac{1}{2}, \tag{4.310}$$

then,

$$\| x_{k+1} - x_k \| \le \frac{(2 q)^{2^k - 1}}{1 - (2 q)^{2^k - 1}} \; \| h_0 \| \tag{4.311}$$

and

$$\| x_k - x^\star \| \le \frac{(2 q)^{2^k - 1}}{1 - (2 q)^{2^k - 1}} \; \| h_0 \| . \tag{4.312}$$

Note that if $F \in C^2[U_0]$ and (4.277) holds, then $\overline{M}$ can replace M in hypothesis (4.282).

Next, we provide a comparison between Theorems 4.5.1 and 4.5.2.

Remark 4.5.1.

(a) We compare sufficient convergence condition (4.286) to (4.301) for $\alpha = \frac{1}{2}$ and $\beta = 2$. We have that

$$\varsigma_0 \le \frac{1}{2} \Longrightarrow p \le \frac{1}{2} \tag{4.313}$$

but not necessarily vice verca, unless if $M_0 = M$. Hence, condition (4.301) can always be replaced by (4.286). That is, the applicability of (DNGSM) has been extended. Note also that (4.278) is not an additional hypothesis, since in practice the computation of M requires that of constant M_0. According to (4.304), only linear convergence is shown provided that hypothesis (4.301) holds. However, in Theorem 4.5.1, we showed quadratic convergence. Theorem 4.5.1 avoids the computation of a given (4.307), which is expensive in general and also uses (4.283) instead of (4.303) which is more difficult to verify. Finally, note that if $p \in [0, \frac{1}{2})$ and $\beta \in [0, 2)$, then condition (4.286) is even weaker than (4.301).

(b) Condition (4.283) is equivalent to $c_{k+1} \le c_0$, since $\| d_k \| = 1$, where

$$c_k = \angle(\nabla F(x_k), d_k),$$

with $\angle(.,.)$ being the angle between two vectors u, v given by:

$$\angle(u, v) = \arccos \frac{u \cdot v}{\| u \| \cdot \| v \|}, \quad u \ne 0, \quad v \ne 0.$$

(c) Condition (4.284) is equivalent to

$$|F(x_k + \theta_k\, d_k) - F(x_k)| \le \alpha\, |F(x_k)|.$$

(d) If $M\,\theta_k \le b\,|\nabla F(x_k) \cdot d_k|$, $b \in [0,1)$, then, condition (4.285) holds for $\beta = 1 - b$. Indeed, we have

$$|\nabla F(x) \cdot d_k - \nabla F(x_k) \cdot d_k| \le M\ \theta_k,$$

so,

$$
\begin{aligned}
|\nabla F(x) \cdot d_k| &\ge |\nabla F(x_k) \cdot d_k| - |\nabla F(x) \cdot d_k - \nabla F(x_k) \cdot d_k| \\
&\ge |\nabla F(x_k) \cdot d_k| - b\,|\nabla F(x_k) \cdot d_k| = \beta\,|\nabla F(x_k) \cdot d_k|.
\end{aligned}
$$

We use now ∞-norm for both vector and matrix. By simply replacing the Euclidean norm by ∞-norm and $|\nabla F(x_k) \cdot d_k|$ by $\|\nabla F(x_k)\|_\infty$ in the proof of Theorem 4.5.1, we arrive at the following corresponding semi-local convergence result for (DMCMCM):

Theorem 4.5.3. *Let $\mathcal{D}_0 \subseteq \mathbb{R}^n$ be a convex set, $F : \mathcal{D}_0 \subseteq \mathbb{R}^n \longrightarrow \mathbb{R}$ be a differentiable mapping. Let $x_0 \in \mathcal{D}_0$, satisfy $F(x_0) \ne 0$ and $\nabla F(x_0) \ne 0$; θ_0 be a given positive parameter, h_0 and x_1 be defined as*

$$h_0 = \frac{\theta_0\, F(x_0)}{F(x_0 + \theta_0\, d_0) - F(x_0)}\, d_0, \qquad x_1 = x_0 + h_0,$$

where $d_0 = e_{m(0)}$ is the $m(0)$th unit vector in $\mathbb{R}^n$. We suppose that for $U_0 \subseteq \mathcal{D}_0$ and $F \in \mathcal{C}^2[U_0]$, there exist constants M_0 and M, such that

$$\|\nabla F(x) - \nabla F(x_0)\|_\infty \le M_0\ \|x - x_0\|_\infty \quad x \in U_0 \quad and \quad M = \sup_{x \in U_0}\ \|F''(x)\|_\infty.$$

Moreover, let the sequence $\{x_k\}$ be generated by (DMCMSM) and further assume direction d_k and positive parameters α, β, θ_k, satisfy for all $k \ge 0$:

$$\|\nabla F(x_k)\|_\infty \le \beta\ \|\nabla F(x) \cdot e_{m(k)}\|_\infty, \qquad x \in (x_k, x_k + \theta_k\, d_k),$$

$$\theta_k \le \alpha\ \|h_k\|_\infty \quad and \quad p = \overline{L}\ \|h_0\|_\infty \le \frac{1}{2},$$

where

$$L_0 = M_0\ \|\nabla F(x_0)\|_\infty^{-1}, \qquad L = \beta\,(1 + \alpha)\,M\ \|\nabla F(x_0)\|_\infty^{-1},$$

$$U_0 = \{x \in \mathbb{R}^n : \|x - x_1\|_\infty \le t^\star\} \subseteq \mathcal{D}_0,$$

with δ and $\overline{L}$ given in Lemma 1.3.1. Then, sequence $\{x_k\}$ generated by (DMCMSM) is well defined, remains in U_0 for all $k \ge 0$ and converges to a solution $x^\star \in U_0$ of equation $F(x) = 0$. Moreover, the following hold

$$\nabla F(x) \ne 0 \quad for\ all \quad x \in U_0 \quad and$$
$$\nabla F(x^\star) \ne 0 \quad unless \quad \|x^\star - x_0\|_\infty = t^\star.$$

Furthermore, the following estimates hold for all $k \ge 0$:

$$\|x_{k+1} - x_k\|_\infty \le t_{k+1} - t_k \le \left(\frac{\delta}{2}\right)^k (2\,p)^{2^k - 1}\ \|h_0\|_\infty$$

and

$$\|x_k - x^\star\|_\infty \le t^\star - t_k \le \left(\frac{\delta}{2}\right)^k \frac{(2\,p)^{2^k - 1}\ \|h_0\|_\infty}{1 - (2\,p)^{2^k}}, \qquad (2\,p < 1),$$

where $t^\star$, $\{t_k\}$ are given in Lemma 1.3.1 and $\eta = \|h_0\|_\infty$, $q = p$.

As in Remark 4.5.1, Theorem 4.5.3 compares favorably to Theorem 3.3 in page 300 of Ref. [20].

So far we showed that the results in [20] can be considerably improved and extended. However, although our approach can be implemented (see Example 4.5.2) there is a difficulty in general verifying condition (4.286) since to calculate h_k, we must first determine θ_k. In the following Section, we show how to overcome this obstacle in case the verification of (4.286) is very difficult or impossible. As before in this Section, we assume now $F : \mathcal{D} \subseteq \mathbb{R}^n \longrightarrow \mathbb{R}$ be a nonlinear differentiable mapping, $x_0 \in \mathcal{D}$, $d_k \in \mathbb{R}^n$ with $\| d_k \| = 1$ and $\theta_0 \in \mathbb{R}$. We shall show the semi-local convergence of (DNGSM) under the following set $(\mathcal{C})$ conditions:

$(\mathcal{C}_1)$ $|F(x_0)| \leq \eta$, $\| \nabla F(x_0) \| \geq \gamma > 0$ and $|\nabla F(x_0) \cdot d_0| \geq \delta \| \nabla F(x_0) \|$, $\delta \in [0, 1]$;

$(\mathcal{C}_2)$ d_k $(k \geq 0)$ satisfies

$\quad (\mathcal{C}_{21})$ $\varrho |\nabla F(x_0) \cdot d_0| \leq |\nabla F(x) \cdot d_0|$, $\quad x \in (x_0, \, x_0 + \theta_0 \, d_0)$, $\quad \varrho \in (0, 1)$,

$\quad (\mathcal{C}_{22})$ $\angle(d_{k+1}, \nabla F(x_{k+1})) \leq \angle(d_k, \nabla F(x_k))$;

$(\mathcal{C}_3)$ θ_k satisfies $\theta_{k+1} \leq \xi \, \theta_k \, \| x_{k+1} - x_k \|$, $\quad \xi \in (0, 1)$;

$(\mathcal{C}_4)$ Mapping ∇F is Lipschitz with constant M on $\mathcal{D}$;

$(\mathcal{C}_5)$ Mapping F has a dd of order one $[x, y; F]$ at the points $(y, y) \in \mathcal{D}^2$ satisfying

$\quad (\mathcal{C}_{51})$ $[x, y; F] \, (x - y) = F(y) - F(x)$ and

$\quad (\mathcal{C}_{52})$ $\| [x, z; F] - [y, z; F] \| \leq N \, \| x - y \|$ for all $(x, y, z) \in \mathcal{D}^3$.

Remark 4.5.2. Note that $(\mathcal{C}_{22})$ implies condition (4.303).

We need two auxiliary lemmas before we show the semi-local convergence of (DNGSM).

Lemma 4.5.3. *Assume* $F : \mathcal{D} \subseteq \mathbb{R}^n \longrightarrow \mathbb{R}$ *is a nonlinear differentiable mapping and sequence* $\{x_n\}$ *generated by (DNGSM) is well defined and remains in* $\mathcal{D}$. *Then, the following hold*

$$|\nabla F(x_k) \cdot d_k| \geq \delta \, \| \nabla F(x_k) \| \quad (k \geq 0) \tag{4.314}$$

and

$$F(x_{k+1}) = ([x_{k+1}, x_k; F] - [v_k, x_k; F]) \, (x_{k+1} - x_k). \tag{4.315}$$

Proof. We shall use induction on k. Estimate (4.314) holds for $k = 0$ by $(\mathcal{C}_1)$. Assume (4.314) holds for all integers j with $j \leq k$. Then, by $(\mathcal{C}_{22})$ and Remark 4.5.2, we have that

$$\frac{\delta \, \| \nabla F(x_{k+1}) \|}{|\nabla F(x_{k+1}) \cdot d_{k+1}|} \leq \frac{\delta \, \| \nabla F(x_k) \|}{|\nabla F(x_k) \cdot d_k|} \leq 1 \Longrightarrow (4.314).$$

Using (DNGSM) and $(\mathcal{C}_{51})$, we have

$$\begin{aligned}
F(x_{k+1}) &= F(x_{k+1}) - F(x_k) - [v_k, x_k; F] \, (x_{k+1} - x_k) \\
&= ([x_{k+1}, x_k; F] - [v_k, x_k; F]) \, (x_{k+1} - x_k).
\end{aligned}$$

That completes the proof of Lemma 4.5.3. $\qquad\square$

It is convenient for us to introduce the initial conditions:

$$a_0 = \frac{1}{\rho\,\gamma\,\delta}, \quad c_0 = a_0\,\eta, \quad b_0 = \frac{M\,\eta}{\rho\,\gamma^2\,\delta},$$

$$N\,a_0\,\eta = r_0, \quad N\,\theta_0 = t_0 \quad \text{and} \quad s_0 = (r_0 + t_0)\,a_0.$$

Then, we can have in turn that

$$\|\,A_0\,\| \leq \left|\frac{\theta_0}{F(v_0) - F(x_0)}\right| \leq \frac{1}{\rho\,\gamma\,\delta} = a_0 < R,$$

for some $R > 0$ to be determined later,

$$\|\,x_1 - x_0\,\| \leq \|\,A_0\,\|\,|F(x_0)| \leq a_0\,\eta, \quad N\,\|\,x_1 - x_0\,\| \leq N\,a_0\,\eta = r_0,$$

$$N\,\|\,x_0 - v_0\,\| \leq N\,\theta_0 = t_0, \quad N\,\|\,A_0\,\|\,\|\,x_0 - v_0\,\| \leq (r_0 + t_0)\,a_0 = s_0,$$

$$M\,\|\,\nabla F(x_0)\,\|^{-1}\,\|\,x_1 - x_0\,\| \leq \frac{M\,\eta}{\gamma^2\,\rho\,\delta} = b_0,$$

so, we get that

$$N\,\|\,x_1 - v_0\,\| \leq N\,\|\,x_1 - x_0\,\| + N\,\theta_0 \leq r_0 + t_0$$

and

$$\|\,F(x_1)\,\| \leq N\,\|\,x_1 - v_0\,\|\,\|\,x_1 - x_0\,\| \leq (r_0 + t_0)\,\|\,x_1 - x_0\,\|.$$

We consider the auxiliary function $f(x) = 1/(1 - x)$ and sequences $(k \geq 1)$

$$r_k = r_{k-1}\,f(b_{k-1})\,s_{k-1}, \quad c_k = c_{k-1}\,f(b_{k-1})\,s_{k-1}, \quad t_k = t_{k-1}\,c_{k-1}\,\xi,$$

$$a_k = a_{k-1}\,f(b_{k-1}), \quad s_k = (r_k + t_k)\,a_k, \quad b_k = b_{k-1}\,f^2(b_{k-1})\,s_{k-1}.$$

Then, using induction, we show the following recurrence relations.

Lemma 4.5.4. *Let us suppose that x_0, v_0, x_k, $v_k \in \mathcal{D}$ $(k \in \mathbb{N})$. If*

$$f^2(b_0)\,s_0 < 1 \quad and \quad \xi\,f(b_0)\,c_0 < 1.$$

Then, the following relations are satisfied for $(k \geq 0)$:

(I_k) $\dfrac{1}{\rho\,|\nabla F(x_k) \cdot d_k|} \leq a_k$ *and* $\|\,A_k\,\| \leq a_k,$

(II_k) $\|\,x_{k+1} - x_k\,\| \leq f(b_{k-1})\,s_{k-1}\,\|\,x_k - x_{k-1}\,\| \leq c_k,$

(III_k) $N\,\|\,A_k\,\|\,\|\,x_k - v_k\,\| \leq s_k,$

(IV_k) $M\,\|\,\nabla F(x_k)\,\|^{-1}\,\|\,x_{k+1} - x_k\,\| \leq b_k,$

(V_k) $\|\,x_{k+1} - x_0\,\| \leq \xi\,\theta_0\,\dfrac{1 - (f(b_0)\,s_0)^{k+1}}{1 - f(b_0)\,s_0}\,\|\,x_1 - x_0\,\|.$

Proof. We first show conditions (I_k)–(V_k) are satisfied for $k = 1$. We have in turn that

$$\| v_1 - x_0 \| \leq \| x_1 - x_0 \| + \theta_1 \leq (1 + \xi\,\theta_0)\ \| x_1 - x_0 \|,$$

$$\| \nabla F(x_1) \| \geq \| \nabla F(x_0) \| - \| \nabla F(x_1) - \nabla F(x_0) \|$$
$$\geq \| \nabla F(x_0) \| - M\ \| x_1 - x_0 \|$$
$$= \| \nabla F(x_0) \|\ (1 - M\ \| \nabla F(x_0) \|^{-1} \| x_1 - x_0 \|) \geq \| \nabla F(x_0) \|\ (1 - b_0),$$

so,

$$\| \nabla F(x_1) \|^{-1} \leq \| \nabla F(x_0) \|^{-1}\ f(b_0),$$

(I_1)

$$\| A_1 \| \leq \left| \frac{\theta_1}{F(v_1) - F(x_1)} \right| \leq \left| \frac{\theta_1}{\nabla F(\lambda_1)\theta_1 \cdot d_1} \right| \leq \frac{1}{\rho\,|\nabla F(x_1) \cdot d_1|}$$
$$\leq \frac{\| \nabla F(x_0) \|}{\rho\,|\nabla F(x_0) \cdot d_0|\ \| \nabla F(x_1) \|} \leq \frac{f(b_0)}{\rho\,|\nabla F(x_0) \cdot d_0|} \leq a_0\,f(b_0) = a_1,$$

(II_2)

$$\| x_2 - x_1 \| \leq \| A_1 \|\ |F(x_1)| \leq a_0\,f(b_0)\,(r_0 + t_0)\ \| x_1 - x_0 \| = f(b_0)\,s_0\ \| x_1 - x_0 \|,$$

$$\| x_2 - x_0 \| \leq \| x_2 - x_1 \| + \| x_1 - x_0 \|$$
$$\leq f(b_0)\,s_0\ \| x_1 - x_0 \| + \| x_1 - x_0 \| \leq (f(b_0)\,s_0 + 1)\ \| x_1 - x_0 \|$$

$$N\ \| x_2 - x_1 \| \leq r_0\,f(b_0)\,s_0 = r_1,$$

$$N\,\theta_1 \leq N\,\xi\,\theta_0\ \| x_1 - x_0 \| \leq N\,\xi\,\theta_0\ \| x_1 - x_0 \| \leq t_0\,\xi\,a_0\,\eta = t_1,$$

(III_1) $N\ \| A_1 \|\ \| x_1 - v_1 \| \leq a_1\,(N\ \| x_2 - x_1 \| + N\,\theta_1) \leq a_1\,(r_1 + t_1) = s_1,$

(IV_1)

$$M\ \| \nabla F(x_1) \|^{-1} \| x_2 - x_1 \| \leq M\ \| \nabla F(x_0) \|^{-1}\ f^2(b_0)\,s_0\ \| x_1 - x_0 \|$$
$$\leq b_0\,f^2(b_0)\,s_0 = b_1,$$

(V_1)

$$\| x_2 - x_0 \| \leq \| x_2 - x_1 \| + \| x_1 - x_0 \|$$
$$\leq (f(b_0)\,s_0 + 1)\ \| x_1 - x_0 \| \leq \xi\,\theta_0\,\frac{1 - (f(b_0)\,s_0)^2}{1 - f(b_0)\,s_0}\ \| x_1 - x_0 \|.$$

The induction proves the result. That completes the proof of Lemma 4.5.4. $\qquad\square$

We can show the following semi-local convergence result for (DNGSM).

Theorem 4.5.4. *Let* $F : \mathcal{D} \subseteq \mathbb{R}^n \longrightarrow \mathbb{R}$ *be a nonlinear differentiable mapping. Suppose that* $(\mathcal{C})$ *conditions,*

$$f^2(b_0)\,s_0 < 1, \quad \xi\,f(b_0)\,c_0 < 1 \quad \text{and} \quad \overline{U}(x_0, R) \subseteq \mathcal{D},$$

where

$$R = \frac{(1 + \xi\,\theta_0)\,a_0\,\lambda}{1 - f(b_0)\,s_0}.$$

Then, sequence $\{x_k\}$ *generated by (DNGSM) is well defined, remains in* $\overline{U}(x_0, R)$ *for all* $k \geq 0$ *and converges to a solution* $x^\star \in \overline{U}(x_0, R)$ *of equation* $F(x) = 0.$

Proof.   We have in turn that

$$\| v_0 - x_0 \| \le \theta_0 < R \implies v_0 \in \mathcal{D} \quad \text{and} \quad \| x_1 - x_0 \| < R \implies x_1 \in \mathcal{D}.$$

Similarly, v_1, $x_2 \in \mathcal{D}$. Assume v_k, $x_{k+1} \in \mathcal{D}$, $k = 1, \cdots, n$. We shall show using induction that v_{k+1}, $x_{k+2} \in \mathcal{D}$. Using the recurrence relations, we get that

$$\| v_{k+1} - x_k \| \le (1 + \xi\,\theta_k) \| x_{k+1} - x_k \| \le (f(b_0)\,s_0)^k (1 + \xi\,\theta_0) \| x_1 - x_0 \|,$$

$$\| x_{k+1} - x_0 \| \le (f(\xi_0)\,s_0)^k (1 + \xi\,\theta_0) \| x_1 - x_0 \| +$$
$$\frac{1 - (f(b_0)\,s_0)^k}{1 - f(b_0)\,s_0} (1 + \xi\,\theta_0) \| x_1 - x_0 \|$$
$$\le (1 + \xi\,\theta_0) \frac{1 - (f(b_0)\,s_0)^{k+1}}{1 - f(b_0)\,s_0} \| x_1 - x_0 \| < R,$$

$$\| x_{k+2} - x_0 \| \le \| x_{k+2} - x_{k+1} \| + \| x_{k+1} - x_0 \|$$
$$\le f(b_k)\,s_k \| y_k - x_k \| + (1 + \xi\,\theta_0) \frac{1 - (f(b_0)\,s_0)^{k+1}}{1 - f(b_0)\,s_0} \| x_1 - x_0 \|$$
$$\le (f(b_0)\,s_0)^{k+1} \| x_1 - x_0 \| + (1 + \xi\,\theta_0) \frac{1 - (f(b_0)\,s_0)^{k+1}}{1 - f(b_0)\,s_0} \| x_1 - x_0 \|$$
$$\le (1 + \xi\,\theta_0) \frac{1 - (f(b_0)\,s_0)^{k+2}}{1 - f(b_0)\,s_0} \| x_1 - x_0 \| < R.$$

Hence, we deduce v_{k+1}, $v_{k+2} \in \mathcal{D}$. Next, we shall show the convergence of sequence $\{x_k\}$. We have that

$$\| x_{k+1} - x_k \| \le (f(b_0)\,s_0)^k \| x_1 - x_0 \|,$$

$$\| x_{k+j} - x_k \| \le \sum_{i=k}^{i=k+j-1} \| x_{i+1} - x_i \| \le \sum_{i=k}^{i=k+j-1} (f(b_0)\,s_0)^i \| x_1 - x_0 \| \tag{4.316}$$
$$\le (f(b_0)\,s_0)^k \frac{1 - (f(b_0)\,s_0)^k}{1 - f(b_0)\,s_0} \| x_1 - x_0 \| .$$

By hypotheses $f(b_0)^2\,s_0 < 1$, we get $f(b_0)\,s_0 < 1$, which together with (4.316) imply that sequence $\{x_k\}$ is Cauchy in a complete space and as such it converges to some $x^\star \in \overline{U}(x_0, R)$ (since $\overline{U}(x_0, R)$ is a closed set).

We shall show that $x^\star$ is a solution of equation $F(x) = 0$. We have that

$$|F(x_{k+1})| \le (r_k + t_k) \| x_{k+1} - x_k \| \le (r_0 + t_0)\,(f(b_0)\,s_0)^k \| x_1 - x_0 \| \longrightarrow 0$$

as $k \longrightarrow \infty$, since $(r_0 + t_0) \| x_1 - x_0 \|$ is bounded and $f(b_0)\,s_0 < 1$. Hence, we deduce $|F(x^\star)| = 0$. That completes the proof of Theorem 4.5.4. $\square$

We can also provide a second semi-local convergence result. The proofs are omitted as almost identical to the corresponding previous ones. First, we need an auxiliary lemma.

Lemma 4.5.5. *Let $\xi \in (0, 1)$ and $k \ge 0$. If $\theta_k \in [0, \tau_k]$, where*

$$\tau_k = \frac{\sqrt{\| \nabla F(x_k) \|^2 + 2\,M\,\xi\,|F(x_k)|} - \| \nabla F(x_k) \|}{M}.$$

Then, we have $\theta_k \le \xi \| x_{k+1} - x_k \| .$

Proof. We have in turn

$$|F(v_k) - F(x_k)| = \left| \int_{x_k}^{v_k} \nabla F(y)\, dy \right|$$

$$= \left| \int_0^1 \left(\nabla F(x_k + \theta\,\theta_k\, d_k)\,\theta_k\, d_k - \nabla F(x_k)\,\theta_k\, d_k + \nabla F(x_k)\,\theta_k\, d_k \right) d\theta \right|$$

$$\leq \theta_k \left(M \int_0^1 \theta\,\theta_k\, d\theta + \| \nabla F(x_k) \| \right)$$

$$= \theta_k \left(\frac{M}{2}\,\theta_k + \| \nabla F(x_k) \| \right) = \frac{M}{2}\,\theta_k^2 + \| \nabla F(x_k) \|\,\theta_k.$$

Hence,

$$\frac{M}{2}\,\theta_k^2 + \| \nabla F(x_k) \|\,\theta_k \leq \xi\,|F(x_k)|$$

provided that $\theta_k \in [0, \tau_k]$. Therefore, we get in turn that

$$\| x_{k+1} - x_k \| = \frac{\theta_k\,|F(x_k)|}{|F(x_k + \theta_k\, d_k) - F(x_k)|} \geq \frac{\theta_k\,|F(x_k)|}{\xi\,|F(x_k)|} = \frac{\theta_k}{\xi}.$$

That completes the proof of Lemma 4.5.5. $\qquad\qquad\qquad\qquad\qquad\square$

We shall assume that the $(\mathcal{C}^\star)$ conditions are: $(\mathcal{C}_1)$, $(\mathcal{C}_2)$, $(\mathcal{C}_4)$, $(\mathcal{C}_5)$ and

$(\mathcal{C}_3^\star)$ $\theta_0 \in [0, \tau_0]$ and $0 < \theta_k \leq \min\{\tau_k, \xi\, \| x_k - x_{k-1} \|\}$ for $k \geq 1$.

Define scalar sequences

$$s_k = f^2(t_{k-1})\,(1 + \xi)\,s_{k-1}^2, \quad s_0 = \frac{N\,\mu^2\,\eta}{\gamma^2}, \quad \mu = \frac{1}{\rho\,\delta},$$

$$t_k = f^2(t_{k-1})\,(1 + \xi)\,s_{k-1}\,t_{k-1}, \quad t_0 = \frac{M\,\mu\,\eta}{\gamma^2}.$$

The corresponding recurrence relations:

$$\| A_k \| \leq \mu\, \| \nabla F(x_k) \|^{-1} \leq \gamma\, \| \nabla F(x_{k-1}) \|^{-1}\, f(t_{k-1}),$$

$$\| x_{k+1} - x_k \| \leq f(t_{k-1})\,(1 + \xi)\,s_{k-1}\, \| x_k - x_{k-1} \|,$$

$$N\, \| A_k \|\, \| x_{k+1} - x_k \| \leq s_k, \quad M\, \| \nabla F(x_k) \|^{-1}\, \| x_{k+1} - x_k \| \leq t_k,$$

$$\| x_{k+1} - x_0 \| \leq (1 + \xi)\,\frac{1 - (f(t_0)\,(1 + \xi)\,s_0)^k}{1 - f(t_0)\,(1 + \xi)\,s_0}\, \| x_1 - x_0 \|$$

and

$$|F(x_k)| \leq (1 + \xi)\,N\, \| x_k - x_{k-1} \|^2.$$

Then, we have the following semi-local convergence result for (DNGSM).

Theorem 4.5.5. *Let $F : \mathcal{D} \subseteq \mathbb{R}^n \longrightarrow \mathbb{R}$ be a nonlinear differentiable mapping. Suppose that $(\mathcal{C}^\star)$ conditions,*

$$f^2(t_0)\,(1 + \xi)\,s_0 < 1 \quad \text{and} \quad \overline{U}(x_0, R) \subseteq \mathcal{D},$$

where

$$R = \frac{1 + \xi}{1 - f(t_0)\,(1 + \xi)\,s_0}\, \| x_1 - x_0 \|.$$

Then, sequence $\{x_k\}$ generated by (DNGSM) is well defined, remains in $\overline{U}(x_0, R)$ for all $k \geq 0$ and converges to a solution $x^\star \in \overline{U}(x_0, R)$ of equation $F(x) = 0$.

We provide now two numerical examples. In the first one, we show that hypotheses of Theorem 4.5.1 are satisfied, but not the corresponding hypotheses of Theorem 4.5.2.

Example 4.5.1. Let $n = 2$. Here, we use the Euclidean inner product and the corresponding norm for both vector and matrix. Choose:

$$x_0 = (1, 1)^T, \quad \mathcal{D}_0 = \{x \in \mathbb{R}^2 : \| x - x_0 \| \leq 1 - b\},$$

for $b \in [0, 1)$ and define mapping F on $\mathcal{D}_0$ by:

$$F(x) = \frac{\lambda_1^3 + \lambda_2^3}{2} - 2\,b, \quad x = (\lambda_1, \lambda_2)^T. \tag{4.317}$$

Then, the gradient ∇F of mapping F is given by

$$\nabla F(x) = \frac{3}{2}\,(\lambda_1^2, \lambda_2^2)^T. \tag{4.318}$$

Using (4.280), (4.278), (4.282) and (4.287), we obtain for $\alpha = \dfrac{1}{2}$, $\beta = 2$:

$$M_0 = \frac{3\sqrt{2}}{2}\,(3 - b), \quad M = 3\sqrt{2}\,(2 - b), \quad L_0 = 3 - b, \quad L = 3\,(2 - b),$$

$$\| \nabla F(x_0) \| = \frac{3\sqrt{2}}{2}, \quad F(x_0) = 1 - \sqrt{2}\,b \quad \text{and} \quad \eta = \frac{\sqrt{2}}{3}\,(1 - 2\,b).$$

We can choose the directions d_k by

$$d_k = \frac{\nabla F(x_k)}{\| \nabla F(x_k) \|},$$

so that conditions (4.283) and (4.303) are satisfied as equalities. Let $b = .4162$. Then, condition (4.301) is violated, since, we have $\eta = .079007397$, $M = 6.71949432$ and

$$\frac{4\sqrt{2}}{3}\,(2 - b)\,(1 - 2\,b) = .500527666 > .5.$$

Hence, there is no guarantee that (DNGSM) starting at x_0 converges to the solution $x^\star$ of equation $F(x) = 0$. Moreover, we obtain that

$$L_0 = 2.5838, \quad L = 4.7514, \quad \overline{L} = 3.259626413,$$

$$\delta = 1.207332475, \quad t^{\star\star} = .199345613 < 1 - b = .5838, \quad \text{and} \quad q = .257534598 < .5.$$

That is condition (4.286) is satisfied and $U_0 \subseteq \mathcal{D}$. The conclusions of Theorem 4.5.1 apply for solving equation $F(x) = 0$. We found $x^\star = (.940684577, .940684577)$.

Example 4.5.2. Let $n = 20$ and the starting point $x_0 = (0.1, 0.1, \ldots, 0.1)$. Here, we use the Euclidean inner product and the corresponding norm for both vector and matrix. We consider the following three nonlinear problems considered in [20]:

$$F_1(x) = \sum_{i=1}^{20} x_i^2\, e^{1-x_i^2}, \tag{4.319}$$

$$F_2(x) = \sum_{i=1}^{20} (\sin x_i)^2 \tag{4.320}$$

and

$$F_3(x) = \sum_{i=1}^{5} (\sin x_i)^2 + \sum_{i=6}^{20} (\tan x_i)^2. \tag{4.321}$$

Note that $x^\star = 0$ is a locally unique solution for each of $F_i(x) = 0$ ($i = 1, 2, 3$). We apply in this example the convergence results previously obtained for (DSM). We also compare (DSM) with the directional Newton method (DNM). The direction d_k ($k \geq 0$) is chosen such that it is sufficiently close to the gradient $\nabla F(x_k)$ of F in each iteration x_k. In the implementation using Maple 13, we consider two cases for direction d_k. In the first case

$$d_k := \nabla F(x_k)/\|\nabla F(x_k)\| \tag{4.322}$$

and in the second case

$$d_k := p_k/\|p_k\|, \tag{4.323}$$

where

$$p_k := \left(\frac{F(x_k + F(x_k)e_1)}{F(x_k)} - 1, \frac{F(x_k + F(x_k)e_2)}{F(x_k)} - 1, \ldots, \frac{F(x_k + F(x_k)e_{20})}{F(x_k)} - 1 \right).$$

e_k ($k = 1, \cdots, 20$) is the kth unit vector of $\mathbb{R}^{20}$. We give three comparison tables (see Tables 4.6–4.8) for error bounds $\|x_k - x^\star\|$ ($k \geq 0$), where the stopping criterion $|F_i(x_k)| < 10^{-12}$ ($i = 1, 2, 3$) is used.

Remark 4.5.3. The results extend in a Hilbert space setting. Indeed, let F be a differentiable operator defined on a convex subset $\mathcal{D}$ of a Hilbert space $\mathcal{H}$ with values in $\mathbb{R}$. Here $x \cdot y$ denotes the inner product of elements x and y in $\mathcal{H}$. We also use $\|x\| = (x \cdot x)^{1/2}$. Moreover, instead of condition (4.283), assume: $\|d_k\| = 1$ and there exists $\kappa \in [0, 1]$, such that $|\nabla F(x_k) \cdot d_k| \geq \kappa \|\nabla F(x_k)\|$. Note that in the case of (4.283), we can set

$$\kappa = \frac{|\nabla F(x_k) \cdot d_k|}{\|\nabla F(x_0)\|} \leq 1.$$

Define

$$L_0 = \frac{M_0}{\kappa\, \|\nabla F(x_0)\|}, \quad L = \frac{\beta\,(1+\alpha)\,M}{\kappa\, \|\nabla F(x_0)\|}, \quad \eta = \|h_0\|$$

in the case of Theorem 4.5.1 and

$$L_0 = \frac{M_0}{\kappa\, \|\nabla F(x_0)\|_\infty}, \quad L = \frac{\beta\,(1+\alpha)\,M}{\kappa\, \|\nabla F(x_0)\|_\infty}, \quad \eta = \|h_0\|_\infty$$

in the case of Theorem 4.5.3. Then, due to the proofs of Theorems 4.5.1 and 4.5.3, the results hold in this more general setting.

Table 4.6   Comparison Table for F_1

k	(DNM) with (4.322)	(DSM) with (4.322)	(DSM) with (4.323)
1	.2213481425	.2334571896	.2330073138
2	.1104022818	.1150173101	.1215974906
3	.05516747919	.05809352097	.06486112837
4	.02757954151	.02908230293	.03635275258
5	.01378924624	.01455220678	.01915900153
6	.006894557562	.007272642285	.009630928543
7	.003447270599	.003636811795	.004812897796
8	.001723634277	.001818963011	.002406148658
9	.0008618170085	.0009092235709	.001203624956
10	.0004309084889	.0004545980359	.0006016686952
11	.0002154542425	.0002273082528	.0003008770080
12	.0001077271212	.0001136578186	.0001504472661
13	.00005386356060	.00005683020130	.00007522944185
14	.00002693178026	.00002841754599	.00003761831663
15	.00001346589015	.00001421126889	.00001881121506
16	.000006732945077	.000007108122620	.000009406895828
17	.000003366472543	.000003556547452	.000004704237278
18	$\sim$	$\sim$	.000002352609769

Table 4.7   Comparison Table for F_2

k	(DNM) with (4.322)	(DSM) with (4.322)	(DSM) with (4.323)
1	.2228584479	.2185022023	.2291197948
2	.1113368951	.1093252604	.1182527578
3	.05565694383	.05469418320	.06314434968
4	.02782703509	.02737622640	.03280017538
5	.01391333797	.01368862309	.01665214627
6	.006956646548	.006841851707	.008367425850
7	.003478320457	.003419925369	.004195078981
8	.001739159878	.001709963795	.002099295662
9	.0008695798963	.0008549568094	.001049898919
10	.0004347899430	.0004274834619	.0005250947457
11	.0002173949714	.0002137417329	.0002625837980
12	.0001086974858	.0001068733876	.0001313046567
13	.00005434874294	.00005343921465	.00006565673218
14	.00002717437149	.00002672207772	.00003283067840
15	.00001358718574	.00001336351914	.00001641606330
16	.000006793592895	.000006684239494	.000008208345478
17	.000003396796451	.000003344597729	.000004104300066

4.6   A Unified Convergence Analysis

A unified local and semi-local convergence analysis for secant-type methods is presented in this section. Our analysis includes the computation of the bounds on the limit points of the majorizing sequences involved. We study the convergence of the secant-type method

$$x_{n+1} = x_n - \mathcal{A}_n^{-1}F(x_n), \; \mathcal{A}_n = \delta F(x_n, y_n) \; \text{ for each } n = 1, 2, \cdots, \qquad (4.324)$$

where x_{-1}, x_0 are initial points, $y_n = \theta_n x_n + (1-\theta_n)x_{n-1}$, $\theta_n \in \mathbb{R}$ and $\mathcal{A}_n \in \mathcal{L}(\mathcal{X}, \mathcal{Y})$.

Table 4.8 Comparison Table for F_3

k	(DNM) with (4.322)	(DSM) with (4.322)	(DSM) with (4.323)
1	.2245686780	.2285067687	.2308497146
2	.1123990871	.1139792812	.1181539085
3	.05621369955	.05721706204	.06282663095
4	.02810861328	.02861766738	.03266887733
5	.01405452689	.01431805650	.01661086509
6	.007027290967	.007160297337	.008356807613
7	.003513648921	.003579879135	.004185849468
8	.001756824890	.001790209649	.002093848982
9	.0008784124981	.0008951552522	.001047627095
10	.0004392062563	.0004476150645	.0005239693784
11	.0002196031293	.0002238285767	.0002619979495
12	.0001098015645	.0001119258299	.0001310142103
13	.00005490078249	.00005597005369	.00006551199164
14	.00002745039116	.00002798994481	.00003275805190
15	.00001372519562	.00001399867952	.00001637986395
16	.000006862597819	.000007002457144	.000008190267678
17	.000003431298914	.000003504042798	.000004095270726

We shall first study some scalar sequences which are related to the secant-type method. Let there be parameters $c \geq 0$, $\nu \geq 0$, $\lambda \geq 0$, $\mu \geq 1$, $l_0 > 0$ and $l > 0$ with $l_0 \leq l$. Define the scalar sequence $\{\alpha_n\}$ by

$$
\begin{cases}
\alpha_{-1} = 0, \ \alpha_0 = c, \alpha_1 = c + \nu \\[2ex]
\alpha_{n+2} = \alpha_{n+1} + \dfrac{l\left(\alpha_{n+1} - \alpha_n + \lambda(\alpha_n - \alpha_{n-1})\right)\left(\alpha_{n+1} - \alpha_n\right)}{1 - l_0\left[\mu(\alpha_{n+1} - c) + \lambda(\alpha_n - c) + c\right]} \quad \text{for each } n = 0, 1, 2, \cdots
\end{cases}
\tag{4.325}
$$

Special cases of the sequence $\{\alpha_n\}$ have been used as majorizing sequences for Secant-type method by several authors. For example:

Case 1 (Secant method) $l_0 = l$, $\lambda = 1$ and $\mu = 1$ has been studied in Refs. [110, 120, 360] and for $l_0 \leq l$ in Refs. [106, 110, 120, 163].

Case 2 (Newton's method) $l_0 = l$, $\lambda = 0$, $c = 0$ and $\mu = 2$ has been studied in Refs. [110, 290] and for $l_0 \leq l$ in Refs. [106, 110, 290].

In this section we shall study the convergence of sequence $\{\alpha_n\}$ by first modifying it. Let

$$
L_0 = \frac{l_0}{1 + (\mu + \lambda - 1)l_0 c} \quad \text{and} \quad L = \frac{l}{1 + (\mu + \lambda - 1)l_0 c}.
\tag{4.326}
$$

Using (4.325) and (4.326), sequence $\{\alpha_n\}$ can be written as

$$
\begin{cases}
\alpha_{-1} = 0, \ \alpha_0 = c, \ \alpha_1 = c + \nu \\[2ex]
\alpha_{n+2} = \alpha_{n+1} + \dfrac{L\left(\alpha_{n+1} - \alpha_n + \lambda(\alpha_n - \alpha_{n-1})\right)\left(\alpha_{n+1} - \alpha_n\right)}{1 - L_0(\mu\alpha_{n+1} + \lambda\alpha_n)} \quad \text{for each } n = 0, 1, 2, \cdots
\end{cases}
\tag{4.327}
$$

Moreover, let

$$
L = b\,L_0 \quad \text{for} \quad \text{some} \quad b \geq 1
\tag{4.328}
$$

and

$$\beta_n = L_0\,\alpha_n. \tag{4.329}$$

Then, we can define sequence $\{\beta_n\}$ by

$$\begin{cases} \beta_{-1} = 0, \quad \beta_0 = L_0 c, \quad \beta_1 = L_0(c+\nu) \\[2mm] \beta_{n+2} = \beta_{n+1} + \dfrac{b\left(\beta_{n+1} - \beta_n + \lambda(\beta_n - \beta_{n-1})\right)\left(\beta_{n+1} - \beta_n\right)}{1 - (\mu\beta_{n+1} + \lambda\beta_n)} \quad \text{for each } n = 0,1,2,\cdots \end{cases} \tag{4.330}$$

Furthermore, let

$$\gamma_n = \frac{1}{\nu + \lambda} - \beta_n \quad \text{for each } n = 0,1,2,\cdots. \tag{4.331}$$

Then, sequence $\{\gamma_n\}$ is defined by

$$\begin{cases} \gamma_{-1} = \dfrac{1}{\mu + \lambda}, \gamma_0 = \dfrac{1}{\mu + \lambda} - L_0 c, \ \gamma_1 = \dfrac{1}{\mu + \lambda} - L_0(c+\nu) \\[2mm] \gamma_{n+2} = \gamma_{n+1} - \dfrac{b\left(\gamma_{n+1} - \gamma_n + \lambda(\gamma_n - \gamma_{n-1})\right)\left(\gamma_{n+1} - \gamma_n\right)}{\mu\gamma_{n+1} + \lambda\gamma_n} \quad \text{for each } n = 0,1,2,\cdots. \end{cases} \tag{4.332}$$

Finally, let

$$\delta_n = 1 - \frac{\gamma_n}{\gamma_{n-1}} \quad \text{for each } n = 0,1,2,\cdots. \tag{4.333}$$

Then, we define the sequence $\{\delta_n\}$ by

$$\begin{cases} \delta_0 = 1 - \dfrac{\gamma_0}{\gamma_{-1}}, \delta_1 = 1 - \dfrac{\gamma_1}{\gamma_0} \\[2mm] \delta_{n+2} = \dfrac{b\delta_{n+1}\left(\lambda\delta_n + (1-\delta_n)\delta_{n+1}\right)}{(1-\delta_n)(1-\delta_{n+1})\left(\mu(1-\delta_{n+1}) + \lambda\right)} \quad \text{for each } n = 0,1,2,\cdots. \end{cases} \tag{4.334}$$

It is convenient for us to define polynomial p by

$$p(t) = \mu t^3 - (\lambda + 3\mu + b)t^2 + (2\lambda + 3\mu + b(\lambda + 1))t - (\mu + \lambda). \tag{4.335}$$

We have that $p(0) = -(\mu + \lambda) < 0$ and $p(1) = b\lambda > 0$. It follows from the intermediate value theorem that p has roots in $(0,1)$. Denote the smallest root by δ. Note that in particular for Newton's method and secant method, respectively, we have that

$$p(t) = (t-1)(2t^2 - (b+4)t + 2)$$

and

$$p(t) = (t-2)(t^2 - (b+2)t + 1).$$

Hence, we obtain, respectively that

$$\delta = \frac{2}{b + 4 + \sqrt{b^2 + 8b}} \quad \text{and} \quad \delta = \frac{2}{b + 2 + \sqrt{b^2 + 4b}}. \tag{4.336}$$

Notice also that
$$p(t) \leq 0 \text{ for each } t \in (-\infty, \delta]. \tag{4.337}$$
We study now the convergence of these sequences starting from $\{\delta_n\}$.

Lemma 4.6.1. *Let $\delta_1 > 0$, $\delta_2 > 0$ and $b \geq 1$ be given parameters. Suppose that*
$$0 < \delta_2 \leq \delta_1 \leq \delta, \tag{4.338}$$
where δ was defined in (4.335). Let $\{\delta_n\}$ be the scalar sequence defined by (4.334). Then, the following assertions hold:

$(\mathcal{A}_1)$ *If*
$$\delta_1 = \delta_2 \tag{4.339}$$
    then,
$$\delta_n = \delta \text{ for each } n = 1, 2, 3, \cdots \tag{4.340}$$

$(\mathcal{A}_2)$ *If*
$$\delta_2 < \delta_1 < \delta \tag{4.341}$$
    then, sequence $\{\delta_n\}$ is decreasing and converges to 0.

Proof. It follows from (4.334) and $\delta_2 \leq \delta_1$ that $\delta_3 > 0$. We shall show that
$$\delta_3 \leq \delta_2. \tag{4.342}$$
In view of (4.334) for $n = 1$, it suffices to show that
$$p_1(\delta_2) = \mu(1 - \delta_1)\delta_2^2 - (1 - \delta_1)(2\mu + \lambda + b)\delta_2 - (\mu + (1 + b)\lambda)\delta_1 + \mu + \lambda \geq 0. \tag{4.343}$$
The discriminant Δ of the quadratic polynomial p_1 is given by
$$\Delta = (1 - \delta_1)\left[(1 - \delta_1)(2\mu + \lambda + b)^2 + 4\mu\lambda b\right] > 0. \tag{4.344}$$
Hence, p_1 has two distinct roots δ_s and δ_l with $\delta_s < \delta_l$. Then, (4.343) shall be true, if $\delta_2 \leq \delta_s$, or since $\delta_2 \leq \delta_1$, if $\delta_1 \leq \delta_s$ or if $p(\delta_1) \leq 0$ which is true by (4.337) and $\delta_1 \leq \delta$. Hence, we showed (4.342). Therefore, relation
$$0 < \delta_{k+1} < \delta_k, \tag{4.345}$$
holds for $k = 2$. Then, we must show that
$$0 < \delta_{k+2} < \delta_{k+1}. \tag{4.346}$$
It follow from (4.334), $\delta_k < 1$ and $\delta_{k+1} < 1$ that $\delta_{k+2} > 0$. Then, in view of (4.334) the right-hand side of (4.346) is true, if
$$\frac{b\delta_{k+1}\left[\lambda\delta_k + (1 - \delta_k)\delta_{k+1}\right]}{(1 - \delta_k)(1 - \delta_{k+1})\left[\lambda + \mu(1 - \delta_{k+1})\right]} \leq \delta_{k+1} \quad \text{or} \quad p(\delta_k) \leq 0,$$
which is true by (4.337) since $\delta_k \leq \delta_1 \leq \delta$. The induction for (4.345) is complete. If $\delta_1 = \delta_2 = \delta$, then it follows from (4.334) for $n = 1$ that $\delta_3 = \delta$ and $\delta_n = \delta$ for $n = 4, 5, \cdots$, which shows (4.340). If $\delta_2 < \delta_1$, the sequence $\{\delta_n\}$ is decreasing, bounded below by 0 and as such it converges to its unique largest lower bound denoted by γ. We then have from (4.334) that
$$\gamma = \frac{b\gamma\left[\lambda\gamma + (1 - \gamma)\gamma\right]}{(1 - \gamma)^2\left[\lambda + \mu(1 - \gamma)\right]} \Rightarrow \gamma = \delta \text{ or } \gamma = 0. \tag{4.347}$$
But $\gamma \leq \delta_1 \leq \delta$. Hence, $\gamma = 0$. The proof of the Lemma is complete. $\qquad\square$

Lemma 4.6.2. *Under the hypothesis* (4.341), *further suppose*

$$l_0 c < 1. \tag{4.348}$$

Then, the sequence $\{\gamma_n\}$ *is decreasingly convergent and sequences* $\{\alpha_n\}$ *and* $\{\beta_n\}$ *are increasingly convergent.*

Proof. Using (4.326), (4.333) and (4.348) we get in turn that

$$\gamma_n = (1 - \delta_n)\gamma_{n-1} = \cdots = (1 - \delta_n)\cdots(1 - \delta_1)\gamma_0$$
$$= (1 - \delta_n)\cdots(1 - \delta_1)\left(\frac{1}{\mu + \lambda} - L_0 c\right) > 0$$

and

$$\gamma_n < \gamma_{n-1},$$

since $\delta_n < 1$. Hence, sequence $\{\gamma_n\}$ converges to its unique largest lower bound denoted by γ^*. We also have that $\beta_n = \frac{1}{\mu + \lambda} - \gamma_n < \frac{1}{\mu + \lambda}$. Thus, the sequence $\{\beta_n\}$ is increasing, bounded from above by $\dfrac{1}{\mu + \lambda}$ and as such it converges to its unique least upper bound denoted by β^*. Then, in view of (4.329) sequence $\{\alpha_n\}$ is also increasing, bounded from above by $\dfrac{L_0^{-1}}{\mu + \lambda}$ and such that it also converges to its unique least upper bound denoted by α^*. The proof is complete. $\qquad\square$

Lemma 4.6.3. *Suppose that* (4.338), (4.339) *and* (4.348) *hold. Then, the following assertions hold for each* $n = 1, 2, \cdots$

$$\delta_n = \delta, \quad \gamma_n = (1 - \delta)^n \gamma_0, \quad \gamma^* = \lim_{n \to \infty} \gamma_n = 0,$$

$$\beta_n = \frac{1}{\mu + \lambda} - (1 - \delta)^n \gamma_0, \quad \beta^* = \lim_{n \to \infty} \beta_n = \frac{1}{\mu + \lambda}$$

and

$$\alpha_n = \frac{1}{L_0}\left[\frac{1}{\mu + \lambda} - (1 - \delta)^n \gamma_0\right], \quad \alpha^* = \lim_{n \to \infty} \alpha_n = \frac{1}{L_0(\mu + \lambda)}.$$

Corollary 4.6.1. *Suppose that the hypotheses of Lemmas 4.6.1 and 4.6.2 hold. Then, sequence* $\{\alpha_n\}$ *defined in* (4.325) *is nondecreasing and converges to*

$$\alpha^* = \beta^* \frac{1 + (\mu + \lambda - 1)l_0 c}{l_0}.$$

We present lower and upper bounds on the limit point α^*.

Lemma 4.6.4. *Suppose that the conditions* (4.341) *and* (4.348) *hold. Then, the following assertion holds*

$$b_1^1 \le \alpha^* \le b_2^1, \tag{4.349}$$

where

$$b_1^1 = \frac{1 + (\mu + \lambda - 1)l_0 c}{l_0}\left[\frac{1}{\mu + \lambda} - \exp\left(-2\left(\frac{\delta_1}{2 - \delta_1} + \frac{\delta_2}{2 - \delta_2}\right)\right)\right],$$

$$b_2^1 = \frac{1 + (\mu + \lambda - 1)l_0 c}{l_0}\left[\frac{1}{\mu + \lambda} - \exp(\delta^*)\right], \qquad (4.350)$$

$$\delta^* = -\left[\frac{1}{1 - \delta_1}\left(\delta_1 + \frac{\delta_2}{1 - r}\right) + \ln\left(\frac{(\mu + \lambda)(1 - (\mu + \lambda - 1)l_0 c)}{1 - l_0 c}\right)\right]$$

and

$$r = b\frac{\lambda\delta_1 + \delta_2(1 - \delta_1)}{(1 - \delta_1)(1 - \delta_2)(\lambda + \mu(1 - \delta_2))}.$$

Proof. Using (4.341) and (4.348) we have that $0 < \delta_3 < \delta_2 < \delta_1$. Let us assume that $0 < \delta_{k+1} < \delta_k < \cdots < \delta_1$. Then, it follows from the induction hypotheses and (4.350) that

$$\delta_{k+2} = \delta_{k+1}b\frac{\delta_k + \delta_{k+1}(1 - \delta_k)}{(1 - \delta_k)(1 - \delta_{k+1})(2 - \delta_{k+1})} < r\delta_{k+1} < r^2\delta_k \le \cdots \le r^{k-1}\delta_3 \le r^k\delta_2.$$

We have that $\gamma^* = \lim\limits_{n\to\infty}\gamma_n = \prod\limits_{i=1}^{\infty}(1 - \delta_n)\gamma_0$. This is equivalent to

$$\ln\left(\frac{1}{\gamma^*}\right) = \sum_{n=1}^{\infty}\ln\left(\frac{1}{1 - \delta_n}\right) + \ln\left(\frac{(\mu + \lambda)(1 + (\mu + \lambda - 1)l_0 c)}{1 - l_0 c}\right),$$

recalling that $\gamma_0 = (1 - l_0 c)/(2(1 + l_0 c))$. We use the following bounds for $\ln t$, $t > 1$:

$$2\left(\frac{t - 1}{t + 1}\right) \le \ln t \le \frac{t^2 - 1}{2t}.$$

First, we shall find an upper bound for $\ln(1/\gamma^*)$. We have that

$$\ln(1/\gamma^*) \le \sum_{n=1}^{\infty}\frac{\delta_n(2 - \delta_n)}{2(1 - \delta_n)} + \ln\left(\frac{(\mu + \lambda)(1 + (\mu + \lambda - 1)l_0 c)}{1 - l_0 c}\right)$$

$$\le \frac{1}{1 - \delta_1}\sum_{n=1}^{\infty}\delta_n + \ln\left(\frac{(\mu + \lambda)(1 + (\mu + \lambda - 1)l_0 c)}{1 - l_0 c}\right)$$

$$\le \frac{1}{1 - \delta_1}(\delta_1 + \delta_2 + \delta_3 + \cdots) + \ln\left(\frac{(\mu + \lambda)(1 + (\mu + \lambda - 1)l_0 c)}{1 - l_0 c}\right)$$

$$\le \frac{1}{1 - \delta_1}(\delta_1 + \delta_2 + r\delta_2 + \cdots + r^n\delta_2 + \cdots) + \ln\left(\frac{(\mu + \lambda)(1 + (\mu + \lambda - 1)l_0 c)}{1 - l_0 c}\right)$$

$$\le \frac{1}{1 - \delta_1}(\delta_1 + \delta_2(r + r^2 + \cdots + r^n + \cdots)) + \ln\left(\frac{(\mu + \lambda)(1 + (\mu + \lambda - 1)l_0 c)}{1 - l_0 c}\right)$$

$$\le \frac{1}{1 - \delta_1}\left(\delta_1 + \frac{\delta_2}{1 - r}\right) + \ln\left(\frac{(\mu + \lambda)(1 + (\mu + \lambda - 1)l_0 c)}{1 - l_0 c}\right) = -\delta^*.$$

As $\beta^* = 1/(\mu + \lambda) - \gamma^*$ and $\alpha^* = L_0^{-1}\beta^*$, we obtain the upper bound in (4.349). Moreover, in order to obtain the lower bound for $\ln(1/\gamma^*)$, we have that

$$\ln(1/\gamma^*) \ge 2\sum_{n=1}^{\infty}\frac{\delta_n}{2 - \delta_n} > 2\left(\frac{\delta_1}{2 - \delta_1} + \frac{\delta_2}{2 - \delta_2}\right),$$

which implies the lower bound in (4.349). The proof of the Lemma is complete. $\square$

From now on we shall denote by (C^1) the hypothesis of Lemmas 4.6.1 and 4.6.2.

Remark 4.6.1.

(a) Let us introduce the notation

$$c^N = \alpha_{N-1} - \alpha_{N-2}, \ \nu^N = \alpha_N - \alpha_{N-1}$$

for some integer $N \geq 1$. Notice that $c^1 = \alpha_0 - \alpha_{-1} = c$ and $\nu^1 = \alpha_1 - \alpha_0 = \nu$. The results in the preceding Lemmas can be weakened even further as follows. Consider the convergence criteria (C_*^N) for $N > 1$: (C^1) with c, ν replaced by c^N, ν^N, respectively

$$\alpha_{-1} < \alpha_0 < \alpha_1 < \cdots < \alpha_N < \alpha_{N+1},$$

$$l_0 \left[\mu(\alpha_{N+1} - c^N) + \lambda(\alpha_N - c^N) + c^N \right] < 1.$$

Then, the preceding results hold with $c, \nu, \delta_1, \delta_2, b_1^1, b_2^1$ replaced, respectively by $c^N, \nu^N, \delta_N, \delta_{N+1}, b_1^N, b_2^N$.

(b) Notice that if

$$l_0 \left[\mu(\alpha_{n+1} - c) + \lambda(\alpha_n - c) + c \right] < 1 \text{ holds for each } n = 0, 1, 2, \cdots, \quad (4.351)$$

then, it follows from (4.325) that sequence $\{\alpha_n\}$ is increasing, bounded from above by $\frac{1+(\mu+\lambda-1)l_0 c}{l_0(\mu+\lambda)}$ and as such it converges to its unique least upper bound α^*. Criterion (4.351) is the weakest of all the preceding convergence criteria for sequence $\{\alpha_n\}$. Clearly all the preceding criteria imply (4.351). Finally, define the criteria for $N \geq 1$

$$(I^N) = \begin{cases} (C_*^N) \\ (4.351) \text{ if criteria } (C_*^N) \text{ fail.} \end{cases} \quad (4.352)$$

We present now the semi-local convergence of the secant-type method using $\{\alpha_n\}$ (defined in (4.325)) as a majorizing sequence. Then, we present the local convergence of the secant-type method in an analogous way. We shall study the secant method for triplets $(\mathcal{F}, x_{-1}, x_0)$ belonging to the class $\mathcal{K} = \mathcal{K}(l_0, l, \nu, c, \lambda, \mu)$ defined as follows.

Definition 4.6.1. Let $l_0, l, \nu, c, \lambda, \mu$ be constants satisfying the hypotheses (I^N) for some fixed integer $N \geq 1$. A triplet $(\mathcal{F}, x_{-1}, x_0)$ belongs to the class $\mathcal{K} = \mathcal{K}(l_0, l, \nu, c, \lambda, \mu)$ if:

$(\mathcal{D}_1)$ $\mathcal{F}$ is a nonlinear operator defined on a convex subset D of a Banach space $\mathcal{X}$ with values in a Banach space $\mathcal{Y}$.

$(\mathcal{D}_2)$ x_{-1} and x_0 are two points belonging to the interior D^0 of D and satisfying the inequality $\|x_0 - x_{-1}\| \leq c$.

$(\mathcal{D}_3)$ There exists a sequence $\{\theta_n\}$ of real numbers and λ, μ such that $|1 - \theta_n| \leq \lambda$ and $1 + |\theta_n| \leq \mu$ for each $n = 0, 1, 2, \cdots$.

$(\mathcal{D}_4)$ $\mathcal{F}$ is Fréchet-differentiable on D^0 and there exists an operator $\delta\mathcal{F} : D^0 \times D^0 \to \mathcal{L}(X, Y)$ such that $\mathcal{A}^{-1} = \delta\mathcal{F}(x_0, y_0)^{-1} \in \mathcal{L}(Y, X)$, $\|\mathcal{A}^{-1}\mathcal{F}(x_0)\| \leq \nu$; for all $x, y, z \in D$, the following hold

$$\|\mathcal{A}^{-1}(\delta\mathcal{F}(x, y) - \mathcal{F}'(z))\| \leq l(\|x - z\| + \|y - z\|)$$

and

$$\|\mathcal{A}^{-1}(\delta\mathcal{F}(x, y) - \mathcal{F}'(x_0))\| \leq l_0(\|x - x_0\| + \|y - x_0\|),$$

where $y_0 = \theta_0 x_0 + (1 - \theta_0)x_{-1}$.

$(\mathcal{D}_5)$ $\overline{U}(x_0, \alpha_0^*) \subseteq D_c = \{x \in D : \mathcal{F} \text{ is continuous at } x\} \subseteq D$, where $\alpha_0^* = (\mu + \lambda - 1)(\alpha^* - c)$ and α^* is given in Lemma 4.6.1.

The semi-local convergence result for the secant method is as follows.

Theorem 4.6.1. *If* $(\mathcal{F}, x_{-1}, x_0) \in \mathcal{K}(l_0, l, \nu, c, \lambda, \mu)$ *then, the sequence* $\{x_n\}$ $(n \geq -1)$ *generated by the secant-type method is well defined, remains in* $\overline{U}(x_0, \alpha_0^*)$ *for each* $n = 0, 1, 2, \cdots$ *and converges to a unique solution* $x^* \in \overline{U}(x_0, \alpha^* - c)$ *of* (1.1). *Moreover, the following assertions hold for each* $n = 0, 1, 2, \cdots$

$$\|x_n - x_{n-1}\| \leq \alpha_n - \alpha_{n-1} \tag{4.353}$$

and

$$\|x^* - x_n\| \leq \alpha^* - \alpha_n, \tag{4.354}$$

where sequence $\{\alpha_n\}$ $(n \geq 0)$ *is given in* (4.325). *Furthermore, if there exists* R *such that*

$$\overline{U}(x_0, R) \subseteq D, \ R \geq \alpha^* - c \ \text{and} \ l_0(\alpha^* - c + R) + \|\mathcal{A}^{-1}(\mathcal{F}^{-1}(x_0) - \mathcal{A})\| < 1, \tag{4.355}$$

then, the solution x^* *is unique in* $\overline{U}(x_0, R)$.

Proof. First, we show that $\mathcal{M} = \delta\mathcal{F}(x_{k+1}, y_{k+1})$ is invertible for $x_{k+1}, y_{k+1} \in \overline{U}(x_0, \alpha_0^*)$. By $(\mathcal{D}_2), (\mathcal{D}_3)$ and $(\mathcal{D}_4)$, we have that .

$$\|y_{k+1} - x_0\| \leq \|\theta_k(x_{k+1} - x_0) + (1 - \theta_{k+1})(x_k - x_0)\|$$

$$\leq |\theta_{k+1}|\|x_{k+1} - x_0\| + |1 - \theta_{k+1}|\|x_k - x_0\| \leq (\mu - 1)(\alpha^* - c) + \lambda(\alpha^* - c) = \alpha_0^*$$

and

$$\begin{aligned}
\|I - \mathcal{A}^{-1}\mathcal{M}\| &= \|\mathcal{A}^{-1}(\mathcal{M} - \mathcal{A})\| \\
&\leq \|\mathcal{A}^{-1}(\mathcal{M} - \mathcal{F}'(x_0))\| + \|\mathcal{A}^{-1}(\mathcal{F}'(x_0) - \mathcal{A})\| \\
&\leq l_0(\|x_{k+1} - x_0\| + \|y_{k+1} - x_0\| + \|x_0 - x_{-1}\|) \\
&\leq l_0(\|x_{k+1} - x_0\| + |\theta_{k+1}|\|x_{k+1} - x_0\| + |1 + \theta_{k+1}|\|x_{k+1} - x_0\| + c) \\
&\leq l_0(\mu(\alpha_{k+1} - c) + \lambda(\alpha_{k+1} - c) + c) < 1.
\end{aligned}$$

$$\tag{4.356}$$

Using the Banach Lemma on invertible operators and (4.356), we deduce that $\mathcal{M}$ is invertible and

$$\|\mathcal{M}^{-1}\mathcal{A}\| \leq (1 - l_0(\mu(\alpha_{k+1} - c) + \lambda(\alpha_{k+1} - c) + c))^{-1}. \tag{4.357}$$

By $(\mathcal{D}_4)$, we have

$$\|\mathcal{A}^{-1}(\mathcal{F}'(u) - \mathcal{F}'(v))\| \le 2l\|u - v\|, \ u, v \in D^0. \tag{4.358}$$

We can write the identity

$$\mathcal{F}'(x) - \mathcal{F}'(y) = \int_0^1 \mathcal{F}'(y + t(x - y))dt(x - y). \tag{4.359}$$

Then, for all $x, y, u, v \in D^0$, we obtain

$$\|\mathcal{A}^{-1}(\mathcal{F}(x) - \mathcal{F}(y) - \mathcal{F}'(u)(x - y))\| \le l(\|x - u\| + \|y - u\|)\|x - y\| \tag{4.360}$$

and

$$\|\mathcal{A}^{-1}(\mathcal{F}(x) - \mathcal{F}(y) - \delta\mathcal{F}(u, v)(x - y))\| \le l(\|x - v\| + \|y - v\| + \|u - v\|)\|x - y\|. \tag{4.361}$$

By a continuity argument (4.358)–(4.361) remain valid if x and/or y belong to D_c. Next, we show (4.353). If (4.353) holds for all $n \le k$ and if $\{x_n\}$ $(n \ge 0)$ is well defined for $n = 0, 1, 2, \cdots, k$, then

$$\|x_n - x_0\| \le \alpha_n - \alpha_0 < \alpha^* - \alpha_0, \ n \le k. \tag{4.362}$$

That is (4.324) is well defined for $n = k + 1$. For $n = -1$ and $n = 0$, (4.353) reduces to $\|x_{-1} - x_0\| \le c$ and $\|x_0 - x_1\| \le \nu$. Suppose (4.353) holds for $n = -1, 0, 1, \cdots, k$ $(k \ge 0)$. By (4.357), (4.361) and

$$\mathcal{F}(x_{k+1}) = \mathcal{F}(x_{k+1}) - \mathcal{F}(x_k) - \mathcal{A}_k(x_{k+1}x_k) \tag{4.363}$$

we obtain in turn the following estimates

$$\begin{aligned}
\|\mathcal{A}^{-1}\mathcal{F}(x_{k+1})\| &= \|\mathcal{A}^{-1}(\delta\mathcal{F}(x_{k+1}, x_k) - \mathcal{A}_k)(x_{k+1} - x_k)\| \\
&\le \left(\|\mathcal{A}^{-1}(\delta\mathcal{F}(x_{k+1}, x_k) - \mathcal{F}'(x_k))\| + \|\mathcal{A}^{-1}(\mathcal{F}'(x_k) - \mathcal{A}_k)\|\right)\|(x_{k+1} - x_k)\| \\
&\le l\left[\|(x_{k+1} - x_k)\| + \|(x_k - y_k)\|\right]\|(x_{k+1} - x_k)\| \\
&\le l(\alpha_{k+1} - \alpha_k + |1 - \theta_k|(t_k - t_{k-1})(t_{k+1} - t_k)
\end{aligned}$$

$$\tag{4.364}$$

and

$$\begin{aligned}
\|x_{k+2} - x_{k+1}\| &= \|\mathcal{A}_{k+1}^{-1}\mathcal{F}(x_{k+1})\| \le \|\mathcal{A}_{k+1}^{-1}\mathcal{A}\|\|\mathcal{A}^{-1}\mathcal{F}(x_{k+1})\| \\
&\le \frac{l(\alpha_{k+1} - \alpha_k + |1 - \theta_k|(\alpha_k - \alpha_{k-1}))}{1 - l_0\left[(1 + |\theta_{k+1}|)(\alpha_{k+1} - c) + c|1 - \theta_{k+1}|(\alpha_k - c) + c\right]}(\alpha_{k+1} - \alpha_k) \\
&\le \alpha_{k+2} - \alpha_{k+1}.
\end{aligned}$$

The induction for (4.353) is complete. It follows from (4.353) and Lemma 4.6.1 that $\{x_n\}$ $(n \ge -1)$ is a complete sequence in a Banach space $\mathcal{X}$ and as such it converges to some $x^* \in \overline{U}(x_0, \alpha^* - c)$ (since $\overline{U}(x_0, \alpha^* - c)$ is a closed set). By letting $k \to \infty$ in (4.364), we obtain $\mathcal{F}(x^*) = 0$. Moreover, estimate (4.354) follows from (4.353) by using standard majoration techniques. Finally, to show the uniqueness in $\overline{U}(x_0, R)$, let $y^* \in \overline{U}(x_0, R)$ be a solution (1.1). Set $\mathcal{T} = \int_0^1 \mathcal{F}'(y^* + t(y^* - x^*))dt$. Using $(\mathcal{D}_4)$ and (4.355), we get in turn that

$$\begin{aligned}
\|\mathcal{A}^{-1}(\mathcal{A} - \mathcal{T})\| &= l_0(\|y^* - x_0\| + \|x^* - x_0\|) + \|\mathcal{A}^{-1}(\mathcal{F}'(x_0) - \mathcal{A})\| \\
&\le l_0\left[(\alpha^* - \alpha_0) + R\right] + \|\mathcal{A}^{-1}(\mathcal{F}'(x_0) - \mathcal{A})\| < 1.
\end{aligned} \tag{4.365}$$

If follows from (4.365) and the Banach lemma on invertible operators that $\mathcal{T}^{-1}$ exists. Using the identity $\mathcal{F}(x^*) - \mathcal{F}'(y^*) = \mathcal{T}(x^* - y^*)$, we deduce that $x^* = y^*$. The proof of the Theorem is complete. $\qquad\qquad\square$

Remark 4.6.2. If follows from the proof of Theorem 4.6.1 that sequences $\{r_n\}$, $\{s_n\}$ defined by

$$\begin{cases} r_{-1} = 0, \ r_0 = c, \ r_1 = c + \nu \\[2mm] r_2 = r_1 + \dfrac{l_0(r_1 - r_0 + |1 - \theta_0|(r_0 - r_{-1}))(r_1 - r_0)}{1 - l_0\left((1 + |\theta_1|)(r_1 - r_0)\right)} \\[4mm] r_{n+2} = r_{n+1} + \dfrac{l(r_{n+1} - r_n + |1 - \theta_n|(r_n - r_{n-1}))(r_{n+1} - r_n)}{1 - l_0\left[(1 + |\theta_{n+1}|)(r_{n+1} - r_0) + (|1 - \theta_{n+1}|)(r_n - r_0)\right]} \end{cases} \tag{4.366}$$

and

$$\begin{cases} s_{-1} = 0, \ s_0 = c, \ s_1 = c + \nu \\[2mm] s_2 = s_1 + \dfrac{l_0(s_1 - s_0 + \lambda(s_0 - s_{-1}))(s_1 - s_0)}{1 - l_0(1 + |\theta_1|)(s_1 - s_0)} \\[4mm] s_{n+2} = s_{n+1} + \dfrac{l(s_{n+1} - s_n + \lambda(s_n - s_{n-1}))(s_{n+1} - s_n)}{1 - l_0(\mu(s_{n+1} - s_0) + \lambda(s_n - s_0))} \end{cases} \tag{4.367}$$

respectively are more precise majorizing sequences for $\{x_n\}$. Clearly, these sequences also converge under the (I^N) hypotheses.

A simple inductive argument shows that if $l_0 < l$ for each $n = 2, 3, \cdots$

$$r_n < s_n < \alpha_n \tag{4.368}$$

$$r_{n+1} - r_n < s_{n+1} - s_n < \alpha_{n+1} - \alpha_n \tag{4.369}$$

and

$$r^* = \lim_{n \to \infty} r_n \leq s^* = \lim_{n \to \infty} s_n \leq \alpha^* = \lim_{n \to \infty} \alpha_n. \tag{4.370}$$

In practice, one must choose $\{\theta_n\}$ so that the best error bounds are obtained. Note also that sequences $\{r_n\}$ or $\{s_n\}$ may converge under even weaker hypotheses. The sufficient convergence criterion (4.338) determines the smallness of c and r. This criterion can be solved for c and r (see for example the h criteria or (4.373) that follow). Indeed, let us demonstrate that in the case of Newton's method, i.e., if $c = 0, \lambda = 0, \mu = 1$. In the case $\kappa_0 = \kappa$ (i.e. $b = 1$), we obtain the famous Kantorovich sufficient convergent criteria. For the secant method, Schmidt in Ref. [397], Potra and Pták in Ref. [360], used the majorizing sequence $\{\alpha_n\}$ for $\theta_n \in [0, 1]$ and $l_0 = l$. That is, he used the sequence $\{t_n\}$ given by

$$\begin{cases} t_{-1} = 0, \ t_0 = c, \ t_1 = c + \nu \\[2mm] t_{n+2} = t_{n+1} + \dfrac{l(t_{n+1} - t_{n-1}) + (t_{n+1} - t_n)}{1 - L(t_n - t_{n+1} + c)} \end{cases} \tag{4.371}$$

whereas our sequence $\{\alpha_n\}$ reduces to

$$\begin{cases} \alpha_{-1} = 0, \ \alpha_0 = c, \ \alpha_1 = c + \nu \\[2mm] \alpha_{n+2} = \alpha_{n+1} + \dfrac{l(\alpha_{n+1} - \alpha_{n-1})(\alpha_{n+1} - \alpha_n)}{1 - l_0(\alpha_{n+1} - \alpha_n + c)} \end{cases} \tag{4.372}$$

Then, in case $l_0 < l$ our sequence is more precise (see also (4.368)–(4.370)). Notice also that the sufficient convergence criterion associated to $\{t_n\}$ is given by

$$lc + 2\sqrt{l\nu} \leq 1. \tag{4.373}$$

Our sufficient convergence criteria can also be weaker in this case (see also the numerical examples). It is worth noting that if $c = 0$ (4.373) reduces to Kantorovich's condition (since $\kappa = 2l$).

Next, we present the local convergence of the secant-type method. Let $x^* \in \mathcal{X}$ be such that $\mathcal{F}(x^*) = 0$ and $\mathcal{F}'(x^*)^{-1} \in \mathcal{L}(\mathcal{Y}, \mathcal{X})$. Using the identities

$$\begin{aligned}
x_{n+1} - x^* = & (A_n^{-1}\mathcal{F}'(x^*))F'(x^*)^{-1}[(\delta\mathcal{F}(x_n, y_n) - F'(x_n)) + \\
& (F'(x_n) - \delta\mathcal{F}(x_n, x^*))](x_n - x^*),
\end{aligned}$$

$$y_n - x_n = (1 - \theta_n)(x_{n-1} - x_n) \quad \text{and} \quad y_n - x^* = \theta_n(x_n - x^*) + (1 - \theta_n)(x_{n-1} - x^*),$$

we easily arrive at:

Theorem 4.6.2. *Suppose that $(\mathcal{D}_1)$ and $(\mathcal{D}_3)$ hold. Moreover, suppose that there exist $x^* \in D, K_0 > 0, K > 0$ such that $\mathcal{F}(x^*) = 0$, $F'(x^*)^{-1} \in \mathcal{L}(\mathcal{Y}, \mathcal{X})$,*

$$\|F'(x^*)^{-1}(\delta\mathcal{F}(x, y) - F'(x^*))\| \leq K_0(\|x - x^*\| + \|y - x^*\|)$$

$$\|F'(x^*)^{-1}(\delta\mathcal{F}(x, y) - F'(z))\| \leq K(\|x - z\| + \|y - z\|) \quad \textit{for each } x, y, z \in D$$

and $\overline{U}(x^*, R_0^*) \subseteq D$, where

$$R^* = \frac{1}{(2\lambda + 1)K + (\lambda + \mu)K_0} \quad \textit{and} \quad R_0^* = (\mu + \lambda - 1)R^*.$$

Then, sequence $\{x_n\}$ generated by the Secant-type method is well defined, remains in $\overline{U}(x^*, R^*)$ for each $n = -1, 0, 1, 2, \cdots$ and converges to x^* provided that $x_{-1}, x_0 \in U(x^*, R^*)$. Moreover, the following estimates hold

$$\|x_{n+1} - x^*\| \leq \hat{e}_n\|x_n - x^*\| \leq \overline{e}_n\|x_n - x^*\| \leq e_n\|x_n - x^*\|,$$

where

$$\hat{e}_n = \frac{\overline{K}(\|x_n - x^*\| + |1 - \theta_n|\|x_{n-1} - x_n\|)}{1 - K_0((1 + |\theta_n|)\|x_n - x^*\| + |1 - \theta_n|\|x_{n-1} - x^*\|)}$$

$$\overline{e}_n = \frac{\overline{K}(\|x_n - x^*\| + \lambda\|x_{n-1} - x_n\|)}{1 - K_0(\mu\|x_n - x^*\| + \lambda\|x_{n-1} - x^*\|)}$$

$$e_n = \frac{\overline{K}(2\lambda + 1)R^*}{1 - K_0(\lambda + \mu)R^*} \quad \textit{and} \quad \overline{K} = \begin{cases} \kappa_0 & \textit{if } n = 0 \\ \kappa & \textit{if } n > 0. \end{cases}$$

Remark 4.6.3. Comments similar to the one given in Remark 4.6.2 can also follow for this case. For example, notice again that in the case of Newton's method $R_* = 2/2\kappa_0 + \kappa$, whereas the one given independently by Rheinboldt in Ref. [383] and Traub in Ref. [418] is given by $R_*^1 = 2/(3\kappa)$. Note that $R_*^1 \leq R_*$. Strict inequality holds in the preceding inequality if $\kappa_0 < \kappa$. Finally, note that κ_0/κ can be arbitrarily small and $R_*/R_*^1 \longrightarrow 3$ as $\kappa_0/\kappa \longrightarrow 0$.

Table 4.9 Comparison Table of $\|\bar{x}_{n+1} - \bar{x}_n\|$ for Example 4.3.4 using various methods

n	$\|x_{n+1} - x_n\|$ $\theta_n = 0$, Newton	$\|x_{n+1} - x_n\|$ $\theta_n = 1$, Secant	$\|x_{n+1} - x_n\|$ $\theta_n = 2$, Kurchatov	$\|x_{n+1} - x_n\|$ $\theta_n = 1/2$, midpoint
1	9.49639×10^{-6}	4.70208×10^{-2}	4.33999×10^{-1}	1.42649×10^{-1}
2	8.18823×10^{-12}	7.77292×10^{-3}	3.28371×10^{-2}	1.51900×10^{-2}
3	5.15077×10^{-24}	5.14596×10^{-5}	2.33370×10^{-3}	1.66883×10^{-4}
4	1.79066×10^{-48}	3.89016×10^{-8}	9.32850×10^{-6}	1.34477×10^{-7}
5	1.95051×10^{-97}	1.77146×10^{-13}	2.214411×10^{-9}	1.03094×10^{-12}
6	2.12404×10^{-195}	5.35306×10^{-22}	1.801201×10^{-15}	5.63911×10^{-21}
ρ	2.00032	1.61815	1.61854	1.61817

Related to the semi-local case we present the following examples.

Example 4.6.1. Consider (1.293) with $a = 0.49$. and we are going to apply the Secant method ($\lambda = 1$, $\mu = 1$, $\theta_n = 0$) to find the solution of (1.293). We take the starting points $x_{-1} = 1.14216 \cdots$, $x_0 = 1$. We consider the domain $\Omega = B(x_0, 2)$. In this case, we obtain that

$$\nu = 0.147967, \quad \nu = 0.14216, \quad l = 2.61119 \quad \text{and} \quad l_0 = 1.74079.$$

Notice that Kantorovich hypothesis $lc + 2\sqrt{l\nu} \leq 1$ is not satisfied, but hypotheses of Theorem 4.6.1 are satisfied, so the convergence of Secant method starting from $x_0 \in B(x_0, 2)$ converges to the solution of (1.293).

For the local case we study the following one.

Example 4.6.2. Let $X = Y = R^3$, $D = U(0, 1)$, $x^* = (0, 0, 0)$ and define function F on D by

$$F(x, y, z) = (e^x - 1, y^2 + y, z). \tag{4.374}$$

We have that for $u = (x, y, z)$

$$F'(u) = \begin{pmatrix} e^x & 0 & 0 \\ 0 & 2y + 1 & 0 \\ 0 & 0 & 1 \end{pmatrix}. \tag{4.375}$$

Using the norm of the maximum of the rows and (4.374), (4.375) we see that since $F'(x^*) = \mathrm{diag}\{1, 1, 1\}$, we can define parameters for Newton's method by

$$K = e/2, \quad K_0 = 1, \quad R^* = \frac{2}{e + 4}, \quad R_0^* = R^*,$$

since $\theta_n = 1$, $\mu = 2$, $\lambda = 0$. Then the Newton's method starting from $x_0 \in B(x^*, R^*)$ converges to a solution of (4.374). Note that using only Lipschitz condition we obtain the Rheinboldt or Traub ball $R_{TR}^* = 2/(3e) < R^*$.

Example 4.6.3. Consider Example 4.3.4. The solution is approximated by an iterative method of (STTM). The computational order of convergence (COC) is shown in the last line of Table 4.9 for various methods.

4.7 Exercises

4.7.1. Introduce the method

$$x_{n+1} = x_n - A_n^{-1} F(x_n) \quad (x_{-1}, x_0 \in D) \quad (n \geq 0)$$

for approximating x^*, where $A_n = [2x_n - x_{n-1}, x_{n-1}]$ $(n \geq 0)$. Let F be a nonlinear operator defined on an open set D of a Banach space X with values in a Banach space Y. Assume that operator F has divided differences of order one and two on D; there exist points x_{-1}, x_0 in D such that $2x_0 - x_{-1} \in D$ and A_0 is invertible on D; There exist constants α, β such that:

$$\|A_0^{-1}([x, y] - [u, v])\| \leq \alpha(\|x - u\| + \|y - v\|),$$

$$\|A_0^{-1}([y, x, y] - [2y - x, x, y])\| \leq \beta\|x - y\|, \text{ for all } x, y, u, v \in D;$$

$$\text{for all} \quad x, y \in D \implies 2y - x \in D.$$

Define constants γ, δ by $\|x_0 - x_{-1}\| \leq \gamma$, $\|A_0^{-1} F(x_0)\| \leq \delta$ and $2\beta\gamma^2 \leq 1$. Moreover define θ, r, h by

$$\theta = \left\{ (\alpha + \beta\gamma)^2 + 3\beta(1 - \beta\gamma^2) \right\}^{1/2},$$

$$r = \frac{1 - \beta\gamma^2}{\alpha + \beta\gamma + \theta} \quad \text{and} \quad h(t) = -\beta t^3 - (\alpha + \beta\gamma)t^2 + (1 - \beta\gamma^2)t.$$

Suppose also that

$$\delta \leq h(r) = \frac{1}{3} \frac{\alpha + \beta\gamma + 2\theta}{1 - 2\beta\gamma^2} r^2 \quad \text{and} \quad U_0 = U(x_0, r_0) \subseteq D,$$

where $r_0 \in (0, r]$ is the unique solution of $h(t) = (1 - 2\beta\gamma^2)\delta$ on interval $(0, r]$. Show:

(a) sequence $\{x_n\}$ $(n \geq -1)$ is well defined, remains in $U(x_0, r_0)$ for all $n \geq -1$ and converges to a solution x^* of equation $F(x) = 0$.

(b) for all $n \geq -1$, we have $\|x_{n+1} - x_n\| \leq t_n - t_{n+1}$, $\quad \|x_n - x^*\| \leq t_n$, where

$$t_{-1} = r_0 + \gamma, \quad t_0 = r_0,$$

$$\gamma_0 = \alpha + 3\beta r_0 + \beta\gamma, \quad \gamma_1 = 3\beta r_0^2 - 2\gamma_0 r_0 - \beta\gamma^2 + 1,$$

(c) for all $n \geq 0$

$$t_{n+1} = \frac{\gamma_0 t_n - (t_n - t_{n-1})^2 \beta - 2\beta t_n^2}{\gamma_1 + 2\gamma_0 t_n - (t_n - t_{n-1})^2 - 3\beta t_n^2} t_n.$$

(d) if D is a convex set and $2\alpha(\gamma + 2r_0) < 1$, then x^* is the unique solution of equation in $\overline{U}(x_0, r_0)$.

Chapter 5

Gauss–Newton Method

There is an extensive literature on the local as well as the semi-local convergence of Gauss–Newton method under Lipschitz and Hölder Lipschitz type conditions. We extend in this chapter the applicability of the Gauss–Newton method for solving nonlinear equations and convex composite optimization problems.

5.1 Regularized Gauss–Newton Method

We provide a local convergence analysis of the iteratively regularized Gauss–Newton method for solving ill-posed problems in a Hilbert space setting. In contrast to earlier studies with Lipschitz conditions, we only use the weaker center Lipschitz conditions. Our analysis expands the applicability of this method. We are concerned with the problem of approximating a solution $x^\dagger$ of $F(x) = y$, where $F : D(F) \subset X \to Y$ is a Fréchet differentiable nonlinear operator between two Hilbert spaces X, Y with domain $D(F)$, range $R(F)$ and y a given element in Y.

Bakushinski [172] introduces the regularized Gauss–Newton method:

$$x^\delta_{k+1} = x^\delta_k - (\alpha_k I + F'(x^\delta_k)^\star F'(x^\delta_k))^{-1}(F'(x^\delta_k)^\star (F(x^\delta_k) - y^\delta) + \alpha_k(x^\delta_k - x_0)) \quad (5.1)$$

to generate a sequence solving the ill-posed problem $F(x) = y$, where $\delta > 0$ is given so that $\|y^\delta - y\| \leq \delta$, $x^\delta_0 = x_0$ is an initial guess of x^+, F' is the Fréchet derivative of operator F, $F'(x)^\star$ is the adjoint of $F'(x)$ and $\{\alpha_k\}$ is a given sequence with

$$\alpha_k > 0, \quad 1 \leq \frac{\alpha_k}{\alpha_{k+1}} \leq r \quad \text{and} \quad \lim_{k\to\infty} \alpha_k = 0 \quad (5.2)$$

for some constant $r > 1$. The generalized discrepancy principle

$$\|F(x^\delta_{n_\delta}) - y^\delta\| \leq \tau\delta < \|F(x_{k_\delta}) - y^\delta\| \quad 0 \leq k \leq n_\delta \quad (5.3)$$

used in [172] is a posteriori rule for choosing the stoping index n_δ, where $\tau > 1$.

It was shown in [172] that if

$$x_0 - x^\dagger = F'(x^\dagger)^\star F'(x^\dagger)^{\frac{v}{2}} w, \quad (5.4)$$

$$\begin{aligned}
&F'(x) = R(x, z)F'(z) + Q(x, z) \\
&\|I - R(x, z)\| \leq C_R \quad x, z \in U(x^\dagger, \delta) \subseteq D(F) \\
&\|Q(x, z)\| \leq C_Q \|F'(x)(x - z)\|
\end{aligned} \quad (5.5)$$

295

then

$$\|x_{n_\delta}^\delta - x^\dagger\| \le O(\delta^{\frac{v}{1+v}}) \quad 0 < v \le 1. \tag{5.6}$$

However, in this case the rate of convergence is not better than $O(\delta^{\frac{1}{2}})$. That is why, an alternative a posteriori rule given by

$$\alpha_{k_\delta}\left(F(x_{k_\delta}^\delta) - y^\delta, (\alpha_{k_\delta}I + F'(x_{k_\delta}^\delta)F'(x_{k_\delta}^\delta)^\star)^{-1}(F(x_{k_\delta}^\delta) - y^\delta) \le \tau^2\delta^2 \tag{5.7}$$

for k_δ being the first integer for which (5.7) is satisfied. It was then shown that if (5.4), $F'(z) = F'(x)R(x, z)$ and

$$\|I - R(x, z)\| \le K_0\|x - z\| \tag{5.8}$$

hold for all $x, z \in U(x^\dagger, \delta)$ and some constant K_0 , then

$$\|x_{k_\delta}^\delta - x^\dagger\| \le C\|w\|^{\frac{1}{1+v}}\delta^{\frac{v}{1+v}} \quad (0 < v \le 2). \tag{5.9}$$

Conditions (5.5) and (5.8) are not the same. In the elegant study [283] some convergence rate results were obtained using the stopping rule (5.7) and a Lipschitz condition on F' under conditions different or weaker than (5.8). In particular the following assumptions were made:

$$U(x^\dagger, \delta) \subset D(F) \quad \text{for some } \delta > 4r\|x_0 - x^\dagger\|, \tag{5.10}$$

$$\|F'(x) - F'(z)\| \le L\|x - z\| \quad \text{for all } x, z \in U(x^\dagger, \delta) \tag{5.11}$$

and

$$\|F'(x)\| \le \alpha_0^{\frac{1}{2}} \quad \text{for all } x \in U(x^\dagger, \delta). \tag{5.12}$$

The sequence also used in these results is the regularized Gauss–Newton method (5.1) corresponding to the noise-free case:

$$x_{k+1} = x_k - (\alpha_k I + F'(x_k)^\star F'(x_k))^{-1}[F'(x_k)^\star(F(x_k) - y) + \alpha_k(x_k - x_0)]. \tag{5.13}$$

Although smallness of $L\|v\|$ has not been specified in [283], the needed smallness conditions appear in the proof. Here, we are motivated by optimization considerations and we show that the quantity $L\|v\|$ can be estimated in a tighter way, under the same or weaker hypotheses. Clearly, this observation leads to a wider choice of initial guesses x_0 and tighter error bounds than before. Such an approach has already been used by us in a Banach space setting [110, 163, 120]. In view of hypothesis (5.11) there exist constants L_1 and L_0 such that

$$\|F'(x) - F'(x^\dagger)\| \le L_1\|x - x^\dagger\| \quad \text{for all } x \in U(x^\dagger, \delta) \tag{5.14}$$

and

$$\|F'(x) - F'(x_0)\| \le L_0\|x - x_0\| \quad \text{for all } x \in U(x_0, \delta_0), \tag{5.15}$$

provided that $\|x_0 - x^\dagger\| + \delta_0 \le \delta$, $\delta_0 \ge 0$ or for all $x \in U(x^\dagger, \delta)$.

It is convenient for us to define some constants and parameters. Let $L_0 > 0$, $L > 0$, $r > 1$, $\tau > 1 + \sqrt{r}$ and $c_0 \in (0, \frac{\tau - 1 - \sqrt{r}}{2\sqrt{r}})$ be given constants. Define constants γ_i, $i = 1, 2, \ldots, 10$ by

$$\gamma_1 = \frac{1}{r}[5 + r^{\frac{1}{2}}(1 + \frac{1}{c_0})]^{-1}, \quad \gamma_2 = 2[r^{\frac{1}{2}}(1 + \frac{1}{c_0})^2]^{-1}, \quad \gamma_3 = \min\{\gamma_1, \gamma_2\},$$

$$\gamma_4 = \frac{1}{3}[r^{\frac{1}{2}}(1 + \frac{1}{c_0})]^{-1}, \quad \gamma_5 = \frac{1}{10r}, \quad \gamma_6 = [\frac{4}{13(1 + r)}]^2,$$

$$\gamma_7 = \min\{\gamma_i, \quad i = 1, 2, \ldots, 6\} \quad \gamma_8 = \frac{2}{5}[r^{\frac{1}{2}}(4 + \frac{1}{c_0})]^{-1}, \quad \gamma_9 = \frac{4}{5r(5 + r^{\frac{1}{2}})}$$

and

$$\gamma_{10} = \min\{\gamma_8, \gamma_9\}.$$

Define also numbers $\varepsilon_0, \varepsilon_1, \varepsilon_2, \varepsilon$ by

$$\varepsilon_0 = \gamma_3, \tag{5.16}$$

$$\varepsilon_1 = \gamma_7, \quad \varepsilon_2 \geq \frac{L}{L_1}\gamma_{10} \quad \text{and} \quad \varepsilon \geq \frac{L}{L_1}\gamma_1.$$

We also need the notations as Refs. [172, 282, 285]:

$$\mathcal{A} = F'(x^\dagger)^\star F'(x^\dagger), \quad \mathcal{A}_k = F'(x_k)^\star F'(x_k), \quad \mathcal{A}_k^\delta = F'(x_k^\delta)^\star F'(x_k^\delta),$$

$$\mathcal{B} = F'(x^\dagger)F'(x^\dagger)^\star, \quad \mathcal{B}_k = F'(x_k)F'(x_k)^\star, \quad \mathcal{B}_k^\delta = F'(x_k^\delta)F'(x_k^\delta)^\star,$$

$$e_k = x_k - x^\dagger \quad \text{and} \quad e_k^\delta = x_k^\delta - x^\dagger.$$

We state a series of results whose proofs are omitted, since they correspond verbatim to the ones in Ref. [283] by simply replacing (5.11) with the actually needed and tighter (5.14).

Lemma 5.1.1. *Assume there exist a solution $x^\dagger$ of equation $F(x) = y$ and $v \in N(F'(x^\dagger)^\star)^\perp$ such that $x_0 - x^\dagger = F'(x^\dagger)v$; conditions (5.2), (5.7), (5.10), (5.12), (5.14) hold and*

$$L_1\|v\| \leq \varepsilon_0, \tag{5.17}$$

where ε_0 is given in (5.16). Then, sequence generated by (GNM) is well defined for all $k \geq 0$. Moreover for all integers $0 \geq k \leq k_\delta$ the following estimates hold:

$$\|e_k^\delta\| \leq r^{\frac{1}{2}}(1 + \frac{1}{c_0})\alpha_k^{\frac{1}{2}}\|v\|.$$

Furthermore, for the integer k_δ determined by (5.7) we have $k_\delta \leq \widetilde{k}_\delta$.

Proposition 5.1.1. *Under the hypotheses of Lemma 5.1.1, but with (5.17) replaced by $L_1\|v\| \leq \varepsilon_1$, the estimate $\|x_{k_\delta}^\delta - x^\dagger\| \leq C\|v\|^{\frac{1}{2}}\delta^{\frac{1}{2}}$ hold.*

Remark 5.1.1. If $L_1 = L$ the results of Lemma 5.1.1 and Proposition 5.1.1 reduce to Lemma 2.2 and Proposition 2.1 in Ref. [283], respectively. Otherwise they constitute an improvement under weaker hypotheses, and less computational cost. The advantages are stated in the introduction of this section. Note that it is easier to compute constant L_1 than L. Condition (5.14) is weaker than (5.11) provided that $L_1 < L$. The rest of the examples in Ref. [283] can be improved along the same lines. However, we leave the details to the motivated reader.

The results obtained here can also be rewritten using the center-Lipschitz condition (5.15), instead of the stronger Lipschitz condition (5.11) (or (5.14)). Simply, replace the estimates (i.e. (5.11) or (5.14)) needing $\|F'(x) - F'(x^\dagger)\|$ by estimates of the form $\|F'(x) - F'(x_0)\| + \|F'(x^\dagger) - F'(x_0)\|$ (i.e. (5.15)), since

$$\|F'(x) - F'(x^\dagger)\| \leq \|F'(x) - F'(x_0)\| + \|F'(x^\dagger) - F'(x_0)\|.$$

5.2 Convex Composite Optimization

A semi-local convergence analysis of the Gauss–Newton method for convex composite optimization is given using the concept of quasi-regularity. Our convergence analysis is presented first under L-average Lipschitz and then under generalized convex majorant conditions. The results extend the applicability of Guass-Newton method under the same computational cost as before. In this section we are concerned with the convex composite optimization problem

$$\min_{x \in \mathbb{R}^l} f(x) := h(F(x)), \tag{5.18}$$

where $h : \mathbb{R}^m \longrightarrow \mathbb{R}$ is convex, $F : \mathbb{R}^l \longrightarrow \mathbb{R}^m$ is Fréchet-differentiable operator and $m, l \in \mathbb{N}^\star$. We assume that the minimum h_{min} of the function h is attained. It is well-known that the study of (5.18) is related to the convex inclusion problem

$$F(x) \in \mathcal{C}, \tag{5.19}$$

where

$$\mathcal{C} = \operatorname{argmin} h \tag{5.20}$$

is the set of all minimum points of h. Recently, in the elegant study by Li and Ng in Ref. [304], the notion of quasi-regularity for $x_0 \in \mathbb{R}^l$ with respect to inclusion (5.19) was used. This notion generalizes the case of regularity studied by Burke and Ferris in Ref. [180] as well as the case when $d \longrightarrow F'(x_0)d - \mathcal{C}$ is surjective (this condition was inaugurated by Robinson in Refs. [385, 386]) (see also Refs. [110, 120, 152, 153]).

A semi-local convergence analysis for Gauss–Newton method (GNM) was presented using the popular algorithm (see Refs. [110, 304]):

Algorithm (GNA) : (ξ, Δ, x_0)

Let $\xi \in [1, \infty)$, $\Delta \in (0, \infty]$ and for each $x \in \mathbb{R}^l$, define $\mathcal{D}_\Delta(x)$ by

$$\mathcal{D}_\Delta(x) = \{d \in \mathbb{R}^l : \| d \| \leq \Delta, \ h(F(x) + F'(x)\,d) \leq h(F(x) + F'(x)\,d') \\ \text{for all } d' \in \mathbb{R}^l \text{ with } \| d' \| \leq \Delta\}. \tag{5.21}$$

Let also $x_0 \in \mathbb{R}^l$ be given. Having $x_0, x_1, \cdots, x_k$ $(k \geq 0)$, determine x_{k+1} by:

If $0 \in \mathcal{D}_\Delta(x_k)$, then STOP;

If $0 \notin \mathcal{D}_\Delta(x_k)$, choose d_k such that $d_k \in \mathcal{D}_\Delta(x_k)$ and

$$\| d_k \| \leq \xi\, d(0, \mathcal{D}_\Delta(x_k)). \tag{5.22}$$

Then, set $x_{k+1} = x_k + d_k$.

Here, $d(x, W)$ denotes the distance from x to W in the finite dimensional Banach space containing W. Note that the set $\mathcal{D}_\Delta(x)$ $(x \in \mathbb{R}^l)$ is nonempty and is the solution of the following convex optimization problem

$$\min_{d \in \mathbb{R}^l, \|d\| \leq \Delta} h(F(x) + F'(x)\,d), \tag{5.23}$$

which can be solved by well-known methods such as the subgradient or cutting plane or bundle methods in Ref. [276].

In Ref. [304], the convergence of (GNA) is based on the generalized Lipschitz conditions inaugurated by Wang in Refs. [427, 428]. In Ref. [152], we presented a finer convergence analysis in a Banach space setting than in Refs. [427, 428, 440, 441] for (GNM), with the advantages $(\mathcal{A})$

(a) Tighter error estimates on the distances involved;
(b) The information on the location of the solution is at least as precise.

These advantages were obtained (under the same computational cost) using same or weaker hypotheses. Here, we provide the same advantages $(\mathcal{A})$ but for (GNA).

Let W be a closed convex subset of $\mathbb{R}^l$ (or $\mathbb{R}^m$). The negative polar of W denoted by $W^\ominus$ is defined as

$$W^\ominus = \{z : <z, w> \leq 0 \quad \text{for each} \quad w \in W\}. \tag{5.24}$$

We need the following notion of generalized Lipschitz condition due to Wang in Refs. [427, 428] (see also Ref. [304]). From now on $L : [0, \infty) \longrightarrow (0, \infty)$ (or L_0) denote a nondecreasing and absolutely continuous function. Moreover, η and α denote given positive numbers.

Definition 5.2.1. Let $\mathcal{Y}$ be a Banach space and let $x_0 \in \mathbb{R}^l$. Let $G : \mathbb{R}^l \longrightarrow \mathcal{Y}$. Then, G is said to satisfy:

(a) The center L_0-average condition on $U(x_0, r)$, if

$$\| G(x) - G(x_0) \| \leq \int_0^{\|x-x_0\|} L_0(u)\, du \quad \text{for all} \quad x \in U(x_0, r). \tag{5.25}$$

(b) The L-average Lipschitz condition on $U(x_0, r)$, if

$$\| G(x) - G(y) \| \leq \int_{\|y-x_0\|}^{\|x-y\|+\|y-x_0\|} L(u)\, du \tag{5.26}$$

for all $x, y \in U(x_0, r)$ with $\| x - y \| + \| y - x_0 \| \leq r$.

Remark 5.2.1. It follows from (5.25) and (5.26) that if G satisfies the L-average condition, then it satisfies the center L_0-Lipschitz condition, but not necessarily vice versa. We have that

$$L_0(u) \leq L(u) \quad \text{for each} \quad u \in [0, r] \tag{5.27}$$

holds in general and L/L_0 can be arbitrarily large.

Definition 5.2.2. Define majorizing function ψ_α on $[0, +\infty)$ by

$$\psi_\alpha(t) = \eta - t + \alpha \int_0^t L(u)\,(t - u)\, du \quad \text{for each} \quad t \geq 0 \tag{5.28}$$

and majorizing sequence $\{t_{\alpha,n}\}$ by

$$t_{\alpha,0} = 0, \quad t_{\alpha,n+1} = t_{\alpha,n} - \frac{\psi_\alpha(t_{\alpha,n})}{\psi'_\alpha(t_{\alpha,n})} \quad \text{for each} \quad n = 0, 1, \cdots . \tag{5.29}$$

$\{t_{\alpha,n}\}$ was used in Ref. [304] as a majorizing sequence for $\{x_n\}$ generated by (GNA).

Sequence $\{t_{\alpha,n}\}$ can also be written equivalently for each $n = 1, 2, \cdots$ and $t_{\alpha,1} = 1$ as

$$t_{\alpha,n+1} = t_{\alpha,n} - \frac{\gamma_{\alpha,n}}{\psi'_\alpha(t_{\alpha,n})} \tag{5.30}$$

where

$$\begin{aligned}
\gamma_{\alpha,n} &= \int_0^1 \int_{t_{\alpha,n-1}}^{t_{\alpha,n-1}+\theta\,(t_{\alpha,n}-t_{\alpha,n-1})} L(u)\, du\, d\theta\, (t_{\alpha,n} - t_{\alpha,n-1}) \\
&= \int_0^{t_{\alpha,n}-t_{\alpha,n-1}} L(t_{\alpha,n-1} + u)\,(t_{\alpha,n} - t_{\alpha,n-1} - u)\, du,
\end{aligned} \tag{5.31}$$

since (see (4.20) in Ref. [304])

$$\psi_\alpha(t_{\alpha,n}) = \frac{\gamma_{\alpha,n}}{\alpha} \quad \text{for each} \quad n = 1, 2, \cdots . \tag{5.32}$$

From now on we show how our convergence analysis for (GNA) is finer than the one in Ref. [304]. Define supplementary majorizing function $\psi_{\alpha,0}$ on $[0, +\infty)$ by

$$\psi_{\alpha,0}(t) = \eta - t + \alpha \int_0^t L_0(u)\,(t - u)\, du \quad \text{for each} \quad t \geq 0 \tag{5.33}$$

and corresponding majorizing sequence $\{s_{\alpha,n}\}$ by

$$s_{\alpha,0} = 0, \quad s_{\alpha,1} = \eta, \quad s_{\alpha,n+1} = s_{\alpha,n} - \frac{\beta_{\alpha,n}}{\psi'_{\alpha,0}(s_{\alpha,n})} \quad \text{for each} \quad n = 0, 1, \cdots, \quad (5.34)$$

where $\beta_{\alpha,n}$ is defined as $\alpha_{\alpha,n}$ with $s_{\alpha,n-1}$, $s_{\alpha,n}$ replacing $t_{\alpha,n-1}$, $t_{\alpha,n}$, respectively.

The results concerning $\{t_{\alpha,n}\}$ are already in the literature (see Refs. [152, 304]), whereas the corresponding ones for sequence $\{s_{\alpha,n}\}$ can be derived in an analogous way by simply using $\psi'_{\alpha,0}$ instead of ψ'_α. First, we need some auxiliary results for the properties of functions ψ_α, $\psi_{\alpha,0}$ and the relationship between sequences $\{s_{\alpha,n}\}$ and $\{t_{\alpha,n}\}$. The proofs of the next four lemmas involving the ψ_α function can be found in Ref. [304], whereas the proofs for function $\psi_{\alpha,0}$ are analogously obtained by simply replacing L by L_0.

Let $r_\alpha > 0$, $b_\alpha > 0$, $r_{\alpha,0} > 0$ and $b_{\alpha,0} > 0$ be such that

$$\alpha \int_0^{r_\alpha} L(u)\, du = 1, \quad b_\alpha = \alpha \int_0^{r_\alpha} L(u)\, u\, du \qquad (5.35)$$

and

$$\alpha \int_0^{r_{\alpha,0}} L_0(u)\, du = 1, \quad b_{\alpha,0} = \alpha \int_0^{r_{\alpha,0}} L_0(u)\, u\, du. \qquad (5.36)$$

Clearly, we have that

$$b_\alpha < r_\alpha \qquad (5.37)$$

and

$$b_{\alpha,0} < r_{\alpha,0}. \qquad (5.38)$$

In view of (5.27), (5.35) and (5.36), we get that

$$r_\alpha \leq r_{\alpha,0} \qquad (5.39)$$

and

$$b_\alpha \leq b_{\alpha,0}. \qquad (5.40)$$

Lemma 5.2.1. *Suppose that $0 < \eta \leq b_\alpha$. Then, $b_\alpha < r_\alpha$ and the following assertions hold:*

(i) ψ_α is strictly decreasing on $[0, r_\alpha]$ and strictly increasing on $[r_\alpha, \infty)$ with $\psi_\alpha(\eta) > 0$, $\psi_\alpha(r_\alpha) = \eta - b_\alpha \leq 0$, $\psi_\alpha(+\infty) \geq \eta > 0$.

(ii) $\psi_{\alpha,0}$ is strictly decreasing on $[0, r_{\alpha,0}]$ and strictly increasing on $[r_{\alpha,0}, \infty)$ with $\psi_{\alpha,0}(\eta) > 0$, $\psi_{\alpha,0}(r_{\alpha,0}) = \eta - b_{\alpha,0} \leq 0$, $\psi_{\alpha,0}(+\infty) \geq \eta > 0$. Moreover, if $\eta < b_\alpha$, then ψ_α has two zeros, denoted by $r_\alpha^\star$ and $r_\alpha^{\star\star}$, such that

$$\eta < r_\alpha^\star < \frac{r_\alpha}{b_\alpha} \eta < r_\alpha < r_\alpha^{\star\star} \qquad (5.41)$$

and if $\eta = b_\alpha$, ψ_α has a unique zero $r_\alpha^\star = r_\alpha$ in (η, ∞); $\psi_{\alpha,0}$ has two zeros, denoted by $r_{\alpha,0}^\star$ and $r_{\alpha,0}^{\star\star}$, such that

$$\eta < r_{\alpha,0}^\star < \frac{r_{\alpha,0}}{b_{\alpha,0}} \eta < r_{\alpha,0} < r_{\alpha,0}^{\star\star},$$

$$r_{\alpha,0}^\star \leq r_\alpha^\star, \qquad (5.42)$$

$$r_{\alpha,0}^{\star\star} \leq r_\alpha^{\star\star} \qquad (5.43)$$

and if $\eta = b_{\alpha,0}$, $\psi_{\alpha,0}$ has a unique zero $r_{\alpha,0}^\star = r_{\alpha,0}$ in (η, ∞).

(iii) $\{t_{\alpha,n}\}$ *is strictly monotonically increasing and converges to* $r_\alpha^\star$.

(iv) $\{s_{\alpha,n}\}$ *is strictly monotonically increasing and converges to its unique least upper bound* $s_\alpha^\star \leq r_{\alpha,0}^\star$.

(v) The convergence of $\{t_{\alpha,n}\}$ *is quadratic if* $\eta < b_\alpha$ *and linear if* $\eta = b_\alpha$.

Lemma 5.2.2. *Let* r_α, $r_{\alpha,0}$, b_α, $b_{\alpha,0}$, ψ_α, $\psi_{\alpha,0}$ *be as defined above. Let* $\overline{\alpha} > \alpha$. *Then, the following assertions hold:*

(i) Functions $\alpha \longrightarrow r_\alpha$, $\alpha \longrightarrow r_{\alpha,0}$, $\alpha \longrightarrow b_\alpha$, $\alpha \longrightarrow b_{\alpha,0}$ *are strictly decreasing on* $[0, \infty)$.

(ii) $\psi_\alpha < \psi_{\overline{\alpha}}$ *and* $\psi_{\alpha,0} < \psi_{\overline{\alpha},0}$ *on* $[0, \infty)$.

(iii) Function $\alpha \longrightarrow r_\alpha^\star$ *is strictly increasing on* $I(\eta)$, *where* $I(\eta) = \{\alpha > 0 : \eta \leq b_\alpha\}$.

(iv) Function $\alpha \longrightarrow r_{\alpha,0}^\star$ *is strictly increasing on* $I(\eta)$.

Lemma 5.2.3. *Let* $0 \leq \lambda < \infty$. *Define functions*

$$\chi(t) = \frac{1}{t^2} \int_0^t L(\lambda + u)(t - u)\, du \quad for\ all \quad t \geq 0 \tag{5.44}$$

and

$$\chi_0(t) = \frac{1}{t^2} \int_0^t L_0(\lambda + u)(t - u)\, du \quad for\ all \quad t \geq 0. \tag{5.45}$$

Then, functions χ *and* χ_0 *are increasing on* $[0, \infty)$.

Lemma 5.2.4. *Define function*

$$g_\alpha(t) = \frac{\psi_\alpha(t)}{\psi_\alpha'(t)} \quad for\ all \quad t \in [0, r_\alpha^\star).$$

Suppose $0 < \eta \leq b_\alpha$. *Then, function* g_α *is increasing on* $[0, r_\alpha^\star)$.

Next, we shall show that sequence $\{s_{\alpha,n}\}$ is tighter than $\{t_{\alpha,n}\}$.

Lemma 5.2.5. *Suppose that hypotheses of Lemma 5.2.1 hold and sequences* $\{s_{\alpha,n}\}$, $\{t_{\alpha,n}\}$ *are well defined for each* $n = 0, 1, \cdots$. *Then, the following assertions hold for* $n = 0, 1, \cdots$

$$s_{\alpha,n} \leq t_{\alpha,n}, \tag{5.46}$$

$$s_{\alpha,n+1} - s_{\alpha,n} \leq t_{\alpha,n+1} - t_{\alpha,n} \tag{5.47}$$

and

$$s_\alpha^\star = \lim_{n \to \infty} s_{\alpha,n} \leq r_\alpha^\star = t_\alpha^\star = \lim_{n \to \infty} t_{\alpha,n}. \tag{5.48}$$

Moreover, if strict inequality holds in (5.27) so does in (5.46) and (5.47) for $n > 1$. *Furthermore, the convergence of* $\{s_{\alpha,n}\}$ *is quadratic if* $\eta < b_\alpha$ *and linear if* $L_0 = L$ *and* $\eta = b_\alpha$.

Proof. We shall first show using induction that (5.46) and (5.47) are satisfied for each $n = 0, 1, \cdots$. These estimates hold true for $n = 0, 1$, since $s_{\alpha,0} = t_{\alpha,0} = 0$ and $s_{\alpha,1} = t_{\alpha,1} = \eta$. Using (5.27) and (5.34) for $n = 1$, we have that

$$s_{\alpha,2} = s_{\alpha,1} - \frac{\beta_{\alpha,1}}{\psi'_{\alpha,0}(s_{\alpha,1})} \leq t_{\alpha,1} - \frac{\gamma_{\alpha,1}}{\psi'_{\alpha}(t_{\alpha,1})} = t_{\alpha,2}$$

and

$$s_{\alpha,2} - s_{\alpha,1} = -\frac{\beta_{\alpha,1}}{\psi'_{\alpha,0}(s_{\alpha,1})} \leq -\frac{\gamma_{\alpha,1}}{\psi'_{\alpha}(t_{\alpha,1})} = t_{\alpha,2} - t_{\alpha,1},$$

since

$$-\psi'_{\alpha,0}(s) \leq -\psi'_{\alpha}(t) \quad \text{for each} \quad s \leq t. \tag{5.49}$$

Hence, estimate (5.46) holds true for $n = 0, 1, 2$ and (5.47) holds true for $n = 0, 1$. Suppose that

$$s_{\alpha,m} \leq t_{\alpha,m} \quad \text{for each} \quad m = 0, 1, 2, \cdots, k+1$$

and

$$s_{\alpha,m+1} - s_{\alpha,m} \leq t_{\alpha,m+1} - t_{\alpha,m} \quad \text{for each} \quad m = 0, 1, 2, \cdots, k.$$

Then, we have that

$$s_{\alpha,m+2} = s_{\alpha,m+1} - \frac{\beta_{\alpha,m+1}}{\psi'_{\alpha,0}(s_{\alpha,m+1})} \leq t_{\alpha,m+1} - \frac{\gamma_{\alpha,m+1}}{\psi'_{\alpha}(t_{\alpha,m+1})} = t_{\alpha,m+2}$$

and

$$s_{\alpha,m+2} - s_{\alpha,m+1} = -\frac{\beta_{\alpha,m+1}}{\psi'_{\alpha,0}(s_{\alpha,m+1})} \leq -\frac{\gamma_{\alpha,m+1}}{\psi'_{\alpha}(t_{\alpha,m+1})} = t_{\alpha,m+2} - t_{\alpha,m+1}.$$

The induction for (5.46) and (5.47) is complete. Finally, estimate (5.48) follows from (5.47) by letting $n \longrightarrow \infty$. The convergence order part for sequence $\{s_{\alpha,n}\}$ follows from (5.47) and Lemma 5.2.1-(v). The proof of Lemma 5.2.5 is complete.$\square$

Remark 5.2.2. If $L_0 = L$, the results in Lemmas 5.2.1–5.2.5 reduce to the corresponding ones in Ref. [304]. Otherwise (i.e. if $L_0 < L$), our results constitute an improvement (see also (5.39)–(5.43)).

We mention some concepts and results on regularities which can be found in Ref. [304]. For a set-valued mapping $T : \mathbb{R}^l \rightrightarrows \mathbb{R}^m$ and for a set A in $\mathbb{R}^l$ or $\mathbb{R}^m$, we denote by

$$D(T) = \{x \in \mathbb{R}^l : Tx \neq \emptyset\}, \quad R(T) = \bigcup_{x \in D(T)} Tx,$$

$$T^{-1}y = \{x \in \mathbb{R}^l : y \in Tx\} \quad \text{and} \quad \| A \| = \inf_{a \in A} \| a \|.$$

Consider the inclusion

$$F(x) \in C, \tag{5.50}$$

where C is a closed convex set in $\mathbb{R}^m$. Let $x \in \mathbb{R}^l$ and

$$\mathcal{D}(x) = \{d \in \mathbb{R}^l : F(x) + F'(x)d \in C\}. \tag{5.51}$$

Definition 5.2.3. Let $x_0 \in \mathbb{R}^l$.

(a) x_0 is quasi-regular point of (5.50) if there exist $R \in (0, +\infty)$ and an increasing positive function β on $[0, R)$ such that

$$\mathcal{D}(x) \neq \emptyset \text{ and } d(0, \mathcal{D}(x)) \leq \beta(\| x - x_0 \|) \, d(F(x), C) \text{ for all } x \in U(x_0, R).$$
(5.52)

$\beta(\| x - x_0 \|)$ is an error bound in determining how for the origin is away from the solution set of (5.50).

(b) x_0 is a regular point of (5.50) if

$$ker(F'(x_0)^T) \cap (C - F(x_0))^{\ominus} = \{0\}.$$
(5.53)

Proposition 5.2.1. *(see Ref. [180]) Let x_0 be a regular point of (5.50). Then, there are constants $R > 0$ and $\beta > 0$ such that (5.52) holds for R and $\beta(\cdot) = \beta$. Therefore, x_0 is a quasi-regular point with the quasi-regular radius $R_{x_0} \geq R$ and the quasi-regular bound function $\beta_{x_0} \leq \beta$ on $[0, R]$.*

Remark 5.2.3.

(a) $\mathcal{D}(x)$ can be considered as the solution set of the linearized problem associated to (5.50)

$$F(x) + F'(x) \, d \in C.$$
(5.54)

(b) If C defined in (5.50) is the set of all minimum points of h and if there exists $d_0 \in \mathcal{D}(x)$ with $\| d_0 \| \leq \Delta$, then $d_0 \in \mathcal{D}_\Delta(x)$ and for each $d \in \mathbb{R}^l$, we have the following equivalence

$$d \in \mathcal{D}_\Delta(x) \Longleftrightarrow d \in \mathcal{D}(x) \Longleftrightarrow d \in \mathcal{D}_\infty(x).$$
(5.55)

(c) Let R_{x_0} denote the supremum of R such that (5.52) holds for some function β defined in Definition 5.2.3. Let $R \in [0, R_{x_0}]$ and $\mathcal{B}_R(x_0)$ denotes the set of functions β defined on $[0, R)$ such that (5.52) holds. Define

$$\beta_{x_0}(t) = \inf\{\beta(t) : \beta \in \mathcal{B}_{R_{x_0}}(x_0)\} \quad \text{for each} \quad t \in [0, R_{x_0}).$$
(5.56)

All functions $\beta \in \mathcal{B}_R(x_0)$ with $\lim_{t \to R^-} \beta(t) < +\infty$ can be extended to an element of $\mathcal{B}_{R_{x_0}}(x_0)$ and we have that

$$\beta_{x_0}(t) = \inf\{\beta(t) : \beta \in \mathcal{B}_R(x_0)\} \quad \text{for each} \quad t \in [0, R).$$
(5.57)

R_{x_0} and β_{x_0} are called the quasi-regular radius and the quasi-regular function of the quasi-regular point x_0, respectively.

Definition 5.2.4.

(a) A set-valued mapping $T : \mathbb{R}^l \rightrightarrows \mathbb{R}^m$ is convex if the following items hold

    (i) $Tx + Ty \subseteq T(x + y)$ for all $x, y \in \mathbb{R}^l$.
    (ii) $T\lambda x = \lambda Tx$ for all $\lambda > 0$ and $x \in \mathbb{R}^l$.
    (iii) $0 \in T0$.

(b) Let $T : \mathbb{R}^l \rightrightarrows \mathbb{R}^m$ a convex set-valued map. The norm of T is defined by $\| T \| = \sup\limits_{x \in D(T)} \{ \| Tx \| : \| x \| \leq 1 \}$. If $\| T \| < \infty$, we say that T is normed.

(c) For two convex set-valued mappings T and $S : \mathbb{R}^l \rightrightarrows \mathbb{R}^m$, the addition and multiplication are defined by $(T + S)x = Tx + Sx$ and $(\lambda T)x = \lambda(Tx)$ for all $x \in \mathbb{R}^l$ and $\lambda \in \mathbb{R}$, respectively.

(d) Let $T : \mathbb{R}^l \rightrightarrows \mathbb{R}^m$, C is closed convex in $\mathbb{R}^m$ and $x \in \mathbb{R}^l$. We define T_x by

$$T_x d = F'(x)d - C \quad \text{for all} \quad d \in \mathbb{R}^l \tag{5.58}$$

and its inverse by

$$T_x^{-1}y = \{ d \in \mathbb{R}^l : F'(x)d \in y + C \} \quad \text{for all} \quad y \in \mathbb{R}^m. \tag{5.59}$$

Note that if C is a cone then T_x is convex. For $x_0 \in \mathbb{R}^l$, if the Robinson condition (see Refs. [385, 386]):

$$T_{x_0} \text{ carries } \mathbb{R}^l \text{ onto } \mathbb{R}^m \tag{5.60}$$

is satisfied, then $D(T_x) = \mathbb{R}^l$ for each $x \in \mathbb{R}^l$ and $D(T_{x_0}^{-1}) = \mathbb{R}^m$.

Remark 5.2.4. Let $T : \mathbb{R}^l \rightrightarrows \mathbb{R}^m$.

(a) T is convex $\iff$ the graph $Gr(T)$ is a convex cone in $\mathbb{R}^l \times \mathbb{R}^m$.
(b) T is convex $\implies T^{-1}$ is convex from $\mathbb{R}^m$ to $\mathbb{R}^l$.

Lemma 5.2.6. *(see Ref. [385]) Let C be a closed convex cone in $\mathbb{R}^m$. Suppose that $x_0 \in \mathbb{R}^l$ satisfies the Robinson condition (5.60). Then we have the following assertions*

(i) $T_{x_0}^{-1}$ is normed.
(ii) If S is a linear operator from $\mathbb{R}^l$ to $\mathbb{R}^m$ such that $\| T_{x_0}^{-1} \| \| S \| < 1$, then the convex set-valued map $\bar{T} = T_{x_0} + S$ carries $\mathbb{R}^l$ onto $\mathbb{R}^m$. Furthermore, $\bar{T}^{-1}$ is normed and

$$\| \bar{T}^{-1} \| \leq \frac{\| T_{x_0}^{-1} \|}{1 - \| T_{x_0}^{-1} \| \| S \|}.$$

The following proposition shows that condition (5.60) implies that x_0 is regular point of (5.50). Using the center L_0-average Lipschitz condition, we also estimate in Proposition 5.2.2 the quasi-regular bound function. The proof is given in an analogous way to the corresponding result in Ref. [304] by simply using L_0 instead of L.

Proposition 5.2.2. *Let C be a closed convex cone in $\mathbb{R}^m$, $x_0 \in \mathbb{R}^l$ and define T_{x_0} as in (5.58). Suppose that x_0 satisfies the Robinson condition (5.60). Then we have the following assertions.*

(i) x_0 is a regular point of (5.50).

(ii) If F' satisfies the center L_0-average Lipschitz condition (5.25) on $U(x_0, R)$ for some $R > 0$. Let $\beta_0 = \parallel T_{x_0}^{-1} \parallel$ and let R_{β_0} such that

$$\beta_0 \int_0^{R_{\beta_0}} L_0(u)\, du = 1. \tag{5.61}$$

Then the quasi-regular radius R_{x_0}, the quasi-regular bound function β_{x_0} satisfy $R_{x_0} \geq \min\{R, R_{\beta_0}\}$ and

$$\beta_{x_0}(t) \leq \frac{\beta_0}{1 - \beta_0 \displaystyle\int_0^t L_0(u)\, du} \quad \text{for each} \quad 0 \leq t < \min\{R, R_{\beta_0}\}. \tag{5.62}$$

Remark 5.2.5. If $L_0 = L$, Proposition 5.2.2 reduces to the corresponding one in Ref. [304]. Otherwise, it constitutes an improvement (see (5.37)–(5.43)).

Assume that the set $\mathcal{C}$ satisfies (5.20). Let $x_0 \in \mathbb{R}^l$ be a quasi-regular point of (5.20) with the quasi-regular radius R_{x_0} and the quasi-regular bound function β_{x_0} (i.e., see (5.56)). Let $\xi \in [1, +\infty)$ and let

$$\eta = \xi\, \beta_{x_0}(0)\, d(F(x_0), \mathcal{C}). \tag{5.63}$$

For all $R \in (0, R_{x_0}]$, we define

$$\alpha_0(R) = \sup\left\{ \frac{\xi\, \beta_{x_0}(t)}{\xi\, \beta_{x_0}(t) \displaystyle\int_0^t L_0(s)\, ds + 1} : \eta \leq t < R\right\}. \tag{5.64}$$

Theorem 5.2.1. *Let $\xi \in [1, +\infty)$ and $\Delta \in (0, +\infty]$. Let $x_0 \in \mathbb{R}^l$ be a quasi-regular point of (5.20) with the quasi-regular radius R_{x_0} and the quasi-regular bound function β_{x_0}. Let $\eta > 0$ and $\alpha_0(R)$ given in (5.63) and (5.64), respectively. Let $0 < R < R_{x_0}$, $\alpha \geq \alpha_0(R)$ be a positive constant and let b_α, r_α as defined in (5.35). Let $\{s_{\alpha,n}\}$ $(n \geq 0)$ and $s_\alpha^\star$ given by (5.34) and (5.48), respectively. Suppose that F' satisfies the L-average Lipschitz and the center L_0-average Lipschitz conditions on $U(x_0, s_\alpha^\star)$. Suppose that*

$$\eta \leq \min\{b_\alpha, \Delta\} \quad \text{and} \quad s_\alpha^\star \leq R. \tag{5.65}$$

Then, sequence $\{x_n\}$ generated by (GNA) is well defined, remains in $\overline{U}(x_0, s_\alpha^\star)$ for all $n \geq 0$ and converges to some $x^\star$ such that $F(x^\star) \in \mathcal{C}$. Moreover, the following estimates hold for each $n = 1, 2, \cdots$

$$\parallel x_n - x_{n-1} \parallel \leq s_{\alpha,n} - s_{\alpha,n-1}, \tag{5.66}$$

$$\parallel x_{n+1} - x_n \parallel \leq (s_{\alpha,n+1} - s_{\alpha,n})\left(\frac{\parallel x_n - x_{n-1} \parallel}{s_{\alpha,n} - s_{\alpha,n-1}}\right)^2, \tag{5.67}$$

$$F(x_n) + F'(x_n)\,(x_{n+1} - x_n) \in \mathcal{C} \tag{5.68}$$

and

$$\parallel x_{n-1} - x^\star \parallel \leq s_\alpha^\star - s_{\alpha,n-1}. \tag{5.69}$$

Proof. By (5.65), (5.66) and Lemma 5.2.1, we have that

$$\eta \le s_{\alpha,n} < s_\alpha^\star \le R \le R_{x_0}. \tag{5.70}$$

Using the quasi-regularity property of x_0, we get that

$$\mathcal{D}(x) \ne \emptyset \ \text{and}\ d(0, \mathcal{D}(x)) \le \beta_{x_0}(\|\,x - x_0\,\|)\, d(F(x), \mathcal{C}) \ \text{for all}\ x \in U(x_0, R). \tag{5.71}$$

We first prove that the following assertion holds

$(\mathcal{T})$ (5.66) holds for all $n \le k - 1 \implies$ (5.67) and (5.68) hold for all $n \le k$.

Denote by $x_k^\theta = \theta\, x_k + (1 - \theta)\, x_{k-1}$ for all $\theta \in [0,1]$. Using (5.70), we have that $x_k^\theta \in U(x_0, s_\alpha^\star) \subseteq U(x_0, R)$ for all $\theta \in [0,1]$. Hence, for $x = x_k$, (5.71) holds, i.e.,

$$\mathcal{D}(x_k) \ne \emptyset \quad \text{and} \quad d(0, \mathcal{D}(x_k)) \le \beta_{x_0}(\|\,x_k - x_0\,\|)\, d(F(x_k), \mathcal{C}). \tag{5.72}$$

We also have that

$$\|\,x_k - x_0\,\| \le \sum_{i=1}^{k} \|\,x_i - x_{i-1}\,\| \le \sum_{i=1}^{k} s_{\alpha,i} - s_{\alpha,i-1} = s_{\alpha,k} \tag{5.73}$$

and

$$\|\,x_{k-1} - x_0\,\| \le s_{\alpha,k-1} \le s_{\alpha,k}. \tag{5.74}$$

Now, we prove that

$$\xi\, d(0, \mathcal{D}(x_k)) \le (s_{\alpha,k+1} - s_{\alpha,k}) \left(\frac{\|\,x_k - x_{k-1}\,\|}{s_{\alpha,k} - s_{\alpha,k-1}} \right)^2 \le s_{\alpha,k+1} - s_{\alpha,k}. \tag{5.75}$$

We show the first inequality in (5.75). We denote by $A_k = \|\,x_{k-1} - x_0\,\|$ and $B_k = \|\,x_k - x_{k-1}\,\|$. We have the following identity

$$\int_0^1 \int_{A_k}^{A_k + \theta\, B_k} L(u)\, du\, d\theta = \int_0^{B_k} L(A_k + u)\left(1 - \frac{u}{B_k}\right) du. \tag{5.76}$$

Then, by the L-average condition on $U(x_0, s_\alpha^\star)$, (5.68) for $n = k - 1$ and (5.72)–(5.76), we get that

$$
\begin{aligned}
\xi\, d(0, \mathcal{D}(x_k)) &\le \xi\, \beta_{x_0}(\|\,x_k - x_0\,\|)\, d(F(x_k), \mathcal{C}) \\
&\le \xi\, \beta_{x_0}(\|\,x_k - x_0\,\|)\, \|\, F(x_k) - F(x_{k-1}) - F'(x_{k-1})(x_k - x_{k-1})\,\| \\
&\le \xi\, \beta_{x_0}(\|\,x_k - x_0\,\|) \int_0^1 \|\,(F'(x_k^\theta) - F'(x_{k-1}))(x_k - x_{k-1})\, d\theta\,\| \\
&\le \xi\, \beta_{x_0}(\|\,x_k - x_0\,\|) \int_0^1 \int_{A_k}^{A_k + \theta\, B_k} L(u)\, du\, B_k\, d\theta \\
&\le \xi\, \beta_{x_0}(\|\,x_k - x_0\,\|) \int_0^{B_k} L(A_k + u)(B_k - u)\, du \\
&\le \xi\, \beta_{x_0}(s_{\alpha,k}) \int_0^{B_k} L(s_{\alpha,k-1} + u)(B_k - u)\, du.
\end{aligned}
\tag{5.77}
$$

For simplicity, we denote $\Xi_{\alpha,k} := s_{\alpha,k} - s_{\alpha,k-1}$. By (5.66) for $n = k$ and Lemma 5.2.3, we have in turn that

$$\frac{\displaystyle\int_0^{B_k} L(s_{\alpha,k-1} + u)(B_k - u)\,du}{B_k^2} \leq \frac{\displaystyle\int_0^{\Xi_{\alpha,k}} L(s_{\alpha,k-1} + u)(\Xi_{\alpha,k} - u)\,du}{\Xi_{\alpha,k}^2}. \tag{5.78}$$

Thus, we deduce that

$$\xi\, d(0, \mathcal{D}(x_k)) \leq \xi\, \beta_{x_0}(s_{\alpha,k})\left(\int_0^{\Xi_{\alpha,k}} L(s_{\alpha,k-1} + u)(\Xi_{\alpha,k} - u)\,du\right)\left(\frac{B_k}{\Xi_{\alpha,k}}\right)^2. \tag{5.79}$$

Using (5.64) and (5.70), we obtain that

$$\frac{\xi\,\beta_{x_0}(s_{\alpha,k})}{\alpha_0(R)} \leq \left(1 - \alpha_0(R)\int_0^{s_{\alpha,k}} L_0(u)\,du\right)^{-1}. \tag{5.80}$$

Note that $\alpha \geq \alpha_0(R)$. By (5.26), we have that

$$\frac{\xi\,\beta_{x_0}(s_{\alpha,k})}{\alpha} \leq \left(1 - \alpha\int_0^{s_{\alpha,k}} L_0(u)\,du\right)^{-1} = -(\psi'_{\alpha,0}(s_{\alpha,k}))^{-1}. \tag{5.81}$$

By (5.29), (5.79)–(5.81), we deduce that the first inequality in (5.75) holds. The second inequality of (5.75) follows from (5.66). Moreover, by (5.65) and Lemma 5.2.5, we have that

$$\Xi_{\alpha,k+1} = -\psi'_{\alpha,0}(s_{\alpha,k})^{-1}\beta_{\alpha,k} \leq -\psi'_{\alpha,0}(t_{\alpha,0})\,\gamma_{\alpha,0} = -\psi'_{\alpha,0}(t_{\alpha,0})\,\psi_\alpha(t_{\alpha,0})$$
$$= \eta \leq \Delta.$$

Hence, (5.75) implies that $d(0, \mathcal{D}(x_k)) \leq \Delta$ and there exists $d_0 \in \mathbb{R}^l$ with $\| d_0 \| \leq \Delta$ such that $F(x_k) + F'(x_k)\,d_0 \in \mathcal{C}$. By Remark 5.2.3, we have that

$$\mathcal{D}_\Delta(x_k) = \{d \in \mathbb{R}^l : \| d \| \leq \Delta \text{ and } F(x_k) + F'(x_k)\,d \in \mathcal{C}\}$$

and

$$d(0, \mathcal{D}_\Delta(x_k)) = d(0, \mathcal{D}(x_k)).$$

We deduce that (5.68) holds for $n = k$ since $d_k = x_{k+1} - x_k \in \mathcal{D}(x_k)$. We also have that

$$\| x_{k+1} - x_k \| \leq \xi\, d(0, \mathcal{D}_\Delta(x_k)) = \xi\, d(0, \mathcal{D}(x_k)).$$

Hence (5.56) holds for $n = k$ and assertion $(\mathcal{T})$ holds. It follows from (5.66) that $\{x_k\}$ is a Cauchy sequence in a Banach space and as such it converges to some $x^\star \in \overline{U}(x_0, s_\alpha^\star)$ (since $\overline{U}(x_0, s_\alpha^\star)$ is a closed set).

We use now mathematical induction to prove that (5.66), (5.67) and (5.68) hold. By (5.63), (5.65) and (5.71), we have that $\mathcal{D}(x_0) \neq \emptyset$ and

$$\xi\, d(0, \mathcal{D}(x_0)) \leq \xi\, \beta_{x_0}(0)\, d(F(x_0), \mathcal{C}) = \eta \leq \Delta.$$

We also have that

$$\| x_1 - x_0 \| = \| d_0 \| \leq \xi\, d(0, \mathcal{D}_\Delta(x_0)) \leq \xi\beta_{x_0}(0)\, d(F(x_0), \mathcal{C}) = \eta = \Xi_{\alpha,0}$$

and (5.66) holds for $n = 1$. By induction argument, we get that

$$\| x_{k+1} - x_k \| \leq \Xi_{\alpha,k+1}\left(\frac{\| x_k - x_{k-1} \|}{\Xi_{\alpha,k}}\right)^2 \leq \Xi_{\alpha,k+1}.$$

The induction is complete. That completes the proof of Theorem 5.2.1. $\square$

Remark 5.2.6.

(a) If $L = L_0$, then Theorem 5.2.1 reduces to the corresponding ones in Ref. [304]. Otherwise, in view of (5.46)–(5.48), our results constitute an improvement. The rest of Ref. [304] paper is improved, since those results are corollaries of Theorem 5.2.1. For brevity, we leave this part to the motivated reader.

(b) In view of the proof of our Theorem 5.2.1, we see that sequence $\{r_{\alpha,n}\}$ given by

$$
\begin{aligned}
&r_{\alpha,0} = 0, \quad r_{\alpha,1} = \eta, \\[2mm]
&r_{\alpha,2} = r_{\alpha,1} - \frac{\alpha \displaystyle\int_0^{r_{\alpha,1}-r_{\alpha,0}} L_0(r_{\alpha,0}+u)\,(r_{\alpha,1}-r_{\alpha,0}-u)\,du}{\psi'_{\alpha,0}(r_{\alpha,1})}, \\[2mm]
&r_{\alpha,n+1} = r_{\alpha,n} - \frac{\alpha \displaystyle\int_0^{r_{\alpha,n}-r_{\alpha,n-1}} L(r_{\alpha,n-1}+u)\,(r_{\alpha,n}-r_{\alpha,n-1}-u)\,du}{\psi'_{\alpha,0}(r_{\alpha,n})}
\end{aligned}
\tag{5.82}
$$

for each $n = 2, 3, \cdots$

is also a majorizing sequences for (GNA). Following the proof of Lemma 5.2.5 and under the hypotheses of Theorem 5.2.1, we get that

$$
r_{\alpha,n} \le s_{\alpha,n} \le t_{\alpha,n},
\tag{5.83}
$$

$$
r_{\alpha,n+1} - r_{\alpha,n} \le s_{\alpha,n+1} - s_{\alpha,n} \le t_{\alpha,n+1} - t_{\alpha,n}
\tag{5.84}
$$

and

$$
r_\alpha^\star = \lim_{n \to \infty} r_{\alpha,n} \le s_\alpha^\star \le r_\alpha^\star.
\tag{5.85}
$$

Hence, $\{r_{\alpha,n}\}$ and $\{s_{\alpha,n}\}$ are tighter majorizing sequences for $\{x_n\}$ than $\{t_{\alpha,n}\}$ used by Li et al. in Ref. [304]. Sequences $\{r_{\alpha,n}\}$ and $\{s_{\alpha,n}\}$ can converge under hypotheses weaker than the ones given in Theorem 5.2.1. Such conditions have already given by us for more general functions ψ and in the more general setting of a Banach space (see Refs. [152, 153]). Therefore, we shall only refer to the popular Kantorovich case as an illustration. Choose $\alpha = 1$, $L(u) = L$ and $L_0(u) = L_0$ for all $u \ge 0$. Then, $\{t_{\alpha,n}\}$ converges under Newton–Kantorovich hypothesis.

(c) There are cases when the sufficient convergence conditions developed in the preceding work are not satisfied. Then, one can use the modified Gauss–Newton method (MGNM). In this case, the majorizing sequence proposed in Ref. [304] is given by

$$
q_{\alpha,0} = 0, \quad q_{\alpha,n+1} = q_{\alpha,n} - \frac{\psi_\alpha(q_{\alpha,n})}{\psi'_\alpha(0)} \quad \text{for each} \quad n = 0, 1, \cdots.
\tag{5.86}
$$

This sequence clearly converges under the hypotheses of Theorem 5.2.1, so that estimates (5.66)–(5.69) hold with sequence $\{q_{\alpha,n}\}$ replacing $\{s_{\alpha,n}\}$. However, according to the proof of Theorem 5.2.1, the hypotheses on $\psi_{\alpha,0}$ can replace the corresponding ones on ψ_α. Moreover, the majorizing sequence is given by

$$p_{\alpha,0} = 0, \quad p_{\alpha,n+1} = p_{\alpha,n} - \frac{\psi_\alpha(p_{\alpha,n})}{\psi'_{\alpha,0}(0)} \quad for \ each \quad n = 0, 1, \cdots . \tag{5.87}$$

Furthermore, we have that

$$\psi_{\alpha,0}(s) \le \psi_\alpha(s) \quad for \ each \quad s \ge 0. \tag{5.88}$$

Hence, we clearly have that for each $n = 0, 1, \cdots$

$$p_{\alpha,n} \le q_{\alpha,n}, \tag{5.89}$$

$$p_{\alpha,n+1} - p_{\alpha,n} \le q_{\alpha,n+1} - q_{\alpha,n} \tag{5.90}$$

and

$$p_\alpha^\star = \lim_{n \to \infty} p_{\alpha,n} \le q_\alpha^\star = \lim_{n \to \infty} q_{\alpha,n}. \tag{5.91}$$

Notice also the advantages (5.37)–(5.43).

In the special case when functions L_0 and L are constants and $\alpha = 1$, we have that the conditions on function ψ_α reduce to Newton–Kantorovich hypothesis, whereas using $\psi_{\alpha,0}$ to

$$h_0 = L_0 \eta \le \frac{1}{2}. \tag{5.92}$$

Notice that

$$\frac{h_0}{h} \longrightarrow 0 \quad as \quad \frac{L_0}{L} \longrightarrow 0. \tag{5.93}$$

Therefore, one can use (MGNM) as a predictor until a cetain iterate x_N for which the sufficient conditions for (GNM) are satisfied. Then, we use x_N as the starting iterate for faster than (MGNM) method (GNM). Such an approach has already been used by us in Ref. [50].

Definition 5.2.5. Let $\mathcal{Y}$ be a Banach space, $x_0 \in \mathbb{R}^l$ and $\alpha > 0$. Let $G : \mathbb{R}^l \longrightarrow \mathcal{Y}$ and $f_\alpha : [0, r) \longrightarrow (-\infty, +\infty)$ be continuously differentiable. Then, G is said to satisfy:

(a) The center-majorant condition on $U(x_0, r)$, if

$$\| G(x) - G(x_0) \| \le \alpha^{-1} \left(f'_\alpha(\| x - x_0 \|) - f'_\alpha(0) \right) \quad \text{for all } x \in U(x_0, r). \tag{5.94}$$

(b) The majorant condition on $U(x_0, r)$, if

$$\| G(x) - G(y) \| \le \alpha^{-1} \left(f'_\alpha(\| x - y \| + \| y - x_0 \|) - f'_\alpha(\| y - x_0 \|) \right) \tag{5.95}$$

for all $x, y \in U(x_0, r)$ with $\| x - y \| + \| y - x_0 \| \le r$.

Clearly, conditions (5.94), (5.95) generalize (5.25) and (5.26), respectively, as well as the ones in Refs. [152, 153] (for $G = F'$ and $\alpha = 1$). Define majorizing sequence $\{t_{\alpha,n}\}$ by

$$t_{\alpha,0} = 0, \quad t_{\alpha,n+1} = t_{\alpha,n} - \frac{f_\alpha(t_{\alpha,n})}{f'_\alpha(t_{\alpha,n})}. \tag{5.96}$$

Moreover, as in (5.64) and for $R > 0$, define (implicitly):

$$\alpha_0(R) := \sup_{\xi \le t < R} -\frac{\eta\,\beta_{x_0}(t)}{f'_{\alpha_0(R)}(t)}. \tag{5.97}$$

Next, we provide sufficient conditions for the convergence of sequence $\{t_{\alpha,n}\}$ corresponding to the ones given in Lemma 5.2.1.

Lemma 5.2.7. *(see Ref. [152, 163]) Let $r > 0$, $\alpha > 0$ and $f_\alpha : [0,r) \longrightarrow (-\infty, +\infty)$ be continuously differentiable. Suppose*

(i) $f_\alpha(0) > 0$, $f'_\alpha(0) = -1$;
(ii) f'_α is convex and strictly increasing;
(iii) equation $f_\alpha(t) = 0$ has positive zeros. Denote by $r^\star_\alpha$ the smallest zero. Define $r^{\star\star}_\alpha$ by

$$r^{\star\star}_\alpha = \sup\{t \in [r^\star_\alpha, r) : f_\alpha(t) \le 0\}. \tag{5.98}$$

Then, sequence $\{t_{\alpha,n}\}$ is strictly increasing and converges to $r^\star_\alpha$. Moreover, the following estimates hold

$$r^\star_\alpha - t_{\alpha,n} \le \frac{D^- f'_\alpha(r^\star_\alpha)}{-2\,f'_\alpha(r^\star_\alpha)}\,(r^\star_\alpha - t_{\alpha,n-1})^2 \quad for\ each \quad n = 1, 2, \cdots, \tag{5.99}$$

where $D^- f'$ is the left directional derivative of f.

We can show the following semi-local convergence result for (GNA) using generalized majorant conditions (5.94) and (5.95).

Theorem 5.2.2. *Let $\xi \in [1, +\infty)$ and $\Delta \in (0, +\infty]$. Let $x_0 \in \mathbb{R}^l$ be a quasi-regular point of (5.20) with the quasi-regular radius R_{x_0} and the quasi-regular bound function β_{x_0}. Let $\eta > 0$ and $\alpha_0(r)$ given in (5.63) and (5.97). Let $0 < R < R_{x_0}$, $\alpha \ge \alpha_0(R)$ be a positive constant and let $r^\star_\alpha$, $r^{\star\star}_\alpha$ as defined in Lemma 5.2.7. Suppose that F' satisfies the majorant condition on $U(x_0, r^\star_\alpha)$, conditions*

$$\eta \le \min\{r^\star_\alpha, \Delta\} \quad and \quad r^\star_\alpha \le R \tag{5.100}$$

hold. Then, sequence $\{x_n\}$ generated by (GNA) is well defined, remains in $\overline{U}(x_0, r^\star_\alpha)$ for all $n \ge 0$ and converges to some $x^\star$ such that $F(x^\star) \in C$. Moreover, the following estimates hold for each $n = 1, 2, \cdots$

$$\| x_n - x_{n-1} \| \le t_{\alpha,n} - t_{\alpha,n-1}, \tag{5.101}$$

$$\| x_{n+1} - x_n \| \le (t_{\alpha,n+1} - t_{\alpha,n})\left(\frac{\| x_n - x_{n-1} \|}{t_{\alpha,n} - t_{\alpha,n-1}}\right)^2, \tag{5.102}$$

$$F(x_n) + F'(x_n)\,(x_{n+1} - x_n) \in C \tag{5.103}$$

and

$$\|\,x_{n-1} - x^\star\,\| \leq r_\alpha^\star - t_{\alpha,n-1}, \tag{5.104}$$

where sequence $\{t_{\alpha,n}\}$ is given by (5.96).

Proof. We use the same notations as in Theorem 5.2.1. We follow the proof of Theorem 5.2.1 until (5.75). Then, using (5.72), (5.95) (for $G = F'$), (5.96), (5.97) and hypothesis $\alpha \geq \alpha_0(R)$, we get in turn that

$$
\begin{aligned}
\xi\,d(0, \mathcal{D}(x_k)) &\leq \xi\,\beta_{x_0}(\|\,x_k - x_0\,\|)\,d(F(x_k), C) \\
&\leq \xi\,\beta_{x_0}(\|\,x_k - x_0\,\|)\,\|\,F(x_k) - F(x_{k-1}) - F'(x_{k-1})\,(x_k - x_{k-1})\,\| \\
&\leq \xi\,\beta_{x_0}(\|\,x_k - x_0\,\|) \int_0^1 \|\,(F'(x_k^\theta) - F'(x_{k-1}))\,(x_k - x_{k-1})\,d\theta\,\| \\
&\leq \xi\,\frac{\beta_{x_0}(t_{\alpha,k})}{\alpha_0(R)} \int_0^1 (f'_\alpha(t_{\alpha,k}^\theta) - f'_\alpha(t_{\alpha,k-1}))\,(t_{\alpha,k} - t_{\alpha,k-1})\,d\theta \\
&\leq \xi\,\frac{\beta_{x_0}(t_{\alpha,k})}{\alpha} \int_0^1 (f'_\alpha(t_{\alpha,k}^\theta) - f'_\alpha(t_{\alpha,k-1}))\,(t_{\alpha,k} - t_{\alpha,k-1})\,d\theta \\
&\leq -f'_\alpha(t_{\alpha,k})^{-1}\,(f_\alpha(t_{\alpha,k}) - f_\alpha(t_{\alpha,k-1}) - f'_\alpha(t_{\alpha,k-1})\,(t_{\alpha,k} - t_{\alpha,k-1})) \\
&= -f'_\alpha(t_{\alpha,k})\,f_\alpha(t_{\alpha,k}),
\end{aligned}
\tag{5.105}
$$

where $t_{\alpha,k}^\theta = \theta\,t_{\alpha,k} + (1 - \theta)\,(t_{\alpha,k} - t_{\alpha,k-1})$ for all $\theta \in [0,1]$. The rest follows as in the proof of Theorem 5.2.1. That completes the proof of Theorem 5.2.2. $\qquad\square$

Remark 5.2.7. In view of condition (5.95), there exists $f_{\alpha,0} : [0, r) \longrightarrow (-\infty, +\infty)$ continuously differentiable such that

$$\|\,G(x) - G(x_0)\,\| \leq \alpha^{-1}\,(f'_{\alpha,0}(\|\,x - x_0\,\|) - f'_{\alpha,0}(0)) \quad \text{for all} \quad x \in U(x_0, r). \tag{5.106}$$

Moreover

$$f'_{\alpha,0}(t) \leq f'_\alpha(t) \quad \text{for all} \quad t \in [0, r] \tag{5.107}$$

holds in general and $\dfrac{f'_\alpha}{f'_{\alpha,0}}$ can be arbitrarily large. These observations motivate us to introduce tighter majorizing sequences $\{r_{\alpha,n}\}, \{s_{\alpha,n}\}$ by

$$
\begin{aligned}
r_{\alpha,0} &= 0, \quad r_{\alpha,1} = \eta = -\frac{f_\alpha(0)}{f'_\alpha(0)}, \\
r_{\alpha,2} &= r_{\alpha,1} - \frac{\alpha\,(f_{\alpha,0}(r_{\alpha,1}) - f_{\alpha,0}(r_{\alpha,0}) - f'_{\alpha,0}(r_{\alpha,0})(r_{\alpha,1} - r_{\alpha,0}))}{f'_{\alpha,0}(r_{\alpha,1})}, \\
r_{\alpha,n+1} &= r_{\alpha,n} - \frac{\displaystyle\int_0^1 (f'_\alpha(r_{\alpha,k}^\theta) - f'_\alpha(r_{\alpha,k-1}))\,(r_{\alpha,k} - r_{\alpha,k-1})\,d\theta}{f'_{\alpha,0}(r_{\alpha,n})}
\end{aligned}
\tag{5.108}
$$

for each $n = 2, 3, \cdots$

and

$$s_{\alpha,0} = 0, \quad s_{\alpha,1} = r_{\alpha,1},$$

$$s_{\alpha,n+1} = s_{\alpha,n} - \frac{\displaystyle\int_0^1 \left(f'_\alpha(s^\theta_{\alpha,k}) - f'_\alpha(s_{\alpha,k-1})\right)\left(s_{\alpha,k} - s_{\alpha,k-1}\right) d\theta}{f'_{\alpha,0}(s_{\alpha,n})} \tag{5.109}$$

for each $n = 0, 1, \cdots$.

We shall provide an example to show that $L_0(t) < L(t)$ for each $u \in [0, R]$ and that hypotheses of Theorem 5.2.1 are satisfied. Then, according to Lemma 5.2.5, new sequence $\{s_{\alpha,n}\}$ is tighter than $\{t_{\alpha,n}\}$ given in Ref. [304], so that estimates (5.46)–(5.48) are satisfied.

Example 5.2.1. Let function $h : \mathbb{R} \longrightarrow \mathbb{R}$ be defined by

$$h(x) = \begin{cases} 0 \text{ if } x \le 0 \\ x \text{ if } x \ge 0. \end{cases}$$

Define function F by

$$F(x) = \begin{cases} \lambda - x + \dfrac{1}{18} x^3 + \dfrac{x^2}{1-x} & \text{if } x \le 1/2 \\ \lambda - \dfrac{71}{144} + 2\,x^2 & \text{if } x \ge 1/2, \end{cases} \tag{5.110}$$

where $\lambda > 0$ is a constant. Then, we have that $C = (-\infty, 0]$,

$$F'(x) = \begin{cases} -2 + \dfrac{1}{(1-x)^2} + \dfrac{x^2}{6} & \text{if } x < 1/2 \\ 4\,x & \text{if } x > 1/2 \end{cases} \tag{5.111}$$

and

$$F''(x) = \begin{cases} \dfrac{2}{(1-x)^2} + \dfrac{x}{3} & \text{if } x < 1/2 \\ 4 & \text{if } x > 1/2. \end{cases} \tag{5.112}$$

We shall first show that F' satisfies the L-average Lipschitz condition on $U(0, 1)$, where

$$L(u) = \frac{2}{(1-u)^3} + \frac{1}{6} \quad \text{for each } u \in [0, 1) \tag{5.113}$$

and the L_0-average condition on $U(0, 1)$, where

$$L_0(u) = \frac{2}{(1-u)^3} + \frac{1}{12} \quad \text{for each } u \in [0, 1). \tag{5.114}$$

It follows from (5.112) that

$$L(u) < L(v) \quad \text{for each} \quad 0 \le u < v < 1 \tag{5.115}$$

and

$$0 < F''(u) < F''(|u|) < L(|u|) \quad \text{for each} \quad \frac{1}{2} \ne u < 1. \tag{5.116}$$

Let $x, y \in U(0,1)$ with $|y| + |x - y| < 1$. Then, it follows from (5.115) and (5.116) that

$$
\begin{aligned}
|F'(x) - F'(y)| &\le |x - y| \int_0^1 F''(y + t\,(x - y))\,dt \\
&\le |x - y| \int_0^1 L(|y| + t\,|x - y|)\,dt.
\end{aligned}
\tag{5.117}
$$

Hence, F' satisfies the L-average Lipschitz condition on $U(0,1)$. Similarily, using (5.111) and (5.114), we deduce that F' satisfies the center L_0-average condition on $U(0,1)$. Notice that

$$
L_0(u) < L(u) \quad \text{for each} \quad u \in [0,1).
\tag{5.118}
$$

We also have $1 = \| T_{x_0}^{-1} \| = \beta_{x_0}(0)$ and T_{x_0} carries $\mathbb{R}$ into $\mathbb{R}$ as $F'(x_0) = -1$. We have that $\xi = 1$, $\alpha = 1$, $F(x_0) = F(0) = \lambda$ and by (5.63)

$$
\eta = \| T_{x_0}^{-1} \| \; d(F(x_0), \mathcal{C}) = \lambda.
$$

Let us choose $\lambda = .05$. By Maple 13 and using (5.28), (5.33), Lemma 5.2.1 (see also Fig. 5.1), we have that

$$
\psi_\alpha(t) = \eta - t + \alpha\left(\frac{1}{12}t^2 - \frac{1}{t-1} - t - 1\right),
$$

$$
\psi_{\alpha,0}(t) = \eta - t + \alpha\left(\frac{1}{24}t^2 - \frac{1}{t-1} - t - 1\right),
$$

$$
r_\alpha^\star = .05322869296 \quad \text{and} \quad r_{\alpha,0}^\star = .05309455977.
$$

Table 5.1 shows that our error bounds $s_{\alpha,n+1} - s_{\alpha,n}$ are finer than $t_{\alpha,n+1} - t_{\alpha,n}$ given in Ref. [304].

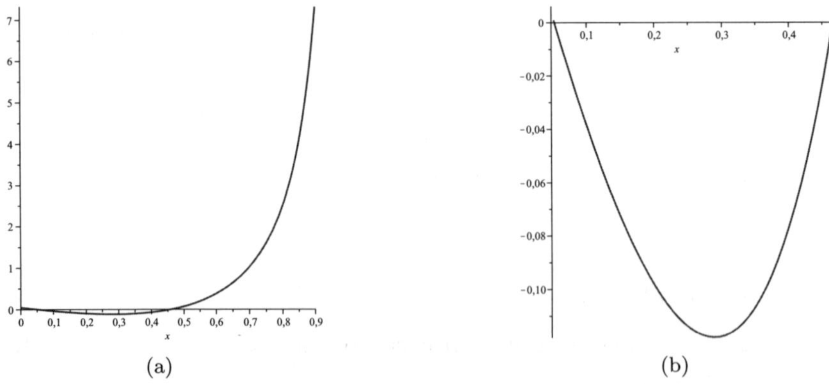

(a)   (b)

Fig. 5.1   (a) Function ψ_α on interval $[0, .9)$. (b) Function $\psi_{\alpha,0}$ on interval $[.052, .47)$.

Table 5.1 Comparison table of majorizing sequences for $\alpha = 1$

n	$s_{\alpha,n}$	$s_{\alpha,n+1} - s_{\alpha,n}$	$t_{\alpha,n}$	$t_{\alpha,n+1} - t_{\alpha,n}$
0	0	.05	0	.05
1	.05	.00308148876	.05	.00321390287
2	.05308148876	.00001307052	.05321390287	.00001479064
3	.05309455928		.05322869351	

5.3 Proximal Gauss–Newton Method

We employ a Gauss–Newton Algorithm (GNA) to find local minimizer of penalized nonlinear least square problems. We develop local convergence analysis of (GNA) from more general Lipschitz-type conditions than used in earlier analysis (see Ref. [389]). Our results, even in special cases, provide a larger attraction ball and tighter error estimates on the distances involved. In this section, we are concerned with the solution of the optimization problem

$$\mathcal{H}(x) := \frac{1}{2} \min_{x \in \mathcal{X}} \|\mathcal{F}(x) - y\|^2 + \mathcal{J}(x), \tag{5.119}$$

where $\mathcal{F}$ is a differentiable operator (Gâteaux or Fréchet) defined on an open and nonempty subset $\mathcal{D}$ of a Hilbert space $\mathcal{X}$ with values in a Hilbert space $\mathcal{Y}$ and $y \in \mathcal{Y}$. Functional $\mathcal{J}$, defined on $\mathcal{X}$ with values in $\mathbb{R} \bigcup \{\infty\}$, is proper, lower semi-continuous and convex (see in Refs. [175, 177, 217]). Diversified problems, arising in applied sciences and in engineering, can be expressed as the optimization problem (5.119). For various applications of the problem (5.119), we refer to the literature (see Refs. [120, 173, 180, 199, 218, 240, 283, 338, 370, 412, 438]). If the functional $\mathcal{J} = 0$, then (5.119) reduces to the classical nonlinear least square problem in Refs. [180, 436] and a popular approach for solving these problems is the Gauss–Newton Method (GNM) (see Refs. [177, 276, 394])

$$x_{n+1} = x_n - \left[\mathcal{F}'(x_n)^\star \mathcal{F}'(x_n)\right]^{-1} \mathcal{F}'(y_n)^\star \Big(\mathcal{F}(x_n) - y\Big). \tag{5.120}$$

The sequence $\{x_n\}$ generated by (GNM) approximate a point x^* such that $\mathcal{F}'(x^*)^\star \mathcal{F}(x^*) = 0$. It is well-known that point x_{n+1} in (5.120) is the minimizer of linearized functional

$$x \longrightarrow \frac{1}{2} \| \mathcal{F}(x_n) + \mathcal{F}'(x_n)(x - x_n) - y \|^2. \tag{5.121}$$

In local analysis, a region of attraction is established from which we pick starting points. So iterative procedures converge to a local minimizer (see Refs. [103, 120, 193, 328]). We note that for the local case we like to have as wide choice of initial guesses and as tight error estimates on distances $\| x_n - x^* \|$ as possible. In Refs. [103, 110, 120] using average Lipschitz conditions, we provided a local convergence analysis with the above stated advantages over earlier works. Sufficient convergence conditions, established in semi-local analysis, must be as weak as possible and estimates $\| x_n - x^* \|$ as tight as possible. In Refs. [120, 152] again using average Lipschitz-type conditions one can find semi-local convergence results with the above advantages over Refs. [199, 240, 299].

Recently, in the elegant work in Ref. [389], Salvo and Villa proposed a generalization of the (GNM) for $\mathcal{J} \neq 0$ which is given by

$$x_{n+1} = \mathsf{prox}_{\mathcal{J}}^{Q(x_n)}\left[x_n - (F'(x_n)^\star \mathcal{F}'(x_n))^{-1}\mathcal{F}'(x_n)^\star\left(\mathcal{F}(x_n) - y\right)\right] \qquad (5.122)$$

where $\mathsf{prox}_{\mathcal{J}}^{Q(x_n)}$ is the proximity operator related to $\mathcal{J}$ (see Refs. [328–330]) and evaluated with respect to the metric defined by the operator $Q(x_n) = \mathcal{F}'(x_n)^\star\mathcal{F}'(x_n)$. Here onwards the algorithm (5.122) will be referred to as (GNA). Apparently, (GNA) is obtained in a way analogous to (GNM) by linearizing $\mathcal{F}$ at the point x_n and computing the minimizer of the corresponding linearized functional

$$x \longrightarrow \frac{1}{2}\,\|\,\mathcal{F}(x_n) + \mathcal{F}'(x_n)(x - x_n) - y\,\|^2 + \mathcal{J}(x). \qquad (5.123)$$

A local convergence analysis for (GNA) was provided in Ref. [120]. Using average Lipschitz type conditions on $\mathcal{F}'$ this analysis reduces to the classical (GNA) (see Refs. [120, 152]). That is we get linear convergence in general and quadratic for zero residual problems. The idea of introducing a proximity operator allows the usage of the rich theory for this kind of operators. Moreover, as already noted in Refs. [120, 152, 180, 302, 306, 431, 437] this approach differs from convex compositive optimization or nonlinear problems with optimization where hypotheses differ as well as the obtained results (see Refs. [177, 287, 291, 301, 394]).

We develop a local convergence analysis with even more general Lipschitz type conditions. Our results not only specialize to the ones in Ref. [389] if average Lipschitz conditions are used but also provide a larger attraction ball and tighter error bounds on $\|\,x_n - x^*\,\|$. These advantages are obtained by using more precise upper bounds on the norms $\mathcal{F}'(x_n)^\dagger$ where $A^\dagger$ is the pseudo-inverse of A and $A^\dagger \in \mathcal{L}(\mathcal{Y}, \mathcal{X})$. Our idea has already produced a finer local as well as semi-local convergence analysis for Newton type methods in a Banach space setting (see Refs. [103, 120, 133, 152, 163]).

Moore–Penrose Generalized inverse (MPG), or pseudo-inverse, are commonly used to find a best-fit or least square solution for a system of linear equations. Here, we recall some useful results about the pseudo-inverse $A^\dagger$ of a linear operator A. Suppose that $A \in \mathcal{L}(\mathcal{X}, \mathcal{Y})$ has a closed range. The MPG of A is the linear operator $A^\dagger \in \mathcal{L}(\mathcal{Y}, \mathcal{X})$ satisfying the following four equations

$$AA^\dagger A = A, \quad A^\dagger AA^\dagger = A^\dagger, \quad (AA^\dagger)^\star = AA^\dagger, \quad (A^\dagger A)^\star = A^\dagger A. \qquad (5.124)$$

Let $P_{N(A)}$ and $P_{R(A)}$ be the orthogonal projectors onto the kernel and the range of A, respectively. From (5.124), we notice that

$$A^\dagger A = I - P_{N(A)}, \quad AA^\dagger = P_{R(A)}. \qquad (5.125)$$

In case A is injective, $N(A) = \{0\}$ and $A^\dagger A = I$, that is $A^\dagger$ is a left inverse of the operator A. Additionally, for each $A \in \mathcal{L}(\mathcal{X}, \mathcal{Y})$ the following conditions are analogous

(1) A is injective and the range of A is closed;

(2) $A^\dagger A$ is invertible in $\mathcal{L}(\mathcal{X}, \mathcal{Y})$.

If one of the above conditions is true, then $A^\dagger = (A^\star A)^{-1}A^T$ and $\|A^\dagger\|^2 = \|(A^T A)^{-1}\|$. The following lemma gives a perturbation bounds for the (MPG) (see Refs. [110, 120, 152]).

Lemma 5.3.1. *Let $A, B \in \mathcal{L}(\mathcal{X}, \mathcal{Y})$ with A injective and $R(A)$ closed. If $\|(B - A)A^\dagger\| < 1$, then B is injective, $R(B)$ is closed and*

$$\|B^\dagger\| \le \frac{\|A^\dagger\|}{1 - \|(B - A)A^\dagger\|}. \tag{5.126}$$

Moreover, the following assertions hold:

$$\|B^\dagger - A^\dagger\| \le \sqrt{2}\,\|A^\dagger\|\|B^\dagger\|\|B - A\| \quad and \quad \|B^\dagger - A^\dagger\| \le \sqrt{2}\,\frac{\|A^\dagger\|^2\|B - A\|}{1 - \|A^\dagger\|\|B - A\|}.$$

We review some useful properties of proximity operators. For a detailed overview about proximity operators, we refer to Refs. [328–330, 389]. Let us consider a positive, continuous and selfadjoint operator $Q \colon \mathcal{X} \to \mathcal{Y}$ which is bounded from below and thus it is invertible. A scalar product on $\mathcal{X}$ (for $x, z \in \mathcal{X}$) is given by

$$\langle x, z \rangle_Q := \langle x, Qz \rangle$$

and, since the following inequalities

$$\frac{\|x\|^2}{\|Q^{-1}\|} \le \| x \|_Q^2 \le \| Q \| \| x \|^2 \tag{5.127}$$

hold thus, the corresponding induced norm $\| \cdot \|_Q$ is equivalent to the given norm on $\mathcal{X}$. The Moreau-Yosida approximation, with respect to the scalar product induced by the operator Q, of a convex and lower semi-continuous function $h \colon \mathcal{X} \to \mathbb{R} \cup \{+\infty\}$ is the function $M_h \colon \mathcal{X} \to \mathbb{R}$ given as

$$M_h(z) = \inf_{x \in \mathcal{X}} \left\{ h(x) + \frac{\|x - z\|_Q^2}{2} \right\}. \tag{5.128}$$

For every $z \in \mathcal{X}$, the infimum in the above equation is attained at a unique $x \in \mathcal{X}$. Thus we define the proximity operator $\mathsf{prox}_h^Q(z)$, for h and with respect to Q, by

$$\mathsf{prox}_h^Q \colon \mathcal{X} \to \mathcal{X}.$$

From the first order optimality condition for (5.128) we obtain

$$p = \mathsf{prox}_h^Q(z) \;\Leftrightarrow\; 0 \in \partial h(p)Q(p - z) \;\Leftrightarrow\; Qz \in (\partial h + Q)(p) \tag{5.129}$$

and

$$\mathsf{prox}_h^Q(z) = (\partial h + Q)^{-1}(Qz).$$

When $Q = I$, then the proximity operator is prox_h.

Lemma 5.3.2. *The proximity operator $\mathsf{prox}_h^Q \colon \mathcal{X} \to \mathcal{X}$ is Lipschitz continuous and the following assertion holds:*

$$\| \mathsf{prox}_h^Q(z_1) - \mathsf{prox}_h^Q(z_2) \| \le \sqrt{\| Q \|\| Q^{-1} \|}\; \| z_1 - z_2 \|. \tag{5.130}$$

In special cases, it can be shown that the computation of the proximity operator with respect to the scalar product induced by Q can be simplified to the computation of the proximity operator with respect to the original norm. The following proposition captures this fact.

Proposition 5.3.1. *Let $A \in \mathcal{L}(\mathcal{X}, \mathcal{Y})$ be injective with closed range and $Q = A^\star A$. Furthermore, let $h\colon \mathcal{X} \to \mathbb{R} \cap \{+\infty\}$ be a proper, convex and lower semi-continuous functional. Then, the following assertion holds*

$$\mathrm{prox}_h^Q = A^\dagger \mathrm{prox}_{h\mathcal{D}A^\dagger} A.$$

Since the norm $\| \cdot \|_Q$, with respect to which the proximity operator is evaluated may vary, the following lemma captures the behavior of the proximity operator when Q varies.

Lemma 5.3.3. *Let Q_1 and Q_2 be bounded from below, continuous and positive selfadjoint operators on $\mathcal{X}$. Then, the following assertion holds for every $z_1, z_2 \in \mathcal{X}$:*

$$\| \mathrm{prox}_h^{Q_1}(z) - \mathrm{prox}_h^{Q_2}(z) \| \leq \| Q_1^{-1} \| \, \| (Q_1 - Q_2)(z - \mathrm{prox}_h^{Q_2}(z)) \| \tag{5.131}$$

and from the assertions (5.130) and (5.132)

$$\begin{aligned}
\| \mathrm{prox}_\mathcal{J}^{Q_1}&(z_1) - \mathrm{prox}_\mathcal{J}^{Q_2}(z_2) \| \\
\leq & \| \mathrm{prox}_\mathcal{J}^{Q_1}(z_1) - \mathrm{prox}_\mathcal{J}^{Q_1}(z_2) \| + \| \mathrm{prox}_\mathcal{J}^{Q_1}(z_2) - \mathrm{prox}_\mathcal{J}^{Q_2}(z_2) \| \\
\leq & \sqrt{\| Q \| \| Q^{-1} \|} \, \| z_1 - z_2 \| + \| Q_1^{-1} \| \, \| (Q_1 - Q_2)(z_2 - \mathrm{prox}_\mathcal{J}^{Q_2}(z_2)) \| .
\end{aligned}$$
$$\tag{5.132}$$

We need the following first order conditions for local minimizers.

Proposition 5.3.2. *Let $x^* \in \mathcal{D}$ be a local minimizer of $\mathcal{H}(x)$ (5.119). Then the following stationary assertion holds $-\mathcal{F}'(x^*)^\star \mathcal{F}(x^*) \in \partial \mathcal{J}(x^*)$. Furthermore, if $\mathcal{F}'(x^*)$ is injective and $R(\mathcal{F}'(x^*))$ is closed, then x^* satisfies the following fixed point equation*

$$x^* = \mathrm{prox}_\mathcal{J}^{Q(x^*)}(x^* - \mathcal{F}'(x^*)^\dagger \mathcal{F}(x^*)) \tag{5.133}$$

where $Q(x^*) = \mathcal{F}'(x^*)^\star \mathcal{F}'(x^*)$.

We develop now local convergence analysis of (GNA) for solving optimization problems (5.119). To accomplish this, we replace (GNA) by an equivalent algorithm by using Proposition 5.3.2. More precisely, we have

Proposition 5.3.3. *Let $\mathcal{F}'(x_n)$ be an injective operator with closed range and moreover let $Q(x_n) = \mathcal{F}'(x_n)^\star \mathcal{F}'(x_n)$. Then,*

$$x_{n+1} = \mathrm{argmin}_{x \in \mathcal{X}} \frac{1}{2} \| \mathcal{F}(x_n) + \mathcal{F}'(x_n)(x - x_n) \|^2 + \mathcal{J}(x) \tag{5.134}$$

is equivalent to

$$x_{n+1} = \mathrm{prox}_\mathcal{J}^{Q(x_n)}(x_n - \mathcal{F}'(x_n)^\dagger \mathcal{F}(x_n)) \tag{5.135}$$

see also Proposition 6 in Ref. [389].

Next we provide the main local convergence result for (GNA). Let $U(x, r)$ be an open ball centered at $x \in \mathcal{D}$ with radius $r > 0$.

Theorem 5.3.1. *Let* $\mathcal{D}_0 \subseteq \mathcal{D}$ *and* $x^* (\in dom(\mathcal{J}) \cap \mathcal{D}_0)$ *be a local minimizer of* $\mathcal{H}$. *Furthermore we suppose:*

(1) $\mathcal{F}'(x^*)$ *is injective with closed range;*
(2) there exist continuous, non-decreasing functions $v, w \colon [0, +\infty) \longrightarrow [0, +\infty)$, *such that for all* $x \in \mathcal{D}_0$, *we have*

$$\| \mathcal{F}'(x) - \mathcal{F}'(x^*) \| \le v(\| x - x^* \|) \| x - x^* \| \tag{5.136}$$

and

$$\| \mathcal{F}(x^*) - \mathcal{F}(x) - \mathcal{F}'(x)(x^* - x) \| \le w(\| x - x^* \|) \| x - x^* \| \tag{5.137}$$

(3)

$$\left[\left(1 + \sqrt{2}\right) \kappa + 1 \right] \alpha \beta^2 v(0) < 1 \tag{5.138}$$

where

$$\alpha = \| \mathcal{F}(x^*) \|, \quad \beta = \| \mathcal{F}'(x^*)^{\dagger} \|, \quad \kappa = \| \mathcal{F}'(x^*)^{\dagger} \| \, \| \mathcal{F}'(x^*) \|.$$

Here, κ *is the condition number of* $\mathcal{F}'(x^*)$.

Define R_0 *and* $q \colon [0, R_0) \longrightarrow [0, +\infty)$ *by*

$$R_0 = \sup \left\{ r \in (0, R) \colon \beta v(r) r < 1 \right\} \tag{5.139}$$

and

$$q(r) = \left(\frac{\beta}{1 - \beta v(r) r} \right)^2 \left[\left(v(r) r + \frac{\kappa}{\beta} \right) \left(w(r) + \left(1 + \sqrt{2}\right) \alpha \beta v(r) \right) + \alpha v(r) \right] r. \tag{5.140}$$

By the definition of functions v, w *and* q, *we notice that the function* q *is continuous and non-decreasing. We define*

$$r^\star = \sup\{ r \in (0, R_0] : q(r) < 1 \}. \tag{5.141}$$

Fix $r \in (0, r^\star]$ *and suppose*

$$U(x^*, r) \subset \mathcal{D}_0. \tag{5.142}$$

Then, for $x_0 \in U(x^*, r)$ *and* $Q(x_n) := \mathcal{F}'(x_n)^\star \mathcal{F}'(x_n)$, *sequence* $\{x_n\}$ *generated by (GNA) is well defined, remains in* $U(x^*, r)$ *for all* $n \ge 0$ *and converges to* x^*. *Moreover, the following assertions hold:*

($\mathcal{A}_1$) $\mathcal{F}'(x_n)$ *is injective with closed range;*
($\mathcal{A}_2$) $\| x_n - x^* \| \le q_0^n \| x_0 - x^* \|$

where for $r_{x_0} = \| x - x_0 \|$, $q_0 = q(r_{x_0}) < 1$.

Next, we rewrite the conditions which have been presented in Proposition 5.3.3 for a local minimizer of $\mathcal{H}$. Let us define G and G_0 by

$$G(x) = x - \mathcal{F}'(x)^\dagger \mathcal{F}(x) \quad and \quad G_0(x) = \mathsf{prox}_{\mathcal{J}}^{Q(x)}(G(x)) \tag{5.143}$$

where $Q(x) = \mathcal{F}'(x)^\star \mathcal{F}'(x)$. Domains of G and G_0 are the subset S of D defined as

$$S := \{x \in \mathcal{D} : \mathcal{F}'(x) \text{ is injective and } R(F'(x)) \text{ is closed}\}. \tag{5.144}$$

If x^* is a local minimizer of $\mathcal{H}$, the fixed point equation of Proposition 5.3.3 can be expressed as a fixed point of G_0

$$x^* = G_0(x^*). \tag{5.145}$$

Proof. Using hypothesis that x^* is a local minimizer of $\mathcal{H}$, the definition of S and G_0, we deduce that x^* is a fixed point of G_0. Then we can write

$$G(x^*) - \mathsf{prox}_{\mathcal{J}}^{Q(x^*)} G(x^*) = G(x^*) - x^* = \mathcal{F}'(x^*)^\dagger \mathcal{F}(x^*). \tag{5.146}$$

For $r \in \mathbb{R}$ with $0 < r \le R$ such that $U(x^*, r) \subseteq \mathcal{D}_0$ and $x \in U(x^*, r)$. Using (5.136) and (5.138), we obtain

$$\| \mathcal{F}'(x) - \mathcal{F}'(x^*) \| \, \big\| \mathcal{F}'(x^*)^\dagger \big\| \le \beta \, v(\| x - x^* \|) \, \| x - x^* \| < \beta \, v(r_{x^*}) \, r_{r_*} \le 1. \tag{5.147}$$

It follows from (5.147) and Lemma 5.3.1 that $\mathcal{F}'(x)$ is injective with closed range and

$$\| \mathcal{F}'(x)^\dagger \| \le \frac{\beta}{1 - \beta \, v(\rho_x) \, \rho_x}. \tag{5.148}$$

Using Lemma 5.3.3 (inequality (5.132)), with $Q_1 = Q(x)$, $Q_2 = Q(x^*)$, $z_1 = G(x)$ and $z_2 = G(x^*)$, and (5.146) we obtain

$$\begin{aligned}
\| G_0(x) - x^* \| &= \| \mathsf{prox}_{\mathcal{J}}^{Q(x)}(G(x)) - \mathsf{prox}_{\mathcal{J}}^{Q(x^*)}(G(x^*)) \| \\
&\le \Big(\| Q(x) \| \| Q(x)^{-1} \| \Big)^{1/2} \| G(x) - G(x^*) \| \\
&\quad + \| Q(x)^{-1} \| \| (Q(x) - Q(x^*))(G(x^*) - \mathsf{prox}_{\mathcal{J}}^{Q(x^*)} G(x^*)) \| \\
&= \Big(\| Q(x) \| \| Q(x)^{-1} \| \Big)^{1/2} \| G(x) - G(x^*) \| \\
&\quad + \| Q(x)^{-1} \| \| (Q(x) - Q(x^*))\mathcal{F}'(x^*)^\dagger \mathcal{F}(x^*) \| .
\end{aligned} \tag{5.149}$$

Additionally

$$\| Q(x) \| = \| \mathcal{F}'(x)^\star \mathcal{F}'(x) \| = \| \mathcal{F}'(x) \|^2$$

$$\| Q(x)^{-1} \| = \Big\| \big[\mathcal{F}'(x)^\star \mathcal{F}'(x) \big]^{-1} \Big\| = \| \mathcal{F}'(x)^\dagger \|^2 .$$

From the properties (5.125) and Lemma 5.3.1, we obtain

$$(Q(x) - Q(x^*))\mathcal{F}'(x^*)^\dagger = \Big(\mathcal{F}'(x)^\star \mathcal{F}'(x) - \mathcal{F}'(x^*)^\star \mathcal{F}'(x^*)\Big)\mathcal{F}'(x^*)^\dagger$$
$$= \mathcal{F}'(x)^\star \mathcal{F}'(x)\mathcal{F}'(x)^\dagger - \mathcal{F}'(x^*)^\star \mathcal{F}'(x^*)\mathcal{F}'(x^*)^\dagger$$
$$= \mathcal{F}'(x)^\star \mathcal{F}'(x)\mathcal{F}'(x)^\dagger - \mathcal{F}'(x)^\star P_{R(\mathcal{F}'(x^*))}$$
$$+ \mathcal{F}'(x)^\star P_{R(\mathcal{F}'(x^*))} - \mathcal{F}'(x^*)^\star P_{R(\mathcal{F}'(x^*))}$$
$$= \mathcal{F}'(x^*)^\star \Big(\mathcal{F}'(x) - \mathcal{F}'(x^*)\Big)\mathcal{F}'(x^*)^\dagger$$
$$\times \Big(\mathcal{F}'(x) - \mathcal{F}'(x^*)\Big)^\star P_{R(\mathcal{F}'(x^*))}$$

which yields

$$\| (Q(x) - Q(x^*))\mathcal{F}'(x^*)^\dagger \| \leq \Big(\| \mathcal{F}'(x) \| \| \mathcal{F}'(x^*)^\dagger \| + 1 \Big) \| \mathcal{F}'(x) - \mathcal{F}'(x^*) \| .$$
$$(5.150)$$

Combining the above inequality with (5.149)

$$\| G_0(x) - x^* \| \leq \| \mathcal{F}'(x) \| \| \mathcal{F}'(x)^\dagger \| \| G(x) - G(x^*) \|$$
$$+ \| \mathcal{F}'(x)^\dagger \|^2 \Big(\| \mathcal{F}'(x) \| \| \mathcal{F}'(x^*)^\dagger \| + 1 \Big) \| \mathcal{F}'(x) - \mathcal{F}'(x^*) \| \| \mathcal{F}(x^*) \| . \quad (5.151)$$

Through the properties (5.125), of the (MPG), and the injectivity of $\mathcal{F}'(x)$ we obtain

$$G(x) - G(x^*) = x - x^* - \mathcal{F}'(x)^\dagger \mathcal{F}(x) + \mathcal{F}'(x^*)^\dagger \mathcal{F}(x^*)$$
$$= \mathcal{F}'(x)^\dagger \Big[\mathcal{F}'(x)(x - x^*) - \mathcal{F}(x) + \mathcal{F}(x^*)\Big] + \Big(\mathcal{F}'(x^*)^\dagger - \mathcal{F}'(x)^\dagger\Big)\mathcal{F}(x^*).$$

Coupling the above equation with Lemma 5.3.1 produces

$$\| G(x) - G(x^*) \| \leq \| \mathcal{F}'(x)^\dagger \| \| \mathcal{F}(x^*) - \mathcal{F}(x) - \mathcal{F}'(x)(x^* - x) \|$$
$$+ \|\mathcal{F}'(x^*)^\dagger - \mathcal{F}'(x)^\dagger\| \| \mathcal{F}(x^*) \|$$
$$\leq \| \mathcal{F}'(x)^\dagger \| \| \mathcal{F}(x^*) - \mathcal{F}(x) - \mathcal{F}'(x)(x^* - x) \|$$
$$+ \sqrt{2} \| \mathcal{F}'(x)^\dagger \| \| \mathcal{F}'(x^*)^\dagger \| \| \mathcal{F}'(x) - \mathcal{F}'(x^*) \| \| \mathcal{F}(x^*) \|$$
$$= \| \mathcal{F}'(x)^\dagger \| \Big[\| \mathcal{F}(x^*) - \mathcal{F}(x) - \mathcal{F}'(x)(x^* - x) \| +$$
$$\sqrt{2} \| \mathcal{F}'(x^*)^\dagger \| \| \mathcal{F}(x^*) \| \| \mathcal{F}'(x) - \mathcal{F}'(x^*) \| \Big]. \quad (5.152)$$

From (5.151), (5.152) and $\| F'(x) \| \leq \| \mathcal{F}'(x) - \mathcal{F}'(x^*) \| + \| \mathcal{F}'(x^*) \|$ we obtain

$$\| G_0(x) - x^* \|$$
$$\leq \| \mathcal{F}'(x)^\dagger \|^2 \Big\{ \Big(\| \mathcal{F}'(x) - \mathcal{F}'(x^*) \| + \| \mathcal{F}'(x^*) \| \Big) \Big[(\| \mathcal{F}(x^*) - \mathcal{F}(x) - \mathcal{F}'(x)(x^* - x) \|$$
$$+ (1 + \sqrt{2}) \| \mathcal{F}'(x^*)^\dagger \| \| \mathcal{F}(x^*) \| \| \mathcal{F}'(x) - \mathcal{F}'(x^*) \|)$$
$$+ \| \mathcal{F}'(x) - \mathcal{F}'(x^*) \| \| \mathcal{F}(x^*) \| \Big] \Big\}$$
$$(5.153)$$

$$\| G_0(x) - x^* \|$$
$$\leq \Big(\frac{\beta}{1 - \beta v(\rho_x)r} \Big)^2 \Big[\Big(v(\rho_x)\rho_x + \frac{\kappa}{\beta} \Big) \Big(w(\rho_x) + (1 + \sqrt{2})\,\alpha\beta v(\rho_x) \Big) + \alpha v(\rho_x) \Big] \rho_x$$
$$\leq q(\| x - x^* \|) \| x - x^* \| .$$

The results now follow from the Banach contraction Mapping principle in Ref. [120] (see also Proposition 7 in Ref. [389]). That completes the proof of the Theorem. $\square$

First we state the following two definitions.

Definition 5.3.1. A set $\mathcal{D}_0$ is called star-shaped with respect to the point $x^* \in \mathcal{D}_0$ if the line segment joining x^* to any other point in $\mathcal{D}_0$ is contained in $\mathcal{D}_0$. We call $\mathcal{D}_0$ star-shaped with respect to the ball $U(x^*, r)$ if D_0 is star-shaped with respect to every $x \in U(x^*, r)$.

Definition 5.3.2. Let $\mathcal{D}_0 \subseteq \mathcal{D}$ be a star-shaped set with respect to $x^* \in \mathcal{D}_0$. Fix $R \in (0, +\infty)$ such that $\sup\limits_{x \in \mathcal{D}_0} \| x - x^* \| \leq R$ and $L\colon [0, R) \longrightarrow \mathbb{R}$ be a positive, continuous and integrable function. Then, the mapping $f\colon D \subseteq \mathcal{X} \longrightarrow \mathcal{Y}$ satisfies the radius-Lipschitz condition of center x^* with L average on $\mathcal{D}_0$ if

$$\| f(x) - f(x^\tau) \| \leq \int_{\tau \| x - x^* \|}^{\| x - x^* \|} L(u) \, du \tag{5.154}$$

where $x^\tau = x^* + \tau(x - x^*)$ and $0 \leq \tau \leq 1$. If a function $\Psi\colon [0, R) \longrightarrow \mathbb{R}$ of L is given as

$$\Psi(u) = \int_0^u L(\omega) \, d\omega,$$

then the inequality (5.154) can be written as

$$\| f(x) - f(x^\tau) \| \leq \Psi(\| x - x^* \|) - \Psi(\tau \| x - x^* \|). \tag{5.155}$$

We compare our results with the corresponding ones in Ref. [389] through the following two special cases.

Case I - Average Lipschitz Conditions

(see Refs. [133, 152, 389, 428, 432]) In view of (5.154) there exists continuous function $L_0\colon [0, R) \longrightarrow [0, +\infty)$ such that

$$\| \mathcal{F}'(x) - \mathcal{F}'(x^*) \| \leq \int_0^{\| x - x^* \|} L_0(u) \, du. \tag{5.156}$$

Clearly,

$$L_0(u) \leq L(u), \quad u \in [0, R) \tag{5.157}$$

holds in general and L/L_0 can be arbitrarily large. Though authors in Ref. [389] have used the case when $L_0(u) = L(u)$, $u \in [0, R)$. However, this is not true in general (see Refs. [120, 133, 152]). Under (5.154), we have

$$\| \mathcal{F}'(x) - \mathcal{F}'(x^*) \| \leq g_0(\| x - x^* \|) \| x - x^* \| \tag{5.158}$$

and

$$\| \mathcal{F}(x^*) - \mathcal{F}(x) - \mathcal{F}'(x)(x^* - x) \| \leq g_1(\| x - x^* \|) \| x - x^* \|^2 \tag{5.159}$$

where

$$g_0(t) = \begin{cases} \dfrac{1}{t} \displaystyle\int_0^t L_0(u)\, du & \text{if} \quad t \in (0, R) \\[2mm] L_0(0) & \text{if} \quad t = 0 \end{cases} \tag{5.160}$$

and

$$g_1(t) = \begin{cases} \dfrac{1}{t^2} \displaystyle\int_0^t u\, L(u)\, du & \text{if} \quad t \in (0, R) \\[2mm] L(0)/2 & \text{if} \quad t = 0 \end{cases} \tag{5.161}$$

are well-defined, continuous, positive and increasing functions. Note that in (5.160), $L_0 = L$ has been used in earlier works in Refs. [389, 301, 302, 304, 306, 431]. Therefore, in this special case according to (5.136), (5.137), (5.158) and (5.159), we can choose

$$v(t) = g_0(t) \tag{5.162}$$

and

$$w(t) = g_1(t)t. \tag{5.163}$$

We define g_0^1 to be g_0 but with L_0 replaced by L. We also define R_0^1, $r_1^\star$ and q^1 (see Ref. [389]) by

$$R_0^1 = \sup \left\{ r \in (0, R) \colon\ \beta\, g_0^1(r)r < 1 \right\}, \tag{5.164}$$

$$r_1^\star = \sup \left\{ r \in (0, R_0^1) \colon\ q^1(r)r < 1 \right\}, \tag{5.165}$$

where

$$q^1(r) = \left(\frac{\beta}{1 - \beta g_0^0(r)r} \right)^2 \left[\left(g_0^0(r)r + \frac{\kappa}{\beta} \right) \left(g_1(r) + (1 + \sqrt{2})\alpha\beta g_0^0(r) \right) + \alpha g_0^0(r) \right]. \tag{5.166}$$

If the equality holds in (5.157), then our results reduce to the ones in Ref. [389]. On the other hand, if strict inequality holds in (5.157), then we have

$$R_0^1 < R_0, \tag{5.167}$$

$$r_1^\star < r^\star \tag{5.168}$$

and

$$q(r) < q^1(r), \quad r \in [0, R). \tag{5.169}$$

Estimates (5.167)–(5.169) justify the claims stated in the introduction of this section. In the special case when L_0 and L are constants and for simplicity we set $\alpha = 0$ then we obtain

$$r_{1,N}^\star = \frac{2}{3L}, \tag{5.170}$$

$$r_N^\star = \frac{2}{2L_0 + L}. \tag{5.171}$$

Convergence radius $r_1^\star$ was given by Ortega et al. in Ref. [338] and Traub in Ref. [418] in connection with the local convergence of Newton's method (NM)

$$z_{n+1} = z_n - \mathcal{F}'(z_n)^{-1}\mathcal{F}(z_n), \tag{5.172}$$

whereas $r^\star$ was given by us in Refs. [110, 133, 152]. Then, again we have

$$r_{1,N}^\star < r_N^\star \quad \text{for} \quad L_0 < L. \tag{5.173}$$

Next, we provide even more general choices for functions v and w than those encountered in Special case I.

Case II - Convex Majorant conditions

(see Refs. [120, 152, 428, 432]) Let $\delta = \sup\{t \in [0, R) : U(x^*, R) \subseteq \mathcal{D}_0\}$ where $\mathcal{D}_0$ is a convex set. Suppose there exist continuously differentiable functions $f_0, f \colon [0, R) \longrightarrow [0, +\infty)$ such that

$$\| \mathcal{F}'(x) - \mathcal{F}'(x^*) \| \le f_0'(\| x - x^* \|) - f_0'(0) \tag{5.174}$$

and

$$\| \mathcal{F}'(x) - \mathcal{F}'(x_\theta) \| \le f'(\| x - x^* \|) - f'(\theta \| x - x^* \|) \tag{5.175}$$

for all $x \in U(x^*, \delta)$, where $x_\theta = x^* + \theta(x - x^*)$, $\theta \in [0, 1]$, $f_0(0) = f(0) = 0$ and $f_0'(0) = f'(0) = -1$, f_0', f' are strictly increasing,

$$f_0(t) \le f(t) \quad \text{and} \quad f_0'(t) \le f'(t). \tag{5.176}$$

We used conditions (5.174)–(5.176) to show that scalar iteration $\{s_n\}$ given by

$$s_0 = \| x_0 - x^* \|, \quad s_{n+1} = \left| \left(s_n - \frac{f(s_n)}{f'(s_n)} \right) f_1'(s_n) \right| \tag{5.177}$$

where

$$f_1(t) = \frac{f'(t)}{f_0'(t)}$$

is null and $\lim\limits_{n \to \infty} x_n = x^*$ in the case of the local convergence of (NM) (see Refs. [120, 133, 152]). Moreover, we showed that functions

$$v(t) = \frac{1}{t}\left(f_0'(t) - f_0'(0) \right), \quad t \in [0, R) \tag{5.178}$$

and

$$w(t) = \frac{1}{t^2}\left(f'(t) - f'(\theta\, t) \right), \quad \theta \in [0, 1], \quad t \in [0, R) \tag{5.179}$$

are well defined, non-negative, continuous and increasing. Hence, v, w, given by (5.178) and (5.179) are another possible choice for functions satisfying (5.136) and (5.137). Moreover, in view of (5.155) these choices of v and w generalize the average Lipschitz conditions given in Special case-I and in particular those given in Ref. [389].

    Finally, we present in this section some numerical examples. The last example involves the solution of minimization problems.

Example 5.3.1. Let $\mathcal{X} = \mathcal{Y} = R^3$, $\mathcal{D} = U(0, 1)$, $x_0 = (0, 0, 0)$ and define function $\mathcal{F}$ on D by

$$\mathcal{F}(x) = (e^{\zeta_1} - 1, \zeta_2^2 + \zeta_2, \zeta_3). \tag{5.180}$$

We obtain the Jacobian matrix

$$\mathcal{F}'(x) = \begin{pmatrix} e^{\zeta_1} & 0 & 0 \\ 0 & 2\zeta_2 + 1 & 0 \\ 0 & 0 & 1 \end{pmatrix}. \tag{5.181}$$

Using (5.180), (5.181) and $\mathcal{F}'(x^*) = \operatorname{diag}\{1, 1, 1\}$, we get

$$\mathcal{F}'(x^*)^{-1} = \operatorname{diag}\{1, 1, 1\}$$

and

$$L_0 = e - 1 < L = e.$$

Table 5.2   Numerical performance of Algorithm (5.135) for test problems

$\mathcal{F}(x)$	a	b	x_0	x^*	average iter.
$\mathcal{F}_1(x)$	$[10{-}4, 10^{-4}]$	$[0.999, +\infty]$	10 randoms	$[0.7573962468236, 0.0213018765882]$	21
$\mathcal{F}_2(x)$	$[2, 3]$	$[0.1, 0.5]$	10 randoms	$[2.5411, 0.2595]$	7
$\mathcal{F}_3(x)$	$[-3, -2]$	$[3, 0.8]$	10 randoms	$[0.89475, 0.800000]$	8

Example 5.3.2. Algorithm (5.135) consists of two steps. The first step constitute the classical Gauss–Newton step ($z_n = x_n - \mathcal{F}'(x_n)^\dagger \mathcal{F}(x_n)$) while in the second step we find a $\mathcal{J}$-projection ($\mathsf{prox}_{\mathcal{J}}^{Q(x_n)}(z_n)$) in the variable metric $Q(x_n) = \mathcal{F}'(x_n)^\star \mathcal{F}'(x_n)$. Here, we consider the problems where $z_n \in \mathcal{X}$. As a result the projection leaves z_n unchanged. Various problems, such as solving system of nonlinear equations or finding an optimal point of a nonlinear multi-variable functions or fitting a nonlinear curve to given data, are formulated as the box constraints optimization problem

$$\min_{x \in \mathbb{R}^n} \| \mathcal{F}(x) - y \|^2 \quad \text{with} \quad a \leq x \leq b.$$

Algorithm (5.135) is implemented in MatLab. If the distance between two consecutive iterates is less that 10^{-12}, i.e. $\| x_{n+1} - x_n \| \leq 10^{-12}$, then the algorithm stops. We consider the following problems:

$$\mathcal{F}_1(x) := \begin{cases} \dfrac{x_1}{1 - x_1} - 5 \ln\left(\dfrac{0.4\,(1 - x_1)}{x_2} \right) + 4.45977 = 0 \\ x_2 - (0.4 - 0.5\,x_1) = 0 \end{cases}$$

with

$$\begin{pmatrix} x_1 \\ x_2 \end{pmatrix}_0 = \begin{pmatrix} 0.9 \\ 0.5 \end{pmatrix},$$

$$\mathcal{F}_2(x) := \begin{cases} \text{Fit the data points :} \\ (\mu_1, \lambda_1) = (1, 3.2939), \ (\mu_2, \lambda_2) = (2, 4.2699), \ (\mu_3, \lambda_3) = (4, 7.1748), \\ (\mu_4, \lambda_4) = (5, 9.3008), \ (\mu_5, \lambda_5) = (8, 20.2590) \\ \text{to the curve :} \ \lambda_i = x_1 \exp\left(x_2\, \mu_i\right) \end{cases}$$

$$\mathcal{F}_3(x) = 100\,(x_2 - x_1^2)^2 + (1 - x_1)^2 \quad \text{for all} \quad x_1, x_2 \in \mathbb{R}.$$

Here, $\mathcal{F}_1(x)$ is a nonlinear system, $\mathcal{F}_2(x)$ is the problem of nonlinear curve fitting to given data and $\mathcal{F}_3(x)$ is the classical Rosenbrock function in two variables. Nonlinear-system $\mathcal{F}_1(x)$ describes conversion in chemical reactor and it is available at the Nonlinear Equations Library at the website: `http://www.polymath-software.com`. During numerical tests, 10 initial points are randomly selected from the box $a \leq x \leq b$. Results are reported in Table 5.2. In Table 5.2: $a, b \in \mathbb{R}^2$. Table 5.2 reports average number of iterations required for 10 randomly selected initializations.

5.4 Inexact Method and Majorant Conditions

We present a local convergence analysis of inexact Gauss–Newton like methods for solving nonlinear equations in a Banach space setting. Using more precise majorant conditions than in earlier studies such as in Ref. [215], we provide a larger radius of convergence; tighter error estimates on the distances involved and a clearer relationship between the majorant function and the associated least squares problem. Moreover, these advantages are obtained under the same computational cost. Let $F : \mathcal{D} \longrightarrow \mathcal{Y}$ be continuously differentiable. We are concerned with the problem of approximating a locally unique solution $x^\star$ of nonlinear least squares problem

$$\min_{x \in \mathcal{D}} \| F(x) \|^2 . \tag{5.182}$$

A solution $x^\star \in \mathcal{D}$ of (5.182) is also called a least squares solution of the equation $F(x) = 0$. We use the inexact Gauss–Newton like method

$$x_{n+1} = x_n + s_n, \quad \mathcal{B}(x_n)\, s_n = -F'(x_n)^\star\, F(x_n) + r_n \quad \text{for each} \quad n = 0, 1, \cdots , \tag{5.183}$$

where $x_0 \in \mathcal{D}$ is an initial point to generate a sequence $\{x_n\}$ approximating $x^\star$. Here, $A^\star$ denotes the adjoint of the operator A, $\mathcal{B}(x) \in \mathcal{L}(\mathcal{X}, \mathcal{Y})$ is an approximation of $F'(x)^\star\, F'(x)$ $(x \in \mathcal{D})$; r_n is the residual tolerance and the preconditioning invertible matrix $\mathcal{P}$ for the linear systems defining the step s_n satisfy

$$\| \mathcal{P}_n\, r_n \| \le \theta_n \, \| \mathcal{P}_n\, F'(x_n)^\star\, F(x_n) \| \quad \text{for each} \quad n = 0, 1, \cdots . \tag{5.184}$$

If $\theta_n = 0$ for each $n = 0, 1, \cdots$, the inexact Gauss–Newton method reduces to Gauss–Newton method. The convergence of these methods requires among other hypotheses that F' satisfies a Lipschitz condition or F'' is bounded in $\mathcal{D}$. Several authors have relaxed these hypotheses (see Refs. [114, 157]). Ferreira et al. in Refs. [210, 217] have used the majorant condition in the local as well as semi-local convergence of Newton-type method. Argyros and Hilout in Refs. [133, 155] have used the majorant condition to provide a tighter convergence analysis. The local convergence of inexact Gauss–Newton method was examined by Ferreira et al. in Ref. [215] using the majorant condition. It was shown that this condition is better than Wang's condition (see Refs. [302, 428]) in some sence. A certain relationship between the majorant function and operator F was established that unifies two previously unrelated results pertaining to inexact Gauss–Newton methods, which are the result for analytical functions and the one for operators with Lipschitz derivative. We are motivated by the elegant work in Ref. [215] and optimization considerations. Using more precise majorant condition and functions, we provide a new local convergence analysis for inexact Gauss–Newton-like methods under the same computational cost and some advantages: larger radius of convergence; tighter error estimates on the distances $\| x_n - x^\star \|$ for each $n = 0, 1, \cdots$ and a clearer relationship between the majorant function and the associated least squares problems (5.182). These advantages are obtained because we use a center-type majorant condition (see (5.185)) for the computation of inverses involved which is more precise than the majorant condition used in Ref. [215]. Moreover, these advantages are obtained under the same computational cost, since as we will see in this section, the computation of the majorant function requires the computation of the center-majorant function. Furthermore, these advantages are very important in computational mathematics, since we have a wider choice of initial guesses x_0 and fewer computations to obtain a desired error tolerance on the distances $\| x_n - x^\star \|$ for each $n = 0, 1, \cdots$.

Let $A : \mathcal{X} \longrightarrow \mathcal{Y}$ be continuous linear and injective with closed image. In this we denote the Moore–Penrose inverse of operator A by A^+ and $\mathcal{I}$ denotes the identity operator on $\mathcal{X}$ (or $\mathcal{Y}$).

Lemma 5.4.1. *(see Refs. [110, 428]) Let $A, E : \mathcal{X} \longrightarrow \mathcal{Y}$ be two continuous linear operators with closed images. Suppose $B = A + E$, A is injective and $\| E A^+ \| < 1$. Then, B is injective.*

Proposition 5.4.1. *(see Ref. [276]) Let $R > 0$. Suppose $g : [0, R) \longrightarrow \mathbb{R}$ is convex. Then, the following holds*

$$D^+ g(0) = \lim_{u \to 0^+} \frac{g(u) - g(0)}{u} = \inf_{u > 0} \frac{g(u) - g(0)}{u}.$$

Proposition 5.4.2. *(see Ref. [276]) Let $R > 0$ and $\theta \in [0, 1]$. Suppose $g : [0, R) \longrightarrow \mathbb{R}$ is convex. Then, $h : (0, R) \longrightarrow \mathbb{R}$ defined by $h(t) = (g(t) - g(\theta t))/t$ is increasing.*

We examine the local convergence of inexact Gauss–Newton-like method. In order for us to show the main Theorem 5.4.1, we need some auxiliary results. The proofs of some of the results are omitted, since these proofs can be found in Ref. [215] by simply replacing function f by f_0. Assume that $x \in \mathcal{D} \longrightarrow F(x)^\star F(x)$ has $x^\star$ as stationarily point. Let $R > 0$, $c = \| F(x^\star) \|$, $\beta = \| F'(x^\star)^+ \|$ and $\kappa = \sup \{t \in [0, R) : U(x^\star, t) \subseteq \mathcal{D}\}$. Suppose that $F'(x^\star)^\star F(x^\star) = 0$, $F'(x^\star)$ is injective and there exist functions $f_0, f : [0, R) \longrightarrow (-\infty, +\infty)$ continuously differentiable, such that the following assumptions hold

$(\mathcal{H}_0)$

$$\| F'(x) - F'(x^\star) \| \leq f_0'(\| x - x^\star \|) - f_0'(0), \tag{5.185}$$

$$\| F'(x) - F'(x^\star + \tau (x - x^\star)) \| \leq f'(\| x - x^\star \|) - f'(\tau \| x - x^\star \|), \tag{5.186}$$

for all $x \in U(x^\star, \kappa)$ and $\tau \in [0, 1]$;

$(\mathcal{H}_1)$ $f_0(0) = f(0) = 0$ and $f_0'(0) = f'(0) = -1$;

$(\mathcal{H}_2)$ f_0', f' are strictly increasing, $f_0(t) \leq f(t)$ and $f_0'(t) \leq f'(t)$ for each $t \in [0, R)$;

$(\mathcal{H}_3)$ $\alpha_0 = \sqrt{2} \, c \, \beta^2 D^+ f_0'(0) < 1$.

Let

$$0 \le \vartheta < 1, \quad 0 \le \omega_2 < \omega_1 \quad \text{such that} \quad \omega_1 \left(\alpha_0 + \alpha_0 \vartheta + \vartheta \right) + \omega_2 < 1, \qquad (5.187)$$

where α_0 is defined in $(\mathcal{H}_3)$. Define parameters ν_0, ρ_0 and r^0 by

$$\nu_0 := \sup\{ t \in [0, R) : \beta \left(f_0'(t) + 1 \right) < 1 \} \qquad (5.188)$$

$$\rho_0 := \sup\{ t \in [0, \nu_0) : (1+\vartheta) \, \omega_1 \, \beta \, \frac{t \, f'(t) - f(t) + \sqrt{2} \, c \, \beta \left(f_0'(t) + 1 \right)}{t \left(1 - \beta \left(f_0'(t) + 1 \right) \right)} + \omega_1 \, \vartheta + \omega_2 < 1 \}$$

$$(5.189)$$

and

$$r^0 := \min\{ \kappa, \rho_0 \}. \qquad (5.190)$$

We provide the following auxiliary lemmas.

Lemma 5.4.2. *Suppose that* $(\mathcal{H}_0)$–$(\mathcal{H}_3)$ *hold. Then, the constant* ν_0 *defined by (5.188) is positive and* $\beta \left(f_0'(t) + 1 \right) < 1$ *for each* $t \in (0, \nu_0)$.

Lemma 5.4.3. *Suppose that* $(\mathcal{H}_0)$–$(\mathcal{H}_3)$ *hold. Then, the following real functions* h_i *($i = 1, 2, 3$) defined on* $(0, R)$ *by*

$$h_1(t) = \frac{1}{1 - \beta \left(f_0'(0) + 1 \right)}, \quad h_2(t) = \frac{t \, f'(t) - f(t)}{t^2} \quad \text{and} \quad h_3(t) = \frac{f_0'(t) + 1}{t}$$

are increasing. Note also that $h_2 \, h_1$ *and* $h_3 \, h_1$ *are increasing on* $(0, R)$.

Lemma 5.4.4. *Suppose that* $(\mathcal{H}_0)$–$(\mathcal{H}_3)$ *hold. Then, the constant* ρ_0 *defined by (5.189) is positive and the following holds for each* $t \in (0, \rho_0)$:

$$(1 + \vartheta) \, \omega_1 \, \beta \, \frac{t \, f'(t) - f(t) + \sqrt{2} \, c \, \beta \left(f_0'(t) + 1 \right)}{t \left(1 - \beta \left(f_0'(t) + 1 \right) \right)} + \omega_1 \, \vartheta + \omega_2 < 1,$$

where ϑ, ω_1 *and* ω_2 *are defined in (5.187).*

Proof. Using $(\mathcal{H}_1)$, we have that

$$\lim_{t \to 0} \frac{t \, f'(t) - f(t)}{t \left(1 - \beta \left(f_0'(t) + 1 \right) \right)} = \lim_{t \to 0} \frac{f'(t) - (f(t) - f(0))/t}{1 - \beta \left(f'(t) + 1 \right)} \frac{1 - \beta \left(f'(t) + 1 \right)}{1 - \beta \left(f_0'(t) + 1 \right)}$$

$$= \frac{1 - \beta \left(f'(0) + 1 \right)}{1 - \beta \left(f_0'(0) + 1 \right)} \lim_{t \to 0} \frac{f'(t) - (f(t) - f(0))/t}{1 - \beta \left(f'(t) + 1 \right)} = 0.$$

By the convexity of f' and f_0' and Proposition 5.4.1, we get that

$$\lim_{t \to 0} \frac{f'(t) + 1}{t \left(1 - \beta \left(f_0'(t) + 1 \right) \right)} = \lim_{t \to 0} \frac{(f_0'(t) - f_0'(0))/t}{1 - \beta \left(f_0'(t) + 1 \right)} = D^+ f_0'(t).$$

We deduce that

$$\lim_{t \to 0} (1 + \vartheta) \, \omega_1 \, \beta \, \frac{t \, f'(t) - f(t) + \sqrt{2} \, c \, \beta \left(f_0'(t) + 1 \right)}{t \left(1 - \beta \left(f_0'(t) + 1 \right) \right)} + \omega_1 \, \vartheta + \omega_2$$

$$= (1 + \vartheta) \, \omega_1 \, \alpha_0 + \omega_1 \, \vartheta + \omega_2.$$

By (5.187), we have that $\omega_1 \left(\alpha_0 + \alpha_0 \vartheta + \vartheta \right) + \omega_2 < 1$. Then, there exists δ_0 such that

$$(1+\vartheta) \, \omega_1 \, \beta \, \frac{t \, f'(t) - f(t) + \sqrt{2} \, c \, \beta \left(f_0'(t) + 1 \right)}{t \left(1 - \beta \left(f_0'(t) + 1 \right) \right)} + \omega_1 \, \vartheta + \omega_2 < 1 \quad \text{for each} \quad t \in (0, \delta_0).$$

The definition of ρ_0 gives that $\delta_0 \le \rho_0$. The proof of Lemma 5.4.4 is complete. $\square$

Lemma 5.4.5. *Suppose that $(\mathcal{H}_0)$–$(\mathcal{H}_3)$ hold. Then, for each $x \in \mathcal{D}$ such that $x \in U(x^\star, \min\{\nu_0, \kappa\})$, $F'(x)^\star F'(x)$ is invertible and the following estimates hold*

$$\| F'(x)^+ \| \leq \frac{\beta}{1 - \beta\left(f_0'(\| x - x^\star \|) + 1\right)}$$

and

$$\| F'(x)^+ - F'(x^\star)^+ \| \leq \frac{\sqrt{2}\,\beta^2\left(f_0'(\| x - x^\star \|) + 1\right)}{1 - \beta\left(f_0'(\| x - x^\star \|) + 1\right)}.$$

In particular, $F'(x)^\star F'(x)$ is invertible in $U(x^\star, r^0)$.

Proof. Since $x \in \mathcal{D}$ such that $x \in U(x^\star, \min\{\nu_0, \kappa\})$, then $\| x - x^\star \| \leq \nu_0$. By Lemma 5.4.2, (5.185) and the definition of β, we have that

$$\| F'(x^\star)^+ \| \, \| F'(x) - F'(x^\star) \| \leq \beta\left(f_0'(\| x - x^\star \|) - f_0'(0)\right) < 1.$$

Consider operators $A = F'(x^\star)$, $B = F'(x)$ and $E = B - A$. Hence, we have that $\| E A^+ \| \leq \| E \| \, \| A^+ \| < 1$. Then, we deduce the desired result by Lemmas 5.4.1 and 5.3.1. That completes the proof of Lemma 5.4.5. $\qquad\square$

Newton's iteration at a point is a zero of the linearization of F at such a point. Hence, we shall study the linearization error at a point in $\mathcal{D}$:

$$E_F(x, y) := F(y) - (F(x) + F'(x)(x - y)) \quad \text{for each} \quad x, y \in \mathcal{D}. \tag{5.191}$$

We shall bound this error by the error in linearization of the majorant function f:

$$e_f(t, u) := f(u) - (f(t) + f'(t)(u - t)) \quad \text{for each} \quad t, u \in [0, R). \tag{5.192}$$

Define also the Gauss–Newton step to the operator F by

$$S_F(x) = -F'(x)^+ F(x) \quad \text{for each} \quad x \in \mathcal{D}. \tag{5.193}$$

Lemma 5.4.6. *Suppose that $(\mathcal{H}_0)$–$(\mathcal{H}_3)$ hold. If $\| x^\star - x \| < \kappa$, then the following assertion holds $\| E_F(x, x^\star) \| \leq e_f(\| x - x^\star \|, 0)$.*

Lemma 5.4.7. *Suppose that $(\mathcal{H}_0)$–$(\mathcal{H}_3)$ hold. Then, for each $x \in \mathcal{D}$ such that $x \in U(x^\star, \min\{\nu_0, \kappa\})$, the following estimate holds*

$$\| S_F(x) \| \leq \frac{\beta\, e_f(\| x - x^\star \|, 0) + \sqrt{2}\,c\,\beta^2\left(f_0'(\| x - x^\star \|) + 1\right)}{1 - \beta\left(f_0'(\| x - x^\star \|) + 1\right)} + \| x - x^\star \|.$$

Proof. Let $x \in \mathcal{D}$ such that $x \in U(x^\star, \min\{\nu_0, \kappa\})$. Using (5.191) and (5.193), we have that

$$\| S_F(x) \|$$
$$= \| F'(x)^+ (F(x^\star) - (F(x) - F'(x)(x^\star - x))) - (F'(x)^+ - F'(x^\star)^+) F(x^\star)$$
$$+ (x^\star - x) \|$$
$$\leq \| F'(x)^+ \| \, \| E_F(x, x^\star) \| + \| F'(x)^+ - F'(x^\star)^+ \| \, \| F(x^\star) \| + \| x^\star - x \|.$$

Then, we deduce the desired result by Lemmas 5.4.5 and 5.4.6. That completes the proof of Lemma 5.4.7. $\qquad\square$

Lemma 5.4.8. *Let parameters ϑ, ω_1 and ω_2 defined by (5.187). Let ν_0, ρ_0 and r^0 as defined in (5.188), (5.189) and (5.190), respectively. Suppose that $(\mathcal{H}_0)$–$(\mathcal{H}_3)$ hold. For each $x \in U(x^\star, r^0) \setminus \{x^\star\}$, define*

$$x_+ = x + s, \quad \mathcal{B}(x)\, s = -F'(x)^\star\, F(x) + r, \tag{5.194}$$

where $\mathcal{B}(x)$ is an invertible approximation of $F'(x)^\star\, F(x)$ satisfying

$$\| \, \mathcal{B}(x)^{-1}\, F'(x)^\star\, F'(x) \, \| \leq \omega_1, \quad \| \, \mathcal{B}(x)^{-1}\, F'(x)^\star\, F'(x) - \mathcal{I} \, \| \leq \omega_2. \tag{5.195}$$

Suppose also that the forcing term θ and the residuals r (as defined in (5.184)) satisfy

$$\| \, \mathcal{P}\, r \, \| \leq \theta \, \| \, \mathcal{P}\, F'(x)^\star\, F(x) \, \| \quad \text{and} \quad \theta \, \mathrm{cond}\, (\mathcal{P} F'(x)^\star\, F'(x)) \leq \vartheta. \tag{5.196}$$

Then, x_+ is well defined and the following estimate holds

$$\| \, x_+ - x^\star \, \| \leq (1 + \vartheta)\, \omega_1\, \beta\, \frac{f'(\| \, x^\star - x \, \|) \, \| \, x^\star - x \, \| - f(\| \, x^\star - x \, \|)}{\| \, x^\star - x \, \|^2\, (1 - \beta\, (f_0'(\| \, x^\star - x \, \|) + 1))}\, \| \, x^\star - x \, \|^2 +$$
$$\left(\frac{(1 + \vartheta)\, \omega_1\, \sqrt{2}\, c\, \beta^2\, (f_0'(\| \, x - x^\star \, \|) + 1)}{\| \, x^\star - x \, \|\, (1 - \beta\, (f_0'(\| \, x - x^\star \, \|) + 1))} + \omega_1\, \vartheta + \omega_2 \right)\, \| \, x^\star - x \, \| .$$
$$\tag{5.197}$$

In particular, $\| \, x_+ - x^\star \, \| < \| \, x^\star - x \, \|$.

Proof. By Lemma 5.4.5 and since $x \in U(x^\star, r^0)$, we have that $F'(x)^\star\, F'(x)$ is invertible. In view of (5.183) and (5.194), we obtain the identity

$$\begin{aligned}
x_+ - x^\star &= x - x^\star - \mathcal{B}(x)^{-1}\, F'(x)^\star\, (F(x) - F(x^\star)) + \mathcal{B}(x)^{-1}\, r + \\
&\quad \mathcal{B}(x)^{-1}\, F'(x)^\star\, F'(x)\, (F'(x^\star)^+ F(x^\star) - F'(x)^+ F(x^\star)) \\
&= \mathcal{B}(x)^{-1}\, F'(x)^\star\, F'(x)\, F'(x)^+\, (F(x^\star) - (F(x) + F'(x)\, (x^\star - x))) + \\
&\quad \mathcal{B}(x)^{-1}\, r + \mathcal{B}(x)^{-1}\, (F'(x)^\star\, F'(x) - \mathcal{B}(x))\, (x - x^\star) + \\
&\quad \mathcal{B}(x)^{-1}\, F'(x)^\star\, F'(x)\, (F'(x^\star)^+ F(x^\star) - F'(x)^+ F(x^\star)).
\end{aligned}$$
$$\tag{5.198}$$

Using (5.191), (5.193), (5.195), (5.196) and (5.198), we get that

$$\begin{aligned}
\| \, x_+ - x^\star \, \| &\leq \omega_1\, \| \, F'(x)^+ \, \|\, \| \, E_F(x, x^\star) \, \| + \| \, \mathcal{B}(x)^{-1}\, r \, \| + \omega_2\, \| \, x^\star - x \, \| + \\
&\quad \omega_1\, \| \, F'(x)^+ - F'(x^\star)^+ \, \|\, \| \, F(x^\star) \, \| \\
&\leq \omega_1\, \| \, F'(x)^+ \, \|\, \| \, E_F(x, x^\star) \, \| + \omega_1\, \vartheta\, \| \, S_F(x) \, \| + \\
&\quad \omega_2\, \| \, x^\star - x \, \| + \omega_1\, \| \, F'(x)^+ - F'(x^\star)^+ \, \|\, \| \, F(x^\star) \, \| .
\end{aligned}$$
$$\tag{5.199}$$

Using (5.192), (5.199) and Lemmas 5.4.5–5.4.7, we deduce that

$$\| \, x_+ - x^\star \, \| \leq (1 + \vartheta)\, \beta\, \omega_1\, \frac{e_f(\| \, x - x^\star \, \|, 0) + \sqrt{2}\, c\, \beta\, (f_0'(\| \, x - x^\star \, \|) + 1)}{1 - \beta\, (f_0'(\| \, x - x^\star \, \|) + 1)} +$$
$$\omega_1\, \vartheta\, \| \, x - x^\star \, \| + \omega_2\, \| \, x^\star - x \, \|$$
$$\leq (1 + \vartheta)\, \beta\, \omega_1\, \frac{f'(\| \, x - x^\star \, \|) \, \| \, x - x^\star \, \| - f(\| \, x - x^\star \, \|) + \sqrt{2}\, c\, \beta\, (f_0'(\| \, x - x^\star \, \|) + 1)}{1 - \beta\, (f_0'(\| \, x - x^\star \, \|) + 1)}$$
$$+ \omega_1\, \vartheta\, \| \, x - x^\star \, \| + \omega_2\, \| \, x^\star - x \, \| .$$
$$\tag{5.200}$$

Hence, (5.197) holds. Note that if we factorize by $\| \, x^\star - x \, \|$ in the right in (5.197), we deduce that $\| \, x_+ - x^\star \, \| < \| \, x^\star - x \, \|$. The proof of Lemma 5.4.8 is complete. $\square$

Next, we provide the main local convergence result for inexact Gauss–Newton-like method.

Theorem 5.4.1. *Let $F : \mathcal{D} \subseteq \mathcal{X} \longrightarrow \mathcal{Y}$ be a continuously differentiable operator. Let parameters ϑ, ω_1 and ω_2 defined by (5.187). Let ν_0, ρ_0 and r^0 as defined in (5.188), (5.189) and (5.190), respectively. Suppose that $(\mathcal{H}_0)$–$(\mathcal{H}_3)$ hold. Then, sequence $\{x_n\}$ generated by (5.183), starting at $x_0 \in U(x^\star, r^0) \setminus \{x^\star\}$ for the the forcing term θ_n, the residual r_n and the invertible preconditioning matrix $\mathcal{P}_n$ satisfying the following estimates for each $n = 0, 1, \cdots$:*

$$\| \mathcal{P}_n \, r_n \| \leq \theta_n \, \| \, \mathcal{P}_n \, F'(x_n)^\star \, F(x_n) \, \|, \quad 0 \leq \theta_n \, \mathrm{cond}\, (\mathcal{P}_n F'(x_n)^\star \, F'(x_n)) \leq \vartheta,$$

$$\| \mathcal{B}(x_n)^{-1} \, F'(x_n)^\star \, F'(x_n) \| \leq \omega_1 \quad \text{and} \quad \| \mathcal{B}(x_n)^{-1} \, F'(x_n)^\star \, F'(x_n) - \mathcal{I} \| \leq \omega_2$$

is well defined, remains in $U(x^\star, r^0)$ for all $n \geq 0$ and converges to $x^\star$. Moreover, the following estimate holds for each $n = 0, 1, \cdots$

$$\| \, x_{n+1} - x^\star \, \| \leq \Xi_n \, \| \, x_n - x^\star \, \|, \tag{5.201}$$

where

$$\Xi_n = (1 + \vartheta)\, \omega_1 \, \beta \, \frac{f'(\| \, x^\star - x_0 \, \|) \, \| \, x^\star - x_0 \, \| - f(\| \, x^\star - x_0 \, \|)}{\| \, x^\star - x_0 \, \|^2 \, (1 - \beta \, (f_0'(\| \, x^\star - x_0 \, \|) + 1))} \, \| \, x^\star - x_n \, \| +$$
$$\frac{(1 + \vartheta)\, \omega_1 \, \sqrt{2} \, c \, \beta^2 \, (f_0'(\| \, x_0 - x^\star \, \|) + 1)}{\| \, x^\star - x_0 \, \| \, (1 - \beta \, (f_0'(\| \, x_0 - x^\star \, \|) + 1))} + \omega_1 \, \vartheta + \omega_2.$$

Proof. By induction argument, Lemmas 5.4.5 and 5.4.8, $\{x_n\}$ starting at $x_0 \in U(x^\star, r^0) \setminus \{x^\star\}$ is well defined in $U(x^\star, r^0)$. By letting $x_+ = x_{n+1}$, $x = x_n$, $r = r_n$, $\mathcal{P} = \mathcal{P}_n$, $\theta = \theta_n$ and $\mathcal{P} = \mathcal{P}_n$ in (5.194)–(5.196), we get that

$$\| \, x_{n+1} - x^\star \, \| \leq$$
$$(1 + \vartheta)\, \omega_1 \, \beta \, \frac{f'(\| \, x^\star - x_n \, \|) \, \| \, x^\star - x_n \, \| - f(\| \, x^\star - x_n \, \|)}{\| \, x^\star - x_n \, \|^2 \, (1 - \beta \, (f_0'(\| \, x^\star - x_n \, \|) + 1))} \, \| \, x^\star - x_n \, \|^2 +$$
$$\left(\frac{(1 + \vartheta)\, \omega_1 \, \sqrt{2} \, c \, \beta^2 \, (f_0'(\| \, x_n - x^\star \, \|) + 1)}{\| \, x^\star - x_n \, \| \, (1 - \beta \, (f_0'(\| \, x_n - x^\star \, \|) + 1))} + \omega_1 \, \vartheta + \omega_2 \right) \| \, x^\star - x_n \, \| .$$

We also have by Lemma 5.4.8 that $\| \, x_n - x^\star \, \| \leq \| \, x_0 - x^\star \, \|$ for each $n = 1, 2, \cdots$. Hence, (5.201) holds. Proposition 5.4.4 implies that $x_{n+1} \in U(x^\star, r^0)$ and $\lim_{n \longrightarrow \infty} x_n = x^\star$. The proof of Theorem 5.4.1 is complete. $\square$

Remark 5.4.1. If $f(t) = f_0(t)$ for each $t \in [0, R)$, then, Theorem 5.4.1 reduces to [Theorem 7] in Ref. [215]. In particular, we have in this case that $\nu = \nu_0$, $\rho = \rho_0$, $\delta = \delta_0$, $\alpha = \alpha_0$, $r = r^0$ and $D^+ f(0) = D^+ f_0(0)$, where ν, ρ, δ, α, r and $D^+ f(0)$ are defined, respectively, as ν_0, ρ_0, δ_0, α_0, r^0 and $D^+ f_0(0)$ by setting $f_0(t) = f(t)$. Otherwise, i.e., if

$$f_0(t) < f(t) \quad \text{and} \quad f_0'(t) < f'(t) \quad \text{for each} \quad t \in [0, R), \tag{5.202}$$

then, we have that

$$\nu \geq \nu_0, \quad \rho \leq \rho_0, \quad \delta \leq \delta_0, \quad \alpha \geq \alpha_0, \quad r \leq r^0 \quad \text{and} \quad D^+ f(0) \geq D^+ f_0(0). \tag{5.203}$$

Note that these advantages are obtained under the same computational cost, since in practice, the computation of function f requires that of f_0. Note also that the local results in Refs. [185, 186, 210, 214] are also extended, since these are special cases of Theorem 5.4.1. In particular, if $\vartheta = 0$ (i.e., if $\theta_n = r_n = 0$ for each $n = 0, 1, \cdots$) in Theorem 5.4.1, we improve the convergence of Gauss–Newton like method under majorant condition, which for $\omega_1 = 1$ and $\omega_2 = 0$ has been obtained in [Theorem 7] of Ref. [213]. These results extend those obtained by Chen and Li in Refs. [185, 186] given only for the case $c = 0$. Moreover, if $c = 0$ and $F'(x^\star)$ is invertible, we extend the convergence of inexact Newton-like methods under majorant condition, which was obtained in [Theorem 4] of Ref. [210]. Furthermore, if $c = \vartheta = \omega_2 = 0$, $\omega_1 = 1$ and $F'(x^\star)$ is invertible in Theorem 5.4.1, we extend the convergence of Newton's method under majorant condition obtained in [Theorem 2.1] of Ref. [210].

We shall show how to choose functions f_0 and f so that (5.202) is satisfied. We present two special cases of Theorem 5.4.1. The first one is based on the center-Lipschitz and Lipschitz conditions (see Refs. [110, 120]). The second one is based on Wang's condition (see Ref. [428]), which generalized Smale's alpha theory for analytic functions (see Ref. [409]).

Remark 5.4.2. Let us define functions f, $f_0 : [0, \kappa] \longrightarrow \mathbb{R}$ by

$$f_0(t) = \frac{L_0\, t^2}{2} - t \quad \text{and} \quad f(t) = \frac{L\, t^2}{2} - t,$$

where L_0 and L are the center-Lipschitz and Lipschitz constants, respectively. We have that $f_0(0) = f(0) = 0$ and $f_0'(0) = f'(0) = -1$. Set also $R = 1/L$. Then, it can easily be seen that Theorem 5.4.1 specializes to the following proposition.

Proposition 5.4.3. *Let* $F : \mathcal{D} \subset \mathcal{X} \longrightarrow \mathcal{Y}$ *be continuously differentiable operator. Let* $x^\star \in \mathcal{D}$, *such that* $F'(x^\star)^\star F(x^\star) = 0$, $F'(x^\star)$ *is injective. Let* $c = \| F(x^\star) \|$, $\beta = \| F'(x^\star)^+ \|$ *and* $\kappa = \sup\{t \geq 0 : U(x^\star, t) \subseteq \mathcal{D}\}$. *Suppose that there exist* $L_0 > 0$, $L > 0$ *such that*

$$\| F'(x) - F'(x^\star) \| \leq L_0 \| x - x^\star \| \quad \text{and} \quad \| F'(x) - F'(y) \| \leq L \| x - y \|,$$

for each $x, y \in U(x^\star, \kappa)$. *Suppose that* $\alpha_0 = \sqrt{2}\, c\, \beta^2\, L < 1$. *Let parameters* ϑ, ω_1 *and* ω_2 *defined by (5.187). Let*

$$r^0 = \min\{\kappa, \frac{2\,(1 - \omega_1\, \vartheta - \omega_2) - 2\,\sqrt{2}\, c\, L_0\, \beta^2\, \omega_1\, (1 + \vartheta)}{\beta\,(L\,(1 + \vartheta)\, \omega_1 + 2\, L_0\,(1 - \omega_1\, \vartheta - \omega_2))}\}.$$

Then, sequence $\{x_n\}$ *generated by (5.183) with* $r_n = 0$ *and* $\mathcal{B}(x_n) = F'(x_n)^\star\, F'(x_n)$, *starting at* $x_0 \in U(x^\star, r^0) \setminus \{x^\star\}$ *for the the forcing term* θ_n *and the invertible preconditioning matrix* $\mathcal{P}_n$ *satisfying the following estimates for each* $n = 0, 1, \cdots$:

$$\| \mathcal{P}_n\, r_n \| \leq \theta_n \| \mathcal{P}_n\, F'(x_n)^\star\, F(x_n) \|, \quad 0 \leq \theta_n\, \text{cond}\,(\mathcal{P}_n F'(x_n)^\star\, F'(x_n)) \leq \vartheta,$$

$$\| \mathcal{B}(x_n)^{-1}\, F'(x_n)^\star\, F'(x_n) \| \leq \omega_1 \quad \text{and} \quad \| \mathcal{B}(x_n)^{-1}\, F'(x_n)^\star\, F'(x_n) - \mathcal{I} \| \leq \omega_2$$

is well defined, remains in $U(x^\star, r^0)$ for all $n \geq 0$ and converges to $x^\star$. Moreover, the following estimate holds for each $n = 0, 1, \cdots$

$$\| x_{n+1} - x^\star \| \leq \Delta_n \| x_n - x^\star \|, \tag{5.204}$$

where

$$\Delta_n = \frac{(1 + \vartheta)\,\omega_1\,\beta\,L}{2\,(1 - \beta\,L_0 \,\| x_0 - x^\star \|)} \,\| x^\star - x_n \| + \frac{(1 + \vartheta)\,\omega_1\,\sqrt{2}\,c\,\beta^2\,L_0}{1 - \beta\,L_0\,\| x_0 - x^\star \|} + \omega_1\,\vartheta + \omega_2.$$

Remark 5.4.3.

(a) If $L_0 = L$, Proposition 5.4.3 reduces to [Theorem 16] in Ref. [215]. Moreover, if $\vartheta = 0$, Proposition 5.4.3 reduces to [Corollary 17] in Ref. [215]. Furthermore, if $c = 0$, then, Proposition 5.4.3 reduces to [Corollary 6.1] in Ref. [186].

(b) If $F(x^\star) = 0$, $F'(x^\star)^+ = F'(x^\star)^{-1}$ and $L_0 < L$, then Theorem 5.4.1 improves the corresponding results on inexact Newton-like methods (see Refs. [246, 279, 305, 311]). In particular for Newton's method. Set $c = \vartheta = \omega_1 = \omega_2 = 0$. Then, we get that $r^0 = \min\{\kappa, \dfrac{2}{\beta\,(2\,L_0 + L)}\}$. This radius is at least as large as the one provided by Traub in Ref. [418], which is given by $r_0^0 = \min\{\kappa, \dfrac{2}{3\,\beta\,L}\}$.

Remark 5.4.4. Let $\gamma \geq \gamma_0$. Let us define functions $f, f_0 : [0, \kappa] \longrightarrow \mathbb{R}$ by

$$f_0(t) = \frac{t}{1 - \gamma_0\,t} - 2\,t \quad \text{and} \quad f(t) = \frac{t}{1 - \gamma\,t} - 2\,t.$$

Then, we have that $f_0(0) = f(0) = 0$, $f_0'(0) = f'(0) = -1$. Set also $R = 1/\gamma$. Note also that

$$f_0'(t) = \frac{1}{(1 - \gamma_0\,t)^2} - 2, \quad f'(t) = \frac{1}{(1 - \gamma\,t)^2} - 2,$$

$$f_0''(t) = \frac{2\,\gamma_0}{(1 - \gamma_0\,t)^3} \quad \text{and} \quad f''(t) = \frac{2\,\gamma}{(1 - \gamma\,t)^3}.$$

Remark 5.4.5. If F is an analytic function, Smale in Ref. [409] used the following choice

$$\gamma = \sup_{n \in \mathbb{N}^\star} \| \frac{F^{(n)}(x^\star)}{n!} \|^{\frac{1}{n}} < +\infty.$$

Using the above definitions and choices of functions (see Remark 5.4.4, Definitions 2.3.1 and 2.3.2), the corresponding specialization of Theorem 5.4.1 along the lines of Proposition 5.4.3 can be obtained. However, we leave this part to the interested reader. Note that clearly if $\gamma_0 = \gamma$, this result reduces to [Theorem 18] in Ref. [215], which in turn reduces to [Example 1] in Ref. [186] if $c = 0$. Otherwise (i.e., if $\gamma_0 < \gamma$), our result is an improvement.

For example where $\gamma_0 < \gamma$ in the case when $F(x^\star) = 0$, $F'(x)^+ = F'(x)^{-1}$ and $c = \vartheta = \omega_1 = \omega_2 = 0$, see section 2.3 of Chapter 1.

5.5　Exercises

5.5.1. Show Lemmas 5.4.1, 5.4.2, 5.4.3, Propositions 5.4.1 and 5.4.2.

5.5.2. Let L and L_0 be non-decreasing functions on $[0, R)$, where $R \in \overline{\mathbb{R}}_+$ and $L_0(t) \leq L(t)$ for all $t \in [0, R)$. Let $\beta > 0$ and $0 \leq \lambda < 1$ be given parameters. Define function $g : [0, R] \longrightarrow \mathbb{R}$ by $g(t) = \beta - t + \int_0^t L_0(u)\,(t - u)\,du$. Define majorizing function $h_\lambda : [0, R] \longrightarrow \mathbb{R}$ corresponding to a fixed pair (λ, L) by $h_\lambda(t) = \beta - (1 - \lambda)\,t + \int_0^t L(u)\,(t - u)\,du$. Define also

$$r_\lambda := \sup\left\{ r \in (0, R) \; : \; \int_0^r L(u)\,du \leq 1 - \lambda \right\},$$

$$b_\lambda := (1 - \lambda)\,r_\lambda - \int_0^{r_\lambda} L(u)\,(r_\lambda - u)\,du \quad \text{and} \quad \Delta = \int_0^R L(u)\,du.$$

(a) show that for all $t \in [0, R]$, we have

$$g'(t) = -1 + \int_0^t L_0(u)\,du, \quad g''(t) = L_0(t),$$

$$h_\lambda'(t) = -(1 - \lambda) + \int_0^t L(u)\,du \quad \text{and} \quad h_\lambda''(t) = L(t).$$

(b) show that $g'(t) \leq h_0'(t)$ for all $t \in [0, R]$.

(c) show that

$$r_\lambda = \begin{cases} R & \text{if } \Delta < 1 - \lambda \\ t_\lambda & \text{if } \Delta \geq 1 - \lambda. \end{cases}$$

(d) show the following inequalities

$$\begin{cases} b_\lambda \geq \displaystyle\int_0^{r_\lambda} L(u)\,u\,du & \text{if } \quad \Delta < 1 - \lambda, \\[4mm] b_\lambda = \displaystyle\int_0^{r_\lambda} L(u)\,u\,du & \text{if } \quad \Delta \geq 1 - \lambda. \end{cases}$$

Assume that $\beta \leq b_\lambda$. show that the following hold

(i) Function h_λ is strictly decreasing on $[0, r_\lambda]$, and has exactly one zero $t_\lambda^\star \in [0, r_\lambda]$, such that $\beta < t_\lambda^\star$.

(ii) Sequence $\{s_{\lambda,n}\}$ is strictly increasing, and converges to $t_\lambda^\star$, where the scalar sequence $\{s_{\lambda,n}\}$ is given by

$$s_{\lambda,0} = 0, \quad s_{\lambda,n+1} = s_{\lambda,n} - \frac{h_\lambda(s_{\lambda,n})}{g'(s_{\lambda,n})}, \quad (n \geq 0).$$

Chapter 6

Halley's Method

Halley's method is also an alternative of Newton's method. The convergence analysis of Halley's method is studied by several authors using Lipschitz and Hölder type conditions. In this chapter, we provide some new results improving earlier studies on this method.

6.1 Semi-local Convergence

We provide in this section a semi-local convergence analysis for Halley's method using convex majorants in order to approximate a locally unique solution of a nonlinear operator equation. Our results reduce and improve earlier ones in special cases. We are concerned with the problem of approximating a locally unique solution $x^\star$ of (1.1). A survey on recent results for Newton-type methods

$$x_{n+1} = x_n - A(x_n)^{-1}F(x_n) \quad (n \geq 0), \quad (x_0 \in \Omega) \tag{6.1}$$

under very general Lipschitz conditions can be found in Ref. [110], and the references there (see also Refs. [100, 98, 290, 368, 368]). Here, $A(x) \in L(X,Y)$, the space of bounded linear operators from X into Y. Note that if $A(x) = F'(x)$ $(x \in \Omega)$, then we obtain Newton's method. If, $A(x) = F'(x)(I - L_F(x))$ $(x \in \Omega)$, where $L_F(x) = \dfrac{1}{2}F'(x)^{-1}F''(x)F'(x)^{-1}F(x)$, then, we obtain Halley's method (see Refs. [43, 110, 206, 261, 290])

$$x_{n+1} = x_n - [I - L_F(x_n)]^{-1}F'(x_n)^{-1}F(x_n) \quad (n \geq 0), \quad (x_0 \in \Omega). \tag{6.2}$$

Although sufficient convergence conditions and error estimates on the distances $\|x_{n+1} - x_n\|$, $\|x_n - x^\star\|$ for Newton's method or Halley's method have been given as special cases of Newton-type methods, a direct approach may yield a finer error analysis. Ferreira and Svaiter in Ref. [217] used the special majorant conditions

$$\| F'(x_0)^{-1}[F'(y) - F'(x)] \| \leq f'(\|y - x\| + \|x - x_0\|) - f'(\|x - x_0\|)$$
$$x, y \in U(x_0, R) \quad \text{for some} \quad R > 0 \tag{6.3}$$

and

$$\|y - x\| + \|x - x_0\| < R$$

335

to provide an elegant Kantorovich-type semi-local convergence analysis for Newton's method. Their corresponding local convergence analysis was given in Ref. [210]. Function $f : [0, R) \to (-\infty, \infty)$ is continuously differentiable satisfying $f(0) > 0$, having zeros in $(0, R)$, whereas f' is convex, strictly increasing with $f'(0) = -1$. Their analysis provides a relationship between the majorizing function f and the nonlinear operator F. Moreover, the assumptions for the Q-quadratic convergence are relaxed.

In Refs. [100, 98, 110, 151] we have shown that under the same hypotheses and computational cost, weaker sufficient conditions and tighter error bounds can be obtained for Newton's method. Indeed, in view of (6.3) there exists a function f_0 such that

$$\|F'(x_0)^{-1}[F'(y) - F'(x_0)]\| \leq f_0'(\|y - x_0\|) - f_0'(0) \tag{6.4}$$

for all $y \in U(x_0, R)$, where function f_0 has the same properties as f but

$$f_0(t) \leq f(t) \quad t \in (0, R] \tag{6.5}$$

holds in general and $f(t)/f_0(t)$ can be arbitrarily large.

Motivated by the above advantages of our technique over earlier ones for Newton's method, we extend our approach in the case of Halley's method (6.2). In a way analogous to (6.3) and (6.4), we consider majorant conditions:

$$\|F'(x_0)^{-1}[F''(y) - F''(x)]\| \leq f''(\|y - x\| + \|x - x_0\|) - f''(\|x - x_0\|), \tag{6.6}$$

for all $x, y \in U(x_0, R)$, where $\|y - x\| + \|x - x_0\| < R$ and $f : [0, R) \to (-\infty, \infty)$ is a twice continuously differentiable function. Here, $F''(z)$ denotes a bilinear operator: $X \times X \to Y$ for each fixed $z \in X$. For simplicity we use $F''(z)z_1z_2$, $F''(z)z_1^2$ instead of the notations $F''(z)(z_1, z_2)$, $F''(z)(z_1, z_1)$, respectively for all z_1 and $z_2 \in X$. Moreover, if $B : X \times X \to Y$ is a bilinear operator and $L : X \to X$ is a linear operator. Then if $x \in X$, by BLx^2 we mean $B(L(x), x)$ (or $BLxx$). Furthermore, LBx^2 is used to denote $LB(x)(x)$. We assume the following conditions hold:

(H_1) $f(0) > 0$, $f''(0) > 0$, $f'(0) = -1$;
(H_2) f'' is convex and strictly increasing in $[0, R)$;
(H_3) f has zeros in $(0, R)$. Denote by $t^\star$ the minimal zero and assume $f'(t^\star) < 0$.

In view of (6.6) there exists a twice continuously differentiable function f_0 satisfying $(H_1) - (H_3)$ (replacing f by f_0 and $t^\star$ by $t_0^\star$) with $t_0^\star \leq t^\star$ such that

$$f_0(t) \leq f(t), \tag{6.7}$$

$$\|F'(x_0)^{-1}[F''(y) - F''(x_0)]\| \leq f_0''(\|y - x_0\|) - f_0''(0) \quad \text{for all } y \in U(x_0, R). \tag{6.8}$$

We shall also assume that

(H_4) $-\dfrac{1}{f_0'(t)} \leq -\dfrac{1}{f'(t)}$ for $t \in (0, t^\star)$;
(H_5) $\|F'(x_0)^{-1}F(x_0)\| \leq f(0)$, $\|F'(x_0)^{-1}F''(x_0)\| \leq f''(0)$

and

(H_6) $U(x_0, R) \subseteq \Omega$.

We present the semi-local convergence analysis of Halley's method (6.2) under conditions $(H_1) - (H_6)$. Special cases and applications are also given in this section. It is convenient for us to define

$$H_F(x) = x - (I - L_F(x))^{-1} F'(x)^{-1} F(x), \quad x \in \Omega,$$

$$H_f(t) = t - \frac{1}{1 - L_f(t)} \frac{f(t)}{f'(t)}, \quad t \in [0, R),$$

$$H_{f_0}(t) = t - \frac{1}{1 - L_{f_0}(t)} \frac{f(t)}{f_0'(t)}, \quad t \in [0, R),$$

where

$$L_f(t) = \frac{f(t) f''(t)}{2(f'(t))^2}, \quad L_{f_0}(t) = \frac{f(t) f''(t)}{2(f_0'(t))^2},$$

$$s_0 = 0, \quad s_{n+1} = H_f(s_n) \ (n \geq 0), \tag{6.9}$$

$$t_0 = 0, \quad t_{n+1} = H_{f_0}(t_n) \ (n \geq 0). \tag{6.10}$$

A simple inductive argument shows

$$t_n \leq s_n, \tag{6.11}$$

$$0 \leq t_{n+1} - t_n \leq s_{n+1} - s_n \tag{6.12}$$

and

$$t_0^\star = \lim_{n \to \infty} t_n \leq t^\star = \lim_{n \to \infty} s_n. \tag{6.13}$$

Inequalities (6.11) and (6.12) are strict if $-\dfrac{1}{f_0'(t)} < -\dfrac{1}{f'(t)}$.

Lemma 6.1.1. *(see Ref. [276]) If $p : [0, R) \to (-\infty, \infty)$ is continuously differentiable and convex, then*

(1) $(1 - \theta) p'(\theta t) \leq \dfrac{p(t) - p(\theta t)}{t} \leq (1 - \theta) p'(t)$ for all $t \in (0, R)$, $\theta \in [0, 1]$;

(2) $\dfrac{p(x) - p(\theta x)}{x} \leq \dfrac{p(y) - p(\theta y)}{y}$ for all $x, y \in [0, R)$, $x < y$, $\theta \in [0, 1]$.

Lemma 6.1.2. *(see Ref. [276]) Let $p : I \subseteq (-\infty, \infty) \to (-\infty, \infty)$ be convex. Then*

(1) For any $z_0 \in int(I)$ there exists left side derivative D^- given by

$$D^- p(z_0) := \lim_{z \to z_0^-} \frac{p(z_0) - p(z)}{z_0 - z} = \sup_{z < z_0} \frac{p(z_0) - p(z)}{z_0 - z};$$

(2) If $x, y, z \in I$ and $x \le y \le z$, then

$$p(y) - p(x) \le [p(z) - p(x)] \frac{y - x}{z - x}.$$

Using the above standard results from convex analysis (see Refs. [110, 276]), Halley's method (see Ref. [110]) and majorant properties (see Refs. [110, 217, 210]), we arrive at the following results on majorizing sequences for Halley's method:

Lemma 6.1.3. *Let f_0, $f : [0, R) \to (-\infty, \infty)$ be twice continuously differentiable functions satisfying $(H_1) - (H_4)$. Then, the following hold:*

(1)(a) f_0', f' are strictly convex and strictly increasing on $[0, R)$;

    (b) f_0, f are strictly convex on $[0, R)$, $f_0(t) > 0$ for $t \in [0, t_0^\star)$, $f(t) > 0$ for $t \in (0, t^\star)$, $f_0(t)$, $f(t)$ have at most one zero on $[t_0^\star, R)$, $[t^\star, R)$, respectively;

    (c) $-1 < f_0'(t) < 0 \; t \in (0, t_0^\star)$, $-1 < f'(t) < 0 \; t \in (0, t^\star)$.

(2) $0 \le L_{f_0}(t) \le \dfrac{1}{4}$ for all $t \in [0, t_0^\star]$, $0 \le L_f(t) \le \dfrac{1}{4}$ for all $t \in [0, t^\star]$, $L_{f_0}(t) \le L_f(t)$ for all $t \in [0, t_0^\star]$.

(3) $t < H_{f_0}(t) < t_0^\star$ for all $t \in [0, t_0^\star)$, $f_0'(t_0^\star) < 0 \Leftrightarrow \exists \, t \in (t_0^\star, R)$ such that $f_0(t) < 0$, $t < H_f(t) < t^\star$ for all $t \in [0, t^\star)$, $f'(t^\star) < 0 \Leftrightarrow \exists \, t \in (t^\star, R)$ such that $f(t) < 0$.

(4)

$$t^\star - H_{f_0}(t) \le g_0(t^\star, t_0^\star)(t^\star - t)^3 \quad t \in [0, t_0^\star),$$

$$t^\star - H_f(t) \le g(t^\star, t^\star)(t^\star - t)^3 \quad t \in [0, t^\star) \quad and \quad g_0(t^\star, t_0^\star) \le g(t^\star, t^\star),$$

where

$$g_0(u, v) = \frac{1}{3} \frac{f''(u)}{f_0'(v)^2} + \frac{2}{9} \frac{D^- f''(u)}{-f_0'(v)} \quad and \quad g(u, v) = \frac{1}{3} \frac{f''(u)}{f'(v)^2} + \frac{2}{9} \frac{D^- f''(u)}{-f'(v)}.$$

Moreover, sequences $\{s_n\}$, $\{t_n\}$ converge to $t^\star$, $t_0^\star$ respectively with Q-cubic order.

Proof.   We show the results for f. Then in an identical way the results for f_0 follow.

(1) Item (a) follows from (H_2) and $f''(0) > 0$ in (H_1). (a) implies that f is strictly convex. By (H_1), (a) and $f(t^\star) = 0$, we get $f(t) = 0$ has at most one zero on $(t^\star, R)$. In view of $f(t^\star) = 0$ and $f(0) > 0$, we obtain $f(t) > 0$ for $t \in [0, t^\star)$. To show (b) we have by Lemma 6.1.1 that

$$f'(t) < \frac{f(t^\star) - f(t)}{t^\star - t}, \quad t \in [0, t^\star).$$

Hence, $0 = f(t^\star) > f(t) + f'(t)(t^\star - t)$. But $f(t) > 0$ on $[0, t^\star)$, so $f'(t) < 0$. Moreover, f' is strictly increasing and $f'(0) = -1$ implies $f'(t) > -1$ for $t \in (0, t^\star)$. That completes the proof for (c).

(2) Define function

$$q(s) = f(t) + f'(t)(s - t) + \frac{1}{2}f''(t)(s - t)^2 \quad s \in [t, t^\star].$$

By $1(b)$, we get $q(t) = f(t) > 0$ and

$$q(t^\star) = f(t) + f'(t)(t^\star - t) + \frac{1}{2}f''(t)(t^\star - t)^2. \tag{6.14}$$

We also have by Taylor's formula

$$\begin{aligned}
f(t^\star) = f(t) &+ f'(t)(t^\star - t) + \frac{1}{2}f''(t)(t^\star - t)^2 \\
&+ \int_0^1 (1 - \theta)[f''(t + \theta(t^\star - t)) - f''(t)](t^\star - t)^2 d\theta.
\end{aligned} \tag{6.15}$$

Since $f(t^\star) = 0$ and function f'' is increasing we have by (6.14), (6.15) $q(t^\star) \leq 0$. Hence there is a zero of q in $[t, t^\star]$. That is the discriminant of q is greater than or equal to zero $\Rightarrow f'(t)^2 - 2f''(t)f(t) \geq 0 \Rightarrow 0 \leq f''(t)f(t)/(f'(t))^2 \leq \frac{1}{2}$. That completes the proof of item (2).

(3) For $t \in [0, t^\star)$ we have by the above $f(t) > 0$, $-1 < f'(t) < 0$ and $0 \leq L_f(t) \leq \frac{1}{4}$ which imply $t < H_f(t)$. Using Lemma 6.1.2 (1) and (H_2) we have $D^- f''(t) > 0$. That is

$$D^- H_f(t) = \frac{f^2(t)[3f''(t)^2 - 2f'(t)D^- f''(t)]}{(f(t)f''(t) - 2f'(t))^2} > 0, \quad t \in (0, t^\star].$$

Hence $H_f(t) < H_f(t^\star) = t^\star$, $t \in [0, t^\star)$. Thus, the first part is shown. For the second part we have that there exists $t \in (t^\star, R)$ such that $f(t) < 0$. Conversely, since $f(t^\star) = 0$, by Lemma 6.1.1, we get that $f(t) > f(t^\star) + f'(t^\star)(t - t^\star)$ for $t \in [t^\star, R) \Rightarrow f'(t^\star) < 0$. By (H_3), $f'(t^\star) < 0 \Rightarrow f(t^{\star\star}) = 0$ for some $t^{\star\star} \in (t^\star, R)$. Moreover, $f(t) < 0$ for some $t \in [t^\star, R)$. A similar result follows for function f_0. Hence, we have that

$$f_0'(t_0^\star) < 0 \Rightarrow \begin{cases} f_0(t_0^{\star\star}) = 0 & t_0^{\star\star} \in (t_0^\star, R) \\ f_0(t) < 0 & t \in (t_0^{\star\star}, R) \end{cases}$$

and

$$f'(t^\star) < 0 \Rightarrow \begin{cases} f(t^{\star\star}) = 0 & t^{\star\star} \in (t^\star, R) \\ f(t) < 0 & t \in (t^{\star\star}, R). \end{cases}$$

(4) From the definition of H_f we have in turn:

$$\begin{aligned}
t^\star - H_f(t) &= \frac{1}{1 - L_f(t)}[(1 - L_f(t))(t^\star - t) + \frac{f(t)}{f'(t)}] \\
&= -\frac{1}{f'(t)(1 - L_f(t))} \int_0^1 [f''(t + \theta(t^\star - t)) - f''(t)](t^\star - t)^2(1 - t)d\theta \\
&\quad + \frac{t^\star - t}{2(1 - L_f(t))} \frac{f''(t)}{f'(t)^2} \int_0^1 f''(t + \theta(t^\star - t))(t^\star - t)^2(1 - \theta)d\theta.
\end{aligned}$$

By the convexity of f'' $t < t^\star$ and Lemma 6.1.2 (2) we get

$$f''(t + \theta(t^\star - t)) - f''(t) \leq [f''(t^\star) - f''(t)]\frac{\theta(t^\star - t)}{t^\star - t}.$$

In view of the fact that f'' is strictly increasing we get

$$t^\star - H_f(t) \leq -\frac{f''(t^\star) - f''(t)}{6f'(t)(1 - L_f(t))}(t^\star - t)^2 + \frac{f''(t^\star)f''(t)}{4(f'(t))^2(1 - L_f(t))}(t^\star - t)^3$$

$$\leq \frac{2}{9}\frac{f''(t^\star) - f''(t)}{-f'(t)}(t^\star - t)^2 + \frac{1}{3}\frac{f''(t^\star)^2}{f'(t^\star)^2}(t^\star - t)^3.$$

$$(6.16)$$

Since $f'(t) < 0$, $f''(0) > 0$, functions f', f'' are increasing on $[0, t^\star)$ and $0 \leq L_f(t) \leq \frac{1}{4}$. We also have

$$\frac{f''(t^\star) - f''(t)}{-f'(t)} \leq \frac{f''(t^\star) - f''(t)}{-f'(t^\star)} = \frac{f''(t^\star) - f''(t)}{-f'(t^\star)(t^\star - t)}(t^\star - t) \leq \frac{D^-f''(t^\star)}{-f'(t^\star)}(t^\star - t)$$

$$(6.17)$$

by Lemma 6.1.2 (1).

The proof is finished by combining (6.16) and (6.17). $\qquad\square$

We also need some lemmas relating F to f_0 and f.

Lemma 6.1.4. *Assume there exists $x_0 \in \Omega$ such that $F'(x_0)^{-1} \in L(Y, X)$, $f_0'(0) = -1$, $-1 < f_0'(t) < 0$ for $t \in (0, t_0^\star)$ and $\|x - x_0\| \leq t < t_0^\star$, where $f_0 : [0, t_0^\star] \to (-\infty, \infty)$ is twice continuously differentiable and satisfies (6.8). Then, $F'(x)^{-1} \in L(Y, X)$ and*

$$\|F'(x)^{-1}F'(x_0)\| \leq -\frac{1}{f_0'(\|x - x_0\|)} \leq -\frac{1}{f_0'(t)}. \qquad (6.18)$$

Proof. Let $x \in \overline{U}(x_0, t)$, then we have by Taylor's formula

$$F'(x) = F'(x_0) + \int_0^1 [F''(x_0 + \theta(x - x_0)) - F''(x_0)](x - x_0)d\theta + F''(x_0)(x - x_0), \quad (6.19)$$

so, in view of (6.8) and (6.19), we obtain in turn that

$$\|F'(x_0)^{-1}[F'(x) - F'(x_0)]\|$$

$$\leq \int_0^1 \|F'(x_0)^{-1}[F''(x_0 + \theta(x - x_0)) - F''(x_0)]\|\|x - x_0\|d\theta$$

$$+ \|F'(x_0)^{-1}F''(x_0)\|\|x - x_0\| \qquad (6.20)$$

$$\leq \int_0^1 [f_0''(\theta\|x - x_0\|) - f_0''(0)]\|x - x_0\|d\theta + f_0''(0)\|x - x_0\|.$$

But $f_0'(0) = -1$ and $-1 < f_0'(t) < 0$ by hypotheses. Hence, we get by (6.20)

$$\|F'(x_0)^{-1}[F'(x) - F'(x_0)]\| \leq f_0'(t) - f_0'(0). \qquad (6.21)$$

It follows from (6.21) and the Banach lemma on invertible operators that $F'(x)^{-1} \in L(Y, X)$ so that (6.18) is satisfied. That completes the proof of the lemma. $\qquad\square$

Remark 6.1.1. Clearly f can replace f_0 in Lemma 6.1.4. However, in view of (H_4), (6.18) is a tighter upper bound for $\|F'(x)^{-1}F'(x_0)\|$ than

$$\|F'(x)^{-1}F'(x_0)\| \leq -\frac{1}{f'(\|x-x_0\|)} \leq -\frac{1}{f'(t)} \quad t \in (0, t^\star). \tag{6.22}$$

This observation leads to more precise majorizing sequence $\{t_n\}$ (see (6.10)) than $\{s_n\}$ (see (6.9)) for Halley's method (6.2). The same observation leads to the advantages already stated in the introduction for Newton's method.

With the exception of the uniqueness part, the following semi-local result for Halley's method (6.2) uses the standard proofs for this method (see Refs. [43, 110, 151, 247, 261]), Lemmas 6.1.3, 6.1.4 and the formulae

$$F(x_{n+1}) = \frac{1}{2}F''(x_n)L_F(x_n)(x_{n+1}-x_n)^2 +$$
$$\int_0^1 [F''(x_n + \theta(x_{n+1}-x_n)) - F''(x_n)](x_{n+1}-x_n)^2 d\theta, \tag{6.23}$$

$$x^\star - x_{n+1} =$$
$$-\Gamma_F(x_n)F'(x_n)^{-1}\int_0^1 [F''(x_n^\theta) - F''(x_n)](x^\star - x_n)^2(1-\theta)d\theta$$
$$+\frac{1}{2}\Gamma_F(x_n)F'(x_n)^{-1}F''(x_n)F'(x_n)^{-1}\int_0^1 F''(x_n^\theta)(x^\star - x_n)^2(1-\theta)d\theta(x^\star - x_n), \tag{6.24}$$

where

$$\Gamma_F = (I - L_F(x))^{-1} \tag{6.25}$$

and

$$x_n^\theta = x_n + \theta(x^\star - x_n), \tag{6.26}$$

to arrive at:

Theorem 6.1.1. *Under conditions* $(H_1) - (H_6)$, *sequence* $\{x_n\}$ *generated by Halley's method (6.2) is well defined, remains in* $U(x_0, t^\star)$ *and converges to a solution* $x^\star \in \overline{U}(x_0, t^\star)$ *of equation* $F(x) = 0$. *Moreover, the following estimates hold for all* $n \geq 0$:

$$F'(x_n)^{-1} \in L(Y, X),$$

$$\|F'(x_n)^{-1}F'(x_0)\| \leq -\frac{1}{f_0'(\|x_n - x_0\|)} \leq -\frac{1}{f_0'(t_n)} \leq -\frac{1}{f'(s_n)},$$

$$\|F'(x_0)^{-1}F''(x_n)\| \leq f''(s_n), \quad \|F'(x_0)^{-1}F(x_n)\| \leq f(s_n),$$

$$(I - L_F(x_n))^{-1} \in L(Y, X),$$

$$\|(I - L_F(x_n))^{-1}\| \leq \frac{1}{1 - L_{f_0}(t_n)} \leq \frac{1}{1 - L_f(s_n)},$$

$$\|x_{n+1} - x_n\| \le t_{n+1} - t_n \le s_{n+1} - s_n,$$

$$\|x_n - x^\star\| \le t^\star - s_n, \quad \|x_{n+1} - x^\star\| \le (t^\star - s_{n+1})\left(\frac{\|x_n - x^\star\|}{t^\star - s_n}\right)^3$$

and

$$\|x_{n+1} - x^\star\| \le g_0(t^\star, t_0^\star)\|x_n - x^\star\|^3 \le g(t^\star, t^\star)\|x_n - x^\star\|^3.$$

Furthermore, $x^\star$ is the unique solution of equation $F(x) = 0$ in $\overline{U}(x_0, t^\star)$. Finally, $x^\star$ is the unique solution of equation $F(x) = 0$ in $U(x_0, \alpha)$, where

$$\alpha = \sup\{t \in [t_0^\star, R) : f_0(t) \le 0\}, \tag{6.27}$$

provided that $x^\star \in \overline{U}(x_0, t_0^\star)$ and (H_5) holds with f_0 replacing f.

Proof. We shall show that $x^\star$ is the unique solution of equation $F(x) = 0$ in $\overline{U}(x_0, t^\star)$. Let $y^\star$ be a solution of equation $F(x) = 0$ in $\overline{U}(x_0, t^\star)$. A simple induction (since $\|y^\star - x_0\| \le t^\star$) shows

$$\|y^\star - x_n\| \le t^\star - s_n \ \ (n \ge 0). \tag{6.28}$$

Indeed, (6.28) holds for $n = 0$, since $s_0 = 0$. Assume (6.28) is true for all $n \le k$. Then, we have from Lemma 6.1.3(4), (6.16), (6.18) and (6.24)–(6.26) that

$$\|y^\star - x_{n+1}\| \le (t^\star - s_{n+1})\left(\frac{\|y^\star - x_n\|}{t^\star - s_n}\right)^3, \tag{6.29}$$

which shows (6.28) for all $n \ge 0$. We get by (6.29) that $\lim_{n\to\infty} x_n = y^\star$ since $\lim_{n\to\infty} s_n = t^\star$. But we have $\lim_{n\to\infty} x_n = x^\star$. Hence, we deduce $x^\star = y^\star$. Finally, we must show F has no zeros in $U(x_0, \alpha) - \overline{U}(x_0, t_0^\star)$. Let $z^\star$ be a solution of equation $F(x) = 0$ such that $t_0^\star < \|z^\star - x_0\| < \alpha$. We have the identity

$$\begin{aligned}
F(z^\star) = {}& F(x_0) + F'(x_0)(z^\star - x_0) + \frac{1}{2}F''(x_0)(z^\star - x_0)^2 \\
& + \int_0^1 [F''(x_0 + \theta(z^\star - x_0)) - F''(x_0)](z^\star - x_0)^2(1 - \theta)d\theta.
\end{aligned} \tag{6.30}$$

Using (6.8) we obtain the estimates

$$\begin{aligned}
& \| \textstyle\int_0^1 F'(x_0)^{-1}[F''(x_0 + \theta(z^\star - x_0)) - F''(x_0)](z^\star - x_0)^2(1 - \theta)d\theta\| \\
& \le \textstyle\int_0^1 [f_0''(\theta\|z^\star - x_0\|) - f_0''(0)]\|z^\star - x_0\|^2(1 - \theta)d\theta \\
& = f_0(\|z^\star - x_0\|) - f_0(0) - f_0'(0)\|z^\star - x_0\| - \frac{1}{2}f_0''(0)\|z^\star - x_0\|^2
\end{aligned}$$

and

$$\|F'(x_0)^{-1}[F(x_0) + F'(x_0)(z^\star - x_0) + \frac{1}{2}F''(x_0)(z^\star - x_0)^2]\|$$

$$\ge \|z^\star - x_0\| - \|F'(x_0)^{-1}F(x_0)\| - \frac{1}{2}\|F'(x_0)^{-1}F''(x_0)\|\|z^\star - x_0\|^2$$

$$\ge \|z^\star - x_0\| - f_0(0) - \frac{1}{2}f_0''(0)\|z^\star - x_0\|^2.$$

It then follows from the estimate

$$f_0(\|z^\star - x_0\|) - f_0(0) + \|z^\star - x_0\| - \frac{1}{2}f_0''(0)\|z^\star - x_0\|^2$$

$$\ge \|z^\star - x_0\| - f_0(0) - \frac{1}{2}f_0''(0)\|z^\star - x_0\|^2,$$

that $f_0(\|z^\star - x_0\|) \ge 0$. But f_0 is strictly positive on $(\|z^\star - x_0\|, R)$ (since f_0 is strictly convex). So, we have $\alpha \le \|z^\star - x_0\|$, which is a contradiction. That completes the proof of the theorem.   □

Remark 6.1.2. Let $f_0 = f$. Moreover, if we assume the Lipschitz condition

$$\|F'(x_0)^{-1}[F''(y) - F''(x)]\| \le L\|y - x\| \text{ for all } x, y \in \Omega,$$

define

$$f(t) = \eta - t + \frac{M}{2}t^2 + \frac{L}{6}t^3,$$

where $\|F'(x_0)^{-1}F(x_0)\| \le \eta = f(0)$ and $\|F'(x_0)^{-1}F''(x)\| \le M = f''(0)$. Then, it follows from Theorem 6.1.1 that the sufficient convergence condition reduces to the Kantorovich-type condition for Halley's method (see Refs. [43, 110, 206, 261, 290])

$$\eta \le \eta_0 := \frac{2(M + 2\sqrt{M^2 + 2L})}{3(M + \sqrt{M^2 + 2L})^2}. \tag{6.31}$$

That is in this case we do not improve (6.31). However, if $f_0 < f$, then again under (6.31) we have convergence for Halley's method but the errors are tighter and the information on the location of the solution at least as precise. Note that a direct study of majorizing iteration $\{t_n\}$ by us in Refs. [110, 151] (see (6.10)) has led to sufficient convergence conditions which can be weaker than (6.31). The results obtained here for special cases of f_0 and f can immediately produce Smale-type (see Refs. [409, 427]) and Nemesirovskii-type (see Ref. [333]) theorems. We leave the details to the motivated reader.

6.2 Local Convergence

A local convergence results for method (6.2) is provided in this section using convex majorants. Kantorovich-type and Smale-type results are considered as applications and special cases. We suppose that operator F has a zero $x^\star$. An interesting problem is to find the radius of convergence for Halley's method (6.2). An open ball $U(x^\star, r) \subset X$ with center $x^\star$ and radius r is called a convergence ball of an iterative method, if the sequence generated by this iterative method starting from any initial values in it converges. The convergence ball of an iterative method is very important because it shows the extent of difficulty for choosing initial guesses for iterative methods.

Recently, we provided a local convergence analysis for Newton's method (see Ref. [116]) using the following majorant conditions:

$$\|F'(x^\star)^{-1}[F'(y) - F'(x)]\| \le h'(\|y - x\| + \|x - x^\star\|) - h'(\|x - x^\star\|), \tag{6.32}$$

$$\|F'(x^\star)^{-1}[F'(y) - F'(x^\star)]\| \le h'_\star(\|y - x^\star\|) - h'_\star(0) \tag{6.33}$$

for all $x, y \in U(x^\star, R)$, $R > 0$, where $\|y - x\| + \|x - x^\star\| < R$ and $h, h_\star : [0, R) \to \mathbb{R}$ are continuously differentiable, convex, strictly increasing functions with $h'_\star(0) = -1$ and $h_\star(R_0) = 0$ for some $R_0 \in (0, R)$. Note that function $h_\star$ is a special case of h,

$$h'_\star(t) \le h'(t), \ t \in [0, R]$$

holds in general and $h'/h'_\star$ can be arbitrarily large.

In the Lipschitz and center-Lipschitz case (6.32) and (6.33) become

$$\|F'(x^\star)^{-1}[F'(y) - F'(x)]\| \le \alpha\|y - x\| \tag{6.34}$$

and

$$\|F'(x^\star)^{-1}[F'(y) - F'(x^\star)]\| \le \alpha_\star\|y - x^\star\|, \tag{6.35}$$

for all $x, y \in U(x^\star, R)$, respectively, provided that

$$h(t) = \frac{\alpha}{2}t^2 - t + \alpha_0, \quad \alpha > 0, \ \alpha_0 > 0 \tag{6.36}$$

and

$$h_\star(t) = \frac{\alpha_\star}{2}t^2 - t + \alpha_1, \quad \alpha_\star > 0, \ \alpha_1 > 0. \tag{6.37}$$

Functions h, $h_\star$ given by (6.36) and (6.37) are continuously differentiable, convex, strictly increasing on $(0, R_0)$ where $R_0 = \dfrac{1}{\alpha_\star}$ and $h'(0) = h'_\star(0) = -1$.

Note that under (6.32) for h given in (6.36) the radius of convergence given in Refs. [217, 211] coincides with the one provided by Traub-Wosniakowski in Ref. [420] and Rheinboldt in Ref. [384]

$$r_{TR} = \frac{2}{3\alpha}. \tag{6.38}$$

However, our radius of convergence is given by

$$r_{AH} = \frac{2}{2\alpha_\star + \alpha}. \tag{6.39}$$

Note that

$$\frac{r_{AH}}{r_{TR}} = \frac{3}{2\dfrac{\alpha_\star}{\alpha} + 1} \to 3 \quad \text{as} \quad \frac{\alpha_\star}{\alpha} \to 0. \tag{6.40}$$

Hence, our radius r_{AH} can be at most three times larger than r_{TR}.

Concerning the error estimates, we have for our case

$$\|x_{n+1} - x^\star\| \le \frac{\alpha\|x_n - x^\star\|^2}{2(1 - \alpha_\star\|x_n - x^\star\|)} \tag{6.41}$$

whereas

$$\|x_{n+1} - x^\star\| \le \frac{\alpha\|x_n - x^\star\|^2}{2(1 - \alpha\|x_n - x^\star\|)} \tag{6.42}$$

is given in Refs. [110, 120, 163]. Therefore if $\alpha = \alpha_\star$, the estimates (6.41) and (6.42) coincide. Otherwise (i.e. if $\alpha_\star < \alpha$) ours are tighter.

In what follows, we extend our ideas to study the local convergence of Halley's method (6.32). Suppose that F is a twice Fréchet-differentiable operator and there exists $x^\star \in D$ such that $F'(x^\star)^{-1} \in L(Y, X)$, the space of bounded linear operators from Y into X. In addition, let $R > 0$ be such that $U(x^\star, R) \subseteq D$ and $h, h_\star : [0, R) \to \mathbb{R}$ be twice continuously differentiable function, we say the operator F'' satisfies the majrant conditions, if

$$\|F'(x^\star)^{-1}[F''(y) - F''(x)]\| \le h''(\|y - x\| + \|x - x^\star\|) - h''(\|x - x^\star\|), \tag{6.43}$$

$$\|F'(x^\star)^{-1}[F''(y) - F''(x^\star)]\| \le h''_\star(\|y - x^\star\|) - h''_\star(0), \tag{6.44}$$

for all $x, y \in U(x^\star, R)$, where $\|y - x\| + \|x - x^\star\| < R$ and the following conditions hold:

(C_1) $h''(0) \geq 0$, $h''_\star(0) > 0$, $h'_\star(0) = -1$;
(C_2) h'', $h''_\star$ are strictly increasing in $[0, R)$, $h''_\star$ is convex on $[0, R)$.
(C_3) $h'_\star$ has zeros in $(0, R)$. Denote by R_0 the minimal zero of $h'_\star$ in $(0, R)$.
(C_4) $\|F'(x^\star)^{-1} F''(x^\star)\| \leq h''_\star(0)$.

Note that if (6.43) holds, then so does (6.44). That is (6.44) is not an additional hypothesis. We also have

$$h''_\star(t) \leq h''(t), \quad t \in [0, R] \tag{6.45}$$

holds in general and $h''/h''_\star$ can be arbitrarily large. We show the cubic convergence of Halley's method under conditions (C_1)–(C_4).

As section 6.1, it is convenient for us to define some functions and parameters. Let us define function g on $[0, R]$ by

$$g(r) = h''_\star(r)(2 + h'_\star(r))r - 2(h'_\star(r))^2.$$

By the (C) conditions g is a continuous function on $[0, R]$, with

$$g(0) = -2(h'_\star(0))^2 = -2 < 0$$

and

$$g(R_0) = h''_\star(R_0)(2 + h'_\star(R_0))R_0 - 2(h'_\star(R_0))^2 = 2h''_\star(R_0)R_0 > 0.$$

It follows from the intermediate value theorem that there exists a zero of function g on $(0, R_0)$. Denote by R_1 the minimal such zero of g on $(0, R_0)$. We then have

$$q(r) = \frac{h''_\star(r)(2 + h'_\star(r))r}{2(h'_\star(r))^2} < 1, \quad G(r) = \frac{1}{1 - q(r)} > 0$$

and

$$P(r) = -\frac{1}{2}G(r)\left(\frac{1}{h'_\star(r)}\right)\left[\frac{1}{r} - \frac{(h''_\star(r))^2}{2h'_\star(r)}\right] > 0$$

are well defined on $[0, R_1]$, since by (C_1) and (C_2), $-1 < h'_\star(r) < 0$. Similarly, we show that there exists a minimal zero $R^\star \in (0, R_1)$ of function

$$Q(r) = P(r)r^2 - 1 \quad \text{and} \quad Q(r) < 0$$

is true for $r \in [0, R^\star)$. Moreover, replace function $h_\star$ by h in conditions $(C_1) - (C_4)$ and in the definition of g, q, G, P and Q. Denote resulting functions by g_0, q_0, G_0, P_0 and Q_0, respectively. Furthermore, denote by r_0, r_1 and $r^\star$ the parameters corresponding to R_0, R_1 and $R^\star$, respectively.

We need a result describing the properties of the h functions.

Lemma 6.2.1. *Let $R > 0$ and let $h_\star : [0, R) \to \mathbb{R}$ be a twice continuously differentiable function which satisfies the conditions $(C_1) - (C_3)$ (for $h_\star$). Then, the following hold*

(a) $h'_\star$ is strictly convex and strictly increasing on $[0, R)$.
(b) $-1 < h'_\star(r) < 0$ for $r \in (0, R_0)$.

Proof.

(a) It follows from (C_2) and $h''_\star(0) > 0$.

(b) By (C_1), (a) and the definition of R_0 we get (b).

That completes the proof of the lemma. $\qquad\qquad\qquad\qquad\qquad\qquad\square$

We need a Banach-type estimate on the inverse of $F'(x)$.

Lemma 6.2.2. *Suppose* $\|x - x^\star\| \leq r < R_0$. *If* $h_\star : [0, R_0) \to \mathbb{R}$ *is twice continuously differentiable and the majorizing function of F at $x^\star$. Then, $F'(x)^{-1} \in L(Y, X)$ and*

$$\|F'(x)^{-1}F'(x^\star)\| \leq -\frac{1}{h'_\star(\|x - x^\star\|)} \leq -\frac{1}{h'_\star(r)}. \qquad (6.46)$$

Proof.   Let $t \in [0, R_0)$ and $x \in \overline{U}(x^\star, R_0)$. We have the identity

$$F'(x) = F'(x^\star) + \int_0^1 [F''(x^\star + \theta(x - x^\star)) - F''(x^\star)](x - x^\star)d\theta + F''(x^\star)(x - x^\star). \quad (6.47)$$

Using (6.44), (C_1), (C_4) and Lemma 6.2.1 (b), we get in turn that

$$\begin{aligned}
&\|F'(x^\star)^{-1}(F'(x) - F'(x^\star))\| \\
&\leq \int_0^1 \|F'(x^\star)^{-1}[F''(x^\star + \theta(x - x^\star)) - F''(x^\star)]\|\|x - x^\star\|d\theta + \\
&\quad \|F'(x^\star)^{-1}F''(x^\star)\|\|x - x^\star\| \\
&\leq \int_0^1 [h''_\star(\theta\|x - x^\star\|) - h''_\star(0)]\|x - x^\star\|d\theta + h''_\star(0)\|x - x^\star\| \\
&= h'_\star(\|x - x^\star\|) - h'_\star(0) \leq h'_\star(r) - h'_\star(0) < 1.
\end{aligned} \qquad (6.48)$$

It follows from (6.48) and the Banach lemma on invertible operators that $F'(x)^{-1} \in L(Y, X)$ so that (6.46) holds. That completes the proof of the lemma. $\qquad\square$

We also need an upper bound on $F''(x)$.

Lemma 6.2.3. *Suppose* $\|x - x^\star\| \leq r < R_0$. *If* $h_\star : [0, R_0) \to \mathbb{R}$ *is twice continuously differentiable and majorizing function to F at $x^\star$. Then, we have*

$$\|F'(x^\star)^{-1}F''(x)\| \leq h''_\star(\|x - x^\star\|) \leq h''_\star(r). \qquad (6.49)$$

Proof.   Using (6.44) and (C_4), we obtain

$$\begin{aligned}
\|F'(x^\star)^{-1}F''(x)\| &\leq \|F'(x^\star)^{-1}[F''(x) - F''(x^\star)]\| + \|F'(x^\star)^{-1}F''(x^\star)\| \\
&\leq h''_\star(\|x - x^\star\|) - h''_\star(0) + h''_\star(0) \\
&= h''_\star(\|x - x^\star\|) \leq h''_\star(r),
\end{aligned}$$

since, $h''_\star$ is strictly increasing. That completes the proof of the lemma. $\qquad\square$

We provide now the main local convergence result for Halley's method (6.2).

Theorem 6.2.1. *Under conditions* $(C_1)-(C_4)$, *sequence* $\{x_n\}$ *generated by Halley's method (6.2) is well defined, remains in* $U(x^\star, R^\star)$ *for all* $n \geq 0$ *and converges to the unique zero* $x^\star$ *of operator* F *in* $U(x^\star, R_0)$, *provided that* $x_0 \in U(x^\star, R^\star)$. *Moreover, the following estimates hold*

$$\|x_{n+1} - x^\star\| \leq P(R^\star)\|x_n - x^\star\|^3. \tag{6.50}$$

In particular, Halley's method (6.2) converges Q-cubically to $x^\star$.

Proof. Using (6.2) and $F(x^\star) = 0$, we get the identity

$$
\begin{aligned}
x_{k+1} - x^\star &= \Gamma_F(x_k)F'(x_k)^{-1}F'(x^\star)F'(x^\star)^{-1}[-F'(x_k)(x^\star - x_k) - F(x_k)] \\
&\quad + \Gamma_F(x_k)L_F(x_k)(x^\star - x_k) \\
&= \Gamma_F(x_k)F'(x_k)^{-1}F'(x_\star)F'(x_\star)^{-1}\int_0^1 (1-\theta)[F''(x_k + \theta(x^\star - x_k)) \\
&\quad - F''(x_k)](x^\star - x_k)^2 d\theta \\
&\quad - \frac{1}{2}\Gamma_F(x_k)F'(x_k)^{-1}F'(x^\star)F'(x^\star)^{-1}F''(x_k)[F'(x_k)^{-1}F'(x^\star)F'(x^\star)^{-1} \\
&\quad \times \int_0^1 (1-\theta)F''(x_k + \theta(x^\star - x_k))(x^\star - x_k)^2 d\theta](x^\star - x_k).
\end{aligned} \tag{6.51}
$$

By (6.49) and the convexity of h'', we get

$$
\begin{aligned}
&h''(\theta\|x_k - x^\star\| + \|x_k - x^\star\|) - h''(\|x_k - x^\star\|) \\
&\leq [h''(\theta R^\star + R^\star) - h''(R^\star)]\frac{\theta\|x^\star - x_k\|}{\theta R^\star}.
\end{aligned} \tag{6.52}
$$

Using (6.46), (6.49) and the definition of operator L_F, we have for $x \in \overline{U}(x^\star, R^\star)$

$$\|L_F(x)\| \leq \frac{1}{2}\|F'(x)^{-1}F'(x^\star)\|^2\|F'(x^\star)^{-1}F''(x)\| \times$$

$$\|\int_0^1 F'(x^\star)^{-1}F'(x^\star + \theta(x - x^\star))d\theta(x - x^\star)\|$$

$$\leq \frac{h''(\|x - x^\star\|)}{2(h'(\|x - x^\star\|))^2}(2 + h'_\star(\|x - x^\star\|))\|x - x^\star\| \leq g(\|x - x^\star\|) \leq g(R^\star) < 1. \tag{6.53}$$

It follows from (6.53) and the Banach lemma on invertible operators that $\Gamma_F \in L(Y, X)$ and

$$\|\Gamma_F\| \leq \frac{1}{1 - g(\|x - x^\star\|)} = G(\|x - x^\star\|). \tag{6.54}$$

Using (6.46), (6.49), (6.51), (6.52) and (6.54) (for $x = x_k$), we get in turn

$$
\begin{aligned}
\|x_{k+1} - x^\star\| &\leq \frac{-G(\|x_k - x^\star\|)\int_0^1 (1-\theta)d\theta\|x_k - x^\star\|^3}{R^\star h'_\star(\|x_k - x^\star\|)} \\
&\quad + \frac{G(\|x_k - x^\star\|)(h''_\star(\|x_k - x^\star\|))^2\|x_k - x^\star\|^3}{4(h'_\star(\|x_k - x^\star\|))^2} = P(R^\star)\|x_k - x^\star\|^3 \\
&= \frac{\|x_k - x^\star\|^3}{R^{\star 2}} \leq \|x_k - x^\star\| < R^\star \quad \text{for all } k \geq 0
\end{aligned} \tag{6.55}
$$

which shows (6.50), $x_{k+1} \in U(x^\star, R^\star)$ and $\lim_{k \to \infty} x_k = x^\star$.

Finally, to show the uniqueness part, let $y^\star \in U(x^\star, R_0)$ be zero of operator F. Define linear operator $L = \int_0^1 F'(y^\star + \theta(x^\star - y^\star))d\theta$. Then, as in (6.46), we get that

$$\|F'(x^\star)^{-1}(L - F'(x^\star))\| \leq h'_\star(\|y^\star - x^\star\|) - h'_\star(0) < 1, \tag{6.56}$$

since, $z = y^\star + \theta(x^\star - y^\star) \in \overline{U}(x^\star, R_0)$. Using (6.56) and the Banach lemma we get $L^{-1} \in L(Y, X)$. Then, in view of the identity $0 = F(x^\star) - F(y^\star) = L(x^\star - y^\star)$, we get $x^\star = y^\star$. That completes the proof of the theorem. $\qquad\square$

According to the discussion above Lemma 6.2.1 if function $h_\star$ is replaced by h in conditions $(C_1) - (C_4)$, Theorem 6.2.1 gives:

Theorem 6.2.2. *Under conditions $(C_1) - (C_4)$ with function h replacing $h_\star$, sequence $\{x_n\}$ generated by Halley's method (6.2) is well defined, remains in $U(x^\star, r^\star)$ for all $n \geq 0$ and converges to the unique zero $x^\star$ of operator F in $U(x^\star, r_0)$, provided that $x_0 \in U(x^\star, r^\star)$. Moreover, the following estimates hold*

$$\|x_{n+1} - x^\star\| \leq P_0(r^\star)\|x_n - x^\star\|^3. \tag{6.57}$$

In particular, Halley's method (6.2) converges Q-cubically to $x^\star$.

Remark 6.2.1. The results obtained here can be used to find zeros $x^\star$ of operators F satisfying autonomous differentiable equations (see Refs. [110, 120, 163]) $F'(x) = T(F(x))$, where $T : Y \to X$ is a known differentiable operator. As

$$F''(x^\star) = F'(x^\star)T'(F(x^\star)) = T(F(x^\star))T'(F(x^\star)) = T(0)T'(0), \tag{6.58}$$

we do not need to know $x^\star$ to verify (6.43), (6.44) and (C_4). Equations where operators F satisfy $F'(x) = T(F(x))$ appear in many branches of applied sciences, such as minimization problems (see Refs. [110, 120, 163]).

We consider two special cases:

I - Kantorovich-type. We define functions h, $h_\star$ by

$$h(r) = \frac{a}{6}r^3 + \frac{b}{2}r^2 - r + a_0, \quad a > 0, \ a_0 > 0 \tag{6.59}$$

and

$$h_\star(r) = \frac{a_\star}{6}r^3 + \frac{b}{2}r^2 - r + a_1 \quad a_\star > 0, \ a_1 > 0 \tag{6.60}$$

for $a > 0$, $a_\star > 0$, $b > 0$ satisfying the conditions

$$\|F'(x^\star)^{-1}[F''(y) - F''(x)]\| \leq a\|y - x\|, \tag{6.61}$$

$$\|F'(x^\star)^{-1}[F''(x) - F''(x^\star)]\| \leq a_\star\|x - x^\star\|, \tag{6.62}$$

for all $x, y \in U(x^\star, R)$ and

$$\|F'(x^\star)^{-1}F''(x^\star)\| \leq b. \tag{6.63}$$

Functions h, $h_\star$ satisfy conditions $(C_1) - (C_4)$, with

$$R_0 = \frac{2}{b + \sqrt{b^2 + 2a_\star}}. \tag{6.64}$$

Table 6.1　Comparison results for the error bound of Halley's method

n	The right side of (6.50)	The right side of (6.57)
0	1.95126e-01	2.36123e-01
1	2.70579e-08	3.27429e-08
2	7.35104e-29	8.89552e-39
3	1.31477e-49	1.59101e-49

We provide a numerical example for case I.

Example 6.1. Returning back to Remark 6.2.1 with $D = U(0,1)$, we get that

$$a = e, \quad a_* = e - 1 \quad \text{and} \quad b = 1.$$

Then, we get that

$$R_0 = 0.6438496678987803, \quad r_0 = 0.5654448141543782,$$
$$R_1 = 0.3203645228203808, \quad r_1 = 0.2878878975673239,$$
$$R^\star = 0.2829778199426982, \quad r^\star = 0.2572415738179739.$$

Choosing $x_0 = 0.25 \in U(0,1)$, we can produce sequence $\{x_n\}$ by using Halley's method (6.2). The comparison results for the error bound by using (6.50) and (6.57) are given in Table 6.1. The table shows that tighter error bounds on the distances $\|x_n - x^\star\|$ can be obtained by using both functions h and h_* instead of by using only function h.

II. Smale-type. We study the local convergence of Halley's method (6.2) assuming that F is analytic and satisfies

$$\|F'(x^\star)^{-1}F^{(n)}(x^\star)\| \leq n!\gamma_*^{n-1} \quad (n \geq 2), \tag{6.65}$$

where

$$\gamma_* = \sup_{n>1} \|\frac{F'(x^\star)^{-1}F^{(n)}(x^\star)}{n!}\|^{1/(n-1)}. \tag{6.66}$$

Note that Smale used (1) (for $x^\star$ replaced by x_0) to study the convergence of (NM) (see Ref. [409]). Define functions h, h_* by

$$h(r) = \frac{\gamma_0 r^2}{1 - \gamma_0 r} - r + \beta_0, \quad \beta_0 > 0, \ 0 \leq r < \frac{1}{\gamma_0} \tag{6.67}$$

and

$$h_*(r) = \frac{\gamma_1 r^2}{1 - \gamma_1 r} - r + \beta_1, \quad \beta_1 > 0, \ 0 \leq r < \frac{1}{\gamma_1}. \tag{6.68}$$

Conditions (6.43), (6.44) using (6.67), (6.68) become respectively

$$\|F'(x^\star)^{-1}[F''(y) - F''(x)]\|$$
$$\leq 2\gamma_0[\frac{1}{(1 - \gamma_0\|y - x\| - \gamma_0\|x - x^\star\|)^3} - \frac{1}{(1 - \gamma_0\|x - x^\star\|)^3}] \tag{6.69}$$

and

$$\|F'(x^\star)^{-1}[F''(x) - F''(x^\star)]\| \le 2\gamma_1\big[\frac{1}{(1 - \gamma_1\|x - x^\star\|)^3} - 1\big]. \qquad (6.70)$$

We have $\gamma_1 \le \gamma_0$ and $\gamma_1 = \gamma_0 = \gamma_\star$ is certainly a choice for γ_1 and γ_0. Condition (6.69) (for $x^\star$, γ replaced by x_0, γ) is the so-called γ-condition introduced by Wang and Han to study the semi-local convergence of Newton's method, see Ref. [427] (see also our works in Refs. [110, 120, 163]). It turns out that there is an easier way ro verify condition (6.43) (or (6.44)).

Proposition 6.2.1. *Let $F : D \subset X \to Y$ be a thrice continuously differentiable operator. Let $h : [0, R) \to \mathbb{R}$ be a thrice continuous differentiable function with convex h''. Then F satisfies (6.43) if and only if*

$$\|F'(x^\star)^{-1}F'''(x)\| \le h'''(\|x - x^\star\|). \qquad (6.71)$$

Proof. If F satisfies (6.71), then we get in turn that

$$
\begin{aligned}
\|F'(x^\star)^{-1}[F''(y) - F''(x)]\| &\le \int_0^1 \|F'(x^\star)^{-1}F'''(x + \theta(y - x))(y - x)\|d\theta \\
&\le \int_0^1 h'''(\|x - x^\star\| + \theta\|y - x\|)\|y - x\|d\theta \\
&= h''(\|y - x\| + \|x - x^\star\|) - h''(\|x - x^\star\|),
\end{aligned}
$$

which shows (6.43). If F satisfies (6.43), then (6.71) certainly holds. That completes the proof of the proposition. $\qquad \square$

Remark 6.2.2. If h is given by (6.67), then (6.71) reduces to

$$\|F'(x^\star)^{-1}F'''(x)\| \le \frac{6\gamma^2}{(1 - \gamma\|x - x^\star\|)^4}, \qquad (6.72)$$

where we set $\gamma_0 = \gamma_1 = \gamma$ for simplicity. As an example, note that if F is an analytic operator and h is given by (6.67), we have

$$
\begin{aligned}
\|F'(x^\star)^{-1}F'''(x)\| &= \|\sum_{n=0}^{\infty} \frac{1}{n!}F'(x^\star)^{-1}F^{(n+3)}(x^\star)(x - x^\star)^n\| \\
&\le \gamma^2 \sum_{n=0}^{\infty}(n + 3)(n + 2)(n + 1)(\gamma\|x - x^\star\|)^n \\
&= \frac{6\gamma^2}{(1 - \gamma\|x - x^\star\|)^4} \quad \text{for} \quad \gamma\|x - x^\star\| < 1,
\end{aligned}
$$

which implies that F satisfies the γ-condition in this case.

6.3   Traub-type Multipoint Method

We present in this section a unified approach to generating majorizing sequences for multipoint iterative procedures to approximate locally the unique solution x^* of (1.1). The semi-local convergence results have the following advantages over earlier work (under the same computational cost): weaker sufficient convergence conditions, more precise error bounds on the distances involved and more precise information on the location of the solution. Special cases are also provided in this section.

Let k denote a fixed natural number or zero, and let $x_{-k}, x_{-k+1}, \ldots, x_{-1}, x_0$ be $k+1$ given points in D. Consider operator $G : X^{k+1} \to Y$, and introduce $(k+1)$-multipoint method (MPM)

$$x_{n+1} = G(x_n, x_{n-1}, \ldots, x_{n-k}) \quad \text{for each} \quad n = 0, 1, 2, \ldots.$$

The operator G is defined in such a way so that if sequence $\{x_n\}$ converges, then $F(x^*) = 0$.

Special choices for the operator G

Case $k = 0$

(i) Let $G(x_n) = x_n - A(x_n)^{-1} F(x_n)$ for each $n = 0, 1, 2, \ldots$, $x_0 \in D$ and $A(x) \in L(X, Y)$. Then, we get the Newton-like method

$$x_{n+1} = G(x_n) = x_n - A(x_n)^{-1} F(x_n), \tag{6.73}$$

(see Refs. [106, 110, 120]).

In particular, if

(ii) $A(x_n) = F'(x_n)$, we obtain the Newton–Kantorovich method

$$x_{n+1} = G(x_n) = x_n - F'(x_n)^{-1} F(x_n), \tag{6.74}$$

(see Refs. [120, 290, 338, 347, 368, 369]).

Case $k = 1$

(iii) For $x_{-1}, x_0 \in D$, let $G(x_n, x_{n-1}) = x_n - [x_n, x_{n-1}; F]^{-1} F(x_n)$ for each $n = 0, 1, 2, \ldots$. Then, we obtain the secant method

$$x_{n+1} = x_n - [x_n, x_{n-1}; F]^{-1} F(x_n). \tag{6.75}$$

Case $k = 2$

(iv) For $x_{-2}, x_{-1}, x_0 \in D$, let

$$G(x_n, x_{n-1}, x_{n-2}) = x_n - ([x_n, x_{n-1}; F] + [x_{n-2}, x_n; F] - [x_{n-2}, x_{n-1}; F])^{-1} F(x_n),$$

for each $n = 0, 1, 2, \ldots$.

To obtain a method considered by Traub and by Potra in the Banach space case:

$$x_{n+1} = x_n - ([x_n, x_{n-1}; F] + [x_{n-2}, x_n; F] - [x_{n-2}, x_{n-1}; F])^{-1} F(x_n). \tag{6.76}$$

Other cases and contraction-type (for G) results can be found in Refs. [110, 120].

In this section, we provide a unified semi-local convergence analysis for (MPM) using majorizing sequences. Our approach has the following advantages over earlier works under the same computational cost:

(a) weaker sufficient convergence conditions,

(b) more precise error bounds on the distances $\|x_{n+1} - x_n\|$, $\|x_n - x^*\|$ for each $n = 0, 1, 2, \ldots$

(c) more precise information on the location of the solution x^*.

We need the definition of general multipoint majorizing iterations.

Definition 6.3.1. Let $N = \{1, 2, 3, \ldots\}$ and set $N_0 = \{0\} \cup N$. Let $k \in N_0$ be fixed and $0 = t_{-k} \leq t_{-k+1} \leq \cdots \leq t_{-1} \leq t_0 \leq t_1$ be $k + 2$ given constants. Set $t_0 = t_{-1} + c$, $t_1 = t_0 + \eta$ for some $c \geq 0$, $\eta \geq 0$. Let also $a_i^1 \geq 0$, $b_i^1 \geq 0$, $c_i^1 \geq 0$, $i = 0, 1, \ldots, k + 1$, $b_{k+2}^1 \in R - \{1\}$. Define multipoint scalar iteration (MSI) $\{t_n\}$ for each $n = 0, 1, 2, \ldots$ by

$$t_{n+2} = t_{n+1} + \frac{\sum_{i=0}^{k} a_i^1 (t_{n+1-i} - t_{n-i}) + \sum_{i=0}^{k+1} b_i^1 t_{n+1-i} + a_{k+1}^1}{1 - \left(\sum_{i=0}^{k+1} c_i^1 t_{n+1-i} + b_{k+2}^1 \right)} (t_{n+1} - t_n). \quad (6.77)$$

Set

$$\alpha_i = \frac{a_i^1}{1 - b_{k+2}^1}, \quad \beta_i = \frac{b_i^1}{1 - b_{k+2}^1}, \quad \gamma_i = \frac{c_i^1}{1 - b_{k+2}^1}. \quad (6.78)$$

Define sequences $\{p_n^1\}$, $\{p_n^2\}$, $\{p_n\}$ by

$$p_n = \frac{p_n^1}{p_n^2} = \frac{\sum_{i=0}^{k} \alpha_i (t_{n+1-i} - t_{n-i}) + \sum_{i=0}^{k+1} \beta_i t_{n+1-i} + \alpha_{k+1}}{1 - \sum_{i=0}^{k+1} \gamma_i^1 t_{n+1-i}}. \quad (6.79)$$

Then, (6.77) can be written as

$$t_{n+2} = t_{n+1} + p_n (t_{n+1} - t_n). \quad (6.80)$$

We shall show (MSI) converges under certain conditions. If

$$0 \leq p_n \leq \delta < 1 \quad \text{for each } n = 1, 2, \ldots \text{ and some } \delta. \quad (6.81)$$

Then, we have that

$$0 \leq t_{n+1} - t_n \leq \delta^n \eta, \quad (6.82)$$

$$0 \leq t_{n+1} \leq \frac{1 - \delta^{n+1}}{1 - \delta} \eta + c \leq \frac{\eta}{1 - \delta} + c = t^{**}, \quad (6.83)$$

$$c + \eta \leq \lim_{n \to +\infty} t_n = t^* \leq t^{**} \quad (6.84)$$

and

$$0 \leq t^* - t_n \leq \frac{\delta^n}{1 - \delta} \eta. \quad (6.85)$$

Define functions f_n, g_k on $[0, +\infty)$ by

$$\begin{aligned}
f_n(t) = & [\alpha_0 t^n + \alpha_1 t^{n-1} + \cdots + \alpha_k t^{n-k}] \eta \\
& + [\beta_0 (1 + t + \cdots + t^n) + \beta_1 (1 + t + \cdots + t^{n-1}) \\
& \quad + \cdots + \beta_{k+1} (1 + t + \cdots + t^{n-1-k})] \eta \\
& + t [\gamma_0 (1 + t + \cdots + t^n) + \gamma_1 (1 + t + \cdots + t^{n-1}) \\
& \quad + \cdots + \gamma_{k+1} (1 + t + \cdots + t^{n-1-k})] \eta \\
& - t + \alpha_{k+1} + (\beta_0 + \gamma_0 + \beta_1 + \gamma_1 + \cdots + \beta_{k+1} + \gamma_{k+1}) c, \quad (6.86)
\end{aligned}$$

$$g_k(t) = \gamma_0 t^{k+2} + \gamma_1 t^{k+1} + \cdots + \gamma_{k+1} t + \beta_0 t^{k+1} + \cdots + \beta_{k+1} t$$
$$+\alpha_0(t^{k+1} - t^k) + \cdots + \alpha_k(t-1). \tag{6.87}$$

Then, we have the following relationship between two consecutive recurrent functions f_n

$$f_{n+1}(t) = f_n(t) + g_k(t)t^{n-k}\eta. \tag{6.88}$$

In view of the intermediate value theorem applied on $[0, t]$, function g_k has a positive zero provided that $\alpha_k > 0$. Denote by δ the minimal such zero.

If

$$\delta \text{ is such that } f_1(t) \leq 0 \text{ for } t \in [0, \delta] \tag{6.89}$$

and

$$0 \leq p_1 \leq \delta < 1, \tag{6.90}$$

then, (6.81)–(6.85) hold. Indeed, according to (6.90), estimates (6.81)–(6.83) hold for $n = 0$. Assume (6.81)–(6.83) hold for all $m \leq n$. Then, instead of (6.81) it suffices to show

$$0 \leq g_n \leq \delta, \tag{6.91}$$

where for $t = \delta$

$$g_n = g_n(t) = \frac{g_n^1(t)}{g_n^2(t)}, \tag{6.92}$$

$$g_n^1(t) = (\alpha_0 t^n + \alpha_1 t^{n-1} + \cdots + \alpha_k t^{n-k})\eta$$
$$+[\beta_0(1 + t + \cdots + t^n) + \beta_1(1 + t + \cdots + t^{n-1})$$
$$+ \cdots + \beta_{k+1}(1 + t + \cdots + t^{n-1-k})]\eta + \alpha_{k+1}$$
$$+(\beta_0 + b_1 + \cdots + \beta_{k+1})c, \tag{6.93}$$

$$g_n^2(t) = 1 - [\gamma_0(1 + t + \cdots + t^n) + \gamma_1(1 + t + \cdots + t^{n-1})$$
$$+ \cdots + \gamma_{k+1}(1 + t + \cdots + t^{n-1-k})$$
$$+(\gamma_0 + \gamma_1 + \cdots + \gamma_{k+1})c]\eta, \tag{6.94}$$

or, instead of (6.91):

$$f_n(\delta) \leq 0. \tag{6.95}$$

But

$$f_n(\delta) = f_1(\delta) \leq 0 \text{ for all } n \text{ (by (6.89))}. \tag{6.96}$$

That is (6.95) (i.e. (6.82) holds for all n). Note also that

$$f_\infty(t) := \lim_{n \to +\infty} f_n(t) \leq \lim_{n \to +\infty} 0 = 0. \tag{6.97}$$

Estimates (6.84) and (6.85) follow from (6.82) and (6.83) by using standard majorization techniques. Hence, we arrived at the following results on majorizing sequences (under the notation of Definition 6.3.1).

Lemma 6.3.1. *If for $\alpha_k > 0$ the minimal zero δ of function g_k satisfies:*

$$0 \leq p_1 \leq \delta < 1 \quad and \quad f_1(t) \leq 0 \quad for\ all \quad t \in [0, \delta],$$

then, (MSI) generated by (6.77), is non-decreasing, bounded from above by t^{**} and converges to its unique least upper bound $t^* \in [0, t^{**}]$ so that estimates (6.82)–(6.85) hold.

We can provide another result on majorizing (MSI):

Lemma 6.3.2. *If there exists δ_0 such that*

$$f_1(\delta_0) \leq 0 \tag{6.98}$$

and

$$0 \leq p_1 \leq \delta_0 \leq \delta < 1, \tag{6.99}$$

then, the conclusions of Lemma 6.3.1 for (MSI) hold with δ_0 replacing δ.

Proof. It suffices to show (6.95) for δ_0 replacing δ. We have by (6.88), (6.98) and (6.99)

$$f_2(\delta_0) = f_1(\delta_0) + g_k(\delta_0)\delta_0^{1-k}\eta \leq 0 + 0 = 0.$$

Assume $f_m(\delta_0) \leq 0$, then

$$f_{m+1}(\delta_0) = f_m(\delta_0) + g_k(\delta_0)\delta_0^{m-k}\eta \leq 0 + 0 = 0,$$

which completes the induction for (6.95). Note also that (6.97) holds. That completes the proof of the Lemma. $\qquad\square$

Remark 6.3.1. Since $f_1(0) \geq 0$ (by (6.86)), condition (6.98) implies (by the intermediate value Lemma) that there exists $\delta_0^* \in [0, \delta]$ such that $f_1(\delta_0^*) = 0$.

Finally, we have that

Lemma 6.3.3. *Assume there exists δ_1 such that*

$$p_1 \leq \delta_1 < 1, \tag{6.100}$$

$$g_k(\delta_1) \geq 0 \tag{6.101}$$

and

$$f_\infty(\delta_1) = \frac{[(\beta_0 + \beta_1 + \cdots + \beta_{k+1}) + \delta_1(\gamma_0 + \gamma_1 + \cdots + \gamma_{k+1})]\eta}{1 - \delta_1} \\ -\delta_1 + \alpha_{k+1} + (\beta_0 + \gamma_0 + \beta_1 + \gamma_1 + \cdots + \beta_{k+1} + \gamma_{k+1})c \leq 0. \tag{6.102}$$

Then, the conclusions of Lemma 6.3.1 for (MSI) hold for δ_1 replacing δ.

Proof. We have by (6.88) and (6.101):

$$f_1(\delta_1) \le f_2(\delta_1) \le \ldots \le f_\infty(\delta_1). \tag{6.103}$$

In view of (6.103) and the proof of Lemma 6.3.1 (see (6.95) for δ_1 replacing δ) we must show $f_\infty(\delta_1) \le 0$, which is true by (6.102). Finally, note that the formula for f_∞ given in (6.102) is obtained from (6.97) and from (6.86) by letting $n \to +\infty$. That completes the proof of the Lemma. $\qquad\square$

We shall now connect the majorizing sequences with (MPM) using the following set of conditions
    (C)

$$\|x_{n+1} - x_n\| \le t_{n+1} - t_n;$$

$$x_n \in \overline{U}(x_0, t^*) \subseteq D;$$

$$t^* = \lim_{n \to +\infty} t_n;$$

there exists a continuous, non-decreasing function $\omega : [0, +\infty) \to [0, +\infty)$ with $\omega(0) = 0$ such that

$$\|F'(x_0)^{-1} F(x_n)\| \le \omega(\|x_n - x_{n-1}\|),$$

$$\|G(u) - G(v)\| \le \xi\|u - v\|$$

for some $\xi \in (0, 1)$, each $u, v \in X^{k+1}$ and $\{t_n\}$ is a Cauchy sequence.

Remark 6.3.2. If sequence $\{t_n\}$ is given by (6.77), then

(1) The function ω can be chosen so that

$$\omega(t_{n+1} - t_n) = p_n^1(t_{n+1} - t_n),$$

    where p_n^1 is given in (6.79).

(2) Under the (C) conditions (MPM) is well defined, remains in $\overline{U}(x_0, t^*)$ and converges to a unique solution x^* of equation $F(x) = 0$ in $\overline{U}(x_0, x^*)$.

Hence, we arrived at the following semi-local convergence result for (MPM).

Theorem 6.3.1. *Under the (C) conditions, further assume hypotheses of Lemma 6.3.1 or Lemma 6.3.2 or Lemma 6.3.3 hold for iteration $\{t_n\}$ given by (6.77). Then, the sequence $\{x_n\}$ generated by (MPM) is well defined, remains in $\overline{U}(x_0, x^*)$ for each $n = 0, 1, 2, \ldots$ and converges to a unique solution x^* of equation $F(x) = 0$ in $\overline{U}(x_0, t^*)$. Moreover, the following estimates hold*

$$\|x_{n+1} - x_n\| \le t_{n+1} - t_n$$

and

$$\|x^* - x_n\| \le t^* - t_n.$$

Remark 6.3.3. The contraction hypothesis for G is only needed to show the uniqueness of the solution x^* in $\overline{U}(x_0, t^*)$. This hypothesis can be dropped if implied by others.

Next, we present a more general majorizing iteration than $\{t_n\}$.

Definition 6.3.2. Let $j, k \in \mathbb{N}_0$ be fixed and $0 = r_{-k} \leq r_{-k+1} \leq \ldots \leq r_{-1} \leq r_0 \leq \ldots \leq r_j \leq r_{j+1}$ be $j + k + 2$ given constants. Set $r_j = r_{j-1} + c$, $r_{j+1} = r_j + \eta$ for some $c \geq 0$, $\eta \geq 0$. Let also $a^1_{j+i} \geq 0$, $b^1_{j+i} \geq 0$, $c^1_{j+i} \geq 0$, $i = 0, 1, 2, \ldots, k+1$, $b_{k+2} \in R - \{1\}$. Define general multipoint scalar iteration (GMSI) for each $n = 0, 1, 2, \ldots$ by

$$r_{j+n+2} = r_{j+n+1} + \frac{\sum_{i=0}^{k} a^1_{j+i}(r_{j+n+1-i} - r_{j+n-i}) + \sum_{i=0}^{k+1} b^1_{j+i} r_{j+n-i} + a^1_{k+1}}{1 - \left(\sum_{i=0}^{k+1} c^1_{j+i} r_{j+n-i} + b^1_{k+2}\right)}$$

$$\times (r_{j+n+1} - r_{j+n}). \tag{6.104}$$

Set

$$\alpha_{j+i} = \frac{a^1_{j+i}}{1 - b_{k+2}}, \quad \beta_{j+i} = \frac{b^1_{j+i}}{1 - b^1_{k+2}}, \quad \gamma_{j+i} = \frac{c^1_{j+i}}{1 - b^1_{k+2}}.$$

Define sequences $\{p^1_{j+n}\}$, $\{p^2_{j+n}\}$, $\{p_{j+n}\}$ by

$$p_{j+n} = \frac{p^1_{j+n}}{p^2_{j+n}} = \frac{\sum_{i=0}^{k} \alpha_{j+i}(r_{j+n+1-i} - r_{j+n-i}) + \sum_{i=0}^{k+1} \beta_{j+i} t_{j+n+1-i} + \alpha_{k+1}}{1 - \sum_{i=0}^{k+1} \gamma^1_{j+i} r_{j+n+1-i}}.$$

Then, we can write

$$r_{j+n+2} = r_{j+n+1} + p_{j+n}(r_{j+n+1} - r_{j+n}).$$

If $0 \leq p_{j+n} \leq \delta < 1$ for each $n = 1, 2, \ldots$, then, we have

$$0 \leq r_{j+n+1} - r_{j+n} \leq \delta^n(r_{j+n} - r_{j+n-1}), \tag{6.105}$$

$$0 \leq r_{j+n+1} \leq \frac{1 - \delta^{n+1}}{1 - \delta}(r_{j+1} - r_j) + (r_j - r_{j+1})$$

$$\leq \frac{r_{j+1} - r_j}{1 - \delta} + r_j - r_{j-1} = r^{**}, \tag{6.106}$$

$$r_{j+1} \leq \lim_{n \to +\infty} r_{j+n} = r^* \leq r^{**} \tag{6.107}$$

and

$$r^* - r_{j+n} \leq \frac{\delta^n}{1 - \delta}(r_{j+1} - r_j). \tag{6.108}$$

Then, Lemma 6.3.1, Lemma 6.3.2 and Lemma 6.3.3 can be written using sequence $\{r_n\}$ instead of $\{t_n\}$. For brevity we present only the result corresponding to Lemma 6.3.2.

Lemma 6.3.4. *Assume that there exist* $\delta_1, k, j \in \mathbb{N}_0$ *such that*

$$p_{j+1} \leq \delta_1 < 1,$$

$$g_{j+k}(\delta_1) \geq 0$$

and

$$f_\infty^j(\delta_1) = \frac{[(\beta_0 + \beta_1 + \cdots + \beta_{k+1}) + \delta_1(\gamma_0 + \gamma_1 + \cdots + \gamma_{k+1})](r_{j+1} - r_j)}{1 - \delta_1}$$

$$-\delta_1 + \alpha_{k+1} + (\beta_0 + \gamma_0 + \beta_1 + \gamma_1 + \cdots + \beta_{k+1} + \gamma_{k+1})(z_j - z_{j-1}) \leq 0.$$

Then, the conclusions (6.105)–(6.108) for (GMSI) $\{r_n\}$ hold with δ_1 replacing δ.

Remark 6.3.4. Clearly hypotheses of Lemma 6.3.4 and iteration $\{r_n\}$ can replace the previous Lemma and iteration $\{t_n\}$ can replace the previous Lemma and iteration $\{t_n\}$ in Theorem 6.3.1.

We return now to special cases (6.73)–(6.76) of (MPM). The advantages (a)–(c) stated in the introduction for these methods can be found in the referred cases for each method.

The (H) conditions that follow together with the corresponding conditions from Lemma 6.3.1 or Lemmas 6.3.2, 6.3.3 imply the (C) conditions (for each case) with the exception of the uniqueness part.

Application 1. Newton-like method (6.73)

(H_1) Let $F : D \subseteq X \to Y$ be Fréchet differentiable and let $A(x) \in L(X, Y)$ be an approximation of $F'(x)$. Assume there exist $x_0 \in D$, a bounded inverse $A(x_0)^{-1}$ of $A(x_0)$ and constants $\eta, L > 0$, $M, L_0, \mu, l \geq 0$ such that for all $x, y \in D$ the following conditions hold:

$$\|A(x_0)^{-1}F(x_0)\| \leq \eta,$$

$$\|A(x_0)^{-1}(F'(x) - F'(y))\| \leq L\|x - y\|,$$

$$\|A(x_0)^{-1}(F'(x) - A(x))\| \leq M\|x - x_0\| + \mu,$$

$$\|A(x_0)^{-1}(A(x) - A(x_0))\| \leq L_0\|x - x_0\| + l,$$

$$\overline{U}(x_0, t^*) \subseteq D$$

and the special case of iteration $\{t_n\}$ (given by (6.77)) (see Refs. [100, 120, 192, 267, 442]) is:

$$\begin{cases} t_0 = 0, \quad t_1 = \eta, \\ \\ t_{n+2} = t_{n+1} + \dfrac{\frac{L}{2}(t_{n+1} - t_n) + (Mt_n + \mu)}{1 - L_0 t_{n+1} - l}(t_{n+1} - t_n). \end{cases}$$

The iteration used in the literature by others in Refs. [290, 442] is given by

$$\begin{cases} s_0 = 0, \quad s_1 = \eta, \\ \\ s_{n+1} = s_n + \dfrac{\frac{\sigma}{2}s_{n+1}^2 - (1 - b)s_{n+1} + \eta}{1 - L_0 s_{n+1} - l}, \quad \sigma = \max\{L, M + L_0\}, \quad b = \mu + l. \end{cases}$$

The comparison, favorable to us, (see (a)–(c) of the introduction) as well as the corresponding to Lemmas 6.3.1–6.3.3 conditions were given in Refs. [100, 120].

Application 2. Newton–Kantorovich method (6.74)

Set $A(x) = F'(x)$ for $x \in D$, $M = \mu = l = 0$ in **Application 1**. Denote by (H_2) the corresponding conditions. Note that $L_0 \leq L$ holds in general and L/L_0 can be arbitrarily large. The sufficient convergence condition given in the Newton–Kantorovich Theorem is given by (1.3). The one given by us (see e.g. hypotheses (6.100)–(6.102) in Lemma 6.3.3) is (1.5), since

$$k = i = 0, \quad a_0^1 = \frac{L}{2}, \quad b_0^1 = a_1^0 = b_2^1 = 0, \quad c_0^1 = L_0, \quad c = 0,$$

$$\alpha_0 = \alpha_0^1 = \frac{L}{2}, \quad b_0 = 0, \quad \gamma_0 = L_0, \quad p_n = \frac{L(t_{n+1} - t_n)}{2(1 - L_0 t_{n+1})},$$

$$\omega(t) = \frac{L}{2}t^2, \quad f_n(t) = \frac{L}{2}t^n \eta + t L_0 (1 + t + \cdots + t^n)\eta - t, \quad f_\infty(t) = \frac{L_0 \eta t}{1 - t} - t,$$

$$g_0(t) = \frac{1}{2}(2L_0 t^2 + Lt - L) \quad \text{and} \quad \delta = \frac{2L}{L + \sqrt{L^2 + 8L_0 L}}.$$

We can do even better if $j = 1$. In this case we have that $\{r_n\}$ reduces to

$$\begin{cases} r_0 = 0, \quad r_1 = \eta, \quad r_2 = r_1 + \dfrac{L_0 \eta^2}{2(1 - L_0 \eta)} \\[2mm] r_{n+2} = r_{n+1} + \dfrac{L(r_{n+1} - r_n)^2}{2(1 - L_0 r_{n+1})}, \quad \text{for each} \quad n = 1, 2, \dots. \end{cases} \tag{6.109}$$

Sequence f_n^1 is given by

$$f_n^1(t) = \frac{L}{2}t^n (r_2 - r_1) + t L_0 (1 + t + \cdots + t^n)(r_2 - r_1) - t.$$

Function f_∞^1 is given by

$$f_\infty^1(t) = \frac{L_0 (r_2 - r_1)t}{1 - t} - t.$$

Then, the hypotheses of Lemma 6.3.4 are satisfied provided that (1.6). Moreover, sequence $\{r_n\}$ given by (6.109) is more precise than $\{t_n\}$ (for $M = \mu = l = 0$) which is more precise than $\{s_n\}$ (for $\sigma = L, b = L = M = \mu = 0$ and $L_0 = L$), if $L_0 < L$ (see also the numerical examples).

Application 3. Secant method (6.75)

Denote by (H_3) the following conditions. Let $F : D \subseteq X \to Y$; $x_{-1}, x_0 \in D$, $\|x_0 - x_{-1}\| \leq c$, $[x_0, x_{-1}; F]^{-1} \in L(X, Y)$, $\|[x_0, x_{-1}; F]^{-1} F(x_0)\| \leq \eta$,

$$\|[x_0, x_{-1}; F]^{-1}([x, y; F] - F'(z))\| \leq L(\|x - z\| + \|y - z\|),$$

for each $x, y, z \in D$,

$$\|[x_0, x_{-1}; F]^{-1}([x, y; F] - F'(x_0))\| \leq L_0(\|x - x_0\| + \|y - x_0\|),$$

for each $x, y \in D$,

$$\overline{U}(x_0, r) \subseteq D$$

and the special case of iteration $\{t_n\}$ is

$$\begin{cases} t_{-1} = 0, t_0 = c, t_1 = c + \eta, \\ t_{n+2} = t_{n+1} + \dfrac{L(t_{n+1} - t_n) + L(t_n - t_{n-1})}{1 - L_0(t_{n+1} - t_0 + t_n)}(t_{n+1} - t_n), \end{cases}$$

(see Refs. [46, 101, 120]). The one used by others in the literature in Refs. [16, 46, 178, 192, 267, 268, 266, 355, 418, 442] is:

$$\begin{cases} s_{-1} = 0, s_0 = c, s_1 = c + \eta, \\ s_{n+2} = s_{n+1} + \dfrac{L(s_{n+1} - s_n) + L(s_n - s_{n-1})}{1 - L(s_{n+1} - s_0 + s_n)}(s_{n+1} - s_n). \end{cases}$$

The comparison, favorable to us, is given in Refs. [46, 101, 120]. Sequence $\{t_n\}$ and the same convergence conditions were also used in Refs. [192, 267, 290, 355, 397, 418, 442] as majorizing for the inverse free Broyden's method

$$x_+ = x - AF(x), \quad y = F(x_+) - F(x), \quad A_+ = A - \frac{AF(x_+) < y, \cdot >}{< y, y >}$$

when X, Y are Hilbert spaces and $< x, y >$ denotes the inner product.

Application 4. Traub–Potra method (6.76)

Denote by (H_4) the following conditions. Let $F : D \subseteq X \to Y$ with divided differences $[\cdot, \cdot; F]$, $[\cdot, \cdot, \cdot; F]$,

$$B_0 = [x_0, x_{-2}; F] + [x_{-1}, x_0; F] - [x_{-2}, x_{-1}; F] \quad \text{is invertible;}$$

$$\|x_0 - x_{-1}\| \le c, \quad \|x_{-1} - x_{-2}\| \le a, \quad \|B_0^{-1} F(x_0)\| \le \eta;$$

$$\|B_0^{-1}([x, y; F] - [u, v; F])\| \le b_0(\|x - u\| + \|y - v\|),$$

for all $x, y, u, v \in D$

$$\|B_0^{-1}([x, y, z; F] - [u, y, z; F])\| \le b_1 \|x - u\|,$$

for each $x, y, z \in D$

$$\overline{U}(x_0, t^*) \subseteq D.$$

For the special case of iteration $\{t_n\}$ (see Refs. [355, 418]). The results obtained here hold for multistep methods (since these methods are converted to multipoint (see Refs. [120, 418]) and also equations containing nondifferentiable terms (since these methods use the same majorizing sequences, see (6.77)).

We provide examples where $L_0 < L$ so we can compare the corresponding convergence conditions.

Table 6.2   Sequences r_n, t_n, s_n

Iteration	r_{n+2}	t_{n+2}	s_{n+2}
0	$0.164706\ldots$	$0.169935\ldots$	$0.173050\ldots$
1	$0.166985\ldots$	$0.173102\ldots$	$0.177334\ldots$
2	$0.166997\ldots$	$0.173126\ldots$	$0.177385\ldots$

Table 6.3   A priori error bounds

Iteration	$|r_{n+2} - r_{n+1}|$	$|t_{n+2} - t_{n+1}|$	$|s_{n+2} - s_{n+1}|$
0	$0.002278\ldots$	$0.003167\ldots$	$0.004284\ldots$
1	$0.000012\ldots$	$0.000024\ldots$	$0.000051\ldots$
2	$3.43829\ldots \times 10^{-10}$	$1.38257\ldots \times 10^{-9}$	$7.24568\ldots \times 10^{-9}$

Table 6.4   Sequences r_n, t_n, s_n

Iteration	r_{n+2}	t_{n+2}	s_{n+2}
0	$0.272629\ldots$	$0.284400\ldots$	$0.290724\ldots$
1	$0.274770\ldots$	$0.287932\ldots$	$0.295915\ldots$
2	$0.274775\ldots$	$0.287946\ldots$	$0.295952\ldots$

Table 6.5   A priori error bounds

Iteration	$|r_{n+2} - r_{n+1}|$	$|t_{n+2} - t_{n+1}|$	$|s_{n+2} - s_{n+1}|$
0	$0.002140\ldots$	$0.003531\ldots$	$0.005190\ldots$
1	$5.21315\ldots \times 10^{-6}$	$0.000014\ldots$	$0.000037\ldots$
2	$3.09166\ldots \times 10^{-11}$	$2.45943\ldots \times 10^{-10}$	$1.94799\ldots \times 10^{-9}$

Example 6.3.1. Let $\mathcal{X} = \mathcal{Y} = \mathbb{R}$, $x_0 = 0$ and $\mathcal{D} = [a, 2 - a]$ for $a \in (0, 1)$. Define the scalar function F on $\mathcal{D}$ by (1.48). We see in Tables 6.2 and 6.3 that $r_{n+2} < t_{n+2} < s_{n+2}$ and $|r_{n+2} - r_{n+1}| < |t_{n+2} - t_{n+1}| < |s_{n+2} - s_{n+1}|$. Then, with the new theory the new error bounds are more precise than the earlier ones. Besides, $r^* = 0.166997\ldots < t^* = 0.173126\ldots < s^* = 0.177385\ldots$. That is more precise information on the location of the solution x^* is obtained.

Example 6.3.2. Consider Example 1.1.4. For $\gamma = 0.1$ and $R_0 = 0.5$ we obtain the corresponding sequences t_n and s_n for the common sequence and for the new sequence defined in this work and compare them. In Tables 6.4 and 6.5 we can see that $r_{n+2} < t_{n+2} < s_{n+2}$, $|r_{n+2} - r_{n+1}| < |t_{n+2} - t_{n+1}| < |s_{n+2} - s_{n+1}|$. Besides, $r^* = 0.274775\ldots < t^* = 0.287946\ldots < s^* = 0.295952$.

Other applications and examples including nonlinear Chandrasekhar-type integral equations appearing in radiative transfer can also be found in Ref. [120].

Example 6.3.3. Consider Example 1.1.5 with ζ replaced by θ. For $\lambda = 0.97548$, $\theta = 0.954585$ we obtain the results appearing in Tables 6.6 and 6.7. We can see that $r_{n+2} < t_{n+2} < s_{n+2}$ and $|r_{n+2} - r_{n+1}| < |t_{n+2} - t_{n+1}| < |s_{n+2} - s_{n+1}|$ for the sequences t_n and s_n constructed in this theory and for the sequence used

Table 6.6   Sequences r_n, t_n, s_n

Iteration	r_{n+2}	t_{n+2}	s_{n+2}
0	$0.314918\ldots$	$0.316731\ldots$	$0.318471\ldots$
1	$0.353020\ldots$	$0.356780\ldots$	$0.362434\ldots$
2	$0.360830\ldots$	$0.365765\ldots$	$0.374947\ldots$

Table 6.7   A priori error bounds

Iteration	$|r_{n+2} - r_{n+1}|$	$|t_{n+2} - t_{n+1}|$	$|s_{n+2} - s_{n+1}|$
0	$0.038102\ldots$	$0.040048\ldots$	$0.043963\ldots$
1	$0.007809\ldots$	$0.008985\ldots$	$0.012513\ldots$
2	$0.000357\ldots$	$0.000501\ldots$	$0.001209\ldots$

in the literature respectively. Besides, we have that $r^* = 0.361189\ldots < t^* = 0.366269\ldots < s^* = 0.376169\ldots$.

6.4   Exercises

6.4.1. Show Lemmas 6.1.1 and 6.1.2.

6.4.2. To solve (1.1), consider the following method

$$y_n = x_n - F'(x_n)^{-1} F(x_n) \quad (n \geq 0),$$

$$\mathcal{B}_n = \mathcal{B}(n, F) = F'(x_n)^{-1} \left(F'(x_n + \frac{2}{3}(y_n - x_n)) - F'(x_n) \right),$$

$$x_{n+1} = y_n - \frac{3}{4} \mathcal{B}_n \left(\mathcal{I} - \frac{3}{2} \mathcal{B}_n \right) (y_n - x_n),$$

where $\mathcal{I}$ is the matrix identity. We suppose that $F : \mathcal{D} \subseteq \mathcal{X} \longrightarrow \mathcal{Y}$ is a thrice differentiable operator. Assume also that there exist $x_0 \in \mathcal{D}$, $L \geq 0$, $M \geq 0$, $N \geq 0$, and $\eta \geq 0$, such that for all $x, y \in \mathcal{D}$, we have the following

$$F'(x_0)^{-1} \in \mathcal{L}(\mathcal{Y}, \mathcal{X}), \; \| F'(x_0)^{-1} F(x_0) \| \leq \eta,$$

$$\| F'(x_0)^{-1} F''(x) \| \leq M, \quad \| F'(x_0)^{-1} F'''(x) \| \leq N,$$

$$\| F'(x_0)^{-1} (F'''(x) - F'''(y)) \| \leq L \| x - y \|,$$

$$M \left(1 + \frac{N}{6 M^2} + \frac{13 L}{36 M^2} \right)^{\frac{1}{3}} \leq K, \quad h = K\eta \leq .46568$$

and

$$\overline{U}(x_0, v^\star) \subseteq \mathcal{D},$$

where $v^\star$ and $v^{\star\star}$ are the zeros of functions $g(t) = \dfrac{K}{2} t^2 - t + \eta$ given by

$$v^\star = \frac{1 - \sqrt{1 - 2h}}{h} \eta, \qquad v^{\star\star} = \frac{1 + \sqrt{1 - 2h}}{h} \eta.$$

Define $\theta = v^\star / v^{\star\star}$ and

$$\Delta_{\theta,\eta} = \frac{(1-\theta)^2\,\eta\,(\sqrt[3]{5}\,\theta)^{4^n-1}}{1 - \dfrac{1}{\sqrt[3]{5}}\,(\sqrt[3]{5}\,\theta)^{4^n}}.$$

Define scalar sequences $\{v_n\}$ and $\{w_n\}$ given by

$$w_n = v_n - g'(v_n)^{-1}\,g(v_n),$$

$$b_n = b(n,g) = g'(v_n)^{-1}\left(g'\!\left(v_n + \frac{2}{3}\,(w_n - v_n)\right) - g'(v_n)\right)$$

and

$$v_{n+1} = w_n - \frac{3}{4}\,b_n\left(1 - \frac{3}{2}\,b_n\right)(w_n - v_n).$$

Show:

(i) $\{v_n\}$ and $\{w_n\}$ are non-decreasing and converge to their common limit $v^\star$.

(ii) $v_n \le w_n \le v_{n+1} \le w_{n+1}$.

(iii) $\{x_n\}$ and $\{y_n\}$ are well defined, remain in $\overline{U}(x_0, v^\star)$ for all $n \ge 0$.

(iv) $\{x_n\}$ and $\{y_n\}$ converge to a unique solution $x^\star \in \overline{U}(x_0, v^\star)$ of (1.1), which is the unique solution of (1.1) in $U(x_0, v^{\star\star})$.

Show also that for all $n \ge 0$, the following estimates hold

(a) $\| y_n - x_n \| \le w_n - v_n$.

(b) $\| x_{n+1} - y_n \| \le w_{n+1} - v_n$.

(c) $\| y_n - x^\star \| \le v^\star - w_n$.

(d) $\| x_n - x^\star \| \le v^\star - v_n \le \Delta_{\theta,\eta}$.

6.4.3. Consider the same notations and assumptions as Exercise 6.4.2.

(a) Show that there exists $M_0 \in [0, M]$, such that

$$\| F'(x_0)^{-1}\,(F'(x) - F'(x_0)) \| \le M_0 \, \| x - x_0 \| \qquad \text{for all } x \in \mathcal{D}.$$

Let g_0 be given by

$$g_0(t) = \frac{M_0}{2}\,t^2 - t + \eta.$$

Consider scalar majorizing sequences $\{\overline{v}_n\}$, $\{\overline{w}_n\}$, defined as follows for initial iterates $\overline{v}_0 = 0$, $\overline{w}_1 = \eta$ by

$$\overline{w}_n = \overline{v}_n - g_0'(\overline{v}_n)^{-1}\,g(\overline{v}_n),$$

$$\overline{b}_n = b(n, g, g_0) = g_0'(\overline{v}_n)^{-1}\left(g'\!\left(\overline{v}_n + \frac{2}{3}\,(\overline{w}_n - \overline{v}_n)\right) - g'(\overline{v}_n)\right)$$

and

$$\overline{v}_{n+1} = \overline{w}_n - \frac{3}{4}\,\overline{b}_n\left(1 - \frac{3}{2}\,\overline{b}_n\right)(\overline{w}_n - \overline{v}_n).$$

Suppose that $M_0 < K$. Show:

(b) $\overline{v}_n < v_n$.

(c) $\overline{w}_n < w_n$.

(d) $\overline{w}_n - \overline{v}_n < w_n - v_n$.

(e) $\overline{v}_{n+1} - \overline{w}_n < v_{n+1} - w_n$.

(f) $\overline{v}^\star \le v^\star$ where $\overline{v}^\star = \lim_{n \to \infty} \overline{v}_n$.

Chapter 7

Chebyshev's Method

Chebyshev's method, or the method of tangent parabola is a cubically convergent method. The analysis of this method is based on the most popular concept of majorizing sequences. In this chapter we present some new results of Chebychev-type methods using more precise majorizing sequence or our new concept of recurrent functions. Our theoretical and numerical results expand the applicability of Chebyshev's method.

7.1 Directional Methods

A semi-local convergence analysis for directional Chebyshev-type methods in m-variables and some applications are provided in this section. Our analysis uses recurrent relations and Newton–Kantorovich-type hypotheses. We are concerned with the problem of approximating a solution $x^\star$ of (1.1) where F is a nonlinear Fréchet-differentiable operator defined in an open convex non-empty subset $\mathcal{D}$ of $\mathbb{R}^m$ (m a natural number) with values in $\mathbb{R}$. In this section we use the Euclidean norms for both vector and matrix. The unit direction d_k is chosen such that

$$d_k \approx \frac{\nabla F(x_k)}{\parallel \nabla F(x_k) \parallel}.$$

The angle between two vectors x, y in a Hilbert space H denoted by $\angle(x, y)$ is given by

$$\angle(x, y) = \arccos \frac{x \cdot y}{\parallel x \parallel \cdot \parallel y \parallel}, \quad x \neq 0, \quad y \neq 0.$$

An and Bai in Ref. [20] used the directional Secant method (DSM)

$$x_{k+1} = x_k + h_k,$$
$$h_k = -\frac{\theta_k \, F(x_k)}{F(x_k + \theta_k \, d_k) - F(x_k)} \, d_k,$$
$$(k \geq 0, \quad x_0 \in \mathbb{R}^n, \quad d_k \in \mathbb{R}^n, \quad \parallel d_k \parallel = 1, \quad \theta_k \geq 0)$$

to generate a sequence $\{x_k\}$ converging to $x^\star$. (DSM) is a useful alternative to (DNM) (see Refs. [116, 176, 300]):

$$x_{k+1} = x_k - \frac{F(x_k)}{\nabla F(x_k) \cdot d_k} \, d_k \quad (k \geq 0),$$

363

where

$$\nabla F(x_k) = \left(\frac{\partial F(x_k)}{\partial x_1}, \frac{\partial F(x_k)}{\partial x_2}, \cdots, \frac{\partial F(x_k)}{\partial x_m} \right),$$

is the gradient of F and d_k is a direction at x_k. (DNM) converges quadratically to $x^\star$, if x_0 is close enough to $x^\star$ (see Ref. [300]). However, as already noted in Ref. [300], the computation of the gradient $\nabla F(x_k)$ may be too expensive as it is the case when the number n of unknowns is large.

In some applications, the mapping F may not be differentiable, or the gradient is impossible to compute. The (DSM) avoids these obstacles. Note that if $m = 1$, (DSM) reduces to the classical Secant method, and (DNM) to Newton's method (see Ref. [100]). The quadratic convergence of (DSM) (see Ref. [20]), and (DNM) (see Ref. [300]) was established for directions d_k sufficiently close to the gradients $\nabla F(x_k)$ and under standard Newton–Kantorovich-type hypotheses. We introduce the directional Chebyshev-type method (DCTM):

$$x_0 \in \mathcal{D},$$
$$y_k = x_k - A_k\, F(x_k), \quad v_k = x_k + \theta_k\, d_k, \quad A_k = \frac{\theta_k}{F(v_k) - F(x_k)} \cdot d_k$$
$$z_k = x_k + a\, (y_k - x_k), \quad a \in [0, 1],$$
$$x_{k+1} = x_k - A_k\, (b\, F(x_k) + c\, F(z_k))$$
$$(k \geq 0, \quad d_k \in \mathbb{R}^m, \quad \| d_k \| = 1, \quad \theta_k \geq 0, \quad b \in [0, 1], \quad c\, (1 - a) = 1 - b)$$

to generate a sequence $\{x_k\}$ approximating the zero $x^\star$.

(DCTM) is a Secant-type analog of the third order Chebyshev-type method (CTM) with efficiency close to Newton's and the same region of accessibility defined on Banach spaces by Ezquerro and Hernández in Ref. [208]:

$$x_0 \in \mathcal{D},$$
$$y_k = x_k - F'(x_k)^{-1}\, F(x_k),$$
$$z_k = x_k + d\, (y_k - x_k), \quad d \in [0, 1],$$
$$x_{k+1} = x_k - \frac{1}{d^2}\, F'(x_k)^{-1}\, ((d^2 + d - 1)\, F(x_k) + F(z_k)), \quad k \geq 0.$$

We need the definition of the first order divided difference for a mapping. We shall use the following conditions:

$(\mathcal{C_1})$

$$|F(x_0)| \leq \lambda, \quad \| \nabla F(x_0) \| \geq \beta > 0$$

and

$$|\nabla F(x_0) \cdot d_0| \geq \alpha\, \| \nabla F(x_0) \|, \quad \alpha \in [0, 1];$$

$(\mathcal{C_2})$ $d_k\ (k \geq 0)$ satisfies

$(\mathcal{C_{21}})$

$$\varrho\, |\nabla F(x_0) \cdot d_0| \leq |\nabla F(x) \cdot d_0|, \quad x \in (x_0, x_0 + \theta_0\, d_0), \quad \varrho \in (0, 1);$$

$(\mathcal{C_{22}})$

$$\angle(d_{k+1}, \nabla F(x_{k+1})) \leq \angle(d_k, \nabla F(x_k));$$

$(\mathcal{C_3})$ θ_k satisfies

$$\theta_{k+1} \leq q\,\theta_k \parallel x_{k+1} - x_k \parallel, \quad q \in (0,1);$$

$(\mathcal{C_4})$ Mapping ∇F is Lipschitz with constant M on $\mathcal{D}$:

$$\parallel \nabla F(x) - \nabla F(y) \parallel \leq M \parallel x - y \parallel \quad \text{for all} \quad x, y \in \mathcal{D};$$

$(\mathcal{C_5})$ Mapping F has a divided difference of order one $[x, y; F]$ at the points $(x, y) \in \mathcal{D}^2$ and

$$\parallel [x, z; F] - [y, z; F] \parallel \leq N \parallel x - y \parallel$$

for all $x, y, z \in \mathcal{D}$.

We need some auxiliary lemmas to establish the semi-local convergence of (DCTM).

Lemma 7.1.1. *(see [3.2.2] in Ref. [338]) Let $\mathcal{D} \subseteq \mathbb{R}^m$ be an open convex non-empty subset and $F : \mathcal{D} \subseteq \mathbb{R}^m \longrightarrow \mathbb{R}$ a differentiable mapping. Then, for any $x, y \in \mathcal{D}$, there exists a vector $\mu \in (x, y)$, such that:*

$$F(y) - F(x) = \nabla F(\mu)\,(y - x),$$

where

$$(x, y) = \{z : z = x + \theta\,y, \quad 0 < \theta < 1\},$$

represents the open straight line between the points x and y.

Remark 7.1.1. Note that $(\mathcal{C_{22}})$ implies

$$\frac{|\nabla F(x_{k+1}) \cdot d_{k+1}|}{\parallel \nabla F(x_{k+1}) \parallel} \geq \frac{|\nabla F(x_k) \cdot d_k|}{\parallel \nabla F(x_k) \parallel} \qquad (k \geq 0). \tag{7.1}$$

Lemma 7.1.2. *Let F be a nonlinear Fréchet-differentiable operator $F : \mathcal{D} \subseteq \mathbb{R}^m \longrightarrow \mathbb{R}$ under the previous conditions $(\mathcal{C_1})$–$(\mathcal{C_5})$ and $\{x_k\}$ the sequence in $\mathcal{D}$ generated by (DCTM).*

Then, the following is satisfied

$$|\nabla F(x_k) \cdot d_k| \geq \alpha \parallel \nabla F(x_k) \parallel \qquad (k \geq 0). \tag{7.2}$$

Proof. Estimate (7.2) holds for $k = 0$ by $(\mathcal{C_1})$. Then, by a simple induction argument and (7.1) we get:

$$\alpha\,\frac{\parallel \nabla F(x_{k+1}) \parallel}{|\nabla F(x_{k+1}) \cdot d_{k+1}|} \leq \alpha\,\frac{\parallel \nabla F(x_k) \parallel}{|\nabla F(x_k) \cdot d_k|} \leq 1.$$

That completes the proof of Lemma 7.1.2. $\qquad\qquad\square$

We need an Ostrowski-type approximation for (DCTM).

Lemma 7.1.3. *Let F be a nonlinear Fréchet-differentiable operator $F : \mathcal{D} \subseteq \mathbb{R}^m \longrightarrow \mathbb{R}$. Assume that sequence $\{x_k\}$ generated by (DCTM) is well defined. Then, the following items hold for all $k \geq 0$:*

$$F(z_k) = (1 - a)\, F(x_k) + a\left([z_k, x_k; F] - [v_k, x_k; F] \right)(y_k - x_k), \tag{7.3}$$

$$x_{k+1} - y_k = -a\, c\, A_k \left([v_k, x_k; F] - [z_k, x_k; F] \right)(y_k - x_k) \tag{7.4}$$

and

$$F(x_{k+1}) = \left([x_{k+1}, x_k; F] - [v_k, x_k; F] \right)(x_{k+1} - x_k) -$$
$$a\, c\left([z_k, x_k; F] - [v_k, x_k; F] \right)(y_k - x_k). \tag{7.5}$$

Proof. Using (DCTM), we obtain in turn that

$$F(z_k) = (1 - a)\, F(x_k) + F(z_k) - F(x_k) + a\,[v_k, x_k; F]\, \frac{\theta_k\, F(x_k)}{F(v_k) - F(x_k)} \cdot d_k$$
$$= (1 - a)\, F(x_k) + F(z_k) - F(x_k) + a\, \frac{F(v_k) - F(x_k)}{F(v_k) - F(x_k)}\, F(x_k)$$

showing (7.3). For $a \neq 1$, we have that

$$x_{k+1} - y_k = -A_k \left(b\, F(x_k) + \frac{1 - b}{1 - a}\, F(z_k) \right) + A_k\, F(x_k)$$
$$= -A_k \left((b - 1)\, F(x_k) + \frac{1 - b}{1 - a}\, F(z_k) \right)$$
$$= -A_k \left((b - 1)\, F(x_k) + (1 - b)\, F(x_k) \right) +$$
$$\frac{a\,(1 - b)}{1 - a} \left([z_k, x_k; F] - [v_k, x_k; F] \right)(y_k - x_k),$$

which is (7.4). We also have that

$$F(x_{k+1}) = F(x_{k+1}) - F(x_k) - [v_k, x_k; F]\, (y_k - x_k)$$
$$= [x_{k+1}, x_k; F]\, (x_{k+1} - x_k) - [v_k, x_k; F]\, (y_k - x_k)$$
$$= \left([x_{k+1}, x_k; F] - [v_k, x_k; F] \right)(x_{k+1} - x_k) +$$
$$[v_k, x_k; F]\, (x_{k+1} - x_k) - [v_k, x_k; F]\, (y_k - x_k)$$
$$= \left([x_{k+1}, x_k; F] - [v_k, x_k; F] \right)(x_{k+1} - x_k) + [v_k, x_k; F]\, (x_{k+1} - y_k)$$
$$= \left([x_{k+1}, x_k; F] - [v_k, x_k; F] \right)(x_{k+1} - x_k) +$$
$$[v_k, x_k; F] \left(\frac{-a\, c\, ([v_k, x_k; F] - [z_k, x_k; F])\, (y_k - x_k)\, \theta_k \cdot d_k}{F(v_k) - F(x_k)} \right)$$
$$= \left([x_{k+1}, x_k; F] - [v_k, x_k; F] \right)(x_{k+1} - x_k) -$$
$$a\, c\left([z_k, x_k; F] - [v_k, x_k; F] \right)(y_k - x_k),$$

which is (7.5). That completes the proof of Lemma 7.1.3. $\square$

It is convenient for us to introduce some notation and initial conditions:

$$a_0 = \frac{1}{\rho\,\alpha\,\beta}, \quad r_0 = N\,a_0\,\lambda,$$

$$t_0 = N\,\theta_0, \quad s_0 = (r_0 + t_0)\,a_0, \quad c_0 = a_0\,\lambda, \quad b_0 = \frac{M\,\lambda}{\rho\,\beta^2\,\alpha}.$$

Then, we have the following relations:

$$\| A_0 \| \leq \left| \frac{\theta_0}{F(v_0) - F(x_0)} \right| \leq \frac{1}{|\nabla F(\mu_0)\cdot d_0|} \leq \frac{1}{\rho\,\alpha\,\beta} = a_0,$$

$$N\,\| y_0 - x_0 \| \leq N\,\| A_0 \|\,|F(x_0)| \leq N\,a_0\,\lambda = r_0,$$

$$N\,\| z_0 - v_0 \| \leq r_0 + t_0,$$

$$N\,\| A_0 \|\,\| z_0 - v_0 \| \leq (r_0 + t_0)\,a_0 = s_0, \quad M\,\| \nabla F(x_0) \|^{-1}\,\| y_0 - x_0 \| \leq \frac{M\,\lambda}{\beta^2\,\rho\,\alpha} = b_0.$$

Then, we get

$$\| x_1 - y_0 \| \leq a\,c\,\| A_0 \|\,\| [v_0, x_0; F] - [z_0, x_0; F] \|\,\| y_0 - x_0 \|$$

$$\leq a\,c\,a_0\,N\,\| z_0 - v_0 \|\,\| y_0 - x_0 \|$$

$$\leq a\,c\,s_0\,\| y_0 - x_0 \|,$$

$$\| x_1 - x_0 \| \leq (1 + a\,c\,s_0)\,\| y_0 - x_0 \| < R,$$

for some $R > 0$ to be determined later,

$$N\,\| x_1 - v_0 \| \leq N\,\| x_1 - x_0 \| + N\,\theta_0$$

$$\leq N\,((1 + a\,c\,s_0)\,\| y_0 - x_0 \| + \theta_0)$$

$$\leq (1 + a\,c\,s_0)\,(r_0 + t_0),$$

and

$$|F(x_1)| \leq N\,\| x_1 - v_0 \|\,\| x_1 - x_0 \| + a\,c\,N\,\| z_0 - v_0 \|\,\| y_0 - x_0 \|$$

$$\leq ((1 + a\,c\,s_0)^2 + a\,c)\,(r_0 + t_0)\,\| y_0 - x_0 \|.$$

We must define auxiliary real functions

$$f(x, y) = \frac{1}{1 - (1 + a\,c\,x)\,y}, \quad g(x) = ((1 + a\,c\,x)^2 + a\,c)\,x,$$

scalar sequences $(k \geq 1)$

$$r_k = r_{k-1}\,f(s_{k-1}, b_{k-1})\,g(s_{k-1}), \quad c_k = c_{k-1}\,f(s_{k-1}, b_{k-1})\,g(s_{k-1}),$$

$$t_k = t_{k-1}\,q\,(1 + a\,c\,s_{k-1})\,c_{k-1}, \quad a_k = a_{k-1}\,f(s_{k-1}, b_{k-1}), \quad s_k = (r_k + t_k)\,a_k$$

and

$$b_k = b_{k-1}\,f^2(s_{k-1}, b_{k-1})\,g(s_{k-1}).$$

Then, we shall show using induction the following recurrence relations.

Lemma 7.1.4. *Let us suppose that* $x_0,\,v_0,\,y_0,\,z_0 \in \mathcal{D}$ *and* $x_k,\,v_k,\,y_k,\,z_k \in \mathcal{D}$ *$(k \in \mathbb{N})$. Assume*

$$f(s_0, b_0)^2\,g(s_0) < 1 \tag{7.6}$$

and

$$q\,(1 + a\,c\,s_0)\,f(s_0, b_0)\,c_0 < 1. \tag{7.7}$$

Then, the following relations are satisfied for $(k \geq 0)$:

(I_k)

$$\frac{1}{\rho\,|\nabla F(x_k)\cdot d_k|} \leq a_k \quad and \quad \|\,A_k\,\| \leq a_k,$$

(II_k)

$$\|\,y_k - x_k\,\| \leq f(s_{k-1},b_{k-1})\,g(s_{k-1})\,\|\,y_{k-1} - x_{k-1}\,\| \leq c_k,$$

(III_k)

$$N\,\|\,A_k\,\|\,\|\,z_k - v_k\,\| \leq s_k,$$

(IV_k)

$$M\,\|\,\nabla F(x_k)\,\|^{-1}\,\|\,y_k - x_k\,\| \leq b_k,$$

(V_k)

$$\|\,x_{k+1} - y_k\,\| \leq a\,c\,s_k\,\|\,y_k - x_k\,\|,$$

(VI_k)

$$\|\,x_{k+1} - x_k\,\| \leq (1 + a\,c\,s_k)\,\|\,y_k - x_k\,\|\,.$$

(VII_k)

$$\|\,x_{k+1} - x_0\,\| \leq (1 + a\,c\,s_0)\,(1 + q\,\theta_0)\,\frac{1 - (f(s_0,b_0)\,g(s_0))^{k+1}}{1 - f(s_0,b_0)\,g(s_0)}\,\|\,y_0 - x_0\,\|\,.$$

Proof. We shall first show conditions (I_k)–(VII_k) are satisfied for $k = 1$. We have in turn:

$$\|\,v_1 - x_0\,\| \leq \|\,x_1 - x_0\,\| + \theta_1 \leq (1 + q\,\theta_0)\,\|\,x_1 - x_0\,\| \leq (1 + a\,c\,s_0)\,(1 + q\,\theta_0)\,\|\,y_0 - x_0\,\|,$$

$$\begin{aligned}
\|\,\nabla F(x_1)\,\| &\geq \|\,\nabla F(x_0)\,\| - \|\,\nabla F(x_1) - \nabla F(x_0)\,\| \\
&\geq \|\,\nabla F(x_0)\,\| - M\,\|\,x_1 - x_0\,\| \\
&\geq \|\,\nabla F(x_0)\,\| - M\,(1 + a\,c\,s_0)\,\|\,y_0 - x_0\,\| \\
&= \|\,\nabla F(x_0)\,\|\,(1 - M\,(1 + a\,c\,s_0)\,\|\,\nabla F(x_0)\,\|^{-1}\,\|\,y_0 - x_0\,\|) \\
&\geq \|\,\nabla F(x_0)\,\|\,(1 - (1 + a\,c\,s_0)\,b_0),
\end{aligned}$$

so, we have that

$$\|\,\nabla F(x_1)\,\|^{-1} \leq \|\,\nabla F(x_0)\,\|^{-1}\,f(s_0,b_0),$$

(I_1)

$$\begin{aligned}
\|\,A_1\,\| &\leq \left|\frac{\theta_1}{F(v_1) - F(x_1)}\right| \leq \left|\frac{\theta_1}{\nabla F(\mu_1)\theta_1\cdot d_1}\right| \\
&\leq \frac{1}{\rho\,|\nabla F(x_1)\cdot d_1|} \leq \frac{\|\,\nabla F(x_0)\,\|}{\rho\,|\nabla F(x_0)\cdot d_0|\,\|\,\nabla F(x_1)\,\|} \\
&\leq \frac{f(s_0,b_0)}{\rho\,|\nabla F(x_0)\cdot d_0|} \leq a_0\,f(s_0,b_0) = a_1,
\end{aligned}$$

(II_2)

$$\| y_1 - x_1 \| \leq \| A_1 \| \, |F(x_1)|$$
$$\leq a_0 \, f(s_0, b_0) \, (r_0 + t_0) \, ((1 + a \, c \, s_0)^2 + a \, c) \, \| y_0 - x_0 \|$$
$$= f(s_0, b_0) \, g(s_0) \, \| y_0 - x_0 \|,$$
$$\| y_1 - x_0 \| \leq \| y_1 - x_1 \| + \| x_1 - x_0 \|$$
$$\leq f(s_0, b_0) \, g(s_0) \, \| y_0 - x_0 \| + (1 + a \, c \, s_0) \, \| y_0 - x_0 \|$$
$$\leq (f(s_0, b_0) \, g(s_0) + 1 + a \, c \, s_0) \, \| y_0 - x_0 \|,$$
$$\| z_1 - x_0 \| \leq a \, \| y_1 - x_1 \| + \| x_1 - x_0 \|$$
$$\leq (f(s_0, b_0) \, g(s_0) + 1 + a \, c \, s_0) \, \| y_0 - x_0 \|,$$
$$N \, \| y_1 - x_1 \| \leq r_0 \, f(s_0, b_0) \, g(s_0) = r_1,$$
$$N \, \theta_1 \leq N \, q \, \theta_0 \, \| x_1 - x_0 \| \leq N \, q \, \theta_0 \, (1 + a \, c \, s_0) \, \| y_0 - x_0 \| \leq t_0 \, q \, (1 + a_0 \, c \, s_0) \, a_0 \, \lambda = t_1,$$

(III_1)

$$N \, \| A_1 \| \, \| z_1 - v_1 \| \leq a_1 \, (N \, \| y_1 - x_1 \| + N \, \theta_1) \leq a_1 \, (r_1 + t_1) = s_1,$$

(IV_1)

$$M \, \| \nabla F(x_1) \|^{-1} \| y_1 - x_1 \| \leq M \, \| \nabla F(x_0) \|^{-1} \, f(s_0, b_0)^2 \, g(s_0) \, \| y_0 - x_0 \|$$
$$\leq b_0 \, f(s_0, b_0)^2 \, g(s_0) = b_1,$$

(V_1)

$$\| x_2 - y_1 \| \leq a \, c \, \| A_1 \| \, \| [v_1, x_1; F] - [z_1, x_1; F] \| \, \| y_1 - x_1 \|$$
$$\leq a \, c \, N \, \| A_1 \| \, \| v_1 - z_1 \| \, \| y_1 - x_1 \| \leq a \, c \, s_1 \, \| y_1 - x_1 \|,$$

(VI_1)

$$\| x_2 - x_1 \| \leq \| x_2 - y_1 \| + \| y_1 - x_1 \| \leq (1 + a \, c \, s_1) \, \| y_1 - x_1 \|,$$

(VII_1)

$$\| x_2 - x_0 \| \leq (1 + a \, c \, s_1) \, \| y_1 - x_1 \| + (1 + a \, c \, s_0) \, \| y_0 - x_0 \|$$
$$\leq ((1 + a \, c \, s_1) \, f(s_0, b_0) \, g(s_0) + 1 + a \, c \, s_0) \, \| y_0 - x_0 \|$$
$$\leq (1 + a \, c \, s_0) \, (1 + f(s_0, b_0) \, g(s_0)) \, \| y_0 - x_0 \|$$
$$\leq (1 + a \, c \, s_0) \, (1 + q \, \theta_0) \, \frac{1 - (f(s_0, b_0) \, g(s_0))^2}{1 - f(s_0, b_0) \, g(s_0)} \, \| y_0 - x_0 \|.$$

The rest follows by a simple induction argument. That completes the proof of Lemma 7.1.4. $\qquad\square$

We can show the following semi-local convergence result for (DCTM).

Theorem 7.1.1. *Let F be a nonlinear Fréchet-differentiable operator under conditions (C_1)–(C_5). We also assume (7.6) and (7.7). Then, if $\overline{U}(x_0, R) \subseteq \mathcal{D}$, where*
$$R = \frac{(1 + a \, c \, s_0) \, (1 + q \, \theta_0) \, a_0 \, \lambda}{1 - f(s_0, b_0) \, g(s_0)}$$
the sequence $\{x_k\}$ generated by (DCTM) starting in x_0, is well defined, remains in $\overline{U}(x_0, R)$ for all $k \geq 0$ and converges to a solution $x^\star \in \overline{U}(x_0, R)$ of equation $F(x) = 0$.

Proof.   Firstly, we have that

$$\| v_0 - x_0 \| \le \theta_0 < R,$$
$$\| y_0 - x_0 \| < R,$$
$$\| z_0 - x_0 \| \le a \, \| y_0 - x_0 \| < R$$

and

$$\| x_1 - x_0 \| \le (1 + a \, c \, s_0) \, \| y_0 - x_0 \| < R.$$

Thus, $v_0, y_0, z_0, x_1 \in \mathcal{D}$ Similarly, we get $v_1, y_1, z_1, x_2 \in \mathcal{D}$. Assume $v_i, y_i, z_i,$ $x_{i+1} \in \mathcal{D}$, $i = 1, \cdots, k$. Then, using Lemma 7.1.3, we prove by induction $v_{k+1},$ $y_{k+1}, z_{k+1}, x_{k+2} \in \mathcal{D}$. We have in turn by the recurrence relations that

$$\begin{aligned}
\| v_{k+1} - x_k \| &\le (1 + q \, \theta_k) \, \| x_{k+1} - x_k \| \\
&\le (1 + a \, c \, s_k)(1 + q \, \theta_k) \, \| y_k - x_k \| \\
&\le (1 + a \, c \, s_0)(1 + q \, \theta_0)(f(s_0, b_0) \, g(b_0))^k \, \| y_0 - x_0 \|,
\end{aligned}$$

$$\begin{aligned}
\| v_{k+1} - x_0 \| &\le (1 + a \, c \, s_0)(f(s_0, b_0) \, g(s_0))^k (1 + q \, \theta_0) \, \| y_0 - x_0 \| + \\
&\quad (1 + a \, c \, s_0) \frac{1 - (f(s_0, b_0) \, g(s_0))^k}{1 - f(s_0, b_0) \, g(s_0)} (1 + q \, \theta_0) \, \| y_0 - x_0 \| \\
&\le (1 + a \, c \, s_0)(1 + q \, \theta_0) \frac{1 - (f(s_0, b_0) \, g(s_0))^{k+1}}{1 - f(s_0, b_0) \, g(s_0)} \, \| y_0 - x_0 \| < R,
\end{aligned}$$

$$\begin{aligned}
\| y_{k+1} - x_0 \| &\le \| y_{k+1} - x_{k+1} \| + \| x_{k+1} - x_0 \| \\
&\le f(s_k, b_k) \, g(s_k) \, \| y_k - x_k \| + \\
&\quad (1 + a \, c \, s_0)(1 + q \, \theta_0) \frac{1 - (f(s_0, b_0) \, g(s_0))^{k+1}}{1 - f(s_0, b_0) \, g(s_0)} \, \| y_0 - x_0 \| \\
&< (1 + a \, c \, s_0)(1 + q \, \theta_0) \frac{1 - (f(s_0, b_0) \, g(s_0))^{k+2}}{1 - f(s_0, b_0) \, g(s_0)} \, \| y_0 - x_0 \| < R,
\end{aligned}$$

$$\| z_{k+1} - x_0 \| \le \| x_{k+1} - x_0 \| + a \, \| y_{k+1} - x_{k+1} \| < R,$$

$$\begin{aligned}
\| x_{k+2} - x_0 \| &\le (1 + a \, c \, s_{k+1}) \, \| y_{k+1} - x_{k+1} \| + \\
&\quad (1 + a \, c \, s_0)(1 + q \, \theta_0) \frac{1 - (f(s_0, b_0) \, g(s_0))^{k+1}}{1 - f(s_0, b_0) \, g(s_0)} \, \| y_0 - x_0 \| \\
&\le (1 + a \, c \, s_0)(f(s_0, b_0) \, g(s_0))^{k+1} \, \| y_0 - x_0 \| + \\
&\quad (1 + a \, c \, s_0)(1 + q \, \theta_0) \frac{1 - (f(s_0, b_0) \, g(s_0))^{k+1}}{1 - f(s_0, b_0) \, g(s_0)} \, \| y_0 - x_0 \| \\
&\le (1 + a \, c \, s_0)(1 + q \, \theta_0) \frac{1 - (f(s_0, b_0) \, g(s_0))^{k+2}}{1 - f(s_0, b_0) \, g(s_0)} \, \| y_0 - x_0 \| < R.
\end{aligned}$$

Hence, we deduce $v_{k+1}, y_{k+1}, z_{k+1}, x_{k+2} \in \mathcal{D}$. Next, we shall show the convergence of sequence $\{x_k\}$ by using recurrence relations:

$$\| x_{k+1} - x_k \| \le (1 + a \, c \, s_k) \, \| y_k - x_k \| \le (1 + a \, c \, s_0)(f(s_0, b_0) \, g(s_0))^k \, \| y_0 - x_0 \|,$$

so,

$$\| x_{k+j} - x_k \| \leq \sum_{i=k}^{i=k+j-1} \| x_{i+1} - x_i \|$$

$$\leq (1 + a\,c\,s_0) \sum_{i=k}^{i=k+j-1} (f(s_0, b_0)\,g(s_0))^i \| y_0 - x_0 \| \tag{7.8}$$

$$\leq (1 + a\,c\,s_0)\,(f(s_0, b_0)\,g(s_0))^k \frac{1 - (f(s_0, b_0)\,g(s_0))^j}{1 - f(b_0)\,g(s_0)} \| y_0 - x_0 \|.$$

It is obvious that $\{x_k\}$ is a Cauchy sequence in a complete space $\mathbb{R}^m$ and as such it converges to some $x^\star \in \overline{U}(x_0, R)$ (since $\overline{U}(x_0, R)$ is a closed set).

We shall show that $F(x^\star) = 0$. Indeed, we have that

$$|F(x_{k+1})| \leq ((1 + a\,c\,s_k)^2 + a\,c)\,(r_k + t_k)\, \| y_k - x_k \|$$

$$\leq ((1 + a\,c\,s_0)^2 + a\,c)(r_0 + t_0)\,(f(s_0, b_0)\,g(s_0))^k \| y_0 - x_0 \|.$$

Thus, by letting $k \to \infty$ it follows $|F(x_k)| \to 0$, since $f(s_0, b_0)\,g(s_0) < 1$ and $((1 + a\,c\,s_0)^2 + a\,c)(r_0 + t_0) \| y_0 - x_0 \|$ is bounded. Hence, we deduce $|F(x^\star)| = 0$. That completes the proof of Theorem 7.1.1. $\qquad\square$

It turns out that in an analogous way a second semi-local convergence result can be given for (DCTM). We first need the lemma.

Lemma 7.1.5. *Let* $q \in (0, 1)$, $M > 0$ *and* $k \geq 0$. *If* $\theta_k \in (0, \Upsilon_k]$, *where*

$$\Upsilon_k = \frac{\sqrt{\| \nabla F(x_k) \|^2 + 2\,M\,q\,|F(x_k)|} - \| \nabla F(x_k) \|}{M}.$$

Then, for all $k \geq 0$ *the following holds* $\theta_k \leq q\, \| y_k - x_k \|$.

Proof. We have in turn that

$$|F(v_k) - F(x_k)| = \left| \int_{x_k}^{v_k} \nabla F(y)\, dy \right|$$

$$= \left| \int_0^1 (\nabla F(x_k + \tau\,\theta_k\,d_k)\,\theta_k\,d_k - \nabla F(x_k)\,\theta_k\,d_k + \nabla F(x_k)\,\theta_k\,d_k)\, d\tau \right|$$

$$\leq \theta_k \left(M \int_0^1 \tau\,\theta_k\,d\tau + \| \nabla F(x_k) \| \right)$$

$$= \theta_k \left(\frac{M}{2}\,\theta_k + \| \nabla F(x_k) \| \right) = \frac{M}{2}\,\theta_k^2 + \| \nabla F(x_k) \|\,\theta_k.$$

Therefore,

$$\frac{M}{2}\,\theta_k^2 + \| \nabla F(x_k) \|\,\theta_k \leq q\,|F(x_k)|$$

provided that $\theta_k \in (0, \Upsilon_k]$. Then, we get that

$$\| y_k - x_k \| = \frac{\theta_k\,|F(x_k)|}{|F(v_k) - F(x_k)|} \geq \frac{\theta_k\,|F(x_k)|}{q\,|F(x_k)|} = \frac{\theta_k}{q}.$$

That completes the proof of Lemma 7.1.5. $\qquad\square$

We shall assume conditions $(\mathcal{C}_1)$, $(\mathcal{C}_2)$, $(\mathcal{C}_4)$, $(\mathcal{C}_5)$ and

$(\mathcal{C}_3^\star)$ $\theta_0 \in (0, \Upsilon_0]$ and $0 < \theta_k \leq \min\{\Upsilon_k, q \parallel y_{k-1} - x_{k-1} \parallel\}$ for $k \geq 1$.

We need to define auxiliary real functions

$$f(x,y) = \frac{1}{1 - (1 + a\,c\,(a + q)\,x)\,y},$$

$$g(x) = (1 + q + a\,c\,(a + q)\,x)\,(1 + a\,c\,(a + q)\,x) + a\,c\,(a + q)$$

and scalar sequences

$$s_k = f(s_{k-1}, t_{k-1})^2\,g(s_{k-1})\,s_{k-1}^2, \quad s_0 = \frac{N\,\gamma^2\,\lambda}{\beta^2}, \quad \gamma = \frac{1}{\rho\,\alpha},$$

$$t_k = f(s_{k-1}, t_{k-1})^2\,g(s_{k-1})\,s_{k-1}\,t_{k-1}, \quad t_0 = \frac{M\,\gamma\,\lambda}{\beta^2}.$$

Then, we obtain exactly as Lemma 7.1.4 the recurrence relations:

$(\widetilde{I_k})$ $\parallel A_k \parallel \leq \gamma \parallel \nabla F(x_k) \parallel^{-1} \leq \gamma \parallel \nabla F(x_{k-1}) \parallel^{-1} f(s_{k-1}, t_{k-1})$,

$(\widetilde{II_k})$ $\parallel y_k - x_k \parallel \leq f(s_{k-1}, t_{k-1})\,g(s_{k-1})\,s_{k-1} \parallel y_{k-1} - x_{k-1} \parallel$,

$(\widetilde{III_k})$ $N \parallel A_k \parallel \parallel y_k - x_k \parallel \leq s_k$,

$(\widetilde{IV_k})$ $M \parallel \nabla F(x_k) \parallel^{-1} \parallel y_k - x_k \parallel \leq t_k$,

$(\widetilde{V_k})$ $\parallel x_{k+1} - y_k \parallel \leq a\,c\,(a + q)\,s_k \parallel y_k - x_k \parallel$,

$(\widetilde{VI_k})$ $\parallel x_{k+1} - x_k \parallel \leq (1 + a\,c\,(a + q)\,s_k) \parallel y_k - x_k \parallel$,

$(\widetilde{VII_k})$ $\parallel x_{k+1} - x_0 \parallel \leq (1 + q + a\,c\,(a + q)\,s_0)\,\dfrac{1 - (f(s_0, t_0)\,g(s_0)\,s_0)^k}{1 - f(s_0, t_0)\,g(s_0)\,s_0} \parallel y_0 - x_0 \parallel$

$(\widetilde{VIII_k})$ $|F(x_k)| \leq N\,g(s_{k-1}) \parallel y_{k-1} - x_{k-1} \parallel^2$.

Hence, we arrive at the second convergence result for (DCTM).

Theorem 7.1.2. *Let F a nonlinear Fréchet-differentiable operator under conditions $(\mathcal{C}_1)$, $(\mathcal{C}_2)$, $(\mathcal{C}_3^\star)$, $(\mathcal{C}_4)$ and $(\mathcal{C}_5)$. We also assume that $f(s_0, t_0)^2\,g(s_0)\,s_0 < 1$. Then, if $\overline{U}(x_0, R) \subseteq \mathcal{D}$, where*

$$R = \frac{1 + q + a\,c\,(a + q)\,s_0}{1 - f(s_0, t_0)\,g(s_0)\,s_0} \parallel y_0 - x_0 \parallel$$

the sequence $\{x_k\}$ generated by (DCTM) is well defined, remains in $\overline{U}(x_0, R)$ for all $k \geq 0$ and converges to a solution $x^\star \in \overline{U}(x_0, R)$ of equation $F(x) = 0$.

Remark 7.1.2.

(a) Condition $(\mathcal{C}_5)$ can be replaced by the stronger, but more popular

$$\parallel [x, y; F] - [u, v; F] \parallel \leq N_1\,(\parallel x - u \parallel + \parallel y - v \parallel)$$

for all x, y, u, $v \in \mathcal{D}$. In this case, we can set $M = 2\,N_1$.

(b) If directions d_k are given by

$$d_k = \frac{\nabla F(x_k)}{\| \nabla F(x_k) \|},$$

the condition $(\mathcal{C}_{21})$ holds for $\alpha = 1$. A possible choice for α can also be

$$\alpha = \frac{|\nabla F(x_0) \cdot d_0|}{\| \nabla F(x_0) \|} \leq 1.$$

(c) Let $d_k = e^{m(k)}$, where $m(k)$ is the index of component of $\nabla F(x_k)$ of maximal modulus:

$$|\nabla F(x_k)\,[m(k)]| := \max_{1 \leq j \leq n} |\nabla F(x_k)\,[j]|.$$

For this choice of d_k, the results obtained here hold, if simply the Euclidean norm is replaced by the infinity norm $\| \cdot \|_\infty$.

(d) Condition $(\mathcal{C}_{22})$ may be too difficult to verify. In this case it can be replaced by weaker

$$\angle(d_k, \nabla F(x_k)) \leq \angle(d_0, \nabla F(x_0)).$$

Similar result can then be obtained using a different technique from recurrence relations. Such technique has been given in a Banach space setting (see Refs. [120, 121]). We leave the details to the motivated reader.

We provide two numerical examples where we show the efficiency of the directional Chebyshev-type methods (DCTM) and we apply the convergence results previously obtained. For this, we compare some directional Chebyshev-type method (DCTM) with the directional Newton method (DNM). In particular, in the first numerical test we compare the iteration number, the computational order of convergence (see Ref. [433])

$$\rho \approx \frac{\ln\left(\|x_n - x^*\|/\|x_{n-1} - x^*\|\right)}{\ln\left(\|x_{n-1} - x^*\|/\|x_{n-2} - x^*\|\right)}$$

and the computational efficiency defined by $\rho^{1/(OC*IN)}$, OC being the operational cost per iteration and (IN) the iteration number of the used method. In this case, if the operator F is such that $F : \mathcal{D} \subseteq \mathbb{R}^m \to \mathbb{R}$, then the operational cost is $4m+3$ for methods (DCTM) and for method (DNM) is $2m+1$. Although for particular cases of the parameters a, b, c the operational cost of methods (DCTM) can be improved. For instance if $a = b = c = 1$, then the operational cost of methods (DCTM) is $3m + 1$.

In the second numerical test, we consider a cubically polynomial equation with $m = 2$. We check that all conditions $(\mathcal{H})$ are satisfied and our Theorem 7.1.1 is applied to solve this equation using (DCTM).

Example 7.1.1. Firstly, we take the following nonlinear problem considered in Ref. [20]:

$$F(x) = \sum_{i=1}^{p} (\sin x_i)^2 + \sum_{i=p+1}^{m} (\tan x_i)^2, \quad p \text{ is a given integer.}$$

Table 7.1 Iteration number (IN) for the directional methods (DCTM) and (DNM)

m	p	(DCTM) IN	(DNM) IN
	5	15	19
20	10	15	19
	15	15	19
	15	16	20
50	25	16	20
	35	16	20
	20	16	20
80	40	17	20
	60	16	20

Table 7.2 The computational order of convergence and the computational efficiency for (DCTM) and (DNM) methods

m	p	(DCTM)		(DNM)	
		COC	CE	COC	CE
20	15	3.47348×10^{-7}	3.79615×10^{-10}	5.89895×10^{-11}	9.59233×10^{-14}
50	15	1.54433×10^{-6}	6.3921×10^{-10}	7.01974×10^{-11}	4.35207×10^{-14}
80	40	1.30575×10^{-6}	3.1871×10^{-10}	1.11893×10^{-11}	3.99680×10^{-15}

From the starting point $x_0 = (0.1, 0.1, \ldots, 0.1)$, we have obtained the iteration number (IN) given in Table 7.1, where the stopping criterion $|F(x_k)| < 10^{-12}$ is used. We have considered the directional Chebyshev-type method (DCTM) with $a = b = c = 1$ and the direction d_k is chosen such that it is sufficiently close to the gradient $\nabla F(x_k)$ of F in each iteration x_k. Notice that if $F(x_k) \neq 0$ and $F(x_k) \approx 0$, then the vector

$$p_k := \left(\frac{F(x_k + F(x_k)e_1)}{F(x_k)} - 1, \frac{F(x_k + F(x_k)e_2)}{F(x_k)} - 1, \ldots, \frac{F(x_k + F(x_k)e_m)}{F(x_k)} - 1 \right)$$

where e_k is the kth unit vector of R^m is near to $\nabla F(x_k)$. Thus, we have chosen $d_k := p_k/\|p_k\|$ in the implementation. Observe in Table 7.1 that the iteration number obtained by the directional Chebyshev-type method (DCTM) with $a = b = c = 1$ is competitive if we compare with the usual directional Newton method (DNM). On the other hand, in Table 7.2 we show the computational order of convergence and the computational efficiency for (DCTM) and (DNM) methods using the logarithmic scale and denoted by (COC) and (CE), respectively. We can see that both (COC) and (CE) for methods (DCTM) are better than that obtained for (DNM). In addition, we observe in both tables that the value of the parameter p has not much influence in the iteration number and neither in the computational order for both directional methods. Finally, in Figure 7.1 we show the computational order of convergence and the computational efficiency for the directional Chebyshev-type method with $a = b = c = 1$ (blue) and for the directional Newton method (red) with $m = 80$ and $p = 40$ and using the logarithmic scale.

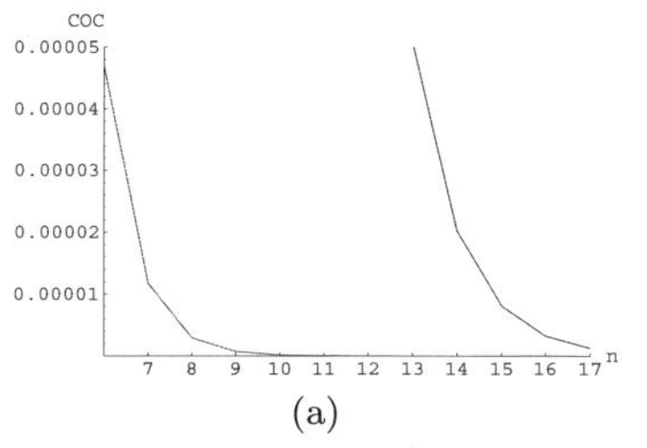

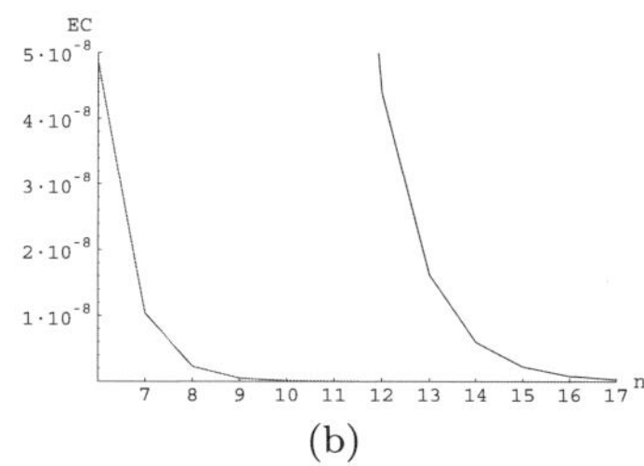

Fig. 7.1 (a) The computational order of convergence for (DCTM) and (DNM) methods. (b) The computational efficiency for (DCTM) and (DNM) methods.

Example 7.1.2. Let $m = 2$. Choose:

$$x_0 = (1,1)^T, \quad \mathcal{D} = \overline{U}(x_0, r) \quad r \in [0, 1/2),$$

and define function F on $\mathcal{D}$ by

$$F(x) = \frac{\varsigma_1^3 + \varsigma_2^3}{2} - 2\, r, \quad x = (\varsigma_1, \varsigma_2)^T.$$

Then, the gradient ∇F of operator F is given by

$$\nabla F(x) = \frac{3}{2}\, (\varsigma_1^2, \varsigma_2^2)^T.$$

Let $x = (\varsigma_1, \varsigma_2)^T$, $y = (\kappa_1, \kappa_2)^T$, $z = (\tau_1, \tau_2)^T$ in $\mathcal{D}$. We have that

$$\nabla F(x) - \nabla F(y) = \frac{3}{2}\, (\varsigma_1^2 - \kappa_1^2, \varsigma_2^2 - \kappa_2^2)^T$$

and

$$[x, z; F] - [y, z; F] = \frac{3}{2}\frac{1}{3}\, (\tau_1\, \varsigma_1 + \varsigma_1^2 - \tau_1\, \kappa_1 - \kappa_1^2, \tau_2\, \varsigma_2 + \varsigma_2^2 - \tau_2\, \kappa_2 - \kappa_2^2)^T$$

$$= \frac{1}{2}\, ((\tau_1 + \varsigma_1 + \kappa_1)\, (\varsigma_1 - \kappa_1), (\tau_2 + \varsigma_2 + \kappa_2)\, (\varsigma_2 - \kappa_2))^T.$$

Consequently,

$$\| \nabla F(x) - \nabla F(y) \| \leq \frac{3}{2}\, \| x + y \| \, \| x - y \|$$

$$\leq \frac{3}{2}\, (1 - r + 1 - r)\, \| x - y \| \leq 3\, (2 - r)\, \sqrt{2}\, \| x - y \|$$

and

$$\| [x, z; F] - [y, z; F] \| \leq \frac{1}{2}\, \| x + y + z \| \, \| x - y \|$$

$$\leq \frac{1}{2}\, (4 - 3\, r)\, \| x - y \| \leq \frac{3}{2}\, (2 - r)\, \sqrt{2}\, \| x - y \|\, .$$

Using the above, we obtain that

$$M = 3\, (2 - r)\, \sqrt{2}, \quad \lambda = 1 - 2\, r, \quad N = \frac{3}{2}\, (2 - r)\, \sqrt{2}, \quad \beta = \frac{3\, \sqrt{2}}{2}$$

and for

$$d_k = \frac{\nabla F(x_k)}{\| \nabla F(x_k) \|},$$

we can choose $\alpha = 1$, so that $(\mathcal{H}_{21})$ is satisfied. Set $a = b = .5$, $c = 1$, $\theta_0 = .5$ and $r = .495$. Then, we get that

$$\lambda = .01, \quad M = 4.515, \quad N = 2.2575, \quad q = .435,$$

$$a_0 = .2357022605, \quad t_0 = 1.12875, \quad r_0 = .005320978531, \quad s_0 = .2673030933$$

and

$$b_0 = .005016666667, \quad c_0 = .002357022605.$$

Conditions (7.6) and (7.7) hold since

$$f(s_0, b_0)^2 \, g(s_0) = .4826545981 < 1$$

and

$$q \, (1 + a \, c \, s_0) \, f(s_0, b_0) \, c_0 = .001168986615 < 1.$$

All conditions $(\mathcal{H})$ are satisfied. That is, Theorem 7.1.1 applies to solve equation $F(x) = 0$ and (DCTM) starting at x_0 converges to $x^\star \in U(x_0, R)$ with $R = .006255089406$. For example $x^\star = (.99999999, .9932883985)^T$ is a solution of $F(x) = 0$.

7.2   Chebyshev-Secant Methods

In this section we give a semi-local convergence theorem for a new family of iterative methods for solving nonlinear equations defined between two Banach spaces. This family is obtained as a combination of the well-known Secant method and Chebyshev method. We give a very general convergence result that allows the application of these methods to non-differentiable problems. The Secant method is described by the following algorithm:

$$x_{n+1} = x_n - [x_{n-1}, x_n; F]^{-1} F(x_n); \quad x_0, \; x_{-1} \text{ given.} \tag{7.9}$$

We have two main aims. Firstly, we consider a multiparametric family of iterative methods that do not use derivatives and generalize method (7.9). Secondly, we obtain a semi-local convergence result for non-differentiable operators. To do this, we change the conditions normally imposed on divided differences. Thus, we relax the requirements that the first order divided difference operator F is Lipschitz or Hölder continuous and just assume the following condition:

$$\|[x, y; F] - [v, w; F]\| \le \omega(\|x - v\|, \|y - w\|); \quad x, y, v, w \in \Omega,$$

where $\omega : R_+ \times R_+ \to R_+$ is a continuous non-decreasing function in its two arguments. It is clear that this condition generalizes the conditions previously indicated.

In fact, when $\omega(u_1, u_2) = k(u_1 + u_2)$, we obtain the Lipschitz continuous case and, when $\omega(u_1, u_2) = k(u_1^p + u_2^p)$, we obtain the (k, p)-Hölder continuous case. Moreover, in general, this condition does not involve the differentiability of the operator F.

The family of iterative methods that we consider in this section arises from the Chebyshev method, a well-known third-order convergence method that has been studied, for instance, in Refs. [181, 10]. Firstly, Hernández in Ref. [263] and later Ezquerro and Hernández in Ref. [208], modify this method by avoiding the computation of the second derivative of F and reducing the number of evaluations of the first derivative of F. Actually, these authors have obtained a modification of the Chebyshev method with order of convergence at least three, which only need to evaluate the first derivative of F. This family of iterative methods is written as follows in Ref. [208]:

$$\begin{cases} x_0 \in \Omega, \\ y_k = x_k - F'(x_k)^{-1} F(x_k), \\ z_k = x_k + p\,(y_k - x_k), \ p \in (0, 1], \\ x_{k+1} = x_k - \dfrac{1}{p^2} F'(x_k)^{-1} \left((p^2 + p - 1)\, F(x_k) + F(z_k)\right), \quad k \geq 0, \end{cases}$$

So, they obtain a uniparametric family of iterative methods which depends only on the first derivative of the operator F, which is evaluated only at one point. Then, to construct a family of iterative methods free of derivatives, as the classical Secant method, we consider an approximation of the first derivative of F from a divided difference of first order; that is, $F'(x_n) \simeq [x_{n-1}, x_n, F]$, where $[x, y; F]$ is a divided difference of first order for the operator F at the points $x, y \in \Omega$. So, we introduce a family of iterative methods that does not use derivatives. We recall these iterative methods as Chebyshev-Secant-type methods and they are written as follows:

$$\begin{cases} x_{-1}, x_0 \in \Omega, \\ y_n = x_n - A_n^{-1} F(x_n), \quad A_n = [x_{n-1}, x_n; F], \\ z_n = x_n + a\,(y_n - x_n), \\ x_{n+1} = x_n - A_n^{-1}\,(b\, F(x_n) + c\, F(z_n)), \quad n \geq 0, \end{cases} \tag{7.10}$$

where a, b, c are non-negative parameters to be chosen so that the sequence $\{x_n\}$ converges to x^*. Note that (7.10) reduces to (7.9) if $b = 1$ and $c = 0$. Even more, if $ac = 0$ and $b > 0$, the Chebyshev-Secant-type method defined in (7.10) becomes the following Secant-type method:

$$x_{n+1} = x_n - b[x_{n-1}, x_n; F]^{-1} F(x_n); \quad x_0, \ x_{-1} \text{ given.} \tag{7.11}$$

This kind of methods has been studied in Refs. [265, 266]. However, if $ac \neq 0$, the methods defined in (7.10) are essentially different from the Secant method (7.9) or Secant-type methods (7.11). We provide a semi-local convergence result for the family of iterative methods (7.10) when they are applied to non-differentiable operators.

First, we analyse the semi-local convergence of the new family of iterative methods given by (7.10). To do this, we use a technique based on proving a system of recurrence relations. Firstly, we suppose that there exists a first-order divided difference $[x, y; F] \in \mathcal{L}(X, Y)$, for all $x, y \in \Omega$. Let us suppose that

(C1) $(1 - b) = (1 - a)c$ and $a \in [0, 1]$,

(C2) $x_{-1}, x_0 \in \Omega$ are such that $\|x_0 - x_{-1}\| \leq \alpha$,

(C3) the linear operator A_0 is invertible, $\|A_0^{-1}\| \leq \beta$ and $\|A_0^{-1}F(x_0)\| \leq \eta$,

(C4) $\|[x, y; F] - [u, v; F]\| \leq \omega(\|x - u\|, \|y - v\|)$, $x, y, u, v \in \Omega$, where $\omega : R_+ \times R_+ \longrightarrow R_+$ is a continuous non-decreasing function in both arguments,

(C5) we denote $m = \max\{p, ac\beta\omega(\alpha, (1 + p)\eta), ac\beta\omega((1 + p)\eta, \eta)\}$, where $p = ac\beta\omega(\alpha, a\eta)$. Suppose equation

$$(1 + \varphi(t))\eta = t\left(1 - \varphi(t)\left(1 + \frac{1}{ac}(1 + \varphi(t))\right)\right), \tag{7.12}$$

where $\varphi(t) = \frac{m}{1 - \beta\omega(\alpha + t, t)}$, has at least one positive root; we denote the smallest positive root of this equation by R and

$$\beta\omega(\alpha + R, R) < 1,$$

(C6) $(1 + M)M\left(1 + \frac{1}{ac}(1 + M)\right) < 1$, where $M = \frac{m}{1 - \beta\omega(\alpha + R, R)}$,

(C7) $B(x_0, R) \subseteq \Omega$.

Next, we give three technical lemmas. In the first one, we give two expressions of the operator F that we need later.

Lemma 7.2.1. *Notice that*

$$F(z_n) = (1 - a)F(x_n) + a(B_n - A_n)(y_n - z_n), \quad n \geq 0,$$

$$F(x_{n+1}) = (A_{n+1} - A_n)(x_{n+1} - x_n) + ac(A_n - B_n)(y_n - x_n), \quad n \geq 0,$$

where $A_n = [x_{n-1}, x_n; F]$, $B_n = [x_n, z_n; F]$ and $(1 - b) = (1 - a)c$.

Lemma 7.2.2. *Let us consider (C1)–(C7). Then*

$$\sum_{i=0}^{n} M^i \left(1 + \frac{1}{ac}(1 + M)\right)^i \eta = (1 + M)\frac{1 - \left(M\left(1 + \frac{1}{ac}(1 + M)\right)\right)^{n+1}}{1 - M\left(1 + \frac{1}{ac}(1 + M)\right)}\eta$$

$$< \frac{(1 + M)\eta}{1 - M\left(1 + \frac{1}{ac}(1 + M)\right)} = R.$$

Lemma 7.2.3. *Let us consider (C1)–(C7). Then, for $n \geq 1$, we have the following recurrence relations:*

[I] There exist A_n^{-1} and $\|A_n^{-1}\| \leq \dfrac{\beta}{1 - \beta\omega(\alpha + R, R)}$,

[II] $\|y_n - x_n\| \le M \left(1 + \frac{1}{ac}(1 + M)\right) \|y_{n-1} - x_{n-1}\|$
$$\le M^n \left(1 + \frac{1}{ac}(1 + M)\right)^n \|y_0 - x_0\|,$$

[III] $\|y_n - x_0\| \le (1 + M) \sum_{i=0}^{n} M^i \left(1 + \frac{1}{ac}(1 + M)\right)^i \|y_0 - x_0\| < R,$

[IV] $\|z_n - x_0\| \le (1 + M) \sum_{i=0}^{n} M^i \left(1 + \frac{1}{ac}(1 + M)\right)^i \|y_0 - x_0\| < R,$

[V] $\|x_{n+1} - x_n\| \le (1 + M)\|y_n - x_n\| \le (1 + M)M^n \left(1 + \frac{1}{ac}(1 + M)\right)^n \|y_0 - x_0\|,$

[VI] $\|x_{n+1} - x_0\| \le (1 + M) \sum_{i=0}^{n} M^i \left(1 + \frac{1}{ac}(1 + M)\right)^i \|y_0 - x_0\| < R.$

Proof. Observe that $\|y_0 - x_0\| \le \eta$, $\|z_0 - x_0\| \le \eta$ and
$\|x_1 - x_0\| \le \left(1 + ac\|A_0^{-1}\|\|A_0 - B_0\|\right) \|y_0 - x_0\| \le (1+p)\|y_0 - x_0\| \le (1+M)\eta < R.$
Now, we can prove that **[I]**–**[VI]** is true for $n = 1$ and assume that **[I]**–**[VI]** are true
for $k = 1, 2, \ldots, n - 1$. Then,

[I]: Since $\|I - A_0^{-1}A_n\| \le \beta\omega(\|x_{n-1} - x_{-1}\|, \|x_n - x_0\|) \le \beta\omega(\alpha + R, R) < 1$, then,
by Banach's lemma, it follows

$$\|A_n^{-1}\| \le \frac{1}{1 - \beta\omega(\|x_{n-1} - x_{-1}\|, \|x_n - x_0\|)} \le \frac{\beta}{1 - \beta\omega(\alpha + R, R)}. \tag{7.13}$$

[II]: $\|y_n - x_n\| \le \|A_n^{-1}\|\|F(x_n)\|$
$$\le \|A_n^{-1}\|\omega(\|x_{n-1} - x_{n-2}\|, \|x_n - x_{n-1}\|)\|x_n - x_{n-1}\|$$
$$+ac\,\omega(\|x_{n-1} - x_{n-2}\|, \|z_{n-1} - x_{n-1}\|)\|y_{n-1} - x_{n-1}\|$$
$$\le M \left(1 + \frac{1}{ac}(1 + M)\right) \|y_{n-1} - x_{n-1}\| \le M^n \left(1 + \frac{1}{ac}(1 + M)\right)^n \|y_0 - x_0\|,$$

[III]: $\|y_n - x_0\| \le \|z_n - x_n\| + \|x_n - x_0\|$
$$\le (1 + M) \sum_{i=0}^{n} M^i \left(1 + \frac{1}{ac}(1 + M)\right)^i \|y_0 - x_0\| < R,$$

[IV]: $\|z_n - x_0\| \le \|z_n - x_n\| + \|x_n - x_0\|$
$$\le (1 + M) \sum_{i=0}^{n} M^i \left(1 + \frac{1}{ac}(1 + M)\right)^i \|y_0 - x_0\| < R,$$

[V]: $\|x_{n+1} - x_n\| \le \left(1 + ac\|A_n^{-1}\|\|A_n - B_n\|\right) \|y_n - x_n\|$
$$\le \left(1 + \frac{m}{1 - \beta\omega(\alpha + R, R)}\right) \|y_n - x_n\| \le (1 + M)\|y_n - x_n\|$$
$$\le (1 + M)M^n \left(1 + \frac{1}{ac}(1 + M)\right)^n \|y_0 - x_0\|$$

[VI]: $\|x_{n+1} - x_0\| \le \|x_{n+1} - x_n\| + \|x_n - x_0\|$
$$\le (1 + M) \sum_{i=0}^{n} M^i \left(1 + \frac{1}{ac}(1 + M)\right)^i \|y_0 - x_0\| < R,$$

so that **[I]**–**[VI]** are true for all positive integers k by mathematical induction. $\square$

We are then ready to prove a semi-local convergence theorem for iterative method (7.10) when they are applied to nondifferentiable operators that satisfy conditions (C1)–(C7).

Theorem 7.2.1. *Let X and Y be two Banach spaces and let $F : \Omega \subseteq X \to Y$ be a nonlinear operator defined on a non-empty open convex domain Ω. We suppose that there exists $[x, y; F] \in \mathcal{L}(X, Y)$, for all $x, y \in \Omega$ and conditions (C1)–(C7) are satisfied. Then, sequence (7.10), starting from x_{-1} and x_0, converges to a unique solution x^* of $F(x) = 0$. Moreover, the solution x^* and the iterates x_n belong to $\overline{U}(x_0, R)$.*

Proof. From Lemma 7.2.3, it follows that (7.10) is a Cauchy sequence, since

$$\|x_{n+k} - x_n\| \le \|x_{n+k} - x_{n+k-1}\| + \|x_{n+k-1} - x_{n+k-2}\| + \cdots + \|x_{n+1} - x_n\|$$

$$\le (1 + M) \left(\|y_{n+k-1} - x_{n+k-1}\| + \|y_{n+k-2} - x_{n+k-2}\| + \cdots + \|y_n - x_n\| \right)$$

$$\le (1 + M) \sum_{i=n}^{n+k-1} \left(M^i \left(1 + \frac{1}{ac}(1 + M) \right)^i \right) \|y_0 - x_0\|$$

$$< (1 + M) M^n \left(1 + \frac{1}{ac}(1 + M) \right)^n \frac{1 - \left(M \left(1 + \frac{1}{ac}(1 + M) \right) \right)^k}{1 - M \left(1 + \frac{1}{ac}(1 + M) \right)} \eta.$$

Consequently, $\{x_n\}$ is convergent and then $\lim_n x_n = x^* \in \overline{U}(x_0, R)$. Observe now that

$$\|F(x_n)\| \le ac\, \omega(\eta, \eta) \left(1 + \frac{1}{ac}(1 + M) \right) \|y_{n-1} - x_{n-1}\|$$

and $\|y_{n-1} - x_{n-1}\| \to 0$ as $n \to \infty$, so that $F(x^*) = 0$.

To prove the uniqueness of the solution x^*, we first assume that y^* is another solution of $F(x) = 0$ in $B(x_0, R)$. Next, we consider the operator $A = [x^*, y^*; F]$, so that if A is invertible, we have $x^* = y^*$, since $A(y^* - x^*) = F(y^*) - F(x^*)$. Indeed,

$$\|A_0^{-1} A - I\| \le \|A_0^{-1}\| \|A - A_0\| \le \|A_0^{-1}\| \| [y^*, x^*; F] - [x_{-1}, x_0; F] \|$$

$$\le \beta \omega(\|y^* - x_{-1}\|, \|x^* - x_0\|) \le \beta \omega(\alpha + R, R) < 1,$$

and the operator A^{-1} exists. $\qquad\qquad\square$

Remark 7.2.1. Using the same information as in Theorem 7.2.1, we can provide at least as tight upper bounds on the distances involved and at least as precise information on the location of the solution x^*. Indeed, let us assume

(C8) $\| [x_{-1}, x_0; F] - [x, y; F] \| \le \omega_0(\|x_{-1} - x\|, \|x_0 - y\|)$, $\quad x, y \in \Omega$, where $\omega_0 : R_+ \times R_+ \longrightarrow R_+$ is a continuous non-decreasing function in both arguments,

(C9) there exists $R_1 \ge R$ such that

$$\beta \, \omega_0(\alpha + R_1, R) < 1$$

and

(C10) $\overline{U}(x_0, R_1) \subseteq \Omega$.

Condition (C8) always follows from (C4) (simply, set $\omega = \omega_0$). Hence, (C8) is not an additional (to (C4)) hypothesis. Note that

$$\omega_0 \leq \omega \tag{7.14}$$

holds in general and $\dfrac{\omega}{\omega_0}$ can be arbitrarily large. It follows from (7.14) and the proof of [**I**] in Lemma 7.2.3 that

$$\|A_n^{-1}\| \leq \frac{1}{1 - \beta\,\omega_0(\|x_{n-1} - x_{-1}\|, \|x_n - x_0\|)} \tag{7.15}$$

which is an at least as tight estimate as (7.13). In particular, if strict inequality holds in (7.14), then (7.15) can replace (7.13), which leads to tighter error bounds on the distances involved. Conditions (C8)–(C9) extend the uniqueness ball for the solution x^*. Indeed, assuming $y^* \in \overline{U}(x_0, R_1)$, the uniqueness proof of Theorem 7.2.1 now gives

$$\|A_0^{-1} A - I\| \leq \beta\,\omega_0(\alpha + R_1, R) < 1.$$

We finish this section with a semi-local convergence theorem for the methods defined in (7.10) in the case $ac = 0$ or, equivalently, for Secant-type methods (7.11) with $b > 0$. We omit the proof because it follows directly from the proof of Theorem 2.1 in Ref. [265].

Theorem 7.2.2. *Let X and Y be two Banach spaces and let $F : \Omega \subseteq X \to Y$ be a nonlinear operator defined on a non-empty open convex domain Ω such that there exists $[x, y; F] \in \mathcal{L}(X, Y)$, for all $x, y \in \Omega$. Let us suppose that*

(c1) The linear operator A_0 is invertible, $\|A_0^{-1}\| \leq \beta$ and $\|A_0^{-1} F(x_0)\| \leq \eta$,
(c2) $x_{-1}, x_0 \in \Omega$ are such that $\|x_0 - x_{-1}\| \leq \alpha$,
(c3) $\|[x, y; F] - [u, v; F]\| \leq \omega(\|x - u\|, \|y - v\|)$, $x, y, u, v \in \Omega$, where $\omega : R_+ \times R_+ \longrightarrow R_+$ is a continuous non-decreasing function in both arguments,
(c4) we denote $m = \max\{\beta\omega(\alpha, \eta), \beta\omega(\eta, \eta)\}$, and suppose that the equation

$$x\left(1 - \frac{bm}{1 - \beta\omega(x + \alpha, x)}\right) - \eta = 0,$$

has at least one positive root; we denote the smallest positive root of this equation by R,
(c5) $\beta\omega(R + \alpha, R) < 1$ and $\dfrac{bm}{1 - \beta\omega(\alpha + R, R)} < 1$,
(c6) $B(x_0, R) \subseteq \Omega$.

Then, sequence (7.11), starting from x_{-1} and x_0, converges to a unique solution x^* of $F(x) = 0$. Moreover, the solution x^* and the iterates x_n belong to $\overline{U}(x_0, R)$.

According to Remark 7.2.1, the results of Theorem 7.2.2 are improved along the same lines. We provide an example to apply our theoretical results. In this example, we solve a system of two equations with two unknowns including a non-differentiable part.

Example 7.2.1. In this example we apply the above result to the following non-linear system:

$$x_1^2 - x_2 + 1 + \tfrac{1}{9}|x_1 - 1| = 0$$
$$x_1 + x_2^2 - 7 + \tfrac{1}{9}|x_2| = 0. \tag{7.16}$$

Observe that system (7.16) is equivalent to $F(x) = 0$, where $F : R^2 \to R^2$, $F = (F_1, F_2)$, $x = (x_1, x_2)$, $F_1(x_1, x_2) = x_1^2 - x_2 + 1 + \tfrac{1}{9}|x_1 - 1|$ and $F_2(x_1, x_2) = x_1 + x_2^2 - 7 + \tfrac{1}{9}|x_2|$. Moreover, for $u, v \in R^2$, $[u, v; F] \in \mathcal{L}(R^2, R^2)$ and

$$[u, v; F]_{i1} = \frac{F_i(u_1, v_2) - F_i(v_1, v_2)}{u_1 - v_1}, \quad [u, v; F]_{i2} = \frac{F_i(u_1, u_2) - F_i(u_1, v_2)}{u_2 - v_2}, \quad i = 1, 2,$$

so that

$$[u, v; F] = \begin{pmatrix} \frac{u_1^2 - v_1^2}{u_1 - v_1} & -1 \\ 1 & \frac{u_2^2 - v_2^2}{u_2 - v_2} \end{pmatrix} + \frac{1}{9} \begin{pmatrix} \frac{|u_1 - 1| - |v_1 - 1|}{u_1 - v_1} & -1 \\ 1 & \frac{|u_2| - |v_2|}{u_2 - v_2} \end{pmatrix}.$$

If we take the Chebyshev norm as vector norm and the matrix norm subordinated to this vector norm, we obtain

$$\|[x, y; F] - [u, v; F]\| \leq \|x - u\| + \|y - v\| + \frac{2}{9},$$

and consequently, from (C5), it follows $\omega(s, t) = s + t + 2/9$.

Now, we apply an iterative method of (7.10) for approximating a solution of (7.16). For example, we choose $a = b = c/2 = 1/2$. We start the corresponding iterative method with $z_{-1} = (0, 0)$ and $z_0 = (1, 1)$. After three iterations applying the iterative method, we obtain

$$z_2 = (1.404149252, 2.249344679) \quad \text{and} \quad z_3 = (1.135872343, 2.358061769),$$

so that if we consider $x_{-1} = z_2$ and $x_0 = z_3$, we can already apply Theorem 7.2.1 to guarantee the convergence of the iterative method to a solution of (7.16). So, we have

$$\alpha = 0.2682\ldots, \quad \beta = 0.4232\ldots, \quad \eta = 0.0215\ldots, \quad m = 0.1088\ldots,$$

the solution of (7.12) is then $R = 0.0469\ldots$, so that $M = 0.1416\ldots$ and

$$(1 + M)M\left(1 + \frac{1}{ac}(1 + M)\right) = 0.5445\ldots < 1.$$

Therefore the hypotheses of Theorem 7.2.1 are satisfied and we can guarantee the convergence of the iterative method to the unique solution

$$x^* = (1.159360850, 2.361824342)$$

of (7.16) in $\overline{U}(x_0, R)$ after four more iterations and using ten significative figures.

Moreover, if we consider the computational order of convergence ρ (see Ref. [239]),

$$\rho \approx \ln\left(\frac{\|x_{n+1} - {}^*\|_\infty}{\|x_n - x^*\|_\infty}\right) \Big/ \ln\left(\frac{\|x_n - x^*\|_\infty}{\|x_{n-1} - x^*\|_\infty}\right), \qquad n \in \mathbb{N},$$

we obtain $\rho = 1.6328\ldots$.

Example 7.2.2. Let $\mathcal{G}(x, t, x(t))$ be a continuous function of its arguments. Let operator F be given by

$$F(x(s)) = x(s) - \int_0^1 \mathcal{G}(x, t, x(t))\, dt.$$

Define divided difference A_n by

$$A_n(s, t) = \frac{\mathcal{G}(s, t, x_n(t)) - \mathcal{G}(s, t, x_{n-1}(t))}{x_n(t) - x_{n-1}(t)}.$$

We refer the reader to Refs. [110, 120, 163, 265] for special choices of function $\mathcal{G}$, so that condition (C4) is satisfied.

7.3 Majorizing Sequences for Chebyshev's Method

We expand in this section the applicability of the Chebyshev method for approximating a locally unique solution of nonlinear equations in a Banach space setting. Our majorizing sequences are finer than the known results in scientific literature as in Refs. [9, 11, 12, 15, 43, 117, 47, 118, 119, 53, 54, 202, 203, 206, 262] and the convergence criteria can be weaker. Numerical examples are also presented which further validate the developed theoretical results. We are concerned with the problem of approximating a locally unique solution $x^\star$ of equation of (1.1). We study the semi-local convergence of the Chebyshev method defined as

$$\begin{aligned} y_n &= x_n - \mathcal{F}'(x_n)^{-1}\mathcal{F}(x_n), \\ x_{n+1} &= y_n - \frac{1}{2}\mathcal{F}'(x_n)^{-1}\mathcal{F}''(x_n)(y_n - x_n)^2 \quad \text{for each} \quad n = 0, 1, 2, \ldots, \end{aligned} \tag{7.17}$$

where $x_0 \in \mathcal{D}$ is an initial point. Here, $\mathcal{F}'(x)$ and $\mathcal{F}''(x)$ denote the first and second Fréchet-derivatives of the operator $\mathcal{F}$, respectively. Note that the Chebyshev method is cubically convergent and it has a long history (see Refs. [250, 262, 434, 181, 288, 343]). The following set of (**C**) conditions have been used

C$_1$. There exists $x_0 \in \mathcal{D}$ such that $F'(x_0)^{-1} \in \mathcal{L}(\mathcal{X}, \mathcal{Y})$.
C$_2$. $\left\|\mathcal{F}'(x_0)^{-1}\mathcal{F}(x_0)\right\| \leq \eta$.
C$_3$. $\left\|\mathcal{F}'(x_0)^{-1}\mathcal{F}''(x)\right\| \leq \mathcal{L}$ for all $x \in \mathcal{D}$.
C$_4$. $\left\|\mathcal{F}'(x_0)^{-1}\left(\mathcal{F}''(x) - \mathcal{F}''(y)\right)\right\| \leq \mathcal{M}\|x - y\|$ for all $x, y \in \mathcal{D}$.

The following sufficient convergence criteria have been given in connection to the (**C**) conditions (see Refs. [9, 12, 15, 120, 202, 203, 206])

$$\eta \le \frac{4\mathcal{M} + \mathcal{L}^2 - \mathcal{L}\sqrt{\mathcal{L}^2 + 2\mathcal{M}}}{3\mathcal{M}(\mathcal{L} + \sqrt{\mathcal{L}^2 + 2\mathcal{M}})} \tag{7.18}$$

or (see Refs. [47, 119])

$$\eta \le \frac{0.485}{\mathcal{K}} \tag{7.19}$$

where

$$\mathcal{K} = \mathcal{L}\sqrt{1 + \frac{\mathcal{M}}{2\mathcal{L}^2}}.$$

However, simple numerical examples can be used to show that conditions (7.18) and (7.19) are not satisfied but the Chebyshev method (7.17) still converges to the solution $x^\star$. As an example, let $\mathcal{X} = \mathcal{Y} = \mathbb{R}$, $x_0 = 1$ and $\mathcal{D} = [\zeta, 2 - \zeta]$ for $\zeta \in (0, 0.5)$. Define function $\mathcal{F}$ on $\mathcal{D}$ by $\mathcal{F}(x) = x^5 - \zeta$. Then, using conditions (**C**), we get that

$$\eta = \frac{(1 - \zeta)}{5}, \quad \mathcal{L} = 4(2 - \zeta)^3, \quad \mathcal{M} = 12(2 - \zeta)^2.$$

Through simple calculations, we may verify that the criteria (7.18) and (7.19) are not satisfied for $\zeta = 0.4$. However, one may see using (7.17) that $\lim_{n \to \infty} x_n = x^\star$. In this section we expand the applicability of the Chebyshev method (7.17), in cases where (7.18) or (7.19) are not satisfied, using the (**C**) conditions together with the following center Lipschitz condition

$C_5.$ $\left\|\mathcal{F}'(x_0)^{-1}\left(\mathcal{F}'(x) - \mathcal{F}'(x_0)\right)\right\| \le \mathcal{L}_0 \|x - x_0\|$ for all $x \in \mathcal{D}$.

We shall refer to (**C$_1$**)-(**C$_5$**) as the (**H**) conditions. Several techniques are usually employed for analyzing the convergence of iterative methods. Among these, the most popular technique is based on the concept of majorizing sequences. In the studies that lead to the convergence conditions (7.18) and (7.19), the computation of the upper bounds on $\|\mathcal{F}'(x_n)^{-1}\mathcal{F}'(x_0)\|$ was based on (**C$_3$**) and the estimate

$$\left\|\mathcal{F}'(x_n)^{-1}\mathcal{F}'(x_0)\right\| \le \frac{1}{1 - \mathcal{L}\|x_n - x_0\|}. \tag{7.20}$$

Instead of (**C$_3$**) we use the more precise and less expensive condition (**C$_5$**) which leads to

$$\left\|\mathcal{F}'(x_n)^{-1}\mathcal{F}'(x_0)\right\| \le \frac{1}{1 - \mathcal{L}_0\|x_n - x_0\|}. \tag{7.21}$$

Note that

$$\mathcal{L}_0 \le \mathcal{L} \tag{7.22}$$

holds in general and $\mathcal{L}/\mathcal{L}_0$ can be arbitrarily large. This change in this section of the semi-local convergence of the Chebyshev method leads to tighter error estimates on the distances $\|y_n - x_n\|$, $\|x_{n+1} - y_n\|$, $\|y_n - x^\star\|$, $\|x_n - x^\star\|$ and weaker convergence criteria.

We study the convergence of scalar sequences that are majorizing for the Chebyshev method (7.17). Let the positive constants be $\mathcal{L}_0 > 0$, $\mathcal{L} > 0$, $\mathcal{M} \geq 0$ and $\eta > 0$. It is convenient for us to define functions γ, α, h_i, $i = 1, 2, 3$ by

$$\gamma(t) = \frac{\mathcal{L}t}{2}, \quad \gamma = \gamma(\eta), \tag{7.23}$$

$$\alpha(t) = \frac{\left(\mathcal{L}\gamma(t)^2 + 2\mathcal{L}\gamma(t) + \dfrac{\mathcal{M}t}{3}\right)}{2(1 - \mathcal{L}_0(1 + \gamma(t))t)}, \quad \alpha = \alpha(\eta) \tag{7.24}$$

$$h_1(t) = \frac{\mathcal{L}^3}{8}t^3 + \frac{1}{2}\left(\mathcal{L}^2 + \frac{\mathcal{M}}{3} + \mathcal{L}_0\mathcal{L}\right)t^2 + \mathcal{L}_0 t - 1 \tag{7.25}$$

$$h_2(t) = \left[\frac{\mathcal{L}}{2}\alpha(t) + \gamma(t)\mathcal{L}_0(1 + \gamma(t))\right]t - \gamma(t) \tag{7.26}$$

and

$$h_3(t) = \frac{\mathcal{M}}{6}t^2 + \left[\frac{\mathcal{L}\gamma(t)(\gamma(t) + 2)}{2} + \mathcal{L}_0(1 + \gamma(t))(1 + \alpha(t))\right]t - 1. \tag{7.27}$$

We denote the minimal positive zeros of the functions h_1, h_2 and h_3 by η_1, η_2 and η_3, respectively. Note that $\alpha(t)$ is well defined on $(0, \eta_1)$ by the choice of η_1. Let us set

$$\eta_0 = \min\{\eta_1, \eta_2, \eta_3\}. \tag{7.28}$$

Then, for all $t \in (0, \eta_0)$ we have that

$$\alpha \in (0, 1), \tag{7.29}$$

$$h_1(t) < 0, \tag{7.30}$$

$$h_2(t) \leq 0 \tag{7.31}$$

and

$$h_3(t) \leq 0. \tag{7.32}$$

We can show the following result on the convergence of majorizing sequences for the Chebyshev method.

Lemma 7.3.1. *Let the positive constants be $\mathcal{L}_0 > 0$, $\mathcal{L} > 0$, $\mathcal{M} \geq 0$ and $\eta > 0$. Suppose that*

$$\eta \begin{cases} \leq \eta_0 & \text{if} \quad \eta_0 \neq \eta_1, \\ < \eta_0 & \text{if} \quad \eta_0 = \eta_1. \end{cases} \tag{7.33}$$

Then, scalar sequence $\{t_n\}$ generated by

$$t_0 = 0, \quad s_0 = \eta, \quad t_{n+1} = s_n + \frac{\mathcal{L}(s_n - t_n)^2}{2(1 - \mathcal{L}_0 t_n)}$$

$$s_{n+1} = t_{n+1} + \frac{\mathcal{L}(t_{n+1} - s_n)^2 + \dfrac{\mathcal{L}^2}{1 - \mathcal{L}_0 t_n}(s_n - t_n)^3 + \dfrac{\mathcal{M}}{3}(s_n - t_n)^3}{2(1 - \mathcal{L}_0 t_{n+1})} \tag{7.34}$$

is increasing, bounded from above by

$$t^{\star\star} = \left(\frac{1 + \gamma}{1 - \alpha}\right)\eta \tag{7.35}$$

and converges to its unique least upper bound $t^\star$ which satisfies

$$\eta \le t^\star \le t^{\star\star}. \tag{7.36}$$

Moreover, the following estimates hold for each $n = 0, 1, 2, \ldots$

$$0 < t_{n+1} - s_n \le \gamma(s_n - t_n) \le \gamma\alpha^n\eta \tag{7.37}$$

and

$$0 < s_{n+1} - t_{n+1} \le \alpha(s_n - t_n) \le \alpha^{n+1}\eta. \tag{7.38}$$

Proof. We use mathematical induction to prove (7.37) and (7.38). Estimates (7.37) and (7.38) hold for $n = 0$ by (7.23), (7.34) and (7.24), since

$$s_1 - t_1 = \frac{\mathcal{L}(t_1 - s_0)^2 + \dfrac{\mathcal{L}^2}{1 - \mathcal{L}_0 t_0}(s_0 - t_0)^3 + \dfrac{\mathcal{M}}{3}(s_0 - t_0)^3}{2(1 - \mathcal{L}_0 t_1)}$$

$$\le \frac{\mathcal{L}\gamma^2(s_0 - t_0)^2 + \dfrac{\mathcal{L}^2}{1 - \mathcal{L}_0 t_0}(s_0 - t_0)^3 + \dfrac{\mathcal{M}}{3}(s_0 - t_0)^3}{2(1 - \mathcal{L}_0 t_1)}$$

$$\le \frac{a(s_0 - t_0)}{2\left[1 - \mathcal{L}_0(s_0 + \gamma(s_0 - t_0))\right]}(s_0 - t_0)$$

$$\le \alpha(s_0 - t_0), \tag{7.39}$$

where

$$a = \mathcal{L}\gamma^2 + 2\gamma\mathcal{L} + \frac{\mathcal{M}}{3}\eta. \tag{7.40}$$

Let us assume that (7.37) and (7.38) hold for all $k \le \eta$. Then, we have that

$$t_{k+1} - s_k \le \gamma(s_k - t_k) \le \gamma\alpha^k\eta, \quad s_{k+1} - t_{k+1} \le \alpha(s_k - t_k) \le \alpha^{k+1}\eta$$

and

$$\begin{aligned}
t_{n+1} &\le s_k + \gamma\alpha^k\eta \le t_k + \alpha^k\eta + \gamma\alpha^k\eta \\
&\le t_{k-1} + \alpha^{k-1}\eta + \alpha^k\eta + \gamma\alpha^{k-1}\eta + \gamma\alpha^k\eta \\
&\le \cdots \le t_2 + (\alpha^2\eta + \alpha^3\eta + \cdots + \alpha^k\eta) + (\gamma\alpha^2\eta + \cdots + \gamma\alpha^k\eta) \\
&\le s_1 + \gamma\alpha\eta + (\alpha^2\eta + \alpha^3\eta + \cdots + \alpha^k\eta) + (\gamma\alpha^2\eta + \cdots + \gamma\alpha^k\eta) \\
&\le t_1 + \alpha\eta + \gamma\alpha\eta + (\alpha^2\eta + \alpha^3\eta + \cdots + \alpha^k\eta) + (\gamma\alpha^2\eta + \cdots + \gamma\alpha^k\eta) \\
&\le \eta + \gamma\eta + \alpha\eta + \gamma\alpha\eta + (\alpha^2\eta + \cdots + \alpha^k\eta) + (\gamma\alpha^2\eta + \cdots + \gamma\alpha^k\eta) \\
&\le \frac{1 - \alpha^{k+1}}{1 - \alpha}(1 + \gamma)\eta < \frac{1 + \gamma}{1 - \alpha}\eta = t^{\star\star}. \tag{7.41}
\end{aligned}$$

Evidently, estimates (7.37) and (7.38) are true provided that

$$\frac{\mathcal{L}(s_k - t_k)}{2(1 - \mathcal{L}_0 t_k)} \leq \gamma \tag{7.42}$$

and

$$\frac{a(s_k - t_k)}{2(1 - \mathcal{L}_0 t_{k+1})} \leq \alpha. \tag{7.43}$$

Estimate (7.42) can be written as

$$\frac{\mathcal{L}\alpha^k \eta}{2} + \gamma \mathcal{L}_0(1 + \gamma)\left(\frac{1 - \alpha^k}{1 - \alpha}\right)\eta - \gamma \leq 0. \tag{7.44}$$

Inequality (7.44) motivates us to define recurrent functions f_k on $[0, 1)$ for each $k = 1, 2, \ldots$ by

$$f_k(t) = \frac{\mathcal{L} t^k \eta}{2} + \gamma \mathcal{L}_0(1 + \gamma)\left(\frac{1 - t^k}{1 - t}\right)\eta - \gamma. \tag{7.45}$$

We need a relationship between two consecutive functions f_k. We have by (7.45) that

$$f_{k+1}(t) = f_k(t) + \frac{\mathcal{L} t^{k+1} \eta}{2} - \frac{\mathcal{L} t^k \eta}{2} + \gamma \mathcal{L}_0(t^k \eta - t^{k-1}\eta + \gamma t^k \eta - \gamma t^{k-1}\eta)$$

$$= f_k(t) + (t - 1)\left[\frac{\mathcal{L}}{2}t + \gamma \mathcal{L}_0(1 + \gamma)\right]t^{k-1}\eta. \tag{7.46}$$

It follows from (7.46) that

$$f_{k+1}(t) \leq f_k(t) \leq \cdots \leq f_1(t). \tag{7.47}$$

In view of (7.44) and (7.47) it suffices to show

$$f_1(\alpha) \leq 0 \tag{7.48}$$

which is true by the choice of η_2, (7.26) and (7.33). Similarly, estimate (7.43) can be written as

$$\frac{a}{2}\alpha^{k-1}\eta + \mathcal{L}_0(1 + \gamma)\left(\frac{1 - \alpha^{k+1}}{1 - \alpha}\right)\eta - 1 \leq 0. \tag{7.49}$$

Define recurrent functions g_k on $[0, 1)$ for each $k = 1, 2, \ldots$ by

$$g_k(t) = \frac{a}{2}t^{k-1}\eta + \mathcal{L}_0(1 + \gamma)\left(\frac{1 - t^{k+1}}{1 - t}\right)\eta - 1. \tag{7.50}$$

Then using (7.50) we get that

$$g_{k+1}(t) = g_k(t) + (t - 1)\left[\frac{a}{2} + \mathcal{L}_0(1 + \gamma)(1 + t)\right]t^{k-1}\eta. \tag{7.51}$$

It follows from (7.51) that

$$g_{k+1}(t) \leq g_k(t) \leq \cdots \leq g_1(t). \tag{7.52}$$

We can show instead of (7.49) that

$$g_1(\alpha) \leq 0, \tag{7.53}$$

which is true by the choice of η_3, (7.27) and (7.33). The induction for (7.37) and (7.38) is complete. Hence, sequence $\{t_n\}$ is an increasing, bounded from above by t^{**} and as such it converges to its unique least upper bound t^*. The proof of the Lemma is complete. $\qquad\square$

We have the following useful and obvious extension of Lemma 7.3.1

Lemma 7.3.2. *Suppose there exists $N \geq 0$ such that*

$$t_0 < s_0 < t_1 < \cdots < t_N < s_N < t_{N+1} < \frac{1}{\mathcal{L}_0} \tag{7.54}$$

and

$$s_N - t_N \begin{cases} \leq \eta_0 & if \quad \eta_0 \neq \eta_1 \\ < \eta_0 & if \quad \eta_0 = \eta_1. \end{cases} \tag{7.55}$$

Then, the conclusions of Lemma 7.3.1 hold for sequence $\{t_n\}$. Moreover, the following estimates hold for each $n = 0, 1, 2, 3, \ldots$

$$0 < t_{N+1+n} - s_{N+n} \leq \gamma_N(s_{N+n} - t_{N+n}) \tag{7.56}$$

and

$$0 < s_{N+1+n} - t_{N+1+n} \leq \alpha_N(s_{N+n} - t_{N+n}) \tag{7.57}$$

where $\gamma_N = \gamma(s_N - t_N)$, $\alpha_N = \alpha(s_N - t_N)$ and $t_N^{\star\star} = \dfrac{1 + \gamma_N}{1 - \alpha_N}(s_N - t_N)$.

Remark 7.3.1.

R1. Note that for $N = 0$, Lemma 7.3.2 reduces to Lemma 7.3.1 with $\alpha_0 = \alpha$ and $\gamma_0 = \gamma$.

R2. Let us define sequences $\{r_n\}$ and $\{v_n\}$ by

$$r_0 = 0, \quad q_0 = \eta, \quad r_1 = q_0 + \frac{\mathcal{K}_0(q_0 - r_0)}{2},$$

$$q_1 = r_1 + \frac{\mathcal{L}_1(r_1 - q_0)^2 + \mathcal{L}_2\mathcal{K}_0(q_0 - r_0)^3 + \dfrac{\mathcal{M}_0}{3}(q_0 - r_0)^3}{2(1 - \mathcal{L}_3 r_1)},$$

$$r_{n+1} = q_n + \frac{\mathcal{L}(q_n - r_n)^2}{2(1 - \mathcal{L}_0 r_n)},$$

$$q_{n+1} = r_{n+1} + \frac{\mathcal{L}(r_{n+1} - q_n)^2 + \dfrac{\mathcal{L}^2(q_n - r_n)^3}{1 - \mathcal{L}_0 r_n} + \dfrac{\mathcal{M}}{3}(q_n - r_n)^3}{2(1 - \mathcal{L}_0 r_{n+1})} \quad (n \geq 1) \tag{7.58}$$

for some $\mathcal{L}_0, \mathcal{L}_1, \mathcal{L}_2, \mathcal{L}_3, \mathcal{K}_0, \mathcal{M}_0$ such that

$$\mathcal{L}_1 \leq \mathcal{L}, \quad \mathcal{L}_2 \leq \mathcal{L}, \quad \mathcal{K}_0 \leq \mathcal{L}, \quad \mathcal{L}_3 \leq \mathcal{L}_0 \quad \text{and} \quad \mathcal{M}_0 \leq \mathcal{M} \tag{7.59}$$

and

$$v_0 = 0, \quad u_0 = \eta, \quad v_{n+1} = u_n + \frac{\mathcal{L}(u_n - v_n)^2}{2(1 - \mathcal{L}v_n)},$$

$$u_{n+1} = v_{n+1} + \frac{\mathcal{L}(v_{n+1} - u_n)^2 + \dfrac{\mathcal{L}^2(u_n - v_n)^3}{1 - \mathcal{L}v_n} + \dfrac{\mathcal{M}}{3}(u_n - v_n)^3}{2(1 - \mathcal{L}_0 v_{n+1})}. \tag{7.60}$$

In view of (7.22), (7.34), (7.58), (7.59), (7.60) a simple inductive argument shows that

$$r_n \le t_n \le v_n, \tag{7.61}$$

$$q_n \le s_n \le u_n, \tag{7.62}$$

$$r_{n+1} - q_n \le s_{n+1} - t_n \le u_{n+1} - v_n, \tag{7.63}$$

$$q_{n+1} - r_{n+1} \le s_{n+1} - t_{n+1} \le u_{n+1} - v_{n+1} \tag{7.64}$$

and

$$r^\star = \lim_{n\to\infty} r_n \le t^\star \le v^\star = \lim_{n\to\infty} v_n. \tag{7.65}$$

Moreover, (7.61)–(7.64) hold as strict inequalities for $n \ge 1$ if (7.22) and (7.59) hold as strict inequalities. Sequence $\{v_n\}$ was shown to be majorizing for the Chebyshev method (7.17) provided that (7.18) or (7.19) hold (see Refs. [9, 12, 15, 43, 117, 47, 118, 119, 53, 54, 202, 203, 206, 262]). We shall prove that tighter sequences $\{r_n\}$ and $\{v_n\}$ are also majorizing for the Chebyshev method (7.17). Then, certainly these majorizing sequences also converge under (7.18) or (7.19). However, these sequences converge under the new convergence criteria given in Lemma 7.3.1 which can be weaker than (7.18) or (7.19). We shall provide the connection of $\mathcal{L}_0$, $\mathcal{L}_1$, $\mathcal{L}_2$, $\mathcal{L}_3$, $\mathcal{K}_0$, $\mathcal{M}_0$ to the equation (1.1) and the Chebyshev method (7.17) so that estimates (7.59) are satisfied.

We need the following Ostrowski-type representation for the Chebyshev method (7.17).

Lemma 7.3.3. *Suppose that the Chebyshev method (7.17) is well defined for each* $n = 0, 1, 2, \ldots$. *Then, the following identity is true for each* $n = 0, 1, 2, \ldots$

$$\mathcal{F}(x_{n+1}) = \int_0^1 \mathcal{F}''(y_n + \theta(x_{n+1} - y_n))(1-\theta)\,d\theta(x_{n+1} - y_n)^2 - $$
$$\frac{1}{2}\int_0^1 \mathcal{F}''(x_n + \theta(y_n - x_n))(y_n - x_n)\,d\theta\,\mathcal{F}'(x_n)^{-1}\mathcal{F}''(x_n)(y_n - x_n)^2$$
$$+ \int_0^1 [\mathcal{F}''(x_n + \theta(y_n - x_n)) - \mathcal{F}''(x_n)](1-\theta)\,d\theta(y_n - x_n)^2. \tag{7.66}$$

Proof. We have that

$$\mathcal{F}(x_{n+1}) = [\mathcal{F}(x_{n+1}) - \mathcal{F}(y_n) - \mathcal{F}'(y_n)(x_{n+1} - y_n)] + \mathcal{F}(y_n) + \mathcal{F}'(y_n)(x_{n+1} - y_n)$$

$$= \int_0^1 \mathcal{F}''(y_n + \theta(x_{n+1} - y_n))(1-\theta)\,d\theta(x_{n+1} - y_n)^2 + \mathcal{F}(y_n) + \mathcal{F}'(y_n)(x_{n+1} - y_n). \tag{7.67}$$

Using the Taylor's expansion

$$\mathcal{F}(y_n) = \mathcal{F}(x_n) + \mathcal{F}'(x_n)(y_n - x_n) + \int_0^1 \mathcal{F}''(x_n + \theta(y_n - x_n))(1-\theta)\,d\theta(y_n - x_n)^2 \tag{7.68}$$

and the definition of the method which gives

$$\mathcal{F}(x_n) + \mathcal{F}'(x_n)(y_n - x_n) = 0,$$

we get that

$$\mathcal{F}(y_n) = \int_0^1 \mathcal{F}''(x_n + \theta(y_n - x_n))(1 - \theta)\, d\theta (y_n - x_n)^2. \tag{7.69}$$

Then, we have that

$$\mathcal{F}(y_n) + \mathcal{F}'(y_n)(x_{n+1} - y_n) = \int_0^1 \mathcal{F}''(x_n + \theta(y_n - x_n))(1 - \theta)\, d\theta (y_n - x_n)^2$$

$$+ \mathcal{F}'(y_n)\left[-\frac{1}{2}\mathcal{F}'(x_n)^{-1}\mathcal{F}''(x_n)(y_n - x_n)^2 \right]$$

$$= \int_0^1 \mathcal{F}''(x_n + \theta(y_n - x_n))(1 - \theta)\, d\theta (y_n - x_n)^2 - \int_0^1 \mathcal{F}''(x_n)(y_n - x_n)^2(1 - \theta)\, d\theta$$

$$- \frac{1}{2}\left(\mathcal{F}'(y_n) - \mathcal{F}'(x_n)\right)\mathcal{F}'(x_n)^{-1}\mathcal{F}''(x_n)(y_n - x_n)^2$$

$$= \int_0^1 \left[\mathcal{F}''(x_n + \theta(y_n - x_n)) - F''(x_n)\right](1 - \theta)\, d\theta (y_n - x_n)^2$$

$$- \frac{1}{2}\int_0^1 \mathcal{F}''(x_n + \theta(y_n - x_n))(y_n - x_n)\, d\theta \, \mathcal{F}'(x_n)^{-1} F''(x_n)(y_n - x_n)^2. \tag{7.70}$$

Identity (7.66) follows from (7.67) and (7.70). The proof of Lemma is complete. $\square$

We can show the main semi-local convergence for the Chebyshev method (7.17) under the (**H**) conditions.

Theorem 7.3.1. *Suppose that the (**H**) conditions and those of Lemma 7.3.1 hold. Moreover, suppose that*

$$\overline{U}(x_0, t^\star) \subseteq \mathcal{D}. \tag{7.71}$$

Then, sequence $\{x_n\}$ generated by the Chebyshev method (7.17) is well defined, remains in $\overline{U}(x_0, t^\star)$ for all $n \geq 0$ and converges to a solution $x^\star \in \overline{U}(x_0, t^\star)$ of equation $\mathcal{F}(x) = 0$. Moreover, the following estimates hold

$$\|y_n - x_n\| \leq s_n - t_n, \tag{7.72}$$

$$\|x_{n+1} - y_n\| \leq t_{n+1} - s_n, \tag{7.73}$$

$$\|x_n - x^\star\| \leq t^\star - t_n \tag{7.74}$$

and

$$\|y_n - x^\star\| \leq t^\star - s_n. \tag{7.75}$$

Furthermore, if there exists $R \geq t^\star$ such that

$$\overline{U}(x_0, R) \subseteq D \tag{7.76}$$

and

$$\frac{\mathcal{L}_0}{2}(t^\star + R) < 1, \tag{7.77}$$

then, the solution $x^\star$ is unique in $\overline{U}(x_0, R)$.

Proof. We shall prove that (7.72) and (7.73) hold using mathematical induction. Using (7.17), ($\mathbf{C_2}$) and (7.34), we get that $\|y_0 - x_0\| = \|\mathcal{F}'(x_0)^{-1}\mathcal{F}'(x_0)\| \le \eta = s_0 - t_0 \le t^\star$. That is (7.72) holds for $n = 0$ and $y_0 \in \overline{U}(x_0, t^\star)$ (by (7.35)). In view of (7.17), (7.34), ($\mathbf{C_3}$) and (7.72), we obtain that

$$\|x_1 - y_0\| \le \frac{1}{2}\|\mathcal{F}'(x_0)^{-1}\mathcal{F}''(x_0)\|\|y_0 - x_0\|^2 \le \frac{\mathcal{L}}{2}(s_0 - t_0)^2 = t_1 - s_0, \qquad (7.78)$$

which shows that (7.73) for $n = 0$. We also get that $\|x_1 - x_0\| \le \|x_1 - y_0\| + \|y_0 - x_0\| \le t_1 - s_0 + s_0 - t_0 = t^\star$, which implies $x_1 \in \overline{U}(x_0, t^\star)$. Let us assume that (7.72), (7.73), $y^\star \in \overline{U}(x_0, t^\star)$ and $x_{k+1} \in \overline{U}(x_0, t^\star)$ hold for all $k \le n$. It follows from the proof of Lemma 7.3.1 and ($\mathbf{C_5}$) that

$$\left\|\mathcal{F}'(x_0)^{-1}\left(\mathcal{F}'(x_{k+1}) - \mathcal{F}'(x_0)\right)\right\| \le \mathcal{L}_0\|x_{k+1} - x_0\| \le \mathcal{L}_0 t_{k+1} < 1. \qquad (7.79)$$

Estimate (7.79) and the Banach Lemma on invertible operators imply that $\mathcal{F}'(x_{k+1})^{-1} \in \mathcal{L}(\mathcal{Y}, \mathcal{X})$ and

$$\left\|\mathcal{F}'(x_{k+1})^{-1}\mathcal{F}'(x_0)\right\| \le \frac{1}{1 - \mathcal{L}_0\|x_{k+1} - x_0\|} \le \frac{1}{1 - \mathcal{L}_0 t_{k+1}}. \qquad (7.80)$$

Then, we have by (7.17), ($\mathbf{C_3}$), (7.34) and (7.80) (for k replacing $k + 1$) and the induction hypothesis that

$$\|x_{k+1} - y_k\| \le \frac{1}{2}\left\|\mathcal{F}'(x_k)^{-1}\mathcal{F}'(x_0)\right\|\left\|\mathcal{F}'(x_0)\mathcal{F}''(x_k)\right\|\|y_k - x_k\|$$

$$\le \frac{\mathcal{L}}{2(1 - \mathcal{L}_0 t_k)}(s_k - t_k)^2 = t_{k+1} - s_k. \qquad (7.81)$$

Using ($\mathbf{C_3}$), ($\mathbf{C_4}$), (7.34), (7.66), (7.80), (7.81) and the induction hypothesis we obtain in turn that

$$\left\|\mathcal{F}'(x_0)^{-1}\mathcal{F}(x_{k+1})\right\| = \left\|\mathcal{F}'(x_0)^{-1}\left\{\int_0^1 \mathcal{F}''(y_k + \theta(x_{k+1} - y_k))(1 - \theta)d\theta\,(x_{k+1} - y_k)^2\right.\right.$$

$$-\frac{1}{2}\int_0^1 \mathcal{F}''(x_k + \theta(y_k - x_k))(y_k - x_k)\,d\theta\,\mathcal{F}'(x_k)^{-1}\mathcal{F}''(x_k)(y_k - x_k)^2$$

$$\left.\left.+\int_0^1 [\mathcal{F}''(x_k + \theta(y_k - x_k)) - \mathcal{F}''(x_k)](1 - \theta)\,d\theta(y_k - x_k)^2\right\}\right\|$$

$$\le \left\|\int_0^1 \mathcal{F}'(x_0)^{-1}\mathcal{F}''(y_k + \theta(x_{k+1} - y_k))(1 - \theta)\,d\theta\right\|\|x_{k+1} - y_k\|^2$$

$$+\frac{1}{2}\left\|\mathcal{F}'(x_0)^{-1}\mathcal{F}''(x_k + \theta(y_k - x_k))\,d\theta\right\|\left\|\mathcal{F}'(x_k)^{-1}\mathcal{F}'(x_0)\right\|\left\|\mathcal{F}'(x_0)^{-1}\mathcal{F}''(x_k)\right\|\|y_k - x_k\|^3$$

$$+\left\|\int_0^1 \mathcal{F}'(x_0)^{-1}[\mathcal{F}''(x_k + \theta(y_k - x_k)) - \mathcal{F}''(x_k)](1 - \theta)\,d\theta\right\|\|y_k - x_k\|^2$$

$$\le \frac{\mathcal{L}}{2}\left[(t_{k+1} - s_k)^2 + \frac{\mathcal{L}^2}{1 - \mathcal{L}_0 t_k}(s_k - t_k)^3 + \frac{\mathcal{M}}{3}(s_k - t_k)^3\right].$$

$$(7.82)$$

Then, by (7.17), (7.34), (7.80) and (7.82), we get that

$$\|y_{k+1} - x_{k+1}\| \le \left\|\mathcal{F}'(x_{k+1})^{-1}\mathcal{F}'(x_0)\right\|\left\|\mathcal{F}'(x_0)^{-1}\mathcal{F}(x_{k+1})\right\|$$

$$\le \frac{\mathcal{L}(t_{k+1} - s_k)^2 + \dfrac{\mathcal{L}^2}{1 - \mathcal{L}_0 t_k}(s_k - t_k)^3 + \dfrac{\mathcal{M}}{3}(s_k - t_k)^3}{2(1 - \mathcal{L}_0 t_{k+1})}$$

$$= s_{k+1} - t_{k+1}. \qquad (7.83)$$

We also have that

$$\|y_{k+1} - x_0\| \leq \|y_{k+1} - x_{k+1}\| + \|x_{k+1} - x_0\|$$
$$\leq s_{k+1} - t_{k+1} + t_{k+1} - t_0 = s_{k+1} \leq t^\star$$

and

$$\|x_{k+2} - x_0\| \leq \|x_{k+2} - y_{k+1}\| + \|y_{k+1} - x_0\|$$
$$\leq t_{k+2} - s_{k+1} + s_{k+1} - t_0 = t_{k+2} \leq t^\star.$$

Hence, y_{k+1} and x_{k+2} belong in $\overline{U}(x_0, t^\star)$. It follows from (7.72), (7.73) and Lemma 7.3.1 that sequence $\{x_n\}$ is complete in a Banach space $\mathcal{X}$ and as such it converges to some $x^\star \in \overline{U}(x_0, t^\star)$ (since $\overline{U}(x_0, t^\star)$ is a closed set). By letting $k \to \infty$ in (7.82) we obtain $\mathcal{F}(x^\star) = 0$. Estimates (7.74) and (7.75) follow from (7.72) and (7.73) by standard majorization techniques. Finally to show the uniqueness part. Let $y^\star \in \overline{U}(x_0, R)$ be a solution of the equation $\mathcal{F}(x) = 0$. Let $Q = \int_0^1 \mathcal{F}'(x^\star + \theta(y^\star - x^\star))\, d\theta$. Using $(\mathbf{C}_5)$, (7.76) and (7.77), we get that

$$\left\|\mathcal{F}'(x_0)^{-1}\left[Q - \mathcal{F}'(x_0)\right]\right\| \leq \int_0^1 \left\|\mathcal{F}'(x_0)^{-1}\left[\mathcal{F}'(x^\star + \theta(y^\star - x^\star)) - \mathcal{F}'(x_0)\right] d\theta\right\|$$

$$\leq \mathcal{L}_0 \int_0^1 \left[(1 - \theta)\|x^\star - x_0\| + \theta\|y^\star - x_0\|\right] d\theta$$

$$\leq \frac{\mathcal{L}_0}{2}(t^\star + R) < 1. \tag{7.84}$$

It follows from (7.84) and the Banach lemma on invertible operators that $Q^{-1} \in \mathcal{L}(\mathcal{Y}, \mathcal{X})$. Then, using the identity $0 = \mathcal{F}(y^\star) - \mathcal{F}(x^\star) = Q(y^\star - x^\star)$, we deduce that $x^\star = y^\star$. The proof of the Theorem is complete. $\qquad\square$

Remark 7.3.2.

(1) The limit point $t^\star$ can be replaced by $t^{\star\star}$ (given in closed form by (7.35)) in Theorem 7.3.1.

(2) The conclusions of Theorem 7.3.1 hold if hypotheses of Lemma 7.3.1 are replaced by those of Lemma 7.3.2.

(3) It follows from the $(\mathbf{H})$ conditions that there exist constants $\mathcal{K}_0$, $\mathcal{L}_1$, $\mathcal{L}_2$, $\mathcal{L}_3$, $\mathcal{M}_0$ satisfying

$$\left\|\mathcal{F}'(x_0)^{-1}\mathcal{F}''(x_0)\right\| \leq \mathcal{K}_0 \tag{7.85}$$

$$\left\|\mathcal{F}'(x_0)^{-1}\mathcal{F}''(y_0 + \theta(x_1 - y_0))\right\| \leq \mathcal{L}_1 \tag{7.86}$$

$$\left\|\mathcal{F}'(x_0)^{-1}\mathcal{F}''(x_0 + \theta(y_0 - x_0))\right\| \leq \mathcal{L}_2 \tag{7.87}$$

$$\left\|\mathcal{F}'(x_0)^{-1}\left[\mathcal{F}'(x_1) - \mathcal{F}'(x_0)\right]\right\| \leq \mathcal{L}_3\|x_1 - x_0\| \tag{7.88}$$

$$\left\|\mathcal{F}'(x_0)^{-1}\left[\mathcal{F}''(x_0 + \theta(y_0 - x_0)) - \mathcal{F}''(x_0)\right]\right\| \leq \mathcal{M}_0\theta\|y_0 - x_0\| \tag{7.89}$$

for all $\theta \in [0, 1]$, where

$$y_0 = x_0 - \mathcal{F}'(x_0)^{-1}\mathcal{F}(x_0) \quad \text{and} \quad x_1 = y_0 - \frac{1}{2}\mathcal{F}'(x_0)^{-1}\mathcal{F}''(x_0)(y_0 - x_0)^2.$$

Table 7.3   Sequences $\{t_n\}$, $\{r_n\}$ and $\{v_n\}$

n	$t_{n+1} - t_n$	$r_{n+1} - r_n$	$v_{n+1} - v_n$
0	$2.136390e - 01$	$1.989000e - 01$	$2.136390e - 01$
1	$6.900971e - 02$	$2.002506e - 02$	$1.003585e - 01$
2	$7.756433e - 03$	$1.464324e - 04$	$6.131980e - 01$
3	$2.508442e - 05$	$7.284484e - 11$	$-3.783329e - 01$
4	$9.987606e - 13$	$8.985923e - 30$	$-1.310068e - 01$
5	$6.307896e - 35$	$1.686765e - 86$	$-3.973089e - 02$
6	$1.589110e - 101$	$1.115655e - 256$	$-5.144345e - 03$
7	$2.540750e - 301$	$3.228164e - 767$	$-3.019584e - 05$
8	$1.038451e - 900$	$0.000000e + 00$	$-7.784198e - 12$
9	$0.000000e + 00$	$0.000000e + 00$	$-1.335738e - 31$

Estimates (7.85)–(7.89) are not additional to the (**H**) conditions, since in practice the verification of (**C**$_2$)–(**C**$_5$) requires the computation of $\mathcal{K}_0$, $\mathcal{L}_1$, $\mathcal{L}_2$, $\mathcal{L}_3$ and $\mathcal{M}_0$. Note that finding these constants only involves computations at the initial data. Moreover, these constants satisfy (7.59). Furthermore, according to the proof of Theorem 7.3.1, $\{r_n\}$ is a majorizing sequence for $\{x_n\}$ which is finer than $\{t_n\}$ and $\{v_n\}$ (see also (7.61)–(7.65) and the Tables).

Example 7.3.1. Let $\mathcal{X} = \mathcal{Y} = \mathbb{R}$ be equipped with the max-norm, $x_0 = 1$, $\mathcal{D} = [a, 2 - a]$. Let us define $\mathcal{F}$ on $\mathcal{D}$ by equation (1.48) with $a \in (0, 1.0)$. Through algebraic manipulations, for the conditions (**C**), we obtain

$$\eta = \frac{1 - a}{3}, \quad \mathcal{L} = 2(2 - a), \quad \mathcal{L}_0 = 3 - a \quad \text{and} \quad \mathcal{M} = 2.$$

Furthermore, we may verify that for $a = 0.49$ the criteria (7.18) and (7.19) are not satisfied. Even though the criteria (7.18) and (7.19) are not fulfilled but the Chebyshev method converges for $a = 0.49$. From the center-Lipschitz condition (**C**$_5$), we get $\mathcal{L}_0 = 3 - a$. Moreover, from Lemma 7.3.1, we obtain

$$\eta_1 = 0.2186578090, \quad \eta_2 = 0.6456538008, \quad \eta_3 = 0.6539443062.$$

From (7.28), we get $\eta_0 = \eta_1 = 0.2186578090$. We can ascertain that (7.33) holds for $a = 0.49$. From (7.85)–(7.89), we obtain that

$$\mathcal{K}_0 = 2, \quad \mathcal{L}_1 = \frac{(4 + 2a)}{3}, \quad \mathcal{L}_2 = 2,$$

$$\mathcal{L}_3 = \frac{\left\| \left(2/3 + 1/3\,a - (-1/3 + 1/3\,a)^2\right)^2 - 1 \right\|}{\left\| -1/3 + 1/3\,a - (-1/3 + 1/3\,a)^2 \right\|} \quad \text{and} \quad \mathcal{M}_0 = 2.$$

We can verify that the conditions (7.59) are fulfilled. Additionally, for the sequences $\{t_n\}$ (given by (7.34)), $\{r_n\}$ (given by (7.58)) and $\{v_n\}$ (given by (7.60)), we produce Table 7.3. In Table 7.3, we observe that the sequence $\{r_n\}$ provides tighter error bounds than the sequence $\{t_n\}$. The convergence of the sequence $\{v_n\}$ is not expected, since (7.18) or (7.19) is not satisfied. Note also that $\{v_n\}$ was essentially used as a majorizing sequence for the Chebyshev method in Refs. [9, 12, 15, 43, 117, 47, 118, 119, 53, 54, 202, 203, 206, 262].

Example 7.3.2. In this example, we provide an application of our results to a special nonlinear Hammerstein integral equation of the second kind. Consider the integral equation

$$x(s) = 1 + \frac{4}{5}\int_0^1 G(s,t)x(t)^3\,dt, \quad s \in [0,1], \tag{7.90}$$

where G is as $\mathcal{G}$ in Example 1.1.2. Let $\mathcal{X} = \mathcal{Y} = \mathbb{C}[0,1]$ and $\mathcal{D}$ be a suitable open convex subset of $\mathcal{X}_1 := \{x \in \mathcal{X} : x(s) > 0, s \in [0,1]\}$, which will be given below. Define $\mathcal{F} : \mathcal{D} \to \mathcal{Y}$ by

$$[\mathcal{F}(x)](s) = x(s) - 1 - \frac{4}{5}\int_0^1 G(s,t)x(t)^3\,dt, \quad s \in [0,1]. \tag{7.91}$$

The first and second derivatives of $\mathcal{F}$ are given by

$$[\mathcal{F}(x)'y](s) = y(s) - \frac{12}{5}\int_0^1 G(s,t)x(t)^2 y(t)\,dt, \quad s \in [0,1], \tag{7.92}$$

and

$$[\mathcal{F}(x)''yz](s) = \frac{24}{5}\int_0^1 G(s,t)x(t)y(t)z(t)\,dt, \quad s \in [0,1], \tag{7.93}$$

respectively. We use the max-norm. Let $x_0(s) = 1$ for all $s \in [0,1]$. Then, for any $y \in \mathcal{D}$, we have that

$$[(I - \mathcal{F}'(x_0))(y)](s) = \frac{12}{5}\int_0^1 G(s,t)y(t)dt, \quad s \in [0,1],$$

which means

$$\|I - \mathcal{F}'(x_0)\| \le \frac{12}{5}\max_{s\in[0,1]}\int_0^1 G(s,t)dt = \frac{12}{5 \times 8} = \frac{3}{10} < 1.$$

It follows from the Banach theorem that $\mathcal{F}'(x_0)^{-1}$ exists and

$$\|\mathcal{F}'(x_0)^{-1}\| \le \frac{1}{1 - \dfrac{3}{10}} = \frac{10}{7}.$$

On the other hand, we have from (7.91) that

$$\|\mathcal{F}(x_0)\| = \frac{4}{5}\max_{s\in[0,1]}\int_0^1 G(s,t)\,dt = \frac{1}{10}.$$

Then, we get $\eta = 1/7$. Note that $\mathcal{F}''(x)$ is not bounded in $\mathcal{X}$ or its subset $\mathcal{X}_1$. Take into account that a solution $x^\star$ of equation (1.1) with $\mathcal{F}$ given by (7.90) must satisfy

$$\|x^\star\| - 1 - \frac{1}{10}\|x^\star\|^3 \le 0, \tag{7.94}$$

i.e., $\|x^\star\| \le \rho_1 = 1.153467305$ and $\|x^\star\| \ge \rho_2 = 2.423622140$, where ρ_1 and ρ_2 are the positive roots of the real equation $z - 1 - z^3/10 = 0$. Consequently, if we look for a solution such that $x^\star < \rho_1 \in \mathcal{X}_1$, we can consider $D := \{x : x \in \mathcal{X}_1 \quad \text{and} \quad \|x\| < r\},$

with $r \in (\rho_1, \rho_2)$, as a nonempty open convex subset of $\mathcal{X}$. For example, choose $r = 1.7$. Using (7.72) and (7.73), we have that for any $x, y, z \in \mathcal{D}$

$$\|[(\mathcal{F}'(x) - \mathcal{F}'(x_0))y](s)\| = \frac{12}{5} \left\| \int_0^1 G(s,t)(x(t)^2 - x_0(t)^2)y(t)dt \right\|$$

$$\leq \frac{12}{5} \int_0^1 G(s,t)\|x(t) - x_0(t)\| \, \|x(t) + x_0(t)\|y(t)dt$$

$$\leq \frac{12}{5} \int_0^1 G(s,t) \, (r+1)\|x(t) - x_0(t)\|y(t)dt, \quad s \in [0,1]$$

and

$$\|(F''(x)yz)(s)\| = \frac{24}{5} \int_0^1 G(s,t)x(t)y(t)z(t)dt, \quad s \in [0,1].$$

Then, we get that

$$\|\mathcal{F}'(x) - \mathcal{F}'(x_0)\| \leq \frac{12}{5}\frac{1}{8}(r+1)\|x - x_0\|$$

$$= \frac{81}{100}\|x - x_0\| \|F''(x)\| \leq \frac{24}{5} \times \frac{r}{8} = \frac{51}{50}$$

and

$$\|[[F''(x) - \mathcal{F}''(\overline{x})]yz](s)\| = \frac{24}{5} \left\| \int_0^1 G(s,t) \, (x(t) - \overline{x}(t)) \, y(t)z(t) \right\| dt$$

$$\leq \frac{24}{5}\frac{1}{8}\|x - \overline{x}\| = \frac{3}{5}\|x - \overline{x}\|.$$

Now we can choose constants as follows:

$$\mathcal{M} = \frac{6}{7}, \quad \mathcal{L} = \frac{51}{35}, \quad \mathcal{L}_0 = \frac{81}{70}, \quad \text{and} \quad \eta = \frac{1}{7}.$$

From Lemma 7.3.1, we obtain

$$\eta_1 = 0.4590282838, \quad \eta_2 = 0.4317611292, \quad \eta_3 = 0.3296230353.$$

Thus

$$\eta_0 = \eta_3 = 0.3296230353.$$

Since, $\eta_0 \neq \eta_1$. Thus from (7.33), we get

$$\frac{1}{7} \leq 0.3296230353.$$

Thus, (7.33) holds. It can be checked that the criteria (7.18) $(0.1428571429 < 0.3070646192)$ and (7.19) $(0.1428571429 < 0.3036094577)$ also hold. Likewise we select the constants

$$\mathcal{K}_0 = \frac{51}{50}, \quad \mathcal{M}_0 = \frac{4}{9}, \quad \mathcal{L}_1 = \frac{51}{50}, \quad \mathcal{L}_2 = \frac{52}{55}.$$

Table 7.4   Sequences $\{t_n\}$, $\{r_n\}$ and $\{v_n\}$

n	$t_{n+1} - t_n$	$r_{n+1} - r_n$	$v_{n+1} - v_n$
0	$1.577259e - 01$	$1.532653e - 01$	$1.577259e - 01$
1	$4.510658e - 03$	$2.327746e - 03$	$4.790202e - 03$
2	$1.612179e - 07$	$2.193287e - 08$	$2.166030e - 07$
3	$7.479352e - 21$	$1.849838e - 23$	$2.042480e - 20$
4	$7.468206e - 61$	$1.109808e - 68$	$1.712530e - 59$
5	$7.434868e - 181$	$2.396576e - 204$	$1.009438e - 176$
6	$7.335743e - 541$	$2.413352e - 611$	$2.067301e - 528$
7	$7.046228e - 1621$	$2.464387e - 1832$	$1.775726e - 1583$
8	$0.000000e + 00$	$0.000000e + 00$	$0.000000e + 00$
9	$0.000000e + 00$	$0.000000e + 00$	$0.000000e + 00$

We can verify that conditions (7.59) are fulfilled. Additionally, for the sequences $\{t_n\}$ (given by (7.34)), $\{r_n\}$ (given by (7.58)) and $\{v_n\}$ (given by (7.60)), we produce Table 7.4. In Table 7.4, we observe that the sequence $\{r_n\}$ provides tighter error bounds than sequences $\{t_n\}$ and $\{v_n\}$. This is also true by (7.61). Concerning the uniqueness balls, let us denote the radii corresponding to (7.84), (7.19) (see Refs. [9, 12, 15, 120, 202, 203, 206]) and (7.18) (see Refs. [47, 119]) by γ_1, γ_2 and γ_3, respectively. These are given as the smallest positive roots of the polynomials

$$p_1(t) = \mathcal{L}_0 t - 1 \quad (\text{for} \quad t^\star = R),$$

$$p_2(t) = \frac{\mathcal{K}}{2} t^2 - t + \eta$$

and

$$p_3(t) = \frac{\mathcal{M}}{3} t^3 + \frac{\mathcal{L}}{2} t^2 - t + \eta$$

respectively. Using the values of $\mathcal{L}_0$, $\mathcal{L}$, $\mathcal{M}$ and η we get

$$\gamma_1 = 0.8641975309, \quad \gamma_2 = 0.1644603931, \quad \gamma_3 = 0.1636113702. \tag{7.95}$$

Here, $\mathcal{K} = 1.597446943$. Note that $\overline{U}(x_0, r - 1) \subseteq \mathcal{D}$, $\mathcal{L}_0 < \mathcal{L}$ and $\gamma_3 < \gamma_2 < \gamma_1$. Therefore, the new approach provides the largest uniqueness ball and since $r - 1 < \gamma_1$, we deduce that $x^\star$ is unique in $\overline{U}(x_0, r - 1) = \overline{U}(1, 0.7) \subseteq \mathcal{D}$.

7.4   Exercises

7.4.1. Consider the super-Halley method for all $n \geq 0$ in the form:

$$F(x_n) + F'(x_n)(y_n - x_n) = 0,$$
$$3F(x_n) + 3F'(x_n)\left[x_n + \tfrac{2}{3}(y_n - x_n)\right](y_n - x_n) + 4F'(y_n)(x_{n+1} - y_n) = 0,$$

for approximating a solution x^* of equation $F(x) = 0$. Let $F : \Omega \subseteq X \to Y$ be a three-times Fréchet-differentiable operator defined on an open convex subset Ω of a Banach space X with values in a Banach space Y. Assume

(1) $\Gamma_0 = F'(x_0)^{-1} \in L(Y, X)$ for some $x_0 \in \Omega$ with $\|\Gamma_0\| \leq \beta$;

(2) $\|\Gamma_0 F(x_0)\| \leq \eta$;

(3) $\|F''(x)\| \leq M \ (x \in \Omega)$;

(4) $\|F'''(x) - F'''(y)\| \leq L \|x - y\| \ (x, y \in \Omega), \ (L \geq 0)$.

Denote by $a_0 = M\beta\eta$, $c_0 = L\beta\eta^3$ and define sequences

$$a_{n+1} = a_n f(a_n)^2 g(a_n, c_n),$$
$$c_{n+1} = c_n f(a_n)^4 g(a_n, c_n)^3,$$

where

$$f(x) = \frac{(1-x)}{x^2 - 4x + 2} \quad \text{and} \quad g(x, y) = \frac{1}{8}\left[\frac{x^3}{(1-x)^2} + \frac{17}{27}y\right].$$

Suppose that $a_0 \in \left(o, \frac{1}{2}\right)$, $c_0 < h(a_0)$, where

$$h(x) = \frac{27(2x - 1)(x - 1)\left(x - 3 + \sqrt{5}\right)\left(x - 3 - \sqrt{5}\right)}{17(1 - x)^2},$$

$$\bar{U}(x_0, R\eta) \subseteq \Omega, \quad R = \left[1 + \frac{a_0}{2(1 - a_0)}\right]\frac{1}{1 - \Delta}$$

and $\Delta = f(a_0)^{-1}$.

Show that iteration $\{x_n\}$ $(n \geq 0)$ is well defined, remains in $\bar{U}(x_0, R\eta)$ for all $n \geq 0$ and converges to a solution x^* of equation $F(x) = 0$. The solution x^* is unique in $U(x_0, \frac{2}{M\beta} - R\eta) \cap \Omega$ and

$$\|x_n - x^*\| \leq \left[1 + \frac{a_0\gamma^{\frac{4^n - 1}{3}}}{2(1 - a_0)}\right]\gamma^{\frac{4^n - 1}{3}}\frac{\Delta^n}{1 - \gamma^{4^n}\Delta}\eta \quad (n \geq 0),$$

where $\gamma = \dfrac{a_1}{a_0}$.

7.4.2. Consider the multipoint iteration method [110]:

$$y_n = x_n - F'(x_n)^{-1} F(x_n)$$
$$G_n = F'(x_n)^{-1}\left[F'\left(x_n + \frac{2}{3}(y_n - x_n)\right) - F'(x_n)\right],$$
$$x_{n+1} = y_n - \frac{3}{4}G_n\left[I - \frac{3}{2}G_n\right](y_n - x_n) \quad (n \geq 0),$$

for approximating a solution x^* of equation $F(x) = 0$. Let $F : \Omega \subseteq X \to Y$ be a three times Fréchet-differentiable operator defined on an open convex subset Ω of a Banach space X with values in a Banach space Y. Assume

(1) $\Gamma_0 = F'(x_0)^{-1} \in L(Y, X)$ exists for some $x_0 \in \Omega$ and $\|\Gamma_0\| \leq \beta$;

(2) $\|\Gamma_0 F(x_0)\| \leq \eta$;

(3) $\|F''(x)\| \leq M \ (x \in \Omega)$;

(4) $\|F'''(x)\| \leq N \ (x \in \Omega)$;

(5) $\|F'''(x) - F'''(y)\| \leq L \|x - y\| \ (x, y \in \Omega), \ (L \geq 0)$.

Denote by $a_0 = M\beta\eta$, $b_0 = N\beta\eta^2$ and $c_0 = L\beta\eta^3$. Define the sequence

$$a_{n+1} = a_n f\left(a_n\right)^2 g\left(a_n, b_n, c_n\right),$$
$$b_{n+1} = b_n f\left(a_n\right)^3 g\left(a_n, b_n, c_n\right)^2,$$
$$c_{n+1} = c_n f\left(a_n\right)^4 g\left(a_n, b_n, c_n\right)^3,$$

where

$$f\left(x\right) = \frac{2}{2 - 2x - x^2 - x^3}$$

and

$$g\left(x, y, z\right) = \frac{1}{216}\left[27x^3\left(x^2 + 2x + 5\right) + 18xy + 17z\right].$$

Suppose that $a_0 \in \left(0, \frac{1}{2}\right)$, $17c_0 + 18a_0b_0 < p\left(a_0\right)$, where

$$p\left(x\right) = 27\left(1 - x\right)\left(1 - 2x\right)\left(x^2 + x + 2\right)\left(x^2 + 2x + 4\right),$$
$$\bar{U}\left(x_0, R\eta\right) \subseteq \Omega, \quad R = \left[1 + \frac{a_0}{2}\left(1 + a_0\right)\right]\frac{1}{1 - \gamma\Delta}, \quad \gamma = \frac{a_1}{a_0}, \quad \Delta = f\left(a_0\right)^{-1}.$$

Show:

(i) Iteration $\{x_n\}$ $(n \geq 0)$ is well defined, remains in $\bar{U}\left(x_0, R\eta\right)$ for all $n \geq 0$ and converges to a solution x^* of equation $F\left(x\right) = 0$, which is unique in $U(x_0, \frac{2}{M\beta} - R\eta) \cap \Omega$.

(ii) The following error bounds hold for all $n \geq 0$:

$$\|x_n - x^*\| \leq \left[1 + \frac{a_0}{2}\gamma^{\frac{4^n - 1}{3}}\left(1 + a_0\gamma^{\frac{4^n - 1}{3}}\right)\right]\gamma^{\frac{4^n - 1}{3}}\frac{\Delta^n}{1 - \gamma^{4^n}\Delta}\eta.$$

7.4.3. Consider the Halley method [110] given by (6.2). Assume

(1) $F'\left(x_0\right)^{-1} \in L\left(Y, X\right)$ exists for some $x_0 \in \Omega$;
(2) $\|F'\left(x_0\right)^{-1} F\left(x_0\right)\| \leq \beta$;
(3) $\|F'\left(x_0\right)^{-1} F''\left(x_0\right)\| \leq \gamma$;
(4) $\|F'\left(x_0\right)^{-1}\left[F''\left(x\right) - F''\left(y\right)\right]\| \leq M\|x - y\|$ $(x, y \in \Omega)$.

Suppose

$$\beta \leq \frac{2[2\sqrt{\gamma^2 + 2M} + \gamma]}{3[\sqrt{\gamma^2 + 2M} + \gamma]^2}, \quad \bar{U}\left(x_0, r_1\right) \subseteq \Omega,$$

$(r_1 \leq r_2)$ where r_1, r_2 are the positive roots of $h\left(t\right) = \beta - t + \frac{\gamma}{2}t^2 + \frac{M}{6}t^3$.
Show:

(i) Iteration $\{x_n\}$ $(n \geq 0)$ is well defined, remains in $\bar{U}\left(x_0, r_1\right)$ for all $n \geq 0$ and converges to a unique solution x^* of equation $F\left(x\right) = 0$ in $\bar{U}\left(x_0, r_1\right)$.

(ii) The following error bounds hold for all $n \geq 0$:

$$\|x^* - x_{n+1}\| \leq (r_1 - t_{n+1}) \left(\frac{\|x^* - x_n\|}{r_1 - t_n} \right)^3,$$

$$\frac{(\lambda_2 \theta)^{3^n}}{\lambda_2 - (\lambda_2 \theta)^{3^n}} (r_2 - r_1) \leq r_1 - t_n \leq \frac{\lambda_1 \theta}{\lambda_1 - (\lambda_1 \theta)^{3^n}} (r_2 - r_1)$$

$$\theta = \frac{r_1}{r_2}, \quad \lambda_1 = \sqrt{\frac{(r_0 - r_2)^2 + r_0 r_2}{(r_0 - r_1)^2 + r_0 r_1}} \leq 1, \quad \lambda_2 = \frac{\sqrt{3}}{2},$$

$-r_0$ is the negative root of h and $t_{n+1} = H(t_n)$, where

$$H(t) = t - \frac{h(t)/h'(t)}{1 - \frac{1}{2} L_h(t)}, \qquad L_h(t) = \frac{h(t)/h''(t)}{h'(t)^2}.$$

Chapter 8

Broyden's Method

A local as well as a semi-local convergence results for Broyden-like methods have been given by several authors using Lipschitz and Hölder type conditions. In this chapter, we present a new result on Broyden's method using ω-type conditions.

8.1 Semi-local Convergence

We present a new semi-local convergence analysis for an inverse free Broyden's method (BM) in order to approximate a locally unique solution of (1.1) in a Hilbert space setting. The operators involved have regularly continuous divided differences. This concept has already been used on Newton's, Secant, Ulm's and other iterative methods. Our approach extends the applicability of (BM) under the same hypotheses as before. Moreover, tighter error bounds and a more precise information on the location of the solution are provided in this section.

Broyden's method (BM)

$$x_+ = x - A F(x), \quad y = F(x_+) - F(x), \quad A_+ = A - \frac{A F(x_+) <y, \cdot >}{<y, y>} \tag{8.1}$$

generates a sequence which approximates $x^\star$ under certain conditions (see Ref. [225]). Here, $A \in \mathcal{L}(\mathcal{Y}, \mathcal{X})$, the space of bounded linear operators from $\mathcal{Y}$ into $\mathcal{X}$ and $< \cdot, \cdot >$ denotes the inner product.

A local as well as a semi-local convergence results for Broyden-like methods have been given by several authors in Refs. [99, 107, 143, 179, 225] (see also, e.g. Refs. [110, 120, 134] and the references therein). Note that (BM) is an inversion-free method. That is, no linear subproblem needs to be solved at each iteration. We are motivated by the elegant work by Galperin in Ref. [225], where a semi-local convergence analysis was provided using regularly continuous divided differences (dd) (see Refs. [110, 267, 313]). Using more precise estimates on the upper bounds of the norms $\| A_+ \|$, we provide under the same hypotheses as in Ref. [225] a new semi-local convergence analysis.

We introduce some definitions and some results on regularly continuous dd. The proofs are omitted and can be found in Refs. [110, 222, 225]. For $x \in \mathcal{X}$, denote

401

by $\mathcal{K}_x$ the subspace of operators vanishing at x

$$\mathcal{K}_x = \{A \in \mathcal{L}(\mathcal{X}, \mathcal{Y}) \; : \; A\,x = 0\}.$$

Let $\mathcal{N}$ be the class of increasing cancave functions $\omega \; : \; \mathbb{R}^+ \longrightarrow \mathbb{R}^+$, with $\omega(0) = 0$. Note that $\mathcal{N}$ contains the functions in the form $\varphi(t) = c\,t^p$, ($c \geq 0$ and $p \in (0, 1]$).

Definition 8.1.1. (see Ref. [110]) An operator $[., .; F]$ belonging in $\mathcal{L}(\mathcal{X}, \mathcal{Y})$ is called the first order divided difference (briefly dd) of F at the points x and y in $\mathcal{X}$ $(x \neq y)$ if the following secant equation holds

$$[x, y; F]\,(y - x) = F(y) - F(x).$$

If F is Fréchet differentiable at x, then $[x, x; F] = \nabla F(x)$. Otherwise, the following limit (if it exists)

$$\lim_{t \searrow 0} [x, x + t\,h; F]\,h = \lim_{t \searrow 0} \frac{F(x + t\,h) - F(x)}{t}$$

varies according to h, with $\| h \| = 1$ and this limit is the Fréchet derivative (or the directional derivative) $\nabla F(x)\,h$ of F in the direction h (i.e., if we suppose that F is Fréchet differentiable at x, then the Fréchet derivative is characterized as a limit of dd in the uniform topology of the space of continuous linear mappings of $\mathcal{X}$ into $\mathcal{Y}$).

Remark 8.1.1.

(a) Let $(x, y) \in \mathcal{X} \times \mathcal{Y}$, the set $\{A \in \mathcal{L}(\mathcal{X}, \mathcal{Y}) \; : \; A\,x = y\}$ constitute an affine manifold in $\mathcal{L}(\mathcal{X}, \mathcal{Y})$.
(b) Let A and A_0 in $\mathcal{L}(\mathcal{X}, \mathcal{Y})$ and $(x, y) \in \mathcal{X} \times \mathcal{Y}$, such that $A_0\,x = A\,x = y$. Then $(A - A_0)\,x = 0$, and $A \in A_0 + \mathcal{K}_x$.

The following result gives some properties of set-valued mapping $\Upsilon_{x,y} \; : \; \mathcal{C}(\mathcal{X}, \mathcal{Y}) \rightrightarrows \mathcal{L}(\mathcal{X}, \mathcal{Y})$ given by $\Upsilon_{x,y}(F) = [x, y; F]$ for the pair $(x, y) \in \mathcal{X}^2$.

Proposition 8.1.1.

(a) $\Upsilon_{x,y}(F) = F$ if and only if F is linear.
(b) $\Upsilon_{x,y}$ is linear, i.e., for F_1, F_2 in $\mathcal{C}(\mathcal{X}, \mathcal{Y})$ and $(\alpha, \beta) \in \mathbb{K}^2$ ($\mathbb{K} = \mathbb{R}$ or $\mathbb{C}$), we have

$$\Upsilon_{x,y}(\alpha\,F_1 + \beta\,F_2) = \alpha\,\Upsilon_{x,y}(F_1) + \beta\,\Upsilon_{x,y}(F_2).$$

(c) If F is a composition of operators F_1 and F_2 (i.e., $F = F_1 \circ F_2$), then

$$\Upsilon_{x,y}(F) = \Upsilon_{F_2(x), F_2(y)}(F_1)\,\Upsilon_{x,y}(F_2).$$

Definition 8.1.2. (see Ref. [222]) The (dd) $[x, y; F]$ is said to be ω-regularly continuous on $\mathcal{D} \subseteq \mathcal{X}$ for $\omega \in \mathcal{N}$ (call it regularity modulus), if the following inequality holds for all $x, y, u, v \in \mathcal{D}$

$$\omega^{-1} \left(\min\{\| [x, y; F] \|, \| [u, v; F] \|\} + \| [x, y; F] - [u, v; F] \| \right)$$

$$-\omega^{-1} \left(\min\{\| [x, y; F] \|, \| [u, v; F] \|\} \right) \leq \| x - u \| + \| y - v \| . \tag{8.2}$$

The (dd) $[x, y; F]$ is said to be regularly continuous on $\mathcal{D}$, if it has a regularity modulus.

In view of (8.2), there exists $\omega_0 \in \mathcal{N}$, such that, for fixed $u = x_0$, $v = x_{-1}$ and all x, y in $\mathcal{D}$

$$\omega_0^{-1} \left(\min\{\| [x, y; F] \|, \| [x_0, x_{-1}; F] \|\} + \| [x, y; F] - [x_0, x_{-1}; F] \| \right)$$

$$-\omega_0^{-1} \left(\min\{\| [x, y; F] \|, \| [x_0, x_{-1}; F] \|\} \right) \leq \| x - x_0 \| + \| y - x_{-1} \| . \tag{8.3}$$

Clearly, we have

$$\omega_0(t) \leq \omega(t) \quad \text{for all} \quad t \in [0, \infty), \tag{8.4}$$

holds in general and $\dfrac{\omega}{\omega_0}$ can be arbitrarily large. If ω_0, ω are linear functions $(\omega(t) = c\, t$ and $\omega_0(t) = c_0\, t)$, then (8.3) and (8.4) become Lipschitz, and center-Lipschitz continuous conditions, respectively, i.e., the following hold respectively for all $(x, y, u, v) \in \mathcal{D}^4$:

$$\| [x, y; F] - [u, v; F] \| \leq c \left(\| x - u \| + \| y - v \| \right) \tag{8.5}$$

and

$$\| [x, y; F] - [x_0, x_{-1}; F] \| \leq c_0 \left(\| x - x_0 \| + \| y - x_{-1} \| \right). \tag{8.6}$$

Estimate (8.4) now gives

$$c_0 \leq c. \tag{8.7}$$

Hypothesis (8.2) is used in Ref. [225] to obtain the upper bounds on the norm $\| A_+ \|$. However, we shall see that the more precise estimate (8.3) is needed. This observation leads to a tighter semi-local convergence analysis with benefits as already stated in the introduction of this section.

We need a result on ω-regular and ω_0-regular continuity of (dd) using conditions (8.2) and (8.3).

Lemma 8.1.1. *(see Ref. [222]) If (dd) $[x, y; F]$ is ω-regularly continuous on $\mathcal{D}$, then*

$$|\omega^{-1}(\| [x, y; F] \|) - \omega^{-1}(\| [u, v; F] \|)| \leq \| x - u \| + \| y - v \|, \quad \forall (x, y, u, v) \in \mathcal{D}^4.$$

Consequently, the following estimate hold for all $(x, y, u, v) \in \mathcal{D}^4$:

$$\omega^{-1}(\|\,[x, y; F]\,\|) \geq (\omega^{-1}(\|\,[u, v; F]\,\|)- \|\,x - u\,\| - \|\,y - v\,\|)^+, \qquad (8.8)$$

where ρ^+ $(\rho \in \mathbb{R})$ *denotes the non-negative part of* ρ: $\rho^+ = \max\{\rho, 0\}$. *In particular, if (dd)* $[x, y; F]$ *is* ω_0-*regularly continuous on* $\mathcal{D}$ *(i.e., condition (8.3) holds), then, (8.8) holds, with* ω_0, x_0 *and* x_{-1} *replacing* ω, u *and* v, *respectively.*

We provide a semi-local convergence analysis of (BM). The proofs are the proper modifications of the ones in Ref. [225], where we use the more precise (8.3) instead of (8.2). First, we denote

$$A_0 = [x_0, x_{-1}; F]^{-1}. \qquad (8.9)$$

As in Ref. [225], for the selected (dd) $[x, y; F]$, such that (8.2) holds with ω modulus, we associate the current iteration (x, A) and we consider $q = (\bar{t}, \bar{\gamma}, \bar{\delta})$, where

$$\bar{t} = \|\,x - x_0\,\|, \quad \bar{\gamma} = \|\,x - x_-\,\|, \quad \bar{\delta} = \|\,x_+ - x\,\| = \|\,A F(x)\,\|.$$

Finally, denote q_+ and $\psi_\omega : \mathbb{R}^{+^2} \longrightarrow \mathbb{R}^+$ by

$$q_+ = (\bar{t}_+, \bar{\gamma}_+, \bar{\delta}_+)$$

and

$$\psi_\omega(u, t) = \omega((u - t)^+ + t) - \omega((u - t)^+), \quad \forall (u, t) \in \mathbb{R}^{+^2},$$

respectively. Note that ψ_ω is not increasing in the first argument, and not decreasing in the second, since ω is concave and increasing.

We provide now a result on q_+ using ω and ω_0-regularity.

Lemma 8.1.2. *Under the hypotheses (8.2) and (8.3), the following estimates hold:*

$$\bar{t}_+ := \|\,x_+ - x_0\,\| \leq \bar{t} + \bar{\delta}, \qquad (8.10)$$

$$\bar{\gamma}_+ := \|\,x_+ - x\,\| = \bar{\delta} \qquad (8.11)$$

and

$$\bar{\delta}_+ \leq \bar{\delta}\, e_{\omega_0, \omega}(q), \qquad (8.12)$$

where

$$e_{\omega_0, \omega}(q) = \frac{\psi_\omega((\omega^{-1}(\|\,A_0^{-1}\,\|) - \bar{\gamma}_0 - 2\bar{t}_- - \bar{\gamma})^+, \bar{\gamma} + \bar{\delta})}{\omega_0(\omega_0^{-1}(\|\,A_0^{-1}\,\|) - \bar{\gamma}_0 - 2\bar{t} - \bar{\delta})}.$$

Proof. (8.10) and (8.11) follow from $\|\,x_+ - x_0\,\| \leq \|\,x_+ - x\,\| + \|\,x - x_0\,\| = \bar{t} + \bar{\delta}$, and the expression of $\bar{\delta}$, respectively. We prove now (8.12). We have that

$$\bar{\delta}_+ \leq \|\,A_+\,\|\,\|\,F(x_+)\,\|.$$

By the Banach lemma on invertible operators and (8.9), we obtain

$$\|\,A_+\,\|^{-1} \geq \|\,A_0\,\|^{-1} - \|\,A_+^{-1} - A_0^{-1}\,\| = \|\,A_0\,\|^{-1} - \|\,[x_+, x; F] - [x_0, x_{-1}; F]\,\|. \qquad (8.13)$$

Using (8.2), we have in turn:

$$\begin{aligned}
&\| [x, y; F] - [u, v; F] \| \leq \\
&\omega(\omega^{-1}(\min\{\| [x, y; F] \|, \| [u, v; F] \|\}) + \| [x, y; F] - [u, v; F] \|) \\
&- \min\{\| [x, y; F] \|, \| [u, v; F] \|\} = \\
&\omega(\min\{\omega^{-1}(\| [x, y; F] \|), \omega^{-1}(\| [u, v; F] \|)\} + \| [x, y; F] - [u, v; F] \|) \\
&- \omega(\min\{\omega^{-1}(\| [x, y; F] \|), \omega^{-1}(\| [u, v; F] \|)\}).
\end{aligned} \tag{8.14}$$

By Lemma 8.2, we have that

$$\omega^{-1}(\| [u, v; F] \|) \geq (\omega^{-1}(\| [x, y; F] \|) - \| x - u \| - \| y - v \|)^+.$$

By (8.14) and the concavity of ω, we get that

$$\begin{aligned}
&\| [x, y; F] - [u, v; F] \| \leq \\
&\omega((\omega^{-1}(\| [x, y; F] \|) - \| x - u \| - \| y - v \|)^+ + \| x - u \| + \| y - v \|) \\
&- \omega((\omega^{-1}(\| [x, y; F] \|) - \| x - u \| - \| y - v \|)^+) = \\
&\psi_\omega(\omega^{-1}(\| [x, y; F] \|), \| x - u \| + \| y - v \|).
\end{aligned} \tag{8.15}$$

Clearly, estimate (8.15) holds with ω_0, x_+, x_0 and x_{-1} replacing ω, y, u and v, respectively. Consequently,

$$\begin{aligned}
&\| [x_+, x; F] - [x_0, x_{-1}; F] \| \\
&\leq \psi_{\omega_0}(\omega_0^{-1}(\| [x_0, x_{-1}; F] \|), \| x_+ - x_0 \| + \| x - x_{-1} \|) \\
&\leq \psi_{\omega_0}(\omega_0^{-1}(\| A_0^{-1} \|), \| x_+ - x_0 \| + \| x - x_0 \| + \| x_0 - x_{-1} \|) \\
&= \psi_{\omega_0}(\omega_0^{-1}(\| A_0^{-1} \|), \overline{t}_+ + \overline{t} + \overline{\gamma}_0).
\end{aligned} \tag{8.16}$$

Hence,

$$\| A_+ \| \leq (\| A_0 \|^{-1} - \psi_{\omega_0}(\omega_0^{-1}(\| A_0^{-1} \|), \overline{\gamma}_0 + \overline{t}_+ + \overline{t}))^{-1} \implies \overline{\gamma}_0 + \overline{t}_+ + \overline{t} < \omega_0^{-1}(\| A_0^{-1} \|) \tag{8.17}$$

since, otherwise

$$\psi_{\omega_0}(\omega_0^{-1}(\| A_0^{-1} \|), \overline{t}_+ + \overline{t} + \overline{\gamma}_0) = \omega_0(\overline{\gamma}_0 + \overline{t}_+ + \overline{t}) \geq \| A_0^{-1} \| \geq \| A_0 \|^{-1}$$

and

$$\| A_0 \|^{-1} - \psi_{\omega_0}(\omega_0^{-1}(\| A_0^{-1} \|), \overline{t}_+ + \overline{t} + \overline{\gamma}_0) \leq 0.$$

We also have that

$$\| A_+ \| \leq \frac{1}{\omega_0(\omega_0^{-1}(\| A_0 \|^{-1}) - \overline{\delta} - 2\overline{t} - \overline{\gamma}_0)}. \tag{8.18}$$

Using (8.18) and since ω_0 is concave and increasing, we deduce

$$\begin{aligned}
\psi_{\omega_0}(\omega_0^{-1}(\| A_0^{-1} \|), \overline{t}_+ + \overline{t} + \overline{\gamma}_0) &= \| A_0 \|^{-1} - \omega_0(\omega_0^{-1}(\| A_0 \|^{-1}) - \overline{t}_+ - \overline{t} - \overline{\gamma}_0) \\
&\leq \| A_0 \|^{-1} - \omega_0(\omega_0^{-1}(\| A_0 \|^{-1}) - \overline{\delta} - 2\overline{t} - \overline{\gamma}_0).
\end{aligned}$$

By (BM), we have the identity

$$\begin{aligned}
F(x_+) &= F(x_+) - F(x) + F(x) \\
&= [x_+, x; F](x_+ - x) - A^{-1}(x_+ - x) \\
&= ([x_+, x; F] - [x, x_-; F])(x_+ - x).
\end{aligned} \tag{8.19}$$

Using (8.15) and (8.19), we obtain

$$\| F(x_+) \| \le \overline{\delta} \, \| \, [x_+, x; F] - [x, x_-; F] \, \|$$
$$\le \psi_\omega(\omega^{-1}(\| A^{-1} \|), \overline{\gamma} + \overline{\delta}).$$

By (8.8), we get that

$$\omega^{-1}(\| A^{-1} \|) = \omega^{-1}([x, x_-; F])$$
$$\ge (\omega^{-1}([x_0, x_{-1}; F]) - \| x - x_0 \| - \| x_- - x_{-1} \|)^+$$
$$\ge (\omega^{-1}(\| A_0 \|^{-1}) - \| x - x_0 \| - \| x_- - x_0 \| - \| x_0 - x_{-1} \|)^+$$
$$\ge (\omega^{-1}(\| A_0 \|^{-1}) - \overline{t} - \overline{t}_- - \overline{\gamma}_0)^+$$
$$\ge (\omega^{-1}(\| A_0 \|^{-1}) - \overline{\gamma} - 2\overline{t}_- - \overline{\gamma}_0)^+.$$

Consequently,

$$\psi_\omega(\omega^{-1}(\| A \|^{-1}), \overline{\gamma} + \overline{\delta}) \le \psi_\omega((\omega^{-1}(\| A_0 \|^{-1}) - \overline{\gamma} - 2\overline{t}_- - \overline{\gamma}_0)^+, \overline{\gamma} + \overline{\delta}),$$

$$\| F(x_+) \| \le \overline{\delta} \, \psi_\omega((\omega^{-1}(\| A_0 \|^{-1}) - \overline{\gamma} - 2\overline{t}_- - \overline{\gamma}_0)^+, \overline{\gamma} + \overline{\delta}) \qquad (8.20)$$

and

$$\overline{\delta}_+ \le \overline{\delta} \, e_{\omega, \omega_0}(q).$$

That completes the proof of Lemma 8.1.2. $\square$

We define the function χ_{ω, ω_0} for all $q = (t, \gamma, \delta)$ by $\chi_{\omega, \omega_0}(q) = q_+ = (t_+, \gamma_+, \delta_+)$ as follows:

$$t_+ = t + \delta, \quad \gamma_+ = \delta, \quad \delta_+ = \delta \, \frac{\psi_\omega((a - 2t + \gamma)^+, \gamma + \delta)}{\omega_0(a_0 - 2t - \delta)}, \qquad (8.21)$$

where

$$a \le \omega^{-1}(\| A_0 \|^{-1}) - \| x_0 - x_{-1} \| \quad \text{and} \quad a_0 \le \omega_0^{-1}(\| A_0 \|^{-1}) - \| x_0 - x_{-1} \| \, .$$

In view of the definitions of a_0 and a, we can certainly assume $a \le a_0$.

Remark 8.1.2.

(a) Since $2t - \gamma \le 2t + \delta < a$, then

$$\psi_\omega((a - 2t + \gamma)^+, \gamma + \delta) = \psi_\omega(a - 2t + \gamma, \gamma + \delta) = \omega(a - 2t + \gamma) - \omega(a - 2t - \delta).$$

Consequently, we can simplify the third component δ_+ in the expression of χ_{ω, ω_0} by:

$$\delta_+ = \delta \, \frac{\omega(a - 2t + \gamma) - \omega(a - 2t - \delta)}{\omega_0(a_0 - 2t - \delta)}.$$

(b) As $\overline{t}_0 = 0$, we can take $t_0 = 0$.

Consider the relation order $\prec$ for $q = (t, \gamma, \delta)$ and $q' = (t', \gamma', \delta')$ by

$$q \prec q' \iff t \le t', \quad \gamma \le \gamma' \text{ and } \delta \le \delta'.$$

Lemma 8.1.3. *Let* $q = (t, \gamma, \delta)$ *and* $q' = (t', \gamma', \delta')$*. Then*

$$0 \prec q \prec q' \implies 0 \prec \chi_{\omega, \omega_0}(q) \prec \chi_{\omega, \omega_0}(q').$$

Proof. We suppose $q \prec q'$. Then, we obtain

$$t \leq t' \text{ and } \delta \leq \delta' \implies t_+ := t + \delta \leq t' + \delta' =: t'_+ \text{ and } \gamma_+ := \delta \leq \delta' =: \gamma'_+.$$

We show now $\delta_+ \leq \delta'_+$. Since ω, ω_0 are concave and increasing and using Remark 8.1.2, we have

$$
\begin{aligned}
\delta_+ &= \delta \, \frac{\omega(a - 2\,t + \gamma) - \omega(a - 2\,t - \delta)}{\omega_0(a_0 - 2\,t - \delta)} \\
&\leq \delta' \, \frac{\omega(a - 2\,t - \delta + (\delta + \gamma)) - \omega(a - 2\,t - \delta)}{\omega_0(a_0 - 2\,t' - \delta')} \\
&\leq \delta' \, \frac{\omega(a - 2\,t' - \delta' + (\delta + \gamma)) - \omega(a - 2\,t' - \delta')}{\omega_0(a_0 - 2\,t' - \delta')} \\
&\leq \delta' \, \frac{\omega(a - 2\,t' - \delta' + (\delta' + \gamma')) - \omega(a - 2\,t' - \delta')}{\omega_0(a_0 - 2\,t' - \delta')} = \delta'_+.
\end{aligned}
$$

That completes the proof of Lemma 8.1.3. $\qquad\square$

Consider the sequence q_n with the initial iterate $q_0 = (t_0, \gamma_0, \delta_0)$ by

$$q_{n+1} = \chi_{\omega, \omega_0}(q_n). \tag{8.22}$$

q_n is a majorizing sequence if $\bar{q}_n \prec q_n$ for all n, where $\bar{q}_n$ is produced by the n-th iteration (x_n, A_n) of (BM).

Lemma 8.1.4. *Let q_0 be an initial iterate for sequence $q_n = (t_n, \gamma_n, \delta_n)$ given by (8.22), such that $\bar{q}_0 \prec q_0$ and $2\,t_n + \delta_n < a$ for all $n \geq 0$. Then, the following hold for all $n \geq 0$:*

(a)

$$\bar{q}_n \prec q_n.$$

(b)

$$\gamma_\infty = \delta_\infty = 0 \quad \text{and} \quad t_n = \sum_{k=0}^{k=n-1} \delta_k \leq .5\,(a - \delta_n),$$

where

$$\gamma_\infty = \lim_{n \to \infty} \gamma_n \quad \text{and} \quad \delta_\infty = \lim_{n \to \infty} \delta_n;$$

(c) The sequence (x_n, A_n) generated by (BM) from the initial iterate (x_0, A_0) converges to a solution $(x^\star, A_\infty)$ of the system:

$$F(x) = 0 \quad \text{and} \quad A\,[x, x; F] = I;$$

(d) The solution $x^\star$ is unique in $U(x_0, a_0 - t_\infty)$;

(e)

$$\| F(x_{n+1}) \| \leq \delta_n \left(\omega(a - 2\,t_n + \gamma_n) - \omega(a - 2\,t_n - \delta_n) \right),$$

$$\| x_n - x_0 \| \leq t_n < t_\infty \leq .5\,a,$$

$$\Delta_n := \| x^\star - x_n \| \leq t_\infty - t_n,$$

$$\| I - A_n \, [x^\star, x^\star; F] \| \leq \frac{\omega(a - 2\,t_n + \gamma_n) - \omega(a - 2\,t_\infty)}{\omega_0(a_0 - 2\,t_n + \gamma_n)}$$

and

$$\frac{\Delta_{n+1}}{\Delta_n} \leq \frac{\omega(a - 2\,t_n + \gamma_n) - \omega(a - t_n - t_\infty)}{\omega_0(a_0 - 2\,t_n + \gamma_n)}.$$

Proof.

(a) We show (a) by induction on n. We suppose that $\overline{q} \prec q$. By Lemma 8.1.2, we have

$$q_+ = (\overline{t}_+, \overline{\gamma}_+, \overline{\delta}_+) \prec (\overline{t} + \overline{\delta}, \overline{\delta}, \overline{\delta}\, e_{\omega_0, \omega}(q)), \tag{8.23}$$

where

$$\omega(\omega^{-1}(\| A_0 \|^{-1}) - \overline{\gamma}_0 - 2\overline{t} - \overline{\delta}) \geq \omega(a - 2\,t - \delta)$$

and

$$\begin{aligned}
\psi_\omega((\omega^{-1}(\| A_0^{-1} \|) - \overline{\gamma}_0 - 2\overline{t}_- - \overline{\gamma})^+, \overline{\gamma} + \overline{\delta}) &\leq \psi_\omega((a - 2\overline{t}_- - \overline{\gamma})^+, \overline{\gamma} + \overline{\delta}) \\
&\leq \psi_\omega((a - 2\,t_- - \gamma)^+, \gamma + \delta) \\
&= \psi_\omega(a - 2\,t + \gamma, \gamma + \delta) \\
&= \omega(a - 2\,t + \gamma) - \omega(a - 2\,t - \delta).
\end{aligned}$$

Then, we get the estimate

$$\overline{\delta}\, e_{\omega_0, \omega}(q) \leq \delta\, \frac{\omega(a - 2\,t + \gamma) - \omega(a - 2\,t - \delta)}{\omega_0(a_0 - 2\,t - \delta)}. \tag{8.24}$$

By (8.23) and (8.24), we deduce that

$$\overline{q}_+ \prec \chi_{\omega, \omega_0}(q) = q_+.$$

The induction is complete.

(b) By hypotheses, $2\,t_n + \delta_n < a$ for all $n \geq 0$, we deduce that $t_n < .5\,(a - \delta_n)$ for all $n \geq 0$. Then, $\{t_n\}$ is increasing and bounded. Thus, $\{t_n\}$ converges to finite limit t_∞. Since $t_{n+1} = t_n + \delta_n$ and $\gamma_{n+1} = \delta_n$, for all $n \geq 0$, we obtain $\gamma_\infty = \delta_\infty = 0$ and $t_\infty \leq .5\,a$.

(c) to (e) First, by (a), we have for $n, m \geq 0$

$$\| x_{n+m} - x_n \| \leq \sum_{k=n}^{k=n+m-1} \| x_{k+1} - x_k \|$$
$$= \sum_{k=n}^{k=n+m-1} \bar{\delta}_k \leq \sum_{k=n}^{k=n+m-1} \delta_k < \sum_{k=n}^{\infty} \delta_k = t_\infty - t_n. \tag{8.25}$$

Then, $\{x_n\}$ is a Cauchy sequence in a Banach space and converges to $x^\star$. By setting $m \longrightarrow \infty$ in (8.25), we deduce for $n \geq 0$ the following estimate

$$\Delta_n = \| x^\star - x_n \| \leq t_\infty - t_n.$$

Moreover, by (8.20), we get

$$\| F(x_{n+1}) \| \leq \bar{\delta}_n \, \psi_w((\omega^{-1}(\| A_0 \|^{-1}) - \bar{\gamma}_0 - \bar{t}_n - \bar{t}_{n-1})^+, \bar{\gamma}_n + \bar{\delta}_n)$$
$$\leq \delta_n \, \psi_w(a - 2 t_n + \gamma_n, \gamma_n + \delta_n)$$
$$= \delta_n \left(\omega(a - 2 t_n + \gamma_n) - \omega(a - 2 t_n - \delta_n) \right),$$

where $\delta_n \longrightarrow \infty$. Then, $F(x^\star) = 0$. Setting $n \longrightarrow \infty$ in equality $A_n [x_n, x_{n-1}; F] = I$, we have $A_\infty [x^\star, x^\star; F] = 0$ and (c) is complete. Replacing A_+ by A_n in (8.18), we have that

$$\| A_n \| \leq \frac{1}{\omega_0(\omega_0^{-1}(\| A_0 \|^{-1}) - \bar{\gamma}_0 - \bar{t}_n - \bar{t}_{n-1})^+}$$
$$\leq \frac{1}{\omega_0(a_0 - t_n - t_{n-1})} = \frac{1}{\omega_0(a_0 - 2 t_n + \gamma_n)}. \tag{8.26}$$

Using (8.15), we have that

$$\| I - A_n [x^\star, x^\star; F] \|$$
$$\leq \| A_n \| \, \| [x_n, x_{n-1}; F] - [x^\star, x^\star; F] \| \tag{8.27}$$
$$\leq \| A_n \| \, \psi_w(\omega^{-1}(\| [x_n, x_{n-1}; F] \|), \| x_n - x^\star \| + \| x_{n-1} - x^\star \|).$$

By Lemma 8.1.1, we get that

$$\omega^{-1}(\| [x_n, x_{n-1}; F] \|) \geq (\omega^{-1}(\| [x_0, x_{-1}; F] \|) - \| x_n - x_0 \| - \| x_{n-1} - x_{-1} \|)^+$$
$$\geq (\omega^{-1}(\| A_0 \|) - \bar{t}_n - \bar{t}_{n-1} - \bar{\gamma}_0)^+$$
$$\geq a - 2 t_n + \gamma_n. \tag{8.28}$$

Using (8.26)–(8.28), we have that

$$\| I - A_n [x^\star, x^\star; F] \| \leq \frac{\omega(a - 2 t_n + \gamma_n) - \omega(a - 2 t_\infty)}{\omega_0(a_0 - 2 t_n + \gamma_n)}. \tag{8.29}$$

By the identity

$$x_{n+1} - x^\star = x_n - x^\star - A_n \, F(x_n)$$
$$= A_n \left([x_n, x_{n-1}; F] - [x_n, x^\star; F] \right) (x_n - x^\star)$$

we obtain similarily as (8.29)

$$\Delta_{n+1} \leq \Delta_n \parallel A_n \parallel \parallel [x_n, x_{n-1}; F] - [x_n, x^\star; F] \parallel$$
$$\leq \Delta_n \frac{\psi_\omega(a - 2\,t_n + \gamma_n, \Delta_{n-1})}{\omega_0(a_0 - 2\,t_n + \gamma_n)}$$
$$\leq \Delta_n \frac{\psi_\omega(a - 2\,t_n + \gamma_n, t_\infty - t_n + \gamma_n)}{\omega_0(a_0 - 2\,t_n + \gamma_n)} \tag{8.30}$$
$$= \Delta_n \frac{\omega(a - 2\,t_n + \gamma_n) - \omega(a - t_n - t_\infty)}{\omega_0(a_0 - 2\,t_n + \gamma_n)}.$$

The proof of (e) is completed.

We prove now (c). Let $y^\star$ be a solution of $F(x) = 0$. Then, $F_0(y^\star) = 0$, where $F_0 = A_0\,F$ and $F_0(y^\star) - F_0(x^\star) = 0 = [y^\star, x^\star; F_0]\,(y^\star - x^\star)$. We deduce by the Banach lemma on invertible operators and Proposition 8.1.1 that $[y^\star, x^\star; F_0] = A_0\,[y^\star, x^\star; F]$ is not invertible and $\parallel I - A_0\,[y^\star, x^\star; F] \parallel \geq 1$. Using (8.27), we have that

$$1 \leq \parallel I - A_0\,[x^\star, x^\star; F] \parallel$$
$$\leq \parallel A_0 \parallel \parallel [x_0, x_{-1}; F] - [y^\star, x^\star; F] \parallel$$
$$\leq \parallel A_0 \parallel \psi_{\omega_0}(\omega_0^{-1}(\parallel [x_0, x_{-1}; F] \parallel), \parallel y^\star - x_0 \parallel + \parallel x_{-1} - x^\star \parallel)$$
$$\leq \parallel A_0 \parallel \psi_{\omega_0}(\omega_0^{-1}(\parallel A_0^{-1} \parallel), \parallel y^\star - x_0 \parallel + \parallel x^\star - x_0 \parallel + \parallel x_0 - x_{-1} \parallel)$$

and

$$\parallel A_0^{-1} \parallel \leq \psi_{\omega_0}(\omega_0^{-1}(\parallel A_0 \parallel^{-1}), \Lambda)$$
$$= \begin{cases} \parallel A_0 \parallel^{-1} - \omega_0(\omega_0^{-1}(\parallel A_0 \parallel^{-1}), \Lambda) & \text{if } \Lambda \leq \omega_0^{-1}(\parallel A_0 \parallel^{-1}) \\ \omega_0(\Lambda) & \text{if } \Lambda \geq \omega_0^{-1}(\parallel A_0 \parallel^{-1}), \end{cases}$$

where $\Lambda = \overline{\gamma}_0 + t_\infty + \parallel y^\star - x_0 \parallel$. It follows that

$$\parallel y^\star - x_0 \parallel \geq \omega_0^{-1}(\parallel A_0 \parallel^{-1}) - \overline{\gamma}_0 - t_\infty \geq a_0 - t_\infty \geq .5\,a.$$

That completes the proof of Lemma 8.1.4. $\qquad\qquad\square$

In the special cases 1 and 2 that follow, we provide more precise estimates than [Section 4] in Ref. [225].

Case 1: Semi-local convergence under (8.5) and (8.6)

Let x_0, x_{-1}, A_0, a, a_0 and q_0 be given, such that

$$\overline{q}_0 \prec q_0 \quad \text{and} \quad 2\,t_n + \delta_n < a \quad \text{for all } n \geq 0. \tag{8.31}$$

Condition (8.31) guarantees the convergence of sequence (x_n, A_n). We denote by $\mathcal{Q}_c$ the set

$$\mathcal{Q}_c = \{(q_0, t_n, \delta_n) : (8.31) \text{ holds}\}.$$

In this subsection $\omega(t) = c\,t$ and $\omega_0(t) = c_0\,t$ with $c_0 \leq c$. The function χ_{ω, ω_0} defined in (8.21) by $\chi_{\omega, \omega_0}(q) = q_+ = (t_+, \gamma_+, \delta_+)$ for all $q = (t, \gamma, \delta)$, is simplified in the following form

$$t_+ = t + \delta, \quad \gamma_+ = \delta, \quad \delta_+ = \delta\,\frac{\overline{c}\,(\gamma + \delta)}{1 - \overline{c}_0\,(\parallel x_0 - x_{-1} \parallel - 2\,t - \delta)}, \tag{8.32}$$

where

$$\bar{c} = c \, \| A_0 \| \quad \text{and} \quad \bar{c}_0 = c_0 \, \| A_0 \| .$$

In [Section 4] of Ref. [225], $c_0 = 0$ and in the denominator of δ_+, function ω, parameter a replace ω_0 and a_0, respectively. Then, (8.32) becomes

$$t_+ = t + \delta, \quad \gamma_+ = \delta, \quad \delta_+ = \delta \, \frac{\gamma + \delta}{a - 2\,t - \delta}. \tag{8.33}$$

Define function Γ for $q = (t, \gamma, \delta)$ by

$$\Gamma(q) = (.5\,a - t)^2 - \delta\,(a - 2\,t + \gamma). \tag{8.34}$$

We present now two results on simplified generator χ_{ω,ω_0} given by (8.33).

Lemma 8.1.5. *(see Ref. [225])*

(a) The function Γ given by (8.34) is an invariant of the generator (8.33):

$$\Gamma(q) = \Gamma(q_+).$$

(b) For all $n \geq 0$,

$$2\,t_n + \delta_n < a \iff \Gamma(0, \gamma_0, \delta_0) \geq 0 \iff 4\,\delta_0\,(a + \gamma_0) \leq a^2$$
$$\iff t_n = .5\,a - \delta_n - \sqrt{\delta_n\,(\gamma_n + \delta_n) + \Gamma(0, \gamma_0, \delta_0)}.$$

Theorem 8.1.1. *Suppose that (8.5) and (8.6) hold. Let x_{-1}, x_0, A_0, a_0, a, γ_0, δ_0, such that*

$$\frac{\| A_0 \|^{-1}}{c_0} - \| x_0 - x_{-1} \| \geq a_0, \qquad \frac{\| A_0 \|^{-1}}{c} - \| x_0 - x_{-1} \| \geq a,$$

$$\| x_0 - x_{-1} \| \leq \gamma_0, \qquad \| A_0\,F(x_0) \| \leq \delta_0$$

and

$$4\,\delta_0\,(a + \gamma_0) \leq a^2. \tag{8.35}$$

Then, the following hold

(a) $(t_n, \gamma_n, \delta_n)$ generated by (8.33) started at $(0, \gamma_0, \delta_0)$ is well defined and converges to $(t_\infty, 0, 0)$, where

$$t_\infty = .5\,(a - \sqrt{a^2 - 4\,\delta_0\,(a + \gamma_0)});$$

(b) The sequence (x_n, A_n) generated by (BM) started at (x_{-1}, x_0, A_0) converges to a solution $(x^\star, A_\infty)$ of the system

$$F(x) = 0 \quad \text{and} \quad A\,[x, x; F] = I;$$

(c) $x^\star$ is the unique solution of (1.1) in $U(x_0, r)$, where

$$r = .5\,(a + \sqrt{a^2 - 4\,\delta_0\,(a + \gamma_0)});$$

(d) For all $n \geq 1$,

$$\| F(x_{n+1}) \| \leq c\,\delta_n\,(\gamma_n + \delta_n), \quad \Delta_n \leq \tilde{t}_\infty - \tilde{t}_n,$$

$$\| x_n - x_0 \| \leq \tilde{t}_n - \tilde{t}_0,$$

$$\| I - A_n\,[x^\star, x^\star; F] \| \leq p_n,$$

and

$$\frac{\Delta_{n+1}}{\Delta_n} \leq q_n,$$

where

$$\tilde{t}_{-1} = 0, \quad \tilde{t}_0 = \| x_0 - x_{-1} \|, \quad \tilde{t}_1 = \tilde{t}_0 + \| A_0\,F(x_0) \|,$$
$$\tilde{t}_n = \tilde{t}_{n-1} + \frac{\overline{c}\,(\tilde{t}_{n-1} - \tilde{t}_{n-3})\,(\tilde{t}_{n-1} - \tilde{t}_{n-2})}{1 - \overline{c}_0\,(-\tilde{t}_{n-1} - \tilde{t}_{n-2} + \tilde{t}_0)}, \quad (n \geq 2),$$

$$p_n = \frac{\overline{c}\,(\gamma_n + 2\,(\tilde{t}_\infty - \tilde{t}_n))}{\overline{c}_0\,(a_0 - 2\,\tilde{t}_n + \gamma_n)}$$

and

$$q_n = \frac{\overline{c}\,(\gamma_n + \tilde{t}_\infty - \tilde{t}_n)}{\overline{c}_0\,(a_0 - 2\,\tilde{t}_n + \gamma_n)},$$

with $\tilde{t}_\infty = \lim\limits_{n \to \infty} \tilde{t}_n \leq t_\infty$.

The estimates in (d) reduce to the corresponding ones in Ref. [225] for $\overline{c} = \overline{c}_0$. Otherwise, they are tighter.

Case 2: Semi-local convergence under regular continuity of (dd)

Suppose that ω is nonlinear and let $q_0 = (t_0, \gamma_0, \delta_0)$. Define the scalar function $F(\cdot/q_0)$ on sequence $\{t_n\}$ $(n \geq 0)$ as follows

$$F(t_n/q_0) = \delta_n\,\omega(a - 2\,t_n + \gamma_n). \tag{8.36}$$

Sequence $F(t_n)$ is decreasing and consequently, function F is invertible on $\{t_n\}$, i.e., for all $n \geq 0$ (see Ref. [225])

$$t_n = F^{-1}(\delta_n\,\omega(a - 2\,t_n + \gamma_n)/q_0). \tag{8.37}$$

We present now two results on generator χ_{ω,ω_0} given by (8.21).

Lemma 8.1.6. *(see Ref. [225])*

(a) The function $F^{-1}(0/q)$ with initial iterate $q = (t, \gamma, \delta)$ is an invariant of the generator (8.21).

(b) For all $n \geq 0$ and $q_0 = (t_0, \gamma_0, \delta_0)$, we have the following equivalence

$$2\,t_n + \delta_n < a \iff F^{-1}(0/q_0) \leq .5\,a.$$

Theorem 8.1.2. *Suppose that (8.2) and (8.3) hold. Let x_{-1}, x_0, A_0, a_0, a, γ_0, δ_0, such that*

$$\omega_0^{-1}(\| A_0 \|^{-1}) - \| x_0 - x_{-1} \| \geq a_0, \quad \omega^{-1}(\| A_0 \|^{-1}) - \| x_0 - x_{-1} \| \geq a,$$

$$\| x_0 - x_{-1} \| \leq \gamma_0, \quad \| A_0 F(x_0) \| \leq \delta_0$$

and

$$F^{-1}(0/(0, \gamma_0, \delta_0)) \leq .5\, a.$$

Then, the following hold

(a) *$(t_n, \gamma_n, \delta_n)$ generated by (8.21) started at $(0, \gamma_0, \delta_0)$ is well defined and converges to $(t_\infty, 0, 0)$, where $t_\infty = F^{-1}(0/(0, \gamma_0, \delta_0))$;*

(b) *The sequence (x_n, A_n) generated by (BM) started at (x_{-1}, x_0, A_0) converges to a solution $(x^\star, A_\infty)$ of the system $F(x) = 0$ and $A[x, x; F] = I$;*

(c) *$x^\star$ is the unique solution of (1.1) in $U(x_0, a_0 - t_\infty)$;*

(d) *For all $n \geq 1$,*

$$\| F(x_{n+1}) \| \leq \delta_n\, (\omega(a - 2\, t_n + \gamma_n) - \omega(a - 2\, t_n - \delta_n)),$$

$$\| I - A_n [x^\star, x^\star; F] \| \leq \frac{\omega(a - 2\, t_n + \gamma_n) - \omega(a - 2\, t_\infty)}{\omega_0(a_0 - 2\, t_n + \gamma_n)}$$

and

$$\frac{\Delta_{n+1}}{\Delta_n} \leq \frac{\omega(a - 2\, t_n + \gamma_n) - \omega(a - t_n - t_\infty)}{\omega_0(a_0 - 2\, t_n + \gamma_n)}.$$

Remark 8.1.3. The results obtained in this section reduce to the corresponding ones in Ref. [225], if equality holds in (8.4). Otherwise, our results provide tighter error bounds than in Ref. [225] (see also the definition of a and a_0). Moreover, the information on the uniqueness of the solution $x^\star$ is more precise, since $a_0 - t_\infty >$ $a - t_\infty$ (see also Lemma 8.1.4-(b)). We shall try to provide weaker conditions of the form of Lemma 8.1.4-(b), Theorem 8.1.1, (8.35) and the main hypothesis in Theorem 8.1.2. The approach will make direct use of the more precise majorant using ω and ω_0 (8.21), not just ω (see $\omega = \omega_0$ in (8.21) used in Ref. [225]). As an example instead of studying (8.33) given in a simpler form

$$s_{-1} = 0, \quad s_0 = \| x_0 - x_{-1} \|, \quad s_1 = s_0 + \| A_0 F(x_0) \|$$

$$s_{n+1} = s_n + \frac{\overline{c}\,(s_n - s_{n-1})(s_n - s_{n-2})}{1 - \overline{c}\,(s_0 - s_n - s_{n-1})},$$

we shall study more precise sequence:

$$\alpha_{-1} = 0, \quad \alpha_0 = \| x_0 - x_{-1} \|, \quad \alpha_1 = \alpha_0 + \| A_0 F(x_0) \|$$

$$\alpha_{n+1} = \alpha_n + \frac{\overline{c}\,(\alpha_n - \alpha_{n-1})(\alpha_n - \alpha_{n-2})}{1 - \overline{c}_0\,(\alpha_0 - \alpha_n - \alpha_{n-1})}.$$

This approach together with our new idea of recurrent functions have already led to weaker sufficient convergence conditions for many iterative methods such as Newton's, Secant and Newton-type methods (under very general conditions, see Refs. [110, 150]). In particular, our work using regularly continuous divided differences can be found in Ref. [137].

8.2   Exercises

8.2.1. Suppose that F', F'' are uniformly bounded by non-negative constants $\alpha < 1$, K respectively, on a convex subset D of X and the ball

$$\bar{U}\left(x_0, r_0 \equiv \frac{2\|x_0 - F(x_0)\|}{1 - \alpha}\right) \subseteq D.$$

Let

$$h_S = \frac{K}{2} \frac{1 + 2\alpha}{1 - \alpha} \frac{\|x_0 - F(x_0)\|}{1 - \alpha} < 1.$$

Consider Stirling's method as follows

$$x_{n+1} = x_n - \left(I - F'(F(x_n))\right)(x_n - F(x_n)) \quad (n \geq 0). \tag{8.38}$$

Show:

(i) Method (8.38) converges to the unique fixed point x^* of F in $\bar{U}(x_0, r_0)$.

(ii) The following estimate holds for all $n \geq 0$:

$$\|x_n - x^*\| \leq h_S^{2^n - 1} \frac{\|x_0 - F(x_0)\|}{1 - \alpha}.$$

Let $F \colon D \subseteq X \to Y$ be analytic. Assume that for $x \in D$

$$\|F'(x)\| \leq \alpha < 1, \quad x_0 \neq F(x_0), \quad x_0 \in D, \quad \frac{\gamma(1 + 2\alpha)\|x_0 - F(x_0)\|}{(1 - \alpha)^2} < 1,$$

$$r_0 < r_1, \quad U(x_0, r_1) \subseteq D \quad \text{and} \quad 0 \neq \gamma \equiv \sup_{\substack{k > 1 \\ x \in D}} \left\|\frac{1}{k!} F^{(k)}(x_0)\right\|^{\frac{1}{k-1}} < \infty,$$

where

$$r_1 = \frac{1}{\gamma}\left[1 - \sqrt[3]{\frac{\gamma(1 + 2\alpha)\|x_0 - F(x_0)\|}{(1 - \alpha)^2}}\right].$$

(iii) Show that $\{x_n\}$ $(n \geq 0)$ generated by Stirling's method (8.38) is well defined, remains in $U(x_0, r_0)$ for all $n \geq 0$ and converges to a unique fixed point x^* of operator F at the rate given before with $K = 2\gamma/(1 - \gamma r_0)^3$.

Let now $X = D = \mathbb{R}$ and define function F on D by

$$F(x) = \begin{cases} -\dfrac{1}{3}x & \text{if} \quad x \leq 3 \\[2mm] \dfrac{1}{45}(x^2 - 7x - 33) & \text{if} \quad 3 \leq x \leq 4 \\[2mm] \dfrac{1}{3}(x - 7) & \text{if} \quad x > 4. \end{cases}$$

(iv) Using Stirling's method for $x_0 = 3$, we obtain the fixed point $x^* = 0$ of F in one iteration, since $x_1 = 3 - (1 + \frac{1}{3})^{-1}(3 + 1) = 0$. Show that Newton's method fails to converge.

Chapter 9

Newton-like Methods

The practice of numerical analysis for finding solutions of nonlinear equations is essentially connected to variants of Newton's method. In this chapter we provide convergence analysis of Newton-like methods.

9.1 Modified Newton Method and Multiple Zeros

In this section, we give an estimate of the convergence radius of the well-known modified Newton's method for multiple zeros, when the involved function satisfies a Hölder and center-Hölder continuity condition. We use iterative methods to find a multiple zero $x^\star$ of multiplicity m $(m > 1)$, i.e., $f^{(j)}(x^\star) = 0$, $j = 0, 1, \ldots, m-1$ and $f^{(m)}(x^\star) \neq 0$, of a nonlinear equation $f(x) = 0$, where $f : D \to \mathbb{R}$ is a smooth function and D is an open interval.

It is known that, the modified Newton method for multiple zeros is given by

$$x_{n+1} = x_n - m\frac{f(x_n)}{f'(x_n)}, \tag{9.1}$$

which converges quadratically (see Ref. [401]). There exists a cubically convergent method for multiple zeros, presented by Hansen and Patrick in Ref. [255]:

$$x_{n+1} = x_n - \frac{f(x_n)}{\frac{m+1}{2m}f'(x_n) - \frac{f(x_n)f''(x_n)}{2f'(x_n)}}, \tag{9.2}$$

which is an extension of the classical Halley method of the third order.

Another cubically convergent method for multiple zeros, is proposed by Traub in Ref. [418]:

$$x_{n+1} = x_n - \frac{m(3-m)}{2}\frac{f(x_n)}{f'(x_n)} - \frac{m^2}{2}\frac{f(x_n)^2 f''(x_n)}{f'(x_n)^3}, \tag{9.3}$$

which is an extension of the well-known Chebyshev method of the third order.

In recent years, a lot of methods for multiple zeros have been presented and many local convergence results have been obtained, see Refs. [189, 190, 197, 198, 219, 278, 334–336, 339, 407, 408, 423] and references

415

therein. In general, these results show that if the initial guess x_0 of the corresponding method is sufficiently close to the zero $x^\star$ of the function involved, then the sequence $\{x_n\}$ generated by this method is well-defined and converges to $x^\star$. But we don't know how close to the zero $x^\star$ the initial guess x_0 should be. That is, these local results give no information on the radius of the convergence ball for the corresponding method.

Here, we say an open ball $B(x^\star, r_\star) \subseteq D$ with center $x^\star$ and radius $r_\star$ is called a convergence ball of an iterative method, if the sequence generated by this iterative method starting from any initial values in it converges. Enlarging the convergence ball of an iterative method is very important, because it shows the extent of difficulty we have to choose iterative initial points. Of course, we will be delighted to have bigger radius for the convergence ball of an iterative method. However, this depends on not only the iterative method we use but also on the conditions of the involved function, such as Lipschitz continuous conditions, Hölder continuous conditions.

It is known that, in the case of a simple zero, many results on the estimate of the radius of the convergence ball have been given for iterative methods, see Refs. [426, 110, 120, 379–382]. Wang in Ref. [426] and Traub in Ref. [420] gave an exact estimate

$$r^\star = \frac{2}{3K} \tag{9.4}$$

on the radius of the convergence ball respectively for Newton's method ($m = 1$ in (9.1)), when function f satisfies the Lipschitz continuous condition:

$$|f'(x^\star)^{-1}(f'(x) - f'(y))| \leq K|x - y|, \ \forall x, y \in D \quad \text{for some} \quad K > 0. \tag{9.5}$$

Argyros in Ref. [95] provided the radius

$$r_\star = \left(\frac{1+p}{K + (1+p)K_0} \right)^{\frac{1}{p}} \tag{9.6}$$

for Newton's method, when function f satisfies the Hölder continuous condition:

$$|f'(x^\star)^{-1}(f'(x) - f'(y))| \leq K|x - y|^p, \ \forall x, y \in D \quad \text{for some} \quad K > 0, \tag{9.7}$$

and the center-Hölder continuous condition:

$$|f'(x^\star)^{-1}(f'(x) - f'(x^\star))| \leq K_0|x - x^\star|^p, \ \forall x \in D \quad \text{for some} \quad K_0 > 0, \tag{9.8}$$

where $0 < p \leq 1$. Note that, when $p = 1$ and $K = K_0$, the radius given by (9.6) reduces to that given by (9.4). Otherwise, since $K_0 < K : r^\star < r_\star$ and the upper bounds on the distance $|x_n - x^\star|(n \geq 0)$ are tighter. This development is very important in computational mathematics, since a wider range of initial guesses x_0 becomes available and at most as few computations are required to obtain a desired error tolerance. Note also that the radio K/K_0 can be arbitrarily large. Moreover, (9.8) is not an additional hypothesis, since in practice the computation of constant K requires that of K_0. Huang in Ref. [280] generalized the Lipschitz continuous condition (9.5) of Newton's method for another type of Hölder continuous condition.

Local results were also given in Refs. [100, 110] using even more general conditions for Newton-like methods. We provide the radius as follows

$$r_\star = \left(\frac{\prod_{i=1}^m (m+p+1-i)}{(m-1)!(K+(m+p)K_0)} \right)^{\frac{1}{p}} \tag{9.9}$$

for the modified Newton's method (9.1) in the case of multiple zeros, when function f satisfies the Hölder continuous condition:

$$|f^{(m)}(x^\star)^{-1}(f^{(m)}(x)-f^{(m)}(y))| \le K|x-y|^p, \ \forall x,y \in D \quad \text{for some} \quad K > 0, \tag{9.10}$$

and the center-Hölder continuous condition:

$$|f^{(m)}(x^\star)^{-1}(f^{(m)}(x) - f^{(m)}(x^\star))| \le K_0|x - x^\star|^p, \ \forall x \in D \quad \text{for some} \quad K_0 > 0, \tag{9.11}$$

where $0 < p \le 1$. Note that, conditions (9.10) and (1.11) are natural generalizations of conditions (9.7) and (9.8) from the case of simple zero($m = 1$) to multiple zero ($m > 1$) and the radius given by (9.9) reduces to the radius given by (9.6) when $m = 1$. We also provide the error analysis, which matches the convergence order of the modified Newton's method (9.1). The advantages of our approach for $m > 1$ are the same as in the case when $m = 1$, which has already been explained above.

We need the definitions of divided differences and their properties.

Definition 9.1.1. (see Ref. [413]) The divided differences $f[a_0, a_1, \ldots, a_k]$, on $k+1$ distinct points $a_0, a_1, \ldots, a_k$ of a function $f(x)$ are defined by

$$\begin{aligned} f[a_0] &= f(a_0), \\ f[a_0, a_1] &= \frac{f[a_0] - f[a_1]}{a_0 - a_1}, \\ &\vdots \\ f[a_0, a_1, \ldots, a_k] &= \frac{f[a_0, a_1, \ldots, a_{k-1}] - f[a_1, a_2, \ldots, a_k]}{a_0 - a_k}. \end{aligned} \tag{9.12}$$

If the function f is sufficiently differentiable, then its divided differences $f[a_0, a_1, \ldots, a_k]$ can be defined if some of the arguments a_i coincide. For instance, if $f(x)$ has k-th derivative at a_0, then it makes sense to define

$$f[\underbrace{a_0, a_0, \ldots, a_0}_{k+1}] = \frac{f^{(k)}(a_0)}{k!}. \tag{9.13}$$

Lemma 9.1.1. *(see Ref. [413]) The divided differences $f[a_0, a_1, \ldots, a_k]$ are symmetric functions of their arguments, i.e., they are invariant to permutations of the $a_0, a_1, \ldots, a_k$.*

Lemma 9.1.2. *(see Ref. [413]) If the function f has $(k+1)$-th derivative, then for every argument x, the following interpolation formula holds*

$$f(x) = f[a_0] + \sum_{i=1}^k f[a_0, a_1, \ldots, a_i] \prod_{j=0}^{i-1} (x-a_j) + f[a_0, a_1, \ldots, a_k, x] \prod_{i=0}^k (x-a_i). \tag{9.14}$$

Lemma 9.1.3. *If the function f has $(m+1)$-th derivative and $x^\star$ is a zero of multiplicity m, then for every argument x, the following formulae hold*

$$f(x) = f[\underbrace{x^\star, x^\star, \ldots, x^\star}_{m}, x](x - x^\star)^m, \tag{9.15}$$

$$f'(x) = f[\underbrace{x^\star, x^\star, \ldots, x^\star}_{m}, x, x](x - x^\star)^m + mf[\underbrace{x^\star, x^\star, \ldots, x^\star}_{m}, x](x - x^\star)^{m-1}. \tag{9.16}$$

Proof. Applying Lemma 9.1.2 to the zero $x^\star$ of multiplicity m and using (9.13), we get (9.15). Differentiating both sides of (9.15) gives (9.16). $\qquad\square$

We also need the following lemma on Genocchi's integral expression formula for divided differences:

Lemma 9.1.4. *(see Ref. [439]) If the function f has continuous k-th derivative, then the following formula holds for any points $a_0, a_1, \ldots, a_k$:*

$$f[a_0, a_1, \ldots, a_k] = \int_0^1 \cdots \int_0^1 f^{(k)}\left(a_0 + \sum_{i=1}^{k}(a_i - a_{i-1})\prod_{j=1}^{i} t_j\right)\prod_{i=1}^{k}(t_i^{k-i} dt_i). \tag{9.17}$$

Let e_n be the error at the nth step, i.e., $e_n = x_n - x^\star$ $(n \geq 0)$. We can show the following local convergence result for the modified Newton method (9.1):

Theorem 9.1.1. *Let $f : D \to \mathbb{R}$ be in $C^m(D)$ for $D \subset \mathbb{R}$, non-empty, open and convex. Let $x^\star$ be a zero of f with multiplicity m $(m \geq 1)$. Suppose both the Hölder continuous condition (9.10) and the center-Hölder continuous condition (1.11) are satisfied. Assume $U(x^\star, r_\star) \subseteq D$, where $r_\star$ is given by (9.9). Then, for any initial point $x_0 \in U(x^\star, r_\star)$, the sequence $\{x_n\}$ generated by the modified Newton method (9.1) is well-defined, remains in $U(x^\star, r_\star)$ for all $n \geq 0$ and converges to $x^\star$ with order at least $1 + p$. Moreover, the following error estimates hold:*

$$|e_{n+1}| \leq \frac{|e_n|^{1+p}}{r_\star{}^p} \quad (n \geq 0). \tag{9.18}$$

Proof. We will show the theorem by induction. In order to simply the proof, we define functions $g(x)$ and $h(x)$ as follows:

$$g(x) = f[\underbrace{x^\star, x^\star, \ldots, x^\star}_{m}, x], \quad h(x) = f[\underbrace{x^\star, x^\star, \ldots, x^\star}_{m}, x, x]. \tag{9.19}$$

Using Lemma 9.1.3, we get that

$$f(x_0) = g(x_0)e_0^m \quad \text{and} \quad f'(x_0) = [h(x_0)e_0 + mg(x_0)]e_0^{m-1}. \tag{9.20}$$

By (9.1) and (9.20), we have that

$$e_1 = e_0 - \frac{mg(x_0)e_0^m}{[h(x_0)e_0 + mg(x_0)]e_0^{m-1}}$$
$$= e_0 - \frac{mg(x_0)e_0}{h(x_0)e_0 + mg(x_0)} = \frac{(mg(x^\star))^{-1}h(x_0)e_0}{(mg(x^\star))^{-1}[h(x_0)e_0 + mg(x_0)]}e_0. \tag{9.21}$$

Here, we suppose that $h(x_0)e_0 + mg(x_0) \neq 0$, which will be shown below.

In fact, by the definition of divided differences, we have that

$$h(x_0)e_0 = f[\underbrace{x^\star, x^\star, \ldots, x^\star}_{m-1}, x_0, x_0] - g(x_0). \tag{9.22}$$

Then, we obtain from (9.13) and (9.22) that

$$|1 - (mg(x^\star))^{-1}[h(x_0)e_0 + mg(x_0)]| = |(mg(x^\star))^{-1}[h(x_0)e_0 + mg(x_0) - mg(x^\star)]|$$
$$= (m-1)!|f^{(m)}(x^\star)^{-1}\big(f[\underbrace{x^\star, x^\star, \ldots, x^\star}_{m-1}, x_0, x_0] - g(x^\star) + (m-1)[g(x_0) - g(x^\star)]\big)|.$$

$$\tag{9.23}$$

Using Lemma 9.1.4 yields

$$f[\underbrace{x^\star, x^\star, \ldots, x^\star}_{m-1}, x_0, x_0] = \int_0^1 \cdots \int_0^1 f^{(m)}\Big(x^\star + e_0 \prod_{i=1}^{m-1} t_i\Big) \prod_{i=1}^m (t_i^{m-i} dt_i), \tag{9.24}$$

$$g(x_0) = \int_0^1 \cdots \int_0^1 f^{(m)}\Big(x^\star + e_0 \prod_{i=1}^m t_i\Big) \prod_{i=1}^m (t_i^{m-i} dt_i), \tag{9.25}$$

$$g(x^\star) = \int_0^1 \cdots \int_0^1 f^{(m)}(x^\star) \prod_{i=1}^m (t_i^{m-i} dt_i). \tag{9.26}$$

Substituting (9.24)–(9.26) into (9.23), using condition (1.11), $x_0 \in U(x^\star, r_\star)$ and the definition of $r_\star$, we get that

$$|1 - (mg(x^\star))^{-1}[h(x_0)e_0 + mg(x_0)]|$$
$$= (m-1)!\Big|\int_0^1 \cdots \int_0^1 \big(f^{(m)}(x^\star)^{-1}(f^{(m)}(x^\star + e_0 \prod_{i=1}^{m-1} t_i) - f^{(m)}(x^\star))\big) \prod_{i=1}^m (t_i^{m-i} dt_i)$$
$$+ (m-1)f^{(m)}(x^\star)^{-1}(f^{(m)}(x^\star + e_0 \prod_{i=1}^m t_i) - f^{(m)}(x^\star)) \prod_{i=1}^m (t_i^{m-i} dt_i))\Big|$$
$$\leq (m-1)!\Big(\int_0^1 \cdots \int_0^1 |f^{(m)}(x^\star)^{-1}(f^{(m)}(x^\star + e_0 \prod_{i=1}^{m-1} t_i) - f^{(m)}(x^\star))| \prod_{i=1}^m (t_i^{m-i} dt_i)$$
$$+ (m-1)\int_0^1 \cdots \int_0^1 |f^{(m)}(x^\star)^{-1}(f^{(m)}(x^\star + e_0 \prod_{i=1}^m t_i) - f^{(m)}(x^\star))| \prod_{i=1}^m (t_i^{m-i} dt_i))$$
$$\leq (m-1)!\Big(K_0|e_0|^p \int_0^1 \cdots \int_0^1 \prod_{i=1}^{m-1} (t_i^{m+p-i} dt_i) dt_m$$
$$+ (m-1)K_0|e_0|^p \int_0^1 \cdots \int_0^1 \prod_{i=1}^m (t_i^{m+p-i} dt_i))$$
$$= (m-1)!\Big(K_0|e_0|^p \prod_{i=1}^{m-1} \frac{1}{m+p+1-i} + (m-1)K_0|e_0|^p \prod_{i=1}^m \frac{1}{m+p+1-i}\Big)$$
$$= (m-1)!(m+p)K_0|e_0|^p \prod_{i=1}^m \frac{1}{m+p+1-i}$$
$$< (m-1)!(m+p)K_0 r_\star^p \prod_{i=1}^m \frac{1}{m+p+1-i} = \frac{(m+p)K_0}{K+(m+p)K_0} < 1.$$

$$\tag{9.27}$$

By Banach's lemma, $h(x_0)e_0 + mg(x_0) \neq 0$ and

$$\left| \left((mg(x^\star))^{-1}[h(x_0)e_0 + mg(x_0)] \right)^{-1} \right|$$

$$\leq \frac{1}{1 - (m-1)!(m+p)K_0|e_0|^p \prod_{i=1}^m \dfrac{1}{m+p+1-i}}$$

$$< \frac{1}{1 - (m-1)!(m+p)K_0 r_\star^p \prod_{i=1}^m \dfrac{1}{m+p+1-i}}. \tag{9.28}$$

On the other hand, using (9.22), (9.24), (9.25) and condition (9.10), we get that

$$|(mg(x^\star))^{-1}h(x_0)e_0| = (m-1)! \left| \int_0^1 \cdots \int_0^1 f^{(m)}(x^\star)^{-1} \left(f^{(m)}(x^\star + e_0 \prod_{i=1}^{m-1} t_i) \right. \right.$$

$$\left. \left. - f^{(m)}(x^\star + e_0 \prod_{i=1}^m t_i) \right) \prod_{i=1}^m (t_i^{m-i} dt_i) \right|$$

$$\leq (m-1)! \int_0^1 \cdots \int_0^1 \left| f^{(m)}(x^\star)^{-1} \left(f^{(m)}(x^\star + e_0 \prod_{i=1}^{m-1} t_i) \right. \right.$$

$$\left. \left. - f^{(m)}(x^\star + e_0 \prod_{i=1}^m t_i) \right) \right| \prod_{i=1}^m (t_i^{m-i} dt_i)$$

$$\leq (m-1)! K |e_0|^p \int_0^1 \cdots \int_0^1 \left(\prod_{i=1}^{m-1} (t_i^{m+p-i} dt_i) \right)(1-t_m)^p dt_m$$

$$= (m-1)! K |e_0|^p \left(\prod_{i=1}^m \frac{1}{m+p+1-i} \right)$$

$$< (m-1)! K r_\star^p \left(\prod_{i=1}^m \frac{1}{m+p+1-i} \right). \tag{9.29}$$

Using (9.21), (9.28), (9.29) and the definition of $r_\star$, we get in turn that

$$|e_1| \leq |e_0| \frac{(m-1)! K |e_0|^p \left(\prod_{i=1}^m \dfrac{1}{m+p+1-i} \right)}{1 - (m-1)!(m+p)K_0|e_0|^p \prod_{i=1}^m \dfrac{1}{m+p+1-i}}$$

$$\leq |e_0| \frac{(m-1)! K r_\star^p \left(\prod_{i=1}^m \dfrac{1}{m+p+1-i} \right)}{1 - (m-1)!(m+p)K_0 r_\star^p \prod_{i=1}^m \dfrac{1}{m+p+1-i}} = |e_0| < r_\star. \tag{9.30}$$

Hence, we deduce $x_1 \in U(x^\star, r_\star)$. Furthermore, using (9.30) and $x_0 \in U(x^\star, r_\star)$, we have that

$$|e_1| \leq |e_0| \frac{(m-1)! K \dfrac{|e_0|^p}{r_\star^p} r_\star^p \left(\prod_{i=1}^m \dfrac{1}{m+p+1-i} \right)}{1 - (m-1)!(m+p)K_0 r_\star^p \prod_{i=1}^m \dfrac{1}{m+p+1-i}} = \frac{|e_0|^{1+p}}{r_\star^p}. \tag{9.31}$$

That is, error estimate (9.18) is true for $n = 0$.

Now we suppose the point x_k generated by the modified Newton's method (9.1) is well-defined and $x_k \in U(x^\star, r_\star)$ holds for some fixed integer $k \geq 1$. Then, all statements of (9.20)–(9.31) will be satisfied if we replace x_0 by x_k, x_1 by x_{k+1}, $|e_0|$ by $|e_k|$ and $|e_1|$ by $|e_{k+1}|$. Therefore, x_{k+1} is well-defined, $x_{k+1} \in U(x^\star, r_\star)$ and

$$|e_{k+1}| \leq \frac{|e_k|^{1+p}}{r_\star^p}. \tag{9.32}$$

By induction, for any initial point $x_0 \in U(x^\star, r_\star)$, the sequence $\{x_n\}$ generated by the modified Newton method (9.1) is well-defined, remains in $U(x^\star, r_\star)$ and the error estimates (9.18) are true. Using (9.18), it is obvious that sequence $\{x_n\}$ converges to $x^\star$ with order $1 + p$ at least. The proof of Theorem 9.1.1 is finished. $\qquad\square$

In particular, by choosing $p = 1$, we have the following corollary for the modified Newton method (9.1):

Corollary 9.1.1. *Let $f : D \to \mathbb{R}$ be in $C^m(D)$ for $D \subset \mathbb{R}$, open and convex. Let $x^\star$ be a zero of f with multiplicity m ($m \geq 1$). Suppose that the Lipschitz continuous condition*

$$|f^{(m)}(x^\star)^{-1}(f^{(m)}(x) - f^{(m)}(y))| \leq K|x - y|, \quad \forall x, y \in D, \text{ for some } K > 0, \tag{9.33}$$

and the center-Lipschitz continuous condition:

$$|f^{(m)}(x^\star)^{-1}(f^{(m)}(x) - f^{(m)}(x^\star))| \leq K_0|x - x^\star|, \forall x \in D, \text{ for some } K_0 > 0, \tag{9.34}$$

are satisfied. Assume:

$$U(x^\star, r_\star) \subseteq D,$$

where

$$r_\star = \frac{(m+1)m}{K + (m+1)K_0}. \tag{9.35}$$

Then, for any initial point $x_0 \in U(x^\star, r_\star)$, the sequence $\{x_n\}$ generated by the modified Newton method (9.1) is well-defined, remains in $U(x^\star, r_\star)$ and converges to $x^\star$ quadratically. Moreover, the following estimates hold:

$$|e_{n+1}| \leq \frac{|e_n|^2}{r_\star} \quad (n \geq 0). \tag{9.36}$$

Remark 9.1.1. For the special case of $m = 1$ and $K = K_0$, Corollary 9.1.1 reduces to the known result of convergence radius of Newton's method given by Wang in Ref. [426] and Traub in Ref. [420], respectively.

Remark 9.1.2. Note that Theorem 9.1.1 reduces to the earlier work in Ref. [95] for the case of $m = 1$.

We give now some examples to show the application of our theoretical results.

Example 9.1.1. Let $D = [-1, 1]$ and define function f on D by

$$f(x) = x(e^x - 1). \tag{9.37}$$

Table 9.1   Comparison Table

n	$e_{n+1}(K_0 = K)$	$e_{n+1}(K_0 \neq K)$
0	0.144975031	0.097792212
1	0.076176259	0.023380448
2	0.021031609	0.001336441
3	0.001603165	0.000004367

We shall show that $x^\star = 0$ is a zero of f with multiplicity $m = 2$. By (9.37), $f(0) = 0$, $f'(x) = (x+1)e^x - 1$, so $f'(0) = 0$, and $f''(x) = (x+2)e^x$ and $f''(0) = 2$. We shall compute constants K_0 and K. Using (1.11) we get that

$$f''(x) - f''(x^\star) = (x+2)e^x - 2 = xe^x + 2(e^x - 1). \tag{9.38}$$

But, we have

$$\begin{aligned}
|e^x - 1| &= |x + \frac{x^2}{2!} + \cdots + \frac{x^n}{n!} + \cdots| \\
&= |(1 + \frac{x}{2!} + \cdots + \frac{x^{n-1}}{n!} + \cdots)x| \\
&\leq |1 + \frac{1}{2!} + \cdots + \frac{1}{n!} + \cdots||x| = (e-1)|x|.
\end{aligned} \tag{9.39}$$

In view of (9.38) and (9.39), we get (1.11) holds for

$$K_0 = \frac{3e - 2}{2} \quad \text{and} \quad p = 1. \tag{9.40}$$

We also have $f'''(x) = (x+3)e^x$. That is,

$$K = 2e. \tag{9.41}$$

Using (9.9), (9.40) and (9.41), we get that

$$r_\star = \begin{cases} \dfrac{3}{2K} = \dfrac{3}{4e} = 0.275909581, & \text{if } K_0 = K, \\[2mm] \dfrac{6}{K + 3K_0} = \dfrac{12}{13e - 6} = 0.409030525, & \text{if } K_0 \neq K. \end{cases} \tag{9.42}$$

Finally, in order for us to compare the error estimates, using (9.18) and (9.42), let us choose $x_0 = 0.2$. Then, we obtain $e_0 = 0.2$. We can provide Comparison Table 9.1.

Example 9.1.2. Let $D = [\frac{5}{4}, \frac{13}{4}]$ and define function f on D by

$$f(x) = \left(\frac{1}{10}x - \frac{1}{15}x^{\frac{3}{2}}\right)^2. \tag{9.43}$$

We shall first show $x^\star = \dfrac{9}{4}$ and $m = 2$. Using (9.43), we obtain that

$$f(x^\star) = 0$$

and

$$f'(x) = 2(\frac{1}{10}x - \frac{1}{15}x^{\frac{3}{2}})'(\frac{1}{10}x - \frac{1}{15}x^{\frac{3}{2}}) = \frac{1}{5}(1 - x^{\frac{1}{2}})(\frac{1}{10}x - \frac{1}{15}x^{\frac{3}{2}}).$$

Hence, we deduce that

$$f'(x^\star) = 0,$$

$$f''(x) = -\frac{1}{10}x^{-\frac{1}{2}}(\frac{1}{10}x - \frac{1}{15}x^{\frac{3}{2}}) + \frac{1}{50}(1 - x^{\frac{1}{2}})^2 = -\frac{1}{20}x^{\frac{1}{2}} + \frac{2x}{75} + \frac{1}{50} \tag{9.44}$$

and

$$f''(x^\star) = \frac{1}{200}. \tag{9.45}$$

Next we shall find an upper bound of the form $K^2|x - y|$ for $|f''(x) - f''(y)|^2$. In view of (9.44), we obtain that

$$|f''(x) - f''(y)|^2 = |-\frac{1}{20}x^{\frac{1}{2}} + \frac{2x}{75} + \frac{1}{50} + \frac{1}{20}y^{\frac{1}{2}} - \frac{2y}{75} - \frac{1}{50}|^2$$
$$= |\frac{1}{20}(y^{\frac{1}{2}} - x^{\frac{1}{2}}) - \frac{2}{75}(y - x)|^2$$

$$\leq (\frac{1}{20})^2|y^{\frac{1}{2}} - x^{\frac{1}{2}}|^2 + (\frac{2}{75})^2|y - x|^2 + 2 \times \frac{2}{75} \times \frac{1}{20}|y - x||y^{\frac{1}{2}} - x^{\frac{1}{2}}|. \tag{9.46}$$

Let

$$\alpha = \frac{\sqrt{5}}{5} \approx 0.447213596. \tag{9.47}$$

Then, for all $x, y \in D$, we have that

$$\alpha|x^{\frac{1}{2}} + y^{\frac{1}{2}}| \geq 1 \Rightarrow |x^{\frac{1}{2}} - y^{\frac{1}{2}}| \leq \alpha|x^{\frac{1}{2}} - y^{\frac{1}{2}}||x^{\frac{1}{2}} + y^{\frac{1}{2}}|$$
$$\Rightarrow |x^{\frac{1}{2}} - y^{\frac{1}{2}}| \leq \alpha|(x^{\frac{1}{2}})^2 - (y^{\frac{1}{2}})^2| \tag{9.48}$$
$$\Rightarrow |x^{\frac{1}{2}} - y^{\frac{1}{2}}| \leq \alpha|x - y|.$$

We also have that

$$|x - y| \leq 2 \quad \text{for all} \quad x, y \in D. \tag{9.49}$$

Using (9.45)–(9.48), we get that

$$|f''(x^\star)^{-1}[f''(x) - f''(y)]|^2$$
$$\leq 40000\,[(\frac{1}{20})^2 \times \frac{1}{5} + (\frac{2}{75})^2 + 2 \times \frac{2}{75} \times \frac{1}{20} \times \frac{1}{\sqrt{5}}]\,|x - y|^2 \tag{9.50}$$
$$\leq K^2\,|x - y|,$$

where

$$K^2 = 80000\,[(\frac{1}{20})^2 \times \frac{1}{5} + (\frac{2}{75})^2 + 2 \times \frac{2}{75} \times \frac{1}{20} \times \frac{1}{\sqrt{5}}].$$

Hence, we deduce that

$$K = 13.86702733 \tag{9.51}$$

and

$$|f''(x^\star)^{-1}[f''(x) - f''(y)]| \leq K\,|x - y|^{\frac{1}{2}} \quad \text{for all} \quad x, y \in D. \tag{9.52}$$

That is,

$$p = \frac{1}{2}.$$
(9.53)

Hence, (9.10) holds for K, p given by (9.51) and (9.53), respectively.
We shall determine constant K_0 in a similar way. Let

$$\gamma = \frac{3 - \sqrt{5}}{2} \approx 0.381966011.$$
(9.54)

We have in turn for all $x \in D$ that

$$\gamma \left| \left(\frac{5}{4}\right)^{\frac{1}{2}} + \left(\frac{9}{4}\right)^{\frac{1}{2}} \right| \geq 1 \Rightarrow \gamma \left| x^{\frac{1}{2}} + \left(\frac{9}{4}\right)^{\frac{1}{2}} \right| \geq 1$$
$$\Rightarrow \gamma \left| x^{\frac{1}{2}} + \left(\frac{9}{4}\right)^{\frac{1}{2}} \right| \left| x^{\frac{1}{2}} - \left(\frac{9}{4}\right)^{\frac{1}{2}} \right| \geq \left| x^{\frac{1}{2}} - \left(\frac{9}{4}\right)^{\frac{1}{2}} \right|$$
$$\Rightarrow \left| x^{\frac{1}{2}} - (x^\star)^{\frac{1}{2}} \right| \leq \gamma |x - x^\star|.$$
(9.55)

We also have that

$$|x - x^\star| \leq \frac{13}{4} - \frac{9}{4} = 1.$$
(9.56)

In view of (9.50) and (9.54)–(9.56), we have that

$$|f''(x^\star)^{-1}[f''(x) - f''(x^\star)]|^2$$
$$\leq 40000 \left[\left(\frac{1}{20}\right)^2 \gamma^2 + \left(\frac{2}{75}\right)^2 + 2 \times \frac{2}{75} \times \frac{1}{20} \gamma \right] |x - x^\star|^2 \leq K_0^2 |x - x^\star|,$$
(9.57)

where

$$K_0^2 = 40000 \left[\left(\frac{1}{20}\right)^2 \gamma^2 + \left(\frac{2}{75}\right)^2 + 2 \times \frac{2}{75} \times \frac{1}{20} \gamma \right].$$

Then, we deduce that

$$K_0 = 9.152493017$$
(9.58)

and

$$|f''(x^\star)^{-1}[f''(x) - f''(x^\star)]| \leq K_0 |x - x^\star|^{\frac{1}{2}}$$
(9.59)

holds for all $x \in D$. Hence, (1.11) holds for p, K_0 given by (9.53) and (9.59), respectively. Using (9.9), we get that

$$r_\star = \begin{cases} 0.005969799 \text{ if } K_0 = K \\ 0.010412606 \text{ if } K_0 \neq K. \end{cases}$$
(9.60)

We also provide an example where only (1.11) holds.

Example 9.1.3. Let $D = \mathbb{R}$. Define function f on D by

$$f(x) = \int_0^x G(t)\, dt,$$
(9.61)

where

$$G(x) = \int_0^x \left(1 + x \sin \frac{\pi}{x}\right) dx.$$
(9.62)

Then, we get for $x^\star$ that $f(x^\star) = 0$,

$$f'(x) = G(x)$$

and

$$f'(x^\star) = G(x^\star) = 0.$$

Moreover, we have that

$$f''(x) = \begin{cases} 1 + x \sin \dfrac{\pi}{x} & \text{if } x \neq 0 \\ 1 & \text{if } x = 0 \end{cases} \tag{9.63}$$

and

$$f''(x^\star) = 1. \tag{9.64}$$

Hence, we conclude $m = 2$. We can then obtain from (9.63) and (9.64) that

$$|f''(x^\star)^{-1}[f''(x) - f''(x^\star)]| = |x \sin \frac{\pi}{x}| \leq |x| = |x - x^\star| \tag{9.65}$$
$$\text{for all} \quad x \in D.$$

That is (1.11) holds for $p = 1$ and $K_0 = 1$.

We shall show that $K > 0$ does not exist. Assume that there exists $K > 0$ such that

$$|f''(x^\star)^{-1}[f''(x) - f''(y)]| \leq |x - y| \quad \text{for all} \quad x, y \in D. \tag{9.66}$$

Let $j \geq 1$ be a fixed integer. Then, for $x = \dfrac{1}{j}$ and $y = \dfrac{2}{2j + 1}$, we get in turn that

$$|f''(x^\star)^{-1}[f''(x) - f''(y)]| = |x \sin \frac{\pi}{x} - y \sin \frac{\pi}{y}|$$
$$= |\frac{1}{j} \sin(j\pi) - \frac{2}{2j + 1} \sin \frac{(2j + 1)\pi}{2}| = \frac{1}{2j + 1} \tag{9.67}$$
$$\leq K|x - y| = K|\frac{1}{j} - \frac{2}{2j + 1}| = K \frac{1}{j(2j + 1)},$$

which implies

$$K \geq j \quad \text{for all} \quad j \geq 1. \tag{9.68}$$

But this is impossible, since $K > 0$ is a constant.

Example 9.1.3 motivates us to find the radius of convergence in the case when only (1.11) holds.

Theorem 9.1.2. *Let $f : D \to \mathbb{R}$ be in $C^m(D)$ for $D \subseteq \mathbb{R}$, non-empty, open and convex. Let $x^\star$ be a zero of f with multiplicity m ($m \geq 1$). Suppose that the center-Hölder continuous condition (1.11) holds and set*

$$r_{\star\star} = \left(\frac{\prod_{i=1}^{m}(m + p + 1 - i)}{K_0(m - 1)!(m + 2p + 2)} \right)^{\frac{1}{p}}. \tag{9.69}$$

Further, assume:

$$U(x^\star, r_{\star\star}) \subseteq D. \tag{9.70}$$

Then, for any initial point $x_0 \in U(x^\star, r_{\star\star})$, the sequence $\{x_n\}$ generated by the modified Newton method (9.1) is well-defined, remains in $U(x^\star, r_{\star\star})$ for all $n \geq 0$ and converges to $x^\star$ with order at least $1+p$. Moreover, the following error estimates hold:

$$|e_{n+1}| \leq \frac{|e_n|^{1+p}}{r_{\star\star}{}^p} \quad (n \geq 0). \tag{9.71}$$

Proof. In view of estimate (9.27), we have that

$$\left| \int_0^1 \cdots \int_0^1 f^{(m)}(x^\star)^{-1} (f^{(m)}(x^\star + e_0 \prod_{i=1}^{m-1} t_i) - f^{(m)}(x^\star)) \right| \prod_{i=1}^m (t_i^{m-i} dt_i)$$

$$\leq K_0 |e_0|^p \prod_{i=1}^{m-1} \frac{1}{m+p+1-i} \tag{9.72}$$

$$= (p+1) K_0 |e_0|^p \prod_{i=1}^m \frac{1}{m+p+1-i}$$

and

$$\left| \int_0^1 \cdots \int_0^1 f^{(m)}(x^\star)^{-1} (f^{(m)}(x^\star + e_0 \prod_{i=1}^m t_i) - f^{(m)}(x^\star)) \right| \prod_{i=1}^m (t_i^{m-i} dt_i)$$

$$\leq K_0 |e_0|^p \prod_{i=1}^m \frac{1}{m+p+1-i}. \tag{9.73}$$

Moreover by adding and subtracting $f^{(m)}(x^\star)$ inside the first parenthesis in (9.29) we obtain by (9.72) and (9.73) that

$$|(mg(x^\star))^{-1} h(x_0) e_0| \leq (m-1)!(p+2) K_0 |e_0|^p \prod_{i=1}^m \frac{1}{m+p+1-i}. \tag{9.74}$$

Using (9.21), (9.28), (9.74) and (9.69), we get in turn that

$$|e_1| \leq |e_0| \frac{(m-1)!(p+2) K_0 |e_0|^p \prod_{i=1}^m \frac{1}{m+p+1-i}}{1 - (m-1)!(m+p) K_0 |e_0|^p \prod_{i=1}^m \frac{1}{m+p+1-i}}$$

$$\leq |e_0| \frac{(m-1)!(p+2) K_0 \alpha^p \prod_{i=1}^m \frac{1}{m+p+1-i}}{1 - (m-1)!(m+p) K_0 \alpha^p \prod_{i=1}^m \frac{1}{m+p+1-i}} = |e_0| < r_{\star\star}. \tag{9.75}$$

Hence, we deduce that $x_1 \in U(x^\star, r_{\star\star})$. Moreover, from (9.69), (9.75) and $x_0 \in U(x^\star, r_{\star\star})$, we have that

$$|e_1| \leq |e_0| \frac{(m-1)!(p+2) K_0 \dfrac{|e_0|^p}{\alpha^p} \alpha^p \prod_{i=1}^m \frac{1}{m+p+1-i}}{1 - (m-1)!(m+p) K_0 \alpha^p \prod_{i=1}^m \frac{1}{m+p+1-i}} = \frac{|e_0|^{1+p}}{r_{\star\star}^p}.$$

That is error estimate (3) is true for $n = 0$. The rest follows as the proof of Theorem 9.1.1. That completes the proof of the theorem. $\qquad\square$

Remark 9.1.3. Returning back to Example 9.1.3 and using (9.69), we deduce that $r_{\star\star} = 1$.

9.2 Weak Convergence Conditions

We provide a semi-local convergence analysis for a Newton-like method under weak conditions for approximating (1.54). In particular, we only assume that the Gateaux derivative of the operator involved is hemicontinuous. We introduce the Newton-like method (NLM):

$$x_{n+1} = x_n - A_n^{-1}(F(x_n) + G(x_n)), \quad (n \geq 0), \ (x_0 \in D) \tag{9.76}$$

to generate a sequence $\{x_n\}$ approximating x^*, where

$$A_n = A(z_n) = F'(z_n) + [z_n, z_{n-1}; G], \quad (n \geq 0), \tag{9.77}$$

where $\{z_n\}(n \geq -1)$ is a sequence in D. Special cases of method (NLM) are

(1) $G = 0$ on D. Then, we obtain Zincenko's iteration (see Refs. [100, 106, 188, 444, 443]).
(2) $z_n = x_n$ $(n \geq -1)$. Then, we obtain Catinas' iteration (see Refs. [100, 106, 182]).
(3) $G = 0$ on D and $z_n = x_n$. Then, we obtain Newton's method (see Refs. [100, 106, 290]).
(4) $F = 0$ on D and $z_n = x_n$. Then, we obtain Secant method (see Refs. [100, 106, 120, 290]).

Other choices have been given in Refs. [100, 106, 369] where semi-local as well as local convergence results have been given, under more general Lipschitz-type conditions on the Fréchet-derivative F' of operator F than in (1)–(4).

In Refs. [94, 97, 120] we provided convergence results for Newton's method, when F' is only continuous on D. This way we extended further the application of Newton's method. In this section, we extend even further the application of Newton-like method (9.76), by assuming that F' is Gateaux differentiable and piecewise hemicontinuous on a subset of D, whereas G is only continuous on D. Finally, we provide a class of equations where earlier results cannot apply to solve equations, but ours can.

We need the following definition of hemicontinuity for an operator.

Definition 9.2.1. An operator $H : D \to Y$ is said to hemicontinuous at each $x \in D$, if for all $\varepsilon > 0$, there exists $\delta > 0$ (δ depending on ε) such that if $|t| \leq \delta$ and $z \in X$, then $\|H(x + tz) - H(x)\| < \varepsilon$.

We can now show the main semi-local convergence result for (NLM):

Theorem 9.2.1. *Let $F : D \subseteq X \to Y$ be a Gateaux-differentiable operator at each point in some neighborhood of $x_0 \in D$ and $G : D \subseteq X \to Y$ a continuous operator with divided difference $[x, y; G]$ satisfying*

$$[x, y; G](x - y) = G(x) - G(y), \quad for \ all \ x, y \in D. \tag{9.78}$$

Moreover, assume that there exist $z_{-1}, z_0 \in D$ and $r > 0$ such that $A_0^{-1} \in L(Y, X)$, $\|A_0^{-1}\| \leq \alpha$,

$$\|F'(x) - F'(z_0)\| \leq \varepsilon_0 \quad \text{for all} \quad x \in U(x_0, r), \tag{9.79}$$

$$\|F'(x) - F'(y)\| \leq \varepsilon \quad \text{for all} \quad x, y \in U(x_0, r), \tag{9.80}$$

$$\|[x, y; G] - [u, v; G]\| \leq \varepsilon_1 \quad \text{for all} \quad x, y, u, v \in U(x_0, r); \tag{9.81}$$

operator $F'(x)$ is piecewise hemicontinuous for each $x \in U(x_0, r)$; for

$$\beta = \frac{\alpha}{1 - \alpha(\varepsilon_0 + \varepsilon_1)}, \quad \eta = \alpha\|F(x_0) + G(x_0)\|, \tag{9.82}$$

$$0 < \gamma = \beta(\varepsilon + \varepsilon_1) < 1, \tag{9.83}$$

$$\frac{\eta}{1 - \gamma} < r \tag{9.84}$$

and

$$\overline{U}(x_0, r) \subseteq D. \tag{9.85}$$

Then, iteration $\{x_n\}$ generated by the (NLM) is well-defined, remains in $U(x_0, r)$ for any sequence $\{z_n\} \subseteq U(x_0, r)$, and converges to a solution x^* of equation $F(x) + G(x) = 0$, which is unique in $\overline{U}(x_0, r)$. Moreover, the following estimates hold for all $n \geq 0$:

$$\|x_n - x^*\| \leq \frac{\gamma^n}{1 - \gamma}\eta. \tag{9.86}$$

Proof.   (**Existence of x^***) We shall show using induction that

$$x_n \in U(x_0, r) \tag{9.87}$$

and

$$\|x_{n+1} - x_n\| \leq \gamma\|x_n - x_{n-1}\|. \tag{9.88}$$

Using (9.76) for $n = 0$, we have by (9.82)

$$\|x_1 - x_0\| \leq \|A_0^{-1}(F(x_0) + G(x_0))\| \leq \eta < r,$$

which implies $x_1 \in U(x_0, r)$. That is estimate (9.87) holds for $n = 1$. We also have in turn that

$$
\begin{aligned}
F(x_1) + G(x_1) &= F(x_1) + G(x_1) - F(x_0) - G(x_0) - A_0(x_1 - x_0) \\
&= F(x_1) - F(x_0) - F'(z_0)(x_1 - x_0) - [z_0, z_{-1}; G](x_1 - x_0) + G(x_1) - G(x_0) \\
&= \int_0^1 [F'(x_0 + t(x_1 - x_0)) - F'(z_0)](x_1 - x_0)\,dt \\
&\quad + ([x_1, x_0; G] - [z_0, z_{-1}; G])(x_1 - x_0).
\end{aligned}
$$

$$\tag{9.89}$$

Hence, by (9.80) and (9.81)

$$\|F(x_1) + G(x_1)\| \leq \|\int_0^1 [F'(x_0 + t(x_1 - x_0)) - F'(z_0)](x_1 - x_0)dt\| \tag{9.90}$$
$$+\|([x_1, x_0; G] - [z_0, z_{-1}; G])(x_1 - x_0)\| \leq (\varepsilon + \varepsilon_1)\|x_1 - x_0\|.$$

Moreover, using (9.79), (9.81) and (9.83), we get

$$\|A_0^{-1}\|\|A_0 - A_1\| \leq \|A_0^{-1}\|\|(F'(z_1) - F'(z_0)) + ([z_1, z_0; G] - [z_0, z_{-1}; G])\| \tag{9.91}$$
$$\leq \alpha(\varepsilon_0 + \varepsilon_1) < 1.$$

It follows from (9.91) and the Banach lemma on invertible operators that A_1^{-1} exists and

$$\|A_1^{-1}\| \leq \beta. \tag{9.92}$$

In view of (9.76) for $n = 1$, we get

$$\|x_2 - x_1\| \leq \|A_1^{-1}(F(x_1) + G(x_1))\| \tag{9.93}$$
$$\leq \|A_1^{-1}\|\|F(x_1) + G(x_1)\| \leq \gamma\|x_1 - x_0\|,$$

which shows (9.88) for $n = 1$.

Let us assume (9.87) and (9.88) hold for all $k \leq n$. We have as in (9.89):

$$F(x_n) + G(x_n) = F(x_n) + G(x_n) - F(x_{n-1}) - G(x_{n-1}) - A_{n-1}(x_n - x_{n-1})$$
$$= F(x_n) - F(x_{n-1}) - F'(z_{n-1})(x_n - x_{n-1}) - [z_{n-1}, z_{n-2}; G](x_n - x_{n-1})$$
$$+G(x_n) - G(x_{n-1})$$
$$= \int_0^1 [F'(x_{n-1} + t(x_n - x_{n-1})) - F'(z_{n-1})](x_n - x_{n-1})dt$$
$$+([x_n, x_{n-1}; G] - [z_{n-1}, z_{n-2}; G])(x_n - x_{n-1}).$$

Hence, we deduce that

$$\|F(x_n) + G(x_n)\| \leq (\varepsilon + \varepsilon_1)\|x_n - x_{n-1}\|. \tag{9.94}$$

We also have that

$$\|A_0^{-1}\|\|A_0 - A_n\| \leq \|A_0^{-1}\|\|(F'(z_n) - F'(z_0)) + ([z_n, z_{n-1}; G] - [z_0, z_{-1}; G])\|$$
$$\leq \alpha(\varepsilon_0 + \varepsilon_1) < 1. \tag{9.95}$$

In view of (9.95) and the Banach lemma A_n^{-1} exists and

$$\|A_n^{-1}\| \leq \beta. \tag{9.96}$$

Using (9.76), (9.94) and (9.96), we get

$$\|x_{n+1} - x_n\| \leq \|A_n^{-1}\|\|F(x_n) + G(x_n)\|$$
$$\leq \beta(\varepsilon + \varepsilon_1)\|x_n - x_{n-1}\| = \gamma\|x_n - x_{n-1}\|,$$

which shows (9.88) for all $n \geq 0$. We also have that

$$\|x_{n+1} - x_0\| \leq \|x_{n+1} - x_n\| + \|x_n - x_{n-1}\| + \cdots + \|x_1 - x_0\|$$
$$\leq (\gamma^n + \gamma^{n-1} + \cdots + 1)\eta = \frac{1 - \gamma^{n+1}}{1 - \gamma}\eta \leq \frac{\eta}{1 - \gamma} < r, \tag{9.97}$$

so, $x_{n+1} \in U(x_0, r)$, which implies (9.87) holds for all $n \geq 0$. The induction for (9.87) and (9.88) is now complete.

Let $n \geq 2$ and $m \geq 0$, then we have in turn by (9.88):

$$\|x_{n+m} - x_n\| \leq \|x_{n+m} - x_{n+m-1}\| + \|x_{n+m-1} - x_{n+m-2}\| + \cdots + \|x_{n+1} - x_n\|$$
$$\leq (\gamma^{n+m-1} + \gamma^{n+m-2} + \cdots + \gamma^n)\eta = \frac{1 - \gamma^m}{1 - \gamma}\gamma^n \eta \leq \frac{\gamma^n}{1 - \gamma}\eta. \tag{9.98}$$

In view of estimate (9.98), sequence $\{x_n\}$ is Cauchy in a Banach space X and as such it converges to some $x^* \in \overline{U}(x_0, r)$ (since $x^* \in \overline{U}(x_0, r)$ is a closed set). Moreover, we have that

$$\|F(x_n) + G(x_n)\| \leq \|A_n\|\|x_{n+1} - x_n\|$$
$$\leq (\|A_n - A_0\| + \|A_0\|)\|x_{n+1} - x_n\| \leq (\|A_0\| + \varepsilon_0 + \varepsilon_1)\|x_{n+1} - x_n\|. \tag{9.99}$$

By letting $n \to \infty$ in (9.99), we obtain $F(x^*) + G(x^*) = 0$.

(**Uniqueness of** x^*) Let y^* be another solution of equation $F(x) + G(x) = 0$ in $U(x_0, r)$. We have in turn

$$\begin{aligned}
x^* - y^* &= x^* - y^* - A_0^{-1}[F(x^*) + G(x^*) - F(y^*) - G(y^*)] \\
&= A_0^{-1}[F(y^*) - F(x^*) - F'(z_0)(y^* - x^*) + G(y^*) - G(x^*) \\
&\quad -[z_0, z_{-1}; G](y^* - x^*)] \\
&= A_0^{-1}\left(\int_0^1 [F'(x^* + t(y^* - x^*)) - F'(z_0)](y^* - x^*)dt \right. \\
&\quad \left. +([y^*, x^*; G] - [z_0, z_{-1}; G])(y^* - x^*) \right).
\end{aligned} \tag{9.100}$$

Hence, we have that $\|x^* - y^*\| \leq \gamma\|x^* - y^*\| < \|x^* - y^*\|$ and we deduce that $x^* = y^*$. That completes the proof of the theorem. $\square$

We shall provide some classes of operators where the results of this section can apply.

Lemma 9.2.1. *Let $q : \mathbb{R} \to \mathbb{R}$ be a continuously differentiable function and q' be a bounded function on $\mathbb{R}$. Define function $Q : I \to I, 1 < p < \infty$ for $I = L^p[a, b]$ by:*

$$Q(y)x = \int_a^x q(y(s))ds.$$

Then, function Q is Gateaux differentiable everywhere and $Q'(y)$ is a hemicontinuous function.

Proof. We shall make use of the dominated convergence theorem and the mean-value theorem. Let $y \in I$ and define operator $P(y) : I \to I$ by

$$P(y)(v)x = \int_a^x v(s)q'(y(s))ds$$

for all $v \in I$. Moreover, for $x \in [a, b]$, $v, y \in I$ and $t \in \mathbb{R} - \{0\}$, define

$$E(y, v, t)x = \frac{1}{t}[q(y(x) + tv(x)) - q(y(x)) - tv(x)q'(y(x))].$$

By hypothesis, there exists $\beta > 0$ such that $\beta = \sup_{x \in [a,b]} |q'(x)|$. Then, we have that $|E(y,v,t)x| \leq 2\beta|v(x)|$ and $E(y,v,t)x \to 0$ as $t \to 0$. It follows by the mean-value theorem that

$$\lim_{t \to 0} \Big(\int_a^b \big(\int_a^x |E(y,v,t)s|ds \big)^p dx \Big)^{\frac{1}{p}} = 0.$$

Moreover, we obtain in turn that

$$|\frac{1}{t}[Q(y+tv) - Q(y) - tP(y)(v)]x| \leq \int_a^x |E(y,v,t)s|ds$$

or

$$\frac{1}{|t|} \Big(\int_a^b |[Q(y+tv) - Q(y) - tP(y)(v)]x|^p dx \Big)^{\frac{1}{p}}$$
$$\leq \Big(\int_a^b \big(\int_a^x |E(y,v,t)s|ds \big)^p dx \Big)^{\frac{1}{p}} \to 0 \quad \text{as} \quad t \to 0,$$

which implies $P(y)$ is the Gateaux derivative of Q at y. Furthermore, we have that

$$\lim_{t \to 0} \|Q'(y+tw) - Q'(y)\|$$
$$= \lim_{t \to 0} \sup_{\|v\| \leq 1} \Big(\int_a^b |\int_a^x v(s)[q'(y(s) + tw(s)) - q'(y(s))]ds|^p dx \Big)^{\frac{1}{p}}$$
$$\leq \lim_{t \to 0} \sup_{\|v\| \leq 1} \Big(\int_a^b \big(\int_a^x |v(s)[q'(y(s) + tw(s)) - q'(y(s))]|ds \big)^p dx \Big)^{\frac{1}{p}}$$
$$\leq \lim_{t \to 0} \sup_{\|v\| \leq 1} \Big(\int_a^b \big(\int_a^x |v(s)|^p ds \big) \big(\int_a^x |[q'(y(s) + tw(s)) - q'(y(s))]|^\mu ds \big)^{\frac{p}{\mu}} dx \Big)^{\frac{1}{p}}$$
$$\leq \lim_{t \to 0} (b-a)^{\frac{2}{p}} \Big(\int_a^b |q'(y(s) + tw(s)) - q'(y(s))|^\mu ds \Big)^{\frac{1}{\mu}} = 0,$$

where $\mu > 0$ is a parameter satisfying $\dfrac{1}{p} + \dfrac{1}{\mu} = 1$. That completes the proof of the lemma. $\qquad\square$

We provide an example of a function which is Gateaux but nowhere Fréchet-differentiable. The result is provided in a Banach space setting.

Lemma 9.2.2. *Assume $q'(x)$ is not a constant function. Then, function Q for $p \in (1,2)$ is everywhere Gateaux but nowhere Fréchet-differentiable.*

Proof. It follows from Lemma 9.2.1 that Q is Gateaux differentiable and the derivative is given by P. Let us define function

$$T(y,w)x = \int_a^x [q(y(s) + w(s)) - q(y(s)) - w(s)q'(y(s))]ds.$$

We shall show that function Q is nowhere Fréchet-differentiable.

Let $y \in I$. We shall show that there exists $w \in I$ such that the Lebesgue measure of the set $S = \{x \in [a,b] : T(y,w)x \neq 0\}$ is positive. Otherwise, the set

$$S_\gamma = \{x \in [a,b] : T(y, w_\gamma) \neq 0\}$$

has measure zero for all $w_\gamma \in I$ with $w_\gamma(x) = \gamma$ on $[a, b]$ and $\gamma \in \mathbb{R}$. That is

$$\int_a^x [q(y(s) + \gamma) - q(y(s))]ds = \int_a^x \gamma q'(y(s))ds$$

almost everywhere on $(a, b]$ and so differentiable with respect to x. We then get

$$q(y(x) + \gamma) - q(y(x)) = \gamma q'(y(x)).$$

Moreover, differentiating with respect to γ, we obtain $q'(y(x) + \gamma) = q'(y(x))$, contradicting the hypothesis that $q'(x)$ is not a constant function. Consequently, there exist $w \in I$, $\varepsilon > 0$ such that $m(S) > 0$ and $m(C) > 0$, where $C = \{x \in [a, b] : |T(y, w)x| > \varepsilon\}$.

Let $\{C_n\}$ be a sequence in C such that $C_{n+1} \subseteq C_n$, $m(C_n) > 0$ $(n \geq 1)$ with $\bigcap_{n \geq 1}^\infty C_n = \emptyset$. Define a sequence $\{\theta_n\}$ of functions in $L^p[a, b]$ by

$$\theta_n(x) = \begin{cases} w(x) & \text{if } x \in C_n; \\ 0 & \text{if } x \notin C_n. \end{cases}$$

There exists $\lambda > 0$ such that $|w(x)| \leq \lambda$ for almost all $x \in [a, b]$ (since, $w \in I$). Moreover, we have that

$$\frac{\|Q(y + \theta_n) - Q(y) - P(y)(\theta_n)\|}{\|\theta_n\|} = \frac{\left(\int_a^b |T(y, \theta_n)x|dx\right)^{\frac{1}{p}}}{\|\theta_n\|} \geq \frac{\varepsilon m(C_n)^{\frac{1}{p}}}{\lambda (m(C_n))^{\frac{1}{p}}} = \frac{\varepsilon}{\lambda} > 0,$$

whereas, $\|\theta_n\| = \left(\int_a^b |\theta_n|^p dx\right)^{\frac{1}{p}} \to 0$ as $n \to \infty$. Hence, Q is nowhere Fréchet differentiable. That completes the proof of the lemma. $\qquad\qquad\square$

By simply replacing (9.81) by (9.102) (see below) and following verbatim the steps of the proof of Theorem 9.2.1 we arrive at:

Theorem 9.2.2. *Let $F : D \subseteq X \to Y$ be a Gateaux-differentiable operator at each point in some neighborhood of $x_0 \in D$, $G : D \subseteq X \to Y$ an operator and*

$$A_n = F'(z_n) \quad (n \geq 0). \tag{9.101}$$

Moreover, assume that there exist $z_0 \in D$ and $r > 0$ such that $A_0^{-1} \in L(Y, X)$, $\|A_0^{-1}\| \leq \alpha$,

$$\|F'(x) - F'(z_0)\| \leq \varepsilon_0, \quad \|F'(x) - F'(y)\| \leq \varepsilon$$

and

$$\|G(x) - G(y)\| \leq \varepsilon_2 \|x - y\| \quad for \ all \quad x, y \in U(x_0, r). \tag{9.102}$$

Operator $F'(x)$ is piecewise hemicontinuous for each $x \in U(x_0, r)$.
    For

$$\beta_0 = \frac{\alpha}{1 - \alpha\varepsilon_0}, \quad \eta = \alpha\|F(x_0) + G(x_0)\|,$$

$$0 < \gamma_0 = \beta_0(\varepsilon + \varepsilon_2) < 1, \quad \frac{\eta}{1 - \gamma_0} < r \quad and \quad \overline{U}(x_0, r) \subseteq D.$$

Then, iteration $\{x_n\}$ generated by the (NLM) is well-defined, remains in $U(x_0, r)$ for any sequence $\{z_n\} \subseteq U(x_0, r)$ and converges to a solution $x^\star$ of equation $F(x) + G(x) = 0$, which is unique in $\overline{U}(x_0, r)$. Moreover, the following estimates hold for all $n \geq 0$

$$\|x_n - x^\star\| \leq \frac{\gamma_0^n}{1 - \gamma_0} \eta.$$

Below, we provide an example for Theorem 9.2.2.

Example 9.2.1. Let $X = Y = L^p[a, b]$. Define operator F on X by

$$F(y)x = y(x) + \int_a^x \sin(y(s))ds$$

almost everywhere $x \in [a, b]$ and $y \in L^p[a, b]$. Set $A(x) = F'(x)$ and further assume that $G : X \to X$ is any continuous operator satisfying (9.102) for some $\varepsilon_2 > 0$. Let also $v, w \in L^p[a, b]$. Then, we have in turn that

$$|(I - F'(v))w(x)| = |\int_a^x \cos(v(s))w(s)ds| \leq \int_a^x |w(s)|ds.$$

Hence, we have in turn that

$$\|(I - F'(v))w\| = \left(\int_a^b |(I - F'(v))w(x)|^p dx\right)^{\frac{1}{p}} \leq (b - a)^{\frac{1}{p}} \|w\|.$$

That is,

$$\|I - F'(v)\| \leq (b - a)^{\frac{1}{p}} = \delta.$$

Assume $\delta \in [0, 1)$. Then $F'(v)^{-1}$ exists and

$$\|F'(v)^{-1}\| \leq \frac{1}{1 - \delta} = \alpha.$$

Let $v_1, v_2 \in U(x_0, r)$ for some r to be specified later. Then we have in turn that

$$|(F'(v_1) - F'(v_2))w(x)| \leq |\int_a^x [\cos(v_1(s)) - \cos(v_2(s))]w(s)ds| \leq 2\int_a^x |w(s)|ds.$$

Consequently, we have that

$$\|F'(v_1) - F'(v_2)\| \leq 2\delta, \quad \text{for all} \quad v_1\, v_2 \in X.$$

Then, we can set $\varepsilon_0 = \varepsilon = 2\delta$. Choose $r > \dfrac{\eta}{1 - \gamma_0}$, where $\eta = \alpha\|F(x_0) + G(x_0)\|$, provided that $0 < \gamma_0 = (2\delta + \varepsilon_2)\alpha < 1$. Then, all hypotheses of Theorem 9.2.2 are satisfied. Therefore, there exists a unique solution y^* in $\overline{U}(x_0, r)$ of equation (1.54), which can be obtained as the limit of (NLM) (9.76).

9.3   Local Convergence for Newton-type Method

We determine in this section the radius of convergence for some Newton-type methods (NTM). A comparison is given between the radii of (NTM) and Newton's method (NM). Another method for approximating $x^\star$ is the cubically convergent Newton-type method (NTM) given by

$$
\begin{aligned}
y_n &= x_n - F'(x_n)^{-1} F(x_n), \quad (n \geq 0), \quad (x_0 \in \mathcal{D}) \\
x_{n+1} &= y_n - F'(x_n)^{-1} F(y_n).
\end{aligned}
\tag{9.103}
$$

(NM) requires the computation of inverse and one function evaluation at each step. There are few situations when one computes the inverse of a matrix. However, (NTM) requires the computation of one inverse and two functions evaluation. Note also that in the case of (NTM), one always computes the solution of a linear system (which is by far more easy to compute than the inverse). We are interested in the local convergence of (NTM). Rheinboldt in Ref. [384], Traub and Woźniakowsi in Ref. [420] used the L-Lipschitz condition ($L > 0$):

$$
\| F'(x^\star)^{-1} (F'(x) - F'(y)) \| \leq L \, \| x - y \| \quad \text{for all} \quad x, y \in \mathcal{D}
\tag{9.104}
$$

to provide the radius of convergence for (NM) given by $r = 2/3\,L$. In view of (9.104), there always exists $L_0 > 0$ such that

$$
\| F'(x^\star)^{-1} (F'(x) - F'(x^\star)) \| \leq L_0 \, \| x - x^\star \| \quad \text{for all} \quad x \in \mathcal{D}.
\tag{9.105}
$$

Note that

$$
L_0 \leq L
\tag{9.106}
$$

holds in general. Argyros in Refs. [100, 110] used a combination of (9.104) and (9.105) to provide the radius of convergence of (NM) by $r_A = 2/(2\,L_0 + L)$. Note that $r \leq r_A$ and this inequality is strict if $L_0 < L$. In Refs. [100, 110], Argyros provided a larger ball using the more precise (9.105) instead of (9.104) to provide a tighter upper bound on the norm $\| F'(x)^{-1} F'(x^\star) \|$ ($x \in \mathcal{D}$). The same approach leads to tighter error bounds on the distances $\| x_{n+1} - x_n \|$, $\| x_n - x^\star \|$ ($n \geq 0$) and at least as precise information on the uniqueness of the solution $x^\star$.

The results under Lipschitz-type conditions are extended in the p-Hölder case ($p \in (0, 1]$) by assuming

$$
\| F'(x^\star)^{-1} (F'(x) - F'(y)) \| \leq L \, \| x - y \|^p \quad \text{for all} \quad x, y \in \mathcal{D}
\tag{9.107}
$$

and

$$
\| F'(x^\star)^{-1} (F'(x) - F'(x^\star)) \| \leq L_0 \, \| x - x^\star \|^p \quad \text{for all} \quad x \in \mathcal{D}.
\tag{9.108}
$$

If $p = 1$, conditions (9.107) and (9.108) are reduced to (9.104) and (9.105), respectively. In the p-Hölder case, the radius of convergence (see Refs. [100, 120]) is

$$
r_p = \left(\frac{1+p}{(1+p)\,L_0 + L} \right)^{1/p}.
$$

In view of (9.106), there exists $c \geq 1$, such that

$$L = c\,L_0. \tag{9.109}$$

Define function f_c on $[0, c]$ by

$$f_c(t) = t\left(((1+p)(1-\tfrac{t}{c})+t)^{1+p} - ((1+p)(1-\tfrac{t}{c}))^{1+p}\right) - (1+p)^{2+p}(1-\tfrac{t}{c})^{3+p} \tag{9.110}$$

for each fixed $c \geq 1$ and $p \in (0, 1]$. We have $f_c(0) = -(1+p)^{2+p} < 0$, $f_c(c) = c^{2+p} \geq 1 > 0$ and f_c is continuous on $[0, c]$. It follows from the intermediate value theorem that there exists a zero of function f_c in $(0, c)$. Denote by t^c the minimal zero of f_c in $(0, c)$. Clearly, we have that

$$t^1 \leq t^c \quad \text{for all} \quad c \geq 1. \tag{9.111}$$

We need an Ostrowski-type representation for (NTM) (see for example Refs. [110, 120, 134]).

Lemma 9.3.1. *If (NTM) is well-defined for all $n \geq 0$, $F(x^\star) = 0$ and $F'(x^\star)^{-1} \in \mathcal{L}(\mathcal{Y}, \mathcal{X})$, then*

$$x_{n+1} - x^\star = (F'(x^\star)^{-1} F'(x_n))^{-1} F'(x^\star)^{-1} \int_0^1 (F'(x_n) - F'(\theta\, y_n + (1-\theta)\, x^\star))\, d\theta \times$$

$$(F'(x^\star)^{-1} F'(x_n))^{-1} F'(x^\star)^{-1} \int_0^1 (F'(x_n) - F'(\theta\, x_n + (1-\theta)\, x^\star))\, d\theta\, (x_n - x^\star) \tag{9.112}$$

holds for all $n \geq 0$.

Proof. Using (9.103) and Taylor's formula, we get in turn that

$$x_{n+1} - x^\star = y_n - x^\star - F'(x_n)^{-1}(F(y_n) - F(x^\star))$$

$$= F'(x_n)^{-1} \int_0^1 (F'(x_n) - F'(\theta\, y_n + (1-\theta)\, x^\star))\, d\theta\, (y_n - x^\star)$$

$$= F'(x_n)^{-1} \int_0^1 (F'(x_n) - F'(\theta\, y_n + (1-\theta)\, x^\star))\, d\theta\, (x_n - x^\star - F'(x_n)^{-1} F(x_n))$$

and since

$$x_n - x^\star - F'(x_n)^{-1} F(x_n)$$

$$= F'(x_n)^{-1} \int_0^1 (F'(x_n) - F'(\theta\, x_n + (1-\theta)\, x^\star))\, d\theta\, (x_n - x^\star),$$

we obtain (9.112). That completes the proof of Lemma 9.3.1. $\square$

We can show the following local convergence result for (NTM).

Theorem 9.3.1. *Let $F : \mathcal{D} \subseteq \mathcal{X} \longrightarrow \mathcal{Y}$ be a Fréchet-differentiable operator. Let $x^\star \in \mathcal{D}$, such that $F(x^\star) = 0$ and $F'(x^\star)^{-1} \in \mathcal{L}(\mathcal{Y}, \mathcal{X})$. Let $p \in (0, 1]$, such that p-Hölder conditions (9.107) and (9.108) hold. Denotes by t^c the minimal zero of f_c in $(0, c)$ and suppose $U(x^\star, r^\star) \subseteq \mathcal{D}$, where*

$$r^\star = \left(\frac{(1+p)\,c}{L}\right)^{1/p}.$$

Then, sequence $\{x_n\}$ generated by (NTM) is well-defined, remains in $\overline{U}(x^\star, r^c)$ for all $n \geq 0$, where

$$r^c = \left(\frac{t^c}{L}\right)^{1/p} \tag{9.113}$$

and $\{x_n\}$ converges to $x^\star \in U(x^\star, r^c)$ solution of (1.1), provided that $x_0 \in U(x^\star, r^c)$. Moreover, for $e_n = \| x_n - x^\star \|$ $(n \geq 0)$, the following estimate holds:

$$e_{n+1} \leq a_{n+1}, \tag{9.114}$$

where scalar iteration $\{a_n\}$ $(n \geq 0)$ is given by

$$a_{n+1} = \frac{L}{(1 - \frac{L}{c} e_n^p)^2} \left(\left(1 + \frac{L e_n^p}{(1+p)(1 - \frac{L}{c} e_n^p)}\right)^{1+p} - 1\right) \frac{e_n^{1+p}}{1+p}.$$

Furthermore, the solution $x^\star$ is unique in $U(x^\star, r^\star)$.

Proof.   Let $x \in U(x^\star, r_0 = (\frac{c}{L})^{1/p})$. Then, using (9.108), we obtain that

$$\| F'(x^\star)^{-1} (F'(x) - F'(x^\star)) \| \leq L_0 \| x - x^\star \|^p < \frac{L}{c} \frac{c}{L} = 1. \tag{9.115}$$

It follows from (9.115) and the Banach Lemma of invertible operators $F'(x)^{-1} \in \mathcal{L}(\mathcal{Y}, \mathcal{X})$ and

$$\| F'(x)^{-1} F'(x^\star) \| \leq \frac{1}{1 - \frac{L}{c} \| x - x^\star \|^p}. \tag{9.116}$$

By hypothesis $x_0 \in U(x^\star, r^c) \subseteq U(x^\star, r_0)$. Hence, (9.116) for $x = x_0$ gives

$$\| F'(x_0)^{-1} F'(x^\star) \| \leq \frac{1}{1 - \frac{L}{c} \| x_0 - x^\star \|^p}. \tag{9.117}$$

Using (9.112) for $n = 0$, (9.107), (9.108) and (9.117), we get that

$$\| x_1 - x^\star \| \leq \frac{L}{(1 - \frac{L}{c} e_0^p)^2} \left(\left(1 + \frac{L e_0^p}{(1+p)(1 - \frac{L}{c} e_0^p)}\right)^{1+p} - 1\right) \frac{e_0^{1+p}}{1+p} \tag{9.118}$$
$$< \| x_0 - x^\star \|,$$

by the definition of t^c, (9.113) and the following estimates

$$\| F'(x^\star)^{-1} \int_0^1 (F'(x_0) - F'(\theta x_0 + (1-\theta) x^\star)) \, d\theta \|$$
$$\leq L \int_0^1 \| x_0 - (1-\theta) x^\star - \theta x_0 \| \, d\theta \tag{9.119}$$
$$= L \| x_0 - x^\star \|^p \int_0^1 (1-\theta)^p \, d\theta = \frac{L \| x_0 - x^\star \|^p}{1+p},$$

$$\| F'(x^\star)^{-1} \int_0^1 (F'(x_0) - F'(\theta\, y_0 + (1-\theta)\, x^\star))\, d\theta \| \le L \int_0^1 \| x_0 - (1-\theta)\, x^\star - \theta\, y_0 \|\, d\theta, \tag{9.120}$$

$$\begin{aligned} x_0 - (1-\theta)\, x^\star - \theta\, y_0 &= x_0 - x^\star - \theta\,(y_0 - x^\star) \\ &= x_0 - x^\star - \theta\, F'(x_0)^{-1} \int_0^1 (F'(x_0) - F'(\theta\, x_0 + (1-\theta)\, x^\star)\, d\theta)\, (x_0 - x^\star) \end{aligned} \tag{9.121}$$

and

$$\begin{aligned} &\| F'(x^\star)^{-1} \int_0^1 (F'(x_0) - F'(\theta\, y_0 + (1-\theta)\, x^\star))\, d\theta \| \\ &\le L\, e_0^p \int_0^1 \left(1 + \frac{\theta\, L\, e_0^p}{(1+p)(1 - \frac{L}{c}\, e_0^p)} \right)^p d\theta \\ &= (1 - \frac{L}{c}\, e_0^p) \left(\left(1 + \frac{L\, e_0^p}{(1+p)(1 - \frac{L}{c}\, e_0^p)} \right)^{1+p} - 1 \right). \end{aligned} \tag{9.122}$$

Estimate (9.118) implies that $x_1 \in U(x^\star, r^c)$ and (9.114) holds for $n = 0$. Let us assume (9.114) and $x_k \in U(x^\star, r^c)$ hold for all $k \le n$. Then, as in (9.116)–(9.122), we have the estimates (with x_n, y_n replacing x_0 and y_0, respectively):

$$\| F'(x_n)^{-1} F'(x^\star) \| \le \frac{1}{1 - \frac{L}{c}\, e_n^p}, \tag{9.123}$$

$$\| F'(x^\star)^{-1} \int_0^1 (F'(x_n) - F'(\theta\, x_n + (1-\theta)\, x^\star))\, d\theta \| \le \frac{L\, e_n^p}{1+p}, \tag{9.124}$$

$$\begin{aligned} &\| F'(x^\star)^{-1} \int_0^1 (F'(x_n) - F'(\theta\, y_n + (1-\theta)\, x^\star))\, d\theta \| \\ &\le (1 - \frac{L}{c}\, e_n^p) \left(\left(1 + \frac{L\, e_n^p}{(1+p)(1 - \frac{L}{c}\, e_n^p)} \right)^{1+p} - 1 \right). \end{aligned} \tag{9.125}$$

Using (9.112), we have that

$$\begin{aligned} &\| x_{n+1} - x^\star \| \\ &\le \| F'(x_n)^{-1} F'(x^\star) \|^2 \int_0^1 \| F'(x^\star)^{-1} (F'(x_n) - F'(\theta\, y_n + (1-\theta)\, x^\star)) \|\, d\theta \times \\ &\quad \int_0^1 \| F'(x^\star)^{-1} (F'(x_n) - F'(\theta\, x_n + (1-\theta)\, x^\star)) \|\, d\theta\, \| x_n - x^\star \|. \end{aligned} \tag{9.126}$$

By (9.123)–(9.126), we obtain

$$e_{n+1} = \| x_{n+1} - x^\star \| \le a_{n+1} < e_n, \tag{9.127}$$

the induction is complete, $x_{n+1} \in U(x^\star, r^c)$ and $\{x_n\}$ converges to $x^\star$.

We shall finally show the uniqueness part. Let $y^\star \in U(x^\star, r^\star)$ be a solution of $F(x) = 0$ and set $Q = \int_0^1 F'(y^\star + \theta\,(x^\star - y^\star))\,d\theta$. Using (9.112), we have in turn that

$$\| A(x^\star)^{-1}\,(Q - F'(x^\star)) \| \leq \| F'(x^\star)^{-1} \int_0^1 (F'(y^\star + \theta\,(x^\star - y^\star)) - F'(x^\star))\,d\theta \|$$

$$\leq \frac{L}{c}\,\| x^\star - y^\star \|^p \int_0^1 \theta^p\,d\theta = \frac{L}{c\,(1+p)}\,\| x^\star - y^\star \|^p < 1.$$

$$(9.128)$$

It follows from (9.128) and the Banach Lemma on invertible operators that Q^{-1} exists. In view of the identity $0 = F(x^\star) - F(y^\star) = Q\,(x^\star - y^\star)$, we conclude that $x^\star = y^\star$. That completes the proof of Theorem 9.3.1. $\qquad\square$

Remark 9.3.1. The condition $U(x^\star, r^\star) \subseteq D$ is needed to show the uniqueness part. It can be replaced by $U(x^\star, r_0) \subseteq D$. However, we then have that $U(x^\star, r_0) \subseteq U(x^\star, r^\star)$.

Next, we shall show that (NTM) converges with order $1 + 2\,p$.

Proposition 9.3.1. *Under the hypotheses of Theorem 9.3.1, we have*

$$e_{n+1} \leq a_{n+1} \leq b_{n+1},\qquad\qquad (9.129)$$

where

$$b_{n+1} = \frac{L^2\left(1 - \dfrac{L}{c}\,e_n^p\right)}{\left(1 - \dfrac{L_0}{c}\,e_n^p\right)^2}\left(\frac{L}{1 - \dfrac{L}{c}\,e_n^p} + \frac{p\,L\,e_n^p}{2\,(1+p)\,\left(1 - \dfrac{L}{c}\,e_n^p\right)^2}\right) e_n^{1+2\,p}.\qquad (9.130)$$

Proof.   Define function $g_{n,c}$ on $[1, \beta]$ by

$$g_{n,c}(t) = t^{1+p},$$

where

$$\beta = 1 + \frac{L\,e_n^p}{(1+p)\,\left(1 - \dfrac{L}{c}\,e_n^p\right)}.$$

Then, we get $g'_{n,c}(t) = (1 + p)\,t^p$ and $g''_{n,c}(t) = (1 + p)\,p\,t^{p-1}$. Taylor's formula guarantees the existence of $\gamma \in [1, \beta]$, such that

$$g_{n,c}(\beta) - g_{n,c}(1) = g'_{n,c}(\beta - 1) + \frac{g''_{n,c}(\gamma)}{2}\,(\beta - 1)^2$$

$$= \frac{L\,e_n^p}{1 - \dfrac{L}{c}\,e_n^p} + \frac{p\,L^2\,e_n^{2\,p}}{2\,\gamma^{1-p}\,(1+p)\,\left(1 - \dfrac{L}{c}\,e_n^p\right)^2} \qquad (9.131)$$

$$\leq L\left(\frac{L}{1 - \dfrac{L}{c}\,e_n^p} + \frac{p\,L\,e_n^p}{2\,(1+p)\,\left(1 - \dfrac{L}{c}\,e_n^p\right)^2}\right) e_n^p,$$

since

$$\frac{1}{\gamma^{1-p}} \le 1 \quad \text{for all} \quad \gamma \in [1, \beta].$$

By the defintion of sequences $\{a_n\}$, $\{b_n\}$ and (9.131), we obtain

$$a_{n+1} \le b_{n+1} \quad \text{for all} \quad n \ge 0.$$

That completes the proof of Proposition 9.3.1. $\qquad\square$

Example 9.3.1. Consider (1.293). Then, for $x^\star = 0$ and using (1.293), we have $F(x^\star) = 0$ and $F'(x^\star) = e^0 = 1$. Moreover, (9.107) and (9.108) hold for $p = 1$, $L = e > L_0 = e - 1$, whereas $c = \dfrac{e}{e-1} = 1.581976707$. We also have that

$$f_c(t) = t\left(\left(2\left(1 - \frac{(e-1)t}{e}\right) + t\right)^2 - 4\left(1 - \frac{(e-1)t}{e}\right)^2\right) - 8\left(1 - \frac{(e-1)t}{e}\right)^4,$$

$$t^c = .6098166918, \qquad r_0 = \left(\frac{c}{L}\right)^{1/p} = .5819767070,$$

$$r^c = \left(\frac{t^c}{L}\right)^{1/p} = .2571658439, \qquad r^\star = 1.163953414.$$

Note that

$$r^c = .2571658439 < r_A = \frac{2}{2\,L_0 + L} = .3249472314.$$

Example 9.3.2. Let $\mathcal{X} = \mathcal{Y} = \mathbb{R}$. Define function F on $\mathcal{D} = [1, 3]$, given by

$$F(x) = \frac{2}{3}\,x^{3/2} - x. \tag{9.132}$$

Then, the zero of F is $x^\star = \dfrac{9}{4} = 2.25$. Using (9.132), $F'(x^\star) = .5$, $L = 2 > L_0 = 1$, $c = 2$ and $p = .5$. Moreover, we have that

$$f_c(t) = t\left(\left(\frac{3}{2}\left(1 - \frac{t}{2}\right) + t\right)^{\frac{3}{2}} - \left(\frac{3}{2}\left(1 - \frac{t}{2}\right)\right)^{\frac{3}{2}}\right) - \left(\frac{3}{2}\right)^{\frac{5}{2}}\left(1 - \frac{t}{2}\right)^{\frac{7}{2}},$$

$$t^c = .6412553390, \qquad r_0 = 1,$$

$$r^c = .5662399399, \qquad r^\star = 1.224744871.$$

Note that

$$r^c = .5662399399 < r_A = \left(\frac{1+p}{(1+p)\,L_0 + L}\right)^{1/p} = .6546536707.$$

Table 9.2   Comparison table (I)

p	.1	.2	.3	.4
t^1	.4368349792	.4429562040	.4486068930	.4538529047
$t^{1.581976707}$	.5466960852	.5556998369	.5640448368	.571820981
t^2	.6027543663	.6134232216	.6233312759	.6325810531

Table 9.3   Comparison table (II)

p	.5	.6	.7	.8
t^1	.4587471202	.4633326601	.4676451584	.4717144081
$t^{1.581976707}$	.5791007272	.5859433703	.5923980761	.5985060824
t^2	.6412553390	.6494219962	.6571373857	.6644488527

Table 9.4   Comparison table (III)

p	.9	1
t^1	.4755655764	.4792201186
$t^{1.581976707}$	.6043023300	.6098166918
t^2	.6713965712	.6780149362

Example 9.3.3. Consider Example 1.1.7. Then, we have that

$$f_c(t) = t\left(\left(2\left(1 - \frac{t}{2}\right) + t\right)^2 - 4\left(1 - \frac{t}{2}\right)^2\right) - 8\left(1 - \frac{t}{2}\right)^4.$$

Using (9.105), (9.106) for $x^\star(x) = 0$ for all $x \in [0,1]$, we get that

$$p = 1, \quad L = 15, \quad L_0 = 7.5, \quad c = 2,$$

$$t^c = .6780149362, \qquad r_0 = .1333333333,$$

$$r^c = .04520099575, \qquad r^\star = .2666666667.$$

Note that

$$r^c = .04520099575 < r_A = \frac{2}{2\,L_0 + L} = .06666666667.$$

Example 9.3.4. In Tables 9.2–9.4, we tabulate values of t^c for $c = 1$, $\dfrac{e}{e-1} = 1.581976707$, 2 and $p = .1, .2, \cdots, 1$. We use Maple 13 for different calculations. It follows from Tables 9.2–9.4 that t^c increases with c and p. To validate this result, see Fig. 9.1(a) (for $p = .5$ and $c \in [1;7]$) and Fig. 9.1(b) (for $c = 2$ and $p \in [.1;1]$). Since we cannot have an explicit solution t^2 in terms of p for function ϕ (Fig. 9.1(b)), we simply plot the function ϕ from the list of points given by Tables 9.2–9.4 for $c = 2$.

9.4   Two-step Newton-like Method

A local convergence analysis is presented for a fast two-step Newton-like method (TSNLM) for solving nonlinear equations in a Banach space setting. (TSNLM)

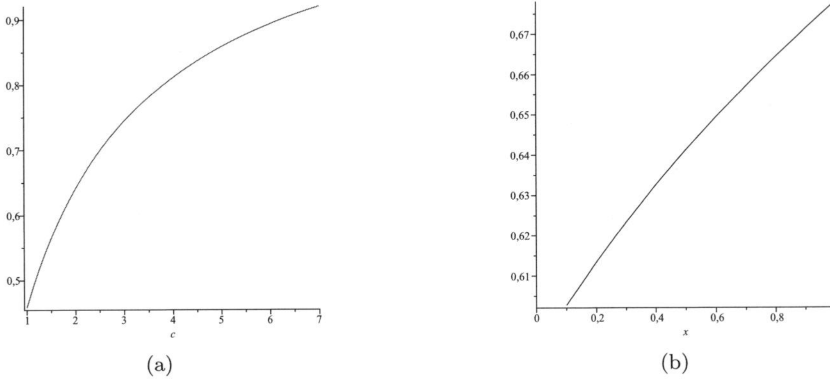

(a)   (b)

Fig. 9.1   (a) Function $\varphi \,:\, c \longrightarrow t^c$ on $[1;7]$. (b) $\phi \,:\, p \longrightarrow t^c$ on $[.1;1]$.

unifies earlier methods such as Newton's, Secant, Newton-like, Chebyshev-Secant, Chebyshev–Newton, Steffensen, Stirling's and other single or multistep methods.

In Ref. [121], we constructed a family of iterative processes free of derivatives as the classic Secant method (SM). Then, we consider an approximation of the first derivative of F from a divided difference of first order. So, we introduced the Chebyshev–Secant-type method (CSTM) given by (4.136) and provided semi-local convergence result for (CSTM). In this section we unify these methods by considering the (TSNLM) as follows:

$$\begin{cases} x_0 \in \mathcal{D}, \\ y_n = x_n - A_n^{-1}\, F(x_n), \quad A_n = A(x_n), \\ z_n = x_n + a\,(y_n - x_n), \\ x_{n+1} = x_n - A_n^{-1}\,(b\,F(x_n) + c\,F(z_n)), \quad n \geq 0, \end{cases}$$

where $A(x) \in \mathcal{L}(\mathcal{X}, \mathcal{Y})$ $(x \in \mathcal{D})$. Clearly, (NM), (SM), (CNTM) and (CSTM) are special cases of (TSNLM).

Semi-local convergence issue is, based on the information around an initial point, to give criteria ensuring the convergence of (TSNLM). The semi-local convergence of (TSNLM) has been given by us in Ref. [138]. We are concerned in this section with the local convergence analysis of (TSNLM). The local convergence analysis issue for (TSNLM) is, based on the information aroud a solution, to find estimates of the radii of convergence balls. Local convergence results were not given for (CNTM) and (CSTM).

The local convergence analysis of (TSNLM) is based on the following conditions:

$(\mathcal{S}_1)$ $F \,:\, \Omega \subseteq \mathcal{X} \longrightarrow \mathcal{Y}$ is Fréchet-differentiable, $A(x) \in \mathcal{L}(\mathcal{X}, \mathcal{Y})$ $(x \in \Omega)$ and there exist $x^\star \in \Omega$, $K > 0$, $L > 0$, $M \geq 0$, $\mu \geq 0$, $\ell \geq 0$ such that for all $x, y \in \mathcal{D}$

$$F(x^\star) = 0, \quad A(x^\star)^{-1} \in \mathcal{L}(\mathcal{Y}, \mathcal{X}), \tag{9.133}$$

$$\| A(x^\star)^{-1}\,(F'(x) - F'(y)) \| \leq K \,\| x - y \|, \tag{9.134}$$

$$\| A(x^\star)^{-1}\left(F'(x) - A(x)\right) \| \leq M \parallel x - x^\star \parallel + \mu, \tag{9.135}$$

$$\| A(x^\star)^{-1}\left(A(x^\star) - A(x)\right) \| \leq L \parallel x - x^\star \parallel + \ell. \tag{9.136}$$

$(\mathcal{S}_2)$

$$U(x^\star, R_1) \subseteq \Omega, \tag{9.137}$$

for some $R_1 > 0$ to be precised later in Lemma 9.4.1 and (9.158).

$(\mathcal{S}_3)$

$$\ell + \mu < 1. \tag{9.138}$$

$(\mathcal{S}_4)$ For $a \in [0,1]$, $b \geq 0$ and $c > 0$ given in (TSNLM), we suppose

$$(1-a)\,c = 1 - b.$$

We need some auxiliary results on the zeros of functions. With the above notations, we have the following lemmas.

Lemma 9.4.1. *Let us define functions ψ_0 on $[0,\infty)$, ψ on $\left[0, \dfrac{1-\ell}{L}\right)$ by*

$$\psi_0(r) = \frac{K}{2}\,r^2 + (M + L)\,r + \ell + \mu - 1 \tag{9.139}$$

and

$$\psi(r) = 4\,\varpi^2(r)\,(1 - ac)\,\rho(r) + 2\,\varpi(r)\,(1 - b)\,K\,a\,(\rho(r) + 2\,\varpi(r))\,r + \\ K\,a^2\,c\,(\rho(r) + 2\,\varpi(r))\,\rho(r)\,r + 2\,a\,c\,\varpi(r)\,\rho(r)^2 - 8\,\varpi(r)^3, \tag{9.140}$$

where ρ, ϖ are functions defined on $[0,\infty)$ by

$$\rho(r) = 2\,M\,r + 2\,\mu + K\,r^2 \quad and \quad \varpi(r) = 1 - (L\,r + \ell). \tag{9.141}$$

Then, the following assertions hold:

(a) Function ψ_0 has a unique positive zero R_0 given by

$$R_0 = \frac{2\,(1 - (\ell + \mu))}{M + L + \sqrt{(M + L)^2 + 2\,K\,(1 - (\ell + \mu))}} \tag{9.142}$$

provided that M, L and K are not all zero.

(b) If

$$(1 - \ell)\,|1 - a\,c|\,\mu + a\,c\,\mu^2 - (1 - \ell)^2 < 0 \tag{9.143}$$

and

$$\psi(R_0) > 0 \tag{9.144}$$

then, function ψ has a zero in $(0, R_0)$. Denote the minimal such zero by R.

Proof.

(a) R_0 is well-defined since M, L, K are not all zero and by (9.138), $R_0 > 0$. Moreover, we have $\psi_0(R_0) = 0$.

(b) We have that

$$\psi(0) = 8\,(1-\ell)\left((1-\ell)\,|1-a\,c|\,\mu + a\,c\,\mu^2 - (1-\ell)^2\right) < 0 \qquad (9.145)$$

by (9.138) and (9.143). Then, the existence of a zero for function ψ follows from intermediate value theorem, (9.144) and (9.145).

This completes the proof of Lemma 9.4.1. $\qquad\square$

It is convenient for us to set

$$e_n = x_n - x^\star, \qquad (9.146)$$

$$u_n = y_n - x^\star, \qquad (9.147)$$

$$v_n = z_n - x^\star, \qquad (9.148)$$

$$G_n = G_n(x^\star, x_n) = \int_0^1 F'(x_n + t\,(x^\star - x_n))\,dt, \qquad (9.149)$$

$$Q_n = Q_n(x^\star, z_n) = \int_0^1 F'(z_n + t\,(x^\star - z_n))\,dt, \qquad (9.150)$$

$$D_n = A_n^{-1}\,(A_n - G_n) \qquad (9.151)$$

and

$$E_n = A_n^{-1}\,(G_n - Q_n). \qquad (9.152)$$

We shall show the following Ostrowski-type representations for (TSNLM).

Lemma 9.4.2. *Assume (TSNLM) is well-defined and there exists $x^\star \in \Omega$, such that $F(x^\star) = 0$, $F(x^\star)^{-1} \in \mathcal{L}(\mathcal{Y}, \mathcal{X})$ and $(\mathcal{S}_4)$ holds. Then, the following assertions hold*

$$F(x_n) = G_n\,e_n, \qquad (9.153)$$

$$F(z_n) = Q_n\,v_n, \qquad (9.154)$$

$$v_n = a\,u_n + (1-a)\,e_n, \qquad (9.155)$$

$$u_n = D_n\,e_n \qquad (9.156)$$

and

$$e_{n+1} = \left((1-a\,c)\,D_n + (1-b)\,E_n + a\,c\,E_n\,D_n + a\,c\,D_n^2\right)e_n. \qquad (9.157)$$

Proof. We have in turn that

$$F(x_n) = F(x_n) - F(x^\star) = G_n\, e_n,$$

$$F(z_n) = F(z_n) - F(x^\star) = Q_n\, v_n,$$

$$v_n - x^\star = e_n + a\,(y_n - x_n) = e_n + a\,(u_n - e_n) = a\,u_n + (1-a)\,e_n,$$

$$u_n = e_n - A_n^{-1}\,F(x_n) = A_n^{-1}\,(A_n - G_n)\,e_n = D_n\, e_n$$

and by (TSNLM), $(\mathcal{S}_4)$, (9.146), (9.153)–(9.156)

$$
\begin{aligned}
e_{n+1} &= e_n - A_n^{-1}\,(b\,F(x_n) + c\,F(z_n)) \\
&= A_n^{-1}\,(A_n\, e_n - (b\,G_n\, e_n + c\,Q_n\, v_n)) \\
&= A_n^{-1}\,(A_n\, e_n - b\,G_n\, e_n - c\,Q_n\,(a\,u_n + (1-a)\,e_n)) \\
&= A_n^{-1}\,(A_n\, e_n - b\,G_n\, e_n - (1-a)\,c\,Q_n\, e_n - a\,c\,Q_n\, u_n) \\
&= A_n^{-1}\,(A_n\, e_n - b\,G_n\, e_n - (1-b)\,Q_n\, e_n - a\,c\,Q_n\, u_n) \\
&= A_n^{-1}\,((A_n - G_n)\, e_n + (1-b)\,Q_n\, e_n + G_n\, e_n - b\,G_n\, e_n - a\,c\,Q_n\, u_n) \\
&= A_n^{-1}\,((A_n - G_n)\, e_n + (1-b)\,(Q_n - G_n)\, e_n + \\
&\quad\ a\,c\,(G_n - Q_n + A_n - G_n) - a\,c\,A_n\, u_n) \\
&= A_n^{-1}\,((A_n - G_n)\, e_n + (1-b)\,(Q_n - G_n)\, e_n + a\,c\,(G_n - Q_n) + \\
&\quad\ a\,c\,(A_n - G_n)) - a\,c\,u_n \\
&= D_n\, e_n + (1-b)\,E_n\, e_n + a\,c\,E_n\, u_n + a\,c\,D_n\, u_n - a\,c\,u_n \\
&= D_n\, e_n + (1-b)\,E_n\, e_n + a\,c\,D_n\, e_n + a\,c\,D_n^2\, e_n - a\,c\,D_n\, e_n \\
&= (1 - a\,c)\,D_n\, e_n + (1-b)\,E_n\, e_n + a\,c\,E_n\, D_n\, e_n + a\,c\,D_n^2\, e_n.
\end{aligned}
$$

This completes the proof of Lemma 9.4.2. $\square$

It is also convenient for us to define sequences on $[0, R_1)$, where

$$R_1 = \min\left\{ R, \frac{1-\ell}{L} \right\}, \tag{9.158}$$

by

$$h_n = \frac{1}{2\,(1 - (L\,\|\,e_n\,\| + \ell))}, \tag{9.159}$$

$$
\xi_n = |1 - a\,c|\, h_n\, \varrho(\|\,e_n\,\|) + (1-b)\,K\,a\,h_n^2(\varrho(\|\,e_n\,\|) + 2\,\varpi(\|\,e_n\,\|))\,\|\,e_n\,\| + \\
K\,a^2\,c\,h_n^2\,(\varrho(\|\,e_n\,\|) + 2\,\varpi(\|\,e_n\,\|))\,\varrho(\|\,e_n\,\|)\,\|\,e_n\,\| + a\,c\,h_n^2\,\varrho(\|\,e_n\,\|)^2 \tag{9.160}
$$

and

$$\xi(r) = \frac{\psi(r) + 8\,\varpi(r)^3}{8\,\varpi(r)^3}, \tag{9.161}$$

where ρ and ϖ are given in (9.140). Note that by the definition of R and R_1, we have that

$$\xi(r) = 1 \tag{9.162}$$

and

$$\xi(R_1) \leq 1. \tag{9.163}$$

We can now show the main local convergence result for (TSNLM).

Theorem 9.4.1. *Assume conditions* $(\mathcal{S}_1)$–$(\mathcal{S}_4)$. *Then, sequence* $\{x_n\}$ *generated by* *(TSNLM) started at* $x_0 \in U(x^\star, R_1)$ *is well-defined, remains in* $U(x^\star, R_1)$ *for all* $n \geq 0$ *and converges to* $x^\star$. *Moreover, the following error estimates hold*

$$\| e_{n+1} \| \leq \xi_n \| e_n \| < \xi(R_1) \| e_n \| \leq \| e_n \|. \tag{9.164}$$

Proof. We shall use induction to show (9.164). We shall first show that (9.164) holds for $n = 0$ and y_0, z_0, $x_1 \in U(x^\star, R_1)$. By hypothesis $x_0 \in U(x^\star, R_1)$. Using (9.136) and the definition of R_1, we get

$$\| A(x^\star)^{-1} (A(x^\star) - A(x_0)) \| \leq L \| x_0 - x^\star \| + \ell < L R_1 + \ell < 1. \tag{9.165}$$

It follows from (9.165) and the Banach Lemma on invertible operators that $A_0^{-1} \in \mathcal{L}(\mathcal{Y}, \mathcal{X})$ and

$$\| A_0^{-1} A(x^\star) \| \leq \frac{1}{1 - (L \| e_0 \| + \ell)} < \frac{1}{1 - (L R_1 + \ell)}. \tag{9.166}$$

Thus, y_0 is well-defined. It follows from (9.156) for $n = 0$, (9.134), (9.135), (9.166) and the definition of R_0 that

$$
\begin{aligned}
\| u_0 \| &\leq \| A_0^{-1} A(x^\star) \| \, \| A(x^\star)^{-1} (A(x_0) - G_0) \| \, \| e_0 \| \\
&\leq \| A_0^{-1} A(x^\star) \| \left(\| A(x^\star)^{-1} (A(x_0) - F'(x_0)) \| + \right. \\
&\quad \left. \| A(x^\star)^{-1} (F'(x_0) - G_0) \| \right) \| e_0 \| \\
&\leq \frac{1}{1 - (L \| e_0 \| + \ell)} \left(M \| e_0 \| + \mu + \frac{K}{2} \| e_0 \|^2 \right) \| e_0 \| \\
&= \frac{\rho(\| e_0 \|)}{2 \varpi(\| e_0 \|)} \| e_0 \| \leq \frac{\rho(R_0)}{2 \varpi(R_0)} \| e_0 \| \leq \| e_0 \| < R_1.
\end{aligned}
\tag{9.167}
$$

That is, $y_0 \in U(x^\star, R_1)$. The vector z_0 is a linear combination of x_0, y_0 and $U(x^\star, R_1) \subseteq \Omega$. Then, using (9.155) for $n = 0$, we get

$$\| v_0 \| \leq a \| u_0 \| + (1 - a) \| e_0 \| \leq \| e_0 \| < R_1. \tag{9.168}$$

That is $z_0 \in U(x^\star, R_1)$. We need to find upper bounds on the norms $\| D_0 \|$ and $\| E_0 \|$. We have in turn using (TSNLM), (9.134), (9.135), (9.166) and (9.151), (9.152) (for $n = 0$)

$$\| D_0 \| \leq \| A_0^{-1} A(x^\star) \| \, \| A(x^\star)^{-1} (A_0 - G_0) \| \leq \frac{\rho(\| e_0 \|)}{2 \varpi(\| e_0 \|)} \tag{9.169}$$

and

$$\| E_0 \| \leq \| A_0^{-1} A(x^\star) \| \, \| A(x^\star)^{-1} (G_0 - Q_0) \|$$

$$\leq \frac{K}{\varpi(\| e_0 \|)} \int_0^1 (1 - t) \| z_0 - x_0 \| \, dt$$

$$\leq \frac{K \, \| z_0 - x_0 \|}{2 \, \varpi(\| e_0 \|)} \leq \frac{K \, a \, \| y_0 - x^\star + x^\star - x_0 \|}{2 \, \varpi(\| e_0 \|)} \leq \frac{K \, a \, (\| u_0 \| + \| e_0 \|)}{2 \, \varpi(\| e_0 \|)}$$

$$\leq K \, a \left(\frac{\rho(\| e_0 \|)}{2 \, \varpi(\| e_0 \|)} + 1 \right) \| e_0 \|$$

$$\leq \frac{K \, a \, M \, \| e_0 \|^2}{2 \, \varpi(\| e_0 \|)} + \frac{K \, a \, \mu \, \| e_0 \|}{2 \, \varpi(\| e_0 \|)^2} + \frac{K^2 \, a \, \| e_0 \|^3}{4 \, \varpi(\| e_0 \|)^2} + \frac{K \, a \, \| e_0 \|}{2 \, \varpi(\| e_0 \|)}$$

$$\leq \frac{K \, a}{4 \, \varpi(\| e_0 \|)^2} \, (\rho(\| e_0 \|) + 2 \, \varpi(\| e_0 \|)) \, \| e_0 \| .$$

$$(9.170)$$

Using (9.169), (9.170) and (9.157), (9.159)–(9.160) (for $n = 0$), we obtain in turn that

$$\| e_1 \| \leq |1 - a \, c| \, \| D_0 \| \, \| e_0 \| + (1 - b) \, \| E_0 \| \, \| e_0 \| + a \, c \, \| E_0 \| \, \| D_0 \| \, \| e_0 \| + a \, c \, \| D_0 \|^2 \, \| e_0 \|$$

$$\leq \frac{|1 - a \, c|}{2 \, \varpi(\| e_0 \|)} \rho(\| e_0 \|) \, \| e_0 \| + \frac{(1 - b) \, K \, a}{4 \, \varpi(\| e_0 \|)^2} \, (\rho(\| e_0 \|) + 2 \, \varpi(\| e_0 \|)) \, \| e_0 \|^2 +$$

$$\frac{a \, c \, K \, a}{4 \, \varpi(\| e_0 \|)^2} \, (\rho(\| e_0 \|) + 2 \, \varpi(\| e_0 \|)) \frac{\rho(\| e_0 \|) \, \| e_0 \|^2}{2 \, \varpi(\| e_0 \|)} +$$

$$\frac{a \, c}{4 \, \varpi(\| e_0 \|)^2} \rho(\| e_0 \|) \, \| e_0 \|$$

$$\leq |1 - a \, c| \, h_0 \, \rho(\| e_0 \|) \, \| e_0 \| + (1 - b) \, K \, a \, h_0^2 \, (\rho(\| e_0 \|) + 2 \, \varpi(\| e_0 \|)) \, \| e_0 \|^2 +$$

$$K \, a^2 c \, h_0^3 (\rho(\| e_0 \|) + 2 \, \varpi(\| e_0 \|)) \rho(\| e_0 \|) \, \| e_0 \|^2 + a \, c \, h_0^2 \rho(\| e_0 \|)^2 \, \| e_0 \|$$

$$= \xi_0 \, \| e_0 \|$$

$$< \frac{1}{8 \, \varpi(R_1)^3} \left(4 \, \varpi^2(R_1) \, |1 - a \, c| \, \rho(R_1) + 2 \, \varpi(R_1) \, (1 - b) \, K \, a \, (\rho(R_1) + \right.$$

$$\left. 2 \, \varpi(R_1)) \, R_1 + K \, a^2 \, c \, (\rho(R_1) + 2 \, \varpi(R_1)) \rho(R_1) \, R_1 + 2 \, a \, c \, \varpi(R_1) \, \rho(R_1)^2 \right) \| e_0 \|$$

$$\leq \xi(R_1) \, \| e_0 \| \leq \| e_0 \| < R_1.$$

$$(9.171)$$

Hence, we deduce $x_1 \in U(x^\star, R_1)$. Suppose $\{x_k\}$ ($0 \leq n \leq k$) is well-defined and $x_n \in U(x^\star, R_1)$, where k is a fixed natural number. By simply exchanging x_0, y_0, z_0, A_0, G_0, Q_0, h_0, H_0, ξ_0 by x_k, y_k, z_k, A_k, G_k, Q_k, h_k, H_k, ξ_k, respectively, we deduce the following results:

(i_k) $A_k \in \mathcal{L}(\mathcal{Y}, \mathcal{X})$ and

$$\| A_k^{-1} A(x^\star) \| \leq \frac{1}{\varpi(\| e_k \|)} < \frac{1}{\varpi(R_1)}; \tag{9.172}$$

(ii_k) y_k is well-defined, $y_k \in U(x^\star, R_1)$ and

$$\| u_k \| \leq \frac{\rho(\| e_k \|) \, \| e_k \|}{2 \, \varpi(\| e_k \|)} < \frac{\rho(R_1)}{2 \, \varpi(R_1)} \, \| e_k \| \leq \| e_k \| < R_1; \tag{9.173}$$

(iii_k) z_k is well-defined, $z_k \in U(x^\star, R_1)$;

(iv_k) x_{k+1} is well-defined and

$$\parallel e_{k+1} \parallel \le \xi_k \parallel e_k \parallel < \xi(R_1) \parallel e_k \parallel \le \parallel e_k \parallel < R_1. \tag{9.174}$$

Hence, (9.164) holds for all n and $x_{k+1} \in U(x^\star, R_1)$. The induction is complete. By letting $k \longrightarrow \infty$ in (9.174), we obtain $\lim\limits_{k \to \infty} x_k = x^\star$. This completes the proof of Theorem 9.4.1. $\qquad\square$

Remark 9.4.1. In the special case of (TSNLM), i.e., if $A(x) = F'(x)$ $(x \in \Omega)$, $M = \mu = \ell = 0$, we have by (9.164)

$$\parallel e_{n+1} \parallel$$
$$\le |1 - a\, c|\, h_n\, K \parallel e_n \parallel^3 + (1-b)\, K\, a\, h_n^2\, (K \parallel e_n \parallel^2 + 2\,(1-L \parallel e_n \parallel)) \parallel e_n \parallel^2$$
$$+ K^2\, a^2\, c\, h_n^3\, (K \parallel e_n \parallel^2 + 2\,(1-L \parallel e_n \parallel)) \parallel e_n \parallel^4 + a\, c\, h_n^2\, K^2 \parallel e_n \parallel^5 .$$
$$\tag{9.175}$$

Moreover, if $a = b = 1$, (9.175) reduces to

$$\parallel e_{n+1} \parallel \le \xi_n^1 \parallel e_n \parallel^3 \le \xi^1(R_1) \parallel e_n \parallel^3, \tag{9.176}$$

where

$$\xi_n^1 = |1 - c|\, h_n\, K + K^2\, c\, h_n^3\, (K \parallel e_n \parallel^2 + 2\,(1-L \parallel e_n \parallel)) \parallel e_n \parallel + c\, h_n^2\, K^2 \parallel e_n \parallel^2$$
$$\le \xi^1(R_1) = |1 - c|\, \frac{K}{2\, \varpi(R_1)} + \frac{K^2\, c}{8\, \varpi(R_3)^3}\, (K\, R_1^2 + 2\,(1 - L\, R_1))\, R_1 + \frac{c\, K^2\, R_1^2}{4\, \varpi(R_1)^2}.$$
$$\tag{9.177}$$

Hence, we deduce that (NTM)

$$y_n = x_n - F'(x_n)^{-1}\, F(x_n)$$
$$x_{n+1} = y_n - c\, F'(x_n)^{-1}\, F(y_n), \quad c > 0$$

is of convergence order three provided that (9.144) holds. Several other choices for A, a, b, c are also possible (see Refs. [110, 120, 163, 208]).

We present now some numerical examples for validating the theoretical results of this section.

Example 9.4.1. Consider (1.293). Then, for $x^\star = 0$, we have $F(x^\star) = 0$ and $F'(x^\star) = e^0 = 1$. Moreover, hypotheses $(\mathcal{S}_1)$–$(\mathcal{S}_4)$ hold for $A(x) = F'(x)$ $(x \in \mathcal{D})$ and

$$K = e = 2.718281828, \quad L = e - 1 = 1.718281828, \quad M = \mu = \ell = 0.$$

Function ψ_0 is given by

$$\psi_0(r) = 1.359140914\, r^2 + 1.718281828\, r - 1.$$

By Lemma 9.4.1 (for example for $a = b = 1$ and $c = 5/6$), function ψ is given by

$$\psi(r) = -69.04763068\, r^2 + 46.67310733\, r^3 - 3.495236830\, r^4$$
$$-4.422853941\, r^5 - 8 + 41.23876387\, r$$

$$R_0 = .4334005735, \quad \psi(R_0) = .5118934862 \quad and \quad R = R_1 = .3612167951.$$

We give in Table 9.5 the values of $\psi(R_0)$, R and R_1 for various values of parameters a, b and c satisfying assumption $(\mathcal{S}_4)$, where $R_0 = .4334005735$.

Table 9.5   Values of $\psi(R_0)$, R and R_1

a	b	c	$\psi(R_0)$	R	R_1
1	1	.8333333333	.5118934862	.3612167951	.3612167951
1	1	.1	.0614272183	.4090368017	.4090368017
.1	.1	1	.1337025358	.3695899743	.3695899743
.1	.5	.5555555556	.0742791866	.3941526077	.3941526077
.2	.6	.5	.1256719449	.3743666097	.3743666097
.4	.7	.5	.2192215268	.3493942691	.3493942691
.45	.868	.24	.1140431053	.3830908277	.3830908277
.8689	.7859	1.633104500	1.021049270	.3030576608	.3030576608
.9999	.9123	877.00	538.7332546	.2708407173	.2708407173

Table 9.6   Values of $\psi(R_0)$, R and R_1

a	b	c	$\psi(R_0)$	R	R_1
.1	1	0	2.9×10^{-11}	.1333333333	.1333333333
.1	.1	1	.1240155518	.04029188762	.04029188762
.1	.5	.5555555556	.06889752882	.07361139239	.07361139239
.2	.6	.5	.1120363027	.04888375116	.04888375116
.4	.7	.5	.1761556092	.01302679070	.01302679070
.45	.868	.24	.08865523446	.06596454142	.06596454142
.8	.7	1.5	.4819296992	.1433430882	.1333333333
.99	.98	2	.3445246534	.08564610417e	.08564610417

Example 9.4.2. Consider Example 1.1.7. All hypotheses of Theorem 9.4.1 hold for $A(x) = F'(x)$ $(x \in \Omega)$, $x^\star = 0$, with $K = 15$, $L = 7.5$ and $M = \mu = \ell = 0$. Function ψ_0 is given by $\psi_0(r) = 7.5\,r^2 + 7.5\,r - 1$. By Lemma 9.4.1 (for example for $a = .2$, $b = .6$ and $c = .5$), function ψ is given by

$$\psi(r) = -990\,r^2 + 2304\,r^3 + 3015\,r^4 - 3.2 + 108\,r - 270\,r^5$$

$$R_0 = .1191391874, \quad \psi(R_0) = .1120363027 \quad \text{and} \quad R = R_1 = .04888375116.$$

We give in Table 9.6 the values of $\psi(R_0)$, R and R_1 for various values of parameters a, b and c satisfying assumption $(\mathcal{S}_4)$, where $R_0 = .1191391874$.

In the last example we are not interested in checking if the hypotheses of Theorem 9.4.1 are satisfied or not, but comparing the numerical behavior of a special case of (TSNLM) with earlier (CSTM).

Example 9.4.3. Consider Example 4.3.4. We shall compare (CSTM) with the Steffensen-type method (STTM) defined by

$$\begin{cases} x_0 \in \Omega, \\ y_n = x_n - [x_n, G(x_n); F]^{-1}\,F(x_k) \\ z_n = x_n + a\,(y_n - x_n) \\ x_{n+1} = x_n - [x_n, G(x_n); F]^{-1}\,(b\,F(x_n) + c\,F(z_n)), \quad n \geq 0, \end{cases}$$

where $G : \mathcal{X} \to \mathcal{X}$. Clearly (STTM) reduces to (TSNLM), if $A(x) = [x, G(x); F]$ $(x \in \Omega)$. Note that (CSTM) and (STTM) are useful alternatives of (CNTM),

Table 9.7 Comparison Table of $\|\overline{x}_{n+1} - \overline{x}_n\|$

n	STTM $(a = b = c = 1)$	CSTM $(a = b = c = 1)$
1	2.49e-01	2.49e-01
2	5.69e-04	6.14e-04
3	3.40e-12	5.76e-07
4	4.34e-37	1.91e-15
5	6.36e-112	4.34e-30
6	1.54e-336	8.04e-62
ρ	2.813536145	2.16697778

especially in cases where the computation of $F''(x_n)$ of $F'(x_n)^{-1}$ is expensive or impossible to compute or their analytic representation is unavailable. The computational order of convergence (COC) is shown in Table 9.7 for various methods. Here (COC) is defined in Ref. [239] (see also section 7.1). The last line in Table 9.7 that the (COC) of (STTM) and (CSTM) is, respectively close to 2.813536145 and 2.16697778. Hence, (STTM) is faster than (CSTM) in this case. Set $G(x) = x$. If we choose $\overline{x}_0 = (1, 1, \ldots, 1)^T$ and $\overline{x}_{-1} = (.99, .99, \ldots, .99)^T$. Assume sequence $\{\overline{x}_n\}$ is generated by (STTM) (or (CSTM)) with different choices of parameters a, b and c. Table 9.7 gives the comparison results for $\|\overline{x}_{n+1} - \overline{x}_n\|$ equipped with the max-norm, which show that (STTM) is faster than (CSTM).

9.5 A Unifying Semi-local Convergence

In this section we unify the semi-local study of Newton-like methods (1.55) under more general Lipschitz-type conditions and majorizing sequences than in earlier studies. The new convergence results for solving (1.54) can be weaker than in earlier studies. An interesting choice of operator A in (1.55) is

$$A(x) := \mathcal{H}(x) = \mathcal{I} - \frac{1}{2} F'(x)^{-1} F''(x) F'(x)^{-1} F(x).$$

Then, if $G(x) = 0$ for each $x \in \mathcal{D}$, (1.55) becomes Halley's method defined by

$$\begin{aligned} &x_0 \in \mathcal{D} \\ &x_{n+1} = x_n - \mathcal{H}(x_n)^{-1} F(x_n) \quad \text{for each} \quad n = 0, 1, 2, \cdots. \end{aligned} \tag{9.178}$$

Majorizing sequences play an important role in the study of iterative methods. We define very general majorizing sequences. We shall first present the motivation for introducing scalar sequences of a certain type appearing in Lemma 9.5.1.

Case 1. Newton's method under Lipschitz continuity conditions.

Let $A(x) = F'(x)$ and $G(x) = 0$ for each $x \in \mathcal{D}$. The following Kantorovich conditions have been used (see Refs. [163, 290, 418])

$$\| F'(x_0)^{-1} F(x_0) \| \leq \eta, \tag{9.179}$$

$$\| F'(x_0)^{-1} (F'(x) - F'(y)) \| \le L \, \| x - y \| \quad \text{for each} \quad x, y \in \mathcal{D} \tag{9.180}$$

together with the majorizing sequence

$$
\begin{aligned}
&t_0 = 0, \quad t_1 = \eta \\
&t_{n+2} = t_{n+1} + \frac{L \, (t_{n+1} - t_n)^2}{2 \, (1 - L \, t_{n+1})} \quad \text{for each} \quad n = 0, 1, \cdots
\end{aligned}
\tag{9.181}
$$

to show the semi-local convergence of Newton's method. We have used (see Refs. [106, 155, 163]), (9.179), (9.180),

$$\| F'(x_0)^{-1} (F'(x) - F'(x_0)) \| \le L_0 \, \| x - x_0 \| \quad \text{for each} \quad x \in \mathcal{D}$$

together with majorizing sequences

$$
\begin{aligned}
&t_0 = 0, \quad t_1 = \eta \\
&t_{n+2} = t_{n+1} + \frac{\dfrac{L}{2} t_{n+1}^2 - t_{n+1} + \eta}{1 - L_0 \, t_{n+1}} \quad \text{for each} \quad n = 0, 1, \cdots
\end{aligned}
\tag{9.182}
$$

or

$$
\begin{aligned}
&t_0 = 0, \quad t_1 = \eta \\
&t_{n+2} = t_{n+1} + \frac{L \, (t_{n+1} - t_n)^2}{2 \, (1 - L_0 \, t_{n+1})} \quad \text{for each} \quad n = 0, 1, \cdots .
\end{aligned}
\tag{9.183}
$$

Iterations (9.182) and (9.183) are tighter than (9.181) since $L_0 \le L$ holds in general and L/L_0 can be arbitrarily large.

Case 2. Newton's method under Hölder continuity conditions.

Let $A(x) = F'(x)$ and $G(x) = 0$ for each $x \in \mathcal{D}$. Suppose that (9.179) holds. The following Hölder-type conditions have been used (see Refs. [163, 290, 418, 444])

$$
\begin{aligned}
&\| F'(x_0)^{-1} (F'(x) - F'(y)) \| \le H \, \| x - y \|^\kappa \\
&\text{for each} \quad x, y \in \mathcal{D} \quad \text{and some} \quad \kappa \in [0, 1]
\end{aligned}
\tag{9.184}
$$

together with the majorizing sequences

$$
\begin{aligned}
&t_0 = 0, \quad t_1 = \eta \\
&t_{n+2} = t_{n+1} + \frac{H \, (t_{n+1} - t_n)^{1+\kappa}}{(1 + \kappa) \, (1 - H \, t_{n+1}^\kappa)} \quad \text{for each} \quad n = 0, 1, \cdots
\end{aligned}
\tag{9.185}
$$

or

$$
\begin{aligned}
&t_0 = 0, \quad t_1 = \eta \\
&t_{n+2} = t_{n+1} + \frac{H \, t_{n+1}^{1+\kappa} + \left(\dfrac{H \, \eta^\kappa}{1 + \kappa} - 1 \right) t_{n+1} - H \, t_{n+1}^\kappa + \eta}{1 - H \, t_{n+1}^\kappa} \\
&\text{for each} \quad n = 0, 1, \cdots
\end{aligned}
$$

to show the semi-local convergence of Newton's method. We have used (see Refs. [110, 163]), (9.179), (9.184),

$$\| F'(x_0)^{-1} (F'(x) - F'(x_0)) \| \le H_0 \, \| x - x_0 \|^\kappa \quad \text{for each} \quad x \in \mathcal{D}$$

together with majorizing sequences

$$t_0 = 0, \quad t_1 = \eta$$

$$t_{n+2} = t_{n+1} + \frac{H \, t_{n+1}^{1+\kappa} + \left(\dfrac{H \, \eta^{\kappa}}{1+\kappa} - 1 \right) t_{n+1} - H \, t_{n+1}^{\kappa} + \eta}{1 - H_0 \, t_{n+1}^{\kappa}} \tag{9.186}$$

for each $n = 0, 1, \cdots$

or

$$t_0 = 0, \quad t_1 = \eta$$

$$t_{n+2} = t_{n+1} + \frac{H \, (t_{n+1} - t_n)^{1+\kappa}}{(1+\kappa) \, (1 - H_0 \, t_{n+1}^{\kappa})} \quad \text{for each} \quad n = 0, 1, \cdots \tag{9.187}$$

in our convergence analysis. We have $H_0 \leq H$ and (9.187) is tighter than (9.186) which in turn is tighter than (9.185). Note also that if $\kappa = 1$ we obtain the Lipschitz case 1.

Case 3. Newton-like method.

The following conditions have been used for all $x, y \in \mathcal{D}$ and some $M \geq 0$, $m \geq 0$, $\ell \geq 0$ and $N \geq 0$

$$\| A(x_0)^{-1} \, (F(x_0) + G(x_0)) \| \leq \eta,$$

$$\| A(x_0)^{-1} \, (F'(x) - F'(y)) \| \leq L \, \| x - y \|$$

$$\| A(x_0)^{-1} \, (F'(x) - A(x)) \| \leq M \, \| x - x_0 \| + m,$$

$$\| A(x_0)^{-1} \, (A(x) - A(x_0)) \| \leq L_0 \, \| x - x_0 \| + \ell,$$

$$\| A(x_0)^{-1} \, (G(x) - G(y)) \| \leq N \, \| x - y \|$$

together with majorizing sequence for $\sigma_0 = \max(L, M + L_0)$ defined by

$$t_0 = 0, \quad t_1 = \eta$$

$$t_{n+2} = t_{n+1} + \frac{\dfrac{\sigma_0}{2} \, (t_{n+1} - t_n) + (M \, t_n + m + N)}{1 - \ell - L_0 \, t_{n+1}} \, (t_{n+1} - t_n) \tag{9.188}$$

for each $n = 0, 1, \cdots$

or

$$t_0 = 0, \quad t_1 = \eta$$

$$t_{n+2} = t_{n+1} + \frac{f(t_{n+1})}{g(t_{n+1})} \quad \text{for each} \quad n = 0, 1, \cdots, \tag{9.189}$$

where for $b = m + \ell + N$, functions f, g are defined by

$$f(s) = \frac{\sigma_0}{2} \, s^2 - (1 - b) \, s + \eta \quad \text{and} \quad g(s) = 1 - L_0 \, s - \ell.$$

We have used the above conditions together with majorizing sequence

$$t_0 = 0, \quad t_1 = \eta$$

$$t_{n+2} = t_{n+1} + \frac{\dfrac{L}{2}(t_{n+1} - t_n) + (M\,t_n + m + N)}{1 - \ell - L_0\,t_{n+1}}(t_{n+1} - t_n) \tag{9.190}$$

for each $n = 0, 1, \cdots$.

We have that sequence (9.190) is tighter than (9.188) which in turn is tighter than (9.189). If $A(x) = F'(x)$ and $G(x) = 0$ for each $x \in \mathcal{D}$, then, conditions and sequences of case 3 reduce to the ones of case 1, since $M = m = N = \ell = 0$. Other cases involving Secant-type and inexact iterative procedures can be found in Refs. [100, 106, 163]. It follows from cases 1–3 (see, in particulier (9.181), (9.183), (9.185), (9.187), (9.188), (9.189)) that the majorizing sequences introduced in the rest of this section are suitable generalizations of majorizing sequences for Newton-like methods. Note also that the majorizing sequences given in cases 1–3 are connected to equation (1.54) through Newton-like method (1.55), the Lipschitz condition and the Ostrowski-type approximation

$$F(x_{n+1}) + G(x_{n+1}) =$$
$$(F(x_{n+1}) - F(x_n) - F'(x_n)(x_{n+1} - x_n)) +$$
$$(F'(x_n) - A(x_n))(x_{n+1} - x_n) + (G(x_{n+1}) - G(x_n)) \quad \text{for each} \quad n = 0, 1, \cdots$$

(see (9.239)–(9.242)).

We need the following results on majorizing sequences for Newton-like method given by (1.55).

Lemma 9.5.1. *Let $\psi : [0, \infty)^2 \longrightarrow [0, \infty)$ be continuous and non-decreasing. Let $\eta > 0$ and $\beta \geq 1$. Define scalar sequence $\{t_n\}$ by*

$$t_0 = 0, \ t_1 = \eta$$
$$t_{n+2} = t_{n+1} + \psi(t_{n+1}, t_n)(t_{n+1} - t_n)^\beta \quad \text{for each} \quad n = 0, 1, \cdots. \tag{9.191}$$

Define functions λ, μ, $t_\beta^{\star\star}$ on interval $(0, +\infty)$ by

$$\lambda(s) = s^{1/(1-\beta)} \quad \text{if } \beta \neq 1, \quad \mu(s) = \frac{\eta}{\lambda(s)} \tag{9.192}$$

and

$$t_\beta^{\star\star}(s) = \begin{cases} \eta + \dfrac{\lambda(s)\,\mu(s)^\beta}{1 - \mu(s)^\beta} & \text{if } \beta \neq 1 \\[2ex] \dfrac{\eta}{1 - s} & \text{if } \beta = 1 \text{ and } s \neq 1 \end{cases} \tag{9.193}$$

for each $s > 0$.

Let $\overline{\psi}(s) = \psi(s, s)$ for each $s \in (0, +\infty)$. Suppose that one of the following sets of conditions is satisfied:

(a) There exists $\alpha > 0$ such that

$$\mu(\alpha) < 1 \tag{9.194}$$

and

$$\overline{\psi}(t_\beta^{\star\star}(\alpha)) \leq \alpha \quad \text{for} \quad \beta \neq 1. \tag{9.195}$$

(b) There exists $\alpha \in (0,1)$ such that

$$\overline{\psi}(t_1^{\star\star}(\alpha)) \leq \alpha.$$

Then, sequence $\{t_n\}$ is non-decreasing, bounded from above by $t_\beta^{\star\star}(\alpha)$ and converges to its unique least upper bound $t_\beta^{\star}(\alpha)$ such that $t_\beta^{\star}(\alpha) \in [0, t_\beta^{\star\star}(\alpha)]$. Moreover, the following estimates hold for each $n = 0, 1, \cdots$

$$t_{n+1} - t_n \leq \begin{cases} \lambda(\alpha)\,\mu(\alpha)^{\beta^n} \leq \lambda(\alpha)\,\mu(\alpha)^{n\,\beta} & if \ \beta \neq 1 \\ \alpha^n\,\eta & if \ \beta = 1 \end{cases} \tag{9.196}$$

and

$$t_\beta^{\star}(\alpha) - t_n \leq \begin{cases} \dfrac{\lambda(\alpha)}{1 - \mu(\alpha)}\,\mu(\alpha)^{\beta^n} \leq \dfrac{\lambda(\alpha)\,\mu(\alpha)^{n\,\beta}}{1 - \mu(\alpha)} & if \ \beta \neq 1 \\ \dfrac{\alpha^n}{1 - \alpha}\,\eta & if \ \beta = 1. \end{cases} \tag{9.197}$$

Proof. We use mathematical induction to prove that

$$\psi(t_{k+1}, t_k) \leq \alpha \quad \text{for each} \quad k = 0, 1, \cdots . \tag{9.198}$$

Case $\beta \neq 1$. Estimate (9.198) holds for $k = 0$ by (9.191)–(9.195). Then, we have by (9.191) and (9.198) that

$$t_2 - t_1 \leq \alpha\,(t_1 - t_0)^\beta.$$

That is (9.196) holds for $n = 0$ and $n = 1$. Assume (9.198) holds for all natural integers $k \leq n+1$. Then, we have that $t_{k+1} - t_k \leq \alpha\,(t_k - t_{k-1})^\beta$. Hence, we deduce that (9.196) holds. We also get for $\beta \neq 1$ that

$$\begin{aligned}
t_{k+1} - t_k &\leq \alpha\,(t_k - t_{k-1})^\beta \\
&\leq \alpha\,(\alpha\,(t_{k-1} - t_{k-2})^\beta)^\beta \\
&= \alpha^{1+\beta}\,(t_{k-1} - t_{k-2})^{\beta^2} \leq \alpha^{(1-\beta^k)/(1-\beta)}\,(t_1 - t_0)^{\beta^k}.
\end{aligned}$$

Hence, we proved the first part in (9.196). Moreover, we obtain that

$$\begin{aligned}
t_{k+1} &\leq t_k + \alpha\,(t_k - t_{k-1})^\beta \\
&\leq t_1 + ((t_1 - t_0)^\beta + (t_2 - t_1)^\beta + \cdots + (t_k - t_{k-1})^\beta) \\
&\leq \eta + \lambda(\alpha)\,(\mu(\alpha)^\beta + \mu(\alpha)^{2\,\beta} + \cdots + \mu(\alpha)^{k\,\beta}) \\
&= \eta + \lambda(\alpha)\,\mu(\alpha)^\beta\,(\mu(\alpha)^\beta + \cdots + \mu(\alpha)^{(k-1)\,\beta}) \\
&= \eta + \lambda(\alpha)\,\mu(\alpha)^\beta\,\frac{1 - \mu(\alpha)^{k\,\beta}}{1 - \mu(\alpha)^\beta} \leq \eta + \frac{\lambda(\alpha)\,\mu(\alpha)^\beta}{1 - \mu(\alpha)^\beta} = t_\beta^{\star\star}(\alpha).
\end{aligned} \tag{9.199}$$

That is, we deduce

$$\psi(t_{k+1}, t_k) \leq \overline{\psi}(t_\beta^{\star\star}(\alpha)) \leq \alpha,$$

by (9.194) and (9.195) since, ψ is non-decreasing. That completes the induction for (9.198) in the case $\beta \neq 1$.

Case $\beta = 1$. Similarly, we obtain that

$$\begin{aligned}
t_{k+1} - t_k &\leq \alpha\,(t_k - t_{k-1}) \\
&\leq \alpha\,(\alpha\,(t_{k-1} - t_{k-2})) = \alpha^2\,(t_{k-1} - t_{k-2}) \leq \alpha^k\,(t_1 - t_0).
\end{aligned}$$

That shows the second part in (9.196). Moreover, we have that

$$t_{k+1} \le t_k + \alpha\,(t_k - t_{k-1})$$

$$\le t_1 + \alpha\,\eta + \alpha^2\,\eta + \cdots + \alpha^k\,\eta = \frac{1 - \alpha^{k+1}}{1 - \alpha} \le t_1^{\star\star}(\alpha).$$

Hence, we get that

$$\psi(t_{k+1}, t_k) \le \overline{\psi}(t_1^{\star\star}(\alpha)) \le \alpha.$$

It follows from (9.192) and (9.196) that sequence $\{t_n\}$ is non-decreasing, bounded from above by $t_\beta^{\star\star}(\alpha)$ and as such it converges to $t_\beta^{\star}(\alpha)$. Finally, estimate (9.197) follows from (9.196) by using standard majorization techniques. The proof of Lemma 9.5.1 is complete. $\qquad\square$

Next, we provide convergence conditions for Newton's majorizing sequence given by (9.183) using Lemma 9.5.1, when $\beta = 2$ and $\beta = 1$, respectively.

Lemma 9.5.2. *Let $\eta > 0$, $L_0 > 0$, $L > 0$ and $\beta = 2$. Define polynomial p on $[0, \infty)$ by*

$$p(s) = L_0\,\eta^3\,s^3 - \left(\frac{L}{2} - L_0\right)\eta^2\,s^2 - (1 - L_0\,\eta)\,s - \frac{L}{2}. \tag{9.200}$$

Denote by α the minimal root of polynomial p on $[0, \infty)$. Suppose that

$$0 < \frac{L}{2\,(1 - L_0\,\eta)} \le \alpha < \min\{\frac{1}{\eta},\, 1\}.$$

Then, the assertions (9.196) and (9.197) hold for sequence $\{t_n\}$ defined by (9.183) for $\lambda = \dfrac{1}{\alpha}$ and $\mu = \alpha\,\eta$. Moreover, the bound $t_2^{\star\star}(\alpha)$ is defined by

$$t_2^{\star\star}(\alpha) = \eta + \frac{\alpha\,\eta^2}{1 - (\alpha\,\eta)^2} + (\alpha_0 - \alpha)\,\eta^2, \tag{9.201}$$

where

$$\alpha_0 = \frac{L}{2\,(1 - L_0\,\eta)}. \tag{9.202}$$

Proof. According to Lemma 9.5.1, we must prove that

$$0 < \psi(t_{k+1}, t_k) \le \alpha, \tag{9.203}$$

where function ψ is given by

$$\psi(t, s) = \frac{L\,(t - s)}{2\,(1 - L_0\,t)}.$$

Estimate (9.203) holds for $k = 1$, since

$$0 < \frac{L}{2\,(1 - L_0\,\eta)} \le \alpha.$$

We can prove instead of (9.203) that

$$\frac{L}{2\,(1 - L_0\,t_{k+1})} \le \alpha,$$

or

$$\frac{L}{2} + \alpha\, L_0\, t_{k+1} - \alpha \leq 0,$$

or

$$\frac{L}{2} + \alpha\, L_0 \left(\eta + \frac{1}{\alpha} \left(1 + (\alpha\,\eta)^2 + \cdots + (\alpha\,\eta)^{2\,(k-1)} \right) (\alpha\,\eta)^2 \right) - \alpha \leq 0. \tag{9.204}$$

Estimate (9.204) motivates us to define recurrent functions f_k for each $k = 0, 1, \cdots$ given by

$$f_k(s) = \frac{L}{2} + (L_0\,\eta - 1)\, s + L_0\, (s\,\eta)^2 \left(1 + (s\,\eta)^2 + \cdots + (s\,\eta)^{2\,(k-1)} \right). \tag{9.205}$$

We need a relationship between two consecutive functions f_k. Notice that

$$\begin{aligned}
f_{k+1}(s) &= \frac{L}{2} + (L_0\,\eta - 1)\, s + L_0\, (s\,\eta)^2 \left(1 + (s\,\eta)^2 + \cdots + (s\,\eta)^{2\,(k-1)} + (s\,\eta)^k \right) \\
&= f_k(s) + L_0\, (s\,\eta)^{2\,(k+1)} \geq f_k(s) \quad \text{for each} \quad k = 0, 1, \cdots .
\end{aligned} \tag{9.206}$$

Define function f_∞ on $[0, \infty)$ by

$$f_\infty(s) = \lim_{k \longrightarrow \infty} f_k(s). \tag{9.207}$$

Using (9.205)–(9.207), we have that (9.204) certainly holds, if

$$f_\infty(\alpha) \leq 0 \quad \text{or} \quad \frac{L}{2} + \alpha\, L_0\,\eta + \frac{\alpha^2\,\eta^2}{1 - \alpha^2\,\eta^2}\, L_0 - \alpha \leq 0,$$

since $f_\infty(\alpha) = \dfrac{L}{2} + \alpha\, L_0\,\eta + \dfrac{\alpha^2\,\eta^2}{1 - \alpha^2\,\eta^2}\, L_0 - \alpha,$
or (by (9.200))

$$p(\alpha) \leq 0. \tag{9.208}$$

But (9.208) is true as equality by the choice of α. The rest follows from the proof of Lemma 9.5.1. Note that (9.201) follows from (9.197), (9.199), (9.202) and the estimate

$$t_2 - t_1 \leq \alpha_0\, (t_1 - t_0)^2 = \alpha\,\eta^2 - \alpha\,\eta^2 + \alpha_0\,\eta^2 = \alpha\,\eta^2 + (\alpha_0 - \alpha)\,\eta^2.$$

The proof of Lemma 9.5.2 is complete. $\qquad\qquad\qquad\qquad\qquad\qquad\qquad\square$

Lemma 9.5.3. *Let $\eta > 0$, $L_0 > 0$, $L > 0$ and $\beta = 1$. Suppose that (1.301) holds, where r_0 is replaced by η. Then, the assertions (9.196) and (9.197) hold for sequence $\{t_n\}$ defined by (9.183), where*

$$\alpha_0 = \frac{L\,\eta}{2\,(1 - L_0\,\eta)}, \quad \alpha = \frac{2\,L}{L + \sqrt{L^2 + 8\,L_0\,L}} \quad \text{and} \quad t_1^{\star\star}(\alpha) = \frac{\eta}{1 - \alpha}.$$

Proof. Let us define function ψ on interval $[0, 1/L_0)$ by

$$\psi(t, s) = \frac{L\,(t - s)}{2\,(1 - L_0\,t)}.$$

Then, we have for each $k = 1, 2, \cdots$ that

$$f_k(s) = \left(\frac{L}{2} + L_0\,(1 + s + \cdots + s^k) \right) \eta - 1,$$

$$f_{k+1}(s) = f_k(s) + \frac{1}{2}\,(2\,L_0\,s^2 + L\,s - L) \quad \text{and} \quad f_\infty(s) = \frac{L_0\,\eta}{1 - s} - 1.$$

In particular, we get that

$$f_k(\alpha) = f_\infty(\alpha) \leq 0 \quad \text{for each} \quad k = 1, 2, \cdots \quad \text{and} \quad \alpha_0 \leq \alpha$$

by the choice of α_0, α, $t_1^{\star\star}(\alpha)$ and (1.301). According to Lemma 9.5.1, the proof of Lemma 9.5.3 is complete. $\square$

Next, we provide more convergence conditions for majorizing sequence $\{t_n\}$.

Lemma 9.5.4. *Let ψ, η, β, λ, μ be as in Lemma 9.5.1. Let $\gamma > 0$ and set $\delta = \gamma - \beta$. Suppose there exists $\alpha \in (0, 1)$ such that*

$$\mu(\alpha) < 1 \tag{9.209}$$

and

$$\text{for} \quad \psi_n := \psi(t_{n+1}, t_n)\,(t_{n+1} - t_n)^\delta, \quad \psi_n \leq \alpha \quad \text{for each} \quad n = 0, 1, \cdots. \tag{9.210}$$

Then, the conclusions of Lemma 9.5.1 hold for sequence $\{t_n\}$ defined by

$$\begin{aligned} &t_0 = 0, \ t_1 = \eta \\ &t_{n+2} = t_{n+1} + \psi(t_{n+1}, t_n)\,(t_{n+1} - t_n)^\gamma \quad \text{for each} \quad n = 0, 1, \cdots. \end{aligned} \tag{9.211}$$

Proof. We use mathematical induction to prove that

$$\psi(t_{k+1}, t_k)\,(t_{k+1} - t_k)^\delta \leq \alpha \quad \text{for each} \quad k = 0, 1, \cdots. \tag{9.212}$$

Estimate (9.212) holds for $k = 0$ by (9.209) and for $k \geq 1$ by (9.210). Then, we have that

$$t_{k+2} - t_{k+1} = \psi(t_{k+1}, t_k)\,(t_{k+1} - t_k)^\gamma \leq \alpha\,(t_{k+1} - t_k)^\beta.$$

The rest follows from the proof of Lemma 9.5.1. The proof of Lemma 9.5.4 is complete. $\square$

Remark 9.5.1. If $\delta \geq 0$ (i.e. if $\gamma \geq \beta$), then (9.210) can be replaced by

$$\overline{\psi}(t_\beta^{\star\star}(\alpha))\,\eta^\beta \leq \alpha,$$

where $t_\beta^{\star\star}(\alpha)$ is given in closed form by (9.193).

We complete this section with a very general result for majorizing sequences $\{t_n\}$ given by (9.183).

Lemma 9.5.5. *Let $\psi : [0,\infty)^2 \longrightarrow [0,\infty)$ be continuous and non-decreasing. Let $\eta > 0$, $\beta \geq 0$ and $\beta \neq 1$. Define the functions for each $s > 0$ by*

$$\lambda(s) = s^{1/(1-\beta)}, \quad \mu(s) = \frac{\eta}{\lambda(s)},$$

$$t^{\star\star}(s) = \eta + \lambda(s)\,\mu(s)^\beta \, \frac{1}{1 - \mu(s)^\beta}, \quad s_n = s_n(s) = \lambda(s)\,\mu(s)^{\beta^n},$$

$$t_{n+1} = t_{n+1}(s) = \eta + s\,(s_0^\beta + s_1^\beta + \cdots + s_{n-1}^\beta),$$

$$t_{n+1}^{\star\star} := t_{n+1}^{\star\star}(s) = \eta + \lambda(s)\,\mu(s)^\beta \, \frac{1 - \mu(s)^{n\,\beta}}{1 - \mu(s)^\beta} \quad \text{and} \quad t^\star(s) = \lim_{n \longrightarrow \infty} t_{n+1}(s).$$

Suppose there exists $\alpha > 0$ such that

$$\mu(\alpha) < 1 \tag{9.213}$$

and one of the following sets of conditions is satisfied:

I. *There exist $\delta \in [0,\alpha]$ and $\alpha_0 > \alpha$ such that*

$$\psi(\eta, 0) \leq \alpha \leq \alpha_0,$$

$$\psi(t_{n+1}(s), t_n(s)) \leq \psi(t_{n+2}(s), t_{n+1}(s)) \quad \text{for each} \quad s \in [\delta, \alpha], \tag{9.214}$$

$$\overline{\psi}(t^\star(\alpha)) \leq \alpha_0, \tag{9.215}$$

or

II. *There exists $\delta \geq \alpha$ such that*

$$\psi(\eta) \leq \alpha,$$

$$\psi(t_{n+2}(s), t_{n+1}(s)) \leq \psi(t_{n+1}(s), t_n(s)) \quad \text{for each} \quad s \in [\alpha, \delta], \tag{9.216}$$

$$\overline{\psi}(t_1(\alpha)) \leq \alpha. \tag{9.217}$$

Then, the conclusions of Lemma 9.5.1 hold for sequence $\{t_n\}$, where $\{t_n\}$ is defined by (9.191) for $\beta \neq 1$.

Proof. **Case I.** We use mathematical induction to prove that

$$\psi(t_{k+1}, t_k) \leq \alpha \quad \text{for each} \quad k = 0, 1, \cdots. \tag{9.218}$$

Estimate (9.218) holds for $k = 0$ by (9.213). Assume (9.218) holds for all $k \leq n$. Then, as in Lemma 9.5.1, we have that $t_{k+1} - t_k \leq s_k(\alpha)$ and $t_{k+1} \leq t_{k+1}(\alpha)$. Evidently, estimate (9.218) is true if

$$\psi(t_{k+1}(\alpha), t_k(\alpha)) \leq \alpha \quad \text{for each} \quad k = 0, 1, \cdots. \tag{9.219}$$

Estimate (9.219) motivates us to define recurrent functions f_k on $[0, \infty)$ by
$$f_k(s) = \psi(t_{k+1}(s), t_k(s)) - s.$$
We need a relationship between two consecutive functions f_k. Notice that
$$f_{k+1}(s) = \psi(t_{k+2}(s), t_{k+1}(s)) - s + f_k(s) - \psi(t_{k+1}(s), t_k(s)) + s.$$
Hence, we obtain that
$$f_{k+1}(s) = f_k(s) + (\psi(t_{k+2}(s), t_{k+1}(s)) - \psi(t_{k+1}(s), t_k(s))). \tag{9.220}$$
Estimate (9.219) can be written as
$$f_k(\alpha) \le 0. \tag{9.221}$$
In view of (9.214), we have that
$$f_{k+1}(\alpha) \ge f_k(\alpha). \tag{9.222}$$
Define function f_∞ on $[0, \infty)$ by
$$f_\infty(s) = \lim_{k \longrightarrow \infty} f_k(s).$$
Then, we have by (9.221) and (9.222) that (9.220) certainly holds, if
$$f_\infty(\alpha_0) \le 0,$$
which is true by (9.215). The rest of the proof follows as in Lemma 9.5.1.
Case II. We proceed as in part **I** but this time we have by (9.216) that
$$f_{k+1}(\alpha) \le f_k(\alpha) \le \cdots \le f_1(\alpha).$$
Evidently, (9.221) holds, if
$$f_1(\alpha) \le 0. \tag{9.223}$$
But (9.223) is true by (9.217).

The proof of Lemma 9.5.5 is complete. $\qquad\square$

Remark 9.5.2. In general, sequence $\{t_n\}$ given by (9.191) certainly converges to $t^\star$ if it is bounded above by some positive number (which is greater than or equal to η). In particular, for the case of Newton's majorizing sequence (9.183), let us assume that there exists σ such that
$$L_0 t_n \le \sigma < 1 \quad \text{for each} \quad n = 0, 1, \cdots. \tag{9.224}$$
Sequence $\{t_n\}$ is then non-decreasing, bounded above by σ/L_0 and as such it converges to some $t^\star$ such that
$$t^\star \le \frac{\sigma}{L_0}. \tag{9.225}$$
Note that we can directly check if (9.224) holds (see Example 9.5.1). Another possibility is given by the following scheme: Let us define
$$L_0 \eta = \sigma_0, \quad b = \frac{1}{1 - \sigma}, \quad b_0 = \frac{1}{1 - \sigma_0},$$
$$c_0 = L_0 b_0, \quad c = L b, \quad \gamma^2 = \frac{c_0}{c}, \quad d = c\gamma\eta, \quad p = d^2$$
$$e_n = c(t_{n+1} - t_n),$$
and
$$s_{n+2} = \eta + c_0 \eta^2 + c c_0^2 \eta^4 + \frac{1}{c}(d^{2^3} + \cdots + d^{2^{k+1}}).$$

Then, we have the following result.

Lemma 9.5.6. *Suppose that*

$$\eta \leq \frac{2}{L_0 + \sqrt{L_0^2 + 4\,L_0\,L}}, \tag{9.226}$$

$$0 < \sigma < 1 - \frac{L_0\,L\,\eta^2}{1 - L_0\,\eta} \tag{9.227}$$

and

$$L_0\,q(\sigma) \leq \sigma, \tag{9.228}$$

where

$$q(t) = \eta + \frac{L_0\,\eta^2}{1 - L_0\,\eta} + \frac{L_0^2}{(1 - L_0\,\eta)^2}\,\frac{L}{1-t}\,\eta^4 + \frac{\dfrac{L^2\,L_0^3\,\eta^6}{(1-t)^2\,(1-L_0\,\eta)^3}}{1 - \dfrac{L\,L_0\,\eta^2}{(1-t)\,(1-L_0\,\eta)}}.$$

Then, sequence $\{t_n\}$ given by (9.183) is non-decreasing, bounded from above by σ/L_0 and converges to its unique upper bound $t^\star$ satisfying (9.225).

Proof. We use mathematical induction to prove (9.224) for each $k = 0, 1, \cdots$. Estimate (9.224) holds for $k = 0, 1$ by (9.191) and (9.226). Note that $d \in (0, 1)$ by (9.226) and (9.227). We have in turn that

$$t_2 - t_1 \leq L_0\,b_0\,(t_1 - t_0)^2 \leq (c_0\,(t_1 - t_0))^2, \quad t_2 \leq t_1 + c_0\,\eta^2,$$

$$t_3 - t_2 \leq c\,(t_2 - t_1)^2, \quad e_2 = c\,(t_3 - t_2) \leq (c\,(t_2 - t_1))^2,$$

$$t_3 \leq t_2 + c\,c_0^2\,\eta^4, \quad e_3 = c\,(t_4 - t_3) \leq (c\,(t_3 - t_2))^2 = e_2^2,$$

$$e_4 = c\,(t_5 - t_4) \leq e_3^2 \leq (e_2^2)^2 = e_2^{2^2}.$$

Hence, we get for $k = 2, 3, \cdots$ that

$$e_{k+1} \leq e_2^{2^{k-1}} \leq ((c\,(t_2 - t_1))^2)^{2^{k-1}} = (c\,(t_2 - t_1))^{2^k}, \quad e_{k+1} \leq d^{2^{k+1}}$$

and

$$\begin{aligned}
t_{k+2} &\leq t_{k+1} + c^{-1}\,d^{2^{k+1}} \leq t_k + c^{-1}\,(d^{2^k} + d^{2^{k+1}}) \leq t_3 + c^{-1}\,(d^{2^3} + \cdots + d^{2^{k+1}}) \\
&\leq t_2 + c\,c_0^2\,\eta^4 + c^{-1}\,(d^{2^3} + \cdots + d^{2^{k+1}}) \\
&\leq \eta + c_0\,\eta^2 + c\,c_0^2\,\eta^4 + c^{-1}\,(d^{2^3} + \cdots + d^{2^{k+1}}) = s_{k+2} \\
&\leq \eta + c_0\,\eta^2 + c\,c_0^2\,\eta^4 + c^{-1}\,(d^{2\times 3} + \cdots + d^{2\,(k+1)}) \\
&\leq \eta + c_0\,\eta^2 + c\,c_0^2\,\eta^4 + c^{-1}\,(p^3 + \cdots + p^{k+1}) \\
&= \eta + c_0\,\eta^2 + c\,c_0^2\,\eta^4 + c^{-1}\,p^3\,(1 + \cdots + p^{k-2}) \\
&= \eta + c_0\,\eta^2 + c\,c_0^2\,\eta^4 + c^{-1}\,p^3\,\frac{1 - p^{k-1}}{1 - p} \leq \eta + c_0\,\eta^2 + c\,c_0^2\,\eta^4 + \frac{p^3}{c\,(1-p)} = q(\sigma).
\end{aligned}$$

Then, (9.224) certainly holds, if $L_0\,q(\sigma) \leq \sigma$. But this estimate is true by (9.228). According to Lemma 9.5.5, the proof of Lemma 9.5.6 is complete. $\qquad\square$

We provide the main semi-local convergence result for Newton-like method.

Theorem 9.5.1. *Let $F : \mathcal{D} \subseteq X \longrightarrow \mathcal{Y}$ be Fréchet-differentiable. Let $G : \mathcal{D} \longrightarrow \mathcal{Y}$ be continuous and let $A(x) \in \mathcal{L}(X, \mathcal{Y})$ be an approximation of $F'(x)$ for each $x \in \mathcal{D}$. Suppose there exist $x_0 \in \mathcal{D}$, $\eta > 0$, $\beta \geq 1$, a bounded inverse $A(x_0)^{-1}$ of $A(x_0)$, functionals $v : \mathcal{D}^2 \longrightarrow [0, \infty)$, $w : \mathcal{D} \longrightarrow [0, 1)$, $z : \mathcal{D} \longrightarrow [0, \infty)$ and $p : \mathcal{D}^2 \longrightarrow [0, \infty)$ such that for all x, $y \in \mathcal{D}$, we have the following*

$$\| A(x_0)^{-1} (F(y) - F(x) - F'(x)(y - x)) \| \leq v(x, y) \| y - x \|^{\beta}, \tag{9.229}$$

$$\| A(x_0)^{-1} (A(x) - A(x_0)) \| \leq w(x), \tag{9.230}$$

$$\| A(x_0)^{-1} (F'(x) - A(x))(y - x) \| \leq z(x) \| y - x \|^{\beta}, \tag{9.231}$$

$$\| A(x_0)^{-1} (G(x) - G(y)) \| \leq p(x, y) \| x - y \|^{\beta}, \tag{9.232}$$

$$\eta \geq \| A(x_0)^{-1} (F(x_0) + G(x_0)) \|; \tag{9.233}$$

$\psi : [0, \infty)^2 \longrightarrow [0, \infty)$ continuous and non-decreasing such that if

$$\psi_{n+1} = \psi(t_{n+1}, t_n) = \frac{q_n}{1 - r_n} \quad for \ each \quad n = 0, 1, \cdots,$$

hypotheses of Lemma 9.5.1 hold and $\overline{U}(x_0, t_{\beta}^{\star}(\alpha)) \subseteq \mathcal{D}$, where

$$r_n = w(x_{n+1}) \quad and \quad q_n = v(x_n, x_{n+1}) + p(x_n, x_{n+1}) + z(x_n).$$

Then, sequence $\{x_n\}$ $(n \geq 0)$ generated by Newton-like method is well defined, remains in $\overline{U}(x_0, t_{\beta}^{\star}(\alpha))$ for each $n = 0, 1, \cdots$ and converges to a solution $x^{\star} \in \overline{U}(x_0, t_{\beta}^{\star}(\alpha))$ of equation (1.54). Moreover, the following estimates hold for each $n = 0, 1, \cdots$

$$\| x_{n+1} - x_n \| \leq t_{n+1} - t_n$$

and

$$\| x_n - x^{\star} \| \leq t_{\beta}^{\star}(\alpha) - t_n. \tag{9.234}$$

Furthermore, suppose that

$$\psi_{n+1}^{\star} = \frac{q_n^{\star}}{1 - r_n} \longrightarrow 0 \quad as \quad n \longrightarrow \infty.$$

Then, limit point $x^{\star}$ is the unique solution of equation (1.54) in $U(x_0, t_{\beta}^{\star}(\alpha))$, where

$$q_n^{\star} = v(x_n, x^{\star}) + p(x_n, x^{\star}) + z(x_n).$$

Proof. We use mathematical induction to prove that sequence $\{x_k\}$ is well-defined,

$$\| x_{k+1} - x_k \| \le t_{k+1} - t_k \quad \text{for all} \quad k = 0, 1, \cdots \tag{9.235}$$

and

$$\overline{U}(x_{k+1}, t_\beta^\star(\alpha) - t_{k+1}) \subseteq \overline{U}(x_k, t_\beta^\star(\alpha) - t_k) \quad \text{for all} \quad k = 0, 1, \cdots . \tag{9.236}$$

Let $x \in \overline{U}(x_1, t_\beta^\star(\alpha) - t_1)$. Then, estimate

$$\| x - x_0 \| \le \| x - x_1 \| + \| x_1 - x_0 \| \le t_\beta^\star(\alpha) - t_1 + t_1 - t_0 = t_\beta^\star(\alpha) - t_0, \tag{9.237}$$

implies that $x \in \overline{U}(x_0, t_\beta^\star(\alpha) - t_0)$. We also have that

$$\| x_1 - x_0 \| = \| A(x_0)^{-1} (F(x_0) + G(x_0)) \| \le \eta = t_1 - t_0. \tag{9.238}$$

Then, (9.238) together with (9.237) show (9.235) and (9.236) for $k = 0$. Let assume $\{x_k\}$ is well-defined for all $k \le n$. We get by

$$\| x_k - x_0 \| \le \sum_{i=1}^{k} \| x_i - x_{i-1} \| \le \sum_{i=1}^{k} (t_i - t_{i-1}) = t_k - t_0 = t_k < t_\beta^\star(\alpha).$$

That is, $x_k \in \overline{U}(x_0, t_\beta^\star(\alpha))$. Using (1.55), we obtain the approximation

$$\begin{aligned}
&F(x_{k+1}) + G(x_{k+1}) \\
&= F(x_{k+1}) + G(x_{k+1}) - F(x_k) - G(x_k) - A(x_k)(x_{k+1} - x_k) \\
&= (F(x_{k+1}) - F(x_k) - F'(x_k)(x_{k+1} - x_k)) + \\
&\quad (F'(x_k) - A(x_k))(x_{k+1} - x_k) + (G(x_{k+1}) - G(x_k)).
\end{aligned} \tag{9.239}$$

Then, it follows from (9.229), (9.231), (9.232) and (9.239), the definition of q_k and the induction hypotheses that

$$\begin{aligned}
&\| A(x_0)^{-1} (F(x_{k+1}) + G(x_{k+1})) \| \\
&\le (v(x_k, x_{k+1}) + p(x_k, x_{k+1}) + z(x_{k+1})) \| x_{k+1} - x_k \| \le q_k (t_{k+1} - t_k)^\beta.
\end{aligned} \tag{9.240}$$

Using (9.230) for $x = x_{k+1}$ and the Banach lemma on invertible operators, we obtain that $A(x_{k+1})^{-1} \in \mathcal{L}(\mathcal{Y}, \mathcal{X})$. Then, we get in turn that

$$\| A(x_{k+1})^{-1} A(x_0) \| \le \frac{1}{1 - w(x_{k+1})} = \frac{1}{1 - r_k}. \tag{9.241}$$

Then, by (1.55), (9.191), (9.240) and (9.241), we have that

$$\begin{aligned}
\| x_{k+2} - x_{k+1} \| &\le \| A(x_{k+1})^{-1} A(x_0) \| \, \| A(x_0)^{-1} (F(x_{k+1}) + G(x_{k+1})) \| \\
&\le \frac{q_k (t_{k+1} - t_k)^\beta}{1 - r_k} = \psi_{k+1} (t_{k+1} - t_k)^\beta = t_{k+2} - t_{k+1}.
\end{aligned}$$
$$\tag{9.242}$$

The induction for (9.235) is complete. Let $x \in \overline{U}(x_{k+1}, t_\beta^\star(\alpha) - t_{k+1})$. Then, we obtain that

$$\| x - x_k \| \le \| x - x_{k+1} \| + \| x_{k+1} - x_k \| \le t_\beta^\star(\alpha) - t_{k+1} + t_{k+1} - t_k = t_\beta^\star(\alpha) - t_k.$$

Hence, (9.236) is true for all k. The induction for (9.235) and (9.236) is now complete. In view of (9.235) and (9.236), sequence $\{x_k\}$ ($k \geq 0$) is Cauchy (since $\{t_n\}$ is Cauchy) in a Banach space $\mathcal{X}$ and as such it converges to some $x^\star \in \overline{U}(x_0, t_\beta^\star(\alpha))$ (since $\overline{U}(x_0, t_\beta^\star(\alpha))$ is a closed set). By letting $k \longrightarrow \infty$ in (9.240), we get $F(x^\star) + G(x^\star) = 0$ (since $A(x_0)^{-1}$, q_k are bounded and $\{t_k\}$ is a Cauchy sequence). Moreover, estimate (9.234) follows from (9.235) by using standard majorization techniques. Finally, to show the uniqueness part, let $y^\star$ be a solution of equation (1.54) in $\overline{U}(x_0, t_\beta^\star(\alpha))$. Using (1.55), we get the approximation

$$
\begin{aligned}
x_{k+1} - y^\star \\
= A(x_0)^{-1} \left(F(y^\star) - F(x_k) - F'(x_k)(y^\star - x_k) \right) + \\
A(x_0)^{-1} \left(F'(x_k) - A(x_k) \right)(y^\star - x_k) + A(x_0)^{-1} \left(G(y^\star) - G(x_k) \right).
\end{aligned}
\tag{9.243}
$$

It follows from (9.178), (9.229)–(9.232), (9.241) and (9.243) that

$$
\| x_{k+1} - y^\star \| \leq \frac{q_n^\star}{1 - r_n} \| x_k - y^\star \| = \psi_{n+1}^\star \| x_k - y^\star \| \longrightarrow 0 \quad \text{as} \quad k \longrightarrow \infty.
$$

Hence, we deduce $x^\star = y^\star$. The proof of Theorem 9.5.1 is complete. $\qquad\square$

Remark 9.5.3.

(a) Note that $t_\beta^{\star\star}(\alpha)$ given in closed form by (9.193) can replace $t_\beta^\star(\alpha)$ in Theorem 9.5.1.

(b) Hypotheses of Lemma 9.5.1 are needed in Theorem 9.5.1, to ensure the convergence of majorizing sequence $\{t_n\}$ given by (9.191). Clearly, these hypotheses can be replaced in Theorem 9.5.1 by the corresponding ones given in the results following Lemma 9.5.1.

We show how to choose sequences and functions introduced in this section for some interesting cases.

Application 9.5.1. Returning back to case 1 in this section, we see that we can choose

$$
q_n = L, \quad r_n = L_0 \quad \text{for each} \quad n = 0, 1, \cdots,
$$

$$
v(x, y) = \left\| \int_0^1 \left(F'(x + t(y - x)) - F'(x) \right) dt \right\|, \quad z(x) = 0, \ \beta = 1,
$$

$$
w(x) = L_0 \| x - x_0 \| \quad \text{and} \quad p(x, y) = 0,
$$

for each x and y in $\mathcal{D}$. Then, we obtain iteration (9.183). Note that in earlier results $L = L_0$ (see Refs. [163, 290, 418]). Hence, our Lipschitz-type conditions and majorizing sequences are more general than in earlier cases such as (see for example Refs. [106, 110, 120]).

Example 9.5.1. Consider now equation (1.54), where $F(x) = (\xi_1^3 - \varrho, \xi_2^3 - \varrho)^T$ and G is given by

$$G(x) = (\epsilon \, |\xi_1 - 1|, \epsilon \, |\xi_2 - 1|), \quad x = (\xi_1, \xi_2)^T. \tag{9.244}$$

Using hypotheses of Theorem 9.5.1, choose $\epsilon = .01$ and $\varrho = .49$, we get that

$$\alpha = .531112045, \quad \eta = .17, \quad L = 3.2 \quad \text{and} \quad L_0 = 2.51.$$

That is Theorem 9.5.1 applies to solve equation (1.54). Moreover, Kantorovich's condition is not satisfied.

Remark 9.5.4. We shall extend our results to include high-order iterative methods along the lines of this section. We are motivated by Traub's theorem for scalar equations $f(t) = 0$ which states (see Ref. [418]).

Theorem 9.5.2. *Iterative procedure given by*

$$s_n = \mathcal{J}(t_n)$$
$$t_{n+1} = s_n - (1 + L_f(t_n) + L_f(t_n)^2 \, g_f(t_n)) \frac{f(s_n)}{f'(t_n)} \tag{9.245}$$
$$\text{for each} \quad n = 0, 2, \cdots$$

has order of convergence $\min\{i+2, \, 2\,i\}$ *for sufficiently smooth function* $g_f(t)$, *where*

$$L_f(s) = f'(s)^{-1} \, f''(s) \, f'(s)^{-1} \, f(s)$$

and i *is the order of* $\mathcal{J}(s)$.

Remark 9.5.5. In the interesting case of Newton's method $\mathcal{J}(s) = s - f'(s)^{-1} f(s)$, the convergence order of $\{t_n\}$ is four, since $i = 2$. Iteration $\{t_n\}$ can also be written as

$$t_{n+1} = m_f(t_n), \tag{9.246}$$

where

$$m_f(t) = t - \frac{1}{f'(t)} \left(f(t) + (1 + L_f(t) + L_f(t)^2 \, g_f(t)) \, f(t - \frac{f(t)}{f'(t)}) \right).$$

The corresponding iterative procedure to (9.246) for solving equation (1.54) is given by

$$x_{n+1} = M_F(x_n), \tag{9.247}$$

where $G_F : \mathcal{D} \subseteq \mathcal{X} \longrightarrow \mathcal{L}(\mathcal{X}, \mathcal{X})$ is a linear operator depending on the operator F and its derivatives. Iterative procedure (9.247) unifies the study of many popular high order iterative procedures such as

▶ Halley's method given by (9.178).
▶ Super-Halley's method

$$x_{n+1} = x_n - (\mathcal{I} + \frac{1}{2}(\mathcal{I} - L_F(x_n))^{-1} L_F(x_n))^{-1} F'(x_n)^{-1} F(x_n).$$

▶ Chebyshev's method

$$x_{n+1} = x_n - (\mathcal{I} + \frac{1}{2} L_F(x_n))\, F'(x_n)^{-1}\, F(x_n).$$

▶ Chebyshev-like methods for $0 \leq \omega_0 \leq 2$

$$x_{n+1} = x_n - (\mathcal{I} + \frac{1}{2} L_F(x_n) + \omega_0\, L_F(x_n)^2)\, F'(x_n)^{-1}\, F(x_n).$$

▶ Two-step method I (convergence order three)

$$y_n = x_n - F'(x_n)^{-1}\, F(x_n)$$
$$x_{n+1} = y_n - F'(x_n)^{-1}\, F(x_n).$$

▶ Two-step method II (convergence order four)

$$y_n = x_n - F'(x_n)^{-1}\, F(x_n)$$
$$x_{n+1} = y_n - (\mathcal{I} + L_F(x_n))\, F'(x_n)^{-1}\, F(y_n).$$

▶ Two-step method III (convergence order five)

$$y_n = x_n - F'(x_n)^{-1}\, F(x_n)$$
$$x_{n+1} = y_n - (\mathcal{I} + L_F(x_n) + \frac{1}{2}(\frac{5}{2}\mathcal{I} - L_{F'}(x_n))\, L_F(x_n)^2)\, F'(x_n)^{-1}\, F(y_n).$$

▶ Two-step method IV (convergence order six)

$$y_n = x_n - (\mathcal{I} + \frac{1}{2} L_F(x_n))\, F'(x_n)^{-1}\, F(x_n)$$
$$x_{n+1} = y_n - (\mathcal{I} + L_F(x_n) + \frac{1}{2}(3\mathcal{I} - L_{F'}(x_n))\, L_F(x_n)^2)\, F'(x_n)^{-1}\, F(y_n).$$

9.6   High Order Traub-type Methods

We propose in this section a collection of hybrid methods combining Newton's method with frozen derivatives and a family of high order iterative schemes. We present semi-local convergence results for this collection to solve (1.1). Using more precise majorizing sequence and under the same or weaker convergence conditions than the ones in earlier studies, we expand the applicability of these iterative procedures. We propose the iterative procedure

$$
\begin{aligned}
&x_0 \in \mathcal{D} \\
&x_n = x_{n-1} - F'(x_0)^{-1}\, F(x_{n-1}) \quad \text{for each} \quad n = 1, 2, \cdots, N_0 \\
&z_0 = x_{N_0} \\
&y_{k-1} = z_{k-1} - F'(z_{k-1})^{-1}\, F(z_{k-1}) \\
&z_k = y_{k-1} - (\mathcal{I} + L_F(z_{k-1}) + L_F(z_{k-1})^2\, G_F(z_{k-1}))\, F'(z_{k-1})^{-1}\, F(y_{k-1}) \\
&\text{for each} \quad k = 1, 2, \cdots
\end{aligned}
\tag{9.248}
$$

to generate a sequence $\{z_k\}$ approximating $x^\star$, where x_0 is a starting point and for $x \in \mathcal{D}$, $L_F(x) = F'(x)^{-1}\, F''(x)\, F'(x)^{-1}\, F(x)$, $G_F(x) : \mathcal{D} \longrightarrow \mathcal{L}(\mathcal{X}, \mathcal{X})$ is a given operator usually depending on the operator F and its derivatives.

Iterative procedure (9.248) unifies many earlier methods, where a slower method is used for a finite number of steps N_0 provided that the starting method converges under certain conditions, then, a faster method takes over from where the starting method is stopped. The last method converges under similar conditions as the first method. We refer the reader to Refs. [106, 104, 111, 116], where the starting method is the modified Newton's method and the last method is a Newton-like method. Other studies where the starting method is Newton's method followed by a faster method can be found in Refs. [9, 12–14, 16–19, 207, 206, 239, 264, 269]. In particular, we expand the applicability of the iterative procedure introduced by Amat et al. in Ref. [9] and given by

$$v_0 = x_0 \in \mathcal{D}$$
$$w_{n-1} = v_{n-1} - F'(v_{n-1})^{-1} F(v_{n-1})$$
$$v_n = w_{n-1} - (\mathcal{I} + L_F(v_{n-1}) + L_F(v_{n-1})^2 G_F(v_{n-1})) F'(v_{n-1})^{-1} F(w_{n-1})$$
$$\text{for each} \quad n = 1, 2, \cdots .$$
$$(9.249)$$

The motivation for introducing iterative procedure (9.249) (or (9.248)) is a result by Traub in Ref. [418] given for scalar nonlinear equation $f(t) = 0$ as follows.

Theorem 9.6.1. *Iterative procedure given by*

$$w_{n-1} = \phi(v_{n-1})$$
$$v_n = w_{n-1} - (\mathcal{I} + L_f(v_{n-1}) + L_f(v_{n-1})^2 G_f(v_{n-1})) \frac{f(w_{n-1})}{f'(v_{n-1})} \qquad (9.250)$$
$$for \ each \quad n = 1, 2, \cdots$$

has order of convergence $\min\{p + 2, 2p\}$*, where* p *is the order of* $\phi(t)$ *for all sufficiently smooth functions* $g_f(t)$*.*

There are reasons for the introduction of iterative procedure (9.248). First, we can recover iterative procedure (9.249) by setting $v_0 = x_{N_0}$. Note that in turn iterative procedure (9.249) generalizes the classical schemes of third order and other higher order methods (see Ref. [9]). Second, the domain of convergence of the modified Newton's method is larger than Newton's method which in turn is larger than other high order methods such as Halley, super-Halley, Chebyshev and other methods. Finally, it is a good practice to start with a method with easier to verify convergence conditions when we are not sufficiently close to the solution (see Refs. [106, 120, 163, 104, 111, 116]).

The following generalized Kantorovich-type conditions were used in Ref. [9] to show semi-local convergence of (9.249).

$(\mathcal{C}_1)$ There exist $x_0 \in \mathcal{D}$ and $\beta > 0$ such that $\Gamma_0 = F'(x_0)^{-1} \in \mathcal{L}(\mathcal{Y}, \mathcal{X})$ and $\| \Gamma_0 \| \leq \beta$;

$(\mathcal{C}_2)$ There exists $\eta > 0$ such that

$$\| \Gamma_0 F(x_0) \| \leq \eta;$$

(C_3) There exists $M > 0$ such that

$$\| F''(x) \| \le M \quad \text{for each} \quad x \in D;$$

(C_4) There exists $K > 0$ such that

$$\| F''(x) - F''(y) \| \le K \| x - y \| \quad \text{for each} \quad x, y \in D;$$

(C_5) There exists a function p defined on an interval $[a, b]$ for $a > 0$ such that $p(a) > 0 > p(b)$, $p'(t) < 0$, $p''(t) > 0$ and $p'''(t) > 0$ in $[a, t^\star]$ where $t^\star$ is the only simple solution of equation $p(t) = 0$.

For $t_0 \in [a, b]$ and $p(t_0) > 0$,

(C_6)

$$\| \Gamma_0 \| \le -1/p'(t_0);$$

(C_7)

$$\| \Gamma_0 F(x_0) \| \le -p(t_0)/p'(t_0);$$

(C_8) for each $x \in D$, $\| x - x_0 \| \le t - t_0 \le t^\star - t_0$

$$\| F''(x) \| \le p''(t);$$

(C_9) for $\| x - y \| \le |u - v|$, $x, y \in D$ and $u, v \in [a, t^\star]$

$$\| F''(x) - F''(y) \| \le |p''(u) - p''(v)|;$$

(C_{10}) for $\| x - x_0 \| \le t - t_0 \le t^\star - t_0$

$$\| L_F(x)^2 \, G_F(x) \| \le L_p(t)^2 \, G_p(t);$$

(C_{11})

$$1 + L_p(t) + L_p(t)^2 \, g_p(t) \ge 0;$$

(C_{12})

$$m'(t) > 0$$

where

$$m(t) = t - \frac{p(t)}{p'(t)} - \left(1 + L_p(t) + L_p(t)^2 \, g_p(t)\right) \frac{p\left(t - \dfrac{p(t)}{p'(t)}\right)}{p'(t)}.$$

We note by (C) the set of conditions (C_1)–(C_{12}).

The following result was given in Ref. [9].

Theorem 9.6.2. *Suppose that for $x_0 \in D$ and $t_0 \in [a, t^\star]$, the (C) conditions hold with*

$$\eta \le \eta_1 = \eta(K, M, \beta) = \frac{4\,K + M^2\,\beta - M\,\beta\,\sqrt{M^2 + 2\,K\,\beta}}{3\,\beta\,K\,(M + \sqrt{M^2 + 2\,K\,\beta})} \tag{9.251}$$

and

$$U(x_0, t^\star - t_0) \subseteq D. \tag{9.252}$$

Then, the following assertions hold

(a) Scalar sequence $\{t_n\}$ generated by

$$t_0 = \eta$$
$$s_{n-1} = t_{n-1} - \frac{p(t_{n-1})}{p'(t_{n-1})}$$
$$t_n = s_{n-1} - \left(1 + L_p(t_{n-1}) + L_p(t_{n-1})^2 \, g_p(t_{n-1})\right) \frac{p(s_{n-1})}{p'(t_{n-1})} \tag{9.253}$$
$$for \ each \quad n = 1, 2, \cdots$$

is increasingly convergent to $t^\star$.

(b) Sequence $\{v_n\}$ generated by iterative procedure (9.249) is well-defined, remains in $U(x_0, t^\star - t_0)$ for each $n \geq 1$ and converges to $x^\star$ which is the unique solution of equation $F(x) = 0$ in $\overline{U}(x_0, t^\star - t_0)$. Moreover, the following estimates hold

$$\| \, v_{n-1} - w_{n-1} \, \| \leq s_{n-1} - t_{n-1} \quad for \ each \quad n = 1, 2, \cdots \tag{9.254}$$

$$\| \, v_n - w_{n-1} \, \| \leq t_n - s_{n-1} \quad for \ each \quad n = 1, 2, \cdots \tag{9.255}$$

and

$$\| \, x^\star - v_n \, \| \leq t^\star - t_n \quad for \ each \quad n = 0, 2, \cdots . \tag{9.256}$$

Remark 9.6.1. A possible choice for p is given by the following polynomial

$$p(t) = \frac{K}{6} t^3 + \frac{M}{2} t^2 - \frac{1}{\beta} t + \frac{\eta}{\beta}. \tag{9.257}$$

If (9.251) holds, then, cubic polynomial p given by (9.257) has two roots $t^\star$ and $t^{\star\star}$ $(t^\star \leq t^{\star\star})$. We can choose a and b such that $0 < a < t^\star$ and $b > b(M, K, \beta) = \dfrac{2}{M\beta + \sqrt{M^2 \beta^2 + 2K\beta}}$ (see Refs. [120, 156, 161, 155, 163]).

Next, we show how to improve the above results under the same conditions $(\mathcal{C})$ and the same computational cost.

It follows from $(\mathcal{C}_3)$ and $(\mathcal{C}_4)$ that the following two conditions hold

$(\mathcal{C}_3^\star)$ There exists $M_0 > 0$ such that

$$\| \, F''(x_0) \, \| \leq M_0$$

and

$(\mathcal{C}_4^\star)$ there exists $K_0 > 0$ such that

$$\| \, F''(x) - F''(x_0) \, \| \leq K_0 \, \| \, x - x_0 \, \| \quad for \ each \quad x \in \mathcal{D}.$$

Note that

$$M_0 \leq M, \tag{9.258}$$

$$K_0 \leq K, \tag{9.259}$$

hold in general and M/M_0, K/K_0 can be arbitrarily large. Conditions $(\mathcal{C}_3^\star)$ and $(\mathcal{C}_4^\star)$ are not additional to $(\mathcal{C}_3)$ and $(\mathcal{C}_4)$, since the computation of M (or K) requires the computation of M_0 (or K_0).

Let us define the cubic polynomial

$$p_0(t) = \frac{K_0}{6} t^3 + \frac{M_0}{2} t^2 - \frac{1}{\beta} t + \frac{\eta}{\beta}. \tag{9.260}$$

Notice that

$$p_0(t) \leq p(t) \tag{9.261}$$

and

$$p_0^{(i)}(t) \leq p^{(i)}(t) \quad \text{for each} \quad i = 1, 2, \cdots, \tag{9.262}$$

where $p^{(i)}$ stands for the i-th derivative of p.

We need the following Banach-type lemma on invertible operators.

Lemma 9.6.1. *Suppose that* $(\mathcal{C}_3^\star)$, $(\mathcal{C}_4^\star)$ *and* $U(x_0, b(M_0, K_0, \beta)) \subseteq \mathcal{D}$. *Then* $F'(x)^{-1} \in \mathcal{L}(\mathcal{Y}, \mathcal{X})$ *and*

$$\| F'(x)^{-1} \| \leq \frac{\beta}{1 - \left(\dfrac{1}{2} \| x - x_0 \|^2 + M_0 \| x - x_0 \| \right)} \leq -\frac{1}{p_0'(t)} \tag{9.263}$$

for all $x \in U(x_0, t)$ *and* $t \leq b(M_0, K_0, \beta)$.

Proof. Let $x \in U(x_0, t)$. Using $(\mathcal{C}_3^\star)$, $(\mathcal{C}_4^\star)$ and the identity

$$\int_{x_0}^x \Gamma_0 \left(F''(y) - F''(x_0) \right) dy = \Gamma_0 F'(x) - \mathcal{I} - \Gamma_0 F''(x_0) (x - x_0),$$

we get in turn that

$$\| \Gamma_0 F'(x) - \mathcal{I} - \Gamma_0 F''(x_0) (x - x_0) \| = \| \int_{x_0}^x \Gamma_0 \left(F''(y) - F''(x_0) \right) dy \|$$
$$\leq \frac{\beta K_0}{2} \| x - x_0 \|^2 .$$

That is

$$\| \Gamma_0 F'(x) - \mathcal{I} \| \leq \beta \left(\frac{K_0}{2} \| x - x_0 \|^2 + M_0 \| x - x_0 \| \right) < 1. \tag{9.264}$$

The result now follows from (9.264) and the Banach lemma on invertible operators. That completes the proof of Lemma 9.6.1. $\qquad\square$

Remark 9.6.2. The more expensive and more difficult to verify conditions $(\mathcal{C}_3)$, $(\mathcal{C}_4)$ were used in Ref. [9] to arrive at the less precise analog of (9.263) given by

$$\| F'(x)^{-1} \| \leq -\frac{1}{p'(t)}. \tag{9.265}$$

Note that

$$-\frac{1}{p_0'(t)} \leq -\frac{1}{p'(t)} \qquad (9.266)$$

holds. Moreover, (9.266) holds as a strict inequality if $M_0 < M$ or $K_0 < K$. The above observations motivate us to introduce majorizing sequence $\{t_n^0\}$ given by

$$t_0^0 = t_0$$
$$s_{n-1}^0 = t_{n-1}^0 - \frac{p(t_{n-1}^0)}{p_0'(t_{n-1}^0)}$$
$$t_n^0 = s_{n-1}^0 - (1 + L_p(t_{n-1}^0) + L_p(t_{n-1}^0)^2 \, g_p(t_{n-1}^0)) \frac{p(s_{n-1}^0)}{p_0'(t_{n-1}^0)} \qquad (9.267)$$

for each $n = 1, 2, \cdots$.

A simple inductive argument shows that

$$s_n^0 \leq s_n, \qquad (9.268)$$

$$t_n^0 \leq t_n, \qquad (9.269)$$

$$s_n^0 - t_n^0 \leq s_n - t_n, \qquad (9.270)$$

$$t_{n+1}^0 - s_n^0 \leq t_{n+1} - s_n \qquad (9.271)$$

and

$$t_0^\star = \lim_{n \to \infty} t_n^0 \leq t^\star. \qquad (9.272)$$

Estimates (9.268)–(9.271) hold as strict inequalities for $n > 1$ if $M_0 < M$ or $K_0 < K$. In view of (9.253) and estimates (9.268)–(9.271), we arrive at the following refinement of Theorem 9.6.2. The proof is obtained from the corresponding one in Theorem 9.6.2 by simply using (9.263) instead of (9.265).

Theorem 9.6.3. *Suppose that hypotheses of Theorem 9.6.2 hold. Then, the following assertions hold*

(a) Scalar sequence $\{t_n^0\}$ generated by (9.267) is increasingly convergent to $t_0^\star$.
(b) The following estimates hold

$$\| v_{n-1} - w_{n-1} \| \leq s_{n-1}^0 - t_{n-1}^0 \leq s_{n-1} - t_{n-1} \quad for \ each \quad n = 1, 2, \cdots \ (9.273)$$

$$\| v_n - w_{n-1} \| \leq t_n^0 - s_{n-1}^0 \leq t_n - s_{n-1} \quad for \ each \quad n = 1, 2, \cdots \quad (9.274)$$

and

$$\| x^\star - v_n \| \leq t_0^\star - t_n^0 \leq t^\star - t_n \quad for \ each \quad n = 0, 1, 2, \cdots . \qquad (9.275)$$

Remark 9.6.3. Theorem 9.6.3 shows that under the same hypotheses of Theorem 9.6.2, more precise estimates on the distances $\| v_n - w_n \|$, $\| v_n - w_{n-1} \|$, $\| x^\star - v_n \|$ for each $n = 1, 2, \cdots$ as well as an at least as precise information on the location of the solution can be obtained. There is a way to weaken the sufficient convergence condition (9.251).

Theorem 9.6.4. *Suppose that $t_0, s_0, t_1, s_1, \cdots, t_{N_0}$ for fixed $N_0 = 0, 1, \cdots$*

$$t_0 \le s_0 \le t_1 \le s_1 \le \cdots \le s_{N_0 - 1} \le t_{N_0} < b(M_0, K_0, \beta) \tag{9.276}$$

and

$$s_{N_0} - t_{N_0} \le \eta_1. \tag{9.277}$$

Moreover, suppose that the (C) *conditions hold with* t_0 *replaced by* t_{N_0}. *Then, the conclusions of Theorems 9.6.2 and 9.6.3 hold. Furthermore, if*

$$s_{N_0}^0 - t_{N_0}^0 \le \eta_1 \tag{9.278}$$

holds instead of (9.277) with t_0 *replaced by* $t_{N_0}^0$. *Then, the conclusions of Theorem 9.6.3 hold.*

Proof.   Define cubic polynomials

$$p_{N_0}(t) = \frac{K}{6} t^3 + \frac{M}{2} t^2 - \frac{1}{\beta} t + \frac{s_{N_0} - t_{N_0}}{\beta} \tag{9.279}$$

and

$$p_{N_0}^0(t) = \frac{K}{6} t^3 + \frac{M}{2} t^2 - \frac{1}{\beta} t + \frac{s_{N_0}^0 - t_{N_0}^0}{\beta}. \tag{9.280}$$

It follows from (9.277) and (9.278), respectively, that p_{N_0} and $p_{N_0}^0$ have roots corresponding to $t^\star$ and $t_0^\star$. Note also that $s_{N_0}^0 - t_{N_0}^0 \le s_{N_0} - t_{N_0}$. That completes the proof of Theorem 9.6.4. $\square$

Next, we show how to expand the convergence domain by considering iterative process (9.248) instead of (9.249). The corresponding iterative process (9.248) to majorizing sequence is given by

$$
\begin{aligned}
r_n &= r_{n-1} + p_0(r_{n-1}) \quad \text{for each} \quad n = 1, 2, \cdots, N_0 \\
t_0^1 &= r_{N_0} \\
s_{k-1}^1 &= t_{k-1}^1 - \frac{p(t_{k-1}^1)}{p_0'(t_{k-1}^1)} \\
t_k^1 &= s_{k-1}^1 - (1 + L_p(t_{k-1}^1) + L_p(t_{k-1}^1)^2 \, g_p(t_{k-1}^1)) \frac{p(t_{k-1}^1)}{p_0'(t_{k-1}^1)}
\end{aligned}
\tag{9.281}
$$

$$\text{for each} \quad k = 1, 2, \cdots.$$

We need a semi-local result for the modified Newton's method $\{x_n\}$.

Theorem 9.6.5. *Suppose that* (C_1), (C_2), $(C_3^\star)$ *and* $(C_4^\star)$ *hold. Let* p_0 *be the polynomial defined in (9.260). Moreover, suppose that*

$$\eta \le \eta_0 = \eta(M_0, K_0, \beta). \tag{9.282}$$

Then

(a) Polynomials p_0 *has two positive roots* r^1, r^2 $(r^1 \le r^2)$ *and a negative root* $-r^0$.

Furthermore, suppose that

$$U(x_0, r^1) \subseteq \mathcal{D}. \tag{9.283}$$

Then

(b) *Sequence $\{x_n\}$ generated by modified Newton's method is well-defined, remains in $U(x_0, r^1)$ for all $n \geq 0$ and converges to a solution $x^\star \in \overline{U}(x_0, r^1)$ of equation $F(x) = 0$. Moreover, the following estimates hold*

$$\| x_{n+1} - x_n \| \leq r_{n+1} - r_n = p_0(r_n) \quad for\ each \quad n = 0, 1, 2, \cdots \tag{9.284}$$

and

$$\| x_n - x^\star \| \leq r^1 - r_n \quad for\ each \quad n = 1, 2, \cdots . \tag{9.285}$$

Furthermore, if $r^1 < r^2$, the solution is unique in $U(x_0, r^2)$. If $r^1 = r^2$, the solution is unique in $\overline{U}(x_0, r^1)$.

Proof.

(a) See Ref. [9].

(b) We shall prove that

$$\| x_{k+1} - x_k \| \leq r_{k+1} - r_k \quad for\ each \quad k = 0, 1, 2, \cdots \tag{9.286}$$

where $\{r_n\}$ is the modified Newton's sequence to solve $p_0(t) = 0$, starting at $r_0 = 0$. Iterate x_1 is well-defined by (9.248) and

$$\| x_1 - x_0 \| = \| \Gamma_0 F(x_0) \| \leq \eta = r_1 - r_0 < r^1 \leq b(M_0, K_0, \beta). \tag{9.287}$$

It follows from (9.287) that $x_1 \in \overline{U}(x_0, r^1)$.

Let $x \in [x_0, x_1]$. Then, we have $x = x_0 + \theta (x_1 - x_0)$ for each $\theta \in [0, 1]$. By Taylor's formula, we get that

$$\Gamma_0 F(x) = \Gamma_0 F(x_0) + (x - x_0) + \frac{1}{2} \Gamma_0 F''(x_0) (x - x_0)^2$$
$$+ \int_0^1 \Gamma_0 \left(F''(x_0 + \theta (x - x_0)) - F''(x_0) \right) (1 - \theta) (x - x_0)^2 \, d\theta. \tag{9.288}$$

It then follows from $(\mathcal{C}_3^\star)$, $(\mathcal{C}_4^\star)$, (9.288), $\theta \eta = t$ and $x - x_0 = \theta (x_1 - x_0) = -\theta \Gamma_0 F(x_0)$ that

$$\| \Gamma_0 F(x) \| = (1 - \theta) \| \Gamma_0 F(x_0) \| + \frac{1}{2} \| \Gamma_0 F''(x_0) \| \, \| x - x_0 \|^2$$
$$+ \int_0^1 \| \Gamma_0 \left(F''(x_0 + \theta (x - x_0)) - F''(x_0) \right) (1 - \theta) (x - x_0)^2 \, d\theta \| \tag{9.289}$$
$$\leq (1 - \theta) \eta + \frac{M_0 \beta}{2} \theta^2 \eta^2 + \frac{K_0}{6} \beta \theta^3 \eta^3 = p_0(t).$$

In particular, for $x = x_1$, we get that $\| x_2 - x_1 \| \leq r_2 - r_1$. In analog way, we obtain $\| x_3 - x_2 \| \leq r_3 - r_2$.

The convergence of $\{r_n\}$ implies that $\{x_n\}$ is a complete sequence in a Banach space $\mathcal{X}$ and as such it converges to some $x^\star \in \overline{U}(x_0, r^1)$ (since $\overline{U}(x_0, r^1)$ is a closed set). By letting $k \to \infty$ in (9.289) for $x = x_k$, we obtain $F(x^\star) = 0$. Estimate (9.286) follows from (9.285) by using standard majorizing techniques. We refer the reader to Refs. [9, 106] for the proof of the uniqueness part.

That completes the proof of Theorem 9.6.5. $\qquad\qquad\qquad\qquad\qquad$ $\square$

Remark 9.6.4.

(a) In the literature, (9.251) is used as a common hypothesis for modified Newton's and Newton's methods. However, note that in view of (9.251), (9.258), (9.259) and (9.282), we have that

$$\eta_1 \le \eta_0. \tag{9.290}$$

Hence, we get that

$$\eta \le \eta_1 \Longrightarrow \eta \le \eta_0 \tag{9.291}$$

but not necessarily vice versa unless if $M_0 = M$ and $K_0 = K$. Note also that

$$\frac{\eta_1}{\eta_0} \longrightarrow \infty \quad \text{as} \quad \frac{M}{M_0} \longrightarrow \infty \quad \text{and} \quad \frac{K}{K_0} \longrightarrow \infty. \tag{9.292}$$

(b) In view of the proof of Theorem 9.6.5 (see (9.289) for $x = x_k$), we get that

$$\| x_{n+1} - x_n \| \le q_n \| x_n - x_{n-1} \| \quad \text{for each} \quad n = 1, 2, \cdots, \tag{9.293}$$

where

$$q_n = \frac{\beta}{2} \left(M_0 + \frac{K_0}{3}\right) \| x_n - x_{n-1} \|. \tag{9.294}$$

(c) According to Theorem 9.6.2, iterative process (9.248) converges provided that the $(\mathcal{C})$ conditions hold for x_{N_0}, t_0^1, respectively, replacing x_0, t_0. In particular, crucial sufficient convergence condition (9.251) must read

$$\| F'(x_{N_0})^{-1} F(x_{N_0}) \| \le \eta_1^{N_0} = \eta(M, K, \beta_{N_0}), \tag{9.295}$$

where

$$\| F'(x_{N_0})^{-1} \| \le \beta_{N_0} \tag{9.296}$$

or

$$\| F'(x_{N_0})^{-1} F'(x_0) \| \, \| F'(x_0)^{-1} F(x_{N_0}) \| \le \eta_1^{N_0}$$

or by (9.248) and (9.293)

$$\| F'(x_{N_0})^{-1} F'(x_0) \| \, q_1^{N_0} \eta \le \eta_1^{N_0}. \tag{9.297}$$

Estimate (9.263) can be used to find upper bounds on $\| F'(x_{N_0})^{-1} F'(x_0) \|$ and $\| F'(x_{N_0})^{-1} \|$ in terms of the initial data.

Set

$$c = \frac{\eta(M, K, c_1)}{c_0 \, c_1 \, \eta},$$

where

$$\| F'(x_{N_0})^{-1} \| \le -\frac{1}{p_0'(r^1)} = c_1 \quad \text{and} \quad \| F'(x_0) \| \le c_0.$$

Then, by (9.297) it suffices to have that

$$q_1^{N_0} \le c. \tag{9.298}$$

Define N_0 by

$$N_0 = 1 + \left[\frac{\ln c}{\ln q_1^{N_0}} \right], \tag{9.299}$$

where $[w]$ denotes the integer part of w.

Similar work leads to even smaller N_0 if (9.276) and (9.277) or (9.276) and (9.278) are verified in an analog way.

Set

$$R^1 = t^\star - t_0^1 \tag{9.300}$$

and

$$\alpha_n = \begin{cases} x_n & \text{if } n \le N_0 \\ z_{n-N_0} & \text{if } n > N_0. \end{cases} \tag{9.301}$$

We can show the following result.

Theorem 9.6.6. *Suppose that the (C) conditions hold. If (9.282) and $U(x_0, r^1 + R^1) \subseteq D$ are satisfied, then iterative process (9.301) starting from α_0 is well-defined and converges to a solution $x^\star$ of equation $F(x) = 0$. Moreover, the solution $x^\star$ and sequence $\{\alpha_n\}$ belong in $\overline{U}(x_0, r^1 + R^1)$.*

Proof. The existence of N_0 has been established in (9.299). We have that $\alpha_0, \alpha_1, \cdots, \alpha_{N_0} \in D$, since they are iterates generated by modified Newton's method. Hence, we have that $\| \alpha_i - \alpha_0 \| \le r^1 < r^1 + R^1$ for each $i = 1, 2, \cdots, N_0$. Then, if we choose $\alpha_{N_0} = x_{N_0}$, the iterates given by (9.248) and $\| \alpha_i - \alpha_{N_0} \| \le r^1 < r^1 + R^1$ for each $i = 1, 2, \cdots, N_0$. Then, if we choose $\alpha_{N_0} = x_{N_0}$, the iterates are given by (9.248) and $\| \alpha_i - \alpha_{N_0} \| \le R^1$ for each $i > N_0$. We have $\| \alpha_i - x_0 \| \le \| \alpha_i - \alpha_{N_0} \| + \| \alpha_{N_0} - x_0 \| < r^1 + R^1$ and $\alpha_i \in U(x_0, r^1 + R^1) \subseteq D$ for each $i > N_0$. That is $\{\alpha_n\} \in D$. Sequence $\{\alpha_n\}$ is complete in D. Indeed, $\{\alpha_n\}$ for $n \le N_0$ is the sequence generated by (9.249) which converges by the choice of N_0 and Theorem 9.6.2. Hence, there exists $x^\star \in \overline{U}(x_0, R^1) \subseteq \overline{U}(x_0, r^1 + R^1)$ such that $F(x^\star) = 0$ and $x^\star = \lim_{n \to \infty} \alpha_n$. That completes the proof of Theorem 9.6.6. $\square$

Next, we establish a uniqueness result for the solution $x^\star$ of equation (1.1).

Theorem 9.6.7. *Under the (C) conditions further suppose that $r^1 + R^1 < \dfrac{2}{M \beta}$. Then the limit point $x^\star$ is the unique solution of equation (1.1) in the region $\Lambda = U(x_0, \dfrac{2}{M \beta} - (r^1 + R^1)) \cap D$.*

Proof. Let us assume that $y^\star$ is a solution of equation (1.1) in Λ. Set $\mathcal{A} = \int_0^1 F'(x^\star + \theta\,(y^\star - x^\star))\,d\theta$. Using $(\mathcal{C}_3)$, we get in turn that

$$
\begin{aligned}
\| F'(x_0)^{-1}\,(\mathcal{A} - F'(x_0)) \| &\leq \| F'(x_0)^{-1} \|\,\| \mathcal{A} - F'(x_0) \| \\
&\leq \beta\,M \int_0^1 \| x^\star + \theta\,(y^\star - x^\star) - x_0 \|\,d\theta \\
&\leq \beta\,M \int_0^1 ((1 - \theta)\,\| x^\star - x_0 \| + \theta\,\| y^\star - x_0 \|)\,d\theta \\
&< \frac{M\,\beta}{2}\,(r^1 + R^1 + \frac{2}{M\,\beta} - (r^1 + R^1)) = 1.
\end{aligned}
\tag{9.302}
$$

It follows from (9.302) and the Banach lemma on invertible operators that $\mathcal{A}^{-1}$ exists. Then, in view of the identity $\mathcal{A}\,(y^\star - x^\star) = F(x^\star) - F(y^\star)$, we deduce that $x^\star = y^\star$. That completes the proof of Theorem 9.6.7. $\square$

Remark 9.6.5.

(a) The domain of existence of solution $U(x_0, r^1 + R^1)$ obtained using iterative process (9.248) is larger than the domains of existence for modified Newton's method and iterative process (9.249) given by $U(x_0, r^1)$ and $U(x_0, R^1)$, respectively. We also note that the domain of uniqueness of solution Λ is subject to the condition $r^1 + R^1 < \dfrac{2}{M\,\beta}$. However, the domain of starting points for iterative process (9.248) is extended with respect to iterative process (9.249). Therefore, we can obtain domains of existence and uniqueness of solutions that cannot be obtained with iterative process (9.249).

(b) Iterative process (9.248) has R-order of convergence at least one (see (9.293)) until iteration N_0 and R-order of convergence at least four (see Theorem 9.6.1) from iteration $N_0 + 1$, since the R-order of convergence of Newton's method at least two. So far we managed to show convergence of iterative process (9.248) provided that (9.282) or (9.276) and (9.277) or (9.276) and (9.278) are satisfied.

Next, we present convergence results in case the above stated conditions are not satisfied. Let us replace conditions $(\mathcal{C}_3^\star)$ and $(\mathcal{C}_4^\star)$ for the remaining of this section by center Lipschitz-condition

$$
\| F'(x) - F'(x_0) \| \leq L_0\,\| x - x_0 \| \quad \text{for each} \quad x \in \mathcal{D} \quad \text{and some} \quad L_0 > 0. \tag{9.303}
$$

Argyros showed in Ref. [104] the following semi-local convergence result for modified Newton's method.

Theorem 9.6.8. *Let $F : \mathcal{D} \subseteq \mathcal{X} \longrightarrow \mathcal{Y}$ be Fréchet differentiable. Suppose that $(\mathcal{C}_1)$, $(\mathcal{C}_2)$ and (9.303) hold. Moreover, suppose that*

$$
h_0 = \beta\,L_0\,\eta \leq \frac{1}{2} \tag{9.304}
$$

and

$$
\overline{U}(x_0, \varrho^1) \subseteq \mathcal{D}, \tag{9.305}
$$

where

$$\varrho^1 = \frac{1 - \sqrt{1 - 2\,h_0}}{\beta\,L_0} \tag{9.306}$$

hold. Then the sequence $\{x_n\}$ generated by the modified Newton's method is well-defined, remains in $U(x_0, \varrho^1)$ for all $n \geq 0$ and converges to a unique solution $x^\star$ of equation $F(x) = 0$ in $\overline{U}(x_0, \varrho^1)$. Moreover, the following estimates hold for all $n \geq 0$

$$\| x_{n+1} - x_n \| \leq q^n \, \| x_1 - x_0 \|, \tag{9.307}$$

$$\| x_n - x^\star \| \leq \frac{q^n}{1 - q}\, \eta \tag{9.308}$$

and

$$p_1(\varrho^1) = 0, \tag{9.309}$$

where

$$q = 1 - \sqrt{1 - 2\,h_0} \quad and \quad p_1(t) = \frac{L_0}{2}\,t^2 - \frac{t}{\beta} + \frac{\eta}{\beta}. \tag{9.310}$$

Remark 9.6.6.

(a) If the convergence conditions for iterative process (9.248), (9.282) or (9.276) and (9.277) or (9.276) and (9.278) do not hold but (9.304) is satisfied. Suppose that

$$-\frac{1}{p_1'(t)} \leq -\frac{1}{p'(t)} \quad for\ each \quad t \in [a, b]. \tag{9.311}$$

Then, it immediately follows from the proofs that p_1, ϱ^1, (9.304), $\dfrac{1}{2\,L_0}$, $\dfrac{1}{L_0}$, q, respectively, can replace p_0, r^1, (9.282), η_0, $b(M_0, K_0, \beta)$, q_1 in all the above stated results in this section.

(b) Concerning the convergence of iterative process (9.249), in a series of Refs. [140, 156, 161, 155], we have provided sufficient convergence conditions for Newton's method that are weaker than the Kantorovich hypothesis. We also have that

$$\| w_n - v_n \| \leq d \, \| w_{n-1} - v_{n-1} \|^2 . \tag{9.312}$$

Then, again if (9.311) holds but (9.251) or (9.276) and (9.277) or (9.276) and (9.278) do not hold but Kantorovich's condition or ours in Refs. [161, 155] hold (call all these the (AHK) conditions), then $-1/p_1'$, ϱ, (AHK), $\dfrac{1}{2\,L_0}$, $\dfrac{1}{L_0}$, $d\eta$, respectively, can replace $-1/p'$, r^1, (9.251), η_1, $b(M, K, \beta)$, q_1 in all the above stated results, where ϱ and d are given in (see Refs. [120, 161, 155, 163]). However, note that in this case

$$s_{n-1} = \begin{cases} t_{n-1} - \dfrac{p(t_{n-1})}{p_1'(t_{n-1})} & \text{if } L_0 = L \\[2ex] t_{n-1} - \dfrac{L\,(t_{n-1} - t_{n-2})^2}{2\,p_1'(t_{n-1})} & \text{if } L_0 < L. \end{cases}$$

Table 9.8   Errors for the approximation of the positive root of $x^2 - 1 = 0$ with $x_0 = 1/4$

n	Newton	Two-step method III
1	$1.12 \cdot 10^0$	$1.12 \cdot 10^0$
2	$2.90 \cdot 10^{-1}$	$2.90 \cdot 10^{-1}$
3	$3.42 \cdot 10^{-2}$	$5.70 \cdot 10^{-4}$
4	$5.64 \cdot 10^{-4}$	0
5	$1.59 \cdot 10^{-7}$	
6	$1.26 \cdot 10^{-14}$	
7	0	

The first step in the family is the modified Newton method, but similar results can be performed taking the classical Newton method or other Newton-type method. The second step consists in the application of a family of high order method depending on a function G_f (see Ref. [9]).

Particular examples of schemes included in this second family with *non-smooth* functions $G_f(x)$ are methods given in Remark 9.5.5. On the other hand, we would like to point out that we cannot increase, in general, the order of convergence of these methods by choosing a suitable function G_f. Indeed we have the following: suppose that r is a simple root of $f(x) = 0$, in other words, $f(r) = 0$ and $f'(r) \neq 0$, then

$$M_f(r) = r\,, \quad M_f'(r) = M_f''(r) = M_f'''(r) = 0\,,$$

but

$$M_f^{(iv)}(r) = -\frac{3f''(r)[2f'(r)f'''(r) - 5f''(r)^2 + 4f''(r)^2 G_f(r)]}{f'(r)^3}\,.$$

If we want this last expression equal to zero, we must impose conditions on the derivatives of higher order of f, which are highly artificial. However, in some particular cases we can find interesting examples. For instance, for equations verifying $f'''(r) = 0$ (as happens with the quadratic equations) we should obtain a fifth order method taking $G_f(x) = \frac{5}{4}$. For quadratic equations this method is in fact two-step method III (see Remark 9.5.5) since in this case $L_{f'}(x) = 0$.

Finally, we conclude with a simple example where all these comments appear numerically. We consider the quadratic equation $x^2 - 1 = 0$ and we approximate the positive root using Newton's method and the hybrid Newton given by two-step method III starting from $x_0 = 1/4$ (see Remark 9.5.5). The hybrid method converges to the positive root and faster than Newton's method, see table 9.8. The different of both schemes start in the third iteration, the two first are the same. With only two extra iterations, due to its fifth order of convergence, our method obtains the convergence. Newton's method needs 5 extra iterations. In this same situation, after 8 iterations the two-step method III converges but to the *negative* root. Clearly, the most efficient method for this problem is our hybrid approach.

9.7 Relaxed Newton's Method

In this section we study the local and semi-local convergence of relaxed Newton's method, that is Newton's method with a relaxation parameter $0 < \lambda < 2$. We give a Kantorovich-like theorem that can be applied for operators defined between two Banach spaces. In fact, we obtain the recurrent sequence that majorizes the one given by the method and we characterize its convergence by a result that involves the relaxation parameter λ. We use a new technique that allows us on the one hand to generalize and on the other hand to extend the applicability of the result given initially by Kantorovich for $\lambda = 1$. We use Newton's method in the scalar case

$$x_{n+1} = x_n - \frac{f(x_n)}{f'(x_n)}, \quad n \geq 0. \tag{9.313}$$

In the beginning, this method was constructed for functions defined on the real line, but after the seminal works of Kantorovich in Ref. [290], Newton's method was extended to general spaces with the aim of solving functional equations in the form (1.1). In this context, the known as *Newton–Kantorovich method* has the following form:

$$x_{n+1} = x_n - \Gamma_n F(x_n), \quad n \geq 0, \tag{9.314}$$

where $\Gamma_n = F'(x_n)^{-1}$ is the inverse of the Fréchet derivative of the nonlinear operator $F(x)$ at the point x_n.

Theorem 9.7.1. (Kantorovich's theorem) *Let us assume that the operator F introduced in (1.1) is defined and Fréchet differentiable in a ball $B(x_0, R)$. Let us assume that the linear operator $F'(x_0)$ is invertible and conditions*

$$\|F'(x_0)^{-1} F(x_0)\| \leq a, \tag{9.315}$$

$$\|F'(x_0)^{-1}[F'(x) - F'(y)]\| \leq b\|x - y\|, \quad x, y \in B(x_0, R), \tag{9.316}$$

hold. In addition, let us suppose,

$$h = ab < 1/2, \quad u_1 = \frac{1 - \sqrt{1 - 2h}}{b} \leq R.$$

Then, Newton–Kantorovich method (9.314) is well-defined and it converges to a solution x^* of (1.1). In addition, x^* is located in $\overline{B}(x_0, u_1)$ and it is the unique solution of (1.1) in the ball $B(x_0, u_2)$, where

$$u_2 = \frac{1 + \sqrt{1 - 2h}}{b}.$$

Finally, the rate of convergence is given by

$$\|x^* - x_n\| \leq \frac{a}{h2^n}(2h)^{2^n}, \quad n \geq 0.$$

That is, the order of convergence is quadratic.

We would like to emphasize the key role of the parameter h defined above in Kantorovich's theory. The previous result has been stated for values $h < 1/2$, but Kantorovich theory can be applied for $h \leq 1/2$. When $h = 1/2$, only the linear convergence can be guaranteed. In addition, in this case, $u_1 = u_2$, and the solution x^* is located and is unique in $\overline{B}(x_0, u_1)$.

Our main goal in this section is to develop a Kantorovich-like theory for a variant of Newton's method that it is called *relaxed Newton's method*:

$$x_{n+1} = x_n - \lambda F'(x_n)^{-1} F(x_n), \quad n \geq 0, \quad 0 < \lambda < 2. \tag{9.317}$$

The main reason to consider this method is because a full Newton step may not be suitable to ensure that convergence of the method is monotone, but introducing the relaxing factor this problem disappears. Some questions about the complex dynamics of this method have been considered in Refs. [201, 248]. Note also that method (9.317) is a special case of Inexact Newton's method defined by:

$$x_{n+1} = x_n - F'(x_n)^{-1} F(x_n) + y_n, \tag{9.318}$$

where y_n is a null residual sequence in X. In particular, if $y_n = (1 - \lambda)F'(x_n)^{-1}F(x_n)$ the Inexact Newton's method reduces to method (9.317). There is a plethora of convergence results for the Inexact Newton's method (see Refs. [110, 120, 163] and the references therein). Hence, the results for Inexact Newton's method can be specialized to provide convergence results for method (9.317). However, we decided in this section to employ a more direct approach which leads to easier to verify sufficient convergence conditions under simpler hypotheses.

In Ref. [159] we can find a semi-local convergence analysis of methods in the form (9.317) but it is related to recurrent sequences and functional spaces. The relevance of our results in this section, compared with the classical Kantorovich theory, is the inclusion of the relaxation parameter λ in the set of sufficient conditions to guarantee the semi-local convergence of (9.317) to a solution of (1.1). To be more precise, throughout this section we assume that the operator F defined in (1.1) is defined and Fréchet differentiable in a ball $B(x_0, R)$.

We will study the local convergence of relaxed Newton's method under center-Lipschitz and Lipschitz conditions. We present the following result.

Theorem 9.7.2. *Let $F : X \to Y$ be Fréchet differentiable in the ball $B(x^*, R) \subseteq X$. Assume that the following conditions hold:*

(i) *There exists a solution x^* of $F(x) = 0$.*
(ii) *There exists $\Gamma = F'(x^*)^{-1}$.*
(iii) $\|\Gamma[F'(x) - F'(y)]\| \leq b\|x - y\|, b > 0, \ x, y \in B(x^*, R).$
(iv) $\|\Gamma[F'(x) - F'(x^*)]\| \leq \beta\|x - x^*\|, \beta > 0, \ x \in B(x^*, R).$
(v) $\rho = \min\{R, \dfrac{2(1 - |1 - \lambda|)}{b + (2 + |1 - \lambda|)\beta}\}, 0 < \lambda < 2.$

Then, the sequence given by relaxed Newton's method (9.317) starting from $x_0 \in B(x^*, \rho)$, remains in $B(x^*, \rho)$ and converges to x^*.

Proof. We need to prove the following conditions

(I) If $x \in B(x^*, \rho)$, then $\beta \|x - x^*\| \leq \beta\rho < 1$, by the choice of ρ given in (v). Hence,

$$\|I - \Gamma F'(x)\| \leq \beta\|x - x^*\| < 1.$$

By the Banach Lemma on inversion of operators, we can ensure that there exists the inverse operator of $\Gamma F'(x)$ and

$$\|F'(x)^{-1} F'(x^*)\| \leq \frac{1}{1 - \beta\|x - x^*\|}.$$

(II) We have the Ostrowski decomposition

$$F(x) = F(x) - F(x^*) - F'(x^*)(x - x^*) + F'(x^*)(x - x^*)$$

$$= \int_{x^*}^{x} [F'(x) - F'(x^*)]\, dx + F'(x^*)(x^* - x)$$

$$= \int_{0}^{1} [F'(x^* + t(x - x^*)) - F'(x^*)]\,(x - x^*)dt + F'(x^*)(x^* - x).$$

By taking norms in the preceding identity and using (iv) we get that

$$\|\Gamma F(x)\| \leq \frac{\beta}{2}\|x - x^*\|^2 + \|x - x^*\|.$$

We also have the following Ostrowski-type decomposition

$$\begin{aligned}
x^* - x_{n+1} &= x^* - x_n + \lambda\Gamma_n F(x_n) - \Gamma_n F(x_n) + \Gamma_n F(x_n) \\
&= x^* - x_n + \Gamma_n F(x_n) + (\lambda - 1)\Gamma_n F(x_n) \\
&= -\Gamma_n \left[F(x^*) - F(x_n) - F'(x_n)(x^* - x_n) + (1 - \lambda)F(x_n)\right] \\
&= -\Gamma_n \left[\int_{x_n}^{x^*} [F'(x) - F'(x_n)]\, dx + (1 - \lambda)F(x_n)\right] \\
&= -\Gamma_n \left[\int_{0}^{1} [F'(x_n + t(x^* - x_n)) - F'(x_n)]\,(x^* - x_n)dt + (1 - \lambda)F(x_n)\right].
\end{aligned}$$

Then, by taking norms we get in turn that

$$\begin{aligned}
\|x^* - x_{n+1}\| &\leq \|\Gamma_n F'(x^*)\| \left[\frac{b}{2}\|x^* - x_n\|^2 + (1 - \lambda)\|\Gamma F(x_n)\|\right] \\
&\leq \frac{1}{1 - \beta\|x^* - x_n\|} \left[\frac{b}{2}\|x^* - x_n\|^2 + (1 - \lambda)\left(\frac{\beta}{2}\|x^* - x_n\|^2 + \|x^* - x_n\|\right)\right] \\
&\leq \frac{1}{1 - \beta\|x^* - x_n\|} \left[\left(\frac{b}{2} + \beta\left(\frac{1 - \lambda}{2}\right)\right)\|x^* - x_n\| + (1 - \lambda)\right]\|x^* - x_n\|.
\end{aligned}$$

Let $a_{n+1} = \dfrac{a_n}{1 - \beta a_n}\left(\left(\dfrac{|1 - \lambda|}{2}\beta + \dfrac{b}{2}\right)a_n + |1 - \lambda|\right)$. Then,

$$\|x^* - x_n\| \leq a_n \quad \forall n \in \mathbb{N}.$$

Let $\gamma = \dfrac{\beta}{b}$ and $\epsilon_n = \beta a_n$

$$\epsilon_{n+1} = \frac{\epsilon_n}{1 - \epsilon_n} \left(\frac{\frac{1}{\gamma} + |1 - \lambda|}{2} \epsilon_n + |1 - \lambda| \right).$$

Using an induction process, we will prove that (ϵ_n) converges to 0. Let us show that $\epsilon_1 < \epsilon_0$:

$$\epsilon_1 = \frac{\epsilon_0}{1 - \epsilon_0} \left(\frac{\frac{1}{\gamma} + |1 - \lambda|}{2} \epsilon_0 + |1 - \lambda| \right) < \epsilon_0$$

or

$$\epsilon_0 < \frac{2(1 - |1 - \lambda|)}{2 + \frac{1}{\gamma} + |1 - \lambda|},$$

which is true by the choice of ρ. Supposing that $\epsilon_n < \epsilon_{n-1} \ \forall n \leq k$. Let us see that it remains valid for $n = k+1$. It's easy to see that $\epsilon_n < \epsilon_{n-1}$ and $1 - \epsilon_n > 1 - \epsilon_{n-1}$. Moreover as $0 < \lambda < 2$, $\frac{1}{\gamma} + |1 - \lambda| > 0$, so $\epsilon_{k+1} < \epsilon_k$. Clearly from the above proof and $x_0 \in B(x^*, \rho)$ it follows that $x_n \in B(x^*, \rho) \ \forall n \geq 0$. The proof is complete. $\quad\square$

Remark 9.7.1. Hypothesis (iv) is not additional to (iii), since (iii) always implies (iv). We also have that $\beta \leq b$ and b/β can be arbitrarily large. Indeed, let us consider as an example $X = Y = R$, $x^* = 0$ and define the function F on X by

$$F(x) = d_0 x - d_1 \sin(1) + d_1 \sin(e^{d_2 x}),$$

where d_0, d_1, d_2 are given prameters. Then, it can easily be seen that for d_2 sufficiently large and d_1 sufficiently small, b/β can be arbitrarily large. Other examples where $\beta < b$ can be found in [110].

If hypothesis (iv) is dropped, then the convergence radius can be defined by

$$\rho_0 = \min\{R, \frac{2(1 - |1 - \lambda|)}{(3 + |1 - \lambda|)b}\}.$$

Notice that $\rho_0 \leq \rho$. The preceding inequality can be strict if $\beta < b$ and

$$\frac{2(1 - |1 - \lambda|)}{(3 + |1 - \lambda|)b} < R.$$

If $\lambda = 1$ (Newton's method case), then ρ, ρ_0 reduce respectively to

$$\rho^A = \min\{R, \frac{2}{2\beta + b}\} \quad \text{and} \quad \rho_0^{TR} = \min\{R, \frac{2}{3b}\}.$$

The radius ρ^A was given by Argyros and Hilout in Refs. [110, 155], whereas ρ^{TR} was given independently by Ortega and Rheinboldt in Ref. [338] and Traub in Ref. [418]. Clearly we have again that $\rho^{TR} \leq \rho^A$. The preceding inequality is strict if $\beta < b$.

Our main purpose is to study the semi-local convergence of relaxed Newton's method defined in (9.317) by means of using similar techniques as Kantorovich did for Newton's method. First of all, we present the following Lemma.

Lemma 9.7.1. *Let $p(r)$ be the polynomial:*

$$p(r) = \frac{b}{2}r^2 - r + a, \tag{9.319}$$

with $a, b > 0$, $ab < 1/2$, $\lambda \in (0,1)$ and let $r_0 = 0$. Then, the sequence given by $r_{n+1} = N_{\lambda,p}(r_n)$ converges to

$$r^* = \frac{1 - \sqrt{1 - 2ab}}{b}, \tag{9.320}$$

where

$$N_{\lambda,p}(r_n) = r_n - \lambda \frac{p(r_n)}{p'(r_n)}. \tag{9.321}$$

Proof. We begin by proving that (r_n) is an increasing sequence. We have that

$$r_1 = r_0 - \lambda \frac{p(r_0)}{p'(r_0)} = \lambda a > 0 = r_0.$$

Let suppose that $r_n > r_{n-1}, \forall n \le k - 1$. We shall show that $r_{n+1} > r_n$. As $p(r) > 0, p'(r) < 0, \forall r \in [0, r^*]$, we deduce that $r_{n+1} > r_n$. So (r_n) is an increasing sequence. Secondly, using an induction process we see that r^* is the upper bound of the sequence (r_n). Indeed, it's immediate that $r^* - r_0 = r^* > 0$. Let's suppose that $r^* - r_k > 0$. We shall prove that the preceding inequality is true by proving that $n = k + 1$

$$r^* - r_{k+1} = N_{\lambda,p}(r^*) - N_{\lambda,p}(r_k).$$

Taking into account the mean value theorem we have that

$$N_{\lambda,p}(r^*) - N_{\lambda,p}(r_k) = N'_{\lambda,p}(\xi)(r^* - r_k), \xi \in (r_k, r^*).$$

As $r^* - r_k > 0$, if $N'_{\lambda,p}(\xi) > 0$ the proof is complete, but:

$$N'_{\lambda,p}(\xi) = (1 - \lambda) + \lambda \frac{p(\xi)p''(\xi)}{p'(\xi)^2}.$$

As $p(z) > 0$ and $p''(z) > 0, \forall \xi \in (0, r^*)$ and $\lambda \in (0,1)$ then $N'_{\lambda,p}(z) > 0$. Hence, (r_n) is bounded and so (r_n) converges to its unique least upper bound $r_0^* < r^*$. Taking limits in (9.321), it's clear that $p(r_0^*) = 0$, then $r_0^* = r^*$. The proof is complete. $\square$

Using this Lemma and majorizing sequences we present the following result:

Theorem 9.7.3. *Let $F : X \to Y$ be an operator defined between two Banach spaces X, Y, and Fréchet differentiable in the ball $B(x_0, R)$. Suppose that, the following conditions are satisfied:*

(i) *There exists the inverse operator of F at the point x_0, moreover we will denote it by $\Gamma_0 = F'(x_0)^{-1}$,*

(ii) $\|\Gamma_0 F(x_0)\| \le a, a > 0,$

(iii) $\|\Gamma_0[F'(x) - F'(y)]\| \le b\|x - y\|, b > 0, \ x, y \in B(x_0, R),$

(iv) $h = ab < 1/2,$

    and

(v) $r^* = \dfrac{1 - \sqrt{1 - 2h}}{b} \le R.$

Then, the sequence defined by:

$$\begin{cases} r_0 = 0 \\ r_{n+1} = r_n - \lambda \dfrac{p(r_n)}{p'(r_n)}, & n \ge 0 \end{cases}$$

where $p(r)$ is defined in (9.319), majorizes the sequence generated by relaxed Newton's method. Moreover sequence (x_n) converges to x^* solution of $F(x) = 0$. Limit point x^* is located in $\overline{B}(x_0, r^*)$, where r^* is defined in (9.320) and is unique in $B(x_0, r^{**})$, where

$$r^{**} = \frac{1 + \sqrt{1 - 2h}}{b}. \tag{9.322}$$

Proof.   First of all taking into account that $p'(r) = br - 1$ we can define the sequence (r_n) as:

$$\begin{cases} r_0 = 0 \\ r_{n+1} = r_n - \lambda \dfrac{\frac{b}{2}r_n^2 - r_n + a}{br_n - 1}, & n \ge 0 \end{cases} \tag{9.323}$$

Lemma 9.7.1 ensures that the sequence (r_n) is increasing and converges to r^*. We shall show that the following conditions are satisfied for every $n \in \mathbb{N}$:

(I_n) $\|F'(x_n)^{-1} F'(x_0)\| \le \dfrac{p'(r_0)}{p'(r_n)}.$

(II_n) $\|\Gamma_0 F(x_n)\| \le -\dfrac{p(r_n)}{p'(r_0)}.$

(III_n) $\|x_{n+1} - x_n\| \le r_{n+1} - r_n.$

If $n = 0$, it's easy to see that

(I_0) $\|\Gamma_0 \Gamma_0^{-1}\| = 1 = \dfrac{p'(r_0)}{p'(r_0)}.$

(II_0) $\|\Gamma_0 F(x_0)\| \le a = -\dfrac{p(r_0)}{p'(r_0)}.$

(III_0) $\|x_1 - x_0\| \le \lambda\|\Gamma_0 F(x_0)\| \le \lambda a = r_1 - r_0.$

Suppose that items (I_n)–(III_n) are true for $n \le k - 1$. We must show that these items are true for $n = k$. Taking into account that $\forall x \in B(x_0, r^*) \subseteq B(x_0, R)$ it's clear that:

$$\|I - \Gamma_0 F'(x)\| \le \|\Gamma_0 [F'(x) - F'(x_0)] \| \le b\|x - x_0\| < br^* \le 1.$$

Consequently, by the Banach Lemma on inversion of operators

$$\|F'(x)^{-1}F'(x_0)\| \le \frac{1}{1 - b\|x - x_0\|}.$$

We also have that

$$\|x_k - x_0\| \le \|x_k - x_{k-1}\| + \cdots + \|x_1 - x_0\| \le r_k - r_0 = r_k < r^*,$$

so $x_k \in B(x_0, r^*)$. Now we consider the following Ostrowski-type decomposition

$$F(x_{k+1}) = F(x_{k+1}) - \lambda F(x_k) - F'(x_k)(x_{k+1} - x_k) - F(x_k) + F(x_k)$$

$$= F(x_{k+1}) - F(x_k) - F'(x_k)(x_{k+1} - x_k) + (1 - \lambda)F(x_k)$$

$$= \int_{x_k}^{x_{k+1}} [F'(x) - F'(x_k)]\, dx + (1 - \lambda)F(x_k)$$

$$= \int_0^1 [F'(x_k + t(x_{k+1} - x_k)) - F'(x_k)] (x_{k+1} - x_k)dt + (1 - \lambda)F(x_k)$$

$$(9.324)$$

and so

$$\|\Gamma_0 F(x_{k+1})\| \le \frac{b}{2}\|x_{k+1} - x_k\|^2 + (1 - \lambda)\|\Gamma_0 F(x_k)\|.$$

Moreover, repeating the process

$$\begin{aligned}
p(r_{k+1}) &= p(r_{k+1}) - \lambda p(r_k) - p'(r_k)(r_{k+1} - r_k) - p(r_k) + p(r_k) \\
&= p(r_{k+1}) - p(r_k) - p'(r_k)(r_{k+1} - r_k) + (1 - \lambda)p(r_k) \\
&= \int_{r_k}^{r_{k+1}} [p'(x) - p'(r_k)]\, dx + (1 - \lambda)p(r_k) \\
&= \int_{r_k}^{r_{k+1}} b(x - r_k)dx + (1 - \lambda)p(r_k) = \frac{b}{2}(r_{k+1} - r_k)^2 + (1 - \lambda)p(r_k).
\end{aligned}$$

Furthermore, we have that

(I_k) $\quad \|F'(x_k)^{-1}F'(x_0)\| \le \dfrac{1}{1 - b\|x_k - x_0\|} \le \dfrac{1}{1 - br_k} \le -\dfrac{1}{p'(r_k)} = \dfrac{p'(r_0)}{p'(r_k)}.$

(II_k) $\quad \|\Gamma_0 F(x_k)\| \le \dfrac{b}{2}\|x_k - x_{k-1}\|^2 + (1 - \lambda)\|\Gamma_0 F(x_{k-1})\| \le$

$\dfrac{b}{2}(r_k - r_{k-1})^2 - (1 - \lambda)\dfrac{p(r_{k-1})}{p'(r_0)} = -\dfrac{1}{p'(r_0)}\dfrac{b}{2}(r_k - r_{k-1})^2 - (1 - \lambda)\dfrac{p(r_{k-1})}{p'(r_0)}$

$= -\dfrac{1}{p'(r_0)}\left[\dfrac{b}{2}(r_k - r_{k-1})^2 + (1 - \lambda)p(r_{k-1})\right] = -\dfrac{p(r_k)}{p'(r_0)}.$

(III_k) $\quad \|x_{k+1} - x_k\| \le \lambda\|\Gamma_k F(x_k)\| \le \lambda\|\Gamma_k F'(x_0)\|\|\Gamma_0 F(x_k)\| \le$

$\lambda \left(\dfrac{p'(r_0)}{p'(r_k)}\right)\left(-\dfrac{p(r_k)}{p'(r_0)}\right) = -\lambda\dfrac{p(r_k)}{p'(r_k)} = r_{k+1} - r_k.$

Now, to obtain the uniqueness region of the solution we assume that $\hat{x}$ is a solution located in the ball $B(x_0, r^{**})$, where r^{**} is defined in (9.322). Then

$$0 = \Gamma_0 [F(\hat{x} - F(x^*)] = W(\hat{x} - x^*), \qquad (9.325)$$

where $W : X \to X$ is the linear operator defined by

$$Wx = \left[\int_0^1 \Gamma_0 F'(x^* + t(\hat{x} - x^*))\, dt \right] x, \quad x \in X.$$

As $\|\hat{x} - x_0\| < r^{**}$ and $\|x^* - x_0\| \leq r^*$, then we have that

$$\|I - W\| = \left\| \int_0^1 \Gamma_0 [F'(x^* + t(\hat{x} - x^*)) - F'(x_0)]\, dt \right\|$$
$$\leq \int_0^1 b\|x^* + t(\hat{x} - x^*) - x_0\|\, dt \leq \int_0^1 b\left((1-t)\|x^* - x_0\| + t\|\hat{x} - x_0\|\right) dt$$
$$< \frac{b}{2}(r^* + r^{**}) = 1.$$

Again by the Banach Lemma on inversion of operators there exists the inverse operator of W. Finally, taking into account (9.325), we deduce that $\hat{x} = x^*$. Hence, x^* is the unique solution of $F(x) = 0$ in the ball $B(x_0, r^{**})$. The proof of the theorem is complete. $\qquad\square$

Furthermore, we present the following corollary.

Corollary 9.7.1. *If $h = 1/2$ the convergence of relaxed Newton's method to a root x^* is also established. However, in this particular case, $t^{**} = t^* = 1/b$. That is the existence and uniqueness domain is the closed ball $\overline{B}(x_0, 1/b)$.*

Once we have proven the previous semi-local convergence results we can present the order of convergence of the method and error bounds. By the Schröder theorem, we can obtain the convergence order of the sequence (9.323). As $r_{n+1} = N_{\lambda,p}(r_n)$ where $N_{\lambda,p}(r_n)$ and $p(r)$ defined in (9.321) and (9.319) respectively:

$$N'_{\lambda,p}(x) = 1 - \lambda(-L_p(x) + 1) = (1 - \lambda) + \lambda L_p(x),$$

so we can ensure that

$$r^* - r_{n+1} = N_{\lambda,p}(r^*) - N_{\lambda,p}(r_n) =$$
$$N_{\lambda,p}(r^*) - \left[N_{\lambda,p}(r^*) + N'_{\lambda,p}(r^*)(r_n - r^*) + O(r_n - r^*)^2 \right]$$
$$= N'_{\lambda,p}(r^*)(r^* - r_n) + O(r_n - r^*).$$

Hence, linear convergence is proven, with asymptotic error constant $N'_{\lambda,p}(r^*) = (1 - \lambda)$. Then the speed of convergence depends on the damping factor λ. This speed is greater when λ is closer to 1 (see Ref. [341]).

Following a technique introduced by Ostrowski, we are going to present upper and lower bounds for the error.

Theorem 9.7.4. *Under the assumptions of Theorem 9.7.3. Then, we have the following error bounds for the sequence (x_n) given by relaxed Newton's method:*

$$\frac{r^*(r^{**} - r^*)(1 - \lambda)^n}{r^{**} - r^*(1 - \lambda)^n} \leq \|x^* - x_n\| \leq \frac{r^*(r^{**} - r^*)Q^n}{r^{**} - r^*Q^n}$$

where $Q = \dfrac{r^* + (1 - \lambda)r^{**}}{r^{**} + (1 - \lambda)r^*}$, r^* and r^{**} are defined in (9.320) and (9.322), respectively.

Proof. Letting $a_n = r^* - r_n$ and $b_n = r^{**} - r_n$. Then, we have that

$$a_{n+1} = r^* - r_{n+1} = r^* - r_n + \lambda \frac{p(r_n)}{p'(r_n)} = a_n - \lambda \frac{a_n b_n}{a_n + b_n} = a_n \frac{a_n + (1 - \lambda)b_n}{a_n + b_n}$$

and

$$b_{n+1} = b_n \frac{b_n + (1 - \lambda)a_n}{a_n + b_n}.$$

It's clear that

$$\frac{a_{n+1}}{b_{n+1}} = \frac{a_n}{b_n} \frac{a_n + (1 - \lambda)b_n}{b_n + (1 - \lambda)a_n}. \tag{9.326}$$

Hence, we have the following

$$\frac{a_n + (1 - \lambda)b_n}{b_n + (1 - \lambda)a_n} = \frac{r^* - r_n + (1 - \lambda)(r^{**} - r_n)}{r^{**} - r_n + (1 - \lambda)(r^* - r_n)} = \frac{r^* + (1 - \lambda)r^{**} - r_n(2 - \lambda)}{r^{**} + (1 - \lambda)r^* - r_n(2 - \lambda)}.$$

Now letting $f(x) = \dfrac{A - Bx}{C - Bx}$ and derivating

$$f'(x) = \frac{-B(C - Bx) + B(A - Bx)}{(C - Bx)^2} = \frac{B(A - C)}{(C - Bx)^2}.$$

In our particular case

$$A - C = r^* + (1 - \lambda)r^{**} - r^{**} - (1 - \lambda)r^* = (r^* - r^{**}) + (1 - \lambda)(r^{**} - r^*) = -\lambda(r^{**} - r^*) < 0.$$

Then, as $0 \le r_n < r^*$, we have that

$$f(0) \ge f(r_n) \ge f(r^*).$$

Letting $q_n = \dfrac{a_n}{b_n}$, we compute the upper bound as follows

$$q_{n+1} \le q_n \frac{r^* + (1 - \lambda)r^{**}}{r^{**} + (1 - \lambda)r^*}.$$

Set $Q = \frac{r^* + (1 - \lambda)r^{**}}{r^{**} + (1 - \lambda)r^*}$, we have that

$$q_{n+1} \le Q q_n \le Q^2 q_{n-1} \le \cdots \le Q^{n+1} q_0.$$

Taking into account, $q_n \le Q^n q_0$, then, we have in turn that

$$a_n \le q_0 b_n Q^n = q_0(r^{**} - r_n)Q^n = q_0(r^{**} - r_n)Q^n = q_0(r^{**} - r^* + a_n)Q^n.$$

In addition, we have that

$$a_n \le \frac{q_0(r^{**} - r^*)Q^n}{1 - q_0 Q^n}$$

since

$$[1 - q_0 Q^n]\, a_n \le q_0(r^{**} - r^*)Q^n.$$

On the other hand, as $Q < 1$, as $r^* + r^{**} - \lambda r^{**} < r^* - \lambda r^* + r^{**}$, we can conclude that

$$a_n \le \frac{r^*(r^{**} - r^*)Q^n}{r^{**} - r^* Q^n}.$$

Let's compute the lower bound. We have that

$$q_{n+1} \geq q_n \frac{r^* + (1-\lambda)r^{**} - (2-\lambda)r^*}{r^{**} + (1-\lambda)r^* - (2-\lambda)r^*} = q_n \frac{r^*(\lambda - 1) + (1-\lambda)r^{**}}{r^{**} - r^*}$$

$$= q_n \frac{(1-\lambda)(r^{**} - r^*)}{r^{**} - r^*} = (1-\lambda)q_n \geq \cdots \geq (1-\lambda)^{n+1}q_0.$$

As $q_n \geq (1-\lambda)^n q_0$, then we have that

$$a_n \geq (1-\lambda)^n b_n q_0 = (1-\lambda)^n (r^{**} - r^* + r^* - r_n)q_0 = (1-\lambda)^n (r^{**} - r^* + a_n)q_0.$$

Consequently, we deduce that

$$a_n \geq \frac{r^*(r^{**} - r^*)(1-\lambda)^n}{r^{**} - r^*(1-\lambda)^n}$$

since $[1 - q_0(1-\lambda)^n] a_n \geq (r^{**} - r^*)q_0(1-\lambda)^n$. This ends the proof. $\qquad\square$

On the other hand, we have that

$$Q = Q(\lambda) = \frac{r^* + (1-\lambda)r^{**}}{r^{**} + (1-\lambda)r^*} = \frac{1 - \dfrac{r^{**}}{r^* + r^{**}}\lambda}{1 - \dfrac{r^*}{r^* + r^{**}}\lambda}.$$

Let

$$A = \frac{r^{**}}{r^* + r^{**}}, \quad B = \frac{r^*}{r^* + r^{**}} \quad \text{and} \quad Q(\lambda) = \frac{1 - A\lambda}{1 - B\lambda}.$$

We have that

$$Q'(\lambda) = \frac{B - A}{(1 - B\lambda)^2}.$$

By $B - A = \dfrac{r^* - r^{**}}{r^* + r^{**}} < 0$, then function $Q(\lambda)$ is decreasing, and consequently the greater the damping factor is the faster method converges to the solution. Lastly, we have the following corollary.

Corollary 9.7.2. *If $\lambda = 1$, from the equation (9.326) we have that $\dfrac{a_{n+1}}{b_{n+1}} = \left(\dfrac{a_n}{b_n}\right)^2$ and so the method has quadratic convergence (see Ref. [341]).*

We present now a different semi-local convergence analysis for relaxed Newton's method based on the center-Lipschitz and Lipschitz condition. First, we need a result on a scalar sequence which will be shown to be majorizing for $\{x_n\}$. The convergence of the sequences in this section is established for $\lambda \in (0, 2)$. Notice that the convergence was established above for $\lambda \in (0, 1)$.

Lemma 9.7.2. *Let $a > 0$, $\beta > 0$, $b > 0$ and $\lambda \in (0, 2)$ be given parameters. Set $M = |1 - \lambda|$ and $\gamma = \beta/b$. Define parameter α by*

$$\alpha = \frac{2(\lambda - 2M\gamma)}{\lambda + \sqrt{\lambda^2 + 8\gamma(\lambda - 2M\gamma)}}. \tag{9.327}$$

Suppose that for $\delta = ab\lambda\gamma$ the following conditions are true

$$2M\gamma < \lambda, \tag{9.328}$$

$$(\lambda - 2)\delta^2 + 2(2 + M)\delta - 2 < 0, \tag{9.329}$$

$$(\frac{\lambda^2}{\delta} - 4M - 4\alpha + 2\alpha\lambda)\delta^2 + 2(\frac{M}{\gamma}\lambda + 4\alpha + 2\alpha M)\delta + 4(M - \alpha) \leq 0 \tag{9.330}$$

and

$$(M + \alpha)(\lambda - 2(1 - \alpha))\delta^2 + 2(M + (3 - 2\alpha)\alpha)\delta + 2(1 - \alpha)(M - \alpha) \leq 0. \tag{9.331}$$

Then, scalar sequence $\{t_n\}$ defined by

$$\begin{cases} t_0 = 0, \ t_1 = \lambda a, \ t_2 = t_1 + \dfrac{1}{2(1 - \gamma b t_1)}[\lambda\gamma b(t_1 - t_0) + 2M](t_1 - t_0) \\[2ex] t_{n+2} = t_{n+1} + \dfrac{1}{2(1 - \gamma b t_{n+1})}[\lambda b(t_{n+1} - t_n) + 2M(1 + \gamma b t_n)](t_{n+1} - t_n), \end{cases}$$
$$\tag{9.332}$$

is well-defined, increasing, bounded from above by

$$t^{**} = \left[1 + \frac{\lambda\gamma b + 2M}{2(1 - \delta)(1 - \alpha)}\right]\lambda a \tag{9.333}$$

and converges to its unique least upper bound t^* which satisfies

$$t^* \in [t_n, t^{**}]. \tag{9.334}$$

Moreover, the following estimates hold for each $n = 1, 2, \cdots$

$$0 < t_{n+2} - t_{n+1} \leq \frac{\lambda\gamma b + 2M}{2(1 - \delta)}\alpha^n n. \tag{9.335}$$

Proof.   We first notice that $\alpha \in (0, 1)$ by (9.327) and (9.328) and $M \in (0, 1)$ since $\lambda \in (0, 2)$. Let

$$\alpha_0 = \frac{\lambda b(t_2 - t_1) + 2(\gamma b t_1 + 1)M}{2(1 - \gamma b t_2)}. \tag{9.336}$$

Using (9.329) and (9.332) we get that

$$\gamma b t_2 < 1. \tag{9.337}$$

Moreover, by (9.327), (9.330) and (9.337) we deduce that

$$0 < \alpha_0 \leq \alpha. \tag{9.338}$$

Next we shall show using induction on the integer $k = 1, 2, \cdots$ that

$$0 < \frac{\lambda b(t_{k+1} - t_k) + 2M(\gamma b t_k + 1)}{2(1 - \gamma b t_2)} \leq \alpha. \tag{9.339}$$

Estimate (9.339) is true for $k = 1$ by (9.338). Then, using (9.332) for $n = 1$ and (9.338) we obtain that

$$0 < t_3 - t_2 \leq \alpha(t_2 - t_1)$$
$$\Rightarrow t_3 \leq t_2 + \alpha(t_2 - t_1) = t_2 + (1 + \alpha)(t_2 - t_1) - (t_2 - t_1) \tag{9.340}$$
$$\Rightarrow t_3 \leq t_1 + \frac{1-\alpha^2}{1-\alpha}(t_2 - t_1) < t^{**}.$$

Suppose that (9.339) holds for each $k = 1, 2, \cdots, m$. Then, we get that

$$0 < t_{k+2} - t_{k+1} \leq \alpha^k(t_2 - t_1) \tag{9.341}$$

and

$$t_{k+2} \leq t_1 + \frac{1 - \alpha^{k+1}}{1 - \alpha}(t_2 - t_1). \tag{9.342}$$

We must show that

$$0 \leq \frac{\lambda b(t_{k+2} - t_{k+1}) + 2M(\gamma b t_k + 1)}{2(1 - \gamma b t_{k+1})} \leq \alpha. \tag{9.343}$$

Evidently, (9.343) holds, if

$$\lambda b(t_{k+2} - t_{k+1}) + 2(\gamma b t_k + 1)M \leq 2\alpha(1 - \gamma b t_{k+1})$$

or using (9.341) and (9.342)

$$\lambda b(t_2-t_1)\alpha^k + 2M\gamma b\frac{1-\alpha^k}{1-\alpha}(t_2-t_1) + 2\alpha\gamma b\frac{1-\alpha^{k+1}}{1-\alpha}(t_2-t_1) + 2(\delta M + \alpha\delta + M - \alpha) \leq 0. \tag{9.344}$$

Estimate (9.344) motivates us to define recurrent functions f_k on $(0, 1)$ by

$$f_k(t) = \lambda b(t_2 - t_1)t^k + 2M\gamma b(t_2 - t_1)(1 + t + \cdots + t^{k-1})$$
$$+ 2\gamma b(t_2 - t_1)t(1 + t + \cdots t^k) + 2((M\gamma b + t\gamma b)\lambda a + M - t). \tag{9.345}$$

We need a relationship between two consecutive functions f_k. Using (9.345) we get that

$$f_{k+1}(t) = f_{k+1}(t) - f_k(t) + f_k(t) = f_k(t) + g(t)(t_2 - t_1)t^k, \tag{9.346}$$

where

$$g(t) = 2\gamma b t^2 + \lambda b t + (2M\gamma - \lambda)b. \tag{9.347}$$

It follows from (9.328) and (9.347) that polynomial g has a unique positive root α given by (9.327). In view of (9.345), estimate (9.344) holds if

$$f_k(\alpha) \leq 0. \tag{9.348}$$

Define function f_∞ on $(0, 1)$ by

$$f_\infty(t) = \lim_{k \to \infty} f_k(t). \tag{9.349}$$

We have from (9.346) and (9.347) that

$$f_{k+1}(\alpha) = f_k(\alpha). \tag{9.350}$$

Then, it follows from (9.349) and (9.350) that (9.348) holds if

$$f_\infty(\alpha) \leq 0. \tag{9.351}$$

By letting $k \to \infty$ and by (9.331) we have that (9.351) is true, since

$$f_\infty(\alpha) = 2\left[M\delta + \alpha\delta + M - \alpha(1-\gamma) + \frac{M\gamma b}{1-\alpha}(t_2 - t_1) + \frac{\gamma b\alpha}{1-\alpha}(t_2 - t_1)\right] \leq 0.$$

The induction for (9.339) is complete. Hence, the sequence $\{t_n\}$ is increasing, bounded from above by t^{**} given in (9.333) and as such it converges to its unique least upper bound t^* which satisfies (9.334). The proof is complete. $\qquad\square$

Remark 9.7.2. Quadratic inequalities (9.329)–(9.331) describe the smallness of a and can be solved for a. However, we decided to leave them as uncluttered as possible, since this representation is too long. It is simple algebra to show that in the interesting case when $\lambda = 1$ (Newton's method) reduce to (see [155]):

$$h_* = La \leq 1/2 \tag{9.352}$$

where

$$L = \frac{1}{8}\left(4b_0 + \sqrt{b_0 b + 8b_0^2} + \sqrt{b_0 b}\right).$$

If $b_0 = b$, then (9.352) reduces to the Newton–Kantorovich hypothesis given in condition (iv) of Theorem 9.7.3. Notice that

$$h \leq 1/2 \Rightarrow h_* \leq 1/2 \tag{9.353}$$

but not necessarily vice versa unless if $b_0 = 0$ and $h_*/h \to 0$ as $b_0/b \to 0$. Hence, the applicability of Newton's method has been extended.

Let us introduce the notation for fixed $N = 0, 1, \cdots$

$$\delta_N = \gamma b(t_{N+1} - t_N).$$

Note that $\delta_0 = \delta$. Next, we present the following weaker than Lemma 9.7.2 result for the convergence of sequence $\{t_n\}$. The proof simply follows from the proof of Lemma 9.7.2 by replacing δ by δ_N.

Lemma 9.7.3. *Suppose that* (9.328) *and* (9.329)–(9.331) *hold with* δ_N *replacing δ for δ_N fixed $N = 0, 1, \cdots$. Moreover, suppose that*

$$0 < t_1 < t_2 < \cdots < t_{N+1} < \frac{1}{\gamma b}.$$

Then, scalar sequence $\{t_n\}$, *given in* (9.332) *is well-defined, increasing, bounded from above by*

$$t_N^{**} = t_{N-1} + \frac{1}{1-\alpha}(t_N - t_{N-1})$$

for $N = 1, 2, \cdots$ and converges to its unique least upper bound t_N^* satisfying

$$t_{N+1} \leq t_N^* \leq t_N^{**}.$$

Moreover, the following estimates hold for each $n = 0, 1, \cdots$

$$0 < t_{N+n} - t_{N+n-1} \leq \alpha^{n-1}(t_{N+1} - t_N).$$

Remark 9.7.3. If $N = 0$, Lemma 9.7.3 reduces to Lemma 9.7.2. Clearly Lemma 9.7.3 is weaker than Lemma 9.7.2.

Next, we present the semi-local convergence of relaxed Newton's method using a center-Lipschitz and a Lipschitz condition as well as $\{t_n\}$ as the majorizing sequence for $\{x_n\}$.

Theorem 9.7.5. *Let $F : \Omega \subset X \to Y$ be Fréchet-differentiable. Suppose that there exist $x_0 \in \Omega$, $\beta > 0$, $b > 0$, $a > 0$, $\lambda \neq 0$ such that there exists $F'(x_0)^{-1}$, in addition*

$$\|F'(x_0)^{-1}F(x_0)\| \leq a,$$

$$\|F'(x_0)^{-1}(F'(x) - F'(x_0))\| \leq \beta\|x - x_0\|, \ \forall x \in \Omega,$$

$$\|F'(x_0)^{-1}(F'(x) - F'(y))\| \leq b\|x - y\|, \ \forall x, y \in \Omega,$$

$$\overline{B}(x_0, t^*) \subseteq \Omega$$

and hypotheses of Lemma 9.7.2 or Lemma 9.7.3 hold. Then, sequence $\{x_n\}$ generated by relaxed Newton's method is well-defined, remains in $\overline{B}(x_0, t^*)$ for each $n = 0, 1, \cdots$ and converges to a solution $x^* \in \overline{B}(x_0, t^*)$ of equation $F(x) = 0$. Moreover, the following estimates hold for each $n = 0, 1, \cdots$

$$\|x_{n+1} - x_n\| \leq t_{n+1} - t_n \quad and \quad \|x_n - x^*\| \leq t_n - t^*,$$

where sequence $\{t_n\}$ is given in (9.332). Furthermore, if there exists $R \geq t^*$ such that $\overline{B}(x_0, t^*) \subseteq D$ and $\beta(t^* + R) < 2$, then, the solution x^* is unique in $\overline{B}(x_0, R)$.

Proof. The proof follows in an analogous way as the proof of Theorem 9.7.3 but we use the Ostrowski decomposition

$$
\begin{aligned}
F'(x_0)^{-1}F(x_{n+1}) = \\
F'(x_0)^{-1} \int_0^1 [F'(x_n + t(x_{n+1} - x_n)) - F'(x_n)](x_{n+1} - x_n)dt \\
+(1 - \tfrac{1}{\lambda})(I + F'(x_0)^{-1}(F'(x_n) - F'(x_0))(x_{n+1} - x_n)
\end{aligned}
\tag{9.354}
$$

instead of (9.324). For the estimates on upper bounds on the norms $\|F'(x_n)^{-1}F'(x_0)\|$ and the uniqueness part we use the more precise and cheaper center-Lipschitz condition instead of the Lispchitz condition. $\qquad\square$

Remark 9.7.4. Note that the definition of t_2 (see (9.332)) is justified from the Ostroswki representation (9.354) for $n = 0$, since then the center-Lipschitz (and not the Lipschitz) condition is used to obtain the estimate on

$$\|F'(x_0)^{-1}\| \left\| \int_0^1 [F'(x_0 + t(x_1 - x_0)) - F'(x_0)]\, dt \right\| \leq \frac{\beta}{2}\|x_1 - x_0\|^2 \leq \frac{\beta}{2}(t_1 - t_0)^2$$

instead of

$$\|F'(x_0)^{-1}\| \left\| \int_0^1 [F'(x_0 + t(x_1 - x_0)) - F'(x_0)]\, dt \right\| \leq \frac{b}{2}\|x_1 - x_0\|^2.$$

We complete this Section with another result for relaxed Newton's method with simpler convergence conditions than in Theorem 9.7.5.

Theorem 9.7.6. *Let $F : \Omega \subset X \to Y$ be Fréchet-differentiable. Suppose that there exist $x_0 \in \Omega$, $\beta > 0$, $b > 0$, $a > 0$, $\lambda \in (0,2)$ such that there exists $F'(x_0)^{-1}$, in addition*

$$\|F'(x_0)^{-1}F(x_0)\| \leq a,$$

$$\|F'(x_0)^{-1}(F'(x) - F'(x_0))\| \leq \beta\|x - x_0\|, \ \forall x \in \Omega,$$

$$\|F'(x_0)^{-1}(F'(x) - F'(y))\| \leq b\|x - y\|, \ \forall x, y \in \Omega.$$

Moreover, suppose that

$$h_1 = \lambda ab \max\{\lambda, \gamma(1 + M)\} \leq \frac{1}{2}(1 - M)^2, \quad and \quad \overline{B}(x_0, s^*) \subseteq \Omega,$$

where M and γ are given in Lemma 9.7.2,

$$s^* = \frac{1 - M - \sqrt{(1 - M)^2 - 2h_1}}{\sigma} \quad and \quad \sigma = b\max\{\lambda, \gamma(1 + M)\}.$$

Then, sequence $\{x_n\}$ generated by relaxed Newton's method is well-defined, remains in $\overline{B}(x_0, s^*)$ for all $n \geq 0$ and converges to a solution $x^* \in \overline{B}(x_0, s^*)$ of equation $F(x) = 0$. Moreover the equation $F(x) = 0$, has a unique solution $x^* \in S$, where

$$S = \begin{cases} \overline{B}(x_0, s^*) \bigcap \Omega & \text{if } h_1 = \frac{1}{2}(1 - M)^2 \\ \overline{B}(x_0, s^{**}) \bigcap \Omega & \text{if } h_1 < \frac{1}{2}(1 - M)^2 \end{cases} \tag{9.355}$$

and

$$s^{**} = \frac{1 - M + \sqrt{(1 - M)^2 - 2h_1}}{\sigma}.$$

Furthermore, the following estimates hold for each $n = 0, 1, \cdots$

$$\|x_{n+1} - x_n\| \leq s_{n+1} - s_n \tag{9.356}$$

and

$$\|x_n - x^*\| \leq s^* - s_n, \tag{9.357}$$

where majorizing sequence $\{s_n\}$ is defined by

$$s_0 = 0, \ s_{n+1} = s_n + \frac{f(s_n)}{g(s_n)},$$

where

$$f(t) = \frac{\sigma}{2}t^2 - (1 - M)t + \lambda a \quad and \quad g(t) = 1 - \gamma bt.$$

Proof. It follows easily from $2\sigma\lambda a \le (1-M)^2$ that function $f(t)$ has two positive roots s^* and s^{**}, $s^* \le s^{**}$ and $s_n \le s_{n+1}$, so that the sequence $\{s_n\}$ converges to s^*. As in Theorem 9.7.5, we obtain $F'(x_n)^{-1} \in L(Y, X)$,

$$\|F'(x_n)^{-1}F'(x_0)\| \le \frac{1}{1 - \beta\|x_n - x_0\|} \le \frac{1}{1 - \beta s_n} \le \frac{1}{1 - b\gamma s_n} \le \frac{1}{g(s_n)}$$

and

$$\|x_{n+1} - x_n\| \le \frac{1}{g(s_n)}\left[\frac{\lambda b}{2}\|x_n - x_{n-1}\| + (M\gamma b\|x_n - x_0\| + M)\right]\|x_n - x_{n-1}\|$$

$$\le \frac{1}{g(s_n)}\left(\frac{\sigma}{2}(s_n - s_{n-1}) + M(\gamma b s_{n-1} + 1)\right)(s_n - s_{n-1})$$

$$= \frac{1}{g(s_n)}\left(\frac{\sigma}{2}(s_n - s_{n-1})^2 + M\gamma b(s_n - s_{n-1})s_{n-1} + M(s_n - s_{n-1})\right)$$

$$-g(s_{n-1})(s_n - s_{n-1}) + f(s_{n-1}))$$

$$= \frac{1}{g(s_n)}\left(\frac{\sigma}{2}s_n^2 - (1-M)s_n + \lambda a - (\sigma - M\gamma b - \gamma b)s_{n-1}(s_n - s_{n-1})\right)$$

$$\le \frac{f(s_n)}{g(s_n)} = s_{n+1} - s_n.$$

Hence, we have for any n that

$$\|x_{n+1} - x_n\| \le s_{n+1} - s_n$$

$$\|F'(x_0)^{-1}(F'(x_{n+1}) - F'(x_0))\| \le \beta\|x_{n+1} - x_0\| \le \beta s_{n+1} \le \gamma b s_{n+1} \le \gamma b s^* < 1$$

and

$$\|x_n - x_0\| \le s_n \le s^*.$$

It follows that: $x_n \in \overline{B}(x_0, s^*)$, sequence $\{x_n\}$ is complete in a Banach space X and as such it converges to some $x^* \in \overline{B}(x_0, s^*)$ (since $\overline{B}(x_0, s^*)$ is a closed set). We also have that

$$\|F'(x_0)^{-1}F(x_n)\| \le \left(\frac{\sigma}{2}(s_n - s_{n-1}) + M(\gamma b s_n + 1)\right)(s_n - s_{n-1})$$

$$\le \left(\frac{\sigma}{2}(s_n - s_{n-1}) + M(\gamma b s^* + 1)\right)(s_n - s_{n-1}) \to 0,$$

as $n \to \infty$, so that $F(x^*) = 0$. Estimate (9.357) follows from (9.356) by using standard majorizing techniques. Finally, for the uniqueness part see the more general proof of Theorem 9.7.7. The proof of the Theorem is complete. $\square$

Remark 9.7.5. The following two conditions are true

(a) Hypothesis $h_1 \le \frac{1}{2}(1-M)^2$ for $0 < \lambda \le 1$ reduces to

$$h_1 = ab\frac{\max\{\lambda, \gamma(2-\lambda)\}}{\lambda} \le \frac{1}{2}$$

which is not weaker than $h = ab \le 1/2$. However, hypothesis $h = ab \le 1/2$ cannot be used as the sufficient convergence condition for relaxed Newton's method for $\lambda \in (1, 2)$. In practice we shall test all the h hypotheses introduced in this section to see if any of them is satisfied.

(b) Results analogous to Lemma 9.7.3 can now follow by exchanging sequence $\{t_n\}$ by $\{s_n\}$.

We present now a semi-local convergence result to solve a nonlinear equation (1.54). Then, relaxed Newton's method for generating a sequence approximating x^* is defined by:

$$x_{n+1} = x_n - \lambda F'(x_n)^{-1}(F(x) + G(x)), \ n \geq 0. \tag{9.358}$$

Clearly if $G = 0$ on Ω, then method (9.358) reduces to method (9.317). We used three different techniques for our semi-local convergence results. However, for the method (9.358) we shall only use the technique of Theorem 9.7.7. For brievity, we leave the analysis of the other two techniques to the motivated reader.

Theorem 9.7.7. *Let* $F : \Omega \subset X \to Y$ *be Fréchet-differentiable and* $G : \Omega \to Y$ *be continuous. Suppose that there exist* $x_0 \in \Omega$, $\beta > 0$, $b > 0$, $a > 0$, $K \geq 0$, $\lambda \in (0, 2)$ *such that there exists* $F'(x_0)^{-1}$, *in addition*

$$\|F'(x_0)^{-1}(F(x_0) + G(x_0))\| \leq a,$$

$$\|F'(x_0)^{-1}(F'(x) - F'(x_0))\| \leq \beta\|x - x_0\|, \ \forall x \in \Omega,$$

$$\|F'(x_0)^{-1}(F'(x) - F'(y))\| \leq b\|x - y\|, \ \forall x, y \in \Omega,$$

$$\|F'(x_0)^{-1}(G(x) - G(y))\| \leq K\|x - y\|, \ \forall x, y \in \Omega.$$

Moreover, suppose

$$\mu = M + \lambda K < 1, \quad h_2 = \lambda ab \max\{\lambda, \gamma(1 + M)\} \leq 1/2(1 - \mu)^2,$$

and

$$\overline{B}(x_0, s^*) \subseteq \Omega,$$

where M *and* γ *are given in Lemma 9.7.2,*

$$s^* = \frac{1 - \mu - \sqrt{(1 - \mu)^2 - 2h_2}}{\sigma} \quad and \quad \sigma = b \max\{\lambda, \gamma(1 + M)\}.$$

Then, the sequence $\{x_n\}$ *generated by relaxed Newton's method (9.358) is well-defined, remains in* $\overline{B}(x_0, s^*)$ *for all* $n \geq 0$ *and converges to a solution* $x^* \in \overline{B}(x_0, s^*)$ *of equation* $F(x) + G(x) = 0$. *Moreover, the equation* $F(x) + G(x) = 0$ *has a unique solution* $x^* \in S$, *where*

$$S = \begin{cases} \overline{B}(x_0, s^*) \bigcap \Omega & if \ h_2 = \frac{1}{2}(1 - \mu)^2 \\ \overline{B}(x_0, s^{**}) \bigcap \Omega & if \ h_2 < \frac{1}{2}(1 - \mu)^2 \end{cases} \tag{9.359}$$

and

$$s^{**} = \frac{1 - \mu + \sqrt{(1 - \mu)^2 - 2h_2}}{\sigma}.$$

Furthermore, the following estimates hold for each $n = 0, 1, \cdots$

$$\|x_{n+1} - x_n\| \leq s_{n+1} - s_n \tag{9.360}$$

and

$$\|x_n - x^*\| \leq s^* - s_n, \tag{9.361}$$

where majorizing sequence $\{s_n\}$ *is defined by*

$$s_0 = 0, \quad s_{n+1} = s_n + \frac{f(s_n)}{g(s_n)},$$

where

$$f(t) = \frac{\sigma}{2}t^2 - (1 - \mu)t + \lambda a \quad and \quad g(t) = 1 - \gamma bt.$$

Proof. Using (9.358) we introduce as in Theorem 9.7.5 the Ostrowski decomposition to obtain

$$F'(x_0)^{-1}(F(x_{n+1}) + G(x_{n+1}))$$
$$= F'(x_0)^{-1} \int_0^1 [F'(x_n + t(x_{n+1} - x_n)) - F'(x_n)] (x_{n+1} - x_n)dt$$
$$+(1 - \frac{1}{\lambda}) \left(I + F'(x_0)^{-1}(F'(x_n) - F'(x_0))\right) (x_{n+1} - x_n)$$
$$+F'(x_0)^{-1}(G(x_{n+1}) - G(x_n)).$$

Then, following the proof of Theorem 9.7.6, we obtain in turn that

$$\|x_{n+1} - x_n\| \leq \frac{1}{g(s_n)} \left[\frac{\lambda b}{2} \|x_n - x_{n-1}\| + M\gamma b \|x_n - x_0\| + \mu \right] \|x_n - x_{n-1}\|$$
$$\leq \frac{1}{g(s_n)} \left(\frac{\sigma}{2}(s_n - s_{n-1})^2 + M\gamma b s_{n-1} + \mu \right) (s_n - s_{n-1})$$
$$\leq \frac{1}{g(s_n)} (\frac{\sigma}{2}(s_n - s_{n-1})^2 + M(s_n - s_{n-1})s_{n-1}$$
$$+\mu(s_n - s_{n-1}) - g(s_{n-1})(s_n - s_{n-1}) + f(s_{n-1}))$$
$$= \frac{1}{g(s_n)} \left(\frac{\sigma}{2}s_n^2 - (1 - \mu)s_n + \lambda a - (\sigma - M\gamma b - \gamma b)s_{n-1}(s_n - s_{n-1}) \right)$$
$$\leq \frac{f(s_n)}{g(s_n)} = s_{n+1} - s_n.$$

The rest follows as in Theorem 9.7.6 until

$$\|F'(x_0)^{-1}(F(x_n) + G(x_n))\| \leq \left(\frac{\sigma}{2}(s_n - s_{n-1})^2 + M\gamma b s^* + \mu \right) (s_n - s_{n-1}) \to 0$$

as $n \longrightarrow \infty$. Hence, $F(x^*) + G(x^*) = 0$. Finally, in order for us to show the uniqueness part, let us assume that $F(y^*) + G(y^*) = 0$, for some $y^* \in \Omega$. Using (9.358) we obtain the identity

$$y^* - x_n = F'(x_0)^{-1}\lambda \int_0^1 [F'(y^* + t(x_n - y^*)) - F'(x_n)] (x_n - y^*)dt$$
$$+(1 - \lambda) \left(I + F'(x_0)^{-1}(F'(x_n) - F'(x_0))\right) (x_n - y^*) + \lambda F'(x_0)^{-1}(G(x_n) - G(y^*)).$$

Then, for $n = 0$ we get that

$$\|y^* - x_1\| \leq \left(\frac{\sigma}{2}\|y^* - x_0\| + \mu\right)\|y^* - x_0\| = \varphi(\xi),$$

where $\xi = \|y^* - x_0\|$. We have that $\|y^* - x_0\| \leq \|y^* - x_1\| + \|x_1 - x_0\| \leq \varphi(\xi) + \lambda a$. That is $f(\xi) \geq 0$. Hence, $y^* \in \overline{B}(x_0, s^*)$. Using induction we shall show that

$$\|y^* - x_k\| \leq s^* - s_n, \ n \geq 0.$$

The preceding estimate is true for $k = 0$, since $y^* \in \overline{B}(x_0, s^*)$. Let $\|y^* - x_k\| \leq s^* - s_k$. Then, we have in turn that

$$
\begin{aligned}
\|y^* - x_k\| &\leq \frac{1}{g(s_k)}\left[\frac{\sigma}{2}\|y^* - x_k\| + M\|x_k - x_0\| + \mu\right]\|y^* - x_k\| \\
&\leq \frac{1}{g(s_k)}\left[\frac{\sigma}{2}s^{*2} + \mu s^* - (\sigma - M\gamma b)s_k(s^* - s_k) - \frac{\sigma}{2}s_k^2 - \mu s_k\right] \\
&= \frac{1}{g(s_k)}\left[s^* - \lambda a - (\sigma - M\gamma b)s_k(s^* - s_k) - \frac{\sigma}{2}s_k^2 - \mu s_k\right] \\
&= s^* - s_k + \frac{1}{g(s_k)}\left[-(s^* - s_k)g(s_k) + s^* - \lambda a - \frac{\sigma}{2}s_k^2 - \mu s_k - (\sigma - M\gamma b)s_k(s^* - s_k)\right] \\
&= s^* - s_k - \frac{1}{g(s_k)}\left[\frac{\sigma}{2}s_k^2 - (1 - \mu)s_k + \lambda a + (\sigma - M\gamma b - \gamma b)s_k(s^* - s_k)\right] \\
&\leq s^* - s_{k+1}.
\end{aligned}
$$

That is, we have that $\lim\limits_{k\to\infty} x_k = y^*$. But, we showed $\lim\limits_{k\to\infty} x_k = x^*$. Hence, we deduce that $x^* = y^*$. The proof of the Theorem is complete. $\qquad\square$

Example 9.7.1. (Extended semi-local convergence) Consider the following nonlinear integral equation of mixed Hammerstein type

$$x(s) = f(s) + \int_A^B G(s, t)\,\alpha(x(t) - f(t))^2 + \beta|x(t) - f(t)|\,dt, \quad s \in [A, B], \quad (9.362)$$

where $x, f \in C[A, B]$, $\alpha, \beta \in R$ and the kernel G is the Green function

$$G(s, t) = \begin{cases} \dfrac{(B - s)(t - A)}{B - A}, & t \leq s, \\[2ex] \dfrac{(s - A)(B - t)}{B - A}, & s \leq t. \end{cases}$$

To simplify the analysis we choose $A = 0$, $B = 1$, $\alpha = \beta = 1/2$ and $f(s) = 0$. To solve (9.362), we transform it into a finite dimensional problem by using a process of discretization. For this, we approximate the integral that appears in (9.362) by the Gauss–Legendre formula

$$\int_0^1 h(t)\,dt \simeq \sum_{i=1}^{8} w_i h(t_i),$$

where the nodes t_i and the weights w_i are known.

Table 9.9   stimates of the error $\|\mathbf{x_n} - \mathbf{x}^*\|$ applying relaxed Newton's method with $\lambda = 0.9$ and $\lambda = 1$ (Newton's method), to solve system (9.363)

n	$\lambda = 0.9, \quad \|\mathbf{x_n} - \mathbf{x}^*\|$	$\lambda = 1, \quad \|\mathbf{x_n} - \mathbf{x}^*\|$
1	$3.409021\ldots \times 10^{-1}$	$2.270807\ldots \times 10^{-1}$
2	$6.788876\ldots \times 10^{-2}$	$2.332819\ldots \times 10^{-2}$
3	$1.240100\ldots \times 10^{-2}$	$2.069182\ldots \times 10^{-3}$
4	$2.221346\ldots \times 10^{-3}$	$1.803860\ldots \times 10^{-4}$
5	$3.964426\ldots \times 10^{-4}$	$1.570110\ldots \times 10^{-5}$
6	$7.0706150\ldots \times 10^{-5}$	$1.366465\ldots \times 10^{-6}$
7	$1.2609058\ldots \times 10^{-5}$	$1.189218\ldots \times 10^{-7}$
8	$2.2485316\ldots \times 10^{-6}$	$1.034962\ldots \times 10^{-8}$

If we denote the approximation of $x(t_i)$ by x_i ($i = 1, 2, \ldots, 8$), then (9.362) is equivalent to the following nonlinear system of equations

$$x_i = \frac{1}{2} \sum_{j=1}^{8} a_{ij} \left(x_j^2 + |x_j| \right), \quad j = 1, 2, \ldots, 8, \tag{9.363}$$

where

$$a_{ij} = \begin{cases} w_j t_j (1 - t_i) & \text{if } j \leq i, \\ w_j t_i (1 - t_j) & \text{if } j > i. \end{cases}$$

System (9.363) is now written as $F(\mathbf{x}) + G(\mathbf{x}) = \mathbf{0}$, where

$$F(\mathbf{x}) \equiv \mathbf{x} - \frac{1}{2} A\mathbf{v_x}, \qquad G(\mathbf{x}) \equiv \frac{1}{2} A\mathbf{w_x} \qquad F : R^8 \longrightarrow R^8,$$

$$\mathbf{x} = (x_1, x_2, \ldots, x_8)^T, \ A = (a_{ij})_{i,j=1}^8, \ \mathbf{v_x} = (x_1^2, x_2^2, \ldots, x_8^2)^T$$

and

$$\mathbf{w_x} = (|x_1|, |x_2|, \ldots, |x_8|)^T.$$

Moreover, $F'(\mathbf{x}) = I - AD(\mathbf{x})$, where $D(\mathbf{x}) = \operatorname{diag}\{x_1, x_2, \ldots, x_8\}$. Choosing as starting point $\mathbf{x_0} = (1, 1, \ldots, 1)^T$ and the max-norm, we obtain $a = 1.13821\ldots$ $b = \beta = 0.140636\ldots$, $k = 0.070318\ldots$. If we take for instance $\lambda = 0.9$ it follows $\mu = M + \lambda k = 0.535159\ldots < 1$ and $h_2 = 0.0400186\ldots$. In consequence, by Theorem 9.7.7, relaxed Newton's method with $\lambda = 0.9$ converges to the trivial solution $\mathbf{x}^* = (0, 0, \ldots, 0)^T$ of system (9.363). The existence of the solution is guaranteed in $\overline{B}(\mathbf{x_0}, 1.36529\ldots)$ and the uniqueness in $B(\mathbf{x_0}, 11.8558\ldots)$. Moreover, In Table 9.9 we can establish the following estimates of the error $\|\mathbf{x_n} - \mathbf{x}^*\|$ that show in Table 9.9.

Example 9.7.2. (Extended semi-local convergence) Consider Example 4.6.1. We take the starting point $x_0 = 1$ and we consider the domain $\Omega = B(x_0, 1)$. In this case, we obtain

$$a = 0.17, \quad b = 3.02, \quad \beta = 2.51 \quad \text{and} \quad \gamma = 0.831126 \cdots.$$

Notice that Kantorovich hypothesis $h \leq 0.5$ is not held. Now, taking $\lambda = 0.99$, we obtain that for $\delta_1 = 0.160253 \cdots$, hypotheses of Lemma 9.7.3 hold. Hence, conditions of convergence for relaxed Newton's method given in Theorem 9.7.5 hold. Hence, relaxed Newton's method starting from $x_0 \in B(x_0, 1)$ converges to the solution $\sqrt[3]{0.49}$.

Example 9.7.3. (Local convergence) Consider Example 4.6.2. We can define parameters

$$b = e, \quad \beta = 2 \quad \text{and} \quad \rho = \min\{1, \frac{2(1 - |1 - \lambda|)}{e + (2 + |1 - \lambda|)2}\}.$$

Then for each $0 < \lambda < 2$, relaxed Newton's method (9.317) starting from $x_0 \in B(x^*, \rho)$. Notice that the radii of this ball is greater than or equal to the one only using Lipschitz condition.

9.8 Exercises

9.8.1. For approximating the solution of (1.1), consider the following method

$$x_{n+1} = x_n - A_n^{-1} F(x_n), \qquad (x_0 \in \mathcal{D}), \quad (n \geq 0), \tag{9.364}$$

where $A_n = [x_n, g(x_n); F]$ for each $n = 0, 1, \cdots$ and $g : \mathcal{D} \longrightarrow \mathcal{X}$ is a continuous operator. Let $\eta, M, \mu \geq 0$, $K, L > 0$ and $l \in [0, 1]$ be given constants. Set $t_0 = (1 - l)/L$. Define functions Δ, A, C on $[0, +\infty)^2$ and B on $[0, +\infty)$ by

$$\Delta(t, \gamma) = \left((K - M) t + M t_0 + \mu\right)^2 - (K - 2\gamma L) t \left((K - 2 M) t + 2(M t_0 + \mu)\right),$$

$$A(t, \gamma) = 2 \left(((K - M) t + M t_0 + \mu) + \sqrt{\Delta(t, \gamma)}\right),$$

$$B(t) = 2 \left((K - 2 M) t + 2(M t_0 + \mu)\right) \quad \text{and} \quad C(t, \gamma) = \frac{B(t)}{A(t, \gamma)}.$$

Assume that the function f_t defined by $f_t(x) = 1 - x - C(t, x)$ has a non-negative zero $\gamma_0 = \gamma(t_0)$ at $t = t_0$, such that

$$\beta = \frac{2 L \eta}{1 - l} \leq \gamma_0 \leq \frac{K}{2 L}; \tag{9.365}$$

or if

$$(K - L)(1 - l) + \mu \geq 0, \quad f_{t_0}(\beta) \geq 0$$

and (9.365) hold; or if function f_t has a non-negative zero γ_0^1 at $t = t_1$ such that

$$\gamma_0^1 = \frac{K}{2 L} \quad \text{and} \quad \beta < \beta_0, \tag{9.366}$$

where

$$\beta_0 = 4\,(1-l)\left[\mu + 2\,(1-l) + \sqrt{(\mu + 2\,(1-l))^2 + 4\,(1-l)^2\,\Big(\frac{K}{2\,L} - 1\Big)}\,\right]^{-1};$$

or if

$$f_{t_1}(\beta_1) \le 0, \qquad \beta_1 = \min\{\beta_0, \frac{K}{2\,L}\}, \quad f_{t_1}(\beta) \ge 0,$$

(9.366) holds. Define scalar sequence $\{t_n\}$ $(n \ge 0)$ by

$$t_1 = t_0 - \eta,$$

$$t_{n+1} = t_n - \frac{\Big(K\,(t_{n-1} - t_n) + 2\,M\,(t_0 - t_{n-1}) + 2\,\mu\Big)\,(t_{n-1} - t_n)}{2\,L\,t_n} \qquad (n \ge 1).$$

Show:

(a) $\{t_n\}$ is well-defined, decreasing and converges to some $t^\star \in [0, t_0]$.

(b) γ_0 (satisfying $f_{t_0}(x) = 0$) is given by

$$\gamma_0 = a + \left(\frac{b}{32^{2/3}\,(c+d)^{1/3}} + \frac{(c+d)^{1/3}}{62^{1/3}}\right)\frac{1}{L\,t_0},$$

where a, b, c and d to be calculated by Maple or Mathematica.

(c) γ_0^1 can be obtained as γ_0 in (b) by simply replacing t_0 by t_1.

Assume there exist $x_0 \in \mathcal{D}_0 \subseteq \mathcal{D}$ ($\mathcal{D}_0$ open and convex) and constants $\eta, p, q, r \ge 0$ and $K, K_0 > 0$, such that for $x, y \in \mathcal{D}_0$, the following conditions hold

$$A_0^{-1} \in \mathcal{L}(\mathcal{X}, \mathcal{X}); \quad \| A_0^{-1}\,F(x_0) \| \le \eta;$$

$$\| A_0^{-1}\,(F'(x) - F'(y)) \| \le K\,\| x - y \|; \quad \| A_0^{-1}\,(F'(x) - F'(x_0)) \| \le K_0\,\| x - x_0 \|;$$

$$\| A_0^{-1}\,(F'(x) - [x, g(x); F]) \| \le p\,\| x - g(x) \|;$$

$$\| A_0^{-1}\,([x, g(x); F] - A_0) \| \le q\,(\| x - x_0 \| + \| g(x) - g(x_0) \|);$$

$$\| g(x) - g(x_0) \| \le r\,\| x - x_0 \|; \quad g(x) \in \mathcal{D}_0 \quad \text{and} \quad \overline{U}(x_0, t_0 - t^\star) \subseteq \mathcal{D}_0.$$

Show:

(d) $\{x_n\}$ generated by (9.364) is well-defined for all $n \ge 0$ and remains in $\overline{U}(x_0, t_0 - t^\star)$.

(e) $\{x_n\}$ converges to a solution $x^\star$ of operator F in $\overline{U}(x_0, t_0 - t^\star)$.

(f) The following inequalities hold

$$\| x_{n+1} - x_n \| \le t_n - t_{n+1} \quad \text{and} \quad \| x_n - x^\star \| \le t_n - t^\star.$$

(g) If $K_0\,(t_0 - t^\star) + K\,\| F(x_0) \| \le 1$, then $x^\star$ is unique in $\overline{U}(x_0, t_0 - t^{\star\star})$.

9.8.2. For approximating (1.54), we consider the two-point and two-step Newton method for initial points $x_0, x_{-1} \in X$ by

$$
\begin{aligned}
y_n &= x_n - A(x_{n-1}, x_n)^{-1}[F(x_n) + G(x_n)] \\
x_{n+1} &= y_n - A(x_n, y_n)^{-1}[F(y_n) + G(y_n)] \quad \text{for each} \quad n = 0, 1, \cdots.
\end{aligned}
\tag{9.367}
$$

Assume there exist $x_{-1} \in X$, $x_0 \in D$ such that $A(x_{-1}, x_0)^{-1} \in L(Y, X)$ and for all $v, w, x, y \in D$ the following conditions hold true:

$$
\|A(x_{-1}, x_0)^{-1}[A(x, y) - A(x_{-1}, x_0)]\| \leq f(\|x - x_{-1}\|, \|y - x_0\|),
$$

$$
\begin{aligned}
&\|A(x_{-1}, x_0)^{-1}[F(y) - F(w) - A(v, w)(x - w) + G(y) - G(w)]\| \\
&\leq g(v, w, x, y)\|x - w\|^p,
\end{aligned}
$$

$$
\begin{aligned}
&\|A(x_{-1}, x_0)^{-1}[F(y) - F(w) - A(x, y)(y - w) + G(y) - G(w)]\| \\
&\leq g_1(x, y, w)\|y - w\|^p,
\end{aligned}
$$

$$
g(v, w, x, y) \leq a \quad \text{and} \quad g_1(x, y, w) \leq a,
$$

where $f \colon [0, \infty) \times [0, \infty) \to [0, \infty)$ is a nondecreasing function, $g \colon D^4 \to [0, \infty)$ and $g_1 \colon D^3 \to [0, \infty)$ for $p \geq 1$, $a > 0$. Define

$$
\|x_0 - y_0\| \leq b_0, \quad \|x_1 - y_0\| \leq b_1, \quad \|x_{-1} - x_0\| \leq c.
$$

For $R > 0$ with $b_0 + b_1 \leq R$, let $r \in [0, R]$, $q = (1/2) \max\{b_0, b_1\}$ and

$$
d = d(r) = \frac{a}{1 - f(r + c, r)}, \quad q_0 = q_0(r) = \frac{1}{2p} d^{\frac{1}{p-1}} \max\{b_0, b_1\}, \quad p \neq 1.
$$

Assume also there exists a minimum point $r^* \in [b_0 + b_1, R]$ such that $\overline{U}(x_0, r^*) \subseteq D$ and

$$
\frac{1}{1 - q_0^p(r^*)} + b_0 \leq r^* \text{ for } p \neq 1 \quad \text{or} \quad \frac{q(r^*)}{1 - d(r^*)} + b_0 \leq r \text{ for } p = 1,
$$

which also satisfies

$$
f(r^* + c, r^*) < 1, \quad q_0(r^*) < 1 \text{ if } p \neq 1 \quad \text{or} \quad d(r^*) < 1 \text{ if } p = 1.
$$

Set

$$
\varepsilon_n^0 = q_0^{p^n}(r^*)r_0^* \text{ if } p \neq 1 \quad \text{and} \quad \varepsilon_n = d^n(r^*)r_0^* \text{ if } p = 1 \quad (n \geq 0),
$$

where

$$
r_0^* = \begin{cases} (1 - q_0^p(r^*))^{-1} & p \neq 1 \\[2mm] \dfrac{q(r^*)}{1 - d(r^*)} & p = 1. \end{cases}
$$

Show

(a) $\{x_n\}$ generated by method (9.367) is well-defined, remains in $\overline{U}(x_0, r^*)$ for all $n \geq 0$ and converges to a solution $x^* \in \overline{U}(x_0, r^*)$ of (1.54).

(b) The following error estimates hold for each $n = 0, 1, \cdots$

$$
\|x_{n+1} - x_n\| \leq q_0^{p^n}, \quad \|x_n - x^*\| \leq \varepsilon_n^0 \quad (p \neq 1)
$$

and

$$
\|x_{n+1} - x_n\| \leq d^n q, \quad \|x_n - x^*\| \leq \varepsilon_n \quad (p = 1).
$$

Chapter 10

Newton–Tikhonov Method for Ill-posed Problems

For solving ill-posed operators problems in Hilbert space, we present in this chapter some new results of convergence analysis for Newton–Tikhonov methods.

10.1 Newton–Tikhonov Method in Hilbert Space

We present in this section a new semi-local convergence analysis of Newton–Tikhonov methods for solving ill-posed operator equations in a Hilbert space setting. Using more precise majorizing sequences and under the same computational cost as in earlier studies such as Refs. [226, 234], we provide: weaker sufficient convergence criteria; tighter error estimates on the distances involved and an at least as precise information on the location of the solution. Applications include Hammerstein nonlinear integral equations.

In this section we are concerned with the problem of approximating a solution of nonlinear ill-posed equation

$$A(x) = y \tag{10.1}$$

where A is a nonlinear operator defined on a subset $D = D(A)$ of a Hilbert space X, and with range $R(A)$ in a Hilbert space Y. Equation (10.1) is ill-posed in the sense that the solution of (10.1) does not depend continuously on the data y. Many regularization methods such as Tikhonov regularization (see Refs. [200, 241, 285, 319, 391, 398]), Gauss–Newton method [174] and other methods (see Refs. [254, 284]) have been used to approximate solution of equation (10.1).

George and colloborators (see Refs. [226, 227, 229–234]) solved equation (10.1) for the special case when A is a Hammerstein-type operator. A Hammerstein-type operator is of the form $A = MF$, where $F : D(F) \subset X \to Z$ is a nonlinear and $M : Z \to Y$ is a bounded linear operator with X, Y and Z are Hilbert spaces. Hence, (10.1) becomes

$$MF(x) = y. \tag{10.2}$$

In particular George and Kunhanandan in Ref. [230] assumed that a solution $x^* \in D(F)$ of (10.2) satisfies

$$\|F(\hat{x}) - F(x_0)\| = \min\{\|F(x) - F(x_0)\| : MF(x) = y, x \in D(F)\}, \tag{10.3}$$

501

and $y^\delta \in Y$ are the available noisy data such that

$$\|y - y^\delta\| \le \delta. \tag{10.4}$$

Then, the Newton–Tikhonov (NT) method defined by, for fixed $\alpha > 0, \delta > 0$

$$x^\delta_{n+1,\alpha} = x^\delta_{n,\alpha} - F'(x^\delta_{n,\alpha})^{-1}(F(x^\delta_{n,\alpha}) - z^\delta_\alpha), \qquad x^\delta_{0,\alpha} = x_0, \tag{10.5}$$

where z^δ_α is an approximation of the solution of the equation $M(z) = y^\delta$, was used to generate a sequence $\{x^\delta_{n,\alpha}\}$ converging quadratically to a solution x^δ_α of the equation

$$F(x) = z^\delta_\alpha \tag{10.6}$$

provided that certain Kantrovich-type criteria are satisfied.

We expand the applicability of (NT) by using more precise majorizing sequence for $\{x^\delta_{n,\alpha}\}$ than the ones given in Ref. [231]. This way we provide a semi-local convergence analysis for (NT) with the advantages (cited in the beginning of this section) over the work in Ref. [230] under the same computational cost. These advantages are obtained, since we use the more precise center-Lipschitz condition instead of the Lipschitz condition for the computation of the upper bounds on the norms $\|F'(x^\delta_{n,\alpha})^{-1}\|$ (see (C3) and (C4)). We also study the semi-local convergence of the simplified Newton–Tikhonov method (SNT) defined by

$$x^\delta_{n+1,\alpha} = x^\delta_{n,\alpha} - F'(x^\delta_{0,\alpha})^{-1}(F(x^\delta_{n,\alpha}) - z^\delta_\alpha), \qquad x^\delta_{0,\alpha} = x_0. \tag{10.7}$$

(SNT) method is used as predictor for (NT) method since the former converges under weaker sufficient convergence criteria than the latter.

We present auxiliary results on scalar sequences which shall be shown to be majorizing for $\{x^\delta_{n,\alpha}\}$.

Definition 10.1.1. Let $L_0 > 0, L > 0, b > 0$ and $n > 0$. Define scalar sequences $\{r^\delta_{n,\alpha}\}, \{s^\delta_{n,\alpha}\}, \{t^\delta_{n,\alpha}\}$ by

$$r^\delta_{0,\alpha} = 0,\ r^\delta_{1,\alpha} = r,\ r^\delta_{2,\alpha} = r^\delta_{1,\alpha} + \frac{3bL_0(r^\delta_{1,\alpha} - r^\delta_{0,\alpha})^2}{2(1 - bL_0 r^\delta_{1,\alpha})},$$

$$r^\delta_{n+2,\alpha} = r^\delta_{n+1,\alpha} + \frac{3bL_0(r^\delta_{n+1,\alpha} - r^\delta_{n,\alpha})^2}{2(1 - bL_0 r^\delta_{n+1,\alpha})}, \forall n = 1, 2, \cdots, \tag{10.8}$$

$$s^\delta_{0,\alpha} = 0,\ s^\delta_{1,\alpha} = r,\ s^\delta_{n+2,\alpha} = s^\delta_{n+1,\alpha} + \frac{3bL(s^\delta_{n+1,\alpha} - s^\delta_{n,\alpha})^2}{2(1 - bL_0 r^\delta_{n+1,\alpha})}, \forall n = 0, 1, 2, \cdots, \tag{10.9}$$

and

$$t^\delta_{0,\alpha} = 0,\ t^\delta_{1,\alpha} = r,\ t^\delta_{n+2,\alpha} = t^\delta_{n+1,\alpha} + \frac{3bL(t^\delta_{n+1,\alpha} - t^\delta_{n,\alpha})^2}{2(1 - bL t^\delta_{n+1,\alpha})}, \forall n = 0, 1, 2, \cdots. \tag{10.10}$$

Then, using a simple inductive argument we obtain the following result where we compare the three scalar sequences.

Proposition 10.1.1. *(see Refs. [110, 116, 120]) Suppose that $L_0 \leq L$ and*

$$t_{n+1,\alpha}^{\delta} < \frac{1}{bL} \tag{10.11}$$

for each $n = 0, 1, 2, \cdots$. Then, sequences $\{r_{n,\alpha}^{\delta}\}, \{s_{n,\alpha}^{\delta}\}$ and $\{t_{n,\alpha}^{\delta}\}$ are well defined, increasing and converge to their unique least upper bounds r^*, s^*, t^* which satisfy for $\gamma = \dfrac{6L}{3L + \sqrt{9L^2 + 24L_0 L}}$,

$$r \leq r^* \leq s^* \leq t^* \leq \frac{1}{bL} \tag{10.12}$$

$r^* \leq \bar{r}^* = r + \dfrac{bL_0 r^2}{2(1 - \gamma)(1 - bL_0 r)}, s^* \leq \bar{s}^* = \dfrac{r}{1 - \gamma}$ *and* $t^* \leq \bar{t}^* \leq 2r$. *Moreover, the following estimates hold for each $n = 1, 2, \cdots$*

$$r_{n,\alpha}^{\delta} \leq s_{n,\alpha}^{\delta} \leq t_{n,\alpha}^{\delta} \tag{10.13}$$

and

$$r_{n+1,\alpha}^{\delta} - r_{n,\alpha}^{\delta} \leq s_{n+1,\alpha}^{\delta} - s_{n,\alpha}^{\delta} \leq t_{n+1,\alpha}^{\delta} - t_{n,\alpha}^{\delta}. \tag{10.14}$$

Furthermore strict inequality holds in (10.13) and (10.14) if $L_0 < L$.

Remark 10.1.1. It follows from (10.12)–(10.14) that if $L_0 < L$ sequence $\{t_{n,\alpha}^{\delta}\}$ is the least tight. This sequence is a majorizing for $\{x_{n,\alpha}^{\delta}\}$ (see Theorem 3.3 in Ref. [231]).

Next we present results respectively for $\{t_{n,\alpha}^{\delta}\}, \{s_{n,\alpha}^{\delta}\}, \{r_{n,\alpha}^{\delta}\}$.

Lemma 10.1.1. *(see Refs. [110, 155]) Suppose that*

$$h = 4Lbr \leq 1. \tag{10.15}$$

Then, sequence $\{t_{n,\alpha}^{\delta}\}$ is increasing convergent to t^*. The convergence is linear if $h = 1$ and quadratic if $h < 1$.

Lemma 10.1.2. *(see Lemma 2.1 in Ref. [116]) Suppose that*

$$h_1 = \frac{b}{4}(3L + 4L_0 + \sqrt{9L^2 + 24L_0 L})r \leq 1. \tag{10.16}$$

Then, $\{s_{n,\alpha}^{\delta}\}$ is increasing convergent to s^*. The convergence is linear if $h_1 = 1$ and quadratic if $h_1 < 1$.

Lemma 10.1.3. *(see Ref. [155]) Suppose that*

$$h_2 = \frac{b}{4}(4L_0 + \sqrt{3L_0 L + 8L_0^2} + \sqrt{3L_0 L})r \leq 1. \tag{10.17}$$

Then, $\{r_{n,\alpha}^{\delta}\}$ is increasing convergent to r^*. The convergence is linear if $h_2 = 1$ and quadratic if $h_2 < 1$.

We also have the following generalization of Lemma 10.1.3.

Lemma 10.1.4. *(see Ref. [155]) Suppose that there exists a minimum integer $N > 1$ such that $r_{i,\alpha}^\delta (i = 0, 1, \cdots, N-1)$ given by (10.14) are well-defined,*

$$r_{i,\alpha}^\delta < r_{i+1,\alpha}^\delta < \frac{1}{bL_0}, \qquad i = 0, 1, \cdots, N-2. \tag{10.18}$$

Then, the following assertions hold

$$bL_0 r_{N,\alpha}^\delta < 1, \tag{10.19}$$

$$r_{N,\alpha}^\delta \le \frac{1}{bL_0}(1 - (1 - bL_0)r_{N-1,\alpha}^\delta \gamma) \tag{10.20}$$

and

$$\gamma_{N-1} = \frac{3L(r_{N+1,\alpha}^\delta - r_{N,\alpha}^\delta)}{2(1 - bL_0 r_{N+1,\alpha}^\delta)} \le \gamma \le 1 - \frac{bL_0(r_{N+1,\alpha}^\delta - r_{N,\alpha}^\delta)}{1 - bL_0 r_{N,\alpha}^\delta}. \tag{10.21}$$

Remark 10.1.2.

(a) Lemma 10.1.4 reduces to Lemma 10.1.3 if $N = 2$ (see also Remark 2.5 in Ref. [155]).
(b) It follows from (10.15), (10.16), (10.17) that

$$h \le 1 \Longrightarrow h_1 \le 1 \Longrightarrow h_2 \le 1 \tag{10.22}$$

but not necessarily vice versa unless if $L_0 = L$. Moreover, we have that

$$\frac{h_4}{h} \to \frac{1}{4}, \frac{h_2}{h} \to 0 \text{ and } \frac{h_2}{h_1} \to 0 \text{ as } \frac{L_0}{L} \to 0. \tag{10.23}$$

Estimates (10.23) show by how many times the applicability of the method is expanded if weaker h_1 or h_2 are used instead of h.
(c) Numerical examples where $L_0 < L$ can be found in Refs. [110, 120].

We present the semi-local convergence of $\{x_{n,\alpha}^\delta\}$. We assume throughout this section that the following (C) conditions hold:

(C1) $F : D(F) \subseteq X \to Y$ is Fréchet differentiable
(C2) There exists $x_0 \in D(F)$ such that $F'(x_0)^{-1} \in L(Y, X)$ and

$$|F'(x_0)^{-1}\| \le b$$

(C3) There exists $L_0 > 0$ such that F' satisfying the center-Lipschitz conditions

$$\|F'(x) - F'(x_0)\| \le L_0\|x - x_0\|$$

for all $x \in D(F)$.
(C4) There exists $L > 0$ such that F' satisfying the Lipschitz condition

$$\|F'(x) - F'(y)\| \le L\|x - y\|$$

for all x and y in $D(F)$.

(C5) $\|F'(x_0)^{-1}(F(x_0) - z_\alpha^\delta)\| \le r$

(C6) $h_2 = bL_2 r \le 1$, where $L_2 = \dfrac{1}{4}(4L_0 + \sqrt{3L_0 L + 8L_0^2} + \sqrt{3L_0 L})$ and

(C7) $\overline{U}(x, r^*) \subseteq D(F)$, where r^* is defined in (10.12).

We present the following semi-local convergence result for $\{x_{n,\alpha}^\delta\}$.

Theorem 10.1.1. *Suppose that the (C) conditions hold. Then, sequence $\{x_{n,\alpha}^\delta\}$ generated by (NT) is well-defined, remains in $\overline{U}(x_0, r^*)$ for all $n \ge 0$ and converges to some $x_\alpha^\delta \in \overline{U}(x_0, r^*)$ such that $F(x_\alpha^\delta) = z_\alpha^\delta$. Moreover, the following estimates hold for each $n = 0, 1, 2 \cdots$*

$$\|x_{n+1,\alpha}^\delta - x_{n,\alpha}^\delta\| \le r_{n+1,\alpha}^\delta - r_{n,\alpha}^\delta \tag{10.24}$$

and

$$\|x_{n+1,\alpha}^\delta - x_\alpha^\delta\| \le r^* - r_{n,\alpha}^\delta. \tag{10.25}$$

Proof. We use mathematical induction to prove that

$$\|x_{k+1,\alpha}^\delta - x_{k,\alpha}^\delta\| \le r_{k+1,\alpha}^\delta - r_{k,\alpha}^\delta \tag{10.26}$$

and

$$\overline{U}(x_{k+1,\alpha}^\delta, r^* - r_{k+1,\alpha}^\delta) \subseteq \overline{U}(x_{k,\alpha}^\delta, r^* - r_{k+1,\alpha}^\delta) \tag{10.27}$$

for each $k = 0, 1, 2, \cdots$. Let $v \in \overline{U}(x_{1,\alpha}^\delta, r^* - r_{1,\alpha}^\delta)$. Then, we obtain

$$\begin{aligned}
\|v - x_{0,\alpha}^\delta\| &\le \|v - x_{1,\alpha}^\delta\| + \|x_{1,\alpha}^\delta - x_{0,\alpha}^\delta\| \\
&\le r^* - r_{1,\alpha}^\delta + r_{1,\alpha}^\delta - r_{0,\alpha}^\delta \\
&= r^* - r_{0,\alpha}^\delta
\end{aligned}$$

which implies $v \in \overline{U}(x_{0,\alpha}^\delta, r^* - r_{0,\alpha}^\delta)$. Note also that

$$\|x_{1,\alpha}^\delta - x_{0,\alpha}^\delta\| = \|F'(x_{0,\alpha}^\delta)^{-1}(F(x_{0,\alpha}^\delta) - z_\alpha^\delta)\| \le r^* - r_{1,\alpha}^\delta + r_{1,\alpha}^\delta - r_{0,\alpha}^\delta \le r = r_{1,\alpha}^\delta - r_{0,\alpha}^\delta.$$

Hence, estimates (10.26) and (10.27) hold for $k = 0$. Suppose that these estimates hold for $n \le k$. Then, we have that

$$\|x_{k+1,\alpha}^\delta - x_{0,\alpha}^\delta\| \le \sum_{i=1}^{k+1} \|x_{i,\alpha}^\delta| - x_{i-1,\alpha}^\delta\| \le \sum_{i=1}^{k+1} (r_{i,\alpha}^\delta - r_{i-1,\alpha}^\delta) \le r_{i+1,\alpha}^\delta \le r^*$$

and

$$\|x_{k,\alpha}^\delta + \theta(x_{k+1,\alpha}^\delta - x_{k,\alpha}^\delta) - x_{0,\alpha}^\delta\| \le r_{k,\alpha}^\delta + \theta(r_{k+1,\alpha}^\delta - r_{k,\alpha}^\delta) - r_{0,\alpha}^\delta \le r^*$$

for each $\theta \in [0, 1]$.

Using (C3), Lemms 10.1.2 (see also Lemma 2.1 in Ref. [116]) and the induction hypotheses we get

$$\begin{aligned}
\|F'(x_{n+1,\alpha}^\delta) - F'(x_{0,\alpha}^\delta)\| &\le L_0 \|x_{k+1,\alpha}^\delta - x_{0,\alpha}^\delta\| \\
&\le L_0 (r_{k+1,\alpha}^\delta - r_{0,\alpha}^\delta) \\
&\le L_0 r_{k+1,\alpha}^\delta < \frac{1}{b}.
\end{aligned} \tag{10.28}$$

It follows from (10.28) and the Banach Lemma on invertible operators that $F'(x_{k+1,\alpha}^\delta)^{-1} \in L(Y,X)$ and

$$\|F'(x_{k+1,\alpha}^\delta)^{-1}\| \leq \frac{b}{1 - bL_0\|x_{k+1,\alpha}^\delta - x_{0,\alpha}^\delta\|} \leq \frac{b}{1 - bL_0(r_{k+1,\alpha}^\delta - r_{0,\alpha}^\delta)}. \qquad (10.29)$$

Using (NT) we obtain the approximation

$$F(x_{k+1,\alpha}^\delta) - z_\alpha^\delta = [F'(x_{k+1,\alpha}^\delta)^{-1}(x_{k+1,\alpha}^\delta - x_{k,\alpha}^\delta) + F(x_{k+1,\alpha}^\delta) - F(x_{k,\alpha}^\delta)$$
$$+ [F'(x_{k,\alpha}^\delta) - F'(x_{k+1,\alpha}^\delta)](x_{k+1,\alpha}^\delta - x_{k,\alpha}^\delta). \qquad (10.30)$$

In view of (C4), (10.8), (10.30), (C3) for $k = 0$ and the induction hypotheses we get

$$\|F(x_{k+1,\alpha}^\delta) - z_\alpha^\delta\| \leq \frac{L}{2}\|x_{k+1,\alpha}^\delta - x_{k,\alpha}^\delta\|^2 + L\|x_{k+1,\alpha}^\delta - x_{k,\alpha}^\delta\|^2$$
$$= \frac{3L}{2}\|x_{k+1,\alpha}^\delta - x_{k,\alpha}^\delta\|^2 \leq \frac{3L}{2}(r_{k+1,\alpha}^\delta - r_{k,\alpha}^\delta)^2.$$

Moreover, by (NT), (10.8), (10.29) and (10.31) we get that

$$\|x_{k+2,\alpha}^\delta - x_{k+1,\alpha}^\delta\| \leq \|F'(x_{k+1,\alpha}^\delta)^{-1}\|\|F(x_{k+1,\alpha}^\delta) - z_\alpha^\delta\|$$
$$\leq \frac{b}{1 - bL_0(r_{k+1,\alpha}^\delta - r_{0,\alpha}^\delta)} \frac{3L}{2}(r_{k+1,\alpha}^\delta - r_{k,\alpha}^\delta)^2$$
$$= r_{k+2,\alpha}^\delta - r_{k,\alpha}^\delta$$

which completes the induction for (10.26). Furthermore, let $w \in \overline{U}(x_{k+2,\alpha}^\delta, r^* - r_{k+2,\alpha}^\delta)$. Then, we have that

$$\|w - x_{k+1,\alpha}^\delta\| \leq \|w - x_{k+2,\alpha}^\delta\| + \|x_{k+2,\alpha}^\delta - x_{k+1,\alpha}^\delta\|$$
$$\leq r^* - r_{k+2,\alpha}^\delta + r_{k+21,\alpha}^\delta - r_{k+1,\alpha}^\delta$$
$$= r^* - r_{k+1,\alpha}^\delta.$$

That is $w \in \overline{U}(x_{k+1,\alpha}^\delta, r^* - r_{k+1,\alpha}^\delta)$. Lemma 10.1.3 implies that $\{r_{k,\alpha}^\delta\}$ is a complete sequence. It then follows from (10.26) and (10.27) that $\{x_{k,\alpha}^\delta\}$ is also complete sequence in the Hilbert space X and as such it converges to some $x_\alpha^\delta \in \overline{U}(x_0, r^*)$ (since $\overline{U}(x_0, r^*)$ is a closed set). By letting $k \to \infty$ in (10.31) we obtain $F(x_\alpha^\delta) = z_\alpha^\delta$. Estimate (10.25) is obtained from (10.24) by using standard majorization techniques. The proof of Theorem is complete. $\qquad \square$

Remark 10.1.3.

(a) If $L_0 = L, r_{n,\alpha}^\delta = t_{n,\alpha}^\delta$ then Theorem 10.1.1 reduces to Theorem 3.3 in Ref. [231] (with corresponding changes). Otherwise, i.e., if $L_0 < L$, according to the above results of this section it constitutes an improvement with advantages as stated in the introduction of this section.

(b) Upper bounds on r^* and s^* in terms of L_0, L, b and n have been given in Refs. [110, 116, 120] (see also (10.12)). These bounds are given in closed form and can certainly replace r^* in (C7).

The rest of the results in Ref. [230] can be improved along the same lines by simply using, respectively, $\bar{r}^*$, L_0 instead of $\bar{t}^*$, L. In order for us to make the section as self-contained as possible we present the proof of one of them and for the proof of the rest we refer the reader to Ref. [230].

Proposition 10.1.2. *Suppose that (10.17), (C3) and (C4) hold. Moreover, suppose that*

$$\|x_0 - x^*\| \le \bar{r}^* < r < \frac{1}{bL_0} = \lambda_0 \tag{10.31}$$

and

$$\overline{U}(x_0, \lambda_0) \subseteq D(F). \tag{10.32}$$

Then, the following assertion holds:

$$\|x^* - x_\alpha^\delta\| \le \frac{b}{1 - bL_0 r}\|F(x^*) - z_\alpha^\delta\|. \tag{10.33}$$

Proof. Using (C3)(instead of (C4)) used in Ref. [230], we get that

$$\begin{aligned}
\|x^* - x_\alpha^\delta\| &= \|x^* - x_\alpha^\delta + F'(x_0)^{-1}[F(x_\alpha^\delta) - F(x^*) + F(x^*) - z_\alpha^\delta]\| \\
&\le \|F'(x_0)^{-1}[F'(x_0)(x^* - x_\alpha^\delta) - (F(x^*) - F(x_\alpha^\delta))]\| \\
&\quad + \|F'(x_0)^{-1}[F(x^*) - z_\alpha^\delta]\| \\
&\le bL_0 r\|x^* - x_\alpha^\delta\| + b\|F(x^*) - z_\alpha^\delta\|,
\end{aligned}$$

which shows (10.33). The proof of the proposition is complete. $\qquad\square$

The following is a consequence of Theorem 10.1.1 and Proposition 10.1.2.

Corollary 10.1.1. *Suppose that hypotheses of Theorem 10.1.1, $\bar{r}^* < r$ and $bL_0 r < 1$ hold. Then, the following assertion holds*

$$\|x^* - x_{n,\alpha}^\delta\| \le \frac{b}{1 - bL_0 r}\|F(x^*) - z_\alpha^\delta\| + r^* - r_{n,\alpha}^\delta$$

for each $n = 0, 1, 2, \cdots$.

Remark 10.1.4. If $L_0 = L$ Proposition 10.1.2 and Corollory 10.1.1 reduce to the corresponding ones in Ref. [230]. Otherwise, i.e., $L_0 < L$ our results constitute an improvement.

We present the semi-local convergence of (SNT).

Theorem 10.1.2. *Suppose that (C1)–(C3), (C5) hold. If in addition*

(C8) $h_0 = \tau bL_0 r < 1$
and
(C9) $\overline{U}(x_0, \rho) \subseteq D(F)$, where $\rho = \dfrac{1 - \sqrt{1 - h_0}}{bL_0}$.

Then, sequence $\{x_{n,\alpha}^{\delta}\}$ generated by (SNT) is well-defined, remain in $\overline{U}(x_0, \rho)$ for all $n \geq 0$ and converges to some $x_{\alpha}^{\delta} \in \overline{U}(x_0, \rho)$ such that $F(x_{\alpha}^{\delta}) = z_{\alpha}^{\delta}$. Moreover the following estimate hold for each $n = 0, 1, 2, \cdots$

$$\|x_{n+2,\alpha}^{\delta} - x_{n+1,\alpha}^{\delta}\| \leq q\|x_{n+1,\alpha}^{\delta} - x_{n,\alpha}^{\delta}\|$$

and

$$\|x_{n,\alpha}^{\delta} - x_{\alpha}^{\delta}\| \leq \frac{q^n r}{1 - q},$$

where $q = 1 - \sqrt{1 - h_0}$.

Proof. Let us define operator T on $\overline{U}(x_0, \rho)$ by $T(x) = x - F'(x_0)^{-1}(F(x) - z_{\alpha}^{\delta})$. Then, we shall show that T is a contraction on $\overline{U}(x_0, \rho)$ and maps $\overline{U}(x_0, \rho)$ into itself. Indeed, we have that for $x, y \in \overline{U}(x_0, \rho)$

$$\begin{aligned} T(x) - T(y) &= x - y - F'(x_0)^{-1}(F(x) - z_{\alpha}^{\delta}) + F'(x_0)^{-1}(F(y) - z_{\alpha}^{\delta}) \\ &= -F'(x_0)^{-1}(F(x) - F(y) - F'(x_0)(x - y)) \end{aligned}$$

and

$$\|T(x) - T(y)\| = \|F'(x_0)^{-1}(F(x) - F(y) - F'(x_0)(x - y))\| \leq bL_0\rho\|x - y\|.$$

But, we have $bL_0\rho < 1$. Hence, T is a contraction operator. Let $x \in \overline{U}(x_0, \rho)$. Then, we have that

$$T(x) - x_0 = T(x) - T(x_0) + T(x_0) - x_0$$

and

$$\|T(x) - x_0\| \leq \|T(x) - T(x_0)\| + \|T(x_0) - x_0\| = \frac{bL_0\rho^2}{2 + r} = \rho$$

by the choice of ρ. The proof of the Theorem is complete. $\square$

Remark 10.1.5.

(a) More subtle arguments show that $h_0 < 1$ can be replaced by $h_0 \leq 1$ (see Refs. [110, 120]).

(b) We have that $h^0 = 2bLr \leq 1 \Rightarrow h_0 \leq 1$ but not vice versa even if $L_0 = L$. Moreover, we have that $\dfrac{h_0}{h^0} \to 0$ as $\dfrac{L_0}{L} \to 0$. Note that the h^0 condition was given in Ref. [231]. Therefore method (SNT) can be used as a predictor until a certain finite iterate N such that $h \leq 1$ holds for $x_{N,\alpha}^{\delta}$, being the initial point of method (NT). Such an approach has been used by us in Ref. [104] for modified Newton and Newton's method.

We present now some applications of the theoretical results of this section. Let us consider the nonlinear Hammerstein operator equation (see Ref. [393])

$$(MFx)(t) = \int_0^1 m(s, t)p(s, x(s))x(s)ds$$

where m is continuous and p is differentiable with respect to the second variable. Define $F : D(F) = H^1(]0,1[) \to L^2(]0,1[)$ by

$$F(x)(s) = p(s, x(s)), \qquad s \in [0,1]$$

and $M : L^2(]0,1[) \to L^2(]0,1[)$ by

$$Mu(t) = \int_0^1 m(s,t)u(s)ds, \qquad t \in [0,1].$$

Then, F is Fréchet differentiable and we have that

$$[F'(x)]u(t) = \partial_2 p(t, x(t))u(t), \qquad t \in [0,1].$$

If in addition $M_1 : H^1(]0,1[) \mapsto H^1(]0,1[)$ is defined by $(M_1 x)(t) := \partial_2 p(t, x(t))$ is locally Lipschitz continuous, one can compute the required constants L_0 and L. Equation (10.2) is equivalent to

$$M[F(x) - F(x_0)] = y - MF(x_0) \tag{10.34}$$

where x_0 is an initial guess. Therefore, the solution x of (10.2) is obtained by first solving

$$Mz = y - KF(x_0) \tag{10.35}$$

for z and then solving

$$F(x) = z + F(x_0) \tag{10.36}$$

for x. Let $\alpha > 0$, $\delta > 0$ be fixed. Then, we consider the regularized solution of (10.35) with y^δ in place of y as

$$z_\alpha^\delta = (M + \alpha I)^{-1}(y^\delta - MF(x_0)) + F(x_0) \tag{10.37}$$

in the case where M (see (10.35)) is positive self adjoint and $Z = Y$. Otherwise, we set

$$z_\alpha^\delta = (M^*M + \alpha I)^{-1}M^*(y^\delta - MF(x_0)) + F(x_0). \tag{10.38}$$

Note that (10.37) is the simplified or Lavrentiev regularization of equation (10.35) and (10.38) is the Tikhonov regularization of (10.35). With thess choices of operators the rest of the results in Ref. [230] involving (SNT) type methods (i.e., using L_0 instead of L) can be improved.

Proposition 10.1.3. *Suppose z_α^δ is given by (10.38) and*

$$\|F(x_0) - F(x^*)\| + \frac{\delta}{\sqrt{\alpha}} < \frac{r}{2b} < \frac{1}{2b^2 L_0}.$$

Then, the following assersion holds

$$\|x_0 - x^*\| \leq \bar{r}^* < r < \frac{1}{bL_0}.$$

Remark 10.1.6.

(a) If $L_0 = L$ Proposition 10.1.3 reduces to Remark 3.4 in Ref. [230]. Otherwise, Proposision 10.1.3 improves Remark 3.4 in Ref. [230].

(b) The rest of the results in the literature (see Refs. [226, 234]) can be extended by simply using (C3) instead of (C4). Note also that there are examples where (C3) holds but not (C4).

Remark 10.1.7. Hereafter we consider z_α^δ as the Tikhonov regularization of (10.35) given in (10.38). All results in the rest of this section are valid for the simplified regularization of (10.35).

In view of the estimate in Corollary 10.1.1, the next task is to find an estimate $\|F(x^*) - z_\alpha^\delta\|$. For this, let us introduce the notation;

$$z_\alpha := F(x_0) + (M^*M + \alpha I)^{-1}M^*(y - MF(x_0)).$$

We may observe that

$$\|F(x^*) - z_\alpha^\delta\| \leq \|F(x^*) - z_\alpha\| + \|z_\alpha - z_\alpha^\delta\| \leq \|F(x^*) - z_\alpha\| + \frac{\delta}{\sqrt{\alpha}}, \tag{10.39}$$

and

$$\begin{aligned}
F(x^*) - z_\alpha &= F(x^*) - F(x_0) - (M^*M + \alpha I)^{-1}M^*M[F(x^*) - F(x_0)] \\
&= [I - (M^*M + \alpha I)^{-1}M^*M][F(x^*) - F(x_0)] \\
&= \alpha(M^*M + \alpha I)^{-1}[F(x^*) - F(x_0)].
\end{aligned} \tag{10.40}$$

Note that for $u \in R(M^*M)$ with $u = M^*Mz$ for some $z \in Z$,

$$\|\alpha(M^*M + \alpha I)^{-1}u\| = \|\alpha(M^*M + \alpha I)^{-1}M^*Mz\| \leq \alpha\|z\| \to 0$$

as $\alpha \to 0$. Now since $\|\alpha(M^*M + \alpha I)^{-1}\| \leq 1$ for all $\alpha > 0$, it follows that for every $u \in \overline{R(M^*M)}$, $\|\alpha(M^*M + \alpha I)^{-1}u\| \to 0$ as $\alpha \to 0$. Thus we have the following theorem.

Theorem 10.1.3. *If* $F(x^*) - F(x_0) \in \overline{R(M^*M)}$, *then* $\|F(x^*) - z_\alpha\| \to 0$ *as* $\alpha \to 0$.

In view of the above theorem, we assume that

$$\|F(x^*) - z_\alpha\| \leq \varphi(\alpha) \tag{10.41}$$

for some positive monotonic increasing function φ defined on $(0, \|M\|^2]$ such that

$$\lim_{\lambda \to 0} \varphi(\lambda) = 0.$$

Suppose φ is a source function in the sense that x^* satisfies a source condition of the form

$$F(x^*) - F(x_0) = \varphi(M^*M)w, \|w\| \leq 1,$$

such that

$$\sup_{0<\lambda<\|M\|^2} \frac{\alpha\varphi(\lambda)}{\lambda+\alpha} \le \varphi(\alpha), \tag{10.42}$$

then the assumption (10.41) is satisfied. Note that if $F(x^*) - F(x_0) \in R((M^*M)^\nu)$, for some ν with, $0 < \nu \le 1$, then by (10.40)

$$\|F(x^*) - z_\alpha\| \le \|\alpha(M^*M + \alpha I)^{-1}(M^*M)^\nu \omega\|$$
$$\le \sup_{0<\lambda\le\|M\|^2} \frac{\alpha\lambda^\nu}{\lambda+\alpha}\|\omega\| \le \alpha^\nu\|\omega\|.$$

Thus in this case $\varphi(\lambda) = \dfrac{\lambda^\nu}{\|\omega\|}$ satisfies the assumption (10.41). Therefore by (10.39) and by the assumption (10.41), we have that

$$\|F(x^*) - z_\alpha^\delta\| \le \varphi(\alpha) + \frac{\delta}{\sqrt{\alpha}}. \tag{10.43}$$

So, we have the following theorem.

Theorem 10.1.4. *Under the assumptions of Corollary 10.1.1 and (10.43),*

$$\|x^* - x_{n,\alpha}^\delta\| \le \frac{b}{1 - b\kappa_0 r}\left(\varphi(\alpha) + \frac{\delta}{\sqrt{\alpha}}\right) + r^* - r_{n,\alpha}^\delta.$$

Note that the estimate $\varphi(\alpha) + \dfrac{\delta}{\sqrt{\alpha}}$ in (10.42) attains minimum for the choice $\alpha := \alpha_\delta$ which satisfies $\varphi(\alpha_\delta) = \dfrac{\delta}{\sqrt{\alpha_\delta}}$. Let $\psi(\lambda) := \lambda\sqrt{\varphi^{-1}(\lambda)}, 0 < \lambda \le \|M\|^2$. Then we have $\delta = \sqrt{\alpha_\delta}\varphi(\alpha_\delta) = \psi(\varphi(\alpha_\delta))$, and

$$\alpha_\delta = \varphi^{-1}(\psi^{-1}(\delta)). \tag{10.44}$$

So the relation (10.43) leads to

$$\|F(x^*) - z_\alpha^\delta\| \le 2\psi^{-1}(\delta).$$

Theorem 10.1.2 and the above observation lead to the following.

Theorem 10.1.5. *Let $\psi(\lambda) := \lambda\sqrt{\varphi^{-1}(\lambda)}, 0 < \lambda \le \|K\|^2$ and the assumptions of Corollary 10.1.1 and (10.41) are satisfied. For $\delta > 0$, let $\alpha_\delta = \varphi^{-1}(\psi^{-1}(\delta))$. If*

$$n_\delta := \min\{n : (r^* - r_{n,\alpha}^\delta) < \frac{\delta}{\sqrt{\alpha_\delta}}\},$$

then

$$\|x^* - x_{\alpha_\delta,n_\delta}^\delta\| = O(\psi^{-1}(\delta)).$$

The error estimate in the above Theorem has optimal order with respect to δ. Unfortunately, an a priori parameter choice (10.44) cannot be used in practice since the smoothness properties of the unknown solution $\hat{x}$ reflected in the function φ are generally unknown. There exist many parameter choice strategies in the literature, for example see Refs. [174, 242, 244, 232, 233, 378, 415].

In Ref. [346], Pereverzev and Schock considered an adaptive selection of the parameter which does not involve even the regularization method in an explicit manner. In this method the regularization parameter α_i are selected from some finite set $\{\alpha_i : 0 < \alpha_0 < \alpha_1 < \cdots < \alpha_N\}$ and the corresponding regularized solution, say $u^\delta_{\alpha_i}$ are studied online. Later George and Nair in Ref. [234] considered the adaptive selection of the parameter for choosing the regularization parameter in Newton–Lavrentiev regularization method for solving Hammerstein-type operator equation. We consider the adaptive method for selecting the parameter α in $x^\delta_{\alpha,n}$. The rest of this section is essentially a reformulation of the adaptive method considered in Ref. [346] in a special context.

Let $i \in \{0, 1, 2, \cdots, N\}$ and $\alpha_i = \mu^{2i}\alpha_0$ where $\mu > 1$ and $\alpha_0 = \delta^2$. Let

$$l := \max\{i : \varphi(\alpha_i) \leq \frac{\delta}{\sqrt{\alpha_i}}\} \tag{10.45}$$

and

$$k := \max\{i : \|z^\delta_{\alpha_i} - z^\delta_{\alpha_j}\| \leq \frac{4\delta}{\sqrt{\alpha_j}}, j = 0, 1, 2, \cdots, i\}. \tag{10.46}$$

The proof of the next theorem is analogous to the proof of Theorem 1.2 in Ref. [346], but for the sake of completeness, we supply its proof as well.

Theorem 10.1.6. *Let l be as in (10.45), k be as in (10.46) and $z^\delta_{\alpha_k}$ be as in (10.38) with $\alpha = \alpha_k$. Then $l \leq k$ and*

$$\|F(x^*) - z^\delta_{\alpha_k}\| \leq (2 + \frac{4\mu}{\mu - 1})\mu\psi^{-1}(\delta).$$

Proof. Note that, to prove $l \leq k$, it is enough to prove that, for $i = 1, 2, \cdots, N$

$$\varphi(\alpha_i) \leq \frac{\delta}{\sqrt{\alpha_i}} \implies \|z^\delta_{\alpha_i} - z^\delta_{\alpha_j}\| \leq \frac{4\delta}{\sqrt{\alpha_j}}, \forall j = 0, 1, 2, \cdots, i.$$

For $j \leq i$, we get in turn

$$\|z^\delta_{\alpha_i} - z^\delta_{\alpha_j}\| \leq \|z^\delta_{\alpha_i} - F(\hat{x})\| + \|F(\hat{x}) - z^\delta_{\alpha_j}\|$$
$$\leq \varphi(\alpha_i) + \frac{\delta}{\sqrt{\alpha_i}} + \varphi(\alpha_j) + \frac{\delta}{\sqrt{\alpha_j}}$$
$$\leq \frac{2\delta}{\sqrt{\alpha_i}} + \frac{2\delta}{\sqrt{\alpha_j}} \leq \frac{4\delta}{\sqrt{\alpha_j}}.$$

This proves the relation $l \leq k$. Now since $\sqrt{\alpha_{l+m}} = \mu^m\sqrt{\alpha_l}$, by using triangle inequality successively, we obtain

$$\|F(\hat{x}) - z^\delta_{\alpha_k}\| \leq \|F(x^*) - z^\delta_{\alpha_l}\| + \sum_{j=l+1}^{k} \frac{4\delta}{\sqrt{\alpha_{j-1}}}$$
$$\leq \|F(\hat{x}) - z^\delta_{\alpha_l}\| + \sum_{m=0}^{k-l-1} \frac{4\delta}{\sqrt{\alpha_l}\mu^m}$$
$$\leq \|F(\hat{x}) - z^\delta_{\alpha_l}\| + (\frac{\mu}{\mu - 1})\frac{4\delta}{\sqrt{\alpha_l}}.$$

Therefore, by (10.42) and (10.45) we have

$$\|F(x^*) - z^\delta_{\alpha_k}\| \le \varphi(\alpha_l) + \frac{\delta}{\sqrt{\alpha_l}} + \left(\frac{\mu}{\mu - 1}\right)\frac{4\delta}{\sqrt{\alpha_l}}$$

$$\le \left(2 + \frac{4\mu}{\mu - 1}\right)\mu\psi^{-1}(\delta).$$

The last step follows from the inequality $\sqrt{\alpha_\delta} \le \sqrt{\alpha_{l+1}} \le \mu\sqrt{\alpha_l}$ and $\dfrac{\delta}{\sqrt{\alpha_\delta}} = \psi^{-1}(\delta)$. This completes the proof. $\qquad\square$

Note that

$$
\begin{aligned}
e_0 = \|x^\delta_{1,\alpha} - x_0\| &= \|F'(x_0)^{-1}(M^*M + \alpha I)^{-1}M^*(y^\delta - MF(x_0))\| \\
&= \|F'(x_0)^{-1}(M^*M + \alpha I)^{-1}M^*(y^\delta - y + y - MF(x_0))\| \\
&\le b(\|(M^*M + \alpha I)^{-1}M^*(y^\delta - y)\| + \\
&\quad \|(M^*M + \alpha I)^{-1}M^*M(F(\hat{x}) - F(x_0))\|) \\
&\le b\left(\omega + \frac{\delta}{\sqrt{\alpha}}\right),
\end{aligned}
$$

so if

$$b\left(\omega + \frac{\delta}{\sqrt{\alpha}}\right) < \frac{1}{bL_2} \tag{10.47}$$

and $r^* < r$. Then hypotheses of Theorem 10.1.1 hold. Again since $\alpha_j = \mu^{2j}\delta^2$, $\dfrac{\delta}{\sqrt{\alpha_k}} = \mu^{-k}$; the condition (10.47) with $\alpha = \alpha_k$ takes the form

$$b\left(\omega + \frac{1}{\mu^k}\right) < \frac{1}{bL_2} \tag{10.48}$$

and $r^* < r$. Then, we have arrived at the algorithm which guarantees

$$\|x^* - x^\delta_{n_k,\alpha_k}\| = O(\psi^{-1}(\delta))$$

with $n_k = \min\{n : r^* - r^\delta_{n,\alpha} \le \mu^{-j}, j = 1, 2, \cdots, i - 1\}$ and improves the corresponding one in Ref. [230].

Note that for $i, j \in \{0, 1, 2, \cdots, n\}$

$$\|z^\delta_{\alpha_i} - z^\delta_{\alpha_j}\| = (\alpha_j - \alpha_i)(M^*M + \alpha_j I)^{-1}(M^*M + \alpha_i I)^{-1}M^*(y^\delta - MF(x_0)).$$

Therefore the adaptive algorithm associated with the choice of the parameter specified in the above theorem is as follows.

```
begin
    i=0
    repeat
       i=i+1
```

$$\text{Solve for } w_i : (M^*M + \alpha_i I)w_i = M^*(y^\delta - MF(x_0))$$

```
        j=-1
        repeat
            j=j+1
```

$$\text{Solve for } z_{i,j} : (M^*M + \alpha_j I)z_{i,j} = (\alpha_j - \alpha_i)w_i$$

$$\text{until}(\ \|z_{i,j}\| \le 4\mu^{-j}\,\text{AND}\ j < i)$$

$$\text{until}(\|z_{i,j}\| \le 4\mu^{-j})$$

k = i-1.

```
    m=0
      repeat
         m=m+1
```

$$\text{until}((r^* - r^\delta_{m,\alpha}) > \frac{1}{\mu^k})$$

$n_k = m$

$$\text{for l=1 to } n_k$$

$$\text{Solve for } u_{l-1} : F'(x^\delta_{l-1,\alpha_k})u_{l-1} = F(x^\delta_{l-1,\alpha_k}) - z^\delta_{\alpha_k}$$

$$x^\delta_{l,\alpha_k} := x^\delta_{l-1,\alpha_k} - u_{l-1}$$

end

Remark 10.1.8. We have considered an iterative regularization method, which is a combination of Newton iterative method with a Tikhonov regularization method, for obtaining approximate solution for a nonlinear Hammerstein-type operator equation $MF(x) = y$, with the available data y^δ in place of the exact data y. If the operator M is a positive self-adjoint bounded linear operator on a Hilbert space, then one may consider Newton Lavrentiev regularization method for obtaining an approximate solution for $MF(x) = y$. It is assumed that the Fréchet derivative $F'(x)$ of the nonlinear operator F has a continuous inverse, in a neighborhood of some initial guess x_0 of the actual solution x^*. The procedure involves solving the equation

$$(M^*M + \alpha I)u^\delta_\alpha = M^*(y^\delta - MF(x_0))$$

and finding the fixed point of the function

$$G(x) = x - F'(x)^{-1}(F(x) - F(x_0) - u^\delta_\alpha)$$

in an iterative manner. For choosing the regularization parameter α and the stopping index for the iteration, we made use of the adaptive method suggested in Ref. [346].

10.2 Two-step Newton–Tikhonov Method in Hilbert Space

Argyros and Hilout used in Ref. [148] a Two-Step Directional Newton Method (TSDNM) to approximate solution of an operator equation in a Hilbert space setting. We use the analogous Two-Step Newton–Tikhonov Method (TSNTM) to solve ill-posed problems under weak conditions.

An equation of the form

$$KF(x) = y, \qquad (10.49)$$

where $F : D(F) \subseteq X \to X$, is nonlinear and $K : X \to Y$ is a bounded linear operator is called a (nonlinear) Hammerstein equation (see Ref. [234]). X and Y denote the Hilbert spaces with inner product $\langle .,. \rangle$ and norm $\|.\|$ respectively.

We assume throughout that $y^\delta \in Y$ are the available noisy data with $\|y - y^\delta\| \leq \delta$ and $\|F'(x)\| \leq M$, $x \in D(F)$ for some $M > 0$. Here and in the following, $F'(.)$ denotes the Fréchet derivative of F. The equation (10.49) is ill-posed in the sense that an arbitrarily small perturbations in the data y may cause an arbitrarily large deviations in the solution x.

Our interest is to find an x_0-minimum norm solution $\hat{x}$ of (10.49), recall that an x_0-minimum norm solution if $\hat{x}$ satisfies (see Refs. [230, 234])

$$\|F(\hat{x}) - F(x_0)\| = \min\{\|F(x) - F(x_0)\| : KF(x) = y, x \in D(F)\}.$$

The element x_0 plays the role of a selection criterion (see Ref. [200]). Because of the nonlinearity of F, the solution $\hat{x}$ may not be unique. Observe that the solution x of (10.49) with y^δ in place of y can be obtained by first solving

$$Kz = y^\delta \qquad (10.50)$$

for z and then solving the nonlinear problem

$$F(x) = z. \qquad (10.51)$$

The above formulation was considered by authors in Refs. [226, 230, 234]. The advantage of the above formulation is that:

- We solve (10.50) and (10.51) separately, to obtain an approximate solution for (10.49). Here one can use any regularization method for linear ill-posed equation for solving (10.50) and any regularization method for solving (10.51). We consider Tikhonov regularization for approximately solving (10.50) and we consider a modified two-step Newton method for solving (10.51).

- The regularization parameter α is chosen according to the adaptive method considered by Pereverzev and Schock in Ref. [346] for the linear ill-posed operator equations and the same parameter α is used for solving the nonlinear operator equation (10.51), so the choice of the regularization parameter does not depend on the nonlinear operator F.

In Ref. [226], George studied an iterative Newton–Tikhonov regularization (NTR) method for approximating (10.49), where z in (10.50) is approximated with z_α^δ;

$$z_\alpha^\delta = (K^*K + \alpha I)^{-1} K^* f^\delta, \qquad \alpha > 0, \qquad \delta > 0.$$

Then (10.51) is solved using the Newton type iteration

$$x_{n+1,\alpha}^\delta = x_{n,\alpha}^\delta - F'(x_0)^{-1}(F(x_{n,\alpha}^\delta) - z_\alpha^\delta)$$

where $x_{0,\alpha}^\delta := x_0$. Local linear convergence was obtained in Ref. [226]. In Ref. [230], George and Kunhanandan used the iteration

$$x_{n+1,\alpha}^\delta = x_{n,\alpha}^\delta - F'(x_{n,\alpha}^\delta)^{-1}(F(x_{n,\alpha}^\delta) - z_\alpha^\delta),$$

where $x_{0,\alpha}^\delta := x_0$ and

$$z_\alpha^\delta = (K^*K + \alpha I)^{-1} K^* (f^\delta - KF(x_0)) + F(x_0) \tag{10.52}$$

for approximately solving (10.49). Local quadratic convergence was established in Ref. [230]. Recall that Ref. [293], a sequence (x_n) in X with $\lim x_n = x^*$ is said to be convergent of order $p > 1$, if there exist positive reals a and b such that, for all $n \in \mathbb{N}$,

$$\|x_n - x^*\| \le a e^{-bp^n}. \tag{10.53}$$

If a sequence (x_n) satisfies $\|x_n - x^*\| \le a l^n$, $0 < l < 1$, then (x_n) is said to be linearly convergent. In Ref. [148], Argyros and Hilout considered a method called Two-Step Directional Newton Method (TSDNM) for approximating a zero x^* of a differentiable function F defined on a convex subset $\mathcal{D}$ of a Hilbert space H with values in $\mathbb{R}$. Motivated by TSDNM we proposed, a Two-Step Newton–Tikhonov Methods (TSNTM) in Ref. [236] for solving (10.49). The convergence of the method for the following regularity classes of the operator F was examined:

(i) **The IFD Class.** $F'(u)^{-1}$ exists and is a bounded operator for all $u \in D(F)$; i.e., $\|F'(u)^{-1}\| \le \beta,\ \forall u \in D(F)$. Consequently, in this situation, the ill-posedness of (10.49) is essentially due to the nonclosedness of the range of the linear operator K (see page 26 in Ref. [374]).

(ii) **The MFD Class.** F is a monotone operator (see Refs. [315, 415]) (i.e., $\langle F(x) - F(y), x - y \rangle \ge 0, \qquad \forall x, y \in D(F)$) and $F'(.)^{-1}$ does not exist. Consequently, in this situation, the ill-posedness of (10.49) is due to the ill-posedness of F as well as the nonclosedness of the range of the linear operator K.

Next we present two examples, first one for IFD class and the second one for MFD class.

Example 10.2.1. Formally a Hammerstein operator A takes the form $A = KF$ where $K : L^2[0,1] \to L^2[0,1]$ is a linear integral operator with kernel $k(s,t)$:

$$Kx(t) = \int_0^1 k(s,t)x(s)ds$$

and $F : D(F) \subseteq L^2[0,1] \to L^2[0,1]$ is the nonlinear superposition operator (see p. 430 in Ref. [295])

$$Fx(s) = f(s, x(s)). \tag{10.54}$$

Let the function f in (10.54) be differentiable with respect to the second variable. Then, it follows that the operator F in (10.54) is Fréchet differentiable with

$$[F'(x)u](t) = \partial_2 f(t, x(t))u(t), \qquad t \in [0,1],$$

where $\partial_2 f(t, s)$ represents the partial derivative of f with respect to the second variable. If, in addition, the existence of a constant $\kappa_1 > 0$ is assumed such that, for all $x \in B_r(x_0)$ and for all $t \in [0,1]$, $\partial_2 f(t, x(t)) \geq \kappa_1$, then $F'(u)^{-1}$ exists and is a bounded operator for all $u \in B_r(x_0)$. So F belongs to the IFD class.

Example 10.2.2. (Example 6.1 in Ref. [332]) Let $F : L^2[0,1] \to L^2[0,1]$ be defined by

$$F(x)(t) = K(x)(t) + f(t), \qquad x, f \in L^2[0,1], \;\; t \in [0,1]$$

where $K : L^2[0,1] \to L^2[0,1]$ is a compact linear operator such that range of K denoted by $R(K)$ is not closed and $\langle Kh, h \rangle \geq 0$ for $h \in L^2[0,1]$. Then, $F(x) = y$ is ill-posed as K is a compact operator with non-closed range. The Fréchet derivative $F'(x)$ of F is given by

$$F'(x)h = Kh, \qquad \forall x, h \in L^2[0,1].$$

Now, since $\langle Kh, h \rangle \geq 0$ for all $h \in L^2[0,1]$, F is monotone. Further $F'(u)^{-1}$ does not exist for any $u \in L^2[0,1]$. Consequently, the operator KF, with K and F as defined above is an example of the MFD Class.

George and Shobha used the popular condition (see Refs. [236, 285, 403])

$(C1)$ There exist a constant L such that for every $x, u \in U(x_0, r)$ and $v \in X$ there exists an element $P(x, u, v) \in X$ such that

$$[F'(x) - F'(u)]v = F'(u)P(x, u, v), \quad \|P(x, u, v)\| \leq L\|v\|\|x - u\|.$$

They used the additional restriction that $0 < L \leq 1$.

There are cases when Lipschitz-type condition (C_1) is violated but the following weaker center-Lipschitz condition is satisfied

$(C1)'$ Let $x_0 \in D(F)$ be fixed. There exists a constant $L_0 > 0, r > 0$ such that for each $x, u \in U(x_0, r) \cup U(\hat{x}, r) \subset D(F)$ and $v \in X$, there exists an element $\Phi(x, u, v) \in X$ such that

$$[F'(x) - F'(u)]v = F'(u)\Phi(x, u, v), \quad \|\Phi(x, u, v)\| \leq L_0\|v\|(\|u - x_0\| + \|x - x_0\|).$$

We note that since $\|u-x\| \le \|u-x_0\| + \|x-x_0\|$ condition $(C1)$ always implies $(C1)'$ with $L_0 = L$ but not necessarily vice versa. If $\|F'(u)^{-1}\| \le \beta, \|F'(x) - F'(u)\| \le \bar{L}\|x-u\|$ and $\|F'(x) - F'(x_0)\| \le \bar{L}_0\|x-x_0\|$ for some $\beta > 0, \bar{L} > 0$ and $\bar{L}_0 > 0$ and $x, u \in D(F)$. Then we can write

$$[F'(x) - F'(u)]v = F'(u)P(x,u,v)$$

where $P(x,u,v) = F'(u)^{-1}(F'(x) - F'(u))v$. Then, the center-Lipschitz condition holds, but the Lipschitz condition does not hold, we have that

$$\begin{aligned}
\|P(x,u,v)\| &= \|F'(u)^{-1}(F'(x) - F'(u))v\| \\
&\le \|F'(u)^{-1}\|\|F'(x) - F'(u)\|\|v\| \\
&\le \|F'(u)^{-1}\|(\|F'(x) - F'(x_0)\| + \|F'(x_0) - F'(u)\|)\|v\| \\
&\le \beta\bar{L}_0(\|x-x_0\| + \|u-x_0\|)\|v\| \\
&= L_0(\|x-x_0\| + \|u-x_0\|)\|v\|
\end{aligned}$$

with $L_0 = \beta\overline{L_0}$. Let $L = \beta\overline{L}$. Then, we have in this case that, under $(C1)'$ we have the following special case:

$(C1)''$ Let $x_0 \in D(F)$ be fixed. There exists a constant $l_0 > 0$ such that for all $w_\theta = x_0 + \theta(\hat{x} - x_0) \in D(F)$ and $v \in X$, there exists an element $P(x_0, w_\theta, v) \in X$ such that

$$[F'(x_0) - F'(w)]v = F'(x_0)\Phi(x_0, w_\theta, v), \qquad \|\Phi(x_0, w_\theta, v)\| \le l_0\|v\|\|x_0 - w_\theta\|.$$

The semi-local convergence analysis in Ref. [236] was based on $(C1)$. Our analysis is based on $(C1)'$. We also show that restrictive condition $(C1)$ can be dropped. Moreover a combination of $(C1)$ and $(C1)'$ can improve the results in Ref. [236].

This section deals with Tikhonov regularized solution z_α^δ of (10.50) and (an a priori and an a posteriori) error estimate for $\|F(\hat{x}) - z_\alpha^\delta\|$. The following assumption is required to obtain the error estimate.

Assumption 1. There exists a continuous, strictly monotonically increasing function $\varphi : (0, a] \to (0, \infty)$ with $a \ge \|K\|^2$ satisfying

$$\lim_{\lambda \to 0} \varphi(\lambda) = 0,$$

$$\sup_{\lambda > 0} \frac{\alpha\varphi(\lambda)}{\lambda + \alpha} \le \varphi(\alpha) \quad \forall \lambda \in (0, a]$$

and there exists $v \in X, \|v\| \le 1$ such that

$$F(\hat{x}) - F(x_0) = \varphi(K^*K)v.$$

Theorem 10.2.1. *(see section 4 in Ref.* [230]*) Let z_α^δ be as in (10.52) and Assumption 1 holds. Then*

$$\|F(\hat{x}) - z_\alpha^\delta\| \le \varphi(\alpha) + \frac{\delta}{\sqrt{\alpha}}. \tag{10.55}$$

Note that the estimate $\varphi(\alpha) + \dfrac{\delta}{\sqrt{\alpha}}$ in (10.55) is of optimal order for the choice $\alpha := \alpha_\delta$ which satisfies $\varphi(\alpha_\delta) = \dfrac{\delta}{\sqrt{\alpha_\delta}}$. Let $\psi(\lambda) := \lambda\sqrt{\varphi^{-1}(\lambda)}, 0 < \lambda \le a$. Then we have $\delta = \sqrt{\alpha_\delta}\varphi(\alpha_\delta) = \psi(\varphi(\alpha_\delta))$ and $\alpha_\delta = \varphi^{-1}(\psi^{-1}(\delta))$. So the relation (10.55) leads to $\|F(\hat{x}) - z_\alpha^\delta\| \le 2\psi^{-1}(\delta)$. We propose in this section to choose the parameter α according to the balancing principle established by Pereverzev and Shock in Ref. [346] for solving ill-posed problems. Let

$$D_N = \{\alpha_i : 0 < \alpha_0 < \alpha_1 < \alpha_2 < \cdots < \alpha_N\}$$

be the set of possible values of the parameter α. The selection of numerical value k for the parameter α according to the balancing principle is performed using the rule

$$l := \max\{i : \varphi(\alpha_i) \le \frac{\delta}{\sqrt{\alpha_i}}\} < N. \tag{10.56}$$

Let

$$k = \max\{i : \alpha_i \in D_N^+\} \tag{10.57}$$

where $D_N^+ = \{\alpha_i \in D_N : \|z_{\alpha_i}^\delta - z_{\alpha_j}^\delta\| \le \dfrac{4\delta}{\sqrt{\alpha_j}}, j = 0, 1, 2, \cdots, i - 1\}$. The following theorem from Ref. [230] is used for our error analysis.

Theorem 10.2.2. *(Ref. [230], Theorem 4.3) Let l be as in (10.56), k be as in (10.57) and $z_{\alpha_k}^\delta$ be as in (10.52) with $\alpha = \alpha_k$. Then $l \le k$ and*

$$\|F(\hat{x}) - z_{\alpha_k}^\delta\| \le (2 + \frac{4\mu}{\mu - 1})\mu\psi^{-1}(\delta).$$

We present the semi-local convergence of (TSNTM), for IFD class under the $(C1)'$ condition, defined by

$$y_{n,\alpha_k}^\delta = x_{n,\alpha_k}^\delta - F'(x_{n,\alpha_k}^\delta)^{-1}(F(x_{n,\alpha_k}^\delta) - z_{\alpha_k}^\delta), \tag{10.58}$$

$$x_{n+1,\alpha_k}^\delta = y_{n,\alpha_k}^\delta - F'(x_{n,\alpha_k}^\delta)^{-1}(F(y_{n,\alpha_k}^\delta) - z_{\alpha_k}^\delta), \tag{10.59}$$

for each $n = 0, 1, 2, \cdots$. We need to introduce some sequences and parameters. Let

$$e_{n,\alpha_k}^\delta := \|y_{n,\alpha_k}^\delta - x_{n,\alpha_k}^\delta\|, \qquad \forall n \ge 0.$$

Suppose that $\delta \in (0, \delta_0]$, where $\delta_0 < (17 - 12\sqrt{2})\dfrac{\sqrt{\alpha_0}}{\beta}$. Let $\|\hat{x} - x_0\| < \rho$. Suppose that

$$\rho < \rho^* = \frac{1}{M}(\frac{17 - 12\sqrt{2}}{\beta} - \frac{\delta_0}{\sqrt{\alpha_0}}), \tag{10.60}$$

$$\gamma_\rho := \beta[M\rho + \frac{\delta_0}{\sqrt{\alpha_0}}]. \tag{10.61}$$

Let

$$r = \frac{1}{L_0} \frac{2\gamma_\rho}{1 - \gamma_\rho + \sqrt{(1 - \gamma_\rho)^2 - 32\gamma_\rho}} \tag{10.62}$$

and

$$q_0 = 2L_0 r, \quad q = 2q_0^2. \tag{10.63}$$

Note that since $\frac{q_0}{2} \in (0,1)$, $q \in (0,1)$, $\gamma_\rho \in (0, 17 - 12\sqrt{2}]$, $r > 0$ is well-defined

and $\frac{1 + L_0 r}{1 - 8L_0^2 r^2} \gamma_\rho = \frac{1 + \frac{p}{2}}{1 - q} \gamma_\rho = L_0 r$. In order for us to simplify the notation, let x_n, y_n and e_n stand respectively, for x_{n,α_k}^δ, y_{n,α_k}^δ and e_{n,α_k}^δ. With the notation introduced so far we can present the semi-local convergence analysis of (TSNTM)(as defined by (10.58) and (10.59)) in the series of results that follow.

Theorem 10.2.3. *Suppose that* $(C1)'$, $\rho < \rho^*$ *for* $\delta \in (0, \delta_0]$ *hold and* r *is given in (10.62). Then, the following assertions hold:*

(a) $\|x_n - y_{n-1}\| \le q_0 \|y_{n-1} - x_{n-1}\| = q_0 e_{n-1},$

(b) $\|x_n - x_{n-1}\| \le (1 + \frac{q_0}{2}) e_{n-1}$ *and*

(c) $e_n \le q e_{n-1}.$

Moreover, if (10.60) holds, then $e_n < 1$.

Proof. Using (TSNTM) we get the identity

$$\begin{aligned}
x_n - y_{n-1} &= y_{n-1} - x_{n-1} - F'(x_{n-1})^{-1}(F(y_{n-1}) - F(x_{n-1})) \\
&= F'(x_{n-1})^{-1}[F'(x_{n-1})(y_{n-1} - x_{n-1}) - (F(y_{n-1}) - F(x_{n-1}))] \\
&= F'(x_{n-1})^{-1} \int_0^1 [F'(x_{n-1}) - F'(x_{n-1} + t(y_{n-1} - x_{n-1}))](y_{n-1} - x_{n-1})dt \\
&= F'(x_{n-1})^{-1} \int_0^1 [F'(x_{n-1}) - F'(x_0) + F'(x_0) - F'(x_{n-1} + t(y_{n-1} - x_{n-1}))] \\
&\quad \times (y_{n-1} - x_{n-1})dt.
\end{aligned}$$

Then by $(C1)'$, we get that

$$\begin{aligned}
\|x_n - y_{n-1}\| &\le \| \textstyle\int_0^1 \Phi(x_{n-1}, x_0, y_{n-1} - x_{n-1})dt \| \\
&\quad + \| \textstyle\int_0^1 \Phi(x_0, x_{n-1} + t(y_{n-1} - x_{n-1}), y_{n-1} - x_{n-1})dt \| \\
&\le L_0[\|x_{n-1} - x_0\| + \textstyle\int_0^1 \|x_{n-1} - x_0 + t(y_{n-1} - x_{n-1})\|dt]\|y_{n-1} - x_{n-1}\| \\
&\le L_0(r + r)\|y_{n-1} - x_{n-1}\| = 2L_0 r \|y_{n-1} - x_{n-1}\| = q_0 e_{n-1},
\end{aligned}$$

which proves (a). Estimate (b) follows from (a) and the triangle inequality;

$$\|x_n - x_{n-1}\| \le \|x_n - y_{n-1}\| + \|y_{n-1} - x_{n-1}\|.$$

To prove (c) we observe that

$$
\begin{aligned}
e_n = \|y_n - x_n\| &\leq \|x_n - y_{n-1} - F'(x_n)^{-1}(F(x_n) - z_\alpha^\delta)\| \\
&\quad + \|F'(x_{n-1})^{-1}(F(y_{n-1}) - z_\alpha^\delta)\| \\
&\leq \|x_n - y_{n-1} - F'(x_n)^{-1}(F(x_n) - F(y_{n-1}))\| \\
&\quad + \|[F'(x_{n-1})^{-1} - F'(x_n)^{-1}](F(y_{n-1}) - z_\alpha^\delta)\| \\
&\leq \|F'(x_n)^{-1}[F'(x_n)(x_n - y_{n-1}) - (F(x_n) - F(y_{n-1}))]\| \\
&\quad + \|[F'(x_{n-1})^{-1} - F'(x_n)^{-1}](F(y_{n-1}) - z_\alpha^\delta)\| \\
&\leq \left\|F'(x_n)^{-1} \int_0^1 [F'(x_n) - F'(y_{n-1} + t(x_n - y_{n-1})]dt(x_n - y_{n-1})\right\| \\
&\quad + \|F'(x_n)^{-1}(F'(x_n) - F'(x_{n-1}))F'(x_{n-1})^{-1}(F(y_{n-1}) - z_\alpha^\delta)\| \\
&\leq \left\|F'(x_n)^{-1} \int_0^1 [F'(x_n) - F'(y_{n-1} + t(x_n - y_{n-1})]dt(x_n - y_{n-1})\right\| \\
&\quad + \|F'(x_n)^{-1}(F'(x_n) - F'(x_{n-1}))(y_{n-1} - x_n)\| \\
&\leq L_0\left[\|x_n - x_0\| + \int_0^1 \|y_{n-1} - x_0 + t(x_n - y_{n-1})\|dt\right]\|x_n - y_{n-1}\| \\
&\quad + L_0[\|x_n - x_0\| + \|x_{n-1} - x_0\|]\|x_n - y_{n-1}\| \\
&\leq 4L_0 r\|y_{n-1} - x_n\| \leq 4L_0 r(2L_0 r)e_{n-1} = qe_{n-1}.
\end{aligned}
$$

From (c) $e_n \leq 1$ if $e_0 \leq 1$. We have that

$$
\begin{aligned}
e_0 = \|y_0 - x_0\| = \|F'(x_0)^{-1}(F(x_0) - z_{\alpha_k}^\delta)\| \\
&\leq \|F'(x_0)^{-1}\|\|(F(x_0) - z_{\alpha_k}^\delta)\| \\
&\leq \beta\|F(x_0) - z_{\alpha_k} + z_{\alpha_k} - z_{\alpha_k}^\delta\| \\
&\leq \beta[\|F(x_0) - F(\hat{x})\| + \|z_{\alpha_k} - z_{\alpha_k}^\delta\|] \\
&\leq \beta\left[\left\|\int_0^1 F'(\hat{x} + t(x_0 - \hat{x}))(x_0 - \hat{x})dt\right\| + \frac{\delta}{\sqrt{\alpha_k}}\right] \\
&\leq \beta\left[M\rho + \frac{\delta}{\sqrt{\alpha_k}}\right] \leq \beta\left[M\rho + \frac{\delta_0}{\sqrt{\alpha_0}}\right] \\
&= \gamma_\rho \leq 17 - 12\sqrt{2} = 0.029437252\cdots < 1.
\end{aligned}
$$

This completes the proof of the Theorem. $\qquad\square$

Theorem 10.2.4. *Suppose that the hypotheses of Theorem 10.2.3 hold and $L_0 \leq 1$. Moreover, suppose that $\overline{U}(x_0, r) \subseteq D(F)$. Then, $x_n, y_n \in U(x_0, r)$ for each $n = 0, 1, 2, \cdots$.*

Proof. Using Theorem 10.2.3 we get that

$$
\|x_1 - x_0\| \leq \left(1 + \frac{q_0}{2}\right)e_0 \leq \left(1 + \frac{q_0}{2}\right)\gamma_\rho \leq r.
$$

Hence, $x_1 \in U(x_0, r)$. Similarly, we obtain that

$$
\begin{aligned}
\|y_1 - x_0\| &\leq \|y_1 - x_1\| + \|x_1 - x_0\| \\
&\leq qe_0 + \left(1 + \frac{q_0}{2}\right)\gamma_\rho \leq \left[q + 1 + \frac{q_0}{2}\right]\gamma_\rho < r,
\end{aligned}
$$

which implies $y_1 \in U(x_0, r)$. Moreover, we have that

$$\|x_2 - x_0\| \leq \|x_2 - x_1\| + \|x_1 - x_0\|$$
$$\leq (1 + \frac{q_0}{2})\|y_1 - x_1\| + (1 + \frac{q_0}{2})\gamma_\rho$$
$$\leq (1 + \frac{q_0}{2})q\gamma_\rho + (1 + \frac{q_0}{2})\gamma_\rho$$
$$= (1 + q)(1 + \frac{q_0}{2})\gamma_\rho < r,$$

which also implies $x_2 \in U(x_0, r)$. Furthermore, we obtain that

$$\|y_2 - x_0\| \leq \|y_2 - x_2\| + \|x_2 - x_0\|$$
$$\leq q\|y_1 - x_1\| + (1 + q)(1 + \frac{q_0}{2})\gamma_\rho$$
$$\leq q^2(1 + \frac{q_0}{2})\gamma_\rho + (1 + q)(1 + \frac{q_0}{2})\gamma_\rho$$
$$\leq (1 + q + q^2)(1 + \frac{q_0}{2})\gamma_\rho < r.$$

Hence, we proved that $y_2 \in U(x_0, r)$. Proceeding in an analogous way we prove that $x_n, y_n \in U(x_0, r)$. That completes the proof of the Theorem. $\qquad\square$

Theorem 10.2.5. *Suppose that the hypotheses of Theorem 10.2.4 hold. Then, sequence $\{x_n\}$ remains in $U(x_0, r)$ for each $n = 0, 1, 2, \cdots$ and converges to a solution $x^\delta_{\alpha_k} \in \overline{U}(x_0, r)$ of (10.51). Moreover, the following estimates hold*

$$\|x_n - x^\delta_{\alpha_k}\| \leq b_0 e^{-\gamma_0 n}, \tag{10.64}$$

where $b_0 = (1 + \frac{q_0}{2})\gamma_\rho$ and $\gamma_0 = -\ln q > 0$.

Proof. Using (b) of Theorem 10.2.3 and (10.60) we get that

$$\|x_{n+m} - x_n\| \leq \sum_{i=0}^{m-1} \|x_{n+i+1} - x_{n+i}\|. \tag{10.65}$$

But, we have

$$\|x_{n+i+1} - x_{n+i}\| \leq (1 + \frac{q_0}{2})q^{n+i} e_0. \tag{10.66}$$

In view of (10.66), inequality (10.65) gives that

$$\|x_{n+m} - x_n\| \leq [1 + q + q^2 + \cdots + q^{m-1}]q^n(1 + \frac{q_0}{2})e_0$$
$$\leq \frac{1 - q^m}{1 - q}(1 + \frac{q_0}{2})q^n e_0. \tag{10.67}$$

It follows from (10.67) that sequence $\{x_n\}$ is complete in a Hilbert space X and as such it converges to some $x^\delta_\alpha \in \overline{U}(x_0, r)$ (since $\overline{U}(x_0, r)$ is a closed set). By letting $m \to \infty$ we obtain (10.64). Finally, to prove $x^\delta_{\alpha_k}$ is a solution of (10.51), note that

$$\|F(x_n) - z^\delta_{\alpha_k}\| = \|F'(x_n)(x_n - y_n)\|$$
$$\leq \|F'(x_n)\|\|(x_n - y_n)\|$$
$$\leq M e_n \leq M q^n \gamma_\rho. \tag{10.68}$$

Now by letting $n \to \infty$ in (10.68) we obtain $F(x^\delta_{\alpha_k}) = z^\delta_{\alpha_k}$. That completes the proof of the Theorem. $\qquad\square$

Remark 10.2.1. Note that $0 < q < 1$ and hence $\gamma_0 > 0$. So by (10.53), sequence (x_n) converges linearly to $x_{\alpha_k}^\delta$.

From now on we suppose that $\rho^* \le r$ holds for x_0 sufficiently close to $\hat{x}$.

Theorem 10.2.6. *(Ref. [234], Theorem 3.3) Suppose that Assumption 1 holds. If in addition $L_0 r < 1$, then*

$$\|\hat{x} - x_{\alpha_k}^\delta\| \le \frac{\beta}{1 - L_0 r} \|F(\hat{x}) - z_{\alpha_k}^\delta\|.$$

The following Theorem is a consequence of Theorems 10.2.5 and 10.2.6.

Theorem 10.2.7. *Suppose that hypotheses of Theorems 10.2.5 and 10.2.6 hold. Then, the following assertion holds*

$$\|\hat{x} - x_n\| \le b_0 e^{-\gamma_0 n} + \frac{\beta}{1 - L_0 r} \|F(\hat{x}) - z_{\alpha_k}^\delta\|$$

where b_0 and γ_0 are as in Theorem 10.2.5.

Now since $l \le k$ and $\alpha_\delta \le \alpha_{l+1} \le \mu \alpha_l$ we have

$$\frac{\delta}{\sqrt{\alpha_k}} \le \frac{\delta}{\sqrt{\alpha_l}} \le \mu \frac{\delta}{\sqrt{\alpha_\delta}} = \mu \varphi(\alpha_\delta) = \mu \psi^{-1}(\delta).$$

This leads to the following theorem,

Theorem 10.2.8. *Let x_n be defined as in (10.59), assumptions in Theorem 10.2.2 and Theorem 10.2.7 hold. Let*

$$n_k := \min\{n : e^{-\gamma_0 n} \le \frac{\delta}{\sqrt{\alpha_k}}\}.$$

Then the following assertion holds

$$\|\hat{x} - x_{n_k}\| = O(\psi^{-1}(\delta)).$$

For an initial guess $x_0 \in X$, $0 < c < \alpha_k$ and for $R(x) := F'(x) + \dfrac{\alpha_k}{c} I$, the TSNTM for MFD class is defined as:

$$\tilde{y}_{n,\alpha_k}^\delta = \tilde{x}_{n,\alpha_k}^\delta - R(\tilde{x}_{n,\alpha_k}^\delta)^{-1}[F(\tilde{x}_{n,\alpha_k}^\delta) - z_{\alpha_k}^\delta + \frac{\alpha_k}{c}(\tilde{x}_{n,\alpha_k}^\delta - x_0)] \tag{10.69}$$

and

$$\tilde{x}_{n+1,\alpha_k}^\delta = \tilde{y}_{n,\alpha_k}^\delta - R(\tilde{x}_{n,\alpha_k}^\delta)^{-1}[F(\tilde{y}_{n,\alpha_k}^\delta) - z_{\alpha_k}^\delta + \frac{\alpha_k}{c}(\tilde{y}_{n,\alpha_k}^\delta - x_0)], \tag{10.70}$$

where $\tilde{x}_{0,\alpha_k}^\delta := x_0$. First we consider a (TSNTM) (i.e., $\tilde{x}_{n,\alpha_k}^\delta$ defined in (10.70)) for approximating the zero x_{c,α_k}^δ of

$$F(x) + \frac{\alpha_k}{c}(x - x_0) = z_{\alpha_k}^\delta \tag{10.71}$$

and then we show that x_{c,α_k}^δ is an approximation to the solution $\hat{x}$ of (10.49). Note that with the above notation

$$\|R(x)^{-1} F'(x)\| \le 1. \tag{10.72}$$

Let

$$\tilde{e}^{\delta}_{n,\alpha_k} := \|\tilde{y}^{\delta}_{n,\alpha_k} - \tilde{x}^{\delta}_{n,\alpha_k}\|, \qquad \forall n \geq 0.$$

Here also for convenience we use the notation $\tilde{x}_n$, $\tilde{y}_n$ and $\tilde{e}_n$ for $\tilde{x}^{\delta}_{n,\alpha_k}$, $\tilde{y}^{\delta}_{n,\alpha_k}$ and $\tilde{e}^{\delta}_{n,\alpha_k}$ respectively. Let $\delta_0 < (17 - 12\sqrt{2})\sqrt{\alpha_0}$,

$$\rho < \rho^* < \frac{1}{M}\left(17 - 12\sqrt{2} - \frac{\delta_0}{\sqrt{\alpha_0}}\right) \tag{10.73}$$

and

$$\tilde{\gamma}_\rho := M\rho + \frac{\delta_0}{\sqrt{\alpha_0}}. \tag{10.74}$$

Let q and q_0 be defined as in (10.63) with ρ as in (10.73).

Theorem 10.2.9. *Let $\tilde{x}_n$ and $\tilde{y}_n$ be defined as in (10.69) and (10.70) respectively. Suppose that $(C1)'$ holds and $\rho < \dfrac{1}{M}\left(17 - 12\sqrt{2} - \dfrac{\delta_0}{\sqrt{\alpha_0}}\right)$ for $\delta \in (0, \delta_0]$ hold. Then, the following assertions hold*

(a) $\|\tilde{x}_n - \tilde{y}_{n-1}\| \leq q_0 \tilde{e}_{n-1}$
(b) $\|\tilde{x}_n - \tilde{x}_{n-1}\| \leq \left(1 + \dfrac{q_0}{2}\right)\tilde{e}_{n-1}$ *and*
(c) $\tilde{e}_{n-1} \leq q\tilde{e}_{n-1}.$

Moreover if (10.73) holds, then $\tilde{e}_n < 1$.

Proof.　Using (TSNTM) we get that identity

$$\tilde{x}_n - \tilde{y}_{n-1} = \tilde{y}_{n-1} - \tilde{x}_{n-1} - R(\tilde{x}_{n-1})^{-1}(F(\tilde{y}_{n-1}) - F(\tilde{x}_{n-1}) + (\alpha_k/c)(\tilde{y}_{n-1} - \tilde{x}_{n-1}))$$
$$= R(\tilde{x}_{n-1})^{-1}[R(\tilde{x}_{n-1})(\tilde{y}_{n-1} - \tilde{x}_{n-1}) - (F(\tilde{y}_{n-1}) - F(\tilde{x}_{n-1})) - (\alpha_k/c)(\tilde{y}_{n-1} - \tilde{x}_{n-1})]$$
$$= R(\tilde{x}_{n-1})^{-1}\int_0^1[F'(\tilde{x}_{n-1}) - F'(\tilde{x}_{n-1} + t(\tilde{y}_{n-1} - \tilde{x}_{n-1}))](\tilde{y}_{n-1} - \tilde{x}_{n-1})dt.$$

Now by (10.72), the proof of (a) and (b) follows as in Theorem 10.2.3. To prove (c) we observe that

$$\tilde{e}_n \leq \|\tilde{x}_n - \tilde{y}_{n-1} - R(\tilde{x}_n)^{-1}(F(\tilde{x}_n) - z^{\delta}_{\alpha_k} + (\alpha_k/c)(\tilde{x}_n - x_0))\|$$
$$+ \|R(\tilde{x}_{n-1})^{-1}(F(\tilde{y}_{n-1}) - z^{\delta}_{\alpha_k} + (\alpha_k/c)(\tilde{y}_{n-1} - x_0))\|$$
$$\leq \|\tilde{x}_n - \tilde{y}_{n-1} - R(\tilde{x}_n)^{-1}(F(\tilde{x}_n) - F(\tilde{y}_{n-1}) + (\alpha_k/c)(\tilde{x}_n - \tilde{y}_{n-1}))\|$$
$$+ \|[R(\tilde{x}_{n-1})^{-1} - R(\tilde{x}_n)^{-1}](F(\tilde{y}_{n-1}) - z^{\delta}_{\alpha_k} + (\alpha_k/c)(\tilde{y}_{n-1} - x_0))\|$$
$$\leq \|R(\tilde{x}_n)^{-1}[R(\tilde{x}_n)(\tilde{x}_n - \tilde{y}_{n-1}) - (F(\tilde{x}_n) - F(\tilde{y}_{n-1})) - (\alpha_k/c)(\tilde{x}_n - \tilde{y}_{n-1})]\|]$$
$$+ \|[R(\tilde{x}_{n-1})^{-1} - R(\tilde{x}_n)^{-1}](F(\tilde{y}_{n-1}) - z^{\delta}_{\alpha_k} + (\alpha_k/c)(\tilde{y}_{n-1} - x_0))\|$$
$$\leq \|R(\tilde{x}_n)^{-1}\int_0^1[F'(\tilde{x}_n) - F'(\tilde{y}_{n-1} + t(\tilde{x}_n - \tilde{y}_{n-1})]dt(\tilde{x}_n - \tilde{y}_{n-1})\|$$
$$+ \|R(\tilde{x}_n)^{-1}(F'(\tilde{x}_n) - F'(\tilde{x}_{n-1}))R(\tilde{x}_{n-1})^{-1}(F(\tilde{y}_{n-1}) - z^{\delta}_{\alpha_k} + (\alpha_k/c)(\tilde{y}_{n-1} - x_0))\|.$$

The remaining part of the proof is analogous to the proof of Theorem 10.2.3.　　$\square$

Theorem 10.2.10. *Suppose that the hypotheses of Theorem 10.2.9 hold. Set $\tilde{r} = \dfrac{1}{L_0}\dfrac{2\tilde{\gamma}_\rho}{1 - \tilde{\gamma}_\rho + \sqrt{(1 - \tilde{\gamma}_\rho)^2 - 32\tilde{\gamma}_\rho}}$, q_0, q are given by (10.63) and $\tilde{\gamma}_\rho$ is given by (10.74). Moreover, suppose that $\overline{U}(x_0, \tilde{r}) \subseteq D(F)$. Then, $\tilde{x}_n, \tilde{y}_n \in U(x_0, \tilde{r})$ for each $n = 0, 1, 2, \cdots$.*

Proof. Analogous to Theorem 10.2.4. $\qquad\square$

Theorem 10.2.11. *Suppose that the hypotheses of Theorem 10.2.10 hold. Then, sequence $\tilde{x}_n$ remains in $U(x_0, \tilde{r})$ for each $n = 0, 1, 2, \cdots$ and converges to a solution $x^\delta_{c,\alpha_k} \in \overline{U}(x_0, \tilde{r})$ which satisfies (10.71). Moreover, the following assertions hold*

$$\|\tilde{x}_n - x^\delta_{c,\alpha_k}\| \leq \tilde{b}_0 e^{-\tilde{\gamma}_0 n},$$

where $\tilde{b}_0 = (1 + \dfrac{q_0}{2})\tilde{\gamma}_\rho$ and $\tilde{\gamma}_0 = -\ln q > 0$.

Proof. Analogous to the proof of Theorem 10.2.5 until the point where we prove that (x^δ_{c,α_k}) solves (10.71). We have

$$\|F(\tilde{x}_n) - z^\delta_{\alpha_k} + \frac{\alpha_k}{c}(\tilde{x}_n - x_0)\| = \|R(\tilde{x}_n)(\tilde{x}_n - \tilde{y}_n)\| \leq \|R(\tilde{x}_n)\|\|(\tilde{x}_n - \tilde{y}_n)\|$$
$$\leq (\|F'(x_n)\| + \frac{\alpha_k}{c})\tilde{e}_n \leq (\|F'(x_n)\| + \frac{\alpha_k}{c})q^n\tilde{e}_0 \leq (\|F'(x_n)\| + \frac{\alpha_k}{c})q^n\tilde{\gamma}_\rho.$$
$$(10.75)$$

Now by letting $n \to \infty$ in (10.75) we obtain $F(x^\delta_{c,\alpha_k}) + \dfrac{\alpha_k}{c}(x^\delta_{c,\alpha_k} - x_0) = z^\delta_{\alpha_k}$. This completes the proof. $\qquad\square$

We need the following assumption in addition to the earlier assumptions for our convergence analysis.

Assumption 2. There exists a continuous, strictly monotonically increasing function $\varphi_1 : (0, b] \to (0, \infty)$ with $b \geq \|F'(x_0)\|$ satisfying $\lim_{\lambda \to 0} \varphi_1(\lambda) = 0$, $\sup_{\lambda > 0} \dfrac{\alpha\varphi_1(\lambda)}{\lambda + \alpha} \leq \varphi_1(\alpha)$ for all $\lambda \in (0, b]$, there exists $v \in X$ with $\|v\| \leq 1$ (see Ref. [332]) such that $x_0 - \hat{x} = \varphi_1(F'(x_0))v$ and for each $x \in B_{\tilde{r}}(x_0)$, there exists a bounded linear operator $G(x, x_0)$ (see Ref. [377]) such that $F'(x) = F'(x_0)G(x, x_0)$ with $\|G(x, x_0)\| \leq k_2$.

Assume that $k_2 < \dfrac{1 - L_0\tilde{r}}{1 - c}$, $\rho \leq \tilde{r} < \dfrac{1}{L_0}$ and for the sake of simplicity assume that $\varphi_1(\alpha) \leq \varphi(\alpha)$ for $\alpha > 0$.

Theorem 10.2.12. *(Ref. [235], Theorem 3.7) Suppose x^δ_{c,α_k} is the solution of (10.71) and Theorem 10.2.2, Assumption 2 and $(C)'$ hold. Then*

$$\|\hat{x} - x^\delta_{c,\alpha_k}\| \leq \frac{\varphi_1(\alpha_k) + (2 + \dfrac{4\mu}{\mu - 1})\mu\psi^{-1}(\delta)}{1 - (1 - c)k_2 - L_0 r} = O(\psi^{-1}(\delta)).$$

Proof. Note that $c(F(x^\delta_{c,\alpha_k}) - z^\delta_{\alpha_k}) + \alpha_k(x^\delta_{c,\alpha_k} - x_0) = 0$, so

$$(F'(x_0) + \alpha_k I)(x^\delta_{c,\alpha_k} - \hat{x}) = (F'(x_0) + \alpha_k I)(x^\delta_{c,\alpha_k} - \hat{x})$$
$$-c(F(x^\delta_{c,\alpha_k}) - z^\delta_{\alpha_k}) - \alpha_k(x^\delta_{c,\alpha_k} - x_0)$$
$$= \alpha_k(x_0 - \hat{x}) - c(F(\hat{x}) - z^\delta_{\alpha_k}) + F'(x_0)(x^\delta_{c,\alpha_k} - \hat{x}) - c[F(x^\delta_{c,\alpha_k}) - F(\hat{x})].$$

Thus

$$\|x^{\delta}_{c,\alpha_k} - \hat{x}\| \leq \|\alpha_k(F'(x_0) + \alpha_k I)^{-1}(x_0 - \hat{x})\| + \|(F'(x_0) + \alpha_k I)^{-1}$$
$$c(F(\hat{x}) - z^{\delta}_{\alpha_k})\| + \|(F'(x_0) + \alpha_k I)^{-1}[F'(x_0)(x^{\delta}_{c,\alpha_k} - \hat{x}) - c(F(x^{\delta}_{c,\alpha_k}) - F(\hat{x}))]\|$$
$$\leq \|\alpha_k(F'(x_0) + \alpha_k I)^{-1}(x_0 - \hat{x})\| + \|F(\hat{x}) - z^{\delta}_{\alpha_k}\| + \Gamma$$

$$(10.76)$$

where $\Gamma := \|(F'(x_0) + \alpha_k I)^{-1} \int_0^1 [F'(x_0) - cF'(\hat{x} + t(x^{\delta}_{c,\alpha_k} - \hat{x})](x^{\delta}_{c,\alpha_k} - \hat{x})dt\|$. So by Assumption 2, we obtain

$$\Gamma \leq \|(F'(x_0) + \alpha_k I)^{-1} \int_0^1 [F'(x_0) - F'(\hat{x} + t(x^{\delta}_{c,\alpha_k} - \hat{x}))]$$
$$\times (x^{\delta}_{c,\alpha_k} - \hat{x})dt\| + (1 - c)\|(F'(x_0) + \alpha_k I)^{-1}F'(x_0)$$
$$\times \int_0^1 G(\hat{x} + t(x^{\delta}_{c,\alpha_k} - \hat{x}), x_0)(x^{\delta}_{c,\alpha_k} - \hat{x})dt\|$$
$$\leq L_0 r \|x^{\delta}_{c,\alpha_k} - \hat{x}\| + (1 - c)k_2 \|x^{\delta}_{c,\alpha_k} - \hat{x}\| \qquad (10.77)$$

and hence by (10.76) and (10.77) we have

$$\|x^{\delta}_{c,\alpha_k} - \hat{x}\| \leq \frac{\|\alpha_k(F'(x_0) + \alpha_k I)^{-1}(x_0 - \hat{x})\| + \|F(\hat{x}) - z^{\delta}_{\alpha_k}\|}{1 - (1 - c)k_2 - L_0 r}$$
$$\leq \frac{\varphi_1(\alpha_k) + (2 + \frac{4\mu}{\mu - 1})\mu\psi^{-1}(\delta)}{1 - (1 - c)k_2 - L_0 r}.$$

That completes the proof of the theorem. $\square$

The following Theorem is a consequence of Theorems 10.2.11 and 10.2.12.

Theorem 10.2.13. *Let $\tilde{x}_n$ be as in (10.70), assumptions in Theorems 10.2.11 and 10.2.12 hold. Then*

$$\|\hat{x} - \tilde{x}_n\| \leq \tilde{b}_0 e^{-\tilde{\gamma}_0 n} + O(\psi^{-1}(\delta))$$

where $\tilde{b}_0$ and $\tilde{\gamma}_0$ are as in Theorem 10.2.11.

Theorem 10.2.14. *Let $\tilde{x}_n$ be as in (10.70), assumptions in Theorem 10.2.2 and Theorem 10.2.13 hold. Let*

$$n_k := \min\{n : e^{-\tilde{\gamma}_0 n} \leq \frac{\delta}{\sqrt{\alpha_k}}\}.$$

Then

$$\|\hat{x} - \tilde{x}_{n_k}\| = O(\psi^{-1}(\delta)).$$

Remark 10.2.2. The results in Ref. [236] can be rewritten without the restrictive condition $L \leq 1$. Indeed, simply use instead of (3.14) in Ref. [236] given by

$$\rho < \frac{1}{M}\left[\frac{17 - 12\sqrt{2}}{\beta} - \frac{\delta_0}{\sqrt{\alpha_0}}\right].$$

The corresponding condition

$$\rho < \min\{\frac{1}{M}(\frac{17 - 12\sqrt{2}}{\beta} - \frac{\delta_0}{\sqrt{\alpha_0}}), \frac{\rho_1}{L}\}$$

where $\rho_1 < \dfrac{0.706442399}{L}$. Then the function

$$g(t) = \frac{L^2}{8}(4 + 3Lt)t^2$$

used there, maps (0,1) into (0,1) without the restriction $L \leq 1$. Using this modification the results in Ref. [236] can be extended in the case when $L > 1$. Moreover, their Theorem 3.7 (see our Theorem 10.2.6) can be improved if $L_0 < L$, since it can be rewritten with L_0 replacing L. Finally their choice (see Remark 3.6 in Ref. [236]) $\rho \leq r < 1/L$ can be replaced by the weaker one where L_0 is replacing L. That is by $\rho \leq r < 1/L_0$.

Note that for $i, j \in \{0, 1, 2, \cdots, N\}$

$$z^\delta_{\alpha_i} - z^\delta_{\alpha_j} = (\alpha_j - \alpha_i)(K^*K + \alpha_j I)^{-1}(K^*K + \alpha_i I)^{-1}[K^*(y^\delta - KF(x_0))].$$

Therefore the balancing principle algorithm associated with the choice of the parameter specified in the above involves the following steps.

- $\alpha_0 = \mu^2\delta^2, \mu > \max\{1, \beta\}$ for IFD class and $\mu > 1$ for MFD class.
- $\alpha_i = \mu^{2i}\alpha_0;$
- solve for w_i: $\quad (K^*K + \alpha_i I)w_i = K^*(y^\delta - KF(x_0));$
- solve for $j < i$, z_{ij}: $\quad (K^*K + \alpha_j I)z_{ij} = (\alpha_j - \alpha_i)w_i;$
- if $\|z_{ij}\| > \dfrac{4}{\mu^{j+1}}$, then take $k = i - 1;$
- otherwise, repeat with $i + 1$ in place of i.
- choose $n_k = \min\{n : e^{-\gamma_0 n} \leq \dfrac{\delta}{\sqrt{\alpha_k}}\}$ in IFD Class and $n_k = \min\{n : e^{-\tilde{\gamma}_0 n} \leq \dfrac{\delta}{\sqrt{\alpha_k}}\}$ in MFD Class
- solve x_{n_k} using the iteration (10.59) or $\tilde{x}_{n_k}$ using the iteration (10.70).

We give an example for MFD class for illustrating the algorithm considered in the above in this section. We apply the algorithm by choosing a sequence of finite dimensional subspace (V_n) of X with dim $V_n = n + 1$. Precisely we choose V_n as the space of linear splines in a uniform grid of $n + 1$ points in $[0, 1]$.

Example 10.2.3. To illustrate the method for MFD class, we consider the same example of nonlinear integral operator as in section 4.3 of Ref. [403]. Consider $KF : L^2(0, 1) \longrightarrow L^2(0, 1)$ where $K : L^2(0, 1) \longrightarrow L^2(0, 1)$ defined by

$$K(x)(t) = \int_0^1 k(t, s)x(s)ds$$

and $F : D(F) \subseteq L^2(0,1) \longrightarrow L^2(0,1)$ defined by

$$F(u) := \int_0^1 k(t,s)u^3(s)ds,$$

where

$$k(t,s) = \begin{cases} (1-t)s \text{ if } 0 \le s \le t \le 1 \\ (1-s)t \text{ if } 0 \le t \le s \le 1. \end{cases}$$

Then for all $x(t), y(t) : x(t) > y(t) :$

$$\langle F(x) - F(y), x - y \rangle = \int_0^1 \left[\int_0^1 k(t,s)(x^3 - y^3)(s)ds \right] (x-y)(t)dt \ge 0.$$

Thus the operator F is monotone. The Fréchet derivative of F is given by

$$F'(u)w = 3 \int_0^1 k(t,s)(u(s))^2 w(s)ds.$$

So for any $u \in B_r(x_0), x_0{}^2(s) \ge k_3 > 0, \forall s \in (0,1)$, we have

$$F'(u)w = F'(x_0)G(u,x_0)w,$$

where $G(u,x_0) = (\dfrac{u}{x_0})^2$. Further observe that

$$[F'(v) - F'(u)]w(s) = 3\int_0^1 k(t,s)u^2(s)ds \frac{k(t,s)(v^2(s) - u^2(s))w(s)ds}{\int_0^1 k(t,s)u^2(s)ds}$$

$$:= F'(u)\Phi(u,v,w),$$

where $\Phi(u,v,w) = \dfrac{k(t,s)(v^2(s) - u^2(s))w(s)ds}{\int_0^1 k(t,s)u^2(s)ds}.$

Note that

$$\Phi(u,v,w) = \frac{k(t,s)(v(s) + u(s))(v(s) - u(s))w(s)ds}{\int_0^1 k(t,s)u^2(s)ds}.$$

Thus F satisfies the $(C)'$ with

$$L_0 \ge \left\| \frac{k(t,s)(v(s) + u(s))ds}{\int_0^1 k(t,s)u^2(s)ds} \right\|.$$

In our computation, we take

$$y(t) = (\frac{1}{18\pi^2})(1-t)(14t - 7 + \cos^3(\pi t)$$

$$+ 6\cos(\pi t))t^2 - (\frac{1}{18\pi^2})t(14t - 7 + \cos^3(\pi t)$$

$$+ 6\cos(\pi t))(1 - t^2) + (\frac{1}{9\pi^2})t(1-t)(14t - 7 + \cos^3(\pi t) + 6\cos(\pi t))$$

and $y^\delta = y + \delta$. Then the exact solution

$$\hat{x}(t) = \cos \pi t.$$

Table 10.1   Comparison Table

n	k	δ	α	$\|\tilde{x}_k - \hat{x}\|$	$\dfrac{\|\tilde{x}_k - \hat{x}\|}{(\delta)^{1/2}}$
8	4	0.1016	0.1094	0.2776	0.8711
16	4	0.1004	0.1069	0.2019	0.6371
32	4	0.1001	0.1063	0.1560	0.4930
64	4	0.1000	0.1061	0.1280	0.4046
128	4	0.1000	0.1061	0.1115	0.3527
256	4	0.1000	0.1060	0.1024	0.3238
512	4	0.1000	0.1060	0.0975	0.3083
1024	4	0.1000	0.1060	0.0950	0.3003

We use

$$
\begin{aligned}
x_0(t) = \cos(\pi t) + 3[&\frac{-1}{4\pi^2}(1 - t + 2\pi t^2 \cos(\pi t) \\
&\times \sin(\pi t) + \pi^2 t^3 + t\cos^2(\pi t) - 2\pi t \cos(\pi t) \\
&\times \sin(\pi t) - \pi^2 t^2 - \cos^2(\pi t)) + \frac{1}{4\pi^2}t \\
&\times(-2\cos(\pi t)\sin(\pi t)\pi - 2\pi^2 t + 2\pi t \cos(\pi t) \\
&\times \sin(\pi t) + \pi^2 t^2 + \cos^2(\pi t) + \pi^2 - \cos^2(\pi t))]
\end{aligned}
$$

as our initial guess, so that the function $x_0 - \hat{x}$ satisfies the source condition

$$
x_0 - \hat{x} = \varphi_1(F'(x_0))1
$$

where $\varphi_1(\lambda) = \lambda$. Thus we expect to have an accuracy of order at least $O(\delta^{1/2})$. We choose $\alpha_0 = (1.3)\delta^2$, $\mu = 1.3$, $\delta = 0.1 = c$, $q = 0.30$ approximately. For all n the number of iteration $n_k = 1$. The results of the computation are presented in Table 10.1. The plots of the exact and the approximate solution obtained are given in Figures 10.1–10.2 and 10.3–10.4.

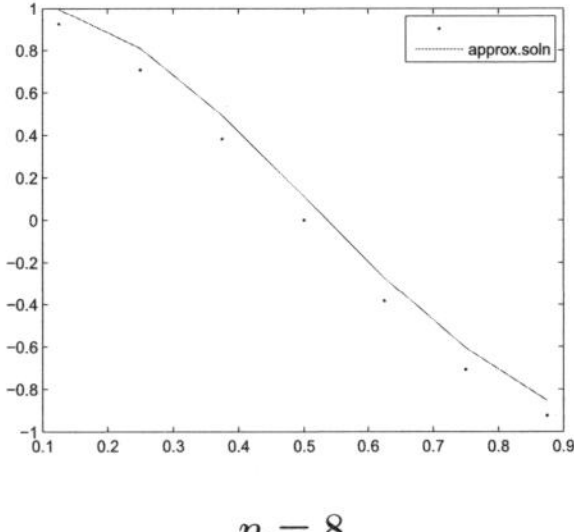

$$n = 8$$

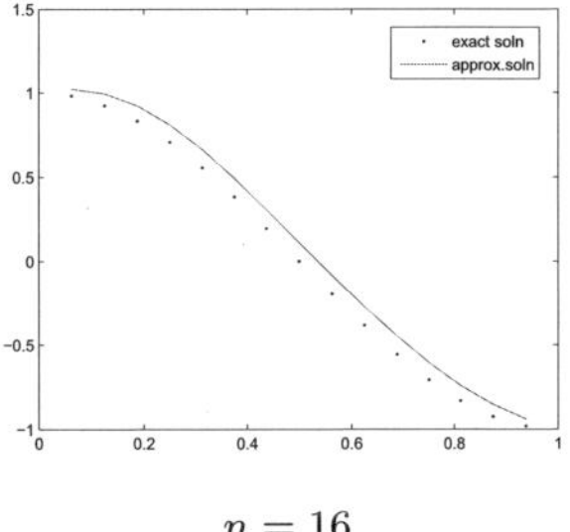

$$n = 16$$

Fig. 10.1   Curve of the exact and approximate solutions

For some examples where $(C1)$ is not satisfied but $(C1)'$ is satisfied, the reader can see the last chapter.

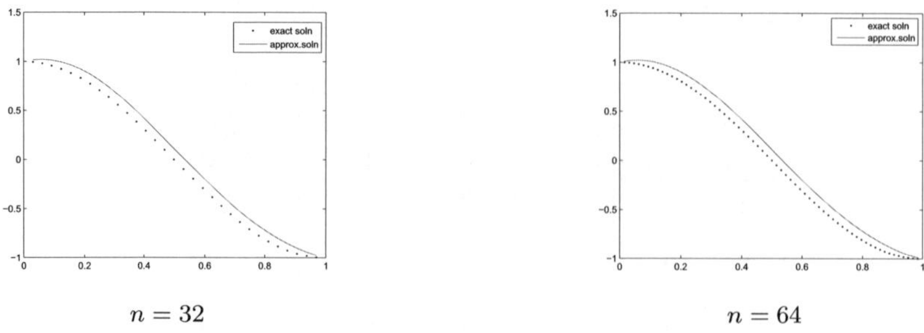

$n = 32$ $n = 64$

Fig. 10.2   Curve of the exact and approximate solutions

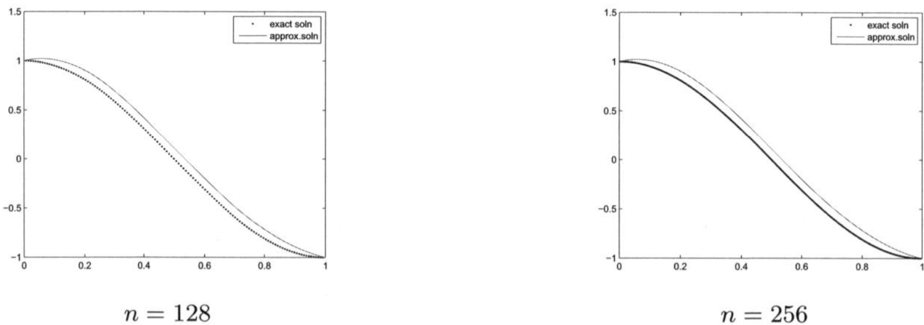

$n = 128$ $n = 256$

Fig. 10.3   Curve of the exact and approximate solutions of MFD class

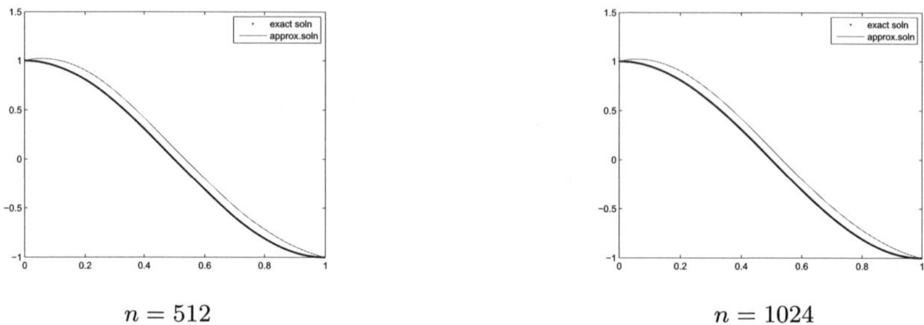

$n = 512$ $n = 1024$

Fig. 10.4   Curve of the exact and approximate solutions of MFD class

10.3   Regularization Methods

An iteratively regularized projection method, which converges quadratically, has been considered for obtaining stable approximate solution to nonlinear ill-posed operator equations $F(x) = y$ where $F : D(F) \subseteq X \longrightarrow X$ is a nonlinear monotone operator defined on the real Hilbert space X. We assume that only a noisy data y^δ with $\|y - y^\delta\| \leq \delta$ are available. Under the assumption that the Fréchet derivative F' of F is Lipschitz continuous, a choice of the regularization parameter using an adaptive selection of the parameter and a stopping rule for the iteration index using a majorizing sequence are presented. We prove that under a general source condition on $x_0 - \hat{x}$, the error $\|x_{n,\alpha}^{h,\delta} - \hat{x}\|$ between the regularized approximation $x_{n,\alpha}^{h,\delta}, (x_{0,\alpha}^{h,\delta} := P_h x_0$ where P_h is an orthogonal projection onto a finite dimensional subspace X_h of X) and the solution $\hat{x}$ is of optimal order. We consider the problem of solving the nonlinear ill-posed operator equation (10.1) approximately when the data y is not known exactly and F replace A. We assume that the operator $F : D(F) \to X$ with domain $D(F) \subseteq X$ is a monotone operator.

Further we assume that $y^\delta \in X$ satisfy condition (10.4) and that (10.1) has a solution $\hat{x}$. Equation (10.1) is ill-posed in the sense that the Fréchet derivative $F'(.)$ is not boundedly invertible (see page 26 in Ref. [374]). Since (10.1) is ill-posed, one has to replace Eq. (10.1) by a nearby equation whose solution is less sensitive to perturbation on the right of y. This replacement is known as regularization. A well-known method for regularizing (10.1), when F is monotone is the method of Lavrentiev regularization (see Ref. [415]). In this method approximation x_α^δ is obtained by solving the singularly perturbed operator equation

$$F(x) + \alpha(x - x_0) = y^\delta. \tag{10.78}$$

In practice one has to deal with some sequence $(x_{n,\alpha}^\delta)$, converging to the solution $\hat{x}$ of (10.1). Many authors considered such sequences (see, Refs. [174, 177, 284, 194, 254]). In Ref. [174] Bakushinsky and Smirnova considered an iteratively regularized Lavrentiev method:

$$x_{k+1}^\delta = x_k^\delta - (A_k^\delta + \alpha_k I)^{-1}(F(x_k^\delta) - y^\delta + \alpha_k(x_k^\delta - x_0)), \tag{10.79}$$

for $k = 0, 1, 2, \cdots$, where $A_k^\delta := F'(x_k^\delta)$ and (α_k) is a sequence of positive real numbers such that $\lim_{k\to\infty} \alpha_k = 0$, as an approximate solution for (10.1). A general discrepancy principle, has been considered in Ref. [174] for choosing the stopping index k_δ and showed that $x_{k_\delta}^\delta \to \hat{x}$ as $\delta \to 0$. However no error estimate for $\|x_{k_\delta}^\delta - \hat{x}\|$ has been given in Ref. [174]. Later in Ref. [314], Mahale and Nair considered the method (10.79) and obtained an error estimate for $\|x_{k_\delta}^\delta - \hat{x}\|$ under weaker assumptions than the assumptions in Ref. [174]. In Ref. [228], George and Elmahdy considered an iterative regularization method;

$$x_{n+1,\alpha}^\delta = x_{n,\alpha}^\delta - (F'(x_0) + \alpha I)^{-1}(F(x_{n,\alpha}^\delta) - y^\delta + \alpha(x_{n,\alpha}^\delta - x_0)), \tag{10.80}$$

where $x_{0,\alpha}^\delta := x_0$ and proved that $(x_{n,\alpha}^\delta)$ converges to the unique solution x_α^δ of (10.78) under the following Assumptions.

Assumption 3. There exists $r_0 > 0$ such that $B_{r_0}(\hat{x}) \subseteq D(F)$ and F is Fréchet differentiable at all $x \in B_{r_0}(\hat{x})$.

Assumption 4. There exists a constant $L > 0$ such that for every $x, u \in B_{r_0}(\hat{x})$ and $v \in X$, there exists an element $\Phi(x, u, v) \in X$ satisfying

$$[F'(x) - F'(u)]v = F'(u)\Phi(x, u, v), \qquad \|\Phi(x, u, v)\| \leq L\|v\|,$$

for all $x, u \in B_{r_0}(\hat{x})$.

Assumption 5. There exists a continuous, strictly monotonically increasing function $\varphi : (0, a] \to (0, \infty)$ with $a \geq \|F'(\hat{x})\|$ satisfying $\lim_{\lambda \to 0} \varphi(\lambda) = 0$ and a vector $v \in X$ with $\|v\| \leq 1$ such that

$$x_0 - \hat{x} = \varphi(F'(\hat{x}))v$$

and

$$\sup_{\lambda \geq 0} \frac{\alpha \varphi(\lambda)}{\lambda + \alpha} \leq \varphi(\alpha), \forall \alpha \in (0, a].$$

The main drawback of the method considered in Ref. [228] is that the initial guess x_0 of the iterative sequence $(x_{n,\alpha}^\delta)$ is highly dependent on l_0 (see Lemma 2.4 and Theorem 2.6 in Ref. [228]), so it is hard to obtain such an initial guess x_0 when l_0 is not small enough. We try to overcome this drawback. We use the following modified form of Assumption 4.

Assumption 6. Let $x_0 \in B_r(\hat{x})$ be fixed. There exists a constant $l_0 > 0$ such that for every $x, x_0 \in B_{r_0}(\hat{x})$ and $v \in X$, there exists an element $\Phi(x, x_0, v) \in X$ satisfying

$$[F'(x) - F'(x_0)]v = F'(x_0)\Phi(x, x_0, v), \quad \|\Phi(x, x_0, v)\| \leq l_0\|v\|\|x - x_0\|,$$

for all $x \in B_{r_0}(\hat{x})$ and $v \in X$.

Note that from Assumption 6 there follows Assumption 4 with $L = l_0\|x - x_0\|$. Hence, the first hypothesis in Assumption 6 is weaker but the second hypothesis is stronger (but more practical) than the corresponding one in Assumption 4. Hence Assumption 6 is stronger than Assumption 4. The autoconvolution problem discussed in Ref. [237] is an example of a nonlinear ill-posed problem satisfying Assumption 4 but not Assumption 6. Further note that $l_0 \leq L$ holds. The results in Ref. [228] really require Assumption 6 not Assumption 4. If $l_0 = L$, the results of this section coincide with the results in Ref. [228]. Otherwise, i.e., if $l_0 < L$, then our convergence results are better, under weaker majorizing sequences. The main advantage of using the stronger Assumption 6 is that, the majorizing sequence we are going to use in this section is independent of the regularization parameter α. Further the majorizing sequence gives an a priori error estimate which can be used to determine the number of iterations needed to achieve a prescribed solution accuracy before actual computation takes place.

Remark 10.3.1. It can be seen that functions $\varphi(\lambda) = \lambda^\nu, \lambda > 0$ for $0 < \nu \leq 1$ and

$$\varphi(\lambda) = \begin{cases} (\ln \frac{1}{\lambda})^{-p}, & 0 < \lambda \leq e^{-(p+1)} \\ 0, & \text{otherwise} \end{cases}$$

for $p \geq 0$ satisfy the above assumption (see Ref. [332]).

Now we shall give an example that satisfies the above assumption.

Example 10.3.1. (see section 4 in Ref. [277]) Let $F : L^2[0,1] \longrightarrow L^2[0,1]$ defined by

$$F(u) := B(u) + (\arctan(u))^3, \tag{10.81}$$

where $B(u) := \int_0^1 e^{-|s-t|}u(t)dt$. Then since the function $u \to \arctan^3(u)$ is increasing on $\mathbb{R}$,

$$\langle (\arctan(u))^3 - (\arctan(v))^3, u - v \rangle \geq 0, \qquad \forall u, v \in L^2[0,1].$$

Again since $e^{-|x|} = \dfrac{1}{\pi} \displaystyle\int_{-\infty}^{\infty} \dfrac{e^{i\lambda x}}{1+\lambda^2} d\lambda$, we have

$$\langle B(u-v), u - v \rangle \geq 0, \qquad \forall u, v \in L^2[0,1].$$

Thus the operator F is monotone. The Fréchet derivative of F is given by

$$F'(u)w = \frac{3(\arctan(u))^2}{1+u^2}w + \int_0^1 e^{-|s-t|}w(t)dt.$$

Note that

$$(F'(v) - F'(x_0))w = \left(\frac{3(\arctan(x_0))^2}{1+u^2}w + \int_0^1 e^{-|s-t|}w(t)dt \right)$$

$$\times \left[\frac{(\dfrac{3(\arctan(v))^2}{1+v^2} - \dfrac{3(\arctan(x_0))^2}{1+u^2})w}{\dfrac{3(\arctan(x_0))^2}{1+x_0^2}w + \displaystyle\int_0^1 e^{-|s-t|}w(t)dt} \right]$$

$$:= F'(x_0)\Phi(v, x_0, w),$$

where

$$\Phi(v, x_0, w) = \frac{(\dfrac{3(\arctan(v))^2}{1+v^2} - \dfrac{3(\arctan(x_0))^2}{1+x_0^2})w}{\dfrac{3(\arctan(x_0))^2}{1+x_0^2}w + \displaystyle\int_0^1 e^{-|s-t|}w(t)dt}.$$

Observe that

$$\Phi(v, x_0, w) = \frac{3(\arctan^2(v)(1+x_0^2) - \arctan^2(x_0)(1+v^2))w}{(1+x_0^2)(1+v^2)(\dfrac{3(\arctan(x_0))^2}{1+x_0^2}w + \displaystyle\int_0^1 e^{-|s-t|}w(t)dt)}$$

$$= \frac{3}{(1+x_0^2)(1+v^2)(\dfrac{3(\arctan(x_0))^2}{1+x_0^2}w + \displaystyle\int_0^1 e^{-|s-t|}w(t)dt)}$$

$$\times [(1+x_0^2)(\arctan(v) + \arctan(x_0))(\arctan(v) - \arctan(x_0))$$

$$+ (x_0 + v)\arctan^2(x_0)(x_0 - v)]w$$

$$:= \Phi_1(v, x_0, w)(v - x_0),$$

where

$$\Phi_1(v, x_0, w) = \cfrac{3}{(1 + x_0^2)(1 + v^2)\left(\dfrac{3(\arctan(x_0))^2}{1 + x_0^2} w + \displaystyle\int_0^1 e^{-|s-t|} w(t)dt\right)}$$

$$\times [(1 + x_0^2)(\arctan(v) + \arctan(x_0))(1 - \frac{1}{3}(x_0^2 + v^2 + x_0 v)$$

$$+ \frac{1}{5}(x_0^4 + x_0^3 v + x_0^2 v^2 + x_0 v^3 + v^4) - \cdots \cdots) - (x_0 + v)\arctan^2(x_0)]w.$$

The above step follows from the relation,

$$\arctan(v) - \arctan(x_0) = (v - x_0)(1 - \frac{1}{3}(x_0^2 + v^2 + x_0 v)$$

$$+ \frac{1}{5}(x_0^4 + x_0^3 v + x_0^2 v^2 + x_0 v^3 + v^4) - \cdots \cdots).$$

So Assumption 6 satisfies with $l_0 \geq \|\Phi_1(v, x_0, w)\|$.

We prove in this section that the sequence $(x_{n,\alpha}^\delta)$ defined in (10.80) converges to x_α^δ as $n \to \infty$ and obtain an error estimate for $\|x_{n,\alpha}^\delta - x_\alpha^\delta\|$. Using an error estimate for $\|x_\alpha^\delta - \hat{x}\|$, we obtained an error estimate for $\|x_{n,\alpha}^\delta - \hat{x}\|$. The error analysis for the order optimal result using an adaptive selection of the parameter α and a stopping rule using a majorizing sequence are also given in this section. Implementation of the adaptive choice of the parameter and the choice of the stoping rule are given.

For proving our results, we consider the sequence $(t_n), n \geq 0$, defined iteratively by $t_0 = 0, t_1 = \eta$,

$$t_{n+1} = t_n + \frac{l_0 \eta}{(1 - r)}(t_n - t_{n-1}), \tag{10.82}$$

where $r \in [0, 1)$, as a majorizing sequence, of the sequence $(x_{n,\alpha}^\delta)$.

The following Lemma is essentially a reformulation of a Lemma in Ref. [228]. For the sake of completeness, we supply its proof as well.

Lemma 10.3.1. *Assume there exist non-negative numbers l_0, η and $r \in [0, 1)$ such that*

$$\frac{l_0}{(1 - r)}\eta \leq r. \tag{10.83}$$

Then the sequence (t_n) defined in (10.82) is increasing, bounded above by $t^{**} := \dfrac{\eta}{1 - r}$, and converges to some t^* such that $0 < t^* \leq \dfrac{\eta}{1 - r}$. Moreover, for $n \geq 0$;

$$0 \leq t_{n+1} - t_n \leq r(t_n - t_{n-1}) \leq r^n \eta \tag{10.84}$$

and

$$t^* - t_n \leq \frac{r^n}{1 - r}\eta. \tag{10.85}$$

Proof. Since the result holds for $\eta = 0, l_0 = 0$ or $r = 0$, we assume that $l_0 \neq 0, \eta \neq 0$ and $r \neq 0$. Observe that $t_1 - t_0 = \eta \geq 0$, assume that $t_{i+1} - t_i \geq 0$, for all $i \leq k$ for some k. Then $t_{k+2} - t_{k+1} = \dfrac{l_0 \eta}{(1-r)}(t_{k+1} - t_k) \geq 0$, so by induction $t_{n+1} - t_n \geq 0$ for all $n \geq 0$. Now since

$$\frac{l_0 \eta}{(1-r)} \leq r,$$

the estimate (10.84) follows from (10.82). Further observe that

$$t_{k+1} \leq t_k + r(t_k - t_{k-1}) \leq \cdots \leq \eta + r\eta + \cdots + r^k \eta$$
$$= \frac{1 - r^{k+1}}{1-r}\eta < \frac{\eta}{1-r}.$$

Hence the sequence $(t_n), n \geq 0$ is bounded above by $\dfrac{\eta}{1-r}$; nondecreasing, so it converges to some $t^* \leq \dfrac{\eta}{1-r}$, and

$$t^* - t_n = \lim_{i \to \infty} t_{n+i} - t_n \leq \lim_{i \to \infty} \sum_{j=0}^{i-1}(t_{n+1+j} - t_{n+j}) \leq \frac{r^n}{1-r}\eta.$$

This completes the proof of the Lemma. $\qquad\square$

To prove the convergence of the sequence $(x_{n,\alpha}^\delta)$ defined in (10.80) we introduce the following notations: Let $R_\alpha(x_0) := F'(x_0) + \alpha I$ and

$$G(x) := x - R_\alpha(x_0)^{-1}[F(x) - y^\delta + \alpha(x - x_0)]. \tag{10.86}$$

Note that with the above notation, $G(x_{n,\alpha}^\delta) = x_{n+1,\alpha}^\delta$ and

$$\|R_\alpha(x_0)^{-1}F'(x_0)\| \leq 1. \tag{10.87}$$

The following Lemma based on Assumption 6 will be used in due course.

Lemma 10.3.2. *For $u, v, x_0 \in B_{r_0}(\hat{x})$*

$$F(v) - F(u) - F'(x_0)(v - u) = F'(x_0) \int_0^1 \Phi(u + t(v - u), x_0, v - u)dt.$$

Proof. Using the Fundamental Theorem of Integration, for $u, v, x_0 \in B_{r_0}(\hat{x})$ we have

$$F(v) - F(u) = \int_0^1 F'(u + t(v - u))(v - u)dt,$$

so by Assumption 6 we have

$$F(v) - F(u) - F'(x_0)(v - u) = F'(x_0) \int_0^1 \Phi(u + t(v - u), x_0, v - u)dt.$$

This completes the proof of the Lemma. $\qquad\square$

Here after we assume that $\|x_0 - \hat{x}\| \leq \rho$ and

$$\frac{l_0}{2}\rho^2 + \rho + \frac{\delta}{\alpha} \leq \eta \leq \min\{\frac{r(1-r)}{l_0}, r_0(1-r)\}. \tag{10.88}$$

Theorem 10.3.1. *Suppose (10.82) holds. Let the assumptions in Lemma 10.3.1 with η as in (10.88) and Assumption 6 be satisfied. Then the sequence $(x_{n,\alpha}^\delta)$ defined in (10.80) is well-defined and $x_{n,\alpha}^\delta \in B_{t^*}(x_0)$ for all $n \geq 0$. Further $(x_{n,\alpha}^\delta)$ is Cauchy sequence in $B_{t^*}(x_0)$ and hence converges to $x_\alpha^\delta \in \overline{U}(x_0, t^*) \subset U(x_0, t^{**})$ and $F(x_\alpha^\delta) + \alpha(x_\alpha^\delta - x_0) = y^\delta$. Moreover, the following estimate holds for all $n \geq 0$,*

$$\|x_{n+1,\alpha}^\delta - x_{n,\alpha}^\delta\| \leq t_{n+1} - t_n \tag{10.89}$$

and

$$\|x_{n,\alpha}^\delta - x_\alpha^\delta\| \leq t^* - t_n \leq \frac{r^n \eta}{(1-r)}. \tag{10.90}$$

Proof.   Let G be as in (10.86). Then for $u, v \in B_{t^*}(x_0)$,

$$
\begin{aligned}
&G(u) - G(v) \\
&= u - v - R_\alpha(x_0)^{-1}[F(u) - y^\delta + \alpha(u - x_0)] + R_\alpha(x_0)^{-1}[F(v) - y^\delta + \alpha(v - x_0)] \\
&= R_\alpha(x_0)^{-1}[R_\alpha(x_0)(u - v) - (F(u) - F(v))] + \alpha R_\alpha(x_0)^{-1}(v - u) \\
&= R_\alpha(x_0)^{-1}[F'(x_0)(u - v) - (F(u) - F(v)) + \alpha(u - v)] + \alpha R_\alpha(x_0)^{-1}(v - u) \\
&= R_\alpha(x_0)^{-1}[F'(x_0)(u - v) - (F(u) - F(v))].
\end{aligned}
$$

Thus by Lemma 10.3.2, Assumption 6, (10.87) and (10.88) we have

$$\|G(u) - G(v)\| \leq l_0 t^* \|u - v\|. \tag{10.91}$$

Now we shall prove that the sequence (t_n) defined in Lemma 10.3.1 is a majorizing sequence of the sequence $(x_{n,\alpha}^\delta)$ and $x_{n,\alpha}^\delta \in B_{t^*}(x_0)$, for all $n \geq 0$. Note that $F(\hat{x}) = y$, so

$$
\begin{aligned}
\|x_{1,\alpha}^\delta - x_0\| &= \|R_\alpha(x_0)^{-1}(F(x_0) - y^\delta)\| \\
&= \|R_\alpha(x_0)^{-1}(F(x_0) - y + y - y^\delta)\| \\
&= \|R_\alpha(x_0)^{-1}(F(x_0) - F(\hat{x}) - F'(x_0)(x_0 - \hat{x}) + F'(x_0)(x_0 - \hat{x}) + y - y^\delta)\| \\
&\leq \|R_\alpha(x_0)^{-1}(F(x_0) - F(\hat{x}) - F'(x_0)(x_0 - \hat{x}))\| \\
&\quad + \|R_\alpha(x_0)^{-1}F'(x_0)(x_0 - \hat{x})\| + \|R_\alpha(x_0)^{-1}(y - y^\delta)\| \\
&\leq \|R_\alpha(x_0)^{-1}F'(x_0) \int_0^1 \Phi(\hat{x} + t(x_0 - \hat{x}), x_0, (x_0 - \hat{x}))dt\| \\
&\quad + \|R_\alpha(x_0)^{-1}F'(x_0)(x_0 - \hat{x})\| + \frac{\delta}{\alpha} \\
&\leq \frac{l_0}{2}\|x_0 - \hat{x}\|^2 + \|x_0 - \hat{x}\| + \frac{\delta}{\alpha} \leq \frac{l_0}{2}\rho^2 + \rho + \frac{\delta}{\alpha} \leq \eta = t_1 - t_0.
\end{aligned}
$$

The last but one step follows from Assumption 6. Assume that

$$\|x_{i+1,\alpha}^\delta - x_{i,\alpha}^\delta\| \leq t_{i+1} - t_i, \qquad \forall i \leq k \tag{10.92}$$

for some k. Then

$$
\begin{aligned}
\|x_{k+1,\alpha}^\delta - x_0\| &= \|x_{k+1,\alpha}^\delta - x_{k,\alpha}^\delta + x_{k,\alpha}^\delta - x_{k-1,\alpha}^\delta + \cdots + x_{1,\alpha}^\delta - x_0\| \\
&\leq \|x_{k+1,\alpha}^\delta - x_{k,\alpha}^\delta\| + \|x_{k,\alpha}^\delta - x_{k-1,\alpha}^\delta\| + \cdots + \|x_{1,\alpha}^\delta - x_0\| \\
&\leq t_{k+1} - t_k + t_k - t_{k-1} + \cdots + t_1 - t_0 = t_{k+1} \leq t^*.
\end{aligned}
$$

So $x_{i+1,\alpha}^{\delta} \in B_{t^*}(x_0)$ for all $i \leq k$, and hence, by (10.91) and (10.92),

$$\|x_{k+2,\alpha}^{\delta} - x_{k+1,\alpha}^{\delta}\| \leq l_0 t^* \|x_{k+1,\alpha}^{\delta} - x_{k,\alpha}^{\delta}\| \leq \frac{l_0 \eta}{(1-r)}(t_{k+1} - t_k) = t_{k+2} - t_{k+1}.$$

Thus by induction $\|x_{n+1,\alpha}^{\delta} - x_{n,\alpha}^{\delta}\| \leq t_{n+1} - t_n$ for all $n \geq 0$ and hence $(t_n), n \geq 0$ is a majorizing sequence of the sequence $(x_{n,\alpha}^{\delta})$. In particular $\|x_{n,\alpha}^{\delta} - x_0\| \leq t_n \leq t^*$, i.e., $x_{n,\alpha}^{\delta} \in B_{t^*}(x_0)$, for all $n \geq 0$. So $(x_{n,\alpha}^{\delta}), n \geq 0$ is a Cauchy sequence and converges to some $x_{\alpha}^{\delta} \in U(x_0, t^*) \subset U(x_0, t^{**})$ and

$$\|x_{\alpha}^{\delta} - x_{n,\alpha}^{\delta}\| \leq t^* - t_n \leq \frac{r^n \eta}{(1-r)}.$$

Now by letting $n \to \infty$ in (10.80) we obtain $F(x_{\alpha}^{\delta}) + \alpha(x_{\alpha}^{\delta} - x_0) = y^{\delta}$. This completes the proof of the Theorem. $\qquad\square$

We will be using the error estimates in the following Proposition, which can be found in Ref. [415], for our error analysis.

Proposition 10.3.1. *(see Proposition 3.1 in Ref. [415]) Let $\hat{x} \in D(F)$ be a solution of (10.1) and let $F : D(F) \subseteq X \mapsto X$ be a monotone operator in X. Let x_{α} be the unique solution of*

$$F(x) + \alpha(x - x_0) = y \tag{10.93}$$

and x_{α}^{δ} be the unique solution of (10.78). Then

$$\|x_{\alpha}^{\delta} - x_{\alpha}\| \leq \frac{\delta}{\alpha} \tag{10.94}$$

and

$$\|x_{\alpha} - \hat{x}\| \leq \|x_0 - \hat{x}\|.$$

To obtain an error estimate for $\|x_{\alpha}^{\delta} - \hat{x}\|$, it is enough to obtain an error estimate for $\|x_{\alpha}^{\delta} - x_{\alpha}\|$ and $\|x_{\alpha} - \hat{x}\|$. Let us introduce the following operators:

$$A := F'(\hat{x}), \tag{10.95}$$

$$M_{\alpha} := \int_0^1 F'(\hat{x} + t(x_{\alpha} - \hat{x}))dt. \tag{10.96}$$

Then by the Mean Value Theorem in integral form we have

$$F(x_{\alpha}) - F(\hat{x}) = M_{\alpha}(x_{\alpha} - \hat{x}). \tag{10.97}$$

The following Theorem gives an estimate for $\|x_{\alpha} - \hat{x}\|$.

Theorem 10.3.2. *Let x_{α} be the unique solution of (10.93) and let the Assumptions 3, 5 and 6 be satisfied. Then we have the following;*

$$\|x_{\alpha} - \hat{x}\| \leq (l_0 r_0 + 1)c_{\varphi}\varphi(\alpha). \tag{10.98}$$

Proof. Since $F(x_\alpha) + \alpha(x_\alpha - x_0) = y$, for any $\alpha > 0$, by (10.97) we have

$$M_\alpha(x_\alpha - \hat{x}) + \alpha(x_\alpha - \hat{x}) = \alpha(x_0 - \hat{x}).$$

Then, we have that

$$
\begin{aligned}
x_\alpha - \hat{x} &= (M_\alpha + \alpha I)^{-1}\alpha(x_0 - \hat{x}) \\
&= [(M_\alpha + \alpha I)^{-1} - (A + \alpha I)^{-1}]\alpha(x_0 - \hat{x}) + \alpha(A + \alpha I)^{-1}(x_0 - \hat{x}) \\
&= (M_\alpha + \alpha I)^{-1}(A - M_\alpha)\alpha(A + \alpha I)^{-1}(x_0 - \hat{x}) + \alpha(A + \alpha I)^{-1}(x_0 - \hat{x}) \\
&= (M_\alpha + \alpha I)^{-1}M_\alpha\Phi(\hat{x}, \hat{x} + t(x_\alpha - \hat{x}), \alpha(A + \alpha I)^{-1}(x_0 - \hat{x})) \\
&\quad + \alpha(A + \alpha I)^{-1}(x_0 - \hat{x}),
\end{aligned}
$$

the last step follows from Assumption 6. Therefore by Proposition 10.3.1, Assumptions 3, 5 and 6 we have,

$$
\begin{aligned}
\|x_\alpha - \hat{x}\| &= \|(M_\alpha + \alpha I)^{-1}M_\alpha\Phi(\hat{x}, \hat{x} + t(x_\alpha - \hat{x}), \alpha(A + \alpha I)^{-1}(x_0 - \hat{x})) \\
&\quad + \alpha(A + \alpha I)^{-1}(x_0 - \hat{x})\| \\
&\leq \|(M_\alpha + \alpha I)^{-1}M_\alpha\Phi(\hat{x}, \hat{x} + t(x_\alpha - \hat{x}), \alpha(A + \alpha I)^{-1}(x_0 - \hat{x}))\| \\
&\quad + \|\alpha(A + \alpha I)^{-1}(x_0 - \hat{x})\| \leq (l_0 r_0 + 1)c_\varphi\varphi(\alpha).
\end{aligned}
$$

This completes the proof of the Theorem. $\qquad\qquad\square$

Combining the estimates in Theorem 10.3.1, (10.94) and (10.98) we obtain the following Theorem.

Theorem 10.3.3. *Let $x_{n,\alpha}^\delta$ be as in (10.80) and let the assumptions in Theorem 10.3.1, (10.94) and (10.98) be satisfied. Then we have the following*

$$\|x_{n,\alpha}^\delta - \hat{x}\| \leq \frac{r^n \eta}{1 - r} + \frac{\delta}{\alpha} + (l_0 r_0 + 1)c_\varphi\varphi(\alpha). \tag{10.99}$$

Let

$$n_\delta := \min\{n : r^n \leq \delta\} \tag{10.100}$$

and let

$$C := \max\{\frac{\eta}{1 - r} + 1, (l_0 r_0 + 1)c_\varphi\}. \tag{10.101}$$

Theorem 10.3.4. *Let $x_{n,\alpha}^\delta$ be as in (10.80) and let the assumptions in Theorem 10.3.3 be satisfied. Let n_δ be as in (10.100) and C be as in (10.101). Then for all $0 < \alpha \leq 1$ we have the following;*

$$\|x_{n_\delta,\alpha}^\delta - \hat{x}\| \leq C(\varphi(\alpha) + \frac{\delta}{\alpha}). \tag{10.102}$$

Note that the error $\varphi(\alpha) + \dfrac{\delta}{\alpha}$ in (10.102) is of optimal order if $\alpha_\delta := \alpha(\delta)$ satisfies, $\alpha_\delta\varphi(\alpha_\delta) = \delta$. Now using the function $\psi(\lambda) := \lambda\varphi^{-1}(\lambda), 0 < \lambda \leq a$ we have $\delta = \alpha_\delta\varphi(\alpha_\delta) = \psi(\varphi(\alpha_\delta))$, so that $\alpha_\delta = \varphi^{-1}(\psi^{-1}(\delta))$. Hence by (10.102) we have the following.

Theorem 10.3.5. *Let $\psi(\lambda) := \lambda\varphi^{-1}(\lambda)$ for $0 < \lambda \leq a$ and let the assumptions in Theorem 10.3.4 hold. For $\delta > 0$, let $\alpha := \alpha_\delta = \varphi^{-1}(\psi^{-1}(\delta))$. Let n_δ be as in (10.100). Then*

$$\|x^\delta_{n_\delta,\alpha} - \hat{x}\| = \bigcirc(\psi^{-1}(\delta)).$$

First we will present a parameter choice rule based on the adaptive method studied in Refs. [324, 346]. In practice, the regularization parameter α is often selected from some finite set

$$D_M(\alpha) := \{\alpha_i = \mu^i\alpha_0, i = 0, 1, \cdots, M\}, \tag{10.103}$$

where $\mu > 1$ and M is such that $\alpha_M < 1 \leq \alpha_{M+1}$. We choose $\alpha_0 := \sqrt{\delta}$, because in general $\varphi(\lambda) = \lambda^\nu, 0 < \nu \leq 1$ and in this case the best possible error estimate is of order $\bigcirc(\sqrt{\delta})$ and from Theorem 10.3.5, it follows that such an accuracy cannot be guaranteed for $\alpha < \sqrt{\delta}$.

Let

$$n_M := \min\{n : r^n \leq \delta\}. \tag{10.104}$$

Then for $i = 0, 1, \cdots, M$, we have

$$\|x^\delta_{n_M,\alpha_i} - x^\delta_{\alpha_i}\| \leq \frac{\delta}{\alpha_i}, \qquad \forall i = 0, 1, \cdots M. \tag{10.105}$$

Let $x_i := x^\delta_{n_M,\alpha_i}$. The parameter choice strategy that we are going to consider here, we selects $\alpha = \alpha_i$ from $D_M(\alpha)$ and operates only with corresponding x_i, $i = 0, 1, \cdots, M$.

Theorem 10.3.6. *Assume that there exists $i \in \{0, 1, 2, \cdots, M\}$ such that $\varphi(\alpha_i) \leq \dfrac{\delta}{\alpha_i}$. Let assumptions of Theorem 10.3.4 and Theorem 10.3.5 hold and let*

$$l := \max\{i : \varphi(\alpha_i) \leq \frac{\delta}{\alpha_i}\} < M,$$

$$k := \max\{i : \|x_i - x_j\| \leq 4C\frac{\delta}{\alpha_j}, j = 0, 1, 2, \cdots, i\}. \tag{10.106}$$

Then $l \leq k$ and

$$\|\hat{x} - x_k\| \leq c\psi^{-1}(\delta),$$

where $c = 6C\mu$.

Proof. To see that $l \leq k$, it is enough to show that, for each $i \in \{1, 2, \cdots, M\}$,

$$\varphi(\alpha_i) \leq \frac{\delta}{\alpha_i} \implies \|x_i - x_j\| \leq 4C\frac{\delta}{\alpha_j}, \qquad \forall j = 0, 1, \cdots, i.$$

For $j \leq i$, by (10.102) we have

$$\|x_i - x_j\| \leq \|x_i - \hat{x}\| + \|\hat{x} - x_j\|$$

$$\leq C(\varphi(\alpha_i) + \frac{\delta}{\alpha_i}) + C(\varphi(\alpha_j) + \frac{\delta}{\alpha_j})$$

$$\leq 2C\frac{\delta}{\alpha_i} + 2C\frac{\delta}{\alpha_j} \leq 4C\frac{\delta}{\alpha_j}.$$

Thus the relation $l \leq k$ is proved. Next we observe that

$$\|\hat{x} - x_k\| \leq \|\hat{x} - x_l\| + \|x_l - x_k\|$$

$$\leq C(\varphi(\alpha_l) + \frac{\delta}{\alpha_l}) + 4C\frac{\delta}{\alpha_l} \leq 6C\frac{\delta}{\alpha_l}.$$

Now since $\alpha_\delta \leq \alpha_{l+1} \leq \mu\alpha_l$, it follows that

$$\frac{\delta}{\alpha_l} \leq \mu\frac{\delta}{\alpha_\delta} = \mu\varphi(\alpha_\delta) = \mu\psi^{-1}(\delta).$$

This completes the proof of the Theorem. $\square$

We finish this section by an algorithm for the determination of a parameter fulfilling the balancing principle (10.106) and also provide a starting point for the iteration (10.80) approximating the unique solution x_α^δ of (10.78). The choice of the starting point involves the following steps:

- Choose $\alpha_0 = \sqrt{\delta}$ and $\mu > 1$.
- Choose $x_0 \in D(F)$ such that $\|x_0 - \hat{x}\| \leq \rho$.
- Choose η satisfying (10.88).

The choice of the stopping index n_M involves the following step:

- Choose n_M such that $n_M = \min\{n : r^n \leq \delta\}$.

Finally the adaptive algorithm associated with the choice of the parameter specified in Theorem 10.3.6 involves the following steps:

Algorithm 10.3.1.

- Set $i \leftarrow 0$.
- Solve $x_i := x_{n_M, \alpha_i}^\delta$ by using the iteration (10.80).
- If $\|x_i - x_j\| > 4C\dfrac{\sqrt{\delta}}{\mu^j}$, $j \leq i$, then take $k = i - 1$.
- Set $i = i + 1$ and return to step 2.

Let H be a bounded subset of positive real numbers such that zero is a limit point of H. Let also $\{P_h\}_{h \in H}$ be a family of orthogonal projections from X into itself. Let

$$\Gamma_h := \|(I - P_h)F'(x_0)\| \tag{10.107}$$

and

$$\gamma_h := \|F'(P_h x_0)(I - P_h)\|. \tag{10.108}$$

We assume that

$$b_h := \|(I - P_h)x_0\| \to 0, \tag{10.109}$$

as $h \to 0$. The above assumption is satisfied if $P_h \to I$ pointwise. Let $(t_{n,h}), n \geq 0$ be defined iteratively by $t_{0,h} = 0, t_{1,h} = \eta_h,$

$$t_{n+1,h} = t_{n,h} + (1 + \frac{\gamma_h}{\alpha})\frac{l_0 \eta_h}{(1 - r_h)}(t_{n,h} - t_{n-1,h}), \tag{10.110}$$

where l_0, α and $r_h \in [0, 1)$ are non-negative numbers, with $(1 + \frac{\gamma_h}{\alpha})\frac{l_0}{(1 - r_h)}\eta_h \leq r_h$. We need the following Lemma, proof of which is analogous to the proof of Lemma 10.3.1, so we ignore the proof.

Lemma 10.3.3. *Assume there exist non-negative numbers l_0, α and $r_h \in [0, 1)$ such that*

$$(1 + \frac{\gamma_h}{\alpha})\frac{l_0}{(1 - r_h)}\eta_h \leq r_h. \tag{10.111}$$

Then the sequence $(t_{n,h})$ defined in (10.110) is increasing, bounded above by $t_h^{**} := \frac{\eta_h}{1 - r_h}$, and converges to some t_h^* such that $0 < \frac{\eta_h}{1 - r_h}$. Moreover, for $n \geq 0$;

$$0 \leq t_{n+1,h} - t_{n,h} \leq r_h(t_{n,h} - t_{n-1,h}) \leq r_h^n \eta_h \tag{10.112}$$

and

$$t_h^* - t_{n,h} \leq \frac{r_h^n}{1 - r_h}\eta_h. \tag{10.113}$$

We considered an iteratively regularized projection method;

$$x_{n+1,\alpha}^{h,\delta} := x_{n,\alpha}^{h,\delta} - (P_h F'(P_h x_0) + \alpha I)^{-1} P_h(F(x_{n,\alpha}^{h,\delta}) - y^\delta + \alpha(x_{n,\alpha}^{h,\delta} - x_0)), \tag{10.114}$$

where $x_{0,\alpha}^{h,\delta} := P_h x_0$, for obtaining a sequence $(x_{n,\alpha}^{h,\delta})$ in a finite dimensional subspace X_h of X. Now we shall prove that the sequence $(t_{n,h})$ is a majorizing sequence of the sequence $(x_{n,\alpha}^{h,\delta})$. Let

$$(1 + \frac{\gamma_h}{\alpha})(\frac{l_0}{2}(b_h + \rho)^2 + b_h + \rho) + \frac{\delta}{\alpha} \leq \eta_h \tag{10.115}$$

$$\leq \min\{\frac{r_h(1 - r_h)}{l_0(1 + \gamma_h/\alpha)}, r_0(1 - r_h)\}.$$

Theorem 10.3.7. *Let the assumptions in Lemma 10.3.3 with η_h as in (10.115) and Assumption 6 be satisfied. Then the sequence $(t_{n,h})$ defined in (10.110) is a majorizing sequence of sequence $(x_{n,\alpha}^{h,\delta})$ defined in (10.114) and $x_{n,\alpha}^{h,\delta} \in B_{t_h^*}(P_h x_0)$ for all $n \geq 0$.*

Proof. Let

$$G(x) = x - R_\alpha(P_h x_0)^{-1}[F(x) - y^\delta + \alpha(x - x_0)],$$

where $R_\alpha(P_h x_0)^{-1} = (P_h F'(P_h x_0)P_h + \alpha P_h)^{-1}$. Then since $R_\alpha(P_h x_0)^{-1} = R_\alpha(P_h x_0)^{-1}P_h = P_h R_\alpha(P_h x_0)^{-1}$; for $u, v \in B_{t_h^*}(P_h x_0)$,

$$\begin{aligned}
G(u) - G(v) &= u - v - R_\alpha(P_h x_0)^{-1}[F(u) - y^\delta + \alpha(u - x_0)] \\
&\quad + R_\alpha(P_h x_0)^{-1}[F(v) - y^\delta + \alpha(v - x_0)] \\
&= R_\alpha(P_h x_0)^{-1}[R_\alpha(P_h x_0)(u - v) - (F(u) - F(v))] + \alpha R_\alpha(P_h x_0)^{-1}(v - u) \\
&= R_\alpha(P_h x_0)^{-1}[F'(P_h x_0)P_h(u - v) - (F(u) - F(v)) + \alpha(u - v)] \\
&\quad + \alpha R_\alpha(P_h x_0)^{-1}(v - u) \\
&= R_\alpha(P_h x_0)^{-1}[F'(P_h x_0)P_h(u - v) - (F(u) - F(v))].
\end{aligned}$$

Now since $G(x_{n,\alpha}^{h,\delta}) = x_{n+1,\alpha}^{h,\delta}$ and $P_h(x_{n,\alpha}^{h,\delta} - x_{n-1,\alpha}^{h,\delta}) = (x_{n,\alpha}^{h,\delta} - x_{n-1,\alpha}^{h,\delta})$ we have

$$\begin{aligned}
(x_{n+1,\alpha}^{h,\delta} - x_{n,\alpha}^{h,\delta}) &= G(x_{n,\alpha}^{h,\delta}) - G(x_{n-1,\alpha}^{h,\delta}) \\
&= R_\alpha(P_h x_0)^{-1}[F'(P_h x_0)(x_{n,\alpha}^{h,\delta} - x_{n-1,\alpha}^{h,\delta}) - (F(x_{n,\alpha}^{h,\delta}) - F(x_{n-1,\alpha}^{h,\delta}))] \\
&= R_\alpha(P_h x_0)^{-1}F'(P_h x_0) \int_0^1 \Phi(x_{n,\alpha}^{h,\delta} + t(x_{n-1,\alpha}^{h,\delta} - x_{n,\alpha}^{h,\delta}), P_h x_0, x_{n-1,\alpha}^{h,\delta} - x_{n,\alpha}^{h,\delta})dt \\
&= R_\alpha(P_h x_0)^{-1}[F'(P_h x_0)P_h + F'(P_h x_0)(I - P_h)] \\
&\quad \times \int_0^1 \Phi(x_{n,\alpha}^{h,\delta} + t(x_{n-1,\alpha}^{h,\delta} - x_{n,\alpha}^{h,\delta}), P_h x_0, x_{n-1,\alpha}^{h,\delta} - x_{n,\alpha}^{h,\delta})dt.
\end{aligned}$$

The last but one step follows from Lemma 10.3.2. So by Assumption 6 and the relation

$$\|R_\alpha(P_h x_0)^{-1}[F'(P_h x_0)P_h + F'(P_h x_0)(I - P_h)]\| \leq 1 + \frac{\gamma h}{\alpha}, \tag{10.116}$$

we have that

$$\|x_{n+1,\alpha}^{h,\delta} - x_{n,\alpha}^{h,\delta}\| \leq (1 + \frac{\gamma h}{\alpha})l_0\|x_{n,\alpha}^{h,\delta} + t(x_{n-1,\alpha}^{h,\delta} - x_{n,\alpha}^{h,\delta}) - P_h x_0\| \|x_{n,\alpha}^{h,\delta} - x_{n-1,\alpha}^{h,\delta}\|. \tag{10.117}$$

Now we shall prove that the sequence $(t_{n,h})$ defined in (10.110) is a majorizing sequence of the sequence $(x_{n,\alpha}^{h,\delta})$ and $x_{n,\alpha}^{h,\delta} \in B_{t_h^*}(P_h x_0)$, for all $n \geq 0$. Note that $F(\hat{x}) = y$, so

$$\begin{aligned}
\|x_{1,\alpha}^{h,\delta} - P_h x_0\| &= \|(P_h F'(P_h x_0) + \alpha I)^{-1}P_h(F(P_h x_0) - y^\delta)\| \\
&= \|(P_h F'(P_h x_0) + \alpha I)^{-1}P_h(F(P_h x_0) - y + y - y^\delta)\|
\end{aligned}$$

$$= \|(P_h F'(P_h x_0) + \alpha I)^{-1} P_h (F(P_h x_0) - F(\hat{x}) + y - y^\delta)\|$$

$$= \|(P_h F'(P_h x_0) + \alpha I)^{-1} P_h (F(P_h x_0) - F(\hat{x}) - F'(P_h x_0)(P_h x_0 - \hat{x})$$

$$+ F'(P_h x_0)(P_h x_0 - \hat{x}) + y - y^\delta)\|$$

$$\leq \|(P_h F'(P_h x_0) + \alpha I)^{-1} P_h (F(P_h x_0) - F(\hat{x}) - F'(P_h x_0)(P_h x_0 - \hat{x}))\|$$

$$+ \|(P_h F'(P_h x_0) + \alpha I)^{-1} P_h F'(P_h x_0)(P_h x_0 - \hat{x})\|$$

$$+ \|(P_h F'(P_h x_0) + \alpha I)^{-1} P_h (y - y^\delta)\|$$

$$\leq \|(P_h F'(P_h x_0) + \alpha I)^{-1} P_h F'(P_h x_0)$$

$$\times \int_0^1 \Phi(\hat{x} + t(P_h x_0 - \hat{x}), P_h x_0, (P_h x_0 - \hat{x})) dt\|$$

$$+ \|(P_h F'(P_h x_0) + \alpha I)^{-1} P_h F'(P_h x_0)(P_h x_0 - \hat{x})\| + \frac{\delta}{\alpha}$$

$$\leq \|(P_h F'(P_h x_0) + \alpha I)^{-1} P_h [F'(P_h x_0) P_h + F'(P_h x_0)(I - P_h)]$$

$$\times \int_0^1 \Phi(\hat{x} + t(P_h x_0 - \hat{x}), P_h x_0, (P_h x_0 - \hat{x})) dt\|$$

$$+ \|(P_h F'(P_h x_0) + \alpha I)^{-1} P_h [F'(P_h x_0) P_h + F'(P_h x_0)(I - P_h)](P_h x_0 - \hat{x})\| + \frac{\delta}{\alpha}$$

$$\leq (1 + \frac{\gamma_h}{\alpha})(\frac{l_0}{2}\|P_h x_0 - \hat{x}\|^2 + \|P_h x_0 - \hat{x}\|) + \frac{\delta}{\alpha}$$

$$\leq (1 + \frac{\gamma_h}{\alpha})(\frac{l_0}{2}(b_h + \rho)^2 + b_h + \rho) + \frac{\delta}{\alpha} \leq \eta_h.$$

The last but one step follows from Assumption 6, (10.116) and the inequality $\|P_h x_0 - \hat{x}\| \leq b_h + \rho$. So $\|x_{1,\alpha}^{h,\delta} - P_h x_0\| \leq t_{1,h} - t_{0,h}$. Assume that

$$\|x_{i+1,\alpha}^{h,\delta} - x_{i,\alpha}^{h,\delta}\| \leq t_{i+1,h} - t_{i,h}, \qquad \forall i \leq k, \tag{10.118}$$

for some k. Then

$$\|x_{k+1,\alpha}^{h,\delta} - P_h x_0\| = \|x_{k+1,\alpha}^{h,\delta} - x_{k,\alpha}^{h,\delta} + x_{k,\alpha}^{h,\delta} - x_{k-1,\alpha}^{h,\delta} + \cdots + x_{1,\alpha}^{h,\delta} - P_h x_0\|$$

$$\leq \|x_{k+1,\alpha}^{h,\delta} - x_{k,\alpha}^{h,\delta}\| + \|x_{k,\alpha}^{h,\delta} - x_{k-1,\alpha}^{h,\delta}\| + \cdots + \|x_{1,\alpha}^{h,\delta} - P_h x_0\|$$

$$\leq t_{k+1,h} - t_{k,h} + t_{k,h} - t_{k-1,h} + \cdots + t_{1,h} - t_{0,h}$$

$$= t_{k+1,h} \leq t_h^*.$$

So $x_{i+1,\alpha}^{h,\delta} \in B_{t_h^*}(P_h x_0)$ for all $i \leq k$, and hence, $x_{k+1,\alpha}^{h,\delta} + t(x_{k,\alpha}^{h,\delta} - x_{k+1,\alpha}^{h,\delta}) \in B_{t_h^*}(P_h x_0)$. Therefore by (10.117) and (10.118) we have

$$\|x_{k+2,\alpha}^{h,\delta} - x_{k+1,\alpha}^{h,\delta}\| \leq l_0(1 + \frac{\gamma_h}{\alpha}) t_h^* \|x_{k+1,\alpha}^{h,\delta} - x_{k,\alpha}^{h,\delta}\|$$

$$\leq l_0(1 + \frac{\gamma_h}{\alpha}) \frac{\eta_h}{(1 - r_h)}(t_{k+1,h} - t_{k,h})$$

$$= t_{k+2,h} - t_{k+1,h}.$$

Thus by induction $\|x_{n+1,\alpha}^{h,\delta} - x_{n,\alpha}^{h,\delta}\| \leq t_{n+1,h} - t_{n,h}$ for all $n \geq 0$ and hence $(t_{n,h}), n \geq 0$ is a majorizing sequence of the sequence $(x_{n,\alpha}^{h,\delta})$. In particular $\|x_{n,\alpha}^{h,\delta} - P_h x_0\| \leq t_{n,h} \leq t_h^*$, i.e., $x_{n,\alpha}^{h,\delta} \in B_{t_h^*}(P_h x_0)$, for all $n \geq 0$. Hence

$$\|x_{n,\alpha}^{h,\delta} - P_h x_0\| \leq t_h^* \leq \frac{\eta_h}{1 - r_h}. \tag{10.119}$$

This completes the proof of the Theorem. $\qquad\square$

Let

$$\tilde{\tilde{r}} := \max\{r, r_h\} \tag{10.120}$$

and

$$q := \frac{1}{2}[2\tilde{\tilde{r}} + l_0 b_h]. \tag{10.121}$$

Note that for $0 < b_h < \dfrac{2(1 - \tilde{\tilde{r}})}{l_0}$, $q < 1$.

Theorem 10.3.8. *Let $x_{n,\alpha}^{h,\delta}$ be as in (10.114) and $x_{n,\alpha}^{\delta}$ be as in (10.80). Let assumptions in Theorems 10.3.1 and 10.3.7 hold. Then we have the following estimate*

$$\|x_{n,\alpha}^{h,\delta} - x_{n,\alpha}^{\delta}\| \leq q^n b_h + \left(\frac{\Gamma_h + l_0\|F'(x_0)\|b_h}{\alpha}\right)\frac{q^n}{(q - r_h)}\eta_h.$$

Proof. Note that

$$
\begin{aligned}
x_{n,\alpha}^{h,\delta} - x_{n,\alpha}^{\delta} &= x_{n-1,\alpha}^{h,\delta} - x_{n-1,\alpha}^{\delta} - (P_h F'(P_h x_0) + \alpha I)^{-1} P_h \\
&\quad \times (F(x_{n-1,\alpha}^{h,\delta}) - y^\delta + \alpha(x_{n-1,\alpha}^{h,\delta} - x_0)) \\
&\quad + (F'(x_0) + \alpha I)^{-1}(F(x_{n-1,\alpha}^{\delta}) - y^\delta + \alpha(x_{n-1,\alpha}^{\delta} - x_0)) \\
&= x_{n-1,\alpha}^{h,\delta} - x_{n-1,\alpha}^{\delta} - [(P_h F'(P_h x_0) + \alpha I)^{-1} P_h - (F'(x_0) + \alpha I)^{-1}] \\
&\quad \times (F(x_{n-1,\alpha}^{h,\delta}) - y^\delta + \alpha(x_{n-1,\alpha}^{h,\delta} - x_0)) \\
&\quad - (F'(x_0) + \alpha I)^{-1}[F(x_{n-1,\alpha}^{h,\delta}) - F(x_{n-1,\alpha}^{\delta}) + \alpha(x_{n-1,\alpha}^{h,\delta} - x_{n-1,\alpha}^{\delta})] \\
&= (F'(x_0) + \alpha I)^{-1}[F'(x_0)(x_{n-1,\alpha}^{h,\delta} - x_{n-1,\alpha}^{\delta}) - (F(x_{n-1,\alpha}^{h,\delta}) - F(x_{n-1,\alpha}^{\delta}))] \\
&\quad - (F'(x_0) + \alpha I)^{-1}[F'(x_0)P_h - P_h F'(P_h x_0)P_h](P_h F'(P_h x_0) + \alpha I)^{-1} \\
&\quad \times P_h[(F(x_{n-1,\alpha}^{h,\delta}) - y^\delta + \alpha(x_{n-1,\alpha}^{h,\delta} - x_0))] \\
&= (F'(x_0) + \alpha I)^{-1}[F'(x_0)(x_{n-1,\alpha}^{h,\delta} - x_{n-1,\alpha}^{\delta}) - (F(x_{n-1,\alpha}^{h,\delta}) - F(x_{n-1,\alpha}^{\delta}))] \\
&\quad - (F'(x_0) + \alpha I)^{-1}[F'(x_0) - P_h F'(x_0) + P_h F'(x_0) - P_h F'(P_h x_0)] \\
&\quad \times (x_{n,\alpha}^{h,\delta} - x_{n-1,\alpha}^{h,\delta}) =: \Gamma_1 - \Gamma_2,
\end{aligned}
$$

where

$$\Gamma_1 = (F'(x_0) + \alpha I)^{-1}[F'(x_0)(x_{n-1,\alpha}^{h,\delta} - x_{n-1,\alpha}^{\delta}) - (F(x_{n-1,\alpha}^{h,\delta}) - F(x_{n-1,\alpha}^{\delta}))]$$

and

$$
\begin{aligned}
\Gamma_2 &= (F'(x_0) + \alpha I)^{-1}[F'(x_0) - P_h F'(x_0) \\
&\quad + P_h F'(x_0) - P_h F'(P_h x_0)](x_{n,\alpha}^{h,\delta} - x_{n-1,\alpha}^{h,\delta}).
\end{aligned}
$$

Note that by Lemma 10.3.2

$$\|\Gamma_1\| = \|(F'(x_0) + \alpha I)^{-1}[F'(x_0)(x_{n-1,\alpha}^{h,\delta} - x_{n-1,\alpha}^{\delta}) - (F(x_{n-1,\alpha}^{h,\delta}) - F(x_{n-1,\alpha}^{\delta}))]\|$$

$$\leq \|(F'(x_0) + \alpha I)^{-1} F'(x_0) \int_0^1 \Phi(x_{n-1,\alpha}^{h,\delta} + t(x_{n-1,\alpha}^{\delta}$$

$$-x_{n-1,\alpha}^{h,\delta}), x_0, x_{n-1,\alpha}^{\delta} - x_{n-1,\alpha}^{h,\delta}) dt\|$$

$$\leq l_0 \int_0^1 \|x_0 - (x_{n-1,\alpha}^{h,\delta} + t(x_{n-1,\alpha}^{\delta} - x_{n-1,\alpha}^{h,\delta}))\| \|x_{n-1,\alpha}^{\delta} - x_{n-1,\alpha}^{h,\delta}\| dt$$

$$\leq l_0 \int_0^1 [t\|x_0 - x_{n-1,\alpha}^{\delta}\| + (1-t)\|P_h x_0 - x_{n-1,\alpha}^{h,\delta}\|$$

$$+(1-t)\|P_h x_0 - x_0\|] \|x_{n-1,\alpha}^{\delta} - x_{n-1,\alpha}^{h,\delta}\| dt$$

$$\leq \frac{l_0}{2}[\frac{\eta}{1-r} + \frac{\eta_h}{1-r_h} + b_h] \|x_{n-1,\alpha}^{h,\delta} - x_{n-1,\alpha}^{\delta}\|$$

$$\leq \frac{1}{2}[r + r_h + l_0 b_h] \|x_{n-1,\alpha}^{h,\delta} - x_{n-1,\alpha}^{\delta}\|$$

$$\leq \frac{1}{2}[2\tilde{\tilde{r}} + l_0 b_h] \|x_{n-1,\alpha}^{h,\delta} - x_{n-1,\alpha}^{\delta}\| \leq q\|x_{n-1,\alpha}^{h,\delta} - x_{n-1,\alpha}^{\delta}\|, \tag{10.122}$$

and by Assumption 6

$$\|\Gamma_2\| = \|(F'(x_0) + \alpha I)^{-1}[(I - P_h)F'(x_0)$$

$$-P_h(F'(P_h x_0) - F'(x_0))](x_{n,\alpha}^{h,\delta} - x_{n-1,\alpha}^{h,\delta})\|$$

$$\leq \|(F'(x_0) + \alpha I)^{-1}(I - P_h)F'(x_0)\|$$

$$+\|(F'(x_0) + \alpha I)^{-1} P_h F'(x_0)\Phi(P_h x_0, x_0, x_{n,\alpha}^{h,\delta} - x_{n-1,\alpha}^{h,\delta})\|$$

$$\leq (\frac{\Gamma_h + l_0\|F'(x_0)\|b_h}{\alpha}) \|x_{n,\alpha}^{h,\delta} - x_{n-1,\alpha}^{h,\delta}\|. \tag{10.123}$$

Therefore by (10.122), (10.122) and (10.123) we have

$$\|x_{n,\alpha}^{h,\delta} - x_{n,\alpha}^{\delta}\| \leq q\|x_{n-1,\alpha}^{h,\delta} - x_{n-1,\alpha}^{\delta}\| + \frac{\Gamma_h + l_0\|F'(x_0)\|b_h}{\alpha} \|x_{n,\alpha}^{h,\delta} - x_{n-1,\alpha}^{h,\delta}\|$$

$$\leq q^n b_h + \frac{\Gamma_h + l_0\|F'(x_0)\|b_h}{\alpha} \eta_h (r_h^{n-1} + q r_h^{n-2} + \cdots + q^{n-1})$$

$$\leq q^n b_h + (\frac{\Gamma_h + l_0\|F'(x_0)\|b_h}{\alpha}) \frac{q^n}{(q - r_h)} \eta_h.$$

This completes the proof of the Theorem. $\qquad\square$

It is known (cf. Proposition 10.3.1) that

$$\|x_\alpha^\delta - x_\alpha\| \leq \frac{\delta}{\alpha} \tag{10.124}$$

and (cf. Theorem 10.3.2) that

$$\|x_\alpha - \hat{x}\| \leq (l_0 r_0 + 1)\varphi(\alpha), \tag{10.125}$$

where x_α is the unique solution of $F(x) + \alpha(x - x_0) = y$.

Combining the estimates in Theorem 10.3.1, Theorem 10.3.8, equation (10.124) and equation (10.125) we obtain the following Theorem.

Theorem 10.3.9. *Let $x_{n,\alpha}^{h,\delta}$ be as in (10.114) and let the assumptions in Theorems 10.3.1 and 10.3.8 be satisfied. Then we have the following;*

$$\|x_{n,\alpha}^{h,\delta} - \hat{x}\| \le q^n b_h + \left(\frac{\Gamma_h + l_0\|F'(x_0)\|b_h}{\alpha}\right)\frac{q^n}{(q-r_h)}\eta_h + \frac{r^n\eta}{1-r} + \frac{\delta}{\alpha} + (l_0 r_0 + 1)\varphi(\alpha).$$

$$(10.126)$$

Let

$$n_\delta := \min\{n : \max\{q^n, r^n\} \le \delta\} \qquad (10.127)$$

and let

$$C_m := \max\{b_h + \frac{\Gamma_h + l_0\|F'(x_0)\|b_h}{(q-r_h)}\eta_h + \frac{\eta}{1-r} + 1, (l_0 r_0 + 1)\}. \qquad (10.128)$$

Theorem 10.3.10. *Let $x_{n,\alpha}^{h,\delta}$ be as in (10.114) and let the assumptions in Theorem 10.3.1 and Theorem 10.3.8 be satisfied. Let n_δ be as in (10.127) and C_m be as in (10.128). Then for all $0 < \alpha \le 1$ we have the following;*

$$\|x_{n_\delta,\alpha}^{h,\delta} - \hat{x}\| \le C_m(\varphi(\alpha) + \frac{\delta}{\alpha}). \qquad (10.129)$$

We observe that the error $\varphi(\alpha) + \dfrac{\delta}{\alpha}$ in (10.129) is of optimal order if $\alpha_\delta := \alpha(\delta)$ satisfies $\alpha_\delta\varphi(\alpha_\delta) = \delta$. Now using the function $\psi(\lambda) := \lambda\varphi^{-1}(\lambda), 0 < \lambda \le a$ we have $\delta = \alpha_\delta\varphi(\alpha_\delta) = \psi(\varphi(\alpha_\delta))$, so that $\alpha_\delta = \varphi^{-1}(\psi^{-1}(\delta))$. Hence by (10.129) we have the following.

Theorem 10.3.11. *Let $\psi(\lambda) := \lambda\varphi^{-1}(\lambda)$ for $0 < \lambda \le a$ and assumptions in Theorem 10.3.10 hold. For $\delta > 0$, let $\alpha =: \alpha_\delta = \varphi^{-1}(\psi^{-1}(\delta))$. Let n_δ be as in (10.127). Then*

$$\|x_{n_\delta,\alpha}^{h,\delta} - \hat{x}\| = O(\psi^{-1}(\delta)).$$

We will present a parameter choice rule based on the adaptive method studied in Refs. [324, 346]. The regularization parameter α is selected from the finite set

$$D_M(\alpha) := \{\alpha_i = \mu^i\alpha_0, i = 0, 1, \cdots, M\}, \qquad (10.130)$$

where $\mu > 1$ and M is such that $\alpha_M < 1 \le \alpha_{M+1}$. We choose $\alpha_0 := \sqrt{\delta}$, because in general $\varphi(\lambda) = \lambda^\nu, 0 < \nu \le 1$ and in this case the best possible error estimate is order $O(\sqrt{\delta})$ and from Theorem 10.3.11, it follows that such an accuracy cannot be guaranteed for $\alpha < \sqrt{\delta}$. Let

$$n_M := \min\{n : \max\{q^n, r^n\} \le \delta\} \qquad (10.131)$$

and let $x_i := x_{n_M,\alpha_i}^{h,\delta}$.

We are going to consider in this section the parameter choice strategy, i.e., we select $\alpha = \alpha_i$ from $D_M(\alpha)$ and operates only with corresponding x_i, $\quad i = 0, 1, \cdots, M$.

Theorem 10.3.12. *(cf. Theorem 10.3.6) Assume that there exists $i \in \{0, 1, 2, \cdots, M\}$ such that $\varphi(\alpha_i) \leq \dfrac{\delta}{\alpha_i}$. Let assumptions of Theorem 10.3.10 and Theorem 10.3.11 hold and let*

$$l := \max\{i : \varphi(\alpha_i) \leq \frac{\delta}{\alpha_i}\} < M,$$

$$k := \max\{i : \|x_i - x_j\| \leq 4C_m \frac{\delta}{\alpha_j}, j = 0, 1, 2, \cdots, i\}. \tag{10.132}$$

Then $l \leq k$ and

$$\|\hat{x} - x_k\| \leq c\psi^{-1}(\delta),$$

where $c = 6C_m\mu$.

In this section we provide an algorithm for the determination of a parameter fulfilling the balancing principle (10.132) and also provide a starting point for the iteration (10.114) approximating the unique solution x_α^δ of (10.78). The choice of the starting point involves the following steps:

- Choose $\alpha_0 = \sqrt{\delta}$, $\mu > 1$ and $q < 1$.
- Choose $x_0 \in D(F)$ such that $\|x_0 - \hat{x}\| \leq \rho$ and η_h satisfying (10.115).

The choice of the stopping index n_M involves the following step:

- Choose n_M such that $n_M = \min\{n : \max\{q^n, r^n\} \leq \delta\}$.

Finally the adaptive algorithm associated with the choice of the parameter specified in Theorem 10.3.12 involves the following steps:

Algorithm 10.3.2.

- Set $i \leftarrow 0$.
- Solve $x_i := x_{n_M, \alpha_i}^{h, \delta}$ by using the iteration (10.114).
- If $\|x_i - x_j\| > 4C_m \dfrac{\sqrt{\delta}}{\mu^j}$, $j \leq i$, then take $k = i - 1$.
- Set $i = i + 1$ and return to step 2.

We consider some simple examples satisfying the assumptions made in this section and present a few computed examples.

We consider the operator $F : L^2[0, 1] \to L^2[0, 1]$ defined by (see Example 6.1 in Ref. [332])

$$F(x)(s) = K^*K(x)(s) + f(s), \qquad x, f \in L^2[0, 1], s \in [0, 1], \tag{10.133}$$

where $K : L^2[0,1] \to L^2[0,1]$ is a compact linear operator such that the range of K denoted by $R(K)$ is not closed in $L^2[0,1]$. Then the equation $F(x) = y$ is ill-posed as K is compact with non-closed range. The Fréchet derivative $F'(.)$ of F is given by

$$F'(x)z = K^*Kz, \qquad \forall x, z \in L^2[0,1]. \tag{10.134}$$

So F is monotone on $L^2[0,1]$. Further for $x, y, z \in L^2[0,1]$

$$[F'(x) - F'(y)]z = 0. \tag{10.135}$$

Hence Assumption 6 holds trivially. Again note that, since $\Phi(x, y, z) = 0 \leq l_0\|z\|\|x - y\|,\ \forall l_0 \geq 0$ we can choose η_h large enough in step 2 of the algorithm. Further, due to (10.134) the iteration $x^{h,\delta}_{m+1,\alpha}$ needs only one step to compute. This can be seen as follows:

$$x^{h,\delta}_{m+1,\alpha} = x^{h,\delta}_{m,\alpha} - (P_h F'(P_h x_0) + \alpha I)^{-1} P_h[F(x^{h,\delta}_{m,\alpha}) - y^\delta + \alpha(x^{h,\delta}_{m,\alpha} - x_0)],$$

i.e.,

$$\begin{aligned}
(P_h F'(P_h x_0) + \alpha I)P_h x^{h,\delta}_{m+1,\alpha} &= (P_h F'(P_h x_0) + \alpha I)P_h x^{h,\delta}_{m,\alpha} \\
&\quad - P_h[F(x^{h,\delta}_{m,\alpha}) - y^\delta + \alpha(x^{h,\delta}_{m,\alpha} - x_0)] \\
&= (P_h K^*K + \alpha I)P_h x^{h,\delta}_{m,\alpha} - P_h[K^*K x^{h,\delta}_{m,\alpha} \\
&\quad + f - y^\delta + \alpha(x^{h,\delta}_{m,\alpha} - x_0)] \\
&= -P_h(f - y^\delta - \alpha x_0). \tag{10.136}
\end{aligned}$$

Now we shall give the details for implementing the algorithm given before in this section. Let (V_n) be a sequence of finite dimensional subspaces of X and let $P_h, h = 1/n$ denote the orthogonal projection on X with range $R(P_h) = V_n$. We assume that $dim V_n = n + 1$, and $\|P_h x - x\| \to 0$ as $h \to 0$ for all $x \in X$. Let $\{v_1, v_2, \cdots, v_{n+1}\}$ be a basis of $V_n, n = 1, 2, \cdots$.

Note that $x^{h,\delta}_{m+1,\alpha} \in V_n$. Thus $x^{h,\delta}_{m+1,\alpha}$ is of the form $\sum_{i=1}^{n+1} \lambda_i v_i$ for some scalars $\lambda_1, \lambda_2, \cdots, \lambda_{n+1}$. It can be seen that $x^{h,\delta}_{m+1,\alpha}$ is a solution of (10.136) if and only if $\bar{\lambda} = (\lambda_1, \lambda_2, \cdots, \lambda_{n+1})^T$ is the unique solution of

$$(M_n + \alpha B_n)\bar{\lambda} = \bar{a}, \tag{10.137}$$

where

$$M_n = ((\langle Kv_i, Kv_j \rangle)), i, j = 1, 2, \cdots, n + 1$$

$$B_n = ((\langle v_i, v_j \rangle)), i, j = 1, 2, \cdots, n + 1$$

and

$$\bar{a} = ((\langle P_h(y^\delta + \alpha x_0 - f), v_i \rangle))^T, i = 1, 2, \cdots, n + 1.$$

Note that (10.137) is uniquely solvable because M_n is a positive definite matrix (i.e., $xM_n x^T > 0$ for all non-zero vector x) and B_n is an invertible matrix. In order to illustrate our method, we consider $X = Y = L^2[0,1]$ and $K : L^2[0,1] \to L^2[0,1]$

as the Fredholm integral operator as defined in Example 10.2.3. We apply the Algorithm in 10.3.2 by choosing V_n as the space of linear splines in a uniform grid of $n+1$ points in $[0, 1]$. Specifically for fixed n we consider $t_i = \dfrac{i-1}{n}, i = 1, 2, \cdots, n+1$ as the grid points. We take the basis function $v_i, i = 1, 2, \cdots, n+1$ of V_n as follows:

$$v_1(t) = \begin{cases} \dfrac{t_2 - t}{t_2} & \text{if } 0 = t_1 \le t \le t_2 \\ 0 & \text{if } t_2 \le t \le t_{n+1} = 1 \end{cases}$$

for $j = 2, 3, \cdots, n$,

$$v_j(t) = \begin{cases} 0 & \text{if } 0 = t_1 \le t \le t_{j-1} \\ \dfrac{t - t_{j-1}}{t_j - t_{j-1}} & \text{if } t_{j-1} \le t \le t_j \\ \dfrac{t_{j+1} - t}{t_{j+1} - t_j} & \text{if } t_j \le t \le t_{j+1} \\ 0 & \text{if } t_{j+1} \le t \le t_{n+1} = 1 \end{cases}$$

and

$$v_{n+1}(t) = \begin{cases} 0 & \text{if } 0 \le t \le t_n \\ \dfrac{t - t_n}{t_{n+1} - t_n} & \text{if } t_n \le t \le t_{n+1}. \end{cases}$$

Let P_h be the orthogonal projection onto V_n. We note that for $x \in C[0, 1]$

$$\|P_h x - x\|_2 = dist(x, R(P_h))$$
$$\le \|\pi_n x - x\|_2 \le \|\pi_n x - x\|_\infty$$

where π_n is the (piecewise linear) interpolatory projection onto V_n. It is known that $\|\pi_n x - x\|_\infty \to 0$ as $n \to \infty$. Therefore using the fact that $C[0, 1]$ is dense in $L^2[0, 1]$, it follows that $\|P_h x - x\|_2 \to 0$ for all $x \in L^2[0, 1]$. The elements $K v_i, i = 1, 2, \cdots, n+1$, the entries of the matrix B_n, M_n and $\bar{a}$ are computed explicitly. For the operator K defined by our operator equation, $\Gamma_h = \gamma_h = \|(I - P_h)F'(x_0)\| = \|(I - P_h)K^*K\| = O(n^{-2})$ (see Ref. [243]).

Example 10.3.2. In this example we take $y = \dfrac{1}{720}(26 + s^6 - 6s^5 + 15s^4 - 36s) + f(s)$ where $f(s) = s^2$ and $x_0 = 0$. Then the exact solution is $\hat{x} = \dfrac{1}{2}(s-1)^2$. Since $\hat{x} - x_0 = \hat{x} = K^*1 \in R(K^*) = R(F'(\hat{x})^{1/2}), \varphi(\lambda) = \lambda^{1/2}$ and hence $\psi^{-1}(\delta) = \varphi(\alpha_\delta) = (\delta)^{1/3}$. According to the theory, we have that $e_k := \|\hat{x} - x_k\| \le c\psi^{-1}(\delta)$, where $c = 6C_m\mu$. Let $y^\delta := y + \delta$. The result are given in Table 10.2 and Figures 10.5–10.8.

Remark 10.3.2. The last column of all Tables show that $e_k = O(\psi^{-1}(\delta))$. During computation we observe that due to the round off error k, e_k remains as constant for large values of n.

Table 10.2   Errors Table for $\delta = 0.001$ and $mu = 1.002$

n	k	e_k	$\dfrac{e_k}{\psi^{-1}(\delta)}$
4	100	0.0377	0.1385
8	99	0.0385	0.1400
16	99	0.0385	0.1399
32	99	0.0385	0.1400
64	99	0.0385	0.1400
128	99	0.0385	0.1400
256	99	0.0385	0.1400
512	99	0.0385	0.1400
1024	99	0.0385	0.1400

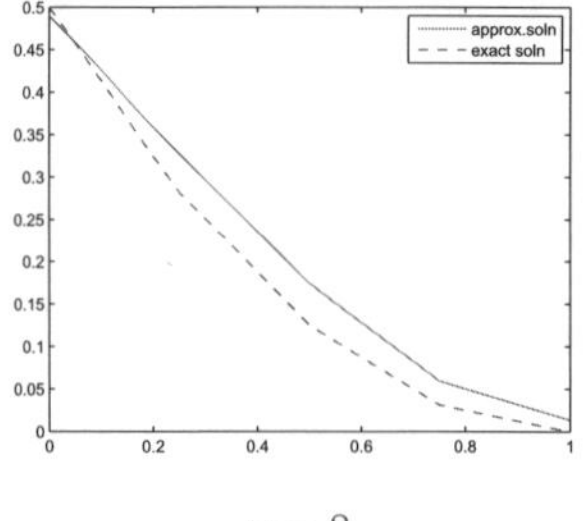

$n = 8$

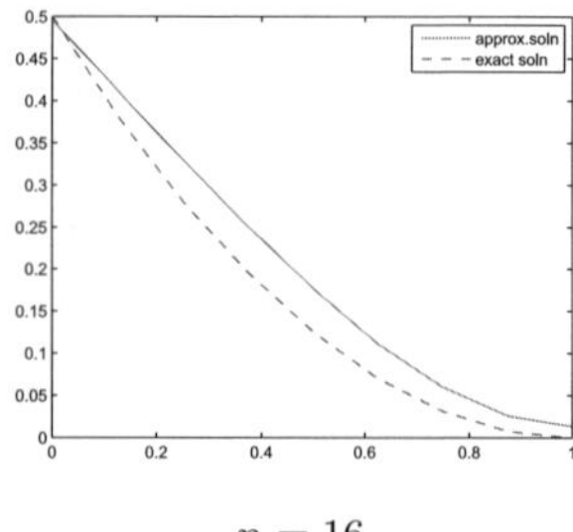

$n = 16$

Fig. 10.5   Curve of the exact and approximate solutions

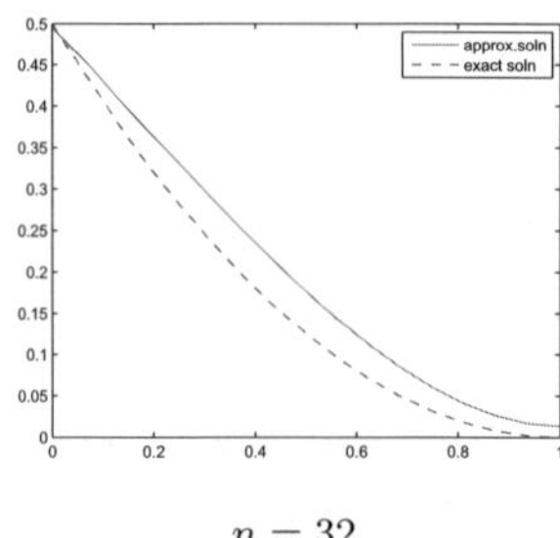

$n = 32$

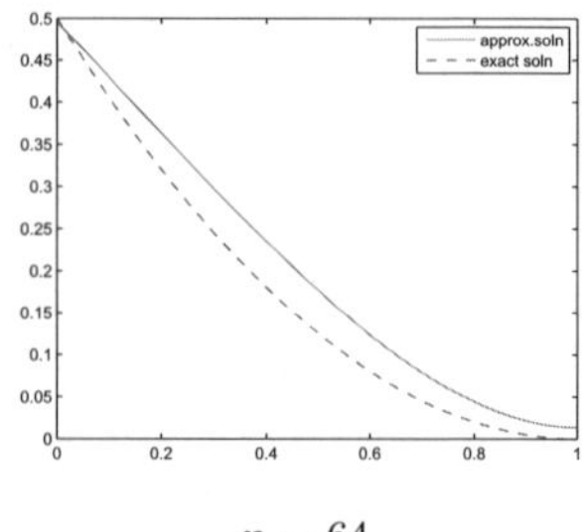

$n = 64$

Fig. 10.6   Curve of the exact and approximate solutions

10.4   Exercises

10.4.1 Let X and Y be two Hilbert spaces. D is a closed subset in X. Consider a modified Newton's method (MNM) in Hilbert

$$\dot{x}(t) = -[F'(x_0)]^{-1}F(x(t)), \quad x(0) = x_0 \in X. \tag{10.138}$$

We associate (10.138) with the autonomous dynamical system

$$\dot{x}(t) = \Phi(x(t)), \quad 0 \le t < +\infty, \quad x(0) = x_0, \tag{10.139}$$

where x_0 is an initial approximation to $x^\star$ and $x(t)$ is the trajectory. Let $F : D \to Y$ and $\Phi : D \to Y$. Assume that

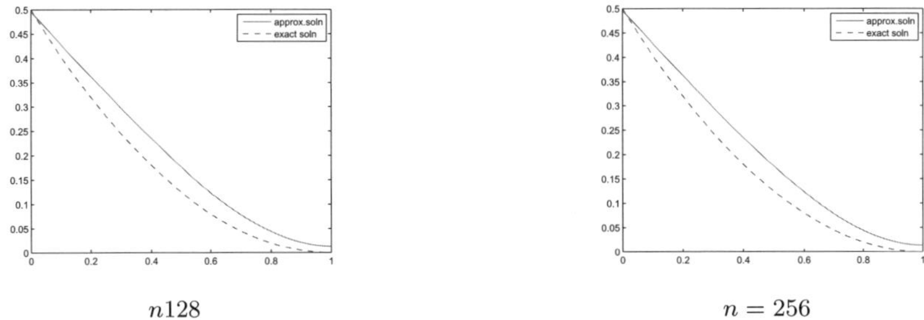

$$n128 \qquad\qquad n = 256$$

Fig. 10.7   Curve of the exact and approximate solutions

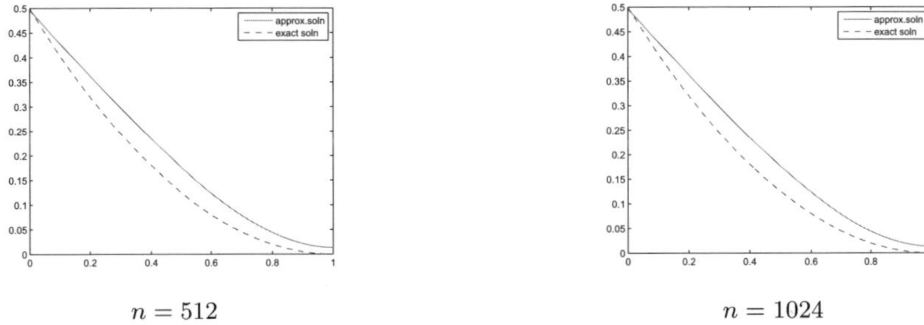

$$n = 512 \qquad\qquad n = 1024$$

Fig. 10.8   Curve of the exact and approximate solutions

(i) there exist $c_1 > 0$ and $c_2 > 0$ such that for

$$r = \frac{c_2\|F(x_0)\|}{c_1};\tag{10.140}$$

(ii) operators F and Φ are Fréchet differentiable on $U(x_0, r)$;

(iii) For all $y \in U(x_0, r)$, the following conditions hold:

$$(F'(y)\Phi(y), F(y)) \le -c_1\|F(y)\|^2,\tag{10.141}$$

$$\|\Phi(y)\| \le c_2\|F(y)\|\tag{10.142}$$

and

$$U(x_0, r) \subseteq D.\tag{10.143}$$

Show:

(a) There exists a global solution $x = x(t)$ to system (10.139) in $U(x_0, r)$, such that

$$\lim_{t\to+\infty} x(t) = x^\star,\tag{10.144}$$

where $x^\star$ is a solution of equation $F(x) = 0$ in $U(x_0, r)$.

(b) The following estimates hold:

$$\|x(t) - x^\star\| \le re^{-c_1 t}\tag{10.145}$$

and

$$\|F(x(t))\| \le \|F(x_0)\|e^{-c_1 t}.\tag{10.146}$$

Bibliography

[1] Absil, P. A., Mahony, R. and Sepulchre, R. (2008). *Optimization Algorithms on Matrix Manifolds*, Princeton University Press, Princeton NJ.

[2] Absil, P. A., Baker, C. G. and Gallivan, K. A. (2007). Trust-region methods on Riemannian manifolds, *Found. Comput. Math.* **7**, pp. 303–330.

[3] Adler, R. L., Dedieu, J. P., Margulies, J. Y., Martens, M. and Shub, M. (2002). Newton's method on Riemannian manifolds and a geometric model for the human spine, *IMA J. Numer. Anal.* **22**, pp. 359–390.

[4] Airapetyan, R. G. (2000). Continuous Newton method and its modification, *Appl. Anal.* **73**, pp. 463–484.

[5] Airapetyan, R. G., Ramm, A. G. and Smimova, A. B. (1999). Continuous analog of Gauss–Newton method, *Math. Models Methods Appl. Sci.* **9**, pp. 463–474.

[6] Airapetyan, R. G., Ramm, A. G. and Smimova, A. B. (2000). Continuous method for solving nonlinear ill-posed problems, Amer. Math. Soc., Providence RI, *Fields. Inst. Commun.* **25**, pp. 111–137.

[7] Airapetyan, R. G. and Ramm, A. G. (2000). Dynamical systems and discrete methods for solving nonlinear ill-posed problems, *Appl. Math. Reviews*, vol. 1, World Sci. Publishers, River Edge, NJ, pp. 491–536.

[8] Alvarez, F., Bolte, J. and Munier, J. (2008). A unifying local convergence result for Newton's method in Riemannian manifolds, *Found. Comput. Math.* **8**, pp. 197–226.

[9] Amat, S., Bermúdez, C., Busquier, S., Legaz, M. J. and Plaza, S. (2012). On a family of high order iterative methods under Kantorovich conditions and some applications, *Abstr. Appl. Anal.*, Art. ID 782170, 14 pp.

[10] Amat, S., Bermúdez, C., Busquier, S. and Mestiri, D. (2009). A family of Halley-Chebyshev iterative schemes for non-Fréchet differentiable operators, *J. Comput. Appl. Math.* **228**, pp. 486–493.

[11] Amat, S., Bermúdez, C., Busquier, S. and Plaza, S. (2010). On a third-order Newton-type method free of bilinear operators, *Numer. Linear Alg. Appl.* **17**, pp. 639–653.

[12] Amat, S. and Busquier, S. (2007). Third-order iterative methods under Kantorovich conditions, *J. Math. Anal. Appl.* **336**, pp. 243–261.

[13] Amat, S., Busquier, S. and Gutiérrez, J. M. (2003). Geometric constructions of iterative functions to solve nonlinear equations, *J. Comput. Appl. Math.* **157**, pp. 197–205.

[14] Amat, S., Busquier, S. and Gutiérrez, J. M. (2006). An adaptive version of a fourth-order iterative method for quadratic equations, *J. Comput. Appl. Math.* **191**, pp. 259–268.

[15] Amat, S., Busquier, S. and Gutiérrez, J. M. (2011). Third-orderiterative methods with applications to Hammerstein equations: A unified approach, *J. Comput. Appl.*

Math. **235**, pp. 2936–2943.

[16] Amat, S., Busquier, S. and Negra, M. (2004). Adaptive approximation of nonlinear operators, *Numer. Funct. Anal. Optim.* **25**, 397–405.

[17] Amat, S., Ezquerro, J. A. and Hernández, M. A. (2012). Approximation of inverse operators by a new family of high-order iterative methods, preprint.

[18] Amat, S., Hernández, M. A. and Romero, N. (2008). A modified Chebyshev's iterative method with at least sixth order of convergence, *Appl. Math. Comput.* **206**, pp. 164–174.

[19] Amat, S., Hernández, M. A. and Romero, N. (2013). Semilocal convergence of a sixth order iterative method for Riccati's equations, *Appl. Numer. Math.*, to appear.

[20] An, H–B. and Bai, Z–Z. (2005). Directional secant method for nonlinear equations, *J. Comput. Appl. Math.* **175**, pp. 291–304.

[21] Appell, J., De Pascale, E., Lysenko, J. V. and Zabrejko, P. P. (1997). New results on Newton–Kantorovich approximations with applications to nonlinear integral equations, *Numer. Funct. Anal. Optim.* **18**, pp. 1–17.

[22] Argyros, I. K. (1977). Improving the rate of convergence of Newton methods on Banach spaces with a convergence structure and applications, *Appl. Math. Lett.* **10**, pp. 21–28.

[23] Argyros, I. K. (1986). On the cardinality of solutions of multilinear differential equations and applications, *Int. J. Math. Math. Sci.* **9**, pp. 757–766

[24] Argyros, I. K. (1987). On the approximation of some nonlinear equations, *Aequat. Math.* **32**, pp. 87–95.

[25] Argyros, I. K. (1987). On polynomial equations in Banach space, perturbation techniques and applications, *Int. J. Math. Math. Sci.* **10**, pp. 69–78.

[26] Argyros, I. K. (1987). Newton-like methods under mild differentiability conditions with error analysis, *Bull. Austral. Math. Soc.* **37**, pp. 131–147.

[27] Argyros, I. K. (1988). On Newton's method and nondiscrete mathematical induction, *Bull. Austral. Math. Soc.* **38**, pp. 131–140.

[28] Argyros, I. K. (1989). On the number of solutions of some integral equations arising in radiative transfer, *Int. J. Math. Math. Sci.* **12**, pp. 297–304

[29] Argyros, I. K. (1990). Improved error bounds for a certain class of Newton-like methods, *J. Approx. Theory Appl.* **61**, pp. 80–98.

[30] Argyros, I. K. (1990). Error founds for the modified secant method, *BIT* **20**, pp. 92–200.

[31] Argyros, I. K. (1990). On the solution of equations with nondifferentiable operators and the Pták error estimates, *BIT* **30**, pp. 752–754.

[32] Argyros, I. K. (1990). On some projection methods for the approximation of implicit functions, *Appl. Math. Lett.* **32**, pp. 5–7.

[33] Argyros, I. K. (1990). The secant method in generalized Banach spaces, *Appl. Math. Comp.* **39**, pp. 111–121.

[34] Argyros, I. K. (1990). A mesh independence principle for operator equations and their discretizations under mild differentiability conditions, *Comput.* **45**, pp. 265–268.

[35] Argyros, I. K. (1991). On the convergence of some projection methods with perturbation, *J. Comp. Appl. Math.* **36**, pp. 255–258.

[36] Argyros, I. K. (1991). On an application of the Zincenko method to the approximation of implicit functions, *Public. Math. Debrecen* **39**, pp. 1–7.

[37] Argyros, I. K. (1991). On an iterative algorithm for solving nonlinear equations, *Beitrage zür Numerischen Math.* **10**, pp. 83–92.

[38] Argyros, I. K. (1992). On a class of quadratic equations with perturbation, *Funct. Approx. Comm. Math.* **XX**, pp. 51–63.

[39] Argyros, I. K. (1992). Improved error bounds for the modified secant method, *Int. J. Computer Math.* **43**, pp. 99–109.

[40] Argyros, I. K. (1992). Some generalized projection methods for solving operator equations, *J. Comp. Appl. Math.* **39**, pp. 1–6.

[41] Argyros, I. K. (1992). On the convergence of generalized Newton-methods and implicit functions, *J. Comp. Appl. Math.* **43**, pp. 335–342.

[42] Argyros, I. K. (1992). On the convergence of inexact Newton-like methods, *Publ. Math. Debrecen* **42**, pp. 1–7.

[43] Argyros, I. K. (1993). On the convergence of a Chebyshev–Halley-type method under Newton–Kantorovich hypothesis, *Appl. Math. Lett.* **5**, pp. 71–74.

[44] Argyros, I. K. (1993). Newton-like methods in partially ordered linear spaces, *J. Approx. Theory Appl.* **9**, pp. 1–10.

[45] Argyros, I. K. (1993). On the solution of undetermined systems of nonlinear equations in Euclidean spaces, *Pure Math. Appl.* **4**, pp. 199–209.

[46] Argyros, I. K. (1993). On the secant method, *Publ. Math. Debrecen* **43**, pp. 223–238.

[47] Argyros, I. K. (1993). On the convergence of an Euler–Chebyshev-type method under Newton-Kantorovich hypotheses, *Pure Math. Appl.* **4**, pp. 369–373.

[48] Argyros, I. K. (1994). A convergence theorem for Newton-like methods under generalized Chen–Yamamato-type assumptions, *Appl. Math. Comput.* **61**, pp. 25–37.

[49] Argyros, I. K. (1994). On the discretization of Newton-like methods, *Inter. J. Computer. Math.* **52**, pp. 161–170.

[50] Argyros, I. K. (1994). On the midpoint method for solving nonlinear operator equations and applications to the solution of integral equations, *Rev. Anal. Numér. Théor. Approx.* **23**, pp. 139–152.

[51] Argyros, I. K. (1995). A unified approach for constructing fast two-step Newton-like methods, *Monatsh. Math.* **119**, pp. 1–22.

[52] Argyros, I. K. (1995). Results on controlling the residuals of perturbed Newton-like methods on Banach spaces with aconvergence structure, *Southwest J. Pure Appl. Math.* **1**, pp. 32–38.

[53] Argyros, I. K. (1996). On the method of tangent hyperbolas, *J. Appr. Theory Appl.* **12**, pp. 78–96.

[54] Argyros, I. K. (1996). On the method of tangent parabolas, *Funct. Approx. Comment. Math.* **24**, pp. 3–15.

[55] Argyros, I. K. (1996). On an extension of the mesh-independence principle for operator equations in Banach space, *Appl. Math. Lett.* **9**, pp. 1–7.

[56] Argyros, I. K. (1996). A generalization of Edelstein's theorem on fixed points and applications, *Southwest J. Pure Appl. Math.* **2**, pp. 60–64.

[57] Argyros, I. K. (1997). Chebyshev–Halley-like methods in Banach spaces, *Korean J. Comp. Appl. Math.* **4**, pp. 83–107.

[58] Argyros, I. K. (1997). Concerning the convergence of inexact Newton methods, *J. Comp. Appl. Math.* **79**, pp. 235–247.

[59] Argyros, I. K. (1997). General ways of constructing accelerating Newton-like iterations on partially ordered topological spaces, *Southwest J. Pure Appl. Math.* **2**, pp. 1–12.

[60] Argyros, I. K. (1997). On a new Newton–Mysovskii-type theorem with applications to inexact Newton-like methods and their discretizations, *IMA J. Num. Anal.* **18**, pp. 43–47.

[61] Argyros, I. K. (1997). On the convergence of two-step methods generated by point-to-point operators, *Appl. Math. Comput.* **82**, pp. 85–96.

[62] Argyros, I. K. (1997). Improved error bounds for Newton-like iterations under Chen–

Yamamoto assumptions, *Appl. Math. Lett.* **10**, pp. 97–100.

[63] Argyros, I. K. (1997). Inexact Newton methods and nondifferentiable operator equations on Banach spaces with a convergence structure, *Approx. Theory Appl.* **13**, pp. 91–104.

[64] Argyros, I. K. (1997). A mesh independence principle for inexact Newton-like methods and their discretizations under generalized Lipschitz conditions, *Appl. Math. Comp.* **87**, pp. 15–48.

[65] Argyros, I. K. (1997). Concerning the convergence of inexact Newton methods, *J. Comput. Appl. Math.* **79**, pp. 235–247.

[66] Argyros, I. K. (1997). Smoothness and perturbed Newton-like methods, *Pure Math. Appl.* **8**, pp. 13–28.

[67] Argyros, I. K. (1997). The asymptotic mesh independence principle for inexact Newton–Galerkin-like methods, *Pure Math. Appl.* **8**, pp. 169–194.

[68] Argyros, I. K. (1997). A new convergence theorem for steffenson's method on Banach spaces and applications, *Southwest J. Pure Appl. Math.* **01**, pp. 23–29.

[69] Argyros, I. K. (1998). *The Theory and Application of Abstract Polynomial Equations*, St.Lucie/CRC/Lewis Publ. Mathematics series, Boca Raton, Florida, U.S.A.

[70] Argyros, I. K. (1998). Sufficient conditions for constructing methods faster than Newton's, Appl. Math. Comp. **93**, pp. 169–181.

[71] Argyros, I. K. (1998). A new convergence theorem for the Jarratt method in Banach spaces, *Comput. Math. Appl.* **36**, pp. 13–18.

[72] Argyros, I. K. (1998). Improving the order and rates of convergence for the Super–Halley method in Banach spaces, *Comp. Appl. Math.* **5**, pp. 465–474.

[73] Argyros, I. K. (1999). Improved error bounds for a Chebyshev–Halley-type method, *Acta Math. Hungarica* **84**, pp. 211–221.

[74] Argyros, I. K. (1999). Convergence domains for some iterative processes in Banach spaces using outer and generalized inverses, *Comput. Anal. Appl.* **1**, pp. 87–104.

[75] Argyros, I. K. (1999). Concerning the convergence of a modified Newton-like method, *J. Anal. Appl. (ZAA)* **18**, pp. 1–8.

[76] Argyros, I. K. (1999). Concerning the radius of convergence of Newton's method and applications, *Korean J. Comp. Appl. Math.* **6**, pp. 451–462.

[77] Argyros, I. K. (1999). Convergence rates for inexact Newton-like methods of singular points and applications, *Appl. Math. Comp.* **102**, pp. 185–201.

[78] Argyros, I. K. (2000). A mesh independence principle for perturbed Newton-like methods and their discretizations, *Korean J. Comp. Appl. Math.* **7**, pp. 139–159.

[79] Argyros, I. K. (2000). Newton methods on Banach spaces with a convergence structure and applications, *Comput. Math. Appl.* **40**, pp. 37–48.

[80] Argyros, I. K. (2000). Semilocal convergence theorems for a certain class of iterative procedures using outer or generalized inverses, *Korean J. Comp. Appl. Math.* **7**, pp. 29–40.

[81] Argyros, I. K. (2000). The effect of rounding errors on a certain class of iterative methods, *Appl. Math.* **27**, pp. 369–375.

[82] Argyros, I. K. (2000). Local convergence of Newton's method for nonlinear equations using outer or generalized inverses, *Czecho. Math. J.* **50**, pp. 603–614.

[83] Argyros, I. K. (2000). On a class of nonlinear implicit quasivariational inequalities, *PanAmer. Math. J.* **10**, pp. 101–109.

[84] Argyros, I. K. (2001). A new semilocal convergence theorem for Newton's method in Banach space using hypotheses on the second Fréchet-derivative, *J. Comput. Appl. Math.* **139**, 369–373.

[85] Argyros, I. K. (2001). On the radius of convergence of Newton's method, *Int. J.*

Comput. Math. **77**, pp. 389–400.

[86] Argyros, I. K. (2001). Semilocal convergence theorems for Newton's method using outer inverses and hypotheses on the second Fréchet-derivative, *Monatshefte fur Mathematik* **132**, pp. 183–195.

[87] Argyros, I. K. (2001). On general auxiliary problem principle and nonlinear mixed variational inequalities, *Nonlinear Funct. Anal. Appl.* **6**, pp. 247–256.

[88] Argyros, I. K. (2001). On an iterative procedure for approximating solutions of quasi variational inequalities, *Adv. Nonlinear Var. Inequa.* **4**, pp. 39–42.

[89] Argyros, I. K. (2001). On generalized variational inequalities, *Adv. Nonlinear Var. Inequa.* **4**, pp. 75–78.

[90] Argyros, I. K. (2001). On a semilocal convergence theorem for a class of quasi variational inequalities, *Adv. Nonlinear Var. Inequa.* **4**, pp. 43–46.

[91] Argyros, I. K. (2002). On the convergence of a Newton-like method based on m-Fréchet-differentiable operators and applications in radiative transfer, *J. Comput. Anal. Appl.* **4**, pp. 141–154.

[92] Argyros, I. K. (2002). A unifying semilocal convergence theorem for Newton-like methods based on center Lipschitz conditions, *Comput. Appl. Math.* **21**, pp. 789–796.

[93] Argyros, I. K. (2002). A semilocal convergence analysis for the method of tangent hyperbolas, *J. Concrete and Appl. Anal.* **1**, pp. 135–144.

[94] Argyros, I. K. (2002). Convergence theorems for Newton-like methods without Lipschitz conditions, *Comm. Appl. Nonlinear Anal.* **9**, pp. 69–68.

[95] Argyros, I. K. (2003). On the convergence and application of Newton's method under weak Hölder continuity assumptions, *Intern. J. Comput. Math.* **5**, pp. 767–780.

[96] Argyros, I. K. (2003). New and generalized convergence conditions for the Newton–Kantorovich method, *J. Appl. Anal.* **9**, pp. 287–299.

[97] Argyros, I. K. (2003). On a theorem of L.V. Kantorovich concerning Newton's method, *J. Comp. Appl. Math.* **155**, pp. 223–230.

[98] Argyros, I. K. (2004). On the Newton–Kantorovich hypothesis for solving equations, *J. Comput. Appl. Math.* **169**, pp. 315–332.

[99] Argyros, I. K. (2004). Convergence of Broyden's method, *Comm. Appl. Nonlinear Anal.* **11**, pp. 77–86.

[100] Argyros, I. K. (2004). A unifying local-semilocal convergence analysis and applications for two-point Newton-like methods in Banach space, *J. Math. Anal. Appl.* **298**, pp. 374–397.

[101] Argyros, I. K. (2005). New sufficient convergence conditions for the Secant method, *Czecho. Math. J.* **55**, pp. 175–187.

[102] Argyros, I. K. (2005). Enlarging the convergence domain of Newton's method under regular smoothness conditions, *Adv. Nonlinear Var. Inequa.* **8**, pp. 121–129.

[103] Argyros, I. K. (2005). On the semilocal convergence of the Gauss–Newton method, *Adv. Nonlinear Var. Inequa.* **8**, pp. 93–99.

[104] Argyros, I. K. (2007). Approximating solutions of equations using Newton's method with a modified Newton's method iterate as a starting point, *Rev. Anal. Numér. Théor. Approx.* **36**, pp. 123–137.

[105] Argyros, I. K. (2007). An improved unifying convergence analysis of Newton's method in Riemannian manifolds, *J. Appl. Math. Comput.* **25**, pp. 345–351.

[106] Argyros, I. K. (2007). *Computational Theory of Iterative Methods.* Series: Studies in Computational Mathematics, 15, Editors: C.K. Chui, and L. Wuytack, Elsevier Publ. Co. New York, U.S.A.

[107] Argyros, I. K. (2007). On the convergence of Broyden-like methods, *Acta Math. Sin. (Engl. Ser.)* **23**, pp. 2087–2096.

[108] Argyros, I. K. (2007). On the local convergence of Newton's method on Lie groups, *PanAmer. Math. J.* **17**, pp. 101–109.

[109] Argyros, I. K. (2008). Concerning the semilocal convergence of Newton's method and convex majorants, *Rend. Circ. Mat. Palermo* **57**, pp. 331–341.

[110] Argyros, I. K. (2008). *Convergence and Applications of Newton-type Iterations*, Springer-Verlag, New York.

[111] Argyros, I. K. (2008). Approximating solutions of equations by combining Newton-like methods, *J. Korea Soc. Educ. Ser. B: Pure and Appl. Math.* **15**, pp. 35–45.

[112] Argyros, I. K. (2009). On a class of Newton-like methods for solving nonlinear equations, *J. Comput. Appl. Math.* **228**, pp. 115–122.

[113] Argyros, I. K. (2009). Newton's method on Lie groups, *J. Appl. Math. Comput.* **31**, pp. 217–228.

[114] Argyros, I. K. (2009). On the semilocal convergence of inexact Newton methods in Banach spaces, *J. Comput. Appl. Math.* **228**, pp. 434–443.

[115] Argyros, I. K. (2010). Local convergence of Newton's method using Kantorovich's convex majorants, *Rev. Anal. Numér. Théor. Approx.* **39**, pp. 97–106.

[116] Argyros, I. K., A semilocal convergence analysis for directional Newton methods, *Math. Comput.* **80** (2011), pp. 327–343.

[117] Argyros, I. K. and Chen, D. (1993). Results on the Chebyshev method in Banach spaces, *Proyecciones* **12**, pp. 119–128.

[118] Argyros, I. K., Chen, D. and Qian, Q. S. (1993). A note on the Halley method in Banach spaces, Appl. *Math. Comput.* **58**, pp. 215–224.

[119] Argyros, I. K., Chen, D. and Qian, Q. S. (1994). Error bound representations of Chebyshev-Halley type methods in Banach spaces, *Rev. Academia de Ciencias Zaragoza* **49**, pp. 57–69.

[120] Argyros, I. K., Cho, Y. J. and Hilout. S. (2012). *Numerical Methods for Equations and its Applications*, CRC Press/Taylor and Francis Group, New York.

[121] Argyros, I. K., Ezquerro, J. A., Gutiérrez, J. M., Hernández, M. A. and Hilout, S. (2011). On the semilocal convergence of efficient Chebyshev-Secant-type methods, *J. Comput. Appl. Math.* **235**, pp. 3195–3206.

[122] Argyros, I. K. and Hilout, S. (2007). Newton's methods for variational inclusions under conditioned Fréchet derivative, *Appl. Math.* **34**, pp. 349–357.

[123] Argyros, I. K. and Hilout, S. (2007). An improved local convergence analysis for Secant-like method, *East Asian Math. J.* **23**, pp. 261–270.

[124] Argyros, I. K. and Hilout, S. (2008). Local convergence of Newton-like methods for generalized equations, *Appl. Math. Comput.* **197**, pp. 507–514.

[125] Argyros, I. K. and Hilout, S. (2008). Multipoint method for generalized equations under mild differentiability conditions, *Functiones et Approximatio* **XXXVIII**, pp. 7–19.

[126] Argyros, I. K. and Hilout, S. (2008). A Fréchet derivative-free cubically convergent method for set-valued maps, *Numer. Algorithms* **48**, pp. 361–371.

[127] Argyros, I. K. and Hilout, S. (2008). Steffensen method for solving generalized equations, *Serdica Math. J.* **34**, pp. 455–466.

[128] Argyros, I. K. and Hilout, S. (2008). On the midpoint method for solving generalized equations, *Punjab Univ. J. Math.* **40**, pp. 63–70.

[129] Argyros, I. K. and Hilout, S. (2008). On the local convergence of Newton-type method in Banach spaces under a gamma-type condition, *Proyecciones J. Math.* **27**, pp. 1–14.

[130] Argyros, I. K. and Hilout, S. (2008). On a Secant-like method for solving generalized equations, *Math. Bohemica* **133**, pp. 313–320.

[131] Argyros, I. K. and Hilout, S. (2008). A cubically convergent method without second order derivative for solving generalized equations, *Internat. J. Modern Math.* **3**, pp. 187–195.

[132] Argyros, I. K. and Hilout, S. (2009). Hummel–Seebeck method for generalized equations under conditioned second Fréchet derivative, *Nonlinear Funct. Anal. Appl.* **14**, pp. 261–269.

[133] Argyros, I. K. and Hilout, S. (2009). On the Local convergence of Gauss–Newton method, *Punjab Univer. Math. J.* **41**, pp. 23–33.

[134] Argyros, I. K. and Hilout, S. (2009). *Efficient Methods for Solving Equations and Variational Inequalities*, Polimetrica Publisher, Milano.

[135] Argyros, I. K. and Hilout, S. (2009). *Aspects of the Computational Theory for Certain Iterative Methods*, Polimetrica Publisher, Milano.

[136] Argyros, I. K. and Hilout, S. (2009). Enclosing roots of polynomial equations and their applications to iterative processes, *Surveys Math. Appl.* **4**, pp. 119–132.

[137] Argyros, I. K. and Hilout, S. (2009). On the convergence of some iterative procedures under regular smoothness, *PanAmer. Math. J.* **19**, pp. 17–34.

[138] Argyros, I. K. and Hilout, S. (2009). On the convergence of two-step Newton-type methods of high efficiency index, *Appl. Math. (Warsaw)* **36**, pp. 465–499.

[139] Argyros, I. K. and Hilout, S. (2009). Newton's method for approximating zeros of vector fields on Riemannian manifolds, *J. Appl. Math. Comput.* **29**, pp. 417–427.

[140] Argyros, I. K. and Hilout, S. (2010). Extending the Newton–Kantorovich hypothesis for solving equations, *J. Comput. Appl. Math.* **234**, pp. 2993–3006.

[141] Argyros, I. K. and Hilout, S. (2010). An improved local convergence analysis for Newton–Steffensen-type method, *J. Appl. Math. Comput.* **32**, pp. 111–118.

[142] Argyros, I. K. and Hilout, S. (2010). A convergence analysis of Newton-like method for singular equations using recurrent functions, *Numer. Funct. Anal. Optim.* **31**, pp. 112–130.

[143] Argyros, I. K. and Hilout, S. (2010). A Kantorovich-type analysis of Broyden's method using recurrent functions, *J. Appl. Math. Comput.* **32**, pp. 353–368.

[144] Argyros, I. K. and Hilout, S. (2010). Convergence conditions for Secant-type methods, *Czecho. Math. J.* **60**, pp. 253–272.

[145] Argyros, I. K. and Hilout, S. (2010). Improved generalized differentiability conditions for Newton-like methods, *J. Complexity* **26**, pp. 316–333.

[146] Argyros, I. K. and Hilout, S. (2010). Inexact Newton methods and recurrent functions, *Appl. Math.* **37**, pp. 113–126.

[147] Argyros, I. K. and Hilout, S. (2010). On Newton-like methods of bounded deterioration using recurrent functions, *Aequat. Math.* **79**, pp. 61–82.

[148] Argyros, I. K. and Hilout, S. (2010). Convergence analysis for directional two-step Newton methods, *Numer. Algorithms* **55**, pp. 503–528.

[149] Argyros, I. K. and Hilout, S. (2011). On the solution of systems of equations with constant rank derivatives, *Numer. Algorithms* **57**, pp. 235–253.

[150] Argyros, I. K. and Hilout, S. (2011). A unifying theorem for Newtons method on spaces with a convergence structure, *J. Complexity* **27**, pp. 39–54.

[151] Argyros, I. K. and Hilout, S. (2011). On the semilocal convergence of the Halley method using recurrent functions, *J. Appl. Math. Comput.* **37**, pp. 221–246.

[152] Argyros, I. K. and Hilout, S. (2011). Extending the applicability of the Gauss–Newton method under average Lipschitz-type conditions, *Numer. Algorithms* **58**, pp. 23–52.

[153] Argyros, I. K. and Hilout, S. (2013). On the semilocal convergence of Newton's method using majorants and recurrent functions, *J. Nonlinear Funct. Anal. Appl.*,

accepted.

[154] Argyros, I. K. and Hilout, S. (2012). On the convergence of inexact two-step Newton-like algorithms using recurrent functions, *J. Appl. Math. Comput.* **38**, pp. 41–61.

[155] Argyros, I. K. and Hilout, S. (2012). Weaker conditions for the convergence of Newton's method, *J. Complexity* **28**, pp. 364–387.

[156] Argyros, I. K. and Hilout, S. (2012). Majorizing sequences for iterative methods, *J. Comput. Appl. Math.* **236**, pp. 1947–1960.

[157] Argyros, I. K. and Hilout, S. (2012). Convergence of Newton's method under weak majorant condition, *J. Comput. Appl. Math.* **236**, pp. 1892–1902.

[158] Argyros, I. K. and Hilout, S. (2012). Traub–Potra-type method for set-valued maps, *Aust. J. Math. Anal. Appl.* **9**, 11 pp.

[159] Argyros, I. K. and Hilout, S. (2012). On the semilocal convergence of damped Newtons method, *Appl. Math. Comput* **5**, 2808–2824.

[160] Argyros, I. K. and Hilout, S. (2012). The majorant method in the theory of Newton–Kantorovich approximations and generalized Lipschitz conditions, submitted.

[161] Argyros, I. K. and Hilout, S. (2013). Estimating upper bounds on the limit points of majorizing sequences for Newton's method, *Numer. Algorithms* **62**, pp. 115–132.

[162] Argyros, I. K. and Hilout, S. (2013). On the convergence of Newton-like methods for solving equations using slantly differentiable operators, *Trans. Math. Programm. Appl.* **1**, pp. 105–115.

[163] Argyros, I. K., Hilout, S. and Tabatabai, M. A. (2011). *Mathematical Modelling with Applications in Biosciences and Engineering*, Nova Publishers, New York, 2011.

[164] Argyros, I. K. and Szidarovszky, F. (1992). Convergence of general iteration schemes, *J. Math. Anal. Appl.* **168**, pp. 42–52.

[165] Argyros, I. K. and Szidarovszky, F. (1992). On the monotone of general Newton-like methods, *Bull. Austral. Math. Soc.* **45**, pp. 489–502.

[166] Argyros, I. K. and Szidarovszky, F. (1993). *The Theory and Applications of Iteration Methods*, C.R.C. Press, Boca Raton, Florida.

[167] Argyros, I. K. and Szidarovszky, F. (1994). On the convergence of modified contractions, *J. Comput. Appl. Math.* **55**, pp. 97–108.

[168] Argyros, I. K. and Verma, R. U. (2000). Semilocal convergence theorems for a certain class of iterative procedures involving m-Fréchet differentiable operators, *Math. Sci. Res. Hot-Line* **4**, 1–12.

[169] Averbuh, V. I. and Smoljanov, O. G. (1967). Differentiation theory in linear topological spaces, *Uspehi Mat. Nauk* **22**, pp. 201–260.

[170] Avila, J. H. (1974). The feasibility of continuation methods for nonlinear equations, *SIAM J. Numer. Anal.* **11**, 102–122.

[171] Bahtin, I. A. and Krasnosel'skii, M. A. (1958). (Russian), *Dokl. Akad. Nauk SSSR* **123**, pp. 17–20.

[172] Bakushinsky, A. (1992). The Problem of the convergence of the iteratively regularized Gauss–Newton method, *Comput. Math. Meth. Phys.* **32**, pp. 1353–1359.

[173] Bakushinsky, A. B. and Kokurin, M. Y. (2004). Iterative Methods for Approximate Solution of Inverse Problems. Mathematics and Its Applications (New York) **577**, Springer, Dordrecht.

[174] Bakushinsky, A. and Smirnova, A. (2005). On application of generalized discrepancy principle to iterative methods for nonlinear ill-posed problems, *Numer. Func. Anal. Optimiz.* **26**, pp. 35–48.

[175] Bellavia, S., Macconi, M. and Morini, B. (2004). STRSCNE: a scaled trust-region solver for constrained nonlinear equations, *Comput. Optim. Appl.* **28**, pp. 31–50.

[176] Ben-Israel, A. and Levin, Y., Maple programs for directional Newton methods, are

available at `ftp://rutcor.rutgers.edu/pub/bisrael/Newton--Dir.mws`

[177] Blaschke, B., Neubauer, A. and Scherzer, O. (1997). On convergence rates for the iteratively regularized Gauss–Newton method, *IMA J. Numer. Anal.* **17**, pp. 421–436.

[178] Bosarge, W. E. and Falb, P. L. (1969). A multipoint method of third order, *J. Optim. Theory Appl.* **4**, pp. 156–166.

[179] Broyden, C. G. (1965). A class of methods for solving nonlinear simultaneous equations, *Math. Comput.* **19**, pp. 577–593.

[180] Burke, J. V. and Ferris, M. C. (1995). A Gauss–Newton method for convex composite optimization, *Math. Programming Ser A.* **71**, pp. 179–194.

[181] Candela, S. and Marquina, A. (1990). Recurrence relations for rational cubic methods. II. The Chebyshev method, *Comput.* **45**, pp. 355–367.

[182] Cătinaş, E. (1994). On some iterative methods for solving nonlinear equations, *Rev. Anal. Numér. Théor. Approx.* **23**, pp. 47–53.

[183] Cătinaş, E. (2005). The inexact, inexact perturbed, and quasi-Newton methods are equivalent models, *Math. Comput.* **74**, pp. 291–301.

[184] Chandrasekhar, S. (1960). *Radiative Transfer*, Dover Publ., New York.

[185] Chen, J. and Li, W. (2005). Convergence of Gauss–Newton's method and uniqueness of the solution, *Appl. Math. Comput.* **170**, pp. 686–705.

[186] Chen J. and Li, W. (2006). Local convergence results of Gauss–Newton's like method in weak conditions, *J. Math. Anal. Appl.* **324**, pp. 1381–1394.

[187] Chen, X. (1990). On the convergence of Broyden-like methods for nonlinear equations with nondifferentiable terms, *Ann. Inst. Statist. Math.* **42**, pp. 387–401.

[188] Chen, X. and Yamamoto, T. (1998). Convergence domains of certain iterative methods for solving nonlinear equations, *Numer. Funct. Anal. Optim.* **10**, pp. 37–48.

[189] Chun, C., Bae, H. J. and Neta, B. (2009). New families of nonlinear third-order solvers for finding multiple roots, *Comput. Math. Appl.* **57**, pp. 1574–1582.

[190] Chun, C. and Neta, B. (2009). A third-order modification of Newton's method for multiple roots, *Appl. Math. Comput.* **211**, pp. 474–479.

[191] Dedieu, J. P. and Nowicki, D. (2005). Symplectic methods for the approximation of the exponential map and the Newton iteration on Riemannian submanifolds, *J. Complexity* **21**, pp. 487–501.

[192] Dennis, J. E. (1971). Toward a Unified Convergence Theory for Newton-like Methods, in *Nonlinear Functional Analysis and Applications* (L.B. Rall, ed.), Academic Press, New York, pp. 425–472.

[193] Dennis, J. E. and Schnabel, R. B. (1996). *Numerical Methods for Unconstrained Optimization and Nonlinear Equations, Classics in Applied Mathematics*, **16**, SIAM, Philadelphia.

[194] Deuflhard, P., Engl, H. W. and Scherzer, O. (1998). A convergence analysis of iterative methods for the solution of nonlinear ill-posed problems under affinely invariant conditions, *Inverse Problems* **14**, pp. 1081–1106.

[195] Deuflhard, P. and Heindl, G. (1979). Affine invariant convergence theorems for Newton's method and extensions to related methods, *SIAM J. Numer. Anal.* **16**, pp. 1–10.

[196] Do Carmo, M. P. (1992). *Riemannian Geometry*, Birkhäuser, Boston, USA.

[197] Dong, C. (1982). A basic theorem of constructing an iterative formula of the higher order for computing multiple roots of an equation, *Math. Numer. Sinica* **11**, pp. 445–450.

[198] Dong, C. (1987). A family of multipoint iterative functions for finding multiple roots of equations, *Int. J. Comput. Math.* **21**, pp. 363–367.

[199] Engl, H. W., Hanke, M. and Neubauer, A. (1996). *Regularization of Inverse Problems*. Mathematics and Its Applications, **375**, Kluwer Academic Publishers Group,

Dordrecht.

[200] Engl, H. W., Kunisch, K. and Neubauer, A. (1989). Convergence rates for Tikhonov regularisation of nonlinear ill-posed problems, *Inverse Problems* **5**, pp. 523–540.

[201] Epureanu B. I. and Greenside, H. S. (1998). Fractal basins on the attraction associated with a the relaxed Newton's method, *SIAM Rev.* **40**, pp. 102–109.

[202] Ezquerro, J. A. (1997). A modification of the Chebyshev method, *IMA J. Numer. Anal.* **17**, pp. 511–525.

[203] Ezquerro, J. A., Gutiérrez, J. M., Hernández, M. A. and Salanova, M. A. (2000). Chebyshev-like methods and quadratic equations, *Rev. Anal. Numér. Théor. Approx.* **28**, pp. 23–35.

[204] Ezquerro, J. A. and Hernández, M. A. (2002). On an application of Newton's method to nonlinear operators with ω-conditioned second derivative, *BIT* **42**, pp. 519–530.

[205] Ezquerro, J. A. and Hernández, M. A. (2002). Generalized differentiability conditions for Newton's method, *IMA J. Numer. Anal.* **22**, pp. 187–205.

[206] Ezquerro, J. A. and Hernández, M. A. (2005). J.A. Ezquerro, M.A. Hernandez, New Kantorovich-type conditions for Halley's method, *Appl. Numer. Anal. Comput. Math.* **2**, pp. 70–77.

[207] Ezquerro, J. A. and Hernández, M. A. (2009). An improvement of the region of accessibility of Chebyshev's method from Newton's method, *Math. Comput.* **78**, pp. 1613–1627.

[208] Ezquerro, J. A. and Hernández, M. A. (2009). An optimization of Chebyshev's method, *J. Complexity*, **25**, pp. 343–361.

[209] Ezquerro, J. M., Hernández, M. A., Romero, N. and Velasco, A. I. (2012). Improving the domain of starting point for secant-like methods, *Appl. Math. Comut.* **219**, pp. 3677–3692.

[210] Ferreira, O. P. (2009). Local convergence of Newton's method from the view point of the majorant principle, *IMA. J. Numer. Anal.* **29**, pp. 746–759.

[211] Ferreira, O. P. (2009). Convergence of Newton's method in Banach space from the view of the majorant principle, *IMA J. Numer. Anal.* **29**, pp. 746–759.

[212] Ferreira, O. P. (2011). Local convergence of Newton's method under majorant condition, *J. Comput. Appl. Math.* **235**, pp. 1515–1522.

[213] Ferreira, O. P. and Goncalves, M. L. N. (2011). Local convergence analysis of inexact Newton-like methods under majorant condition, *Comput. Optim. Appl.* **48**, pp. 1–21.

[214] Ferreira, O. P., Gonçalves, M. L. N and Oliveira, P. R. (2011). Local convergence analysis of Gauss–Newton like methods under majorant condition, *J. Complexity* **27**, pp. 111–125.

[215] Ferreira, O. P., Gonçalves, M. L. N and Oliveira, P. R. (2012). Local convergence analysis of inexact Gauss–Newton like methods under majorant condition, *J. Comput. Appl. Math.* **236**, pp. 2487–2498.

[216] Ferreira, O. P. and Svaiter, B. F. (2002). Kantorovich's theorem on Newton's method in Riemannian manifolds, *J. Complexity* **18**, pp. 304–329.

[217] Ferreira, O. P. and Svaiter, B. F. (2009). Kantorovich's majorants principle for Newton's method, *Comput. Optim. Appl.* **42**, pp. 213–229.

[218] Floudas, C. A. and Pardalos, P. M. (1990). *A Collection of Test Problems for Constrained Global Optimization Algorithms*, Lecture Notes in Computer Science **455**, Springer, Berlin.

[219] Frontini, M. and Sormani, E. (2003). Modified Newton's method with third-order convergence and multiple roots, *J. Comput. Appl. Math.* **156**, pp. 345–354.

[220] Galantai, A. (2000). The theory of Newton's method, *J. Comput. Appl. Math.* **124**, pp. 25–44.

[221] Galperin, A. (2005). On convergence domains of Newton's and modified Newton methods, *Numer. Funct. Anal. Optim.* **26**, pp. 385–405.

[222] Galperin, A. (2006). Secant method with regularity continuous divided differences, *J. Comput. Appl. Math.* **193**, pp. 574–575.

[223] Galperin, A. (2009). Optimal iterative methods for nonlinear equations, *Numer. Funct. Anal. Optim.* **30**, pp. 499–522.

[224] Galperin, A. (2009). Ulm's method without derivatives, *Nonlinear Anal.* **71**, pp. 2094–2113.

[225] Galperin, A. (2011). The second Broyden's method for operators with regularly continuous divided differences, preprint.

[226] George, S. (2006). Newton–Tikhonov regularization of ill-posed Hammerstein operator equation, *J. Inverse and Ill-Posed Problems* **2**, pp. 135–146.

[227] George, S. (2006). Newton–Lavrentiev regularization of ill-posed Hammerstein type operator equation, *J. Inverse and Ill-Posed Problems* **6**, pp. 573–582.

[228] George, S. and Elmahdy, A. I. (2010). An analysis of Lavrentiev regularization for nonlinear ill-posed problems using an iterative regularization method, *Int. J. Comput. Appl. Math.* **5**, pp. 369–381.

[229] George, S. and Elmahdy, A. I. (2012). A quadratic convergence yielding iterative method for nonlinear ill-posed operator equations, *Comput. Methods Appl. Math.* **12**, pp. 32–45.

[230] George, S. and Kunhanandan, M. (2009). An iterative regularization method for Ill-posed Hammerstein type operator equation, *J. Inv. Ill-Posed Problems* **17**, pp. 831–844.

[231] George, S. and Kunhanandan, M. (2010). Iterative regularization methods for ill-posed Hammerstein type operator equation with monotone nonlinear part, *Int. J. Math. Analysis* **4**, pp. 1673–1685.

[232] George, S. and Nair, M. T. (1993). An a posteriori parameter choice for simplified regularization of ill-posed problems, *Integr. Equat. Oper. Th.* **16**, pp. 392–399.

[233] George, S. and Nair, M. T. (1998). On a generalized Arcangeli's method for Tikhonov regularization with inexact data, *Numer. Funct. Anal. Optim.* **19**, pp. 773–787.

[234] George, S. and Nair, M. T. (2008). A modified Newton–Lavrentiev regularization for nonlinear ill-posed Hammerstein-type operator equation, *J. Complexity* **24**, pp. 228–240.

[235] George, S. and Shobha, M. E. (2012). Two-step Newton–Tikhonov method for Hammerstein-type equations: finite-dimensional realization, *ISRN Applied Mathematics* **1**, pp. 65–78.

[236] George, S. and Shobha M. E. (2012). Two-step Newton–Tikhonov methods for Hammerstein equations, submitted.

[237] Gorenflo, R. and Hofmann, B. (1994). On autoconvolution and regularization, *Inverse Problems* **10**, pp. 353–373.

[238] Gragg, W. B. and Tapia, R. A. (1974). Optimal error bounds for the Newton–Kantorovich theorem, *SIAM J. Numer. Anal.* **11**, 10–13.

[239] Grau, M. and Noguera, M. (2004). A variant of Cauchy's method with accelerated fifth-order convergence, *Appl. Math. Lett.* **17**, pp. 509–517.

[240] Groetsch, C. W. (1977). *Generalized Inverses of Linear Operators: Representation and Approximation.* Monographs and Textbooks in Pure and Applied Mathematics **37**, Marcel Dekker, New York.

[241] Groetsch, C. W. (1984). *Theory of Tikhonov Regularization for Fredholm Equation of the First Kind*, Pitman Books, London.

[242] Groetsch, C. W. and Guacaneme, J. E. (1987). Arcangeli's method for Fredhom

equations of the first kind, *Proc. Amer. Math. Soc.* **99**, pp. 256–260.

[243] Groetsch, C. W., King, J. T. and Murio, D. (1982). Asymptotic analysis of a finite element method for Fredholm equations of the first kind. In: Treatment of Integral Equations by Numerical Methods, Edited by: C. T. Baker and G. F. Miller, Academic Press, London, England, 1–11.

[244] Guacaneme, J. E. (1990). A parameter choice for simplified regularization, *Rostock Math. Kolloq.* **42**, pp. 59–68.

[245] Gupta, D. K. and Parhi, S. K. (2009). A third order method for fixed points in Banach spaces, *J. Math. Anal. Appl.* **359**, pp. 642–652.

[246] Guo, X. (2007). On semilocal convergence of inexact Newton methods, *J. Comput. Math.* **25**, pp. 231–242.

[247] Gutiérrez, T. M., Hernández, M. A. and Salanova, M. A. (1997). A family of the Chebyshev–Halley type methods in Banach spaces, *Bull. Austral. Math. Soc.* **55**, pp. 113–130.

[248] Haeseler, F. V. and Kriete, H. (1993). Surgery for Relaxed Newton's method, *Complex Var., Theor. Appl.: Inter. J.* **22**, pp. 129–143.

[249] Hall, B. C. (2003). *Lie Groups, Lie Algebras and Representations. An elementary introduction*, Graduate Texts in Mathematics **222**, Springer-Verlag, New York.

[250] Halley, E. (1694). A new, exact, and easy method of finding roots of any equations generally, and that without any previous reduction, *Philos. Trans. Roy. Soc. London* **18**, pp. 136–145.

[251] Han, D. and Wang, X. (1997). The error estimates of Halley's method, *Numer. Math. J. Chinese Univ.* (English Ser.) **6**, pp. 231–240.

[252] Han, D. F. and Wang, X. (1998). Convergence of a deformed Newton method, *Appl. Math. Comput.* **94**, 65–72.

[253] Han, S., Pang, J. and Rangaraj, N. (1992). Globally convergent Newton methods for nonsmooth equations, *Math. Oper. Res.* **17**, pp. 586–607.

[254] Hanke, M. Neubauer, A. and Scherzer, O. (1995). A convergence analysis of Landweber iteration of nonlinear ill-posed problems, *Numer. Math.* **72**, pp. 21–37.

[255] Hansen, E. and Patrick, M. (1977). A family of root finding methods, *Numer. Math.* **27**, pp. 257–269.

[256] Harker, P. T. and Pang, J. S. (1990). Finite dimensional variational inequality and nonlinear complementarity problems: a survey of theory, algorithms and applications, *Math. Program.* **48**, pp. 161–220.

[257] Häubler, W. M. (1986). A Kantorovich-type convergence analysis for the Gauss–Newton method, *Numer. Math.* **48**, pp. 119–125.

[258] Helgason, S. (1978). *Differential Geometry, Lie Groups and Symmetric Spaces.* Pure and Applied Mathematics, 80, Academic Press, Inc. Harcourt Brace Jovanovich, Publishers, New York-London.

[259] Helgason, S. (1982). *Differential Geometry, Lie Groups and Symmetric Spaces*, Pergamon Press, Oxford.

[260] Hellinger, E. and Toeplitz, O. (1953). *Integralgleichungen und Gleichungen mit Unendlichvielen Unbekannten*, (German), Chelsea Publishing Company, New York, pp. 1335–1616.

[261] Hernández, M. A. (1991). A note on the Halley method, *Numer. Math.* **59**, pp. 273–276.

[262] Hernández, M. A. (2000). Second-derivative-free variant of the Chebyshev method for nonlinear equations, *J. Optim. Theory Appl.* **104**, pp. 501–515.

[263] Hernández, M. A. (2001). Chebyshev's approximation algorithms and applications, *Comput. Math. Appl.* **41**, pp. 433–445.

[264] Hernández, M. A. and Romero, N. (2005). On a characterization of some Newton-like methods of R-order at least three, *J. Comput. Appl. Math.* **183**, pp. 53–66.

[265] Hernández, M. A. and Rubio, M. J. (2002). The Secant method for nondifferentiable operators, *Appl. Math. Lett.* **15**, pp. 395–399.

[266] Hernández, M. A. and Rubio, M. J. (2002). An uniparametric family of iterative processes for solving nondifferentiable equations, *J. Math. Anal. Appl.* **275**, pp. 821–834.

[267] Hernández, M. A., Rubio, M. J. and Ezquerro, J. A. (2000). Secant-like methods for slving nonlinear integral equations of the Hammerstein type, *J. Comput. Appl. Math.* **115**, pp. 245–254.

[268] Hernández, M. A., Rubio, M. J. and Ezquerro, J. A. (2005). Solving a special case of conservative problems by Secant-like method, *Appl. Math. Cmput.* **169**, pp. 926–942.

[269] Hernández, M. A. and Salanova, M. A. (2000). Modification of the Kantorovich assumptions for semilocal convergence of the Chebyshev methods, *J. Comput. Appl. Math.* **126**, pp. 131–143.

[270] Hilout, S. (2007). Steffensen-type methods on Banach spaces for solving generalized equations, *Adv. Nonlinear Var. Inequa.* **10**, pp. 105–113.

[271] Hilout, S. (2007). An uniparametric Newton–Steffensen-type methods for perturbed generalized equations, *Adv. Nonlinear Var. Inequa.* **10**, pp. 115–124.

[272] Hilout, S. (2007). Superlinear convergence of a family of two-step Steffensen-type methods for generalized equations, *Internat. J. Pure Appl. Math.* **40**, pp. 1–10.

[273] Hilout, S. (2007). A two-step Steffensen-type method for nonsmooth variational inclusions, *Comm. Appl. Nonlinear Anal.* **14**, pp. 27–34.

[274] Hilout, S. (2008). Convergence analysis of a family of Steffensen-type methods for generalized equations, *J. Math. Anal. Appl.* **339**, pp. 753–761.

[275] Hilout, S. (2008). An uniparametric Secant-type method for nonsmooth generalized equations, *Positivity* **12**, pp. 281–287.

[276] Hiriart-Urruty, J. B. and Lemaréchal, C. (1993). *Convex Analysis and Minimization Algorithms (two volumes): I. Fundamentals II. Advanced Theory and Bundle Methods*, Grundlehren der Mathematischen Wissenschaften, Vols. **305** and **306**, Springer-Verlag, Berlin.

[277] Hoang, N. S. and Ramm, A. G. (2010). Dynamical systems methosd for solving nonlinear equations with monotone operators, *Math. Comput.* **79**, pp. 239–258.

[278] Homeier, H. (2009). On Newton-type methods for multiple roots with cubic convergence, *J. Comput. Appl. Math.* **231**, pp. 249–254.

[279] Huang, Z. A. (2003). Convergence of inexact Newton method, *J. Zhejiang Univ. Sci. Ed.* **30**, pp. 393–396.

[280] Huang, Z. D. (2004). The convergence ball of Newton's method and the uniqueness ball of equations under Hölder-type continuous derivatives, *Comput. Math. Appl.* **25**, pp. 247–251.

[281] Jin, Q. N. (2000). Error estimates of some Newton-type methods for solving nonlinear inverse problems in Hilbert scales, *Inverse Problems* **16**, pp. 187–197.

[282] Jin, Q. N. (2000). On the iteratively regularized Gauss–Newton method for solving nonlinear ill-posed problems, *Math. Comput.* **69**, pp. 1603–1623.

[283] Jin, Q. N. (2008). A convergence analysis of the iteratively regularized Gauss–Newton method under the Lipschitz condition, *Inverse Problems* **24**, pp. 1–24.

[284] Jin, Q. N. and Hou, Z. Y. (1996). Finite-dimensional approximations to the solutions of nonlinear ill-posed problems, *Appl. Anal.* **62**, pp. 253–261.

[285] Jin, Q. N. and Hou, Z. Y. (1999). On an a posteriori parameter choice strategy for Tikhonov regularization of nonlinear ill-posed problems, *Numer. Math.* **83**, pp.

139–159.

[286] Kaltenbacher, B. (1998). On Broyden's method for the regularization of nonlinear ill-posed problems, *Numer. Funct. Anal. Optim.* **19**, pp. 807–833.

[287] Kaltenbacher, B. and Hofmann, B. (2010). Convergence rates for the iteratively regularized Gauss–Newton method in Banach spaces, *Inverse Problems* **26**, no. 3, 035007, 21 pp.

[288] Kanno, S. (1992). Convergence theorems for the method of tangent hyperbolas, *Math. Japon.* **37**, pp. 711–722.

[289] Kantorovich, L. V. (1948). On Newton's method for functional equations (Russian), *Dokl. Akad. Nauk. SSSR* **59**, pp. 1237–1240.

[290] Kantorovich, L. V. and Akilov, G. P., *Functional Analysis*, Pergamon Press, Oxford, 1982.

[291] Kanzow, C. (2001). An active set-type Newton method for constrained nonlinear systems, *Appl. Optim.* **50**, pp. 179–200.

[292] Kelley, C. T. (1986). A Shamanskii-like acceleration scheme for nonlinear equations at singular roots, *Math. Comp.* **47**, pp. 609–623.

[293] Kelley, C. T. (1995). *Iterative Methods for Linear and Nonlinear Equations*, (with separately available software), Frontiers in Applied Mathematics, **16**, SIAM, Philadelphia.

[294] Kou, J., Li, Y. and Wang, X. (2007). A modification of Newton method with third-order convergence, *Appl. Math. Comput.* **181**, pp. 1106–111.

[295] Krasnoselskii, M. A., Zabreiko, P. P., Pustylnik, E. I. and Sobolevskii, P. E. (1976). Integral operators in spaces of summable functions, Translated by T. Ando, Noordhoff International publishing, Leyden.

[296] Krishnan, S. and Manocha, D. (1997). An efficient surface intersection algorithm based on lower-dimensional formulation, *ACM Trans. on Graphics* **16**, pp. 74–106.

[297] Laasonen, P. (1969). Ein überquadratisch konvergenter iterativer algorithmus, *Ann. Acad. Sci. Fenn. Ser. I* **450**, pp. 1–10.

[298] Lang, S. (2002). *Introduction to Differentiable Manifolds*, Springer-Verlag, New York.

[299] Langer S. (2010). Investigation of preconditioning techniques for the iteratively regularized Gauss–Newton method for exponentially ill-posed problems, *SIAM J. Sci. Comput.* **32**, pp. 2543–2559.

[300] Levin, Y. and Ben-Israel, A. (2002). Directional Newton methods in n variables, *Math. Comput.* **71**, pp. 251–262.

[301] Lewis, A. S. and Wright, S. J. (2008). A proximal method for composite minimization, arXiv:0812.0423v1 [math.OC]

[302] Li, C., Hu, N. and Wang, J. (2010). Convergence bahavior of Gauss–Newton's method and extensions to the Smale point estimate theory, *J. Complexity* **26**, pp. 268–295.

[303] Li, C., López, G. and Martín-Márquez, V. (2009). Monotone vector fields and the proximal point algorithm on Hadamard manifolds, *J. Lond. Math. Soc.* **79**, pp. 663–683.

[304] Li, C. and Ng, K. F. (2007). Majorizing functions and convergence of the Gauss–Newton method for convex composite optimization, *SIAM J. Optim.* **18**, pp. 613–642.

[305] Li, C. and Shen, W. P. (2008). Local convergence of inexact methods under the Hölder condition, *J. Comput. Appl. Math.* **222**, pp. 544–560.

[306] Li, C. and Wang, X. H. (2002). On convergence of the Gauss–Newton method for convex composite optimization, *Math. Program. Ser A.* **91**, pp. 349–356.

[307] Li, C. and Wang, J. (2005). Convergence of the Newton method and uniqueness of

zeros of vector fields on Riemannian manifolds, *Sci. China Ser. A* **48**, pp. 1465–1478.

[308] Li, C. and Wang, J. (2006). Newton's method on Riemannian manifolds: Smale's point estimate theory under the γ-condition, *IMA J. Numer. Anal.* **26**, pp. 228–251.

[309] Li, C. and Wang, J. (2008). Newton's method for sections on Riemannian manifolds: Generalized covariant α-theory, *J. Complexity* **24**, pp. 423–451.

[310] Li, C., Wang, J. H. and Dedieu, J. P. (2009). Smale's point estimate theory for Newton's method on Lie groups, *J. Complexity* **25**, pp. 128–151.

[311] Li, C., Zhang, W-H. and Jin, X-Q. (2004). Convergence and uniqueness properties of Gauss–Newton's method, *Comput. Math. Appl.* **47**, pp. 1057–1067.

[312] Lukács, G. (1989). *The Generalized Inverse Matrix and the Surface-surface Intersection Problem.* Theory and practice of geometric modeling (Blaubeuren, 1988), pp. 167–185, Springer, Berlin.

[313] Magaril-Il'yaev, G. G. and Tikhomirov, V. M. (2003). *Convex Analysis: Theory and Applications*, Translated from the 2000 Russian edition by Dmitry Chibisov and revised by the authors, Translations of Mathematical Monographs **222**, AMS, Providence, RI.

[314] Mahale, P. and Nair, M. T. (2009). Iterated Lavrentiev regularization for nonlinear ill-posed problems, *ANZIAM J.* **51**, pp. 191–217.

[315] Mahale, P. and Nair, M. T. (2009). A Simplified generalized Gauss–Newton method for nonlinear ill-posed problems, *Math. Comp.* **78**, pp. 171–184.

[316] Mahony, R. E. (1994). *Optimization Algorithms on Homogeneous Spaces: with Applications in Linear Systems Theory*, Ph.D. Thesis, Department of Systems Engineering, University of Canberra, Australia.

[317] Mahony, R. E. (1996). The constrained Newton method on a Lie group and the symmetric eigenvalue problem, *Linear Algebra Appl.* **248**, pp. 67–89.

[318] Mahony, R. E., Helmke, U. and Moore, J. B. (1993). Pole placement algorithms for symmetric realisations, *Proceedings of 32nd IEEE Conference on Decision and Control*, San Antonio, TX, pp. 1355–1358.

[319] Mair, B. A. (1994). Tikhonov regularization for finitely and infinitely smoothing operators, *SIAM J. Math. Anal.* **25**, pp. 135–147.

[320] Maĭstrovskiĭ, G. D. (1972). The optimality of the Newton method (Russian), *Dokl. Akad. Nauk SSSR* **204**, pp. 1313–1315.

[321] Mann, W. R. (1953). Mean value methods in iteration, *Proc. Amer. Math. Soc.* **4**, pp. 506–510.

[322] Marcotte, P. and Wu, J. H. (1995). On the convergence of projection methods, *J. Optim. Theory Appl.* **85**, pp. 347–362.

[323] Mărușter, S. (1976). Quasi-nonexpansivity and two classical methods for solving nonlinear equations, *Proc. Amer. Math. Soc.* **62**, pp. 119–123.

[324] Mathe, P. and Perverzev, S. V. (2003). Geometry of linear ill-posed problems in variable Hilbert scales, *Inverse Problems* **19**, pp. 789–803.

[325] Meyer, P. W. (1984). Das modifizierte Newton-Verfahren in verallgemeinerten Banach–Räumen, *Numer. Math.* **43**, pp. 91–104.

[326] Meyer, P. W. (1987). Newton's method in generalized Banach spaces, *Numer. Funct. Anal. Optim.* **9**, pp. 244–259.

[327] Meyer, P. W. (1992). A unifying theorem on Newton's method, *Numer. Funct. Anal. Optim.* **13**, pp. 463–473.

[328] Moreau, J. J. (1962). Fonctions convexes duales et points proximaux dans un espace hilbertien, *C. R. Acad. Sci. Paris* **255**, pp. 2897–2899.

[329] Moreau, J. J. (1963). Propriétés des applications prox, *C. R. Acad. Sci. Paris* **256**, pp. 1069–1071.

[330] Moreau, J. J. (1965). Proximité et dualité dans un espace Hilbertien, *Bull. Soc.*

Math. Fr. **93**, pp. 273–299.

[331] Moudafi, A. (2007). On finite and strong convergence of a proximal method for equilibrium problems, *Numer. Funct. Anal. Optim.* **28**, pp. 1347–1354.

[332] Nair, M. T. and Ravishankar, P. (2008). Regularized versions of continuous newton's method and continuous modified newton's method under general source conditions, *Numer. Funct. Anal. Optim.* **29**, pp. 1140–1165.

[333] Nesterov, Y. and Nemirovskii, A. (1994). Interior-point polynomial algorithms in convex programming, *SIAM Studies Appl. Math.* **13**, Philadelphia, PA.

[334] Neta, B. (2008). New third order nonlinear solvers for multiple roots, *Appl. Math. Comput.* **202**, pp. 162–170.

[335] Neta, B. (2010). Extension of Murakami's High order nonlinear solver to multiple roots, *Int. J. Comput. Math.* **87**, pp. 1023–1031.

[336] Neta, B. and Johnson, A. N. (2008). High order nonlinear solver for multiple roots, *Comput. Math. Appl.* **55**, pp. 2012–2017.

[337] Ng, K. F. and Zheng, X. Y. (2004). Characterizations of error bounds for convex multifunctions on Banach spaces, *Math. Oper. Res.* **29**, pp. 45–63.

[338] Ortega, L. M. and Rheinboldt, W. C. (1970). *Iterative Solution of Nonlinear Equations in Several Variables*, Academic press, New York.

[339] Osada, N. (1994). An optimal multiple root-finding method of order three, *J. Comput. Appl. Math.* **51**, pp. 131–133.

[340] Ostrowski, A. M. (1940). Sur la convergence et l'estimation des erreurs dans quelques procédés de résolution des équations numériques. (French), Memorial volume dedicated to D. A. Grave [Sbornik posvjaščenii pamjati D. A. Grave], 213–234. publisher unknown, Moscow.

[341] Ostrowski, A. M. (1966). *Solution of Equations and Systems of Equations*, 2nd ed., Academic Press, New York.

[342] Ostrowski, A. M. (1971). La méthode de Newton dans les espaces de Banach, *C. R. Acad. Sci. Paris* Sér. A–B **272**, pp. 1251–1253.

[343] Ostrowski, A. M. (1973). *Solution of Equations in Euclidean and Banach Spaces*, Academic Press, New York.

[344] Owren, B. and Welfert, B. (2000). The Newton iteration on Lie groups, *BIT* **40**, pp. 121–145.

[345] Penrose, R. (1955). A generalized inverse for matrices, *Proc. Cambridge Philos. Soc.* **51**, pp. 406–413.

[346] Pereverzev, S. and Schock, E. (2005). On the adaptive selection of the parameter in regularization of ill-posed problems, *SIAM. J. Numer. Anal.* **43**, pp. 2060–2076.

[347] Petković, M. S., Herceg, D. and Ilić, S. M. (1971). *Point Estimation Theory and its Applications*, University of Novi Sad, Institute of Mathematics, Novi Sad.

[348] Potra, F. A. (1980). A characterisation of the divided differences of an operator which can be represented by Riemann integrals, *Rev. Anal. Numér. Théor. Approx.* **9**, pp. 251–253.

[349] Potra, F. A. (1981). An application of the induction method of V. Pták to the study of regula falsi. With a loose Russian summary, *Aplikace Matematiky* **26**, pp. 111–120.

[350] Potra, F. A. (1981). The rate of convergence of a modified Newton's process. With a loose Russian summary, *Aplikace Matematiky* **26**, pp. 13–17.

[351] Potra, F. A. (1981). An application of the induction method of V. Pták to the study of Regula Falsi, *Aplikace Matematiky* **26**, pp. 111–120.

[352] Potra, F. A. (1982). *On the Convergence of a Class of Newton-like Methods*, Iterative Solution of Nonlinear Systems of Equations, (Oberwolfach, 1982), Lecture Notes in Math. **953**, Springer, Berlin–New York, pp. 125–137.

[353] Potra, F. A. (1981/1982). An error analysis for the secant method, *Numer. Math.* **38**, pp. 427–445.

[354] Potra, F. A. (1984). On the a posteriori error estimates for Newton's method, *Beiträge Zur Numer. Math.* **12**, pp. 125–138.

[355] Potra, F. A. (1984/1985). On an iterative algorithm of order 1.839... for solving nonlinear equations, *Numer. Funct. Anal. Optim.* **7**, pp. 75–106.

[356] Potra, F. A. (1985). Sharp error bounds for a class of Newton-like methods, *Libertas Math.* **5**, pp. 71–84.

[357] Potra, F. A. and Pták, V. (1980). Nondiscrete induction and double step secant method, *Math. Scand.* **46**, pp. 236–250.

[358] Potra, F. A. and Pták, V. (1980). On a class of modified Newton processes, *Numer. Func. Anal. Optim.* **2**, pp. 107–120.

[359] Potra, F. A. and Pták, V. (1980). Sharp error bounds for Newton's process, *Numer. Math.* **34**, pp. 63–72.

[360] Potra, F. A. and Pták, V. (1984). *Nondiscrete Induction and Iterative Processes*, Research Notes in Mathematics **103**, Pitman Advanced Publishing program, Boston.

[361] Pták, V. (1975). Concerning the rate of convergence of Newton's process, *Comment. Math. Univ. Carolinae* **16**, pp. 699–705.

[362] Pták, V. (1976). A modification of Newton's method, *Časopis Pěst. Mat.* **101**, pp. 188–194.

[363] Pták, V. (1976). Nondiscrete mathematical induction and iterative existence proofs, *Linear Algebra Appl.* **13**, pp. 223–238.

[364] Pták, V. (1976). The rate of convergence of Newton's process., *Numer. Math.* **25**, pp. 279–285.

[365] Pták, V. (1976). Nondiscrete mathematical induction. General topology and its relations to modern analysis and algebra, IV (*Proc. Fourth Prague Topological Sympos.*, Prague), Part A, 166–178, Lecture Notes in Math., Vol. **609**, Springer, Berlin.

[366] Pták, V. (1977). What should be a rate of convergence?, *RAIRO Anal. Numér.* **11**, pp. 279–286.

[367] Pták, V. (1979). Stability of exactness. Special issue dedicated to Władysław Orlicz on the occasion of his seventy-fifth birthday, Comment. Math., Special Issue **2**, pp. 283–288.

[368] Proinov, P. D. (2009). General local convergence theory for a class of iterative processes and its applications to Newton's method, *J. Complexity* **25**, pp. 38–62.

[369] Proinov, P. D. (2010). New general convergence theory for iterative processes and its applications to Newton–Kantorovich type theorems, *J. Complexity* **26**, pp. 3–42.

[370] Ramlau, R. and Teschke, G. (2005). Tikhonov replacement functionals for iteratively solving nonlinear operator equations, *Inverse Problems* **21**, pp. 1571–1592.

[371] Ramm, A. G. (2001). Linear ill-posed problems and dynamical systems, *J. Math. Anal. Appl.* **258**, pp. 448–456.

[372] Ramm, A. G. (2002). Acceleration of convergence of a continuous analog of the Newton method, *Appl. Anal.* **81**, pp. 1001–1004.

[373] Ramm, A. G. (2004). Dynamical systems method for solving operator equations, *Commun. Nonlinear. Sci. Numer. Simul.* **9**, pp. 383–402.

[374] Ramm, A. G. (2005). *Inverse Problems. Mathematical and Analytical Techniques with Applications to Engineering*. With a foreword by Alan Jeffrey. Mathematical and Analytical Techniques with Applications to Engineering. Springer, New York.

[375] Ramm, A. G. and Smimova, A. B. (2001). On stable numerical differentiation, *Math. Comput.* **70**, pp. 1131–1153.

[376] Ramm, A. G. and Smimova, A. B. (2002). Continuous regularized Gauss–Newton-

type algorithm for nonlinear ill-posed equations with simultaneous updates of inverse derivative, *Int. J. Pure. Appl. Math.* **2**, pp. 23–34.

[377] Ramm, A. G., Smimova, A. B. and Favini, A. (2003). Continuous modified Newton's-type method for nonlinear operator equations, *Annali di Math.* **182**, pp. 37–52.

[378] Raus, T. (1984). On the discrepancy principle for the solution of ill-posed problems, *Acta Comment. Univ. Tartuensis* **672**, pp. 16–26.

[379] Ren, H. (2006). On the local convergence of a deformed Newton's method under Argyros-type condition, *J. Math. Anal. Appl.* **321**, pp. 396–404.

[380] Ren, H. and Wu, Q. B. (2006). The convergence ball of the Secant method under Hölder continuous divided differences, *J. Comput. Appl. Math.* **194**, pp. 284–293.

[381] Ren, H. and Wu, Q. B. (2006). Mysovskii-type theorem for the Secant method under Hölder continuous Fréchet derivative, *J. Math. Anal. Appl.* **320**, pp. 415–424.

[382] Ren, H., Wu, Q. B. (2007). Convergence ball of a modified secant method with convergence order $1.839\ldots$, *Appl. Math. Comput.* **188**, pp. 281–285.

[383] Rheinboldt, W. C. (1968). A unified convergence theory for a class of iterative processes, *SIAM J. Numer. Anal.* **5**, pp. 42–63.

[384] Rheinboldt, W. C. (1977). An adaptive continuation process for solving systems of nonlinear equations, Polish Academy of Science, *Banach Ctr. Publ.* **3**, pp. 129–142.

[385] Robinson, S. M. (1972). Extension of Newton's method to nonlinear functions with values in a cone, *Numer. Math.* **19**, pp. 341–347.

[386] Robinson, S. M. (1975). Stability theory for systems of inequalities. I. Linear systems, *SIAM J. Numer. Anal.* **12**, pp. 754–769.

[387] Rockafellar, R. T. (1967). *Convex Analysis*, Princeton University Press, Princeton.

[388] Romero, N. (2006). *Familias paramétricas de procesos iterativos de alto orden de convergencia*, Ph.D. Dissertation, Logroño, Spain.

[389] Salzo, S. and Villa S. (2012). Convergence analysis of a proximal Gauss–Newton method, *Comput. Optimiz. Appl.* **53**, pp. 557–589.

[390] Scherzer, O. (1989). The use of Tikhonov regularization in the identification of electrical conductivities from overdetermined problems, *Inverse Problems* **5**, pp. 227–238.

[391] Scherzer, O. (1993). A parameter choice for Tikhonov regularization for solving nonlinear inverse problems leading to optimal rates, *Appl. Math.* **38**, pp. 479–487.

[392] Scherzer, O., Engl, H. W. and Anderssen, R. S. (1993). Parameter identification from boundary measurements in parabolic equation arising from geophysics, *Nonlinear Anal.* **20**, pp. 127–156.

[393] Scherzer, O., Engl, H. W. and Kunisch, K. (1993). Optimal a posteriori parameter choice for Tikhonov regularization for solving nonlinear ill-posed problems, *SIAM. J. Numer. Anal.* **30**, pp. 1796–1838.

[394] Scherzer, O., Grasmair, M., Grossauer, H., Haltmeier, M. and Lenzen, F. (2009). Variational Methods in Imaging, *Applied Mathematical Sciences* **167**, Springer, New York.

[395] Schmidt, J. W. (1963). Eine übertragung der Regula Falsi and gleichungen in Banachräumen, I, *Z. Angew. Math. Mech.* **43**, pp. 97–110.

[396] Schmidt, J. W. and Schwetlick, H. (1968). Ableitungsfreie Verfahren mit hohere Konvergenzgeschwindigkeit, *Comput.* **3**, pp. 215–226.

[397] Schmidt, J. W. (1978). Untere Fehlerschranken fur Regula–Falsi Verhafren, *Period. Hungar.* **9**, pp. 241–247.

[398] Schock, E. (1984). On the asymptotic order of accuracy of Tikhonov regularization, *J. Optim. Th. Appl.* **44**, pp. 95–104.

[399] Schock, E. (1985). Integral equations of the first kind, *Studia Mathematica* **LXXX1**,

1–11.

[400] Schröder, J. (1956). Nichtlineare majoranten beim verfahren der schrittweisen näherung, *Arch. Math.* **7**, pp. 471–484.

[401] Schröer, E. (1870). Üer unendlich viele Algorithmen zur Auflöung der Gleichungen, *Math. Ann.* **2**, pp. 317–365.

[402] Schwetlick, H. (1979). Numerische Lösung nichtlinearer Gleichungen, R. Oldenburg Verlag, Munchen Wien.

[403] Semenova, E. V. (2010). Lavrentiev regularization and balancing principle for solving ill-posed problems with monotone operators, *Comput. Methods Appl. Math.* **4**, pp. 444–454.

[404] Shamanskii, V. E. (1967). A modification of Newton's method, *Ukrain. Mat. Zh.* **19**, 133–138.

[405] Shen, W. and Li, C. (2009). Kantorovich-type convergence criterion for inexact Newton methods. *Appl. Numer. Math.* **59**, pp. 1599–1611.

[406] Shen, W. and Li, C. (2010). Smale's α-theory for inexact Newton methods under the γ-condition, *J. Math. Anal. Appl.* **369**, pp. 29–42.

[407] Shengguo, L., Housen, L. and Lizhi, C. (2009). A new fourth-order iterative method for finding multiple roots of nonlinear equations, *Appl. Math. Comput.* **215**, pp. 1288–1292.

[408] Shengguo, L., Housen, L. and Lizhi, C. (2009). Some second-derivative-free variants of Halley's method for multiple roots, *Appl. Math. Comput.* **215**, pp. 2192–2198.

[409] Smale, S. (1986). *Newton Method Estimates from Data at One Point*. The merging of disciplines: new directions in pure, applied and computational mathematics (R. Ewing, K. Gross, C. Martin, eds), Springer-Verlag, New York, pp. 185–196.

[410] Spivak, M. (2005). *A Comprehensive Introduction to Differential Geometry*, Vol. **I**, third ed., Publish or Perish Inc., Houston, Texas.

[411] Spivak, M. (2005). *A Comprehensive Introduction to Differential Geometry*, Vol. **II**, third ed., Publish or Perish Inc., Houston, Texas.

[412] Stewart, G. W. (1969). On the continuity of the generalized inverse, *SIAM J. Appl. Math.* **17**, pp. 33–45.

[413] Stoer, J. and Bulirsch, K. (1976). *Introduction to Numerical Analysis*, Springer-Verlag.

[414] Tapia, R. A. (1971). The Kantorovich theorem for Newton's method, *Amer. Math. Monthly* **11**, pp. 10–13.

[415] Tautenhahn, U. (2002). On the method of Lavrentiev regularization for nonlinear ill-posed problems, *Inverse Problems* **18**, 191–207.

[416] Taylor, A. E. (1957). *Introduction to Functional Analysis*, Wiley, New York.

[417] Tikhonov, A. N. (1963). Regularizations of incorrectly posed problems, *Soviet Mathematics Doklady* **4**, pp. 1624–1627.

[418] Traub, J. F. (1964). *Iterative Methods for the Solution of Equations*, Englewood Cliffs, New Jersey: Prentice Hall.

[419] Traub, J. F. (editor). (1975). *Analytic Computational Complexity*, Academic Press.

[420] Traub, J. F. and Woźniakowsi, H., (1979). Convergence and complexity of Newton iteration for operator equations, *J. Assoc. Comput. Mach.* **26**, pp. 250–258.

[421] Varadarajan, V. S. (1984). *Lie Groups, Lie Algebras and their Representations*. Reprint of the 1974 edition. Graduate Texts in Mathematics **102**, Springer-Verlag, New York.

[422] Vandergraft, J. S. (1967). Newton's method for convex operators in partially ordered spaces, *SIAM J. Numer. Anal.* **4**, pp. 406–432.

[423] Victory, H. D. and Neta, B. (1983). A higher order method for multiple zeros of

nonlinear functions, *Int. J. Comput. Math.* **12**, pp. 329–335.

[424] Wang, J. H. and Li, C. (2006). Uniqueness of the singular points of vector fields on Riemannian manifolds under the γ-condition, *J. Complexity* **22**, pp. 533–548.

[425] Wang, J. H. and Li, C. (2011). Kantorovich's theorems for Newton's method for mappings and optimization problems on Lie groups, *IMA J. Numer. Anal.* **31**, pp. 322–347.

[426] Wang, X. H. (1980). *The Convergence Ball on Newton's Method*, (in Chinese), Chinese Science Bulletin, A Special Issue of Mathematics, Physics, Chemistry **25**, pp. 36–37.

[427] Wang, X. H. (1999). Convergence of Newton's method and inverse function theorem in Banach space, *Math. Comput.* **68**, pp. 169–186.

[428] Wang, X. H. (2000). Convergence of Newton's method and uniqueness of the solution of equations in Banach space, *IMA J. Numer. Anal.* **20**, pp. 123–134.

[429] Wang, X. H. and Han, D. F. (1990). On dominating sequence method in the point estimate and Smale theorem, *Sci. China Ser. A* **33**, pp. 135–144.

[430] Wang, X. H. and Han, D. F. (1997). Criterion α and Newton's method under weak conditions, (Chinese) Math. Numer. Sin., 19(1) (1997), 103–112; translation in Chinese *J. Numer. Math. Appl.* **19**, pp. 96–105.

[431] Wang, X. H. and Li, C. (2003). Convergence of Newton's method and uniqueness of the solution of equations in Banach spaces, II, *Acta Math. Sin. (Engl. Ser.)* **19**, pp. 405–412.

[432] Wedin, P. (1973). Perturbation theory for pseudo-inverses, *BIT Numer. Math.* **13**, pp. 217–232.

[433] Weerakoon, S. and Fernando, T. G. I. (2000). A variant of Newton's method with accelerated third-order convergence, *Appl. Math. Lett.* **13**, pp. 87–93.

[434] Werner, W. (1981). *Some Improvements of Classical Iterative Methods for the Solution of Nonlinear Equations*, Numerical Solution of Nonlinear Equations Lecture Notes in Mathematics **878**, pp. 426–440.

[435] Wolfe, M. A. (1978). Extended iterative methods for the solution of operator equations, *Numer. Math.* **31**, pp. 153–174.

[436] Womersley, R. S. (1985). Local properties of algorithms for minimizing nonsmooth composite functions, *Math. Program. Ser. A* **32**, pp. 69–89.

[437] Womersley, R. S. and Fletcher, R. (1986). An algorithm for composite nonsmooth optimization problems, *J. Optim. Theory Appl.* **48**, pp. 493–523.

[438] Xu, C. (2009). *Nonlinear Least Squares Problems.* In: Floudas, C.A., Pardalos, P.M. (eds.) Encyclopedia of Optimization, Springer, New York, pp. 2626–2630.

[439] Xu, L. and Wang, X. (1983). *Topics on Methods and Examples of Mathematical Analysis* (in Chinese), High Education Press.

[440] Xu, X. B. and Li, C. (2007). Convergence of Newton's method for systems of equations with constant rank derivatives, *J. Comput. Math.* **25**, pp. 705–718.

[441] Xu, X. B. and Li, C. (2008). Convergence criterion of Newton's method for singular systems with constant rank derivatives, *J. Math. Anal. Appl.* **345**, pp. 689–701.

[442] Yamamoto, T. (1987). A convergence theorem for Newton-like methods in Banach spaces, *Numer. Math.* **51**, pp. 545–557.

[443] Zabrejko, P. P. and Nguen, D. F. (1987). The majorant method in the theory of Newton–Kantorovich approximations and the Pták error estimates, *Numer. Funct. Anal. Optim.* **9**, pp. 671–684.

[444] Zinčenko, A. I. (1963). Some approximate methods of solving equations with non–differentiable operators, (Ukrainian), *Dopovidi Akad. Nauk Ukraïn. RSR*, pp. 156–161.